Offshore Operation Facilities

Offshore Operation Facilities
Equipment and Procedures

Huacan Fang and Menglan Duan

Offshore Oil/Gas Research Center,
China University of Petroleum,
Beijing, China.

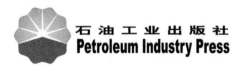

石 油 工 业 出 版 社
Petroleum Industry Press

ELSEVIER

AMSTERDAM • BOSTON • HEIDELBERG • LONDON
NEW YORK • OXFORD • PARIS • SAN DIEGO
SAN FRANCISCO • SYDNEY • TOKYO
Gulf Professional Publishing is an imprint of Elsevier

G P
P

Gulf Professional Publishing is an imprint of Elsevier
225 Wyman Street, Waltham, MA 02451, USA
The Boulevard, Langford Lane, Kidlington, Oxford, OX5 1GB, UK

Permissions may be sought directly from Elsevier's Science & Technology Rights Department in Oxford, UK:
phone (+44) (0) 1865 843830; fax (+44) (0) 1865 853333; email: permissions@elsevier.com. Alternatively
you can submit your request online by visiting the Elsevier web site at http://elsevier.com/locate/permissions,
and selecting Obtaining permission to use Elsevier material

Notice
No responsibility is assumed by the publisher for any injury and/or damage to persons or property as a matter
of products liability, negligence or otherwise, or from any use or operation of any methods, products,
instructions or ideas contained in the material herein

Library of Congress Cataloging-in-Publication Data
Fang, Huacan.
  Offshore operation facilities : equipment and procedures / Huacan Fang, Menglan Duan.
    pages cm
  Includes bibliographical references and index.
  ISBN 978-0-12-396977-4 (hardback)
1. Offshore oil well drilling. I. Duan, Menglan. II. Title.
  TN871.3.F35 2014
  622'.33819—dc23
                        2014021033

British Library Cataloguing in Publication Data
A catalogue record for this book is available from the British Library

ISBN: 978-0-12-396977-4

For information on all Gulf Professional Publishing
publications visit our web site at http://store.elsevier.com

Working together
to grow libraries in
developing countries

www.elsevier.com • www.bookaid.org

Contents

About the Author

Huacan Fang, Professor at China University of Petroleum (Beijing), was born on March 9, 1930, in Dingyuan, Anhui Province. As a member of the Communist party of China, Professor Fang graduated from the Department of Mechanical Engineering of Beiyang University (now Tianjin University) and has worked successively as a teacher at the Department of Petroleum at Tsinghua University, the Associate Head of the Department of Mechanical Engineering at the Beijing Petroleum Institute, the Head of the Department of Mechanical Engineering at the East-China Petroleum Institute (name changed due to relocation from Beijing to Shandong), Vice President of East-China Petroleum Institute since 1984, Vice President of University of Petroleum (East China) since 1988, and the Deputy Director of the Council of China University of Petroleum (Beijing) in 1992. He retired in 1996.

During the years 1996 to 2008, he served as Team Lead of the Teaching Expert Panel at the Department of Electrical and Mechanical Engineering at China University of Petroleum (Beijing). In 2009, he took the role of Leader at the university-level Teaching Expert Panel; now Professor Fang is also a consultant at the Equipment Industry Association of CNPC and Sinopec. Having cultivated more than 40 postgraduate students, he created 11 courses for both undergraduate and postgraduate students, including Oil Drilling Machinery, Offshore Oil Drilling Equipment and Structures, Mechanical Vibrations, Theoretical Foundation of Offshore Oil Drilling Equipmen, Reliability Engineering, and others.

He has obtained more than 10 research project grants including The Research into the Safety and Reliability of the Offshore Structures. Because of this project, he won the Second Prize of the National Prize for Progress in Science and Technology, and the First and Second Provincial and Ministerial Prizes for Progress in Science and Technology Progress 10 times and obtained two Chinese patents. In addition, Professor Fang has published nearly 200 papers, of which more than 40 were published abroad and some 30 have been included in SCI and EI. He has published more than 14 books, including *Theoretical Foundation of Offshore Oil Drilling Equipment, The Fatigue Life of Offshore Oil Steel Structure, The Reliability Analysis of Offshore Structures In Ice Areas, The Safety Reliability Analysis of Oil and Gas Pipeline, Mechanism Design by English, Design of Mechanical Elements by English, About the Classroom Teachings of Higher Education.* In addition, Professor Fang established a new branch of "Fuzzy Probability Fracture Mechanics" used in offshore oil engineering and published his book *Fuzzy Probability Fracture Mechanics (for Ocean Engineering)* in 1999. In 1992, he was awarded National Outstanding Expert status for his significant contributions and was granted government subsidies.

Menglan Duan, born in 1966, held a Bachelor's degree (1987) in petroleum drilling engineering at the Department of Petroleum Engineering, Jianghan University of Petroleum (Yangtze University since 2003), a Master's degree (1990) in offshore structural engineering and a Ph.D. degree (1993) in petroleum mechanical engineering at the Department of Mechanical Engineering, University of Petroleum of China. He was later a postdoctoral fellow both in the Division of Solid Mechanics at the Institute of Mechanics and at the, Chinese Academy of Sciences, and in the Department of Mechanical Engineering, University of Rochester, New York, USA. In 1995, he conducted research as a senior engineer in the offshore engineering department of the Headquarters of China Classification Society. Then, he was promoted to be a research professor in the CCS Research and Development Center. From 2001 to 2002, he headed for the establishment of the joint venture of risk management services between CCS and ABS, and from 2002 to 2006, he was the chief engineer of CCS Industrial Corp. He is now the head of the Offshore Oil/Gas Research Center, China University of Petroleum and the director of China-Brazil Joint Research Center for Subsea Engineering.

Prof. Menglan Duan has a wide range of research areas and has conducted over 140 projects related to fixed and mobile platforms, subsea pipelines and risers, subsea production systems, in the following areas: Arctic Mechanics and Offshore Engineering; Subsea Engineering; Fatigue and Fracture of Materials and Structures; Dynamics of Offshore Engineering Structures; Mechanical Behaviour of Materials; Risk and Reliability in Offshore and Ocean Engineering. He has published over 300 technical papers in many international journals and conferences. His academic titles include the following international organizations: Chair of China Branch of Society for Underwater Technology; Technical Committee Member of the 14th and 15th International Ship and Offshore Structures Congress (ISSC); Technical Program Member of the International Society of Offshore and Polar Engineers (ISOPE); Committee Member of the 11th International Symposium on Practical Design of Ships and other Floating Structures; International Scientific Committee Member of the 7th International Conference on Thin-Walled Structures; International Scientific Committee Member of the 3th International Conference on Fracture Fatigue and Wear; Committee Member of the 7th International Workshop on Ship Hydrodynamics; Secretary of 2012 SUT International Conference on Subsea Technologies and Deepwater Engineering. He is also the Editor-in-Chief of Development and Application of Oceanic Engineering and Frontiers in Fossil Fuel Engineering while simultaneously an editorial board member of many international journals, such as Ships and Offshore Structures, Ocean Systems Engineering, Ocean Systems Engineering, International Journal of Energy Engineering, Thermal Energy and Power Engineering, Petroleum Science, China Ocean Engineering. He is the chairperson of the following international conferences: MARIENV'95, ISOPE'2000, ISSC'2000, ISSC'2003, PACOMS'2006, ISOPE'2007, APAC'2007, ISOPE'2008, ISOPE'2009, IDOT'2009, SUT'2010, ISOPE'2010, PRADS'2010, IWSH'2011, ISOPE'2011, ISOPE'2012, ISOPE'2013, SUTTC'2013.

Foreword

Professor Fang Huacan was the former Vice-Chancellor of University of Petroleum. He was born in March 1930. At the age of 80, he accomplished the writing of *Offshore Operation Facilities* monograph with 1.2 million words; it includes two volumes published by Petroleum Industry Press. Recently, I heard that this voluminous book will come out after several years of effort. It is dedicated to his 80th birthday. I am sincerely happy. So, I would like to address my sincere congratulations for his birthday and the forthcoming *Offshore Operation Facilities*.

Since 1952, Professor Fang was a teacher at Tsinghua University, Beijing Petroleum Institute, East China Petroleum Institute, Petroleum University (renamed in 1988). He taught undergraduate and graduate students many subjects: oil drilling machinery, petroleum mining machinery, oil drilling machinery reliability design, offshore oil drilling equipment and structures, offshore oil drilling equipment theoretical foundation, to name but a few. What is more, he has rich teaching experience and writing capacity.

Professor Fang Huacan has been engaged in teaching and research work from the beginning of onshore oil field equipment. In answering the call of the Ministry of Petroleum Industry in 1964, he changed his research and teaching field from oil field equipment to offshore oil drilling equipment and structures, which enabled him to be the frontier teacher and researcher in offshore oil engineering.

Professor Fang has done a lot of teaching and research on the subject of offshore oil engineering. For instance, he has trained many undergraduate and nearly 40 graduate students. He published a number of books, including *Theoretical Foundation of Offshore Oil Drilling Equipment, Fatigue Life of Offshore Oil Steel Structures, Reliability Analysis of Petroleum Steel Structures in Offshore Ice Region, Safety Reliability Analysis of Oil and Gas Long-Distance Pipeline, Fuzzy Probability Fracture Mechanics (Ocean Engineering)*, and other monographs. He has been engaged in more than 10 research projects including "Security and Reliability of Maritime Structures". Professor Fang at one time won the second prize of the National Science and Technology Progress Award, ten provincial and ministerial level scientific and technological progress prizes and holds two national patents. His 200 papers have been published, among which, more than 40 papers were published aboard and 30 papers published in SCI and EI. It can be said that Professor Fang Huacan has made a great harvest both in teaching and scientific research on offshore oil engineering. Indeed, he is a master in the field of offshore oil engineering.

Starting with The Environment and Environmental Load of Offshore Oil Engineering (Chapter 1) *Offshore Operation Facilities*, written by Professor Fang Huacan not only gives a thorough discussion on Offshore Oil and Gas Drilling Engineering And Equipment (Chapter 2), Marine Petroleum (Gas) Engineering and Equipment (Chapter 3), Submarine Pipelines and Pipeline Cable Engineering (Chapter 6) and Safety System Engineering for Offshore Oil (Chapter 7), but also delivers special analysis on Special Problems of Deep-Sea Oil and Gas Engineering (Chapter 4) and Special Problems in Sea Petroleum Engineering for Beaches and Shallow Sea Areas (Chapter 5). Comprehensive, systematic, rich, thorough, the book encompasses marine and petroleum technology and equipment, theory and practice. I highly recommended this book, whether for content as a reference or as a science and technology text. Therefore, it will be helpful for teaching, scientific research, and engineering staff. Though at ripe old age, Professor Fang Huacan does not forget his cause in offshore oil engineering. Instead, he still has a strong will to finish this book. This kind of high professionalism, indomitable will, perseverance, and rigorous scholarship is worth learning. As his former classmate and colleague, I pay tribute to him once again and wish him a happy birthday, health, and longevity.

Former Deputy Commander and Chief Engineer of Daqing Oil Field
Former Deputy Director and Chief Engineer of Offshore
Oil Exploration Bureau of Ministry of Petroleum Industry
Former Deputy Commander of Tarim Oil Field
Senior Engineer and Professor

Bingcheng Wang
July, 2009

Preface

After Dagang Oil Field was discovered in 1964, which was called Plant 641 at that time, the Chinese offshore oil industry started in 1965 when the Ministry of Petroleum Industry decided to "go to the sea." Although offshore oil began 70 years later in China than in the rest of the world, through decades of hard work, China's offshore oil and gas exploration and development has witnessed astonishing progress and thus has made great achievements that attract worldwide attention. China has discovered six big oil and gas basins in the $130*10$ m^2 offshore continental shelf of the Bohai Sea, South Yellow Sea, East China Sea, South Sea Pearl River Mouth, Yinggehai in Beibu Gulf, and the shallow water of Taiwan Strait. It is estimated that China has an abundance in offshore oil and gas resources with $(150 \sim 200) \times 10^8$ t of oil resources and 14×10^{12} m^3 of natural gas resources.

By creating the Foreign Cooperative Exploitation of Offshore Petroleum Resources Act in 1982, China established China National Offshore Oil Corporation (CNOOC, for short). Against this backdrop, China's offshore oil industry took a new step. Over the past 20 years, China has been following the principle of foreign cooperation while maintaining independent management; at the same time, China has insisted on the innovation of systems and technology, as well as following the path of internationalized development. In terms of offshore oil and gas development and production, China has achieved continuously high-speed and high-efficiency performance. Oil production developed from an annual output of 8×104 t in 1982 to 4293×104 t in 2008 (equivalent to total oil and gas production in that year). Take the CNOOC as an example: Its gross assets reached CNY 428.5 billion with its net assets reaching CNY 291.6 billion, and its sales volume, gross profit, and taxes paid that year reached CNY 198.3 billion, 67.7 billion, and 56.9 billion, respectively. This made a significant contribution to the national economy.

Moreover, most of the 20 oil and gas fields that CNOOC has put into production have made substantial growth in recoverable reserves. While under high and rapid exploration, it maintains the growth in oil and gas production, realizing better development indicators of oil and gas fields than the designed ones. In brief, in China the past decades of offshore oil development is a history of technological innovation. It is during that period that a scientific innovation system was established, the key and supporting technologies of production and management were obtained, and a large group of technology and management talents were nurtured, laying a solid foundation for further future development in the offshore oil industry.

Since so many achievements and experiences related to the offshore oil industry have accumulated, it is highly in need of a systematic and comprehensive book that encompasses both scientific theories and practice. From 1965, I transferred my work from in-land oil to offshore oil, teaching and researching into offshore oil engineering. I have a strong feeling that I am obliged to contribute by writing *Offshore Operation Facilities* to express my passion for it over the years.

Offshore oil engineering is a systematic engineering, made up of well drilling, oil extraction, oil and gas production and processing, oil and gas storage and transportation, and so forth. Therefore, from a systematic point of view, this book comprehensively explains offshore oil engineering as a whole. Chapter One discusses the computing theories and solutions of the structure of offshore oil engineering bearing the oil environmental loads, based on the characteristics of various wind, waves, currents, and sea ice on land. Chapters Two and Three focus on the unique ocean technologies of towing and positioning, floating well drilling, wellhead tieback completion, floating oil and gas production and oil and gas gathering and transportation, and drill string compensator, a dynamic positioning system, subsea production system, and offshore oil and gas production and processing facilities. The aim of these chapters is to tackle the high priority tasks of the offshore oil industry.

Chapter Four introduces the technologies and devices used for deepwater drilling, oil- and gas-gathering according to the demand for oil and gas from the deep sea such as dual gradient drilling, tension leg platform, and so on. Chapter Five is about beach sea petroleum engineering and special problems such as the artificial island and so on. Chapter Six is about subsea pipeline engineering, introducing specific design and construction methods, such as the stability designing calculation during and after laying, as well as pipe-laying vessels, and other specifics that are different from on- land work. Chapter Seven introduces safety topics on how to ensure the normal operation of offshore oil and gas fields, focusing on computing theories and solutions that analyze the safety and reliability of offshore oil engineering structures, fault tree analysis of offshore oil and gas fields, and theories and methods in risk analysis of offshore oil engineering. It took me two years to finish this book. During the long period of writing, I mainly paid attention to the following aspects.

First was the combination of oil and sea. Objective matters are interdependent. Various disciplines of natural science are also inextricable. Therefore, the cross-disciplinary overlapping has drawn wide attention. *Offshore Operation Facilities*, as a cross-disciplinary subject and a major combination of oil and marine science, is related to subjects such as ocean, petroleum, mechanics, civil engineering, and so on. Thus, this book must solve the problem of the organic integration of the two. For example, when talking about the vibration induced by sea ice, this book tries to combine the requirements for reducing ice-induced vibration in oil engineering with ice freezing and melting in marine science, ice mechanics in kinematics and mechanics, vibration theory in mechanical engineering, analysis of the vibration mitigation efficacy of the platform with ice cones in offshore oil engineering.

Second was the combination of the universal and the particular. Objective matters have both universality and particularity, so if we do not study the particularity of its contradiction, it is impossible to determine the special nature of an object that distinguishes itself from others. As a result, the main task of this book is to reveal the unique nature of offshore oil engineering that differs from onshore petroleum engineering. It is for this reason that the characteristics of the ocean have to be highlighted. For example, when introducing drilling engineering, the text emphasizes preparatory work before drilling that is different from onshore drilling; these include towing and placing, offshore floating drilling operating procedures, technology and equipment for underwater completions, refind wellhead and tieback operations for abandoned offshore wells to highlight marine features in general drilling engineering.

Third was the combination of technology and equipment. Generally speaking, the equipment is the method, which is supposed to serve for technology from being designed, used and even repaired. It can be said that if isolated from technology, equipment will lose its objective and its targeted service. As a Chinese saying goes: "with the skin gone, to what can the hair attach itself?" Therefore, the disadvantages of technology and equipment are separated. From this point of view, this book combines technologies in offshore oil drilling, production, transportation, oil- and gas-gathering and transportation, and offshore oil and gas production and processing, with their corresponding equipment.

Fourth was the combination of theories and practice. We all know that theory comes from practice while practice is the only criterion to test theory. Therefore, how to solve the problem of combining theory with practice was one of the important elements to consider while writing this book. When discussing engineering technologies, the onsite staff were consulted to make sure what is here is practical and feasible.

When it involved engineering equipment, great efforts were made to collect as much relevant information as possible so that the text could be up to date about future products; when it comes to theoretical results, detailed application procedures, methods, application examples are given so that the results can be put into use. In short, I have tried to make new progress by combining theory and practice.

Huacan Fang
March 2009

Acknowledgments

This book cites both domestic and foreign references in the chapters, and the authors of various works have made great contributions to the completion of it. I want to give my sincere gratitude to all of them. As some of the documents and postgraduate theses references are not published, they were not listed as citations. Here, I want to give them my apologies and thank the authors of them; they are: Doctors Chen Guoming, Duan Menglan, Luo Yanscheng, Yue Qianjin, Duan Zhongdong, among others; Masters Yu Xiongying, Xu Fayan, Xu Xingping, Sui Xinzhong, Zhao Xuenian, Jia Xinglan, and others. My thanks also go to enterprises and institutions such as the China National Offshore Oil Corporation, the Bohai Branch Corporation, Sea Petroleum Branch, Oil Nanhai West Branch, Offshore Engineering Inc., Chinese Petroleum Group Economic Research Institute, and CNOOC Limited Research Center in Beijing. Thanks are also due to senior engineers, such as Yu Xiongying, who provided me with precious field experience and various references for this book.

The recommendations of Gu Xinyi, Academician of the Chinese Academy of Engineering and Professor Zhao Guozhen promoted the publication of this book; it also received great support from the leadership of China University of Petroleum (Beijing), including from Professor Zhang Laibin, Professor Wu Xiaolin, and the leadership of the Department of Mechanical and Electrical Engineering such as Professor Zhang Hong, Professor Wang Deguo, and also Lu Xiumei, Liang Wei, as well as Professor Duan Menglan of Offshore Oil and Gas Research Center, the leadership of China University of Petroleum (East China), Professor Tong Xinghua, Professor Qi Mingxia, along with the Department of Mechanical and Electrical Engineering and the Marine Oil Drilling Equipment Laboratory. In addition, thanks to Editor He Li from Petroleum Industry Press. Many thanks to all of them!

Later, Elsevier Inc agreed with the Petroleum Industry Press and initiated translating it into in English as *Offshore Operations Facilities: Equipmet and Procedures*. The Offshore Oil and Gas Research Center at China University of Petroleum (Beijing) paid careful attention to the process of translation, and put in a lot of hard work. At the same time, I also want to express sincere thanks to Elsevier Inc.

March 2009

Synopsis

This book discusses offshore oil engineering systematically and comprehensively from the feature perspective of the ocean. It is composed of seven chapters. The main contents include the environment of offshore oil and gas engineering and environmental loads, offshore oil and gas drilling and production engineering and equipment, special problems in deepwater and beach-shallow sea oil and gas engineering, subsea pipeline engineering, and safety system engineering.

This book is written for scientific and technical personnel and the management who work in petroleum engineering, offshore oil engineering, and the related industries; it can also be used as a reference by teachers and students of other higher education specialties.

The Environment and Environmental Load of Offshore Oil Engineering

Although collectively referred to as the "sea," seawater with depth less than $\sim$ 2000–3000 meters is commonly known as the sea, while that with a depth more than 3000 meters is known as the ocean. Ocean is the core of the sea, and accounts for approximately 89% of the total area of seawater. The total volume of seawater nowadays is about 13.7×10^8 m^3, making up around 97% of the total water on the Earth, and the total area of seawater across the world occupies 78.9% of the Earth's surface, the area of which is 5.11×10^8 km^2. The ocean is rich in natural resources, among which oil and natural gas are some of the most important strategic resources. The world's offshore oil production began in the 1940s. In the 1960s, offshore oil production reached 1 million barrels per day. In 2005, it reached 25 million barrels per day (almost 1.25 billion tons per year), which accounts for one-third of world crude oil production. At present, more than 2000 offshore oil and gas fields are found all over the world. According to the statistics by the Cambridge Energy Research Associates, offshore oil production accounted for 33% of global oil production in 2009, and this proportion will rise to 35% in 2020. Offshore gas production accounted for 31% of global gas production in 2009, and this is estimated to rise to 41% in 2020.

China is a large country in terms of both land and sea. China has a coastline of more than 18,000 km, and the island coastline of Hainan, Taiwan, etc. reaches 2,300 km. According to the United Nations Convention on the Law of the Sea, the sea area under China's jurisdiction is around 488×10^4 km^2, and the amount of oil and gas reserves reaches about $(40 \sim 50) \times 10^8$ t. The continental shelf is the extended perimeter of each continent. After preliminary investigation, more than 300 sedimentary basins available for exploration (covering around 450×10^4 km^2) have been found in the immense area of the continental shelves of the China Bohai Sea, China Yellow China Sea, and East China Sea, of which the marine sedimentary layers occupy an area of about 250×10^4 km^2. About 10 large-scale offshore oil and gas basins have been found in China, namely, Bohai Basin, North Yellow Sea Basin, South Yellow Sea Basin, East China Sea Basin, West Taiwan Basin, Pearl River Mouth Basin of South China Sea, the southeast part of the basin near the Ryukyu Islands, the Beibu Gulf Basin, Yinggehai Basin, and Taiwan Shoal Basin. According to the third petroleum resource assessment results in 2008, China's offshore oil resources measure 24.6 billion tons, which accounts for 30% of the total oil resources in China. And the offshore natural gas resources are 16 trillion cube meters, which accounts for 30% of the total gas in China. In the above-mentioned offshore oil and gas resources in China, 70% of them are in the deep-sea area.

Though only since the late 1950s has China begun shifting oil and gas research and exploration from the land to the sea, the country has seen rapid progress in this process. By the end of 2002, oil and gas investigations in all seas and waters and the evaluation of oil and gas resources in the major basins

Offshore Operation Facilities: Equipment and Procedures. http://dx.doi.org/10.1016/B978-0-12-396977-4.00001-9

have been completed. Now, 235×10^4 km seismic lines have been acquired (with the three-dimensional lines being 153×10^4 km long, and the two-dimensional lines 82×10^4 km); 817 wildcat and appraisal wells have been completed (with 513 wildcat wells and 304 appraisal wells); 207 geological structures containing oil and gas have been discovered; and 52 oil and gas fields have been found. As of the end of 2008, there are more than 80 oil and gas fields in China's sea area. In 2008, the annual oil production was 189 million tons and the annual nature gas production was 74.49 billion cubic meters. To the end of 2008, cumulative oil production is 5.02 billion tons and cumulative nature gas production is 940 billion cubic meters. It is predicted that by 2020, China's offshore oil production will reach 37 to 41 million tons.

Offshore oil engineering consists of a process that starts with offshore exploration for oil and gas, including exploration, development, and production, and continues through to the transport of the oil and gas to land. The environment of offshore oil engineering can vary greatly. It is influenced not only by the climate (similar to the land), but also the severe environmental conditions of the sea, for instance, waves, currents, sea ice, typhoons, monsoons, etc.

Since offshore oil engineering is carried out in the harsh conditions of the marine environment, engineering structures, such as drilling platforms, oil extraction platforms, floating production platforms, production storage and offloading vessels, single point mooring devices, and submarine pipelines and cables, have to withstand the marine environment load coming from the wind, waves, currents, sea ice, and even earthquakes, tsunamis, etc., either in its construction or in use. These marine environmental loads may cause enormous damage, and even affect the normal operation of offshore oil engineering and the production in offshore oil and gas fields. For example, the wind load that hurricanes on the sea exert on offshore platforms can severely affect the security of the platforms. On November 25, 1979, China's "Bohai Sea II" drilling platform was capsized due to the wind load during towage for well relocation. Similarly, on December 25, 1983, the "Java Sea" drillship rented by the U.S. Atlantic Richfield Company was turned over in the South China Sea during its operation because of a typhoon. In the Gulf of Mexico, the hurricane in October 1964 destroyed more than 30 platforms and the hurricane in October 1967 led to many cases of platform collapse. Moreover, wave loads can have a destructive impact on offshore platforms. In August 1980, four drilling platforms in the Mexico Gulf were destroyed by great tidal waves; in November 1989, the American drillship "Seacrest" was also overturned by huge waves. By 1989, statistics showed that more than 50 offshore drilling platforms and ships worldwide have been engulfed by the sea. Based on past experience, the marine environment conditions of where the platform settles in has an important impact on the fatigue of the structure. For instance, the wind and waves are generally stable in the Gulf of Mexico, but hurricane can occur during a certain part of the season, so the low cycle fatigue should be taken into consideration. And in the North Sea, the marine environment condition is difficult, so the high cycle fatigue should be considered. In 1966, when periodically checking the four-legged fixed platform that settled in the North Sea after 6 years, it was 600 mm lower than the low-tide level waterline, and the pole was entirely out of the chord, which is caused by high-cycle, low-stress fatigue crack in the section observed. At the same time, there was a large amount of corrosion products attached to the pole, which made the out diameter of the pole change from 300 mm to 700 mm (with a wall thickness of 12 mm). Consequently, the resistance force was greatly increased. Also, the other attachments on the platform increased the resistance. In addition, the destructive force of the sea ice load cannot be ignored. The China Bohai Sea is an ice zone, and its ice-frozen period lasts approximately three months in winter. During the wintertime, the minimum temperature is about $-22\ ^{\circ}$C, and the maximum wind speed is about 32 m/s. Approximately every 10 years, there will be a normal/severe ice regime in the China Bohai Sea. Its sea ice thickness is normally ~20–40 cm, and sometimes it can be ~40–60 cm, and even up to 80 cm.

In the China Bohai Sea, there have been profound lessons learned from the ice load acting on China's offshore oil engineering structures. Take the old "No.2" oil platform as an example. This platform is composed of a life platform, an equipment platform, and a production platform. In 1969, the China Bohai Sea was severely frozen. In February of that year, the life platform was the first to be overturned by the ice, and then the equipment and production platforms had both been turned over by March 8th. In 1974 a beacon tower in the China Bohai Sea was torn down by ice. Also, in 1979, one of the legs of a self-elevating drilling platform was broken by the ice load, resulting in a serious accident. All the above-mentioned incidents illustrate the significant influence of ice load on engineering structures.

In conclusion, obviously it is of great necessity and importance to study the environment and the environmental load of offshore oil engineering. Therefore, based on this topic, the first chapter is intended to introduce the combinations of wind and wind load; tide, current, and current force; wave and wave force; sea ice and ice load; and earthquake force and the combination of it with other wave forces.

1.1 Environment of Offshore Oil Engineering

The following sections give some background on the environment of offshore oil engineering.

1.1.1 Natural Environmental Conditions

Generally, the natural environmental conditions, such as wind, wave, tide, current, earthquake, and tsunami, may affect offshore oil engineering during construction. In addition, we must consider rain, snow, fog, frost, temperature, foundation of soil, seawater corrosion, and marine life.

1.1.2 Designed Environmental Conditions

1. Working environmental conditions
   Working environmental conditions refer to the natural environments that often appear during the production and operation of offshore oil engineering. However, it should be noted that these conditions should be selected to ensure normal construction and operations.
2. Extreme environmental conditions
   Extreme environmental conditions refer to natural conditions that are extremely serious and rarely arise in the service life of offshore oil engineering after putting the system into production. But the extreme environmental conditions should be carefully considered to ensure the safety of the project.

1.1.3 Recurrence Interval

The recurrence interval refers to the length of time between two occurrences of extreme environmental conditions. For offshore oil engineering, the recurrence interval is generally not less than 50 years.

However, the 50-year recurrence interval refers to the recurrence interval considering all extreme environmental conditions except the earthquake. In addition, for some small structures of offshore oil engineering, such as unattended wellhead platforms, the recurrence interval should be two to three times the designed service life of structures, but should not be less than 30 years.

1.1.4 Marine Environment Loads

Marine environment loads refer to the loads that the natural marine environment place on offshore oil engineering.

1.1.5 Classification of the Loads on Offshore Oil Engineering

The loads on offshore oil engineering can be divided into three categories:

1. Environment loads
   Environmental loads refer to the loads mainly caused by marine natural environmental conditions, such as wind, wave, current, ice, earthquake, etc.
2. Operation loads
   For offshore oil engineering in the period of service, all other loads except environmental loads are operation loads. These usually include:
   - Fixed loads, such as the static load caused by gravity
   - Live loads, the static load of increased or decreased weight
   - Dynamic load
3. Construction loads

Construction loads refer to the loads on offshore oil engineering during construction. These are often temporary and unique.

1.1.6 The Combination of Marine Environment Loads

A combination load is the superposition of a variety of loads from marine environmental conditions. Generally, for offshore oil engineering, it is the combination for the worst possible case. But it does not count the earthquake load, because it is so occasional.

When combining loads, specific analysis must be done based on the actual circumstances, as follows:

1. Different stages
   For example, during the construction stage or the working season after the project is put into operation, the worst natural marine environmental conditions that might appear are not the same, so the loads should be combined according to the actual situation.
2. Different objects
   For different designs of offshore oil engineering, there are also different load combinations. For example, when designing an artificial island or designing local components of a leg of an oil platform, the load combinations of the marine environment loads considered should be different.

1.2 Wind and Wind Load

The force that atmosphere exerts on a unit area of the Earth's surface is called the atmospheric pressure, or air pressure. Due to the uneven distribution of air pressure in the horizontal direction, the air moves from high-pressure areas to low-pressure areas; this air movement is called wind. The reason for the uneven distribution of pressure in the horizontal direction is generally the changes in the air density and the region of air flowing in is greater than the outflow.

1.2.1 Wind Characteristics

The wind usually is characterized by two quantities, the wind direction and wind velocity.

1.2.1.1 Wind Velocity

Wind velocity is the distance of air flow in unit time, generally with m/s or km/h as the unit. According to the measurement of the wind velocity, it is often divided into 12, known as the Beaufort wind scale. Later, a few levels are added and it becomes a common wind scale, as shown in Table 1-1.

Table 1-1 Universal Wind Scale

Wind Scale	Wind Name	State of the Sea	Sea Wave Height, m		Equivalent Wind Velocity		
			General	Highest	n mile/h	km/h	m/s
0	No wind	As a mirror			<1	<1	0
1	No wind	Wavelet	0.1	0.1	1~3	1~5	0.3~1.5
2	Breeze	Small wave	0.2	0.3	4~6	6~11	1.6~3.3
3	Gentle breeze	Small wave	0.3	1.0	7~10	12~19	3.4~5.4
4	Moderate breeze	Light waves	1.0	1.5	11~16	20~28	5.5~7.9
5	Cool breeze	Waves	2.0	2.5	17~21	29~38	8.0~10.7
6	Strong breeze	Big Wave	3.0	4.0	22~27	39~49	10.8~13.8
7	Moderate breeze	Billow	4.0	5.5	28~33	50~61	13.9~17.1
8	Gale	Wild waves	5.5	7.5	34~40	62~74	17.2~20.7
9	Strong gale		7.0	10.0	41~47	75~88	20.8~24.4
10	Whole gale	Turbulent	9.0	12.5	48~55	89~102	24.5~28.4
11	Storm wind	Can't imagine	11.0	16.0	56~63	103~117	28.5~32.6
12	Hurricane		14.0	>16.0	64~71	118~133	32.7~36.9
13			>14.0		72~80	134~149	37.0~41.4
14					81~89	150~166	41.5~46.1
15					90~99	167~183	46.2~50.9
16					100~108	184~201	51.0~56.0
17					109~118	202~220	56.1~61.2

Note: 1 n mile = 1852 m.

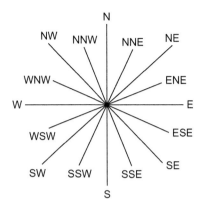

FIGURE 1-1

Orientation diagram for wind direction.

1.2.1.2 *Wind Direction*

The wind direction is that from where the wind comes. In meteorology, 16 azimuths are usually used to describe wind directions: N (North), NNE (North of Northeast), NE (Northeast), ENE (East of Northeast), E (East), ESE (East of Southeast), SE (Southeast), SSE (South of Southeast), S (South), SSW (South of Southwest), SW (Southwest), WSW (West of Southwest), W (West), the WNW (West of Northwest), NW (Northwest), and NNW (North of Northwest), as shown in Figure 1-1.

1.2.1.3 *Wind Rose Graphics*

The image that shows the wind direction, wind velocity (wind strength), and the percentage frequency of the wind in a region is called a wind rose graphic, as shown in Figure 1-2.

1. Wind direction
   Draw 16 lines in the wind rose graphic, with each lines representing one of the 16 wind directions, as shown in Figure 1-2 (for 8 directions).
2. Wind strength
   The width of the rectangle along the 16 sections of wind rose represents the strength of the wind. The wind strength can be shown by the wind velocity, and the Beaufort wind scale. The rectangle width expresses the graduated scale of the wind strength, as the scale shown in Figure 1-2, which represents a Beaufort wind scale from 0 to 17. The rectangular length represents the wind velocity (m/s) at 10 m above the sea surface, marked on the right of the rectangle.
3. The frequency percentage of the wind
   The equally spaced concentric circles in the wind rose graphic can represent the percentage frequency of the wind. Starting from the center of the circle, each interval is 5%. Figure 1-2 uses four numbers (5, 10, 15, 20) represented by concentric circles to display the wind frequency percentage, and the number marked out near the center of the first concentric circles represents the percentage frequency of wind in the eight directions corresponding to the eight lines.

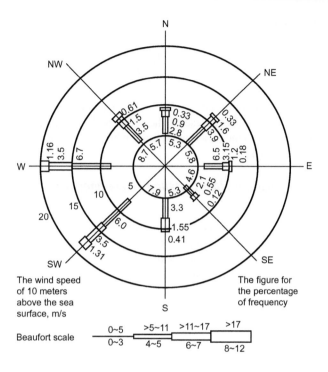

FIGURE 1-2

Wind rose graphic.

1.2.2 Design Wind Velocity

The observed value of the wind velocity, which is calculated according to the design specification standard, is called the design wind velocity.

1.2.2.1 Prescriptive Observation Height

Usually the height of the design wind velocity at the observation point is 10 meters above the sea surface. Therefore, if the height observation point is any height Z, it should convert observed average wind velocity $\bar{u}_z$ into the average velocity $\bar{u}_{10}$ that is 10 m above the sea surface. Each of the following two equations generally can be applied.

1. Logarithmic conversion equation

$$\bar{u}_z = \bar{u}_{10}\left[1 + \frac{(c_{10})^{\frac{1}{2}}}{k}\ln\frac{Z}{10}\right] \tag{1-1}$$

2. Index conversion equation

$$\bar{u}_z = \bar{u}_G \left[\frac{Z}{Z_G}\right]^{\alpha} \tag{1-2}$$

where K represents parameters, generally $K = 0.4$; C_{10} represents a coefficient, generally $C_{10} = (0.80 + 0.114\bar{u}_{10}) \times 10^{-3}$; α represents the index and for the open sea $\alpha = 0.16$; Z_G represents the height from the sea to the observation point where the average wind velocities close to a constant, generally $Z_G = 300\ m$.

1.2.2.2 Prescriptive Observation Time-Distance (Time Interval)

1. Time interval
Time interval refers to the length of time between each observation. There are the following two time interval standards for offshore oil engineering in the specification:
  - 1 min time interval: Mainly used in the local component design of offshore oil engineering, and the observed average wind velocity is $\bar{u}_1$.
  - 10 min time interval: Mainly used in the overall design of offshore oil engineering, and the observed average wind velocity is $\bar{u}_{10}$.
2. The mean wind velocity
As the wind velocity is observed in a certain time interval, the,wind velocity is changing, so either $\bar{u}_1$ or $\bar{u}_{10}$ are average values. The average value is collectively referred as $\bar{u}$, so

$$\bar{u} = \frac{1}{T} \int_{t_0+\frac{T}{2}}^{t_0+\frac{T}{2}} u dt \tag{1-3}$$

where T represents time interval; t_0 represents half of the time interval; and u represents the wind velocity at an instant dt.

1.2.2.3 Prescriptive Number of Observations
In China, the number of observations of wind velocity is based on the following two conditions:

1. Timed observations
Generally these can be taken 3 times, 4 times, or even 24 times a day.
2. Automatic observations
This refers to recording observed values automatically, and in this situation it is continuous day and night.

1.2.2.4 Prescriptive Maximum Average Wind Velocity
In accordance with the prescriptive time interval and the number of observations, take the maximum average wind velocity values with different numbers as the maximum average wind velocity. Obviously, if the specified number of observations is more, the maximum average wind velocity value received from the timing observation is closer to automatically recorded values.

1.2.2.5 Prescriptive Maximum Sustained Wind Velocity

For the offshore oil platform structure, the design wind velocity is the maximum sustained wind velocity in accordance with the technical specifications of the China Classification Society. When the duration of one wind velocity reaches the prescriptive time interval and it is the maximum value in the automatically recorded long-term observations, this wind velocity is taken as the maximum sustained wind velocity. If lacking of long-term recording data of observed wind velocity, the maximum sustained wind velocity of 2 min for four-time daily observations must be converted to the maximum sustained wind velocity of 10 min for annual automatic recording observations of 10 min. Then the designed wind velocity can be obtained by mathematical statistics.

1.2.2.6 Prescriptive Conversion of Onshore to Offshore Design Wind Velocity

When there is a lack of long-term observation data of offshore wind velocity, we can use the neighboring onshore long-term wind velocity data, multiplied by the wind velocity increase coefficient for offshore wind velocity. This increase coefficient is the ratio of offshore wind velocity and onshore wind velocity, as shown in Table 1-2.

1.2.3 Wind Pressure

For one measure of wind velocity (average velocity), the pressure (N/m^2), which is perpendicular to the plane of wind direction, is the air pressure corresponding to the wind velocity. The wind pressure is one of the fundamental measures for the design of offshore oil engineering. During design we must consider the tractive force including air pressure and overturning torque on the fixed end.

1.2.3.1 Steady Wind Pressure and Fluctuating Wind Pressure

Because the air movement over a large range on the ocean is very complex, the effect of wind pressure on the surface of offshore oil engineering changes along with wind direction, wind velocity, and the wind's turbulence structure, which makes it difficult to choose wind pressure value in the design. In order to solve this problem while engineering, we need a way to simplify the complex process, such as taking long-term changes of the wind as a quasi-static load, or taking the short-period changes with little energy as a quasi-dynamic load. The former air pressure is called the stable air pressure, and the latter is known as the fluctuating wind pressure.

Table 1-2 Land and Sea Wind Relationship

Distance to the Coastline, km	Wind Velocity Ratio of Sea and Coastal Onshore
<2	<1.10
2~30	1.10~1.14
30~50	1.14~1.22
50~100	1.22~1.30
>100	According to the measured or survey data

1.2.3.2 Calculation of Steady Wind Pressure

The steady wind pressure is the pressure acting on the surface of the engineering structure within a certain time interval when taking the wind velocity and direction as not changing over time, calculated as follows:

1. Basic wind pressure

   For an air current with a certain wind velocity, the dynamic pressure in the flow process is known as the basic wind pressure, represented by p_0. The air in the flow process obeys the law of energy conservation. That is, the sum of the pressure energy composed by static pressure p and the kinetic energy ($\frac{1}{2}mu^2$) composed by wind velocity u and mass m should be a constant (Bernoulli equation). Obviously, if the velocity of flow in kinetic energy decreases, then the dynamic pressure is reduced, and the static pressure increases. Until all the dynamic pressure is converted to static pressure, the maximum static pressure of the fluid is called the total pressure, expressed by p_t. In this way, the total pressure can be written as the sum of static pressure and dynamic pressure, that is

$$p + \frac{1}{2}\frac{w}{g}u_t^2 = p_t \tag{1-4}$$

   Therefore, basic wind pressure p_0 is the dynamic pressure of air in the flow process, which should be the difference between the total pressure p_t and the static pressure p:

$$p_0 = p_t - p = \frac{1}{2}\frac{w}{g}u_t^2 \tag{1-5}$$

   where u_t represents the design wind velocity with t minutes time interval, m/s; w represents the weight density of the air (gravity); and generally $w = 12.015\text{N/m}^3$.

   If we set g in Equation (1-5) to 9.8m/s^2, and take $\alpha = \frac{1}{2}\frac{w}{g} = \frac{1}{2} \times \frac{12.015}{9.8} = 0.613$ as the wind pressure coefficient, the basic wind pressure calculation equation is

$$p_0 = \alpha u_t^2 \tag{1-6}$$

2. The wind pressure on the surface of the engineering structure

   When we calculate the wind pressure acting on the engineering structure, we should consider the effect of other factors. Considering that the air pressure changes with height, we should multiply the basic wind pressure P_0 by the wind pressure height variation coefficient K_g (as shown in Table 1-3). Considering the effect of the size, scale, and shape factors of engineering structures, we should multiply the basic wind pressure by the shape coefficient k of the structure (as shown in Table 1-4). Thus, the air pressure P_w on the surface of engineering structures can be written as

$$p_w = k_g k p_0 = k_g k \frac{1}{2}\frac{w}{g}u_t^2 = k_g k \alpha\, u_t^2 \tag{1-7}$$

1.2.3.3 Calculation of the Fluctuating Wind Pressure

Fluctuating wind pressure, also known as wind-induced vibration, is the dynamic response of engineering structures to the vibration caused by the action of fluctuating wind. For a towering engineering structure, because of its lower stiffness, the fundamental natural vibration period is longer. If it is under

Table 1-3 Sea Wind Pressure Coefficient

Height above Sea Level, m	≤2	5	10	15	20	30	40
K_g	0.64	0.84	1.00	1.10	1.18	1.29	1.37

Height above sea level, m	50	60	70	80	90	100	150
K_g	1.43	1.49	1.54	1.58	1.62	1.64	1.79

Table 1-4 Wind Pressure Coefficient (k)

Form Factor	k
Cylindrical	0.5
Hull (surface type)	1.0
Deckhouse	1.0
The independent structural shape (cranes, angle, channel, beam)	1.5
Below deck area (smooth surface)	1.0
Below deck area (exposed beams and trusses)	1.3
Rig derrick (each side)	1.25

the action of fluctuating winds, especially strong winds, its pulsating displacement obviously increases. Therefore, for these towering engineering structures, in addition to considering the stable air pressure generated by the average wind velocity, we should also consider the fluctuating wind pressure generated by the fluctuating wind velocity.

Due to the complexity of the wind problem, although the fluctuating wind pressure acting on the towering engineering structures is simplified as a quasi-dynamic load, it is still not completely improved and remains to be further studied. So today in design we more often turn it into an equivalent static load by a method using a so-called dynamic coefficient. This method is simple and practical.

The first-order free vibration period of towering engineering structures with $T \geq 0.5s$ is

$$p_w = \beta p_0 \tag{1-8}$$

In this equation, β is the wind vibration factor, and it can be determined from Table 1-5 depending on the natural vibration period T; p_w and p_0 have the same meaning as before. Equation (1-8) shows that the design wind pressure should be β times than the basic wind pressure when considering wind-induced vibration for a towering engineering structure.

The natural vibration period of the important towering engineering structures is $0.25\ s \leq T \leq 0.5\ s$.

In this case, although the natural vibration period is shorter, we should also consider wind vibration coefficient β. When $T = 0.25\ s$, take $\beta = 1.25$; the β value shall be determined by an interpolation method when $0.25\ s \leq T \leq 0.5\ s$.

Table 1-5 Wind Vibration Coefficient

	Vibration Period T, s					
	0.5	1.0	1.5	2.0	3.5	5
Type of Structure						
Steel structure	1.45	1.55	1.62	1.65	1.70	1.75
Reinforced concrete structure	1.40	1.45	1.48	1.50	1.55	1.60

1.2.4 Wind Load

The wind load is horizontal tractive force of the wind pressure that acts on engineering structures, including the overturning torque on the fixed end.

1.2.4.1 Calculation of Wind Load

Set the wind area of the offshore oil engineering structure (the projected area of structure contour that is perpendicular to the wind) to A; then the horizontal tractive force (wind load) F_w on engineering structures caused by air pressure is

$$F_w = K_g K p_0 A = K_g K \alpha \, u_t^2 A \tag{1-9}$$

1.2.4.2 Calculation of the wind area

Because an offshore oil engineering structure is complex, we should classify the calculation of the wind area. Taking an offshore oil platform as an example, according to the different local geometric shapes of the structure, the calculating method can be divided into the following four categories:

1. The first category includes the drilling derrick and lifting parts. They can be regarded as being composed of small diameter cylinder components. The wind area of each cylindrical member should be calculated according to the projection area. A derrick structure can be seen as a space truss structure; the wind area is 60% of the total area.
2. The second category mainly includes the living area and office structures, and they can be regarded as composed of the surface of the rectangular area. For a group of houses, the wind area is calculated as a superposition of the projection area.
3. The third group includes the deck and helipad, and they can be regarded as consisting of the end edge of a smooth horizontal plane. The wind area should be calculated by the area of the end edge.
4. The fourth category includes the support structures of the platform. Generally, the calculation of the wind area is in accordance with the projected area of the support structure.

1.2.4.3 Recommended Value of the Design Wind Velocity

To calculate the wind load with Equation (1-9), we need to know the design wind velocity value u_t. The "Offshore Drilling Vessel Design Specification" in China lists the recommended values of design wind velocity mainly for the sea of China, as shown in Figure 1-3. Figure 1-3(a) is the classification number of the sea area. Figure 1-3(b) is the design wind velocity under extreme conditions with a time interval of 1 min. Figure 1-3(c) is the design wind velocity under normal conditions with a time interval of 10

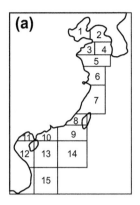 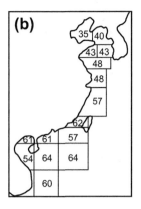 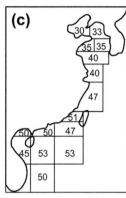

FIGURE 1-3

The recommended value of the design wind velocity in the main sea area of China: (a) the classification and labeling of the sea area; (b) the design wind velocity under extreme conditions; (c) the design wind velocity under normal conditions.

minutes. They all give the average maximum wind velocity of 10 meters above the sea surface, of which the recurrence interval is 30 years.

1.3 Tides, Currents, and Current Force

The rhythm of the sea tide is variability and periodic. It has an important influence on the planning and the construction of offshore oil engineering and the design, construction, and operation of engineering structures. Current is one of the main dynamic phenomena in the sea. Like wind and waves, it acts on the offshore oil engineering structure directly and has an important influence on the strength and stability of the structure. Therefore, this section will briefly introduce these conditions of the natural marine environment.

1.3.1 Tides

We will start with a review of tides.

1.3.1.1 Basic Concepts

1. Tides and tidal currents
   The phenomenon of the cyclical lifting movement formed by the sea surface that changes over time is called tides. The tidal current is the movement of water in the tidal wave, that is, the cyclical current of seawater in the horizontal direction that is simultaneous with the rhythm of the tide under the action of the horizontal tidal force. Where tide appears, there must be tidal current. Tides and tidal currents are more obvious in coastal and shallow water areas.
2. High tide and low tide
   In every period of tide lift and drop, the highest water level is known as the high tide; conversely, the lowest water level is called the low tide.

3. Tide hour and tide level
The time when tide occurs is called tide hour, such as high tide hour and low tide hour. The height from the tide surface to the tidal height datum plane at a certain moment is called tide level (e.g., high tide level and low tide level).

4. Tidal range
The water head of adjacent high and low tides is called the tidal range. The water head from low tide to high tide is called the rising tidal range; the water head from high tide to low tide is called the ebb tidal range.

5. The datum plane and characteristics of tide level

a. The datum plane
In order to determine the water depth, tidal level, and topographic survey, we need a starting surface to measure elevation, and this starting surface is called the datum plane. In China, the unified elevation datum plane is stipulated to be the mean sea level of the China Yellow Sea.

b. The mean sea level
The average height of the sea in a certain period is called the mean sea level. The calculation method is to take the arithmetic mean after getting the sum of hourly tide levels for a certain period. Mean sea level is generally divided into daily mean sea level, monthly mean sea level, annual mean sea level, and multiyear mean sea level. Because of the tide with change period long as 18.61a, when calculating the mean sea level, it should take datas of tidal levels recorded for 19a, at least 9a. When comparing the mean sea level calculated by the continuous 30-day tide information with the long-term mean sea level, there is often a difference of about ±50 cm.

c. Characteristic tide level
Characteristics tide level refers to the statistical characteristics of the tide. Usually in engineering, these are the highest tide level and lowest tide level (historically observed), the mean highest tide level and the mean lowest tide level (average annual maximum tide level and the lowest value), mean high tide and mean low tide of spring tide (mean value for years of high tide and low tide of spring tide and neap tide, which occurs two times a month) and mean high tide and mean low tide of neap tide.

1.3.1.2 Tide-Generating Force and Tidal Origin

1. Tide-generating force
The tide-generating force is the force that attracts the water covering the Earth's surface and results in a tide. It consists of an attractive and a centrifugal force.

a. Attractive force
This is the force that acts on each particle of water covering the Earth's surface and is caused by the attraction of celestial bodies, mainly the Moon. In accordance with the laws of gravity, the size of this attractive force is proportional to the mass of the Moon and the mass of the particle on Earth, and is inversely proportional to the square of the distance between the moon and the particle.

b. Centrifugal force
This refers to the inertial centrifugal force generated by rotation around a common barycenter of the Earth and the Moon. The centrifugal force acting on each particle on the Earth's surface

is equal, and the direction and the size of it is the same as the inertial centrifugal force acting on the particle at the center of the Earth.

The resultant force of the above-mentioned attractive force and centrifugal force is called the tide-generating force.

2. Tidal ellipse

The spheroid of homogeneous seawater covering the Earth's surface that is under the tide-generating force is called the tidal ellipse. The formation of the tidal ellipse is the result of the balance of gravity, tide-generating force, and pressure difference at sea level. If a homogeneous water layer covered the Earth's surface, under the tide-generating force, the water would have no inertia and wouldn't be affected by friction forces and eccentric force caused by the Earth's rotation. This hydrosphere will become the tidal ellipse, as shown in Figure 1-4.

During the rotation of Earth, a point on the surface is at the locations A → B → C → D → A shown in Figure 1-4 at a different time within a day. So, there are two high and two low tides in a day.

1.3.1.3 The Design Tidal Level

High and low tide level are important pieces of hydrological data in the design of offshore oil engineering. This not only directly affects the determination of the elevation of the engineering structure and depth of the construction sea area, but also affects the selection and design calculation of the engineering structure type. Therefore, it should be considered an important parameter.

1. The design high level and design low level

The design high level and design low level refer to the high level and the low level of the offshore oil engineering or its structure under normal conditions. According to the related technical code for engineering in China, the following provisions have been made:

a. The design high level, provisions said that it should regard the water level that cumulative frequency of high level is 10% as a design high level.

b. The design low level, provisions said that it should regard the water level that cumulative frequency of low level is 90% as a design low level.

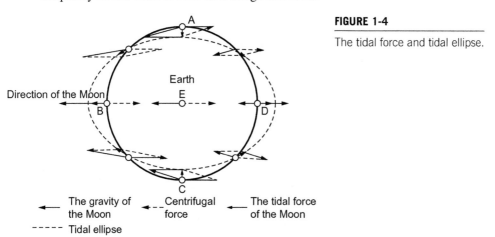

FIGURE 1-4

The tidal force and tidal ellipse.

That is to say, for the design one does not need to know the high tide and low tide in a certain day. The design water level can be confirmed using the statistical law of change of tide. Generally, in accordance with the statistics data of high tide level and low tide level in one or more years, the difference of the maximum and minimum values is divided at intervals of 10 cm. Then get a frequency and corresponding cumulative frequency for a water level. The cumulative frequency (%) is obtained according to the corresponding cumulative frequency and the total number of the values. Last, determine the designed high level and designed low level according to the 10% and 90% levels of the cumulative frequency. Because the difference between the statistical data for a year and years is less than 10 cm, it is usually the statistical data for a year is used.

2. Checking high level and low level

This involves checking the high or low water level of the offshore oil engineering or its structure under abnormal operating conditions. This water level is usually not caused by astronomical factors alone, but the combination of astronomical tide and the increase or decrease of water level caused by cold waves, typhoons, low pressure, earthquakes, and tsunami. It is used as a safety check in the design of engineering structures under extreme conditions.

According to the provisions of relevant technical specifications in China, the annual frequency statistical method is usually used to determine the appropriate water level to check. That is, in the sea where there are high tide level and low tide level with recurrence interval of more than 20 years, a frequency analysis method is often used to calculate high tide level and low tide level with a recurrence interval of 50 years and take the result as the checking level.

It should be pointed out that a water level with a recurrence interval of 50 years is not just once each 50 years, but over a long period time, the average probability of an appearance of that water level over 50 years is 1.

3. The reference value of the design tide level in each sea area of China

When the measured tidal level data of a year or years is missing and someone needs the rough level for an emergency, the following sea level in each sea area along the coast of China can be taken as a design high tide level.

   - The China Bohai Sea: In the Yingkou area, it is 3.0 m. In the Qinhuangdao area, it is 0.9 m. In the Laizhou Bay and Bohai Strait area, it is 1.5 to 2.5 m. In the center of the China Bohai Sea, it is generally less than or equal to 2.0 m.
   - The China Yellow Sea: The sea level is less than 2 m.
   - East China Sea: In Zhejiang and along the northern coast of Fujian, it is ~4–5 m. Generally along the coast of Fujian, it is 5 m. Near Wenzhou, the maximum of is about 8 m.
   - The South China Sea: Along the east and west coast of Taiwan and in the Guangdong Jia Zi area, it is ~1.5–2.5 m. From Shantou, Guangdong to Dapeng Bay, the average is ~1–1.5m. From Dapeng Bay to Zhanjiang, it is ~2–3 m. From the Qiongzhou Strait westward to the Northern Gulf, it is ~4–5.5m. In the southeast of Hainan Island, it is ~1–2.0m, and it is ~3–4.0 m in the northwest.

1.3.2 Currents

In the sea, the transfer of a large-scale mass of seawater from one area to another area in the horizontal or vertical direction is known as the current. Currents are comprehensively formed by the complex overlap of short-term, occasional, and nonperiodic seawater transfer. A current's velocity and direction is constantly changing with the transfer of time and space. The following will describe the classification of ocean currents, the characteristics of the currents in China, and the calculation of current velocity.

1.3.2.1 Classification of Currents

Currents can be classified according to their origin, length of time, the temperature, the salinity value of current water, etc. We will briefly introduce current classification here.

1. Wind current

   The movement generated by the shear stress acting on the sea surface when wind blows on the sea is called wind currents. It can be divided into the following categories:

   a. Drift current

   This refers to the constant current (blow current) caused by the prevailing wind in a large-scale spatial-temporal range. Drift current is the steady constant movement of surface waters in the ocean. The formation of the constant movement is the result of the balance of the angular force caused by the Earth's rotation, the friction force between the still lower water and the upper moving water, and the shear stress of sea surface wind.

   In the deep ocean far away from the coast, the seabed has no influence on the drift movement, so the drift current is called the unlimited deep-sea drift current. On the other hand, in the inshore waters the seabed will have an impact on the drift movement, so the drift current is called the limited deep-sea drifting current.

   b. Local wind currents

   In the large-scale spatial-temporal range, currents are formed by the prevailing winds during a short-term local weather process or discontinuous local gust, known as the local wind currents.

2. Gradient current

   Gradient current is the permanent current that is the result of the equilibrium of the horizontal pressure gradient force and the deflecting force of the Earth's rotation when the isobaric surface (such free-flowing sea) tilts. Then, water moves from the high-pressure area to the low-pressure area. Therefore, the tilt of the isobaric surface and the difference of horizontal distribution of the pressure are important factors in seawater movement. According to the different reasons for tilt of an isobaric surface, the gradient current can be divided into the following:

   a. Barogradient current

   The barogradient current is the gradient current caused by the tilted sea surface formed by uneven distribution of atmospheric pressure over the ocean. If seawater density distribution is uniform, the pressure decreases on sea surface with high atmospheric pressure and increases on seas surface with low atmospheric pressure. The sea surface is inclined, and at the same time each isobaric surface below the sea surface will tilt and become parallel with the sea surface. This will generate an equal horizontal pressure gradient force from the surface to the bottom, so water movement occurs.

   b. Density gradient current

   The density gradient current is the current considering the tilted isobaric surface caused by uneven seawater density distribution only, while not considering the action of external forces. Obviously, when the water density distribution is not uniform, the distance between isobaric surfaces is large in low-density areas and small in high-density areas. So the isobaric surface will tilt and produce the horizontal pressure gradient force in the horizontal plane, leading to water movement.

3. Tidal current

   The tidal current is the movement of water particles in tides. It is the cyclical flow of seawater in the horizontal direction under the action of the horizontal tidal force, simultaneously with the tide. Therefore, the character of the tidal current in various sea areas is basically consistent with the character of the tides. It can also be divided into three streams: half-day tide, mixed tide, and full-day tide. Therefore, according to the form of the movement of the tidal current, it can also be divided into the following two categories:

a. Reversing current

Reversing current generally occurs in the inshore waters, estuaries, bay mouths, waterways, straits, and narrow bays. In these areas, due to the limitations of the terrain conditions, the flow direction of the tidal current is cyclical and changes as the direction goes positive and negative, so it is called the reversing current. There is a big difference between the maximum and minimum velocity, and the minimum velocity is close to or equal to zero, so it is also known as slack water.

b. Rotating current

Rotating current is a common form of tidal current. It mostly occurs in the open sea area. The direction of it changes gradually over time, rather than simply dividing into positive and negative direction. The difference of velocity is not large. Due to the action of the deflecting force of the Earth's rotation, the direction of the rotating current is clockwise in the ocean in the Northern Hemisphere, and is counterclockwise in the Southern Hemisphere.

4. Wave current

The wave current is the seawater current caused by waves in inshore areas.

5. Runoff into the sea

River runoff, which is significantly increased with melting ice or heavy precipitation, continues to extend outside of estuary or coastal waters into the sea. This will lead to the movement of water. This movement is known as runoff into the sea.

6. Steady current, periodic current, and short-term current

According to the variation of current with time, the tidal current can be divided into the following categories:

a. Steady current

The steady current basically does not change with time. Theoretically, after the various forces that impact currents reach equilibrium, the direction, velocity, and strength should not change. Generally the ocean surface environment controlled by the prevailing winds on Earth is called the steady current.

b. Periodic current

The currents with a cyclical variation, which repeat within a certain time frame, are called periodic currents. Examples include the monsoon currents and tidal currents, etc.

c. Short-term current

The brief currents caused by the accidental change of external conditions are called short-term currents. For example, the pressure gradient currents or the wind currents caused by cyclone belong to the short-term current category.

7. Warm current and the cold current

According to the temperature of the current and surrounding seawater, currents can be divided into the following:

a. Warm current

The temperature of the current is higher than the surrounding seawater and the energy is transmitted and transporting from the current to the area external to it. This is called the warm current.

b. Cold current

A cold current occurs when temperature of the current is lower than the surrounding seawater.

8. Brine current and freshwater current
   According to the difference between the salinity of currents and the salinity of surrounding seawater, currents can be divided into the following:
   a. Brine current
      This is seawater current with higher salinity than the surrounding water.
   b. Freshwater current
      This is seawater current with lower salinity than the surrounding water.
9. Compensation current

   The compensation current is named according to the characteristics of the current. Because of the movement of water, a seawater deficit will inevitably form in some sea areas with a seawater concentration in other areas. Water will flow from the latter to the former area according to the incompressibility and continuity of water. This water current is compensation current.

   In addition to the nine kinds of current mentioned above, currents can, for example, also be divided into linear current and circulation in accordance with the approach of current. Currents can also be divided into surface current, deep current ,and coastal current in accordance with the position of the currents in the ocean.

   It should be noted that the observed actual currents in the ocean often are the result of the combined effects of the variety of different currents mentioned above. For example, in the inshore area, wind currents, barogradient currents, and wave currents usually exist at the same time.

1.3.2.2 Characteristics of the Littoral Environment in China

1. Characteristics of the littoral environment in the China Bohai Sea, China Yellow Sea, and East China Sea
   Since these three water areas are located in the shallow edge on the same continental shelf, the offshore circulation system is mainly constituted of the Chinese coastal currents and a branch of the Kuroshio current, as shown in Figure 1-5. The so-called Kuroshio current is the circulation of the North Equatorial Current that reaches the east coast of Asia by the obstruction of a land barrier split into two branches turned to the north and south near the Philippines and Taiwan, as shown in Figure 1-5.
2. Characteristics of the environment in the South China Sea
   The South Sea is a typical deep-sea basin, so the environment there, which belongs to the character as drifting, is mainly dominated by monsoons.

1.3.2.3 Current Velocity

Current velocity is a vector. Its direction refers to the flow direction and it takes degrees (°) as a unit, with the north as zero and clockwise measurement. The size of the current velocity is expressed by cm/s or kn. 1 kn means 1mile per hour, namely 51.44 cm/s, which is approximately equal to 50 cm/s.

   Current velocity is usually actually measured in several ways such as single station continuous observation, multistation synchronization continuous observation, and large current observation. After observation, the tidal current and nontidal current (residual current) should be taken separately. Assuming that the direction and velocity of the nontidal current in a relatively short period is a constant value, one should respectively count the frequency of different velocities and directions in each small area, and draw out the sea current velocity roses. As this is very cumbersome,

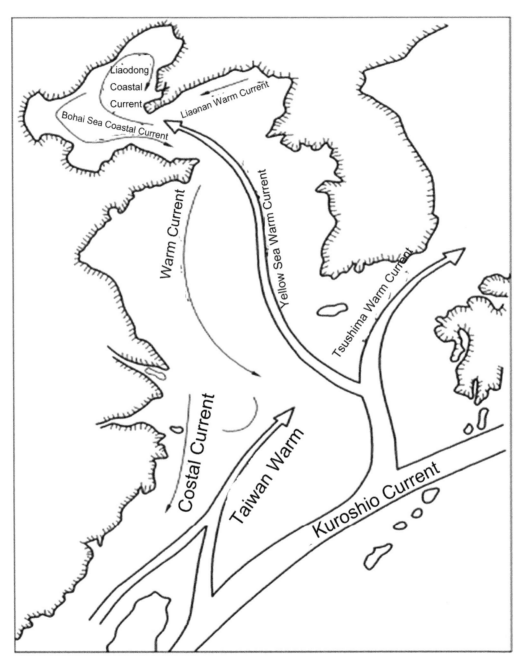

FIGURE 1-5

The coastal environment of the China Bohai Sea, China Yellow Sea, and East China Sea.

currently the method is to make a simple calculation based on the measured data, which we introduce in the following.

1. Estimation of the velocity of wind currents in the inshore area

   For an inshore area with a significant tide, wind current often composes the main part of the residual current. Therefore, according to the observed data for wind, we can estimate the wind current velocity by the following equations:

   $$u_{cw} = K_w u_w \tag{1-10}$$

   $$\theta_{cw} = \beta_c \tag{1-11}$$

   where u_{cw} represents the value of wind current velocity, m/s; θ_{cw} represents the direction of the wind current velocity (°); u_w represents the wind velocity, m/s; K_w represents the wind factor, $0.024 \leq K_w \leq 0.050$ (generally $K_w = 0.030$, in the China Bohai Sea, $0.025 \leq K_w \leq 0.030$); and β_c represents the direction along the isobaths (°).

2. Calculation of average maximum velocity of tidal current

   For a sea area with small nontidal current, the measured velocity can be approximately regarded as the velocity of the tidal current. That is, from the rate of the mean ranges R_{mg}, R_{mm}, and R_{ms} in the large, medium, and small tide and the tidal range R_{td} in an observational day, and from the maximum velocity vector u_{td} measured in an observational day, the maximum average velocity vectors u_{mg}, u_{mm}, and u_{ms} during the large, medium, and small tide can be calculated by the following equations:

   $$u_{mg} = \frac{R_{mg}}{R_{td}} u_{td} \tag{1-12}$$

   $$u_{mm} = \frac{R_{mm}}{R_{td}} u_{td} \tag{1-13}$$

   $$u_{ms} = \frac{R_{ms}}{R_{td}} u_{td} \tag{1-14}$$

3. Calculation of the maximum possible current velocity

   The maximum possible current velocity should be calculated using a statistical relationship based on the measured current. In the inshore area, the maximum possible velocity $u_{c\ max}$ of the current should be the vector sum of the maximum possible velocity $u_{ct\ max}$ of the tidal current and the maximum possible velocity $u_{cw\ max}$ of the wind current, namely

   $$u_{c\ max} = u_{ct\ max} + u_{cw\ max} \tag{1-15}$$

   According to Equation (1-10), $u_{cw\ max}$ in Equation (1-15) can be replaced by the maximum wind velocity $u_{w\ max}$ when calculating, namely

   $$u_{cw\ max} = K_w u_{w\ max} \tag{1-16}$$

4. Calculation of velocity of current changing with depth

   For some engineering structures in offshore oil engineering, where the depth under the sea surface is $d(m)$ and the height above the seabed is $Z(m)$, if on the sea surface the velocity of the tidal current u_{ct} (m/s) and the velocity of the wind current u_{cw} (m/s) are known, we can calculate the velocity of current u_{cz} with the height of Z:

$$u_{cz} = u_{ct}\left(\frac{Z}{d}\right)^{\frac{1}{7}} + u_{cw}\left(\frac{Z}{d}\right) \tag{1-17}$$

1.3.2.4 Current Force

Currents and tides not only scour the offshore oil engineering piles and the seabed around the foundation, but also affect structures through the action of the current force. Because the velocity of the tide and currents (mainly wind currents) changes slowly with time, one often takes the tide and currents as a steady flow the engineering design. Thus, the force acting on the structure or on its foundation can be regarded as a drag force (resistance), now specifically analyzed in the following.

1. Only consider the action of currents or tides
   According to the specification of the China Classification Society, the calculation is as follows:
   a. The current force f_{D_c} acting on the unit length of the circular member:

   $$f_{D_c} = \frac{1}{2}C_D\rho Au_c^2 \tag{1-18}$$

   where f_{D_c} represents the drag force, which is proportional to the kinetic energy currents or the tide only, N/m; ρ represents the mass density of seawater, kg/m^3; u_c represents the velocity of the current or tide, considering generally the wind currents, [calculated according to Equation (1-16)], m/s; A represents the projected area of a member per unit length, which is perpendicular to the direction of the currents or the tide, m^2/m (if the member is circular cross-section, $A = D \times 1 = D$, m^2; D represents the diameter of circular component, m; and C_D represents the drag coefficient (which shall be measured according to the appropriate test). When lacking test data, take 0.65 for a circular member with a smooth surface; and take 1.05 for a circular member with a rough surface.
   b. The current force F_{D_c} acting on the whole circular member:

   $$F_{D_c} = \int_0^d \frac{1}{2}C_D\rho Au_{cz}dz \tag{1-19}$$

   where u_{cz} represents the velocity of currents at the height Z above the seabed [shown in Equation (1-17)], which is a function of Z, m/s; and d represents the seawater depth, m:
2. Consideration of the combined effects of currents and waves
   a. The force f_{Dcw} acting on the unit length of the circular member:

   $$f_{Dcw} = \frac{1}{2}\rho C_D D(u_c + u)^2 \tag{1-20}$$

   where D represents the diameter of circular component, m; and u represents the horizontal velocity of water particles in a wave, m/s.

b. The force F_{Dcw} acting on the whole circular member:

$$F_{Dcw} = \int_0^{d+\eta} \frac{1}{2}\rho C_D D(u_c + u)^2 dz \qquad (1\text{-}21)$$

where F_{Dcw} represents the synthetic drag force acting on the whole component (shown in Figure 1-6), N; η represents the wave surface (its height from the still water level), m; and Z represents the height from seabed to the tiny section. The other symbols are the same meaning as before.

1.3.2.5 Von Karman eddy current

When the tides or currents and waves move uniformly along the direction perpendicular to the axis of the circular members, there will be a Von Karman eddy current around the member. These eddy currents generate a variable force. When the frequency of the force is the same or close to the natural frequency of the engineering structures, there will be a resonance resulting in serious fatigue damage on structures.

If we set the firing frequency of vortex caused by this eddy current to f, then

$$f = S\frac{u}{D} \qquad (1\text{-}22)$$

where u represents the velocity of the tide or current and wave that is perpendicular to the axis of the component, m/s; D represents the diameter of the component, m; S represents the Strouhal number (it can be acquired from Figure 1-7 according to the Reynolds number, Re).

The Reynolds number Re is calculated as follows:

$$R_e = \frac{uD}{v} \qquad (1\text{-}23)$$

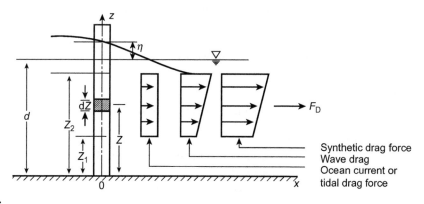

FIGURE 1-6

The drag force of the current and/or wave acting on the circular member.

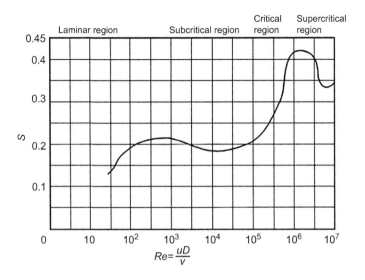

FIGURE 1-7

The relationship between the Strouhal number and the Reynolds number.

where v is the kinematic viscosity coefficient of seawater. Thus, for sea water

$$R_e = 0.9 \times 10^9 uD \tag{1-24}$$

1.4 Waves and the Random Wave Spectrum

Offshore oil engineering structures could be affected by waves at any moment. The power of waves is amazing. The resultant force of the wind pressure and wave pressure on an offshore oil platform in water of approximately 100 m depth can be over 1×10^4 t, which shows the potential destructiveness of a wave. Therefore, in offshore oil engineering, it is necessary to recognize waves and their load on the structure. This section introduces basic features of waves.

1.4.1 Brief Description

In the following, we present a brief discussion of the basics of waves.

1.4.1.1 Stormy Waves and Flush

As the proverb goes, "there is no smoke without fire, there are no waves without wind." Therefore, the wind is the external cause for the formation of waves. A wave is generally a waving phenomenon caused by all kinds of forces, such as the wind force. But a sea wave generally means stormy waves and flush.

1. Stormy waves

   The waving phenomenon in the ocean directly caused by the wind is called stormy waves. The direction of stormy waves is consistent with the direction of the wind.

According to fluid mechanics, the air is compressible fluid and the free surface is the interface between two kinds of fluid (air and water). Therefore, when there is relative movement between the two contiguous fluids, waves are formed on the interface. In other words, when air flows on the sea, waves will be formed.

2. Flush

When the stormy waves are in an area of no wind or little, they still can spread; this phenomenon is called flush. The characteristics of flush involve inerratic shape, a smooth wave surface, and a long period. When the sea is calm, the sudden wave is flush.

1.4.1.2 Growth of Waves

The growth of waves is connected with the following three factors:

1. Wind velocity. This is the main factor in the growth of waves.

2. Wind duration. Under the condition that the velocity and direction of wind are constant, the time that wind acts on the sea surface is called the wind duration. It is also one of the important factors for the growth of waves. For example, although the strength of a gust may be very large, it is difficult to form billow because of the limited energy captured by water during the short wind duration.

3. Wind field. The distance touched by the wind is known as the wind field. It affects the accumulation of the energy of water; thus it is also one of the factors in the growth of waves. For example, despite the strength of the wind velocity and wind duration, a billow can't occur in the small pond, which is limited by the size of the wind field.

1.4.1.3 Disappearance of Waves

When a wave is out of the wind field, it becomes a flush in an area with little or no wind. It continues to spread, relying on inertia. However, when a wave spreads from a deep to a shallow area of the sea, wave energy decreases because of the disturbance caused by waves spread from the surface to the bottom of the sea. When the energy loss is greater than the energy accepted by the fluctuating water surface, the wave height first decreases. Then, because most of the wave energy is concentrated in shallow water, the wave height increases gradually until there is a complete break and loss of the original shape. Usually, the wave breaks when it is transmitted to a certain depth. Energy is whittled down by a crushing phenomenon, and the wave height lowers. After that, if wave can adapt to the new water depth, this breaking will stop. Otherwise, this breaking will continue until it spreads to the coastline. The relationship between the depth d_b and the wave height H_b of the breaking wave is $d_b = 1.25H_b$.

1.4.2 Basic Parameters of a Wave

The basic parameters of a wave are parameters that quantify the characteristics of waves, such as the height, length, period, and steepness of waves. The most important parameters among them are the height and period, which are also called wave factors.

1.4.2.1 Wave Height

1. Height of an independent wave

An independent wave (shown in Figure 1-8) has a regular wave profile that changes with time t and space. With the wave profile as the vertical ordinate $\varepsilon(t)$ and the still water surface coinciding

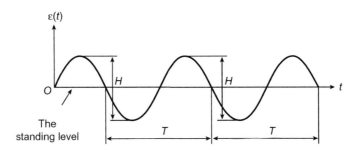

FIGURE 1-8

Independent wave.

with the abscissa, the average displacement of the wave is zero from the still water surface in a period T:

$$\int_{0}^{t+T} \varepsilon(t)dt = 0 \tag{1-25}$$

As shown in Figure 1-8, the height (theoretical significance) of a regular independent wave is the vertical height from the crest to the adjacent trough, which is generally expressed by H.

2. Statistical wave height

The wave profiles actually measured on the sea are irregular, as shown in Figure 1-9. They vary in amplitude, period, etc. Therefore, wave height is defined according to the characteristic values. In other words, wave height is represented by statistics. They are as follows:

a. The average wave height $\overline{H}$

Generally, during 20 min or so of observation, the arithmetic mean value of the sum of the wave heights at the static observation point divided by the wave number N is known as the average wave height:

$$\overline{H} = \frac{1}{N}(H_1 + H_2 + H_3 + \cdots + H_N) \tag{1-26}$$

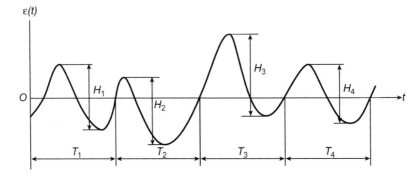

FIGURE 1-9

The measured wave curve.

b. Root-mean-square wave height H_{rms}

This defines wave height based on the characteristic values, so it is called the root-mean-square wave height. Assume the observed wave heights are H_i, $i = 1, 2, 3, \ldots , N$. Then

$$H_{rms} = \frac{1}{N}(H_1^2 + H_2^2 + H_3^2 + \cdots H_N^2)^{1/2} \tag{1-27}$$

c. Characterization of wave height (effective wave height) H_S

Generally, during a 20 min or so observation, the arithmetic mean of the highest part among the observed waves at static observation points is the wave height of statistical significance and it is called the characterization of wave height, or effective wave height, expressed by H_S. If the proportion of the highest wave height is 1/3, it is called the average height of the highest one-third of waves, $H_{1/3}$; if the proportion of the highest wave height is 1/10, it is called the average height of the highest one-tenth of waves, $H_{1/10}$. Their mathematical expressions are

$$H_{1/3} = \frac{1}{N/3}\left(H_1^* + H_2^* + \cdots + H_{N/3}^*\right) \tag{1-28}$$

$$H_{1/10} = \frac{1}{N/10}\left(H_1^* + H_2^* + \cdots + H_{N/10}^*\right) \tag{1-29}$$

where H_i^* ($i = 1, 2\ldots, H_{N/3}, H_{N/10}$) is the highest wave height during the observation time.

d. Wave height corresponding to a certain cumulative probability

Through actual observation, the probability density curve of wave height can be obtained, as shown in Figure 1-10. The vertical coordinate represents the probability density $p(H)$ of wave height H, and the horizontal coordinate represents wave height H. According to Figure 1-10, the cumulative probability is $P(H)$ and the cumulative probability of a wave height between H_1 and H_2 is

$$P(H) = \int_{H_1}^{H_2} P(H)dH \tag{1-30}$$

If $H_1 = 0$, then the cumulative probability $P(H)$ obtained is the percentage of the possible wave height that is not more than H_2. As this percentage can ensure that the wave height is not larger

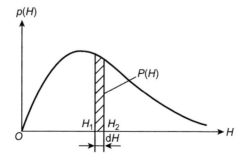

FIGURE 1-10

Probability density curve of wave height.

than wave height H_2, we call it the factor of assurance. H_2 is the wave height corresponding to a certain cumulative rate.

3. The probability distribution of wave height

   At fixed observation points, the measured wave height changes over time and is unsystematic. However, it obeys a certain statistical distribution. At present, the common probability distributions of wave height are as follows:

   a. The probability distribution of wave height for a short period

   Generally, for waves with a narrowband spectrum counted over a short period of time, the amplitude or wave height obeys Rayleigh distribution. The distribution curve of the probability density $p(H)$ is shown in Figure 1-11 and $P(H)$ is

   $$P(h) = 1 - \exp\left(-\frac{H^2}{4\sigma_H^2}\right) \tag{1-31}$$

   In this equation, σ_H^2 is the statistical variance value of the wave height, and the relationship between it and the characterization wave height $H_{1/3}$ is approximately the following:

   $$H_{1/3} = \sqrt{8\sigma_H^2} \tag{1-32}$$

   Therefore, Equation (1-31) can also be

   $$P(H) = 1 - \exp\left(-2\left(\frac{H}{H_{1/3}}\right)^2\right) \tag{1-33}$$

   b. Probability distribution of long-term wave height

   When dealing with months and years observation data for waves, generally the synthesis of a large amount of measured data for a short period during a stable process is used. Thus, statistical results show that the broadband spectrum waves obey a normal distribution and the two-parameter obey a Weibull distribution, as follows:

   $$P(H) = \exp\left(-\frac{\pi}{4}\frac{H^2}{\overline{H}^2}\right) \text{(Normal distribution)} \tag{1-34}$$

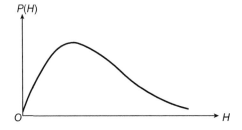

FIGURE 1-11

The probability density curve showing that wave height obeys Rayleigh distribution.

$$P(H) = 1 - \exp\left[-\left(\frac{H}{\delta}\right)^{\varepsilon}\right] \text{(Weibull distribution)} \tag{1-35}$$

where $\overline{H}$ is average wave height, namely the mathematical expectation of wave height (generally, $\overline{H} = 0.89H_{rms}$), and σ and ε are constant.

4. The relationship among different wave heights

   a. The relationship between average wave height $\overline{H}$ and root-mean-square wave height H_{rms}

   According to the definition in mathematical statistics, the the σ_H^2 should be the square of root-mean-square H_{rms}, i.e., $\sigma_H^2 = (H_{rms})^2$; the standard deviation σ_H should be equal to H_{rms}, i.e., $\sigma_H = H_{rms}$. The average wave height $\overline{H}$ is the mathematical expectation of wave height $E(H)$. When the probability density of wave height is $p(H)$,

$$\overline{H} = E(H) = \int_0^\infty Hp(H)dH \tag{1-36}$$

Obviously, replace the probability density function $p(H)$ in Equation (1-36) by the Rayleigh distribution:

$$p(H) = \frac{H}{\sigma_H^2} \exp\left(-\frac{1}{2}\frac{H^2}{\sigma_H^2}\right) \tag{1-37}$$

Replace σ_H^2 by $(H_{rms})^2$, and the equation obtained from Equation (1-36) is:

$$\overline{H} = 0.89\, H_{rms} \tag{1-38}$$

   b. The relationship between characterization wave height $H_{1/3}$ and root-mean-square wave height H_{rms}

   According to the definition of wave height $H_{1/3}$, set the wave height corresponding to the cumulative probability of 33.33% as H_u, divide Equation (1-36) by a third, and replace $\overline{H}$ by $H_{1/3}$:

$$H_{1/3} = \int_{H_u}^\infty \frac{Hp(H)dH}{1/3} \tag{1-39}$$

With the common relation $H_u = 1.05H_{rms}$, if the probability density function $p(H)$ in Equation (1-39) obeys the Rayleigh distribution and is substituted into Equation (1-37), it can be concluded that the relationship of $H_{1/3}$ and H_{rms} is

$$H_{1/3} = \sqrt{2H_{rms}} \tag{1-40}$$

1.4.2.2 Wave Period

1. The definition of the period

   a. Period across the zero axis

   The observation curve of ocean waves is shown in Figure 1-12 and it can be seen in the figure that not all the crests or troughs of the wave cross the zero axis. However, the period of a wave

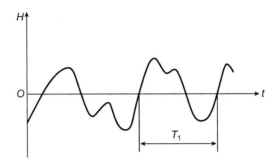

FIGURE 1-12

The measured period of the waves.

usually refers to the time interval when the adjacent crests and troughs cross the zero axis, as shown in Figure 1-9. We must then think about how to deal with the crests and troughs not crossing the zero axis. Generally in engineering, the rule is that only the peaks and troughs that cross the zero axis count and peaks or troughs that do not cross the zero axis are ignored. So the time interval when adjacent peaks and troughs cross the zero axis is defined as the period across the zero axis, as represented by the T_i in Figure 1-12.

b. Apparent period

The time interval during which two adjacent significant crests go through a fixed observation point is called the apparent period.

c. Characterization period

Over a specified observation time, the arithmetic mean value of the period of the highest one-third wave height is called the characterization period or significant period.

d. Average period

The mathematical expectation of the period according to statistics is the average period.

2. The probability distribution of a period

Statistics shows that the apparent period T_v generally obeys the normal distribution; thus its probability density function $p(T_v)$ should be

$$p(T_v) = \frac{1}{\sigma_{T_v}\sqrt{2\pi}}\exp\left(-\frac{T^2_v}{2\sigma^2_{Tv}}\right) \tag{1-41}$$

where σ_{Tv} refers to the statistical characteristic value of the apparent period, T_v refers to the standard deviation, and σ^2_{Tv} is its variance.

If $p(T_v)$ is known, then the cumulative probability and the average period can be respectively calculated.

a. The cumulative probability of period $P(T_v)$

$$P(T_v) = \int_{T_v}^{\infty} P(T_v)dT_v = \frac{1}{\sigma_{T_v}\sqrt{2\pi}}\int_{T_v}^{\infty}\exp\left(-\frac{T^2_v}{2\sigma^2_{Tv}}\right)dT_v \tag{1-42}$$

In the equation, $P(T_v)$ is the percentage of the apparent period that is not less than T_v, and T_v is the apparent period corresponding to the cumulative probability $P(T_v)$.

b. Calculation of average period $\overline{T}$

According to the definition of mathematical expectation $E(T_v)$ in statistics, the average period can be obtained as

$$\overline{T} = E(T_v) = \int_0^\infty T_v p(T_v) dT_v \tag{1-43}$$

Substituting Equation (1-41), we obtain

$$\overline{T}_v = \int_0^\infty T_v \frac{1}{\sigma_T \sqrt{2\pi}} \exp\left(-\frac{T_v^2}{2\sigma^2_{T_v}} \right) = \left(\frac{4}{\pi} \right)^4 \Gamma\left(\frac{5}{4} \right) \sigma_{T_v} \tag{1-44}$$

where

$$\Gamma(n) = \int_0^\infty x^{n-1} e^{-x} dx \tag{1-45}$$

c. The maximum probable period

Generally, the period with maximum probability is the maximum probable period T_m. Obviously, at the peak of the probability density curve of period T_v, i.e. $\frac{dp(Tv)}{dTv} = 0$, probability density $p(T_v)$ will reach its maximum value, and T_v should be the maximum probable period. Then, according to $\frac{dp(Tv)}{dTv} = 0$, substitute the probability density function $P(T_v)$ of Equation (1-41) in, and the maximum probable period T_m is

$$T_m = \left(\frac{3}{4} \right)^{\frac{3}{4}} \cdot \frac{\overline{T}}{\Gamma\left(\frac{5}{4} \right)} \approx \overline{T} \tag{1-46}$$

1.4.2.3 Other Wave Parameters

In addition to wave height and wave period, there are basic parameters such as wavelength, wave steepness, and wave velocity.

1. Wavelength

The horizontal distance between the two adjacent (across the zero axis) crests is referred to as the wavelength, generally expressed by λ. The definition and statistics for the wavelength are similar to that which we reviewed with the wave height and period, so it is not covered in detail here.

2. Wave steepness

For independent waves, the ratio of its wave height H and wavelength λ is known as the wave steepness and it is expressed by δ:

$$\delta = \frac{H}{\lambda} \tag{1-47}$$

3. Wave velocity

   Usually the ratio of wavelength and apparent period T_v is regarded as wave velocity, C:

$$C = \frac{\lambda}{T_v} = \frac{\omega}{K} \tag{1-48}$$

   where ω refers to circular frequency of the waves, $\omega = \frac{2\pi}{T_v}$, and K refers to wave number, $K\lambda = 2\pi$.

1.4.3 Characteristics of Basic Wave Parameters in China

In this section, we discuss the characteristics of wave direction, height, and period in China

1.4.3.1 Wave Direction

The wave direction is the propagation direction of waves, and it mainly depends on the direction of the wind (except flush). As China is a typical monsoon zone with a primary north wind in winter and primary south wind in summer, the wave direction of Chinese sea area corresponds with those characteristics.

1.4.3.2 Wave Height

1. In winter

   As the north wind is strong and has a long duration, the wave height along the coast is large. For example, a wave height more than 8 m can appear in the China Bohai Sea in winter.

2. In summer

   Because of the frequent typhoon sin the East China Sea and the South China Sea, waves are larger. Generally in the China Bohai Sea, China Yellow Sea, and East China Sea wave height can be as high as 5 to 6 m, and in the Xisha Islands of the South China Sea wave height sometimes is up to 10 m.

3. In spring

   In the Bohai Bay and northern coast of Shandong Peninsula, there are waves above grade 5.

4. In autumn

   In the China Bohai Sea, China Yellow Sea, and East China Sea, usually waves up to grades 5 to 6 appear.

1.4.3.3 Period

General observation shows that the period of waves in the north is smaller than in the south from the China Bohai Sea and China Yellow Sea to the East China Sea, and the period of waves along the coast of the South China Sea is lower.

1. In winter

   From south of Shandong Peninsula to Fujian, the average period increases from roughly 4 to 5 s to 6 to 7 s. The maximum is about 11 s and the average period of the East China Sea coast is only roughly 3 to 4 s; the maximum is about 10 s.

2. During the spring and autumn

   The average period in each area is the same. The average period of the China Bohai Sea is less than 4 s with the maximum at about 8 s. From North of Jiangsu to the Zhejiang and Fujian area, the average period is less than 5 s with the maximum about 11 s. The average period in the South China Sea is more than 5 s and the maximum can reach 12 s.

3. In summer

The average period from Liaodong Bay to the Yangtze River estuary area is not more than 5 s, and the maximum is about 13 s. Along the coast of Zhejiang and Fujian, the average period is above 5 s, and the maximum is 19 s or so. The average period is 5 s on the west coast of Guangdong and the maximum is about 11 s. Along the coasts of Xisha, Hainan Island, and north of the Beibu Gulf, the average period is 4 s and the maximum is about 11 s.

1.4.4 Random Wave Spectrum

The waves of the sea are a series of fluctuations that is complex, irregular, and changes with time and location, composed of a large number of simple harmonic waves with different wave height (amplitude), frequency, transmission direction, and phase clutter, so we call it a random wave. This random phenomenon is inseparable from the wind, because the changes of wind velocity and pressure on the sea surface related to the position and time is extremely complex and random. As waves are random phenomena, the characteristics and basic parameters of ocean waves observed also have different values that change with time and position. The value involves random variables and thus it can be regarded as a random process.

Because it is difficult to describe random sea waves with the liquid fluctuation theory and deterministic function used nowadays in marine engineering, the random wave spectrum theory is used to describe the random waves. The random wave spectrum is the energy spectrum and frequency spectrum of a random ocean wave that is used to describe the distribution of energy related to each part in the frequency domain. It defines how much energy distributes in the high-frequency part and how much energy distributes in the low-frequency part. The next sections will introduce the theoretical basis for this topic, the creation method, and the steps of the random wave spectrum.

1.4.4.1 Theoretical Basis

1. Mathematic expression of the random ocean wave surface

Random waves are considered to be composed of many simple harmonic waves with different amplitudes (a_n), frequencies (ω), and phases. Therefore, wave surface $\varepsilon(t)$ changing with time can be expressed by a Fourier series in mathematics as follows:

$$\varepsilon(t) = a_0 + \sum_{n=-\infty}^{\infty} a_n \cos n\omega t + \sum_{n=1}^{\infty} b_n \sin n\omega t \qquad (1\text{-}49)$$

If written in the form of a complex number, this is

$$\varepsilon(t) = \sum_{n=-\infty}^{\infty} C_n e^{in\omega t} \qquad (1\text{-}50)$$

Using knowledge of the Fourier series, the coefficient C_n should be

$$C_n = \frac{1}{T} \int_{-\frac{T}{2}}^{\frac{T}{2}} \varepsilon(t) e^{-in\omega t} dt = \frac{\omega}{2\pi} \int_{-\frac{T}{2}}^{\frac{T}{2}} \varepsilon(t) e^{-in\omega t} dt \qquad (1\text{-}51)$$

The coefficient C_n is substituted into Equation (1-51), and then Equation (1-50) can be converted into

$$\varepsilon(t) = \sum_{n=-\infty}^{+\infty} \frac{\omega}{2\pi} \int_{-\frac{T}{2}}^{\frac{T}{2}} \varepsilon(t) e^{-in\omega t} dt \cdot e^{in\omega t} \tag{1-52}$$

The superposition of discrete values $n\omega$ in Equation (1-52) is written as integral $d\omega$. When $d\omega \to 0$, $T \to \infty$. Then in Equation (1-52), replace the index $n\omega$ integral limit $-T/2, T/2$, and ω by $\omega, -\infty, \infty$, and $d\omega$ respectively, then

$$\varepsilon(t) = \int_{-\infty}^{+\infty} \left[\frac{1}{2\pi} \varepsilon(t) e^{-i\omega t} dt \right] e^{i\omega t} d\omega \tag{1-53}$$

In Equation (1-53)

$$\varepsilon(\omega) = \frac{1}{2\pi} \int_{-\infty}^{+\infty} \varepsilon(t) e^{-i\omega t} dt \tag{1-54}$$

Equation (1-53) can be written as

$$\varepsilon(t) = \int_{-\infty}^{\infty} \varepsilon(\omega) e^{i\omega t} d\omega \tag{1-55}$$

The relationship between $\varepsilon(\omega)$ in Equation (1-54) and $\varepsilon(t)$ in Equation (1-55) is a mutual Fourier transform in mathematics. Moreover, Equation (1-55) is the mathematical expression of the wave surface of random sea waves expressed by the circular frequency ω.

2. Mathematic expression of the energy of a random ocean wave
   For phenomenon of motion in nature, the energy or power of the movement all can be expressed by $\varepsilon(t)^2$, which is the square of the function $\varepsilon(t)$ changing with time t. For example, for electron motion, the alternating current (i) is the function $\varepsilon(t)$ changing with time. When electricity resistance R is 1, voltage $E = i \cdot R = \varepsilon(t) \cdot 1 = \varepsilon(t)$, then electric power $P_e = E \cdot i = \varepsilon(t) \cdot \varepsilon(t) = \varepsilon(t)^2$. Obviously, the power or energy can be expressed by the square of $\varepsilon(t)$.
   If the average total power (energy) per unit water particle in random waves is $\overline{E}$, in accordance with the above principle, it can be written as the function of the square $\varepsilon(t)^2$ of wave surface, as follows:

$$\overline{E} = \lim_{T \to \infty} \frac{1}{T} \int_{-\frac{T}{2}}^{\frac{T}{2}} \varepsilon(t)^2 dt \tag{1-56}$$

Replace $\varepsilon(t)^2$ in Equation (1-56) by $\varepsilon(t) \cdot \varepsilon(t)$, and substitute each $\varepsilon(t)$ with Equation (1-55), and we can obtain

$$\bar{E} = \lim_{T \to \infty} \frac{1}{T} \int_{-\frac{T}{2}}^{\frac{T}{2}} \varepsilon(t) \left[\int_{-\infty}^{\infty} \varepsilon(\omega) e^{i\omega t} d\omega \right] dt = \int_{-\infty}^{\infty} \varepsilon(\omega) \left[\lim_{T \to \infty} \frac{1}{T} \left[\int_{-\infty}^{\infty} \varepsilon(t) \varepsilon(\omega) e^{i\omega t} dt \right] \right] d\omega \quad (1\text{-}57)$$

As $\varepsilon(\omega) = \int_{-\infty}^{\infty} \varepsilon(t) e^{-i\omega t} dt$ and $\varepsilon^*(\omega) = \int_{-\infty}^{\infty} \varepsilon(t)^{i\omega t} dt$ are mutual conjugates, substitute $\varepsilon^*(\omega)$ into Equation (1-57):

$$\bar{E} = \int_{-\infty}^{\infty} \varepsilon(\omega) \lim_{T \to \infty} \frac{1}{T} \varepsilon^*(\omega) d\omega = \int_{-\infty}^{\infty} \left[\lim_{T \to \infty} \frac{1}{T} |\varepsilon(\omega)^2| \right] d\omega \quad (1\text{-}58)$$

3. Mathematic expression of the power spectral density of random waves
   Take $S_{\varepsilon\varepsilon}(\omega)$ as the power spectral density of random waves, and the average total power $\bar{E}$ should be

$$\bar{E} = \int_{-\infty}^{\infty} S_{\varepsilon\varepsilon}(\omega) d\omega \quad (1\text{-}59)$$

Substitute Equation (1-58) into Equation (1-59) to get

$$\int_{-\infty}^{\infty} \left[\lim_{T \to \infty} \frac{1}{T} |\varepsilon(\omega)^2| \right] d\omega = \int_{-\infty}^{\infty} S_{\varepsilon\varepsilon}(\omega) d\omega \quad (1\text{-}60)$$

So, we can be obtain the following from Equation (1-60):

$$S_{\varepsilon\varepsilon}(\omega) = \lim_{T \to \infty} \frac{1}{T} |\varepsilon(\omega)^2| \quad (1\text{-}61)$$

4. Relationship between power spectral density function and wave autocorrelation function
   Set $R_{\varepsilon\varepsilon}(\tau)$ as the autocorrelation function of a random sea wave surface $\varepsilon(t)$ and by the definition of the autocorrelation function in engineering mathematics, we get

$$R_{\varepsilon\varepsilon}(\tau) = \lim_{T \to \infty} \frac{1}{T} \int_{0}^{T} \varepsilon(t) \varepsilon(T + t) dt \quad (1\text{-}62)$$

In Equation (1-62), τ is the time interval of the autocorrelation for random ocean waves. If $\varepsilon(t + \tau)$ is written as a Fourier transform equation according to the Equation (1-55), then

$$\varepsilon(t + \tau) = \int_{-\infty}^{\infty} \varepsilon(\omega) e^{i\omega(t+\tau)} d\omega \quad (1\text{-}63)$$

Then substitute Equation (1-63) into Equation (1-62):

$$R_{\varepsilon\varepsilon}(\tau) = \lim_{T \to \infty} \frac{1}{T} \int_{-\frac{T}{2}}^{\frac{T}{2}} \varepsilon(T) \int_{-\infty}^{\infty} \varepsilon(\omega) e^{i\omega(t+\tau)} d\omega dt$$

$$= \int_{-\infty}^{\infty} \lim_{T \to \infty} \frac{1}{T} \left[\varepsilon(\omega) \int_{-\frac{T}{2}}^{\frac{T}{2}} \varepsilon(t) e^{i\omega\tau} dt \right] e^{i\omega\tau} d\omega \tag{1-64}$$

$$= \int_{-\infty}^{\infty} \lim_{T \to \infty} \frac{1}{T} \varepsilon(\omega) \varepsilon^*(\omega) e^{i\omega\tau} d\omega = \int_{-\infty}^{\infty} \lim_{T \to \infty} \frac{1}{T} |\varepsilon(\omega)^2| e^{i\omega\tau} d\omega$$

$$= \int_{-\infty}^{\infty} S_{\varepsilon\varepsilon}(\omega) e^{i\omega\tau} d\omega$$

Comparing Equation (1-64) with Equation (1-55), it can be found that the relationship between the autocorrelation function $R_{\varepsilon\varepsilon}(\tau)$ of a wave surface and the power spectral density function $S_{\varepsilon\varepsilon}(\omega)$ of random wave is a mutual Fourier transform. From Equations (1-55) and (1-54), we get

$$\left. \begin{array}{l} R_{\varepsilon\varepsilon}(\tau) = \displaystyle\int_{-\infty}^{\infty} S_{\varepsilon\varepsilon}(\omega) e^{i\omega\tau} d\omega \\[4mm] S_{\varepsilon\varepsilon}(\omega) = \dfrac{1}{2\pi} \displaystyle\int_{-\infty}^{\infty} R_{\varepsilon\varepsilon}(\tau) e^{i\omega\tau} d\tau \end{array} \right\} \tag{1-65}$$

Equation (1-65) shows that the power spectral density $S_{\varepsilon\varepsilon}(\omega)$ can be obtained according to the random wave autocorrelation function $R_{\varepsilon\varepsilon}(\tau)$ subsequent to acquisition for the curve (time history) of the wave surface of a random wave over time at the observation points, which is the theoretical basis of the production of the random wave power spectrum.

1.4.4.2 Creation Method

The random wave spectrum in the usual sense actually is the power spectral density $S_{\varepsilon\varepsilon}(\omega)$ of a random ocean wave with a distribution of the average total power of a unit water particle for a random wave in the frequency domain. Usually there are two kinds of creation methods: the direct method and indirect method, as follows.

1. Direct method

   This is the method that directly obtains the power spectral density of random ocean waves using mathematical methods on the basis of the large number of observed sample curves of the change of the wave over time and in accordance with the theoretical basis of the random ocean wave spectrum. The production methods and steps are as follows:

a. Wave observation time

Here we obtain a wave record at a fixed point of observation and it is advisable to take about 10 to 20 min for each segment, as shown in Figure 1-13. The abscissa of curve is "Time, 0.125s" and the ordinate is "Wave height, m".

b. Centralization of the wave ordinate

Because the ordinate value of the observation is not relative to the mean water level, it should be centralized. According to the theory that the mathematical expectation of the wave height is zero when the variance is equal to the mean square value, take the average value of the ordinate as the neutral axis (center) of the coordinate axis.

c. Determination of the time interval and the total number of intervals

The time interval Δt is the step length of the abscissa corresponding to two adjacent ordinate values in the wave record, as shown in Figure 1-14. The time interval is less than the minimum period of the sample curve in principle. Moreover, when calculating the average value for the autocorrelation function of the sample curve, the total number m of intervals needed generally is taken as $30 < m < 60$.

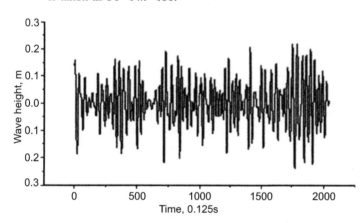

FIGURE 1-13

Time-history curve of actual observed wave surface.

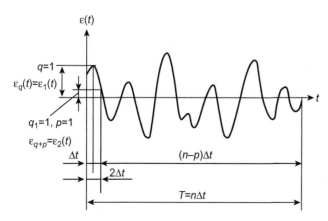

FIGURE 1-14

Wave analysis of random wave spectrum created by direct method.

d. Determine the total number of $\varepsilon(t)$

According to the total time T of the observed sample curve, the total number of $\varepsilon(t)$ can be determined:

$$n = \frac{T}{\Delta t} + 1 \tag{1-66}$$

where Δt is the time interval.

e. Take the adjacent ordinate value

In order to calculate the autocorrelation function, take enough number of the adjacent wave surfaces $\varepsilon(t)$ and $\varepsilon(t + \tau)$ in turn. Take the ordinate value of q as $\varepsilon_q(t)$ in the sample curve, and then take the adjacent ordinate value $\varepsilon_{q+p}(t)$ after a time interval Δt. Then from Figure 1-14, when $q = 1$, the ordinate value is $\varepsilon_q(t) = \varepsilon_1(t)$, while another adjacent ordinate value is $\varepsilon_q(t) = \varepsilon_1(t) = \varepsilon_{1+1}(t) = \varepsilon_2(t)$ due to $q = 1$, $p = 1$. Therefore, it can take enough number of wave surface by such analogy method.

f. Calculate the autocorrelation function

According to the definition of the autocorrelation function, Equation (1-62) gives a continuous function expressed by the integral. But when making a spectrum the discrete value R_p for the autocorrelation function of $\varepsilon_q(t)$ and $\varepsilon_{q+p}(t)$ is necessary. So when calculating the autocorrelation function, the expression after discretization should be provided. Take the wave surface $\varepsilon_q(t)$ in q and then take $\varepsilon_{q+p}(t)$ after $p\Delta t$. As shown in Figure 1-14, this is

$$R_p = \frac{2}{n-p} \sum_{q-1}^{n-p} \varepsilon_q(t) \cdot \varepsilon_{q+p}(t) \tag{1-67}$$

R_p in Equation (1-67) is the approximate average value of the autocorrelation function discretization subsequent to $p\Delta t$. $\frac{2}{n-p}$ in the equation represents the percentage of the weight with a weighted average, because the ratio of the time interval $2\Delta t$ between $\varepsilon_q(t)$ and $\varepsilon_{q+p}(t)$ to the total time $((n-p)\Delta t)$ is $\frac{2}{n-p}$.

g. Calculate discretized spectral values

Equation (1-65) provides the continuous function of the power spectral density:

$$S_{\varepsilon\varepsilon}(\omega) = \frac{1}{2\pi} \int_{-\infty}^{+\infty} R_{\varepsilon\varepsilon}(\tau) e^{-i\omega\tau} d\tau \tag{1-68}$$

Therefore, when making the random wave spectrum, it is necessary to calculate the discretized power spectral density values. The calculation equations are as follows:

From Euler's identity:

$$e^{-i\omega\tau} = \cos\omega\tau + i\sin\omega\tau \tag{1-69}$$

If we take only the real component of Equation (1-69), then the discretization power spectral density values S_{hp} in Equation (1-68) can be

$$S_{hp} = \frac{1}{2\pi} R_p \cos\omega(\Delta t) \tag{1-70}$$

Equation (1-70) is the result of replacing the $R_{\varepsilon\varepsilon}(\tau)$ in Equation (1-68) with the discretization autocorrelation function R_p and replacing $e^{-i\omega\tau}$ with $\cos\omega\Delta t$ (τ is the time interval Δt).

If the ordinal number of waves across the zero axis is h, the period of waves is T, the total number of time intervals required for determining the autocorrelation function value R_p is m, and the frequency (circular frequency) corresponding to the autocorrelation function value R_p is ω, from Figure 1-14 we can get

$$\omega = \frac{2\pi}{T} = \frac{2\pi}{m\Delta t/\frac{h}{2}} = \frac{\pi h}{m\Delta t} \tag{1-71}$$

Substitute Equation (1-71) into Equation (1-70), and the expression of the discretization power spectral density values S_{hp} corresponding to the discretization autocorrelation function value R_p that is calculated from Equation (1-67) is

$$S_{hp} = \frac{1}{2\pi} R_p \cos \frac{\pi h}{m\Delta t} \Delta t \tag{1-72}$$

h. Draw power spectral density plot

From Equation (1-72), a circular frequency value ω ($\omega = \pi h/m\Delta t$) can be calculated from Equation (1-71) according to h when $m\Delta t$ is elapsed. The autocorrelation function R_p can also be calculated from Equation (1-67), so the corresponding power spectral density values S_{hp} can be obtained. Therefore if we take circular frequency ω as the abscissa and power spectral density as the ordinate, several ω values and the power spectrum density of discrete values can be calculated according to sample curves of a random wave. Draw these values on the coordinate diagram and we can get many of the points as the discretization spectral values seen in Figure 1-15. Then connect them to the ladder histogram in Figure 1-15 and this ladder histogram can be expressed by the approximate integral equation

$$S_h = \frac{2}{\pi} \left(\frac{1}{2} R_0 + \sum_{p-1}^{m-1} R_p \cos \frac{\pi hp}{m} + \frac{1}{2} R_m \cos \pi h \right) \Delta t \tag{1-73}$$

In Equation (1-73), R_0 and R_m separately are discrete values of the autocorrelation function when $p = 0$ and $p = m$. The other symbols are the same as those Equation (1-71): S_h is the power

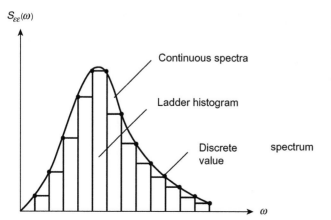

$S_{\varepsilon\varepsilon}(\omega)$

Continuous spectra

Ladder histogram

Discrete value

spectrum

ω

FIGURE 1-15

Direct system as a power spectral density plot.

spectral density function of the ladder histogram. Connect each endpoint of the ladder in the ladder histogram into a continuous curve as shown in Figure 1-15, and we get the continuous spectrum diagram. If the continuous power spectrum density function is expressed by the ordinate $S_{\varepsilon\varepsilon}(\omega)$, we can obtain the mathematical expressions of power spectral density function $S_{\varepsilon\varepsilon}(\omega)$ of a random wave corresponding to a continuous curve through fitting.

Figure 1-16 shows the power spectrum density function curve of random waves that is drawn by the direct method according to the sample curves of observed random wave for a period of time (Figure 1-13). The statistical characteristics of wave height and wave period from these sample curves are as shown in Table 1-6.

2. The indirect method

Using existing expressions of the random wave spectrum, we substitute some coefficients in these expressions calculated using actually measured sea wave data in the engineering sea area. Then, after appropriate modification of these expressions, we determine the random wave power spectral density expressions. This is the indirect method. Let's take the South China Sea as the sea area of petroleum engineering and create the random wave spectrum by the indirect method in accordance with the actually measured wave data of the South China Sea. The methods and steps are as follows.

a. Select the existing random wave spectrum

Considering that the condition of the South China Sea is better than Europe's North Sea, choose the existing P–M (Pierson-Moscowitz) spectrum as an expression of the random ocean wave spectrum of South China Sea, as follows:

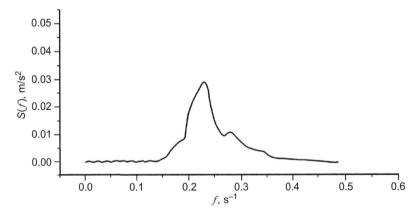

FIGURE 1-16

Random ocean wave spectrum according to the measured data using the direct method.

Table 1-6 Sample of Random Wave Statistical Values

$\overline{H}$, m	1.09	$\overline{T}$, s	3.555
$\overline{H}_{\frac{1}{10}}$, m	1.60	$\overline{T}_{\frac{1}{10}}$, s	3.29
$\overline{H}_{\frac{1}{3}}$, m	1.41	$\overline{T}_{\frac{1}{3}}$, s	3.84

$$S_{\varepsilon\varepsilon}(\omega) = \frac{\alpha g^2}{\omega^5} \exp\left[-\beta\left(\frac{g}{U\omega}\right)^4\right] \tag{1-74}$$

where ω represents the circular frequency; g represents the gravitational acceleration; U represents $U_{19.5}$, the wind velocity.5m above the sea surface; and α, β are coefficients.

b. Determine the related coefficients that should be calculated

In Equation (1-74), α and β are the calculated coefficients according to the conditions of the South China Sea. The expressions of the minimum P-M are as follows:

$$\alpha = 4\pi^3\left(\frac{H_{\frac{1}{3}}}{g\overline{T}^2}\right)^2 \tag{1-75}$$

$$\beta = 16\pi^3\left(\frac{u}{g\overline{T}}\right)^4 \tag{1-76}$$

In these equations, $\overline{T}$ is the average period, $H_{\frac{1}{3}}$ is the weighted average height of the highest one-third wave; and u is the wind velocity at 19.5 m high. According to Equation (1-1), u should be

$$u_{19.5} = u_{10}\left[1 + \frac{(C_{10})^{\frac{1}{2}}}{K}\ln\frac{Z}{10}\right] \tag{1-77}$$

The coefficient K can be 0.4 and C_{10} should be calculated as follows:

$$C_{10} = (0.80 + 0.114u_{10}) \times 10^{-3} \tag{1-78}$$

c. The statistical data of peaks and troughs

According to the actually observed sample curve of the random waves from the South China Sea by the University of Petroleum, and counting using the first crest counting method, the frequency and probability of crests and troughs has been calculated, as shown in Table 1-7.

d. Statistical data of period across zero

According to the same sample curves, we count the period across the zero axis, and find out the probability on the basis of the frequency of their occurrence, as shown in Table 1-8.

e. Compute coefficients in the spectrum expression

We compute the coefficient α according to Equation (1-75):

$$H_{\frac{1}{3}} = H_{C\frac{1}{3}} - H_{t\frac{1}{3}} = \frac{3}{104}(4 \times 13.4 + 10 \times 10.05 + 12 \times 6.70 + 9 \times 3.35)$$

$$-\frac{3}{116}(-11.73 \times 2 - 8.38 \times 5 - 5.03 \times 14 - 1.68 \times 18) = 12m$$

$$\overline{T} = \frac{\sum T_i}{N} = \frac{1}{66}(2 \times 34 + 24 + 28 + 10 + 18 + 35 + 16 + 9 + 11 + 13 + 15)\frac{1}{0.17} = 23s$$

$$\alpha = 4\pi^3\left(\frac{H_{\frac{1}{3}}}{g\overline{T}^2}\right)^2 = 4\pi^3\left[\frac{12}{9.8(23)^2}\right] = 6.65 \times 10^{-4}$$

where $H_{C\frac{1}{3}}$, $H_{t\frac{1}{3}}$ represent the height values for which the proportion of the number of waves with higher wave crest or trough is 1/3; T_i represents the period across the zero axis; and N represents

Table 1-7 The Observed Data of the Peaks and Troughs from the South China Sea

	Class Number	Class Interval, m	Value of Class Level	Value of Class Value	Frequency, n_i	Cumulative Frequency, $\sum n_i$	Probability, $\varphi_i = \frac{n_i}{n}$	Cumulative Probability, $\sum \varphi_i$
Peak value	1	-3.5~-2.5	-3	-10.5	1	104	0.010	1.000
	2	-2.5~-1.5	-2	-6.7	5	103	0.048	0.990
	3	-1.5~-0.5	-1	-3.35	13	98	0.125	0.942
	4	-0.5~-0.5	0	0	22	85	0.212	0.807
	5	0.5~1.5	1	3.45	37	63	0.356	0.605
	6	1.5~2.5	2	6.70	12	26	0.115	0.249
	7	2.5~3.5	3	10.05	10	14	0.090	0.134
	8	3.5~4.5	4	13.40	4	4	0.038	0.038
Valley value	1	-4~-13	-3.5	-11.73	2	116	0.017	1.000
	2	-3~-2	-2.5	-8.38	5	114	0.043	0.983
	3	-2~-1	-1.5	-5.03	14	109	0.120	0.940
	4	-1~0	-0.5	-1.68	26	95	0.226	0.820
	5	0~1	0.5	1.68	38	69	328	0.576
	6	1~2	1.5	5.03	21	31	0.181	0.268
	7	2~3	2.5	8.38	9	10	0.078	0.087
	8	3~4	3.5	11.73	1	1	0.009	0.007

Table 1-8 The Observed Data of the Cross Zero Axis Period from the South China Sea

Class No.	2	3	4	5	6	7	8	9	10	11	13	15
Frequency	34	8	7	2	3	5	2	1	1	1	1	1
Cumulative frequency	66	32	24	17	15	12	7	5	4	3	2	1
Probability	0.515	0.122	0.106	0.030	0.046	0.076	0.030	0.015	0.015	0.015	0.015	0.015
Cumulative probability	1.000	0.485	0.36	0.257	0.227	0.181	0.105	0.075	0.060	0.045	0.030	0.015

the total number of periods across zero [namely the cumulative frequency in Table (1-8)]. We compute β from Equation (1-76):

$$\beta = 16\pi^3 \left(\frac{u}{gT}\right)^4 = \left(\frac{36.7}{9.8 \times 23}\right) 16\pi^3 = 0.35$$

where the actual wind velocity u_{10} is 33.4 m/s. So C_{10} computed from Equation (1-78) is

$$C_{10} = (0.80 + 0.114u_{10}) \times 10^{-3} = (0.80 + 0.114 \times 33.4) \times 10^{-3} = 4.067 \times 10^{-3}$$

Therefore, the wind velocity with a height of 19.5 meters is as follows:

$$u = u_{19.5} = \left[1 + \frac{(C_{10})^{\frac{1}{2}}}{K} \ln \frac{Z}{10}\right] u_{10} = \left[1 + \frac{(4.067 \times 10^{-3})}{0.4} \ln \frac{195}{10}\right] \times 33.4 = 36.7m/s$$

f. Expression of random wave spectrum

Take the calculated α, β, and u and gravitational acceleration g = 9.8 m/s^2 into Equation (1-74):

$$S_{\varepsilon\varepsilon}(\omega) = \frac{0.0638}{\omega^5} e^{\left(-\frac{0.0018}{\omega^4}\right)} \tag{1-79}$$

For different circular frequencies ω, different values of $S_{\varepsilon\varepsilon}(\omega)$ can be determined. Take ω and $S_{\varepsilon\varepsilon}(\omega)$ as the abscissa and ordinate, respectively, and Equation (1-79) can be expressed by a continuous curve. The power spectral density function curve of random waves is as seen in Figure 1-17. From the figure, the spectral peaks are concentrated in a narrow range of the frequency domain where the energy of ocean waves is the largest, which is worth considering in the construction of offshore oil engineering in the South China Sea.

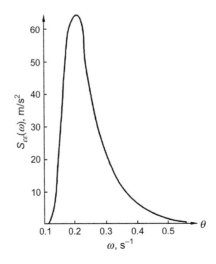

FIGURE 1-17

Random wave spectra by the indirect system.

1.4.4.3 *Several Accepted Random Wave Spectrums*

1. Mathematical expressions of spectrums
   Within the mathematical expressions of random wave spectrums accepted at home and aboard presently, most of them have the form of the expression given by Newman in 1952, as follows:

   $$S_{\varepsilon\varepsilon}(\omega) = \frac{A}{\omega^p}\exp\left(-B\frac{1}{\omega^q}\right) \qquad (1\text{-}80)$$

   where p is roughly 5 to 6 and q is roughly 2 to 4 generally; the coefficients A and B are the parameters of the wave factors (wave height, period, etc.).
   The advantage of this spectrum expression is that it is simple in structure and convenient . The $1/\omega^p$ on the right of Equation (1-80) has a larger influence on the high-frequency part of the spectrum and decreases along with the frequency; the exponential part of the function changes rapidly from 0 to 1, which makes the value of spectrum 0 when ω is 0. Therefore, multiplying the two parts, the spectrum increases sharply in the beginning, and then decreases sharply tending to zero gradually, just as shown in Figure 1-18. The value of p, q, A, and B in Equation (1-80) can be adjusted according to the different conditions of various sea area.
   However, this expression of the Newman spectrum is a type of empirical equation with an inadequate theoretical basis. Also, there is no high order torque exist for this expression, which is harmful for lucubrate. Because of these disadvantages, expressions of the spectrum internationally accepted are obtained by taking a large number of observations of different sea areas in the world, accumulating statistical data, and making theoretical generalization on the basis of Newman spectrum. Here are these expressions.

   a. Improved JONSWAP spectrum
      The JONSWAP (Joint North Sea Wave Project) spectrum was jointly developed by the countries related to oil and gas in Europe's North Sea in 1968–1969. After years of improvment, the power spectrum density function $S_{\varepsilon\varepsilon}(\omega)$ is as follows:

      $$S_{\varepsilon\varepsilon}(\omega) = \beta_j H_{\frac{1}{3}}^2 \frac{\omega_p^4}{\omega^5}\exp\left[-\frac{5}{4}\left(\frac{\omega_p}{\omega}\right)^4\right]\gamma^{\exp\left[\frac{-(\omega-\omega_p)^2}{2\sigma^2\omega_p^2}\right]}$$

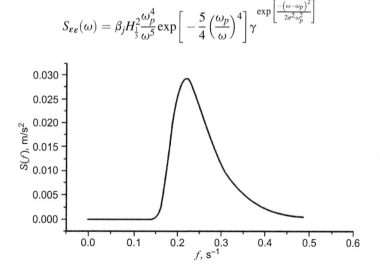

FIGURE 1-18

The curve of Newman's wave spectrum expression.

where $T_p = \dfrac{T_{\frac{1}{3}}}{1-0.132(\gamma+0.2)^{-0.559}}$ and

$$\beta_j = \dfrac{T_{\frac{1}{3}}}{0.23 + 0.0336\gamma - 0.185(1.9+\gamma)^{-1}}(1.094 - 0.01915 \ln \gamma) \tag{1-82}$$

where γ is the rising factor for the spectral peak (generally the observation value is about 1.5 to 6.0, and the average is 3.3); σ is the peak shape factor (when $\omega \le \omega_p$ take $\sigma = 0.07$, and when $\omega_p \le \omega$, take $\sigma = 0.09$); and $\omega_p = 2\pi/T_p$, where T_p is the peak spectral period.

b. B-M (Bretschneider-Mitsuyasu) spectrum

This was developed in 1970, and its expression is as follows:

$$S_{\varepsilon\varepsilon}(\omega) = A\dfrac{(2\pi)^4}{\omega^5}\exp\left[-B\dfrac{(2\pi)^4}{\omega^4}\right] \tag{1-83}$$

where $A = 0.257\dfrac{H_{\frac{1}{3}}^2}{T_{\frac{1}{3}}^4}; B = \dfrac{1.03}{T_{\frac{1}{3}}^4};$ and T_p is the peak spectral period, using $T_p = 1.05T_{1/3}$, i.e., $T_{1/3} = T_p/1.05$.

c. The improved Wallops spectrum

The Wallops spectrum was created in 1987, and its expression is as follows:

$$S_{\varepsilon\varepsilon}(\omega) = \beta_\omega H_{\frac{1}{3}}^2 \dfrac{\omega_p^{m-1}}{\omega^m}\exp\left[-\dfrac{m}{4}\left(\dfrac{\omega_p}{\omega}\right)^4\right] \tag{1-84}$$

where

$$\beta_\omega = \dfrac{0.0628m^{\frac{m-1}{4}}}{4^{\frac{m-1}{4}}\Gamma[(m-1)]}\left[1 + 0.7458(m+2)^{-1.057}\right]$$

$$T_p = \dfrac{T_{\frac{1}{3}}}{1-0.238(m-1.5)^{-0.684}}; \omega_p = \dfrac{2\pi}{T_p}$$

where the coefficient m is 5 in deep water, and is roughly 3 to 4 in shallow water.

d. P-M (Pierson-Moscowitz) spectrum

This was developed in 1964, and the expression is as follows:

$$S_{\varepsilon\varepsilon}(\omega) = \dfrac{\alpha g^2}{\omega^5}\exp\left[-\beta\left(\dfrac{g}{u\omega}\right)^4\right] \tag{1-85}$$

The calculation of the coefficient α and β and the wind velocity u in Equation (1-85) is as noted in Equations (1-75), (1-76), and (1-77).

In additions, there also is the ISSC spectrum created in 1967 and the spectrum developed by Professor Shengchang Wen in 1990.

2. Comparison of several random wave spectrums

Figure 1-19 shows the power spectral density curves of four random wave spectrums, where B represents the B-M spectrum, C represents the spectrum developed by Shengchang Wen, D represents the JONSWAP spectrum, and E represents the spectrum put forward by the Chinese Academy of Sciences. Their element values are all determined through actual observation, and

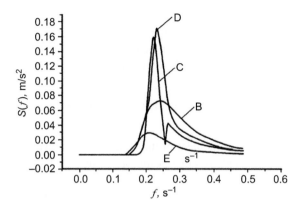

FIGURE 1-19

Comparison of the random ocean wave spectrums.

their characterization wave heights and periods are, respectively, $H_{1/3} = 1.60$ m and $T_{1/3} = 3.84$ s. From the figure, we can see that

- The differences in spectral values given by different spectrum expressions are great, and in particular the differences of the peak values are the most obvious.
- Because of the different sea conditions in different sea areas, the measured data is quite different, which is the primary source of the difference of spectrum values.
- It is not realistic to make a spectrum expression suitable for random sea waves in all sea areas of the world. It is desirable to use a special spectrum aimed at a specific area such as a direct method aimed at a certain area or anindirect method with an accepted spectrum.

1.4.4.4 The Relationship between Random Wave Power Spectrums and the Elements of Waves

1. The relationship between a random wave power spectrum and the average wave height of random waves $\overline{H}$

We can write the expression that the probability distribution of random wave heights obeys a Rayleigh distribution, as seen in Equation (1-37),s ubstitute the variance σ_H^2 into m_0 in Equation (1-87), and according to the definition of mathematical expectation (average) write the average wave height $\overline{H}$ as $\overline{H} = \int_0^\infty Hp(H)dH$[from Equation (1-36)]. By derivation we will get

$$\overline{H} = \sqrt{2\pi m_0} = \sqrt{2\pi \sigma_H^2} = \sqrt{2\pi} \left[\int_{-\infty}^{\infty} S_{\varepsilon\varepsilon}(\omega)d\omega \right]^{1/2} \tag{1-86}$$

2. The relationship between a random wave power spectrum and the variance of wave height,σ_H^2
Multiply both sides of Equation (1-61) by $\int_{-\infty}^{\infty} d\omega$ and we will get

$$\int_{-\infty}^{\infty} S_{\varepsilon\varepsilon}(\omega)d\omega = \int_{-\infty}^{\infty} \left[\lim_{T \to \infty} |\varepsilon(\omega)^2| \right] d\omega \tag{1-87}$$

In Equation (1-87), the left side is the power spectrum of random waves and the right side is the mean square value of the wave surface $\varepsilon(\omega)$ or the variance σ_H^2 (statistical characteristic variance of wave height) in mathematical statistics. Therefore, the relationship betweein them is

$$\int_{-\infty}^{\infty} S_{\varepsilon\varepsilon}(\omega)d\omega = \text{the area of the density function curve in the power spectrum} = \sigma_H^2 = m_0 \quad (1\text{-}88)$$

where m_0 is called the zero order moment of the random wave power spectrum.

3. The relationship between a random wave power spectrum and the average height of the highest one-third wave $H_{\frac{1}{3}}$

In accordance with the concept that the wave height probability distribution of random waves obeys the Rayleigh distribution and the definition of $H_{\frac{1}{3}}$, we write the expression of weighted average, and replace σ_H^2 in the Rayleigh distribution expression by m_0:

$$H_{\frac{1}{3}} = 4.0\sqrt{m_0} = 4.0 \left[\int_{-\infty}^{\infty} S_{\varepsilon\varepsilon}(\omega)d\omega \right]^{\frac{1}{2}} \quad (1\text{-}89)$$

4. The relationship between a random wave power spectrum and the average height of the highest one-tenth wave $H_{\frac{1}{10}}$

After derivation with the principle just above, we have

$$H_{\frac{1}{10}} = 5.1\sqrt{m_0} = 5.1 \left[\int_{-\infty}^{\infty} S_{\varepsilon\varepsilon}(\omega)d\omega \right]^{\frac{1}{2}} \quad (1\text{-}90)$$

5. The relationship between a random wave power spectrum and the average period of a random wave $\overline{T}$

The probability $P[\varepsilon(t)]$ of wave height $\varepsilon(t)$ and the probability $P[d\varepsilon/dt]$ of the velocity change $\frac{d\varepsilon}{dt}$ of a wave (namely the slope of the wave curve) can be regarded as Gaussian distributions. Take the wave crossing the zero axis each time as the wave height being zero $[\varepsilon(t) = 0]$, which is the product of the probability of mutually independent events obeying Gaussian distribution. Therefore, we can get the expression of the average total number N_0 of waves across the zero axis in unit time, $N_0 = P[\varepsilon(t), d\varepsilon/dt]/dt$. Then divide it by 2 (because a period means the wave crosses zero twice), and we will get the mean frequency, $N_0/2$, whose reciprocal is the average period $\overline{T}$. Deduce the expression $\overline{T} = \frac{2}{N_0}$:

$$\overline{T} = \left[\frac{\int_{-\infty}^{\infty} S_{\varepsilon\varepsilon}(\omega)d\omega}{\int_{-\infty}^{\infty} \omega^2 S_{\varepsilon\varepsilon}(\omega)d\omega} \right]^{\frac{1}{2}} = \left(\frac{m_0}{m_2} \right)^{\frac{1}{2}} \quad (1\text{-}91)$$

In Equation (1-91), m_2 is the secondary moment of the random wave power spectrum.

From the above, as long as the random wave power spectrum can be created, the elements of random waves, such as wave height, period, and other statistical characteristic data, will be obtained. So it is convenient to choose these required statistical characteristic values when calculating the wave loads on the sea oil engineering structure.

1.4.4.5 Random Wave Direction Spectrum

1. Expressions

   In the previous study of random waves, the random wave surface $\varepsilon(t)$ is regarded as composing of series of simple harmonic waves $\varepsilon(t) = A sin(Kx\text{-}\omega t)$ with different amplitudes A, frequencies ω, and initial phases ε. However, we did not consider determining the propagation direction of a random wave.

   Set the angle consisting of random waves and the x axis in the x–y plane as θ, and take θ as the direction of propagation of random waves. Therefore, the expression of a random wave $\varepsilon(x,y,t)$ that is composed of a series of simple harmonic waves and takes the variability element θ into condition in addition to amplitude A, frequency ω, and initial phase ε is

$$\varepsilon(x, y, t) = \sum_{n-1}^{\infty} A_n \cos\left(\frac{\omega_n^2}{g} \cos\theta_n + \frac{\omega_n^2 y}{g} \sin\theta_n - \omega_n t + \varepsilon_n \right) \tag{1-92}$$

   where

   a. ε_n represents the initial phase comprehensively considering various factors.

   b. For $\frac{\omega_n^2}{g}$, wave velocity C is generally expressed by $\sqrt{\frac{g}{k}}$. But from Equation (1-48) $C = \frac{\omega_n}{K}$, so $\frac{\omega_n}{K} = \sqrt{\frac{g}{K}}$. Then $\frac{\omega^2 n}{g} = K$, and K is wave number, thus $\frac{\omega^2 n}{g}$ means wave number.

   c. $x\cos\theta_n$ represents the component on the x axis of the distance (length) that the random wave propagates forward along the angle θ, i.e., wavelength λ_x

   d. $y\sin\theta_n$ represents the component on the y axis of wave propagation distance, i.e., wavelength λ_y.

   e. For $\frac{\omega^2 x}{g}\cos\theta_n$, according to the above analysis, we should have $K\lambda_x$. But $K\lambda_x = \omega t$ from Equation (1-48), so $\frac{\omega^2 x}{g}\cos\theta_n$ is the phase after the consideration of the changes of wave propagation direction.

   f. $\omega_n t$ represents the phase only considering the change of amplitude, frequency, and initial phase.

   g. $\left(\frac{\omega^2 n}{g} x \cos\theta_n + \frac{\omega}{g} y \cos\theta_n - \omega_n t + \varepsilon_n \right)$ obviously represents the phase, that is the phase when considering in totality the change of amplitude, frequency, and initial phase, and the direction of propagation

   h. A_n represents different amplitudes.

   All in all, Equation (1-92) is the mathematical expression describing random waves that are composed of simple harmonic waves with different amplitudes A_n, frequencies ω_n, initial phases ε_n, directions of propagation θ_n, and phases.

   If the power spectral density obtained according to the above is $S_{\varepsilon\varepsilon}(\omega,\theta)$, the integral form of the directional spectrum in consideration with the change of propagation direction should be

$$\int_{0}^{\infty} \int_{-\pi}^{\pi} S_{\varepsilon\varepsilon}(\omega, \theta)d\omega d\theta \tag{1-93}$$

Equation (1-93) is the expression of the direction spectrum of a random wave, and its power spectral density function curve is shown in Figure 1-20.

2. Power spectral density function curve

According to the theory of creating a random wave spectrum by the direct method, the power spectrum density function curve of a random wave direction spectrum can also obtained, as shown in Figure 1-20. It shows that, after taking the change of the propagation direction of random waves into consideration, the energy is obviously concentrated in roughly $-90°$ to $90°$, and the peak of the energy in the positive direction is at $0°$. It is for this reason that waves are vertical to winds without component and offset, and the max energy occurs.

3. The relationship between the direction spectrum and the random wave spectrum without consideration of changes of wave propagation direction

If the average total energy per unit water particle in a random wave is E, the expression for a random wave without consideration of the changes of the propagation direction is

$$E = \int_{-\infty}^{\infty} S_{\varepsilon\varepsilon}(\omega)d\omega = \sigma_H^2 = m_0 \tag{1-94}$$

But the direction spectrum with consideration of changes of propagation direction should be

$$E = \int_{0}^{\infty} \int_{-\pi}^{\pi} S_{\varepsilon\varepsilon}(\omega, \theta)d\omega d\theta = \sigma_H^2 = m_0' \tag{1-95}$$

As the average total energy is E, make Equation (1-94) equal to Equation (1-95). Then

$$\int_{0}^{\infty} S_{\varepsilon\varepsilon}(\omega)d\omega = \int_{0}^{\infty} \int_{-\pi}^{\pi} S_{\varepsilon\varepsilon}(\omega, \theta)d\omega d\theta$$

$$S_{\varepsilon\varepsilon}(\omega) = \int_{-\pi}^{\pi} S_{\varepsilon\varepsilon}(\omega, \theta)d\theta \tag{1-96}$$

FIGURE 1-20

Power spectral density function curve.

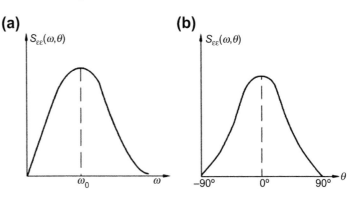

(a) $S_{\varepsilon\varepsilon}(\omega, \theta)$ ω_0 ω

(b) $S_{\varepsilon\varepsilon}(\omega, \theta)$ $-90°$ $0°$ $90°$ θ

Equation (1-96) is the relationship between the power spectrum density function $S(\omega, \theta)$ of the direction spectrum and the power spectral density function $S_{\varepsilon\varepsilon}(\omega)$ without consideration of changes of wave propagation direction.

Furthermore, from Equation (1-94) and Equation (1-95), we can also obtained that $\sigma_H^2 = \sigma_{HD}^2$ and $m_0 = m_0'$. They show that the variance of the wave height and the zero order moment (the area under the power spectral density curve) of these two spectrums are equal.

1.5 **Wave Theory and the Wave Force of Ocean Wave**

For offshore oil engineering, the main purpose of studying the ocean wave is to analyze the interaction between ocean waves and ocean engineering structure and calculate the wave loads acting on the structure. But we nees to know the rules of the changing of velocity and pressure for all the points in the waves to achieve this goal. Thus, we can determine the distribution of the wave energy of ocean waves and the size of the forces that the waves are putting on the engineering structure. In short, we need to study the calculation of the wave force and wave theory. This section will introduce these topics.

1.5.1 **Fluid Mechanics Basics**

The basic concepts, moment, and the dynamics equation in fluid mechanics will be used to analyze the wave theory of waves.

1.5.1.1 *Basic Concepts*

1. The velocity field and the velocity potential

   Seawater can be regarded as a "chunk object" and the motion state of its particle is different along with the position in space (x, y, z) and the time (t). The velocity is a physical quantity both with size (scalar) and direction (vector) to describe motion (state). Therefore, the physical phenomena that describe the size and direction of the velocity of a seawater particle ("chunk object") changing with position in space (x, y, z) and time (t) is called a physical field (velocity). This physical phenomenon can be described with a mathematical function, which is called the mathematical field (velocity). Using vector v to present the mathematics field of seawater particle velocity in the ocean, this function is

$$v = v(x, y, z, t) = iu + jv + kw \tag{1-97}$$

   In Equation (1-97), $i, j,$ and k respectively are unit vectors along the $x, y,$ and z coordinate axes; $u, v,$ and w are the three velocity components on the $x, y,$ and z axes. They can be expressed as in the following functions:

$$\left.\begin{array}{l} u = u(x, y, z, t) \\ v = v(x, y, z, t) \\ \omega = \omega(x, y, z, t) \end{array}\right\} \tag{1-98}$$

If the velocity field describes the change of the size of its velocity value (scalar), not the change of direction, then it can be expressed by a scalar math function. This function is called the velocity potential; we often use Φ to express this, namely

$$\Phi = \Phi(x, y, z, t) \tag{1-99}$$

Obviously, the velocity field, which is a vector, and the velocity potential, which is a scalar, are different.

2. Divergence, gradient, and rotation

    a. Divergence of a vector field

        Suppose we are using div v to represent the divergence of the velocity field v; it should be defined as

$$\mathrm{div}v = \nabla.v = \left(i\frac{\partial}{\partial x} + j\frac{\partial}{\partial y} + k\frac{\partial}{\partial z} \right)v$$

$$= i\frac{\partial v}{\partial x} + j\frac{\partial v}{\partial y} + k\frac{\partial v}{\partial z} = \left|\frac{\partial}{\partial x}\right|v\cos\alpha + \left|\frac{\partial}{\partial x}\right|v\cos\beta + \left|\frac{\partial}{\partial z}\right|v\cos\gamma = \frac{\partial u}{\partial x} + \frac{\partial v}{\partial y} + \frac{\partial w}{\partial z} \tag{1-100}$$

In Equation (1-100), α, β, and γ respectively represent the angle between the direction of the velocity v and the coordinates of the x, y, and z axes. The products of v and the cosine of these angles are respectively the three velocity components u, v, and w. ∇ is spelled as "deal"; it is a sign of operation, expressed as the following equation:

$$\nabla = i\frac{\partial}{\partial x} + j\frac{\partial}{\partial y} + k\frac{\partial}{\partial z} \tag{1-101}$$

From Equation (1-100), we see that the divergence is the synthesis of the degree of change along the three axes x, y, and z (change rate $\frac{\partial u}{\partial x} + \frac{\partial v}{\partial y} + \frac{\partial w}{\partial z}$) of the value of the components of velocity field v on the three axes x, y, and z.

2. Gradient of a scalar field

Use grad Φ to represent the gradient of velocity potential Φ (scalar); we can use the operator ∇ to define it as

$$\mathrm{grad}\Phi = \nabla\Phi = \left(i\frac{\partial}{\partial x} + j\frac{\partial}{\partial y} + k\frac{\partial}{\partial z} \right)\Phi$$

$$= i\frac{\partial\Phi}{\partial x} + j\frac{\partial\Phi}{\partial y} + k\frac{\partial\Phi}{\partial z} \tag{1-102}$$

But if the velocity potential exists, $\nabla\phi$ is equal to v according to engineering mathematics, so it also can be concluded from Equation (1-102) that

$$\nabla\Phi = v = iu + jv + kw = i\frac{\partial\Phi}{\partial x} + j\frac{\partial\Phi}{\partial y} + k\frac{\partial\Phi}{\partial z} \tag{1-103}$$

Therefore, from Equation (1-103) we can obtain

$$u = \frac{\partial \Phi}{\partial x}, v = \frac{\partial \Phi}{\partial y}, \omega = \frac{\partial \Phi}{\partial z} \tag{1-104}$$

Equation (1-103) shows that the gradient is the degree (rate) of change of the size and direction of the velocity changing along the axes x, y, and z. It is obtaining a vector $i\frac{\partial \Phi}{\partial x} + j\frac{\partial \Phi}{\partial y} + k\frac{\partial \Phi}{\partial z}$ from a scalar Φ, while the divergence is obtaining a scalar $\frac{\partial u}{\partial x} + \frac{\partial v}{\partial y} + \frac{\partial \omega}{\partial z}$ from a vector v. This is why the gradient is different from the divergence.

3. Rotation of the vector field

Suppose the product of the velocity field vector v and the operator ∇, which is regarded as a vector, is represented by $\nabla \times v$ (it is a scalar product without the multiplication sign), describe it as the rotation, and represent it with rot$v$; then

$$\text{rot}v = \nabla \times v = \left(i\frac{\partial}{\partial x} + j\frac{\partial}{\partial y} + k\frac{\partial}{\partial z} \right) \left(iu + jv + k\omega \right)$$

$$= ii\frac{\partial u}{\partial x} + ij\frac{\partial u}{\partial y} + ik\frac{\partial u}{\partial z} + ij\frac{\partial v}{\partial x} + ij\frac{\partial v}{\partial y} + kj\frac{\partial v}{\partial z} + ik\frac{\partial \omega}{\partial x} + kj\frac{\partial \omega}{\partial y} + kk\frac{\partial \omega}{\partial z}$$

$$= 0\frac{\partial u}{\partial x} + (-k)\frac{\partial u}{\partial y} + (+j)\frac{\partial u}{\partial z} + (-k)\frac{\partial u}{\partial x} + 0\frac{\partial v}{\partial y} + (-i)\frac{\partial v}{\partial z} + (-j)\frac{\partial \omega}{\partial x} + (+i)\frac{\partial \omega}{\partial y} + 0\frac{\partial \omega}{\partial z}$$

$$= i\begin{vmatrix} \frac{\partial}{\partial y} & \frac{\partial}{\partial z} \\ v & \omega \end{vmatrix} - j\begin{vmatrix} \frac{\partial}{\partial x} & \frac{\partial}{\partial z} \\ u & \omega \end{vmatrix} + k\begin{vmatrix} \frac{\partial}{\partial x} & \frac{\partial}{\partial y} \\ u & v \end{vmatrix}$$

$$= \begin{vmatrix} i & j & k \\ \frac{\partial}{\partial x} & \frac{\partial}{\partial y} & \frac{\partial}{\partial z} \\ u & v & \omega \end{vmatrix} = \begin{vmatrix} i & j & k \\ \frac{\partial}{\partial x} & \frac{\partial}{\partial y} & \frac{\partial}{\partial z} \\ -y\omega & x\omega & 0 \end{vmatrix} = \begin{vmatrix} i & j & k \\ \frac{\partial}{\partial x} & \frac{\partial}{\partial y} & \frac{\partial}{\partial z} \\ \omega r & \omega r & 0 \end{vmatrix} = f(\omega) \tag{1-105}$$

In Equation (1-105), because the linear velocities u, v, and w can be represented by the product of angular velocity ω and radius r (x or y coordinates), the u, v, and w in the determinant are transformed correspondingly.

Equation (1-105) shows that the rotation rot$v$ is a function $f(\omega)$ of rotating angular velocity ω. Therefore, if $\omega = 0$, rot$v = 0$ means that we have irrotational motion; if $\omega \neq 0$, rot$v \neq 0$ means that we have rotational movement. So, rot$v$ represents the degree of rotation, and thus it is called the rotation.

4. Steam line and the steam function

a. Steam line

The steam line is a group of curves, but all the tangent directions of curves in the group should be the same as the direction of the velocity of the water particles at the same point.

As shown in Figure 1-21, suppose the point $M(x, y)$ is a point on the line, the tiny tangent at this point is dr, and the velocity vector at this point is v. Then the characteristic of the spatial steam line of the stationary flowing fluid is

$$dr \times v = (idx + jdy + kdz)(iu + jv + k\omega) = 0 \tag{1-106}$$

b. Steam function

Normally when fluid moves as a stationary current, in addition to the velocity potential Φ, a stream function also exists. The steam function is a scalar that describes the motion characteristics of a steam line with mathematical functions; generally it is written in the form of $\Psi\ (x, y)$:

$$\Psi(x, y) = c \tag{1-107}$$

That is,

$$\frac{\partial \Psi}{\partial x} dx + \frac{\partial \Psi}{\partial y} dy = 0 \tag{1-108}$$

5. The relationship between the steam function and the potential function

From Figure 1-21, we can see that the following geometric relationships exist:

$$\frac{u}{|v|} = \frac{dx}{dr}, \frac{v}{|v|} = \frac{dy}{dr} \tag{1-109}$$

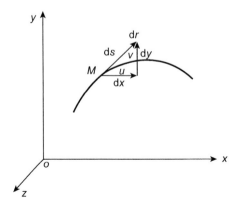

FIGURE 1-21

Diagram of streamline and current function.

That is,

$$\frac{dx}{u} = \frac{dr}{|v|} = \frac{dy}{|u|} \tag{1-110}$$

It can be concluded that

$$udy - vdx = 0 \tag{1-111}$$

Substitute the expressions for u and v in Equation (1-104) into Equation (1-111). Then

$$\frac{\partial \Phi}{\partial x} dy - \frac{\partial \Phi}{\partial x} dx = 0 \tag{1-112}$$

If we set Equation (1-112) equal to Equation (1-108), the following relationships can be obtained:

$$\frac{\partial \Phi}{\partial x} = \frac{\partial \psi}{\partial y}, \frac{\partial \Phi}{\partial y} = \frac{\partial \psi}{\partial x} \tag{1-113}$$

Equation (1-113) showcases the relationship between the stream function Ψ and potential function Φ.

6. Complex potential and complex velocity
   a. Complex potential
   If the potential function $\Phi(x, y)$ is taken as the real part of complex function $F(Z)$, and the stream function $\Psi(x, y)$ is taken as its imaginary part, it can be written

$$F(Z) = \Phi(x, y) + i\Psi(x, y) \tag{1-114}$$

   $F(Z)$ in Equation (1-114) is known as the complex potential.
   b. Complex velocity
   The first order partial derivative of complex potential function $F(Z)$ is called the complex velocity, and the mathematic expression of it is as follows:

$$\frac{dF(Z)}{dZ} = \frac{\partial F(Z)}{\partial Z}\frac{\partial Z}{\partial x} = \frac{\partial F(Z)}{\partial x} = \frac{\partial \Phi}{\partial x} + i\frac{\partial \psi}{\partial x} = u - iv \tag{1-115}$$

7. Stationary current and unsteady current
   When the velocity field of the fluid is a stability field, namely the velocity only changes with the location of the field point (x, y, z), and does not change with time t, then the fluid movement is called stationary current. Velocity field v can be written as

$$v = v(x, y, z) \tag{1-116}$$

On the contrary, the fluid velocity field is instable, namely the velocity not only changes with the location of the sites (x, y, z), but also changes with time t; thus the fluid movement is called unsteady current, and the expression of v is

$$v = v(x, y, z, t) \tag{1-117}$$

1.5.1.2 Basic Equations of Kinematics

1. Continuity equation

   The movement of fluid must follow the law of conservation of mass; namely it must obey the continuity equation.

   Now pick out a fixed geometry space surrounded by a closed surface S in the fluid and suppose its volume is τ, so in the unit time and in the space taken, the increase of fluid quality should be $\int_\tau \int \int \frac{\partial \rho}{\partial \tau} d\tau$. At the same time, fluid quality that moves through the boundary into the space should be $-\int \int_S \int \rho v_n ds = 0$. This is because, at this time, velocity components of the fluid particle along the outer normal direction of the boundary are v_n and the mass density of the fluid is ρ. Then, in accordance with the principle of conservation of mass, we have

   $$\int_\tau \int \int \frac{\partial \rho}{\partial t} d\tau + \int \int_S \int \rho v_n ds = 0 \tag{1-118}$$

   For and ideal fluid, because $\rho = $ constant, namely $\frac{\partial \rho}{\partial t} = 0$, Equation (1-118) can be written as

   $$\rho \left(\frac{\partial u}{\partial x} + \frac{\partial v}{\partial y} + \frac{\partial \omega}{\partial z} \right) = 0 \tag{1-119}$$

   It can be concluded using Equation (1-119) and Equation (1-100) that

   $$\frac{\partial u}{\partial x} + \frac{\partial v}{\partial y} + \frac{\partial \omega}{\partial z} = \nabla v = 0 \tag{1-120}$$

   Equation (1-120) is usually called the continuity equation of fluid motion.

2. The Laplace equation

   From Equation (1-120), we can determine that $\Delta v = 0$, but from Equation (1-103), $v = \Delta \Phi$, so we can write

   $$\nabla(\nabla \Phi) = \nabla^2 \Phi = \frac{\partial^2 \Phi}{\partial x^2} + \frac{\partial^2 \Phi}{\partial y^2} + \frac{\partial^2 \Phi}{\partial z^2} = 0 \tag{1-121}$$

   Because the second-order differential equation of the velocity potential Φ given by Equation (1-121) satisfies the Laplace equation in engineering mathematics, this second-order differential equation of velocity potential of fluid movement is called the Laplace equation.

3. Lagrange and Bernoulli equations

   a. Bernoulli equation describing steady flow

      The Bernoulli equation can be used to describe the conservation of energy for steady flow, namely

      $$\frac{1}{2} v^2 + \Omega + p = C \tag{1-122}$$

Here

$$\Omega = gZ$$

$$P = \frac{p}{\rho}$$

Also, $\frac{1}{2}v^2$ represents the kinetic energy for a unit mass; Ω represents the potential energy of a unit mass; Z represents height; P represents the pressure energy for a unit mass; ρ represents the quality density of seawater; and p represents pressure.

b. Lagrange equations describing the unsteady flow

For unsteady flow, the velocity field not only changes with the site's position (x, y, z), but also changes with time t. Therefore, the constant C' on the right of the energy conservation equation should be replaced by the function $F(t)$ changing with time. On the left of Equation (1-122), in addition to velocity energy term $\frac{1}{2}v^2$, there should also be a work $\frac{1}{\rho}\rho\frac{\partial\Phi}{\partial t} = \frac{\partial\Phi}{\partial t}$ acting on a unit mass by the inertia force $\rho\frac{\partial\Phi}{\partial t}$, which is produced by velocity acceleration $\frac{\partial\Phi}{\partial t}$ changing over time. Therefore, in accordance with the principle of conservation of energy as seen in Equation (1-122), we can write

$$\frac{\partial\Phi}{\partial t} + \frac{1}{2}v^2 + \Omega + p = F(t) \tag{1-123}$$

Equation (1-123) is the Lagrange equation describing the energy conservation of unsteady flow.

1.5.1.3 Fundamental Equations of Dynamics (Euler Equation)

We now analyze the basic equations using the Euler method to describe fluid motion with a velocity field.

1. Force analysis

Take out a small unit (dx, dy, dz) from the moving fluid, as shown in Figure 1-22. The volume of this small unit is $dxdydz$. The pressure that this small unit bears is p, the gravitational acceleration along the x, y, and z axis directions respectively is $X = g_x$, $Y = g_y$, and $Z = g_z = -g$. Then, the force acting on this small unit can be described as follows.

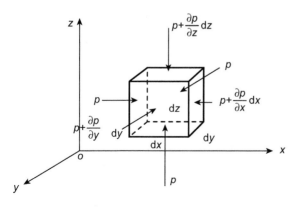

FIGURE 1-22

Analysis diagram of fluid dynamic equation.

a. The surface force, that is, the product of area and pressure. If we take the surface force along the x axis as an example, it should be

$$pdydz - \left(p + \frac{\partial p}{\partial x}dx\right)dydz$$

Similarly, the surface force along the y axis and z axis respectively is

$$pdxdz - \left(p + \frac{\partial p}{\partial y}dy\right)dxdz$$

$$pdxdy - \left(p + \frac{\partial p}{\partial z}dz\right)dxdy$$

b. The body force, for which the gravity shall be the product of mass and the gravitational acceleration. So the body forces acting on the small volume $dxdydz$ along the x, y, and z axes are

$$\rho dxdydzX; \rho dxdydzY; \rho dxdydzZ$$

c. The inertia force is the product of mass and acceleration. If the acceleration along the x, y, and z axes are $\frac{du}{dt}, \frac{dv}{dt}$, and $\frac{dw}{dt}$, the inertia force along this three axes are

$$-\rho dxdydz\frac{du}{dt}; -\rho dxdydz\frac{dv}{dt}; -\rho dxdydz\frac{dw}{dt}$$

2. Balance of forces along the x, y, and z axes
Take the direction along the x axis as an example; if $\Sigma F_x = 0$, then

$$\sum F_x = pdydz - \left(p + \frac{\partial p}{\partial x}dx\right)dydz + \rho dxdydzX - \rho dxdydz\frac{du}{dt} = 0$$

or

$$\sum F_x = -\frac{1}{\rho}\frac{\partial p}{\partial x} + X - \frac{du}{dt} = 0 \qquad (1\text{-}124)$$

In these equations, F_x represents the force along the x axis.
Similarly, the superposition of force F_y along the y axis and F_z along the z axis is zero (that is, $\sum F_y = 0$ and $\sum F_z = 0$). So, we can write

$$\left.\begin{array}{l} X - \dfrac{1}{\rho}\dfrac{\partial p}{\partial x} = \dfrac{du}{dt} \\[2mm] pXY - \dfrac{1}{\rho}\dfrac{\partial p}{\partial y} = \dfrac{dv}{dt} \\[2mm] Z - \dfrac{1}{\rho}\dfrac{\partial p}{\partial z} = \dfrac{dw}{dt} \end{array}\right\} \qquad (1\text{-}125)$$

This is known as the fundamental equation of dynamics.

3. Simplify the expression

   The X, Y, and Z in Equation (1-125) are represented in a unified fashion with F. And the gradient of pressure $\frac{\partial p}{\partial x}$, $\frac{\partial p}{\partial y}$ and $\frac{\partial p}{\partial z}$ is expressed in a unified fashion by Δp. Then the derivative of velocity components u, v, and w to time $\frac{du}{dt}$, $\frac{dv}{dt}$ and $\frac{dw}{dt}$ is expressed in a unified fashion by $\frac{dv}{dt}$. Equation (1-125) can be simplified into the following equation:

$$F = -\frac{1}{\rho}\Delta p = \frac{dv}{dp} \tag{1-126}$$

1.5.1.4 Boundary Conditions

1. Kinematic boundary conditions
   a. Boundary conditions on the seabed

   The seabed is one boundary of seawater. The origin of the coordinates is at sea level. The axis that is perpendicular to sea level is the z axis and the direction towards the seabed along the z axis is the negative direction. The depth of sea is h, and the boundary is where the coordinate $z = -h$. Here, the particle of water can't cross through the seabed and only moves along the tangent direction of the boundary; thus the velocity component w along the z axis of water particle should be zero, i.e.,

$$\omega\Big|_{z=-h} = \frac{\partial \Phi}{\partial z}\Big|_{z=-h} = 0 \tag{1-127}$$

   b. Free surface boundary conditions

   The free surface is the sea surface, which is the other boundary of seawater. The z coordinate there is $z = 0$. If the velocity component along the z axis is w, and the wave surface of the sea wave on the sea surface is ε, we get

$$\omega\Big|_{z=0} = \frac{\partial \Phi}{\partial z}\Big|_{z=0} = \frac{\partial \varepsilon}{\partial t}\Big|_{z=0} \tag{1-128}$$

   From this we can obtain the boundary conditions.
2. Dynamic boundary conditions
   a. Free surface boundary conditions

   For unsteady flow, according to the Lagrange equation, if the pressure on the sea surface is p_0, then the boundary condition of the free surface can be written as

$$\frac{\partial \Phi}{\partial t} + \frac{1}{2}v^2 + gz + \frac{p}{\rho} = F(t) = \frac{p_0}{\rho}$$

   At the sea surface, the z coordinate is $z = 0$, where pressure is $p = p_0$ and the velocity vector $v \approx 0$, so from the equation above, it can be concluded that

$$gz = -\frac{\partial \Phi}{\partial t}$$

So, the boundary conditions can be written as

$$z|_{z=0} = \varepsilon|_{z=0} = -\frac{1}{g}\frac{\partial \Phi}{\partial t}\bigg|_{z=0} \tag{1-129}$$

b. Seabed boundary conditions

On the seabed the z coordinate is $z = -h$, the pressure $p = p$, and $v \approx 0$, so for the unsteady flow, the Lagrange equation can be written as

$$\frac{\partial \Phi}{\partial t} + gh + \frac{p}{\rho} = F(t) = \frac{p_0}{\rho} \tag{1-130}$$

So the boundary conditions can be written as

$$-\frac{\partial \Phi}{\partial t}\bigg|_{z=-h} = \frac{p - p_0}{\rho} + gh\bigg|_{z=-h|} \tag{1-130'}$$

1.5.2 The Undulation Theory of Slight Waves

Wave theory is a system theory that uses mathematics to describe the fluctuation of water particles in waves. General wave theory includes the wave equation, velocity potential, each component of wave velocity, the pressure distribution, and a finite amplitude wave. Different hypotheses can be divided into a variety of wave theory such as the slight wave and the finite amplitude wave. Here we introduce the slight undulation theory first. The main hypothesis of a slight wave is that the amplitude of the fluctuation of waves is regarded as infinitely small relative to wavelength, which means we analyzed it as small amplitude fluctuations. The assumption of undulation theory of slight waves has significantly differences with the method that is applied in engineering design, which involves large wave heights. But this theory on the basis of a finite amplitude wave can express the steps and method of fluctuation characteristics. So we take undulation theory of slight waves as an example here, and introduce the method and steps of building a wave theory of ocean waves systematically.

We use an analysis of unsteady flow to analyze the fluctuation theory of slight waves here. But in order to facilitate analysis, we consider the two-dimensional fluctuations only, and ignore the three-dimensional complex fluctuations. Now we will introduce the steps and method to build the undulation theory of slight waves.

1.5.2.1 Basic Assumptions

The undulation theory of slight waves was established and based on the following assumptions:

1. The amplitude of a wave is infinitely small related to wavelengths.
2. Wave motion is irrational, but it does have unsteady flow.
3. Gravity is the sole external force apart from the viscous force.
4. The pressure of the sea surface is constant.

1.5.2.2 Export the Velocity Potential Function

1. Preliminary: Set the velocity potential expression
   From the dynamic boundary conditions at the sea surface shown in Equation (1-129), the wave surface is $\varepsilon|_{z=0} = -\frac{1}{g}\frac{\partial \Phi}{\partial t}\Big|_{z=0}$. We can thus conclude that

$$\int -d\Phi = \int \varepsilon g dt = \int gA \cos(kx - \omega t) dt$$

That is,

$$\Phi = \frac{Ag}{\omega} \sin(kx - \omega t) \tag{1-131}$$

Hence, $Z(z)$ represents the function of amplitude changing with the z coordinates on the Z axis. Following Equation (1-131), the velocity potential function can be set preliminarily as follows:

$$\Phi = Z(z)\cos(kx - \omega t) \tag{1-132}$$

2. Second derivative of velocity potential
   The expression of the second derivative is

$$\left.\begin{aligned}
\frac{\partial^2 \Phi}{\partial x^2} &= Z(z)\big[-k^2 \cos(kx - \omega t)\big] \\
\frac{\partial^2 \Phi}{\partial z^2} &= Z''(z)[\cos(kx - \omega t)]
\end{aligned}\right\} \tag{1-133}$$

3. Establish the Laplace equation
   From our knowledge of the Laplace equation in Equation (1-121), when the two-dimensional fluctuations occur, $\frac{\partial^2 \Phi}{\partial y^2} = 0$. Using this and Equation (1-133) we can write

$$\frac{\partial^2 \Phi}{\partial x^2} + \frac{\partial^2 \Phi}{\partial z^2} = Z''(z) - k^2[Z(z)] = 0 \tag{1-134}$$

4. Solve the differential equation for the velocity potential
   We want to solve the second-order constant coefficient linear homogeneous differential equation of Equation (1-134); we can get $Z(z)$ through the following steps.
   a. Write out the general solution

$$Z(z) = C_1 e^{kz} + C^2 e^{-kz} \tag{1-135}$$

   b. State the first derivative

$$Z'(z) = C_1 k e^{kz} + C^2 k e^{-kz} \tag{1-136}$$

c. Get the boundary condition of movement
When $z = -h$ under the sea, from Equation (1-127), use Equation (1-132) to calculate the boundary condition of movement:

$$\frac{\partial \Phi}{\partial z}\bigg|_{z=-h} = Z'(z)\cos(kx - \omega t) = 0 \tag{1-137}$$

d. Determine the constants C_1 and C_2
Substitute Equation (1-136) into Equation (1-137), and make it equal to $\frac{C}{2}$. Then

$$C_1 e^{-kh} = C_2 e^{kh} = \frac{C}{2} \tag{1-138}$$

Multiply Equation (1-138) by $e^{k(z+h)}$ and $e^{-k(z+h)}$, respectively:

$$\left.\begin{array}{l} \frac{1}{2}C e^{k(z+h)} = C_1 e^{-kh}\left[e^{k(z+h)}\right] = C_1 e^{hz} \\[2mm] \frac{1}{2}C e^{-k(z+h)} = C_2 e^{kh}\left[e^{-k(z+h)}\right] = C_2 e^{-kz} \end{array}\right\} \tag{1-139}$$

e. State the solution of the differential equations
Substitute Equation (1-139) into Equation (1-135), and the solutions $Z(z)$ of the differential equation, Equation (1-134), can be obtained:

$$Z(z) = C_1 e^{kz} + C^2 e^{-kz} = \frac{1}{2}C\left[e^{k(z+h)} + e^{-k(z+h)}\right]$$

$$= C\left[\frac{e^{k(z+h)} + e^{-k(z+h)}}{2}\right] = C \cosh k(z+h) \tag{1-140}$$

f. State the expression for velocity potential
Substitute Equation (1-140) into Equation (1-132). It can be concluded that the expression of velocity potential Φ is as follows:

$$\Phi = Z(z)\cos(kx - \omega t) = C \cosh k(z+h)\cos(kx + \omega t) \tag{1-141}$$

1.5.2.3 The Wave Surface Expression

Set the wave surface of two-dimensional fluctuation of seawater, which is unsteady flow, to ε. Then using Equation (1-141) and the dynamic boundary condition on the sea surface in Equation (1-129), the expression for a wave surface can be obtained as follows:

$$\varepsilon(x, z, t) = \varepsilon|_{z=0} = -\frac{1}{g}\frac{\partial \Phi}{\partial t}\bigg|_{z=0}$$

$$= -\frac{1}{g}[C \cosh k(z+h)\omega \sin(kx - \omega t)] \tag{1-142}$$

In Equation (1-142)

$$-\frac{1}{g}C\omega\cosh(z+h) = A \tag{1-143}$$

The expression for a wave surface in the Equation (1-142) can be written as

$$\varepsilon(x,z,t) = A\sin(kx - \omega t) \tag{1-144}$$

Because of the small amplitude at the sea surface, $z \approx 0$, the A in Equation (1-144) should be

$$A = -\frac{\omega}{g}C\cosh kh \tag{1-145}$$

and

$$C = -\frac{-Ag}{\omega\cosh kh} \tag{1-146}$$

Obviously, A and C shall be determined by the period (frequency ω), wave number (k), and depth (h) in different sea areas.

1.5.2.4 Determine the Velocity Component

We study a two-dimensional wave here. We just involve the horizontal velocity component u and vertical velocity component w. So, we can determine u and w respectively according to Equation (1-104) and Equation (1-141):

$$u = \frac{\partial \Phi}{\partial x} = -ck\cosh k(z+h)\sin(kx - \omega t)$$
$$= \frac{Agk}{\omega}\frac{\cosh k(z+h)}{\cosh kh}\sin(kx - \omega t) \tag{1-147}$$

$$\omega = \frac{\partial \Phi}{\partial z} = \frac{Agk}{\omega}\frac{\sinh k(z+h)}{\cosh kh}\cos(kx - \omega t) \tag{1-148}$$

For a sine wave, the amplitude A is seen as half of the wave height H, namely $A = H/2$. And $k = \frac{2\pi}{\lambda}$, $\omega = \frac{2\pi}{T}$, the wavelength $\lambda = T\sqrt{\frac{g\lambda}{2\pi}}$ is given in Equation (1-148). Then, $\frac{Agk}{\omega}$ in Equation (1-147) and Equation (1-148) can be changed to

$$\frac{Agk}{\omega} = \frac{H}{2}\cdot\frac{T}{2\pi}\cdot\frac{2\pi}{\lambda}g = \frac{H}{2}\cdot\frac{T}{2\pi}\cdot\frac{1}{T^2\left(\frac{g}{2\pi}\right)}\cdot 2\pi g = \frac{\pi H}{T} \tag{1-149}$$

If $\frac{Agk}{\omega}$ in Equation (1-147) and Equation (1-148) has been changed to Equation (1-149) we can conclude that

$$u = \frac{\pi H}{T}\frac{\cosh k(z+h)}{\cosh kh}\sin(kx - \omega t) \tag{1-150}$$

$$\omega = \frac{\pi H}{T}\frac{\sinh k(z+h)}{\cosh kh}\cos(kx - \omega t) \tag{1-151}$$

1.5.2.5 Determine the Pressure Distribution

According to the Lagrange equation from Equation (1-130) or the dynamic boundary conditions at the seabed given by Equation (1-130), we can determine that the pressure distribution is as follows:

$$\frac{p - p_0}{\rho} = -\frac{\partial \Phi}{\partial t} - gz = -[C \cosh k(z + h)\omega \sin(kx - \omega t)] - gz$$

$$= -C \cosh k(z + h)\omega \sin(kx - \omega t) - gz \tag{1-152}$$

In Equation (1-152), z represents the vertical axis (the z axis) from the sea surface, namely the distance from the surface down to the seabed; p is the pressure distribution; and p_0 is the surface pressure. If $z = -h, p$, which is the pressure distribution of the sea floor, can be obtained from Equation (1-130).

1.5.2.6 Determine the Wave Velocity

Wave velocity C is the propagation velocity of a wave. If the wavelength is λ and the period is T, you can write the wave velocity C as follows:

$$C = \frac{\lambda}{T} = \frac{2\frac{\pi}{k}}{2\frac{\pi}{\omega}} = \frac{\omega}{k} \tag{1-153}$$

From the dynamic boundary conditions of the sea surface ($z = 0$) we know that

$$\frac{\partial \Phi}{\partial z}\bigg|_{z=0} = \frac{\partial \varepsilon}{\partial t}\bigg|_{z=0} = -\frac{1}{g}\frac{\partial^2 \Phi}{\partial t^2}\bigg|_{z=0}$$

That is,

$$g\left(\frac{\partial \Phi}{\partial z}\right)\bigg|_{z=0} = -\frac{\partial^2 \Phi}{\partial t^2}\bigg|_{z=0} \tag{1-154}$$

According to the expression of velocity potential Φ in Equation (1-141), take $\frac{\partial \Phi}{\partial z}$ and $\frac{\partial^2 \Phi}{\partial t^2}$. We conclude that

$$gk \sinh k(z + h)|_{z=0} = \omega^2 \cosh k(z + h)|_{z=0}$$

So, we will get

$$\omega^2 = gk \tanh h|_{z=0}, \omega|_{z=0} = \sqrt{gk \tanh h} \tag{1-155}$$

Take Equation (1-155) into Equation (1-153), and we can calculate wave velocity C as

$$C = \frac{\omega}{k} = \frac{\sqrt{gk \tanh h|}}{k} \tag{1-156}$$

or

$$C = \frac{\lambda}{T} = \frac{\lambda}{\frac{2\pi}{\omega}} = \frac{\lambda}{2\pi}\sqrt{gk \tanh h|} = \sqrt{\frac{\lambda^2}{(2\pi)^2}g\frac{2\pi}{\pi}\tanh h}$$

$$= \sqrt{\frac{g\lambda}{2\pi}\tanh h} \tag{1-157}$$

For deep water, $h \to \infty$, $\tanh h \to 1$, so C can be

$$C = \sqrt{\frac{g\lambda}{2\pi}} \qquad (1\text{-}158)$$

1.5.2.7 Determine the Wave Energy

From fluctuating seawater, take a small volume of wavelength λ of unit length along the y axis, small section dx along the x axis for small section dx, and wave height $\varepsilon = dz$ along the z axis, as shown in Figure 1-23. Then the volume of this small piece should be $dz \times 1 \times dx$. Now the kinetic energy and potential energy are as follows.

1. **Kinetic energy**

   We consider a two-dimensional velocity component that only consists of $u = \frac{\partial \Phi}{\partial x}$ and $\omega = \frac{\partial \Phi}{\partial z}$. If $\left(\frac{\partial \Phi}{\partial x} + \frac{\partial \Phi}{\partial z}\right)^2$ is regarded approximately as $\left(\frac{\partial \Phi}{\partial x}\right)^2 + \left(\frac{\partial \Phi}{\partial z}\right)^2$, the kinetic energy $K.E$ of fluctuating seawater with a wavelength λ can be written as

$$K \cdot E = \frac{1}{2}\rho \iint (dz \times 1 \times dx) \left[\left(\frac{\partial \Phi}{\partial x}\right)^2 + \left(\frac{\partial \Phi}{\partial z}\right)^2\right]$$

$$= \frac{1}{2}\rho \int_0^\lambda \int_{-\infty}^0 (u^2 + \omega^2) dx dz \qquad (1\text{-}159)$$

   Take the expressions of u and ω in Equation (1-147) and Equation (1-148) into Equation (1-159), and change ω to the approximate expression $\omega^2 = kg$ in Equation (1-155) for arithmetic operations. Then you can write the $K.E$ as follows:

$$K \cdot E = \frac{1}{4}\rho g A^2 \lambda \qquad (1\text{-}160)$$

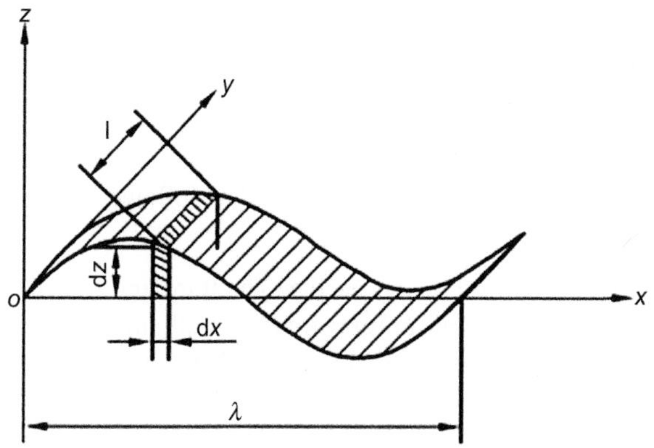

FIGURE 1-23

Energy diagram of fluctuating seawater.

2. Potential energy

Assume we have a small volume with $dz = \varepsilon$, and take as a position the mean value of the height of the wave surface, namely, $1/2\varepsilon$. The potential energy $P.E$ of fluctuations of seawater with wavelength λ can be written as

$$P.E = \int_0^\lambda \rho(dx \times 1 \times \varepsilon)g\frac{1}{2}\varepsilon = \rho g \int_0^\lambda \varepsilon^2 dx \qquad (1\text{-}161)$$

Then take the expression of wave surface ε in Equation (1-144) into Equation (1-161), so $\omega^2 = kg$. The potential energy $P.E$ of a wavelength is then as follows:

$$P.E = \frac{1}{4}\rho g\, A^2\lambda = \frac{1}{4}\rho g\left(\frac{H}{2}\right)^2\lambda = \frac{1}{16}\rho g H^2\lambda \qquad (1\text{-}162)$$

3. Total energy

The total energy of the fluctuation of water with wavelength E should be the sum of the kinetic energy $K.E$ and potential energy $P.E$, namely

$$E = K.E + P.E = \frac{1}{16}\rho g H^2\lambda + \frac{1}{16}\rho g H^2\lambda = \frac{1}{8}\rho g H^2\lambda \qquad (1\text{-}163)$$

In Equation (1-163), ρ is the density of seawater, g is the acceleration of gravity, H is the wave height, and λ is the wavelength.

In short, the above takes the slight wave as an example, and introduces the steps and method of building the sea wave theory of the system including wave expressions; the velocity potential function; calculation equations for the velocity components, pressure distribution, and wave propagation velocity; and calculation equations for the energy of the wave. This slight undulation theory is known as the sine wave theory by the side of the characteristics of the wave equation, and it is also called the Airy wave theory after the main creator of this theory.

1.5.3 The Finite Amplitude Undulation Theory

The slight undulation theory mentioned above takes wave height and wavelength (or water depth) as infinitesimal. Next, we will remove the assumption that the wave height is infinitesimal to study water wave theory. This is the finite amplitude undulation theory.

1.5.3.1 Main Characteristic of the Theory

The characteristics of finite amplitude undulation theory mainly can be described as follows.

1. Basic assumptions

Four assumptions were provided to explain slight undulation theory earlier. The third and fourth assumptions are still applicable. But the first assumption about infinitesimally small amplitudes should be removed, and the second one that takes seawater as unsteady current does not apply to the finite amplitude wave theory. So, the finite amplitude wave theory takes seawater as stationary current.

2. Wave shape

In finite amplitude undulation theories, such as the Stokes wave theory, the effect of wave steepness is considered. It proves that the wave shape is not a simple cosine wave, but rather that the shape of the crest is narrow and the trough wide, similar to the shape of the cycloid, as shown in Figure 1-24(b). The wave shape deduced by the slight undulation theory is shown in Figure 1-24(a), with small wave heights and wave steepness. There is the elliptical cosine wave theory suitable for shallow waves in addition to Stokes wave theory in finite amplitude undulation theory. It can reflect the wave steepness and influence related to the wave height and the wave shape shown in Figure 1-24(c). In addition, there is the wave theory that describes the limit state of an elliptical cosine wave where the depth tends to be small, known as the solitary wave theory. The wave surface deduced by this theory distributes entirely on quiet water, and the wavelength tends to infinitely great, as shown in Figure 1-24(d).

3. Analysis

There are differences in analysis between the finite amplitude undulation theory and slight undulation theory when establishing wave theory.

a. "Make" water a stationary current

Finite amplitude undulation theory assumes that when an advanced wave spreads from left to right with wave velocity C and an unchanged waveform, all seawater moves towards the opposite direction with velocity C, making the water move as a result of the resultant motion of the wave. Related to a fixed point in space, the status does not change with time, and thus the synthetic movement of seawater is "made" a stationary current, so we can analysis it with stationary current theory.

b. The water moves according to plane parallel motion

If we assume a fixed plane for a stationary current of water, every water particle on the same straight line vertical to this plane must move forward along the synthetic and parallel track with the same velocity and acceleration. So, according to the definition of plane parallel motion, we only need to study the movement of water particles in a plane parallel to a fixed plane when analyzing all water areas. The movement of water particles not changing over time

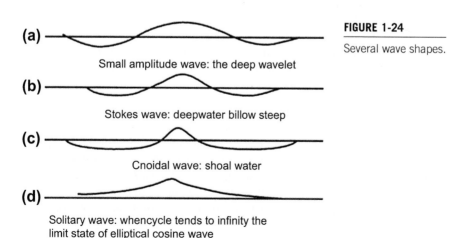

(a)

Small amplitude wave: the deep wavelet

(b)

Stokes wave: deepwater billow steep

(c)

Cnoidal wave: shoal water

(d)

Solitary wave: whencycle tends to infinity the limit state of elliptical cosine wave

FIGURE 1-24

Several wave shapes.

FIGURE 1-25

Water movement at different depths.

is called stationary planar motion in this plane. In finite amplitude undulation theory seawater is considered to be moving in plane parallel motion; commonly this method is referred to as the stationary planar motion analysis method.

Using this analysis method, as shown in Figure 1-25, the surface contour line of water is the current line. The deeper the water, the more straight the current line, and the current line turns into a straight line eventually. When the current line fully changes t a straight line, we can adopt the method of assuming stationary current to make the water move left at velocity C. Because all seawater is at a constant velocity (C) for translation and no new mechanics relations are brought in, the pressure along the free surface of synthetic movement stays constant. Namely the pressure on the surface contour line is $p = p_a$.

4. Function description

Finite amplitude undulation theory adopts the analysis method described above. It can be summed up as solving for the stationary current of seawater through mathematics, with the surface contour line being the current line, and the pressure defined as $p = p_a$. The current line gradually straightens with the changing of the water depth, and becomes a straight line eventually. So, for stationary current and water of irrational movement, using mathematics there is the current function $\Psi\ (x, z)$ in addition to a velocity potential function $\Phi(x, z)$. Therefore we need to describe a complex potential function $\Phi + iz$, that is, we also need to use an analytic function $x + iz$ with complex variables to describe this situation. Obviously, the difference in the description of the mathematical functions compared to slight undulation theory composes an important feature of finite amplitude undulation theory.

1.5.3.2 Methods and Steps to Build the Theory

As mentioned above, in relation to the different wave shapes, finite amplitude undulation theory can be divided into Stokes wave theory, elliptical cosine wave theory, solitary wave theory, etc. Here we take Stokes wave theory as an example below to introduce the methods and steps to build a theory.

As is known to all, the basic idea of establishment of wave theory is taking the complex wave as the composition of an unlimited number of cosine waves with different amplitude range, frequency range, direction, and phase. So as the basic idea for the establishment of Stokes wave finite amplitude wave theory, we use a finite number of simple cosine waves with proportional frequency to approach a finite amplitude wave with a single period.

1. Derive the expression for the complex potential
   As mentioned above, the description of finite amplitude wave theory for water movement is expressed by complex potential in mathematics, as follows.

   a. Complex potential of water parallel current at a constant velocity
   In order to "make" seawater into a stationary current, we need to make the seawater move in the opposite direction to the wave propagation with velocity C. As is shown in Figure 1-25, if the undulation does not exist, which is mean that the free interface of sea water is ox axis, then water flow only at constant velocity C along the ox minus axial. The complex potential of this parallel current at a constant velocity can be expressed by the analytic complex function $x + iz$, so the complex potential ω_1 can be written as follows:

   $$\omega_1 = -C(x + iz) \tag{1-164}$$

   b. Complex potential of periodic motion of water particles
   The velocity components u and ω of water particles have been given by Equation (1-147) and Equation (1-148). If integrate these two equation, the trajectory of the movement of water particle will be obtained as a closed ellipse curve, as shown in Figure 1-26, and the horizontal half axis of the ellipse is $\alpha_0 = aCthkh$. Consequently as shown in figure, the elliptical horizontal half axis of the trajectory decreases with the depth of the water h, so that when at the bottom of the sea, it just moves as a horizontal vibration, and the elliptical vertical half axis β_0 (that is, the amplitude of A) tends to zero.
   Because of the cosh $k(z + h)/\cosh kh$ in Equation (1-147) and thesinh $k(z + h)/\cosh kh$ in Equation (1-148), these terms can be expressed by e^{kz} in deep water, so the complex potential w_2 of the periodic motion of water particles can be expressed the with analytic function $i\beta ce^{-ik(x+iz)}$ of complex variable $x + iz$. Then the complex potential ω_2 is as follows:

   $$w_2 = i\beta Ce^{-ik(x+iz)} \tag{1-165}$$

   c. Complex potential of the synthesis of two kinds of motion
   Set w as the complex potential of the synthesis motion. Taking the sum of Equation (1-164) and Equation (1-165), we can obtain

   $$\begin{aligned}
   \omega = w_1 + w_2 &= -C(x + iz) + i\beta Ce^{-ik(x+iz)} \\
   &= -C(x + iz) + i\beta Ce^{-ik}(\cos kx - i \sin kx)
   \end{aligned} \tag{1-166}$$

   The synthesis of the complex potential of Equation (1-166) is made up of thevelocity potential Φ and stream function Ψ, i.e.,

   $$\omega = \Phi + i\Psi \tag{1-167}$$

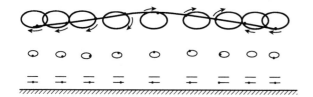

FIGURE 1-26

The motion of water particles.

So if we take the real part of Equation (1-166) for Φ, and imaginary part for Ψ, we will obtain

$$\Phi = -Cx + \beta Ce^{kz} \sin kx \qquad (1\text{-}168)$$

$$\Psi = -Cz + \beta Ce^{kz} \cos kx \qquad (1\text{-}169)$$

In the equations above, C is the wave velocity and β is a constant.

2. Establish expression for the wave surface

From Equation (1-107), we know that when the water is represented as a two-dimensional planar stationary current, the stream function shall be a constant. For convenience, take the constant as zero, namely $\Psi = 0$. Then as Equation (1-169) is equal to zero, we get

$$z = \beta e^{kz} \cos kx \qquad (1\text{-}170)$$

Equation (1-170) is the mathematical expression given by the Stokes wave theory for the displacement z along the z axis.

From Equation (1-170) is not hard to see that for a wave where the amplitude is infinitely small compared to the wavelength (slight undulation theory), namely $kz \rightarrow 0$, then $e^{kz} \rightarrow 1$, and we get

$$z = \beta \cos kx \qquad (1\text{-}171)$$

Obviously, Equation (1-170) is the expression of the wave surface of the slight undulation theory given by Equation (1-144), and it has been shown that the constant β is the amplitude of the fluctuations.

But for a finite amplitude wave, e^{kz} cannot be equal to 1, e^{kz} should develop according to the Maclaurin series. As long as $|kz| \langle 1$, the series e^{kz} converges, and the form is

$$e^{kz} = 1 + \frac{kz}{1!} + \frac{1}{2!}(kz)^2 + \dots + \frac{1}{n!}(kz)^n \qquad (1\text{-}172)$$

So if substitute this series expansion into Equation (1-170), we can obtain

$$z = \beta \cos kx + \beta kx \cos kx + \frac{1}{2}\beta k^2 z^2 \cos kx + \dots$$

$$= \beta \left[1 + kz + \frac{1}{2}(kz)^2 + \frac{1}{6}(kz)^3 + \dots \right] \cos kx \qquad (1\text{-}173)$$

Obviously, both sides of the equal sign of Equation (1-173) have z, so a solution cannot be obtained directly. Usually, we solve it with the stepwise approximation method.

a. Wave function of the third-order Stokes wave theory

For the third-order wave, we can adopt the step approximation method:

$$z = \beta z_0 + \beta^2 z_1 + \beta^3 z_2 \qquad (1\text{-}174)$$

Take Equation (1-174) into Equation (1-173), then divide each term by β, and we can obtain

$$z_0 + \beta z_1 + \beta^2 z_2 = \left[1 + k\left(\beta z_0 + \beta^2 z_1 + \beta^3 z_2\right) + \frac{1}{2}k^2 \left(\beta z_0 + \beta^2 z_1 + \beta^3 z_2\right)^2 \right.$$

$$\left. + \frac{1}{6}k^3 \left(\beta z_0 + \beta^2 z_1 + \beta^3 z_2\right)^3 \right] \cos kx \qquad (1\text{-}175)$$

Unfold the right side of Equation (1-175), and remove each term higher than the second order of β. Then we can write

$$z_0 + \beta z_1 + \beta^2 z_2 = \left[1 + \beta k z_0 + \beta^2 k z_1 + \beta^2 \left(\frac{1}{2} k^2 z_0^2 \right) \right] \cos kx$$

or

$$z_0 + \beta z_1 + \beta^2 z_2 = \cos kx + \beta(kz_0 \cos kx) + \beta^2 \left[\left(kz_1 + \frac{1}{2} k^2 z_0^2 \right) \cos kx \right] \tag{1-176}$$

Set the coefficients of β with the same power equal in Equation (1-176) on both sides. Then

$$
\left.
\begin{aligned}
z_0 &= \cos kx \\[4pt]
z_1 &= kz_0 \cos kx = k\cos^2 kx = \frac{1}{2}k + \frac{1}{2}k \cos 2\,kx \\[4pt]
z_2 &= \left(kz_1 + \frac{1}{2}k_2 z_0^2 \right) \cos kx = \frac{3}{2} k^2 \cos^3 kx \\[4pt]
&= \frac{9}{8} k^2 \cos kx + \frac{3}{8} k^2 \cos 3\,kx
\end{aligned}
\right\}
\tag{1-177}
$$

Substitute z_0, z_1, and z_2 into Equation (1-177) and Equation (1-174), and we can obtain

$$z = \frac{1}{2}k\beta^2 + \beta \left(1 + \frac{9}{8} k^2 \beta^2 \right) \cos kx + \frac{1}{2} k\beta^2 \cos 2\,kx + \frac{3}{8} k^2 \beta^3 \cos 3\,kx \tag{1-178}$$

Set $\beta \left(1 + \frac{9}{8} k^2 \beta^2 \right) = A$ in Equation (1-178), so $\beta = A - \frac{9}{8} k^2 \beta^3 \approx A$. Now we can write

$$z = \frac{1}{2}kA^2 + A \cos kx + \frac{1}{2}kA^2 \cos 2\,kx + \frac{3}{8} k^2 A^3 \cos 3\,kx \tag{1-179}$$

If we translate the coordinate ox axis until it is coincident with the still water level, and consider waveform advances along the direction of position x, then take the initial phase ωt into consideration, and change z in Equation (1-179) into wave surface ε, the wave equation is as follows:

$$\varepsilon = A \cos(kx - \omega t) + \frac{1}{2}kA^2 \cos 2(kx - \omega t) + \frac{3}{8} k^2 A^3 \cos 3(kx - \omega t) \tag{1-180}$$

Equation (1-180) is the function of the wave surface for the third-order Stokes wave theory.

b. The wave function for the fourth-order Stokes wave theory

Take another term (item 4) when e^{kz} in Equation (1-172) unfolds according to the Maclaurin series expansion, and then remove each term that is higher than the third order of β only in Equation (1-175), and assume the z in Equation (1-174) is $z = \beta z_0 + \beta^2 z_1 + \beta^3 z_2 + \beta^3 z_3$.

Adopting the same methods and steps as above, we can obtain the wave function z for the fourth-order Stokes wave theory:

$$z = \frac{1}{2}k\beta^2 + k^3\beta^3 + \beta\left(1 + \frac{9}{8}k\beta^2\right)\cos kx + \left(\frac{1}{2}k\beta^2 + \frac{11}{6}k^3\beta^4\right)\cos 2kx$$
$$+ \frac{3}{8}k^3\beta^3\cos 3kx + \frac{1}{3}k^3\beta^4\cos 4kx \tag{1-181}$$

Similarly, use $A = \beta\left(1 + \frac{9}{8}k\beta^2\right)$ and we get

$$z = A\cos kx + \left(\frac{1}{2}kA^2 + \frac{17}{24}k^3A^4\right)\cos 2kx + \frac{3}{8}k^2A^3\cos 3kx + \frac{1}{3}k^3A^4\cos 4kx \tag{1-182}$$

3. Other parameter expressions

Now we will take the fifth-order Stokes wave theory as an example, and list the mathematical expressions of characteristic parameters commonly used in engineering.

a. Velocity potential ϕ

$$\frac{k\phi}{C} = \sum_{n=1}^{5}\phi_n\cosh[nk(z+h)]\sin n(kx - \omega t) \tag{1-183}$$

b. Wave velocity C

$$\frac{C^2}{gh} = \frac{\tanh(kh)}{kh}\left[1 + \lambda^2 C_1 + \lambda^4 C_2\right] \tag{1-184}$$

c. Wave surface ε

$$k\varepsilon = \sum_{n=1}^{5}\varepsilon_n'\cos n(kx - \omega t) \tag{1-185}$$

d. Horizontal velocity u of a water particle

$$\frac{u}{C} = \sum_{n=1}^{5}n\phi_n'\cosh[nk(z+h)]\cos n(kx - \omega t) \tag{1-186}$$

e. Vertical velocity w of a water particle

$$\frac{w}{C} = \sum_{n=1}^{5}n\phi_n'\sinh[nk(z+h)]\sin n(kx - \omega t) \tag{1-187}$$

f. Pressure p

$$\frac{p}{\rho gh} = 1 - \left(\frac{z+h}{h}\right) - \frac{C^2}{gh}\left\{\frac{\partial\phi/\partial t}{C^2} + \frac{1}{2}\left[\left(\frac{u}{C}\right)^2 + \left(\frac{w}{C}\right)^2\right]\right\} \tag{1-188}$$

g. Time derivative $\frac{\partial \phi}{\partial t}$ of velocity potential ϕ

$$\frac{\partial \phi / \partial t}{C^2} = -\sum_{n=1}^{5} n\phi'_n \cosh[nk(z+h)]\cos n(kx - \omega t) \qquad (1\text{-}189)$$

In all the equations above, h is the depth of the water; ω is circular frequency; k is the wave number; λ is wavelength; z is the coordinate value along the z axis; t is time; and ϕ'_n and ε'_n are as in following equations:

$$\phi'_1 = \lambda A_{11} + \lambda^3 A_{13} + \lambda^5 A_{15};$$
$$\phi'_2 = \lambda^2 A_{22} + \lambda^4 A_{24};$$
$$\phi'_3 = \lambda^3 A_{33} + \lambda^5 A_{35};$$
$$\phi'_4 = \lambda^4 A_{44};$$
$$\phi'_5 = \lambda^5 A_{55};$$
$$\varepsilon'_1 = \lambda;$$
$$\varepsilon'_2 = \lambda^2 B_{22} + \lambda^4 B_{24};$$
$$\varepsilon'_3 = \lambda^3 B_{33} + \lambda^5 B_{35};$$
$$\varepsilon'_4 = \lambda^4 B_{44};$$
$$\varepsilon'_5 = \lambda^5 B_{55};$$

When calculating the characteristic parameters above, the wavelength λ and wave number k must be used from Equation (1-183) to Equation (1-189). These two parameters must be determined by the following two equations:

$$\frac{1}{kh}\left[\lambda + B_{33}\lambda^3 + (B_{35} + B_{55})\lambda^5\right] = \frac{H}{2h} \qquad (1\text{-}190)$$

$$kh\tanh(kh)\left[1 + C_1\lambda^2 + C_2\lambda^4\right] = 4\pi^2\frac{h}{gT^2} \qquad (1\text{-}191)$$

In Equation (1-190) and Equation (1-191), h is the depth of the water, T is the wave period, and g is the acceleration of gravity. In this manner, using $\frac{h}{gT^2}$ and $\frac{H}{h}$, λ and k can be calculated using the stepwise approximation method through Equation (1-190) and Equation (1-191).

The coefficients A, B, and C are a function of kh. They have been given by Skjelbreia and Hendrickson, and they can be obtained from the relevant manuals or reference books.

For the commonly used second-order Stokes wave theory, the velocity potential ϕ is as follows:

$$\phi = \frac{H\lambda}{T}\frac{\cosh(z+h)}{\sinh kh}\sin(kx - \omega t) + \frac{3\pi H}{16T}\frac{\cosh(z+h)}{\sinh^4 kh}\sin 2(kx - \omega t) \qquad (1\text{-}192)$$

The horizontal velocity u and verticality w of a water particle are, respectively,

$$u = \frac{\pi H}{T}\frac{\cosh k(z+h)}{\sinh kh}\cos(kx - wt) + \frac{3}{4}\left(\frac{\pi H}{T}\right)\left(\frac{\pi H}{\lambda}\right)\frac{\cosh 2 k(z+h)}{\sinh^4 kh}\cos 2(kx - \omega t) \quad (1\text{-}193)$$

$$w = \frac{\pi H}{T}\frac{\sinh k(z+h)}{\sinh kh}\sin(kx - wt) + \frac{3}{4}\left(\frac{\pi H}{T}\right)\left(\frac{\pi H}{\lambda}\right)\frac{\sinh 2 k(z+h)}{\sinh^4 kh}\sin 2(kx - \omega t) \quad (1\text{-}194)$$

1.5.3.3 Division of Wave Theory into Different Types

It was pointed out when discussing the characteristics of finite amplitude undulation theory above that different wave shapes and finite amplitude wave theory can be divided into the following: a Stokes wave theory suitable for deep water; a solitary wave theory suitable for sea areas with shallow depth; and an elliptical cosine wave theory suitable for both deep water and shallow water. Now, we introduce these theories.

1. Second-, third-, and fifth-order Stokes wave theory
   This theory was first proposed by Stokes and later by Rayleigh and others who did in-depth study, and it is widely used in ocean engineering. As for the methods and steps of establishment of this wave theory and expression of some characteristic parameters, we presented a detailed introduction earlier, so we don't restate that here.
2. Eliptical cosine wave theory
   This was proposed by Korteweg and De Vries, and further improved by Coulaigan and Patterson and Laidon. Then, the second-order and third-order solution was given by Laidon and Chapuygives, and the fifth-order solution was given by Fington; these have been applied in ocean engineering. The characteristic of this wave theory is that the wave surface is expressed by an elliptical cosine function C_n, and it belongs to shallow water waves; the following will introduce the expressions of the characteristic parameters.
   a. Velocity potential ϕ
      Similar to Stokes wave theory, it is assumed that the solution of velocity potential ϕ is expressed with the following form of power series:

      $$\phi = \sum_n Z^n \phi_n(x,t), \phi_1(x,t) = 0 \qquad (1\text{-}195)$$

      In Equation (1-195), Z is a wave of the y coordinate.
   b. First-order wave equation ε
      The wave equation given by Coulaigan and Patterson, which is the first order approximate solution, is as follows:

      $$\varepsilon = Z_s - h \qquad (1\text{-}196)$$

      In Equation (1-196), Z_s is the distance from the bottom of the water to the wave surface; Z_s is calculated as

      $$Z_s = Z_t + HC_n^2\left[2K(k)\left(\frac{x}{\lambda} - \frac{t}{T}\right), t\right] \qquad (1\text{-}197)$$

      In Equation (1-197), Z_t is the distance from the bottom of the water to the bottom of the trough:

      $$Z_t = h - H + \frac{16h^3}{3\lambda^2}K(k)[K(k) - E(k)] \qquad (1\text{-}198)$$

In the equations above, h is the depth of the water, and $K(k)$ and $E(k)$ are the first and second type of complete elliptic integrals, respectively:

$$K(k) = \int_0^{\frac{\pi}{2}} \frac{d\theta}{1 - k^2 \sin^2 \theta} \tag{1-199}$$

$$E(k) = \int_0^{\frac{\pi}{2}} \sqrt{1 - k^2 \sin^2 \theta} d\theta \tag{1-200}$$

In these equations, k is the modulus of the elliptic integral, the value of which is about 0 to 1. In Equation (1-197), $C_n()$ is the Jacobian elliptical cosine function. According to the nature of the elliptical cosine function, $C_n^2 \left[2K(k) \left(\frac{x}{\lambda} - \frac{t}{T} \right), k \right]$ with a period of $2K(k)$. Thus Equation (1-197) describes the periodic wave as well.

From Equation (1-197) to Equation (1-200), we can see that different moduli k have different wave curve shapes, and between the modulus and elements of a wave exist an elliptical cosine wave dispersion relation, that is,

$$\frac{16}{3}[K \cdot K(k)]^2 = \left(\frac{\lambda}{h} \right)^2 \left(\frac{H}{h} \right) \tag{1-201}$$

In Equation (1-201), $\left(\frac{\lambda}{h} \right)^2 \left(\frac{H}{h} \right) = U$ is called Eritrea's number; λ and H represent wavelength and wave height, respectively; and h is the depth of the water.

c. Wave velocity C

$$C = \sqrt{gh} \left\{ 1 + \frac{H}{h} \left[-1 + \frac{1}{k^2} \left(2 - 3 \frac{E(k)}{K(k)} \right) \right] \right\}^{\frac{1}{2}} \tag{1-202}$$

d. Wave period T

With wavelength λ in Equation (1-201) and wave velocity C in Equation (1-202), the expression of wave period T can be obtained as follows:

$$T = \frac{4h}{\sqrt{3gH}} \left\{ \frac{K \cdot K(k)}{\sqrt{1 + \frac{H}{h} \left[-1 + \frac{1}{k^2} \left(2 - 3 \frac{E(k)}{K(k)} \right) \right]}} \right\} \tag{1-203}$$

Above all, as long as the wave parameter λ, H and water depth h, or wavelength $\frac{\lambda}{h}$ and wave height H/h are known, the modulus K and second kind of ellipse integral $K(k)$ can be calculated through Equation (1-201) by using the iterative method. Then, the wave shape of the elliptical cosine wave can be completely determined according to Equation (1-198), Equation (1-197), and Equation (1-196), as shown in Figure 1-24(c).

e. Wave equation when modulus $k \to 0$

When $k \to 0$, the second kind of ellipse integral $K(k)$ becomes

$$K(k) = \int_0^{\frac{\pi}{2}} d\theta = \frac{\pi}{2} \tag{1-204}$$

The Jacobian elliptical cosine function $C_n(r,k)$ is as follows:

$$C_n(r,k) = \cos r \tag{1-205}$$

So, the wave equation ε of an epsilon cosine wave should be as follows:

$$\begin{aligned} \varepsilon &= H\cos^2\left[\pi\left(\frac{x}{\lambda} - \frac{t}{T}\right)\right] - h + Z_t \\ &= \frac{H}{2} + \frac{H}{2}\cos(kx - \omega t) - h + Z_t = \frac{H}{2}\cos(kx - \omega t) \end{aligned} \tag{1-206}$$

Obviously, wave equation ε in Equation (1-206) is consistent with the wave equation [Equation (1-144)] deduced by slight undulation theory. It suggests that the shallow cosine wave derived by slight undulation theory is a limiting case of nonlinear elliptic cosine wave when $k = 0$.

f. Wave equation when modulus $k = 1$

When $k = 1$, $K(k) \to \infty$, $C_n(r, 1) = \sec h(r)$. Substitute them into Equation (1-197) and Equation (1-196); the wave surface ε is

$$\varepsilon = H \sec h^2\left[\sqrt{\frac{3H}{4h}}\left(\frac{x}{h} - \frac{Ct}{h}\right)\right] \tag{1-207}$$

In Equation (1-207), H is wave height, h is the depth of the water, C is wave velocity, t is time, and X is the x coordinate along the x axis.

The expression of wave epsilon ε in Equation (1-207) is derived by solitary wave theory; therefore, we can see that solitary wave is another limiting case of the elliptical cosine wave when the modulus $k = 1$. The wavelength and wave period [see Equation (1-201) and Equation (1-203)] tend to infinity; namely they form the wave shape of a solitary wave, as shown in Figure 1-24(d). In a word, elliptical cosine wave theory is quite comprehensive and covers slight wave and solitary wave theory. However, the expression of the characteristic parameters is rather complex, and application is inconvenient. In order to simplify the calculation, Wiegel provided a series of calculation curves and graphs that can be obtained from reference books such as "The Coast Protection Manual," but some special calculating programs also have be used in recent years.

3. Solitary wave theory

We have already mentioned that the solitary wave with only one peak or trough belongs to a kind of aperiodic fluctuation. At first, there is no practical significance to using solitary wave theory to describe waves; however, when the waves enter shallow waters, the trough becomes smoother and crest becomes steeper. Finally, the wavelength gradually tends to infinity. Therefore, the wave at this moment can be regarded as a series of single solitary waves. In this way, because the waveform of solitary wave [see Figure 1-24(d)] is similar to a wave in inshore shallow water, and its mathematical description is simple, in some offshore engineering, especially in the inshore area, the solitary wave has been widely used in practice.

The existence of the solitary wave was first observed by Russell; later Poohsing, Rayleigh, MoCowen, and others obtained important achievements through theoretical research; until the 1960s the different high-order approximate solutions of solitary wave system theory were gradually put forward by Coulaiga and Patterson as well as Keller and Leiden and others. The following will give some characteristic parameters of solitary wave theory.

a. Wave function ε

We have an exception foor the elliptical cosine wave mentioned above; when $k = 1$, $K(k) \to \infty$, and $C_n(r, 1) = \sec h(r)$, ε is

$$\varepsilon = H \sec h^2 \left[\sqrt{\frac{3H}{4h} \left(\frac{x}{h} - \frac{Ct}{h} \right)} \right]$$

This equation is the wave function of a solitary wave given by Equation (1-207).

b. Wavelength λ

Because $K(k) \to \infty$ when the modulus $k = 1$, and the period of function $C_n^2 \left[2K(k) \left(\frac{x}{\lambda} - \frac{t}{T} \right), k \right]$ is $2K(k) \to \infty$, the wavelength of a solitary wave is infinite.

c. Wave velocity C

$$C = \sqrt{g(h + H)} \tag{1-208}$$

d. Wave pressure p

Similar to the front equation (such as Equation (1-188)), it can be determined from the Lagrange equation that p is

$$p = \rho g \left((h - z) + H \sec h^2 \left[\sqrt{\frac{3H}{4h} \left(\frac{x}{h} - \frac{Ct}{h} \right)} \right] \right) \tag{1-209}$$

e. Horizontal velocity u of water particles

According to the solitary wave theory proposed by McCormick, which corresponds relatively well to to the measured value, the horizontal velocity u of water particles is as follows:

$$u = NC \frac{\left[1 + \cos\left(\frac{Mz}{h}\right) \cosh\left(\frac{Mx}{h}\right) \right]}{\left[\cos\left(\frac{Mz}{h}\right) + \cosh\left(\frac{Mx}{h}\right) \right]^2} \tag{1-210}$$

f. Vertical velocity ω of water particles

$$w = NC \frac{\sin\left(\frac{Mz}{h}\right) \sinh\left(\frac{Mx}{h}\right)}{\left[\cos\left(\frac{Mz}{h}\right) + \cosh\left(\frac{Mx}{h}\right) \right]^2} \tag{1-211}$$

In Equation (1-210) and Equation (1-211), h is the depth of the water; z and x are the coordinate values on the z and x axes, respectively; C is wave velocity; and M and N are the two coefficients that are obtained through getting a joint solution of the following two equations according to the relative wave height $\frac{H}{h}$:

$$\frac{H}{h} = \frac{N}{M}\tan\frac{1}{2}\left[M\left(1 + \frac{H}{h}\right)\right] \tag{1-212}$$

$$N = \frac{2}{3}\sin^2\left[M\left(1 + \frac{2}{3}\frac{H}{h}\right)\right] \tag{1-213}$$

We should point out that the solitary wave is a translation wave, namely water particles move along the direction of wave propagation rather than backward. Before the arrival of the wave crest, water particles far away from the high crest ($x = 10h$) are almost in a stationary state. As the crest arrives, water particles move upward (rate of $+w$) and forward (rate of $+u$). When the crest goes through the moment ($x = 0$), horizontal velocity u reaches the maximum and vertical velocity w is zero while the upward displacement of the particle is at a maximum. As the wave goes through, water particles begin to decline and the particle velocity u of a water particle also gradually slows down, finally returning to the original depth position of the water particles, but the water particles are moving with a net forward displacement in the horizontal direction.

Various wave theories were introduced before. There are differences in the basic assumptions, simplified approximate methods, and results and each wave theory has its own applicable scope. So, there is not a unified and general wave theory. Many scholars have done a lot of research on restricted conditions and the applicable scopes of different wave theories. There is research presented by Dean that is shown in Figure 1-27; another research result is proposed by Miywattl is shown in Figure 1-28(a). Figure 1-28(b) is the applicable scope given by the standards of the China Classification Society.

In Figure 1-27 and Figure 1-28 (a), the ordinate is h/gT^2 where h is wave height, T is period, and g is the gravitational acceleration; the horizontal ordinate is H/gT^2 where H is the depth of the water and T is the same as above. In Figure 1-27, the critical breaking point is $h = h_b$, in which h is wave height and h_b is broken wave height. In these two figures, h represents the wave height, h_s represents symptomatic wave height, λ represents wavelength, H represents the water depth, and U_r represents the relative water depth.

1.5.4 Effect of Small Scale Structures on Wave Force

The theories of ocean waves that can be used to calculate the velocity and acceleration components along the x, y, and z axes for seawater particle motion have been analyzed.

The size of the structure has an important impact on the wave force acting on engineering structures, so it needs to be studied as well. The following will first introduce the wave force acting on a small-scale structure.

1.5.4.1 Morison Equation and Its Application

For small-scale offshore structures, the Morison equation is usually used to calculate the wave force. It can be assumed that the structure does not exist in the ocean if it is a small-scale vertical isolated pile,

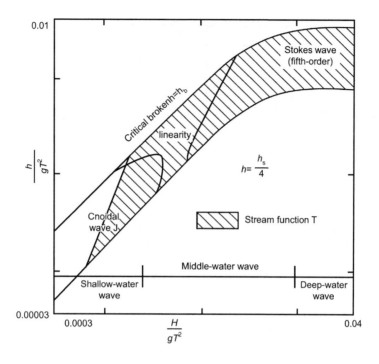

FIGURE 1-27

Applicable scope of various wave theories presented by Dean.

and when the pile is not inserted in the wave, it is the equivalent of using the horizontal velocity u and acceleration $\frac{\partial u}{\partial t}$ of waving water particles at the center of a pile to respectively calculate the horizontal drag force and inertial force that waves exert on a pile. The Morison equation pointed out that the wave force acting on a unit length of the vertical isolated cylinder on the sea can be regarded as the superposition of resistance and inertial force, in which the drag force (resistance) is caused by the velocity of the water when the water goes through the cylindrical point and the inertial force is caused by the acceleration of water particles.

We assume that we have an isolated vertical cylindrical pile with a diameter of D fixed in the water waves at depth d, as shown in Figure 1-29. Now take a small section of dz at any height z on the pile. In accordance with the Morison equation, the wave force df acting on the small section can be represented as

$$df = \left[\frac{1}{2} C_D \rho D |u| u + C_M \rho \frac{\pi}{4} D^2 \frac{\partial u}{\partial t} \right] dZ \qquad (1\text{-}214)$$

In the equation, C_D and C_M respectively represent the coefficient of drag force (resistance) and inertial force; ρ is the quality of seawater density; u can be instead by $|u|u$ with due to sometimes positive and sometimes negative is horizontal velocity of the water point.

We often use the sum of the drag force and inertial force, which are assumed to be unrelated, to express the wave force acting on an isolated vertical cylindrical pile, as shown in Equation (1-214), which is called the Morison equation.

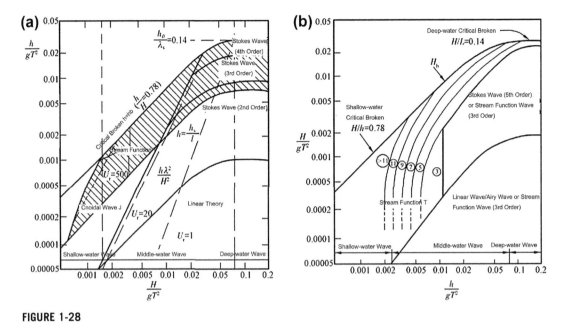

FIGURE 1-28

Applicable scope of wave theories given by Miywattl and the China Classification Society.

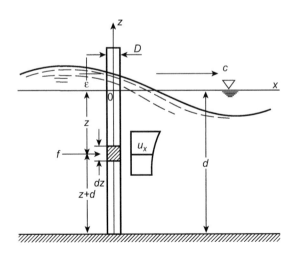

FIGURE 1-29

Wave force acting on an isolated vertical cylindrical pile.

Although the assumptions of the Morison equation have some issues, it has been widely used in ocean engineering, because of its simple expression and convenient operation. The main applications are the following:

1. Calculation of wave force acting on vertical cylindrical structure
   a. Total wave force F on the whole pile
      As shown in Figure 1-29, integrate Equation (1-214) along the pile from $-d$ to $+\varepsilon$, namely,

$$F = \int_{-d}^{\varepsilon} df = \int_{-d}^{\varepsilon} \frac{1}{2} C_D \rho D |u| u \, dZ + \int_{-d}^{\varepsilon} C_{MP} \frac{\pi}{4} D^2 \frac{\partial u}{\partial t} \, dZ \qquad (1\text{-}215)$$

   b. Torque M of the total wave force acting on the pile

$$M = \int_{-d}^{\varepsilon} df \cdot (z + d) \qquad (1\text{-}216)$$

2. Calculation of the wave force acting on a tilted single cylindrical structure
   Among ocean engineering structures, in addition to the vertical cylindrical structures, there are tilted cylinder components acting as support bars. The Morison equation can be applied to calculate the effect of wave force, but it is complex.
   a. Vector analysis of velocity
      As shown in Figure 1-30, when the cylinder is tilt, at any particle on the pile, the velocity which is orthogonal to the pile is vector v_n and the acceleration is $\dot{v}_n$, but usually v_n and $\dot{v}_n$ is not in a straight line. Such as, in the Morison equation, the composition of the drag force constituted of velocity and the inertia force constituted of acceleration must be written in form of vector, therefore, first do vector analysis of velocity.

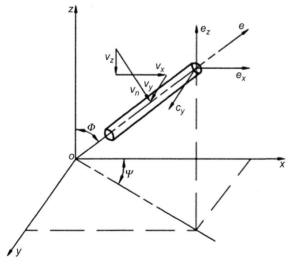

FIGURE 1-30

The wave force acting on an isolated tilted cylindrical structure.

Assume the projection of v_n on three axes respectively is v_x, v_y, v_z; it will be represented as

$$v_n = i v_x + j v_y + k v_z \qquad (1\text{-}217)$$

If the velocity of a water particle is u and the projection on three axes respectively is u_x, u_y, u_z (for two dimensions, $u_y=0$), then

$$u = i u_x + j u_y + k u_z \qquad (1\text{-}218)$$

And if the unit vector along the axis of the cylinder is set to l and its projections along three axes respectively are l_x, l_y, l_z we have

$$l = i l_x + j l_y + k l_z \qquad (1\text{-}219)$$

From the angles ϕ and ψ shown in Figure 1-30, we have

$$\begin{aligned} l_x &= \sin\phi \cos\psi \\ l_y &= \sin\phi \sin\psi \\ l_z &= \cos\psi \end{aligned} \qquad (1\text{-}220)$$

But from vector analysis, we know that the velocity vector $\dot{v}_n$, which is orthogonal to the axis of the cylinder, can be obtained according to a triple vector product as follows:

$$\mathbf{v}_n = \mathbf{l} \times (\mathbf{u} \times \mathbf{l}) \qquad (1\text{-}221)$$

Equation (1-219) and Equation (1-218) can be substituted into Equation (1-211); then

$$\mathbf{v} = i\left[u_x\left(1 - l_x^2\right) - u_z l_z l_x\right] + j\left[-u_x l_x l_y - u_z l_z l_y\right] + k\left[-u_x l_x l_z + u_z\left(1 - l_z^2\right)\right] \qquad (1\text{-}222)$$

If we know that Equation (1-217) is equal to Equation (1-222), it can be concluded that

$$v_x = \left[u_x\left(1 - l_x^2\right) - u_z l_z l_x\right] = u_x - l_x(l_x u_x + l_z u_z) \qquad (1\text{-}223)$$

$$v_y = \left[-u_x l_x l_y - u_z l_z l_y\right] = -l_y(l_x u_x + l_z u_z) \qquad (1\text{-}224)$$

$$v_z = \left[-u_x l_x l_z + u_z\left(1 - l_z^2\right)\right] = u_z - l_z(l_z u_z + l_x u_x) \qquad (1\text{-}225)$$

If the modulus of v_n is $|v_n|$, it's known that

$$\begin{aligned} |\mathbf{v}_n| &= (\mathbf{v}_v.\mathbf{v}_n)^{\frac{1}{2}} = \left(v_x^2 + v_y^2 + v_z^2\right)^{\frac{1}{2}} \\ &= \left[u_x^2 + u_z^2 - (l_x u_x + l_z u_z)^2\right]^{\frac{1}{2}} \\ &= \left[u_x^2 + u_z^2 - (\sin\phi \cos\psi u_x + \cos\phi u_z)^2\right]^{\frac{1}{2}} \end{aligned} \qquad (1\text{-}226)$$

b. Vector analysis of acceleration

Analyzed in the same way, it can be concluded that the projection of acceleration $\mathbf{v}_n$ on three axes respectively is $\dot{v}_x$, $\dot{v}_y$, $\dot{v}_z$:

$$\dot{v}_x = \left(1 - l_x^2\right)\frac{\partial u_x}{\partial t} - l_y l_x \frac{\partial u_z}{\partial t} \tag{1-227}$$

$$\dot{v}_y = -l_x l_y \frac{\partial u_x}{\partial t} - l_x l_y \frac{\partial u_z}{\partial t} \tag{1-228}$$

$$\dot{v}_z = -l_x l_z \frac{\partial u_x}{\partial t} + \left(1 - l_z^2\right)\frac{\partial u_z}{\partial t} \tag{1-229}$$

In these equations, $\frac{\partial u_x}{\partial t}$ and $\frac{\partial u_z}{\partial t}$ respectively are the acceleration component of a water particle along the x axis and z axis. They can be calculated by the horizontal velocity component u_x and vertical velocity component u_z of a water particle given in the wave theory; the l_x and l_y in these equation can also be expressed by Equation (1-220).

c. Wave force acting on unit length of a tilted cylinder

The component of this wave force acting along the direction of each coordinate axis are f_x, f_y, and f_z, respectively; then, in accordance with the Morison equation, we can write

$$\begin{bmatrix} f_x \\ f_y \\ f_z \end{bmatrix} = \frac{1}{2} C_D \rho D |v_n| \begin{bmatrix} v_x \\ v_y \\ v_z \end{bmatrix} + C_M \rho \frac{\pi D^2}{4} \begin{bmatrix} \dot{v}_x \\ \dot{v}_y \\ \dot{v}_z \end{bmatrix} \tag{1-230}$$

In this way, $|\mathbf{v}_n|$, v_x, v_y, v_z and $\dot{v}_x$, $\dot{v}_y$, $\dot{v}_z$ can be obtained in accordance with the above equations and then substituted into Equation (1-230); then, f_x, f_y, f_z can be calculated. According to the method given by Equation (1-215) and Equation (1-216), we can calculate the total wave force and torque acting on the tilted cylindrical structure.

3. Calculation of the wave force on an isolated cylinder in relative motion

In offshore oil engineering, sometimes in addition to the wave action, there is also the action of the current. We may have a floating structure that also moves in waves, so, for the isolated cylinder components (including vertical and tilted), we need to calculate the acting force when it is in relative motion. For the calculation, we also need to apply Morison equation but also need to calculate the drag force and inertial force.

a. Drag force

Set the wave force acting on a unit length of the cylinder components to f_D, and the relative velocity vertical to the cylinder axis to vector $\mathbf{v}_{nr}$; then

$$f_D = \frac{1}{2} C_D \rho D |\mathbf{v}_{nr}| \mathbf{v}_{nr} \tag{1-231}$$

In accordance with the velocity vector analysis and Equation (1-221), $|\mathbf{v}_{nr}|$ follows:

$$\mathbf{v}_{nr} = \mathbf{l} \times (\mathbf{u}_r \times \mathbf{l}) \tag{1-232}$$

In this equation, $\mathbf{u}_r$ represents the relative velocity vector of the water particle, which is relative to the cylinder component. Assume there exists a velocity vector $\mathbf{u}_c$ for the current and a velocity vector $\mathbf{u}_m$ in the center of cylinder when the structure moves in the wave (for a fixed structure, $\mathbf{u}_m = 0$), and a forward velocity vector $\mathbf{u}_a$ which represent the migration of all floating structures in seawater. Then

$$\mathbf{u}_r = (\mathbf{u}_h + \mathbf{u}_c) - \mathbf{u}_m - \mathbf{u}_a \tag{1-233}$$

In the equation, $\mathbf{u}_h$ represents the velocity vector of water particles, which is caused by sea waves at the center of small-scale cylinder components.

In this way, using the method introduced in Equation (1-232) and carrying out the value of $|\mathbf{v}_{nr}|$ and $\mathbf{v}_{nr}$ and then substituting them into Equation (1-231), the force f_D can be calculated.

b. Inertial force

Assume that the inertial force caused by waves on the unit length cylinder components is f_i; then, according to the Murray-Lin equation, we can write

$$f_i = C_M \rho \frac{\pi D^2}{4} \dot{\mathbf{v}}_{nr} \tag{1-234}$$

In the equation, $\dot{\mathbf{v}}_{nr}$ represents the relative acceleration vector, which is perpendicular to the axis of the cylinder component. Depending on Equation (1-221), we can write

$$\dot{\mathbf{v}}_{nr} = \mathbf{l}_x (\dot{\mathbf{u}}_r \times \mathbf{l}) \tag{1-235}$$

In the equation, $\dot{\mathbf{u}}_r$ represents the relative acceleration vector of water particles, which is relative to the center of component. If at this center, the acceleration vector of water particles is $\dot{\mathbf{u}}_h$, and the acceleration vector of the component which is in the absolute motion of the waves is $\dot{\mathbf{u}}_m$, then

$$\dot{\mathbf{u}}_r = \dot{\mathbf{u}}_h - \dot{\mathbf{u}}_m \tag{1-236}$$

In this way, Equation (1-234) can be applied to obtain the inertial force f_i. After f_i and f_D are obtained, according to the method of Equation (1-215) and Equation (1-216), the total wave force and torque of it acting on the component can be calculated. As shown in Figure 1-30, when the angle $\phi = 0$, the wave force acting on a vertical isolated cylinder can be obtained.

1.5.4.2 Linearization of the Morison Equation

As shown in Equation (1-214), the resistance item in the Morison equation is the nonlinear quantity $|u|u$. It is difficult to calculate; therefore, the Morison equation needs to be linearized. Usually we use the method of equivalent root-mean-square (RMS), that is, take the statistical regularity of velocity u of water particles of seawater as obeying the normal distribution and get the value of its RMS u_{rms}, then make $u_{rms}\sqrt{\frac{8}{\pi}}$ approximately equivalent with u. So the $|u|u$ in the Morison equation can be replaced by $u_{rms}\sqrt{\frac{8}{\pi}}$ and u. Thus, the wave force f acting on a unit length of the vertical isolated cylindrical structure can be written as

$$f = \frac{1}{2}\rho C_D D u_{rms}\sqrt{\frac{8}{\pi}} + \rho C_M \frac{\pi}{4} D^2 \frac{\partial u}{\partial t} \tag{1-237}$$

1.5.4.3 Sphere of Application of the Morison Equation

The premise of the Morison equation is the hypothesis that the structure does not exist in seawater and does not affect the velocity distribution of a water particle. Obviously, when the size of structure is very small, namely $\frac{D}{\lambda} \leq 0.2$, we can satisfy the conditions of the hypothesis. Therefore, generally the range that the Morison equation applies to for small-size structures is defined like this:

$$\frac{D}{\lambda} \leq 0.2 \tag{1-238}$$

In the equation, λ is wavelength and D is the diameter of the structure.

1.5.4.4 Problems When Applying the Morison Equation

1. Problem with selection of wave theory

As shown in Equation (1-237), when using the Morison equation to calculate the wave force acting on the structure, we must use the horizontal velocity component u and acceleration $\frac{\partial u}{\partial t}$ of seawater particles. At this moment the calculation equation of the horizontal velocity component u must be used, which is given when analyzing various wave theories above. But the expression of calculation of the velocity component given by various wave theories is different. Therefore, in accordance with the characteristics of the sea area where engineering structures are located and the characteristic parameters of the working water depth and waves such as wave height and wave period (refer to the curves in Figure 1-27 and Figure 1-28 given above), we must reasonably select a suitable wave theory. This is the first problem that should be noticed.

2. Problem with selection of related coefficient

From Equation (1-237), we can see that the coefficient of the drag force (resistance) C_D and the coefficient of inertial force (quality coefficient) C_M in the Morison equation have a direct impact on the size of the drag force and inertial force acting on the structure; that is, they are directly related to the economics of ocean engineering design. Usually the two coefficients should be determined by experiment, but sometimes the test data are fragmented and these should be determined prudently. Table 1-9 and Table 1-10, respectively, list several common values of the coefficient of the drag force and the inertial force for various structures.

Table 1-9 Drag Force Coefficient for Several Typical Structures

The Shape of the Object (when the axis is perpendicular to the paper)	The Reference Area[①] (cylinder per unit length)	The Drag Force Coefficient (*l* — the Length of the Column)
Cylinder	D	1.0 ($l \gg D$)
Square column	D	2.0 ($l \gg D$)

Continued

Table 1-9 Drag Force Coefficient for Several Typical Structures—cont'd

The Shape of the Object (when the axis is perpendicular to the paper)	The Reference Area[①] (cylinder per unit length)	The Drag Force Coefficient (I − the Length of the Column)
Square column	$\sqrt{2}D$	1.55 $(I >> D)$
Plate	D	2.01 $(I >> D)$
Circular plate	$\frac{\pi}{4}D^2$	1.2
Sphere	$\frac{\pi}{4}D^2$	0.5
Cube	D^2	1.05

① *The projection area of objects in the direction of the flow.*

Table 1-10 Inertial Force Coefficient of Several Typical Structures

The Shape of the Object (when the axis perpendicular to the paper)	The Reference Area (cylinder per unit length)	Coefficient of Inertial Force (l — the length of the column)
Cylinder	$\frac{\pi}{4}D^2$	2.0 $(l >> D)$
Square column	D^2	2.19 $(l >> D)$
Plate	$\frac{\pi}{4}D^2$	1.0 $(l >> D)$
Sphere	$\frac{\pi}{6}D^3$	1.5
Cube	D^3	1.67

Table 1-11 Value of Coefficient N When Calculating Drag Force

Degree of Adhesion	Relative Roughness ε/D	N
General degree of attachment	$\varepsilon/D \leq 0.02$	1.15
Moderate attachment	$0.02 < \varepsilon/D < 0.04$	1.25
Severe attachment	$\varepsilon/D \geq 0.04$	1.40

For cylinder structures commonly used in offshore oil engineering, the specifications of the China Classification Society recommend that C_D be 0.6 ~ 1.0, C_D be 0.65 for components with smooth surfaces, and C_D be 1.05 for components with rough surfaces. As for the coefficient of inertial force, C_M is recommended as 1.3~2.0 and we take $C_M = 1.6$ for components with smooth surfaces, and $C_M = 1.2$ for components with rough surfaces.

In addition, if there are sea creatures adhering to components, the diameter D of components in the Morison equation should be calculated with the actual diameter. At the same time, the drag force acting on structures should be multiplied by the corresponding coefficient n; the value of n should be selected from Table 1-11 according to different degrees of biological adhesion. The ε in the table represents the average thickness of the attaching ocean organism, m; D is the diameter of the components in m.

3. Problem with selection of design wave height

To apply the Morison equation to calculate the wave force acting on the structure, as mentioned above, we need to calculate the velocity of water particles. Taking the horizontal velocity u in Equation (1-150) as an example, we require the design wave height H and the period T. In accordance with the regulations of the China Classification Society rules, the period T can take the average period, but the design wave height H should be the smaller value of the possible value of the maximum wave height u (H_{max}) and the critical wave height H_b of the broken wave.

a. Possible value of maximum wave height u (H_{max})

The China Classification Society recommends selecting the value according to Table 1-12. In the table, $\overline{H}$, which uses meters for a unit, is the average value of the wave height. N is the wave number; generally the corresponding wave number should be chosen between 100~1000. Take

Table 1-12 Possible Values for Design Wave Height

N	$\overline{H}/h$										
	0.00	0.05	0.10	0.15	0.20	0.25	0.30	0.35	0.40	0.45	0.50
10	1.78	1.745	1.705	1.670	1.635	1.600	1.565	1.530	1.495	1.460	1.420
20	2.01	1.955	1.900	1.845	1.795	1.745	1.690	1.640	1.590	1.540	1.490
50	2.27	2.195	2.10	2.050	1.980	1.920	1.850	1.780	1.715	1.650	1.580
100	2.45	2.360	2.270	2.190	2.110	2.030	1.950	1.870	1.795	1.720	1.650
200	2.62	2.520	2.415	2.320	2.220	2.130	2.035	1.950	1.860	1.770	1.690
500	2.83	2.710	2.595	2.480	2.270	2.260	2.155	2.050	1.955	1.860	1.770
1000	2.98	2.850	2.720	2.590	2.470	2.350	2.235	2.125	2.015	1.910	1.810

$N = 1000$ for the South China Sea and the East China Sea, take $N = 100 \sim 1000$ for the China Yellow Sea and the China Bohai Sea, and take $N = 100 \sim 500$ for shallow waters.

b. Critical wave height of broken waves H_b

The related specification of China Classification Society recommends selecting the value of this from Figure 1-31. In this figure, L_0, is deepwater wavelength, in m; d_b is the breakage depth, in m; and T is the wave period, in s.

1.5.5 The Random Wave Force

As mentioned above, the actual waves in the ocean are random waves, thus the wave force acting on ocean engineering structures should be random wave forces. So, how to calculate the random wave force? Here we use a small-scale structure as an example, and introduce the calculation method for random wave forces.

1.5.5.1 Calculation Methods

There are four methods to calculate the wave force of random waves.

1. Characteristic wave method

This method is based on the probability distribution of random wave heights. We determine the wave height when the cumulative probability of a certain number N of wave height is equal to 1, then take this wave height as the characteristic wave height and substitute it into the calculation of velocity components of water particles, thereby calculating the velocity and acceleration components and using them to calculate the wave force acting on the structure according to the

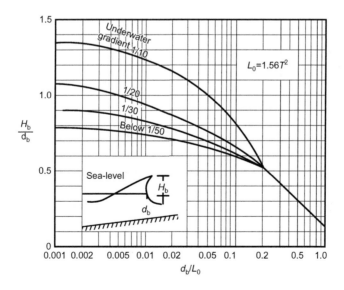

FIGURE 1-31

The value of the critical wave height of broken waves H_b.

Morison equation. For example, if the average wave height is $\overline{H}$ and the cumulative probability is $P(H)$ when the random wave heights obey the normal distribution, then

$$P(H) = e^{-\frac{\pi}{4}\frac{H^2}{\overline{H}^2}} = 1 \tag{1-239}$$

When $P(H) = 1$, the characteristic wave height H can be calculated to determine wave forces. When introducing the problems of applying the Morison equation above, we put forward that for the wave height one might choose the possible value of the maximum wave height; essentially this is using characteristic wave method to calculate wave forces.

2. Probability distribution method

This method takes the probability distribution of the wave force of a random wave acting on the structure as the same as the probability distribution of the wave height of random waves. In this way, the maximum value from the probability of the random wave force can be directly determined. We calculate the wave force F when $P(F) = 1$. For example, the wave force, the same as wave height, also obeys the normal distribution and the mean of it is $\overline{F}$. Then F can be obtained from the following equation:

$$P(F) = 1 = e^{-\frac{\pi}{4}\frac{F}{\overline{F}}} \tag{1-240}$$

Certainly, as with Equation (1-239), no matter the wave height or the wave force, the probability $P(H)$ or $P(F)$ all need a certain number N for the wave number.

3. Simulation method

The simulation method is the method that in accordance with the statistical characteristic variances σ_A (amplitude variance) and σ_H (wave height variance) of wave height calculated from the random wave spectrum, calculates the wave force by simulating wave height with the variance. Because of the random wave spectrum $\int_{-\infty}^{\infty} S_{\varepsilon\varepsilon}(\omega)d\omega = \sigma_A^2 = m_0$ known from Equation (1-87) and $H_{1/3} = 4.0\sqrt{m_0}$ known from Equation (1-88) and Equation (1-89), $H_{1/10} = 5.1\sqrt{m_0}$.

4. Spectrum method

This is the method that uses the random wave force spectrum acting on the structure, which is directly calculated from random wave spectrum to describe the wave force. This method is commonly used and will be described below.

1.5.5.2 The Commonly Used Wave Force Spectrum Method

1. The wave force spectrum per unit length

Here was are still using the Morison equation to calculate the drag force and inertial force of wave action for a small size of ocean engineering structure.

a. The drag force spectrum $S_{f_D f_D}(\omega)$ of wave action per unit length

The drag force on a unit length of the structure that is changing over time is $f_D(t)$ and we express the drag force $\frac{1}{2}\rho C_D D$ with ϕ_D:

$$f_D(t) = \phi_D u_{rms}\sqrt{\frac{8}{\pi}}u = A\omega e^{kz}\sin(kx - \omega t)\left(\phi_D u_{rms}\sqrt{\frac{8}{\pi}}\right)$$

$$= \left(\omega e^{kz}\phi_D u_{rms}\sqrt{\frac{8}{\pi}}\right)A\sin(kx - \omega t) = \left|T_{f_D f_D}(\omega)\right|\varepsilon(t) \tag{1-241}$$

Take the square of both sides of the equals sign in Equation (1-241) and multiply by $\lim\limits_{T \to \infty} \frac{1}{T}$; we can then write

$$\lim_{T \to \infty} \frac{1}{T} f_D^2(t) = \lim_{T \to \infty} \frac{1}{T} |T_{f_D f_D}(\omega)|^2 \varepsilon^2(t) = |T_{f_D f_D}(\omega)|^2 \lim_{T \to \infty} \frac{1}{T} \varepsilon^2(t) \tag{1-242}$$

But from Equation (1-61), we know that $\lim\limits_{T \to \infty} \frac{1}{T} |\varepsilon^2(t)| = S_{\varepsilon\varepsilon}(\omega)$. In this way, $\lim\limits_{T \to \infty} \frac{1}{T} f_D^2(t)$ should also be the power spectral density $S_{f_D f_D}(\omega)$ of the drag force f_D on a unit length; then Equation (1-242) can be written as

$$S_{f_D f_D}(\omega) = |T_{f_D f_D}(\omega)|^2 S_{\varepsilon\varepsilon}(\omega) = \left(\omega e^{kz} \phi_D u_{rms} \sqrt{\frac{8}{\pi}} \right)^2 S_{\varepsilon\varepsilon}(\omega) \tag{1-243}$$

b. The inertial force spectrum of wave action per unit length

Set the inertial force on a unit length of the structure that is changing with time as $f_i(t)$ and replace $\rho C_M \frac{\pi}{4} D^2$ which express the inertial force with ϕ_M. Then get the derivative of the expression of horizontal velocity u of the water particle to u and give the expression of acceleration $\frac{\partial u}{\partial t}$. Follow the same method to deal with Equation (1-241). From this we can obtain

$$\lim_{T \to \infty} \frac{1}{T} f_i^2 = \left(\phi_M \omega^2 e^{kz} \right)^2 \lim_{T \to \infty} \frac{1}{T} \varepsilon^2(t) = |T_{f_i f_i}(\omega)|^2 \lim_{T \to \infty} \frac{1}{T} \varepsilon^2(t) \tag{1-244}$$

Similarly, $\lim\limits_{T \to \infty} \frac{1}{T} f_i^2(t)$ can be replaced by the power spectral density $S_{f_i f_i}(\omega)$ of the inertial force of wave action per unit length of the structure; then Equation (1-244) is as follows:

$$S_{f_i f_i}(\omega) = |T_{f_i f_i}(\omega)|^2 S_{\varepsilon\varepsilon}(\omega) = \left(\phi_M \omega^2 e^{kz} \right)^2 S_{\varepsilon\varepsilon}(\omega) \tag{1-245}$$

c. The total wave forces spectrum $S_{ff}(\omega)$ per unit length

According to the Morrison equation, the total wave force is the superposition of the drag force and the inertial force, so from Equation (1-243) and Equation (1-245), we obtain

$$
\begin{aligned}
S_{ff}(\omega) &= S_{f_D f_D}(\omega) + S_{f_i f_i}(\omega) \\
&= \left[|T_{f_D f_D}(\omega)|^2 + |T_{f_i f_i}(\omega)|^2 \right] S_{\varepsilon\varepsilon}(\omega) \\
&= |T_{ff}(\omega)|^2 S_{\varepsilon\varepsilon}(\omega)
\end{aligned}
\tag{1-246}
$$

2. The wave force spectrum $S_{FF}(\omega)$ on the total length of a structure immersed in water

Assume that the length of the submerged structure is S, then take a small section of dz along the vertical axis z and the integral from zero to S; we get

$$
\begin{aligned}
S_{FF}(\omega) &= \int_0^S S_{f_D f_D}(\omega) dz + \int_0^S S_{f_i f_i}(\omega) dz \\
&= \left\{ \left[\int_0^S T_{f_D f_D}(\omega) dz \right]^2 + \left[\int_0^S T_{f_i f_i}(\omega) dz \right]^2 \right\} \times S_{\varepsilon\varepsilon}(\omega) \\
&= |T_{FF}(\omega)|^2 S_{\varepsilon\varepsilon}(\omega)
\end{aligned}
\tag{1-247}
$$

3. The transfer function of the random wave force spectrum
The random wave spectrum can be converted into a random wave force spectrum by multiplying the square of a function called the transfer function.
  a. Transfer function of random wave force spectrum on unit length of a structure
  The $T_{ff}(\omega)$ in Equation (1-246) is the transfer function, thus

$$\left|T_{ff}(\omega)\right|^2 = \left|T_{f_D f_D}(\omega)\right|^2 + \left|t_{f_i f_i}(\omega)\right|^2 \tag{1-248}$$

  In the equation, $T_{f_D f_D}(\omega)$ and $T_{f_i f_i}(\omega)$ are respectively called the transfer function of the drag force and the inertial force of random wave forces on a unit length structure:

$$T_{f_D f_D}(\omega) = \omega e^{kz}\phi_D u_{rms}\sqrt{\frac{8}{\pi}} \tag{1-249}$$

$$T_{f_i f_i}(\omega) = \omega^2 \phi_M e^{kz} \tag{1-250}$$

  b. Transfer function of random wave force spectrum on a submerged structure.

  The $T_{FF}(\omega)$ in Equation (1-247) is the transfer function; then

$$[|T_{FF}(\omega)|]^2 = \left[\int_0^S T_{f_D f_D}(\omega)dz\right]^2 + \left[\int_0^S T_{f_i f_i}(\omega)dz\right]^2$$

$$= \left(\int_0^S \omega\phi_D e^{kz} u_{rms}\sqrt{\frac{8}{\pi}}dz\right)^2 + \left(\int_0^S \omega^2 \phi_M e^{kz}dz\right)^2 \tag{1-251}$$

From the above analysis, we see that the random wave force spectrum on the total length of the submerged structure can be determined after calculating the transfer function of the random ocean wave spectrum.

1.5.6 Wave Force on a Large-Scale Structure

The premise of the research on the wave force acting on a small-scale structure assumes that the structure does not exist in seawater. But in fact, all ocean engineering structures are in seawater and have a certain scale. This brings complexity to the stress analysis of a large-scale ($D/\lambda > 0.2$) structure.

1.5.6.1 Characteristics of Wave Force Acting on Large-Scale Structures

1. Relative scale effect
Due to that the fact that the scale of structures present in the sea is large, there will be incident waves that touch the structure; this will produce scattered waves and this is called the scattering effect. Such scattering waves and incident waves interfere with each other and will cause changes in the current field around the structure. This effect is called the relative scale effect.

2. Free surface effect

Due to the increase of the characteristic length of the structure, the length is in the same order of magnitude as the wavelength of ocean waves. In this way, no matter whether the structure is completely submerged in water or partly above the water, it will produce the effect of waves on a free surface because of the existence of structures in the water; this effect is known as the free surface effect.

3. Viscous effect

Because we considering structures existing in seawater, the size of resistance of waves on the structure is related to the the scale of structures; this effect is called the viscous effect.

1.5.6.2 Calculation Principles of the Wave Force on Large-Scale Structures

For calculation of the wave force acting on large-scale structures, because we need to consider the effects of wave scattering on incident waves, we will need to use the diffraction theory for analysis. The characteristics of this calculation principle are as follows: first, determine the velocity potential ϕ_i of the incident waves of structures; second, calculate the velocity potential ϕ_m of the scattering waves reflected back after waves touch the structure. We can calculate the velocity potential ϕ of the current field with the superposition of ϕ_i and ϕ_m; in turn, we use the synthetic velocity potential ϕ to determine the dynamic pressure distribution on the surface of the structure with Cauchy integral equation and then calculate the wave force and torque acting on the structure.

The details of applying the diffraction theory to calculate the wave force acting on large-scale structure will be described in detail in Chapter 4.

1.5.6.3 The Scope of Diffraction Theory

As introduced above, the Morison equation is composed of the drag force and the inertial force. The drag force that is created by viscous effects mainly depends on the ratio of the size of the trajectory of water particles' motion and the size of the structure. Usually when the water depth and wavelength are constant, this ratio is expressed by the ratio of wave height H and characteristic length of the structure $2a$, namely $H/2a$. Whether we need to consider the viscous effect, what is whether the drag force should be accepted or rejected, is determined by the size of this ratio $H/2a$. In addition, the ratio of the scale of the structure $2\pi a$ and wavelength λ, as the relative scale indicators of the structure, is directly related to the application of the Morison equation, that is, the question of whether we need to apply the diffraction theory or not. So, taking $H/2a$ as the ordinate and $\frac{2\pi a}{\lambda}$ as abscissa, we can use Figure 1-32 to determine whether the diffraction theory, the viscous effects, and the Morison equation are suitable or not. Figure 1-32 can be divided into the following three areas.

a. $\frac{H}{2a} < 1$

At this moment, because the value of $H/2a$ is small, the relative displacement value of the water particle is very small. Thus the water near structures will still stick on the surface of structures and we do not see the phenomenon of flow separation, which is caused by the diffraction around the structure, so the viscous effect is insignificant and can be completely ignored.

b. $\frac{2\pi a}{\lambda} < 0.2$

At this moment, because the value of $\frac{2\pi a}{\lambda}$ is small and the diffraction effect insignificant and thus can be completely ignored, the Morison equation can be applied to all $H/2a$.

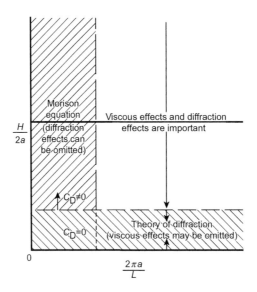

FIGURE 1-32

Scope of application of diffraction theory.

c. The area that $\frac{2\pi a}{\lambda}$ and $\frac{H}{2a}$ all is small

At this moment, as shown in the cross area in the lower left corner of Figure 1-32, because $\frac{H}{2a}$ is small and viscous effect is too small to be considered, the drag coefficient C_D is tending to zero and the drag force can be ignored. Thus, the wave force is dominated by inertial force and the shape and volume of the structure are the principal factor in deciding the size of the wave force.

1.6 Sea Ice and Ice Load

For the China Bohai Sea in the north of China and the north of the China Yellow Sea, because they are in the north and the cold air invades in winter, we see a varying degree of icing phenomenon every year. Though the ice conditions in the China Sea area is not as severe as that in the cold region, there are still years where the ice is heavy. For example, the China Bohai Sea was seriously frozen in February 1936; the coast of Laizhou Bay was frozen for a stretch reaching 30 ~ 40 km wide; the accumulation of ice was up to 27 m thick. Under serious ice conditions, offshore oil production facilities can't normally produce because of supply disruption, and sometimes some of the offshore oil engineering structures will be destroyed, or even be pushed over. For example, in February to March 1969, serious ice conditions occurred in the China Bohai Sea that hadn't happened in six decades; under the action of the alternating ice load, "The Old No.2" oil production platform was damaged, leading to the platform being completely pushed over by the sea ice. Thus, for the design and construction of oil engineering structures in the Chinese sea ice area, we should attach great importance to the ice load. In addition, the velocity of the floating ice that is formed with the ocean current or wind can not be ignored for its impact on offshore engineering structures. In fact, the sea ice is far more serious in its impact on a structure than the sea waves in the China Bohai Sea. Therefore, ice load is often cited as the control load of the engineering structure design.

1.6.1 General Situation

1.6.1.1 The Classification of Sea Ice

The characteristics of the ice conditions in China Sea area are often compartmentalized into speed, class, and form.

1. Speed

   This can be divided into two classes as noted below according to the movement state of sea ice:
   a. Floating ice, namely the ice drifting with the wind, current, and waves.
   b. Fast ice, namely the ice freezing near the coast and islands; although it cannot flow, it can lift up and down with the tide.

2. Class

   This can be divided according to the ice generation, development process of ice, and consideration of its thickness. It can be divided into new ice (C), pancake ice (I) (it's like a disc pie under 3 m in diameter, with a thickness of less than 5 cm), ice rind (P) (it is bigger than pancake ice and its thickness is greater than 5 cm), plate ice (B) (flat surface, its thickness is about 5 to 15 cm), grey-white ice (H) (gray, its thickness is 15 ~ 30 cm).

3. Form

   This divided according to the characteristics of the shape and physical appearance of the sea ice and the size of the ice.
   a. Flat ice (N): The ice surface is smooth.
   b. Rafted ice (Z): Ice overlaps with ice and the gradation is distinct.
   c. Hummocked ice (D): Under the action of wind, wave, and current, ice promiscuously overlaps with ice on the ice.

   The classification of the ice in the China Sea is shown in Table 1-13.

1.6.1.2 The Age of Sea Ice

Usually ice age is the time interval from its appearance to its disappearance. There is a process of forming, development, and disappearance for sea ice like other natural phenomena. So, according to the laws, the age of ice can be divided into three periods.

1. Freezing period (development)

   The time after the ice's appearance and before its pleniglacial period is about a bit more than one month (generally it is from the middle of November to the end of December for the China Bohai Sea). During this time, the mount of ice, which is thin, crisp, fragile, and has high salinity, is generally 30% ~ 40% of the total sea area, and there are few incidents of ice layers overlapping.

2. Pleniglacial period

   This period lasts about 50 ~ 70 days; for the China Bohai Sea this is generally from mid-to-late January and February. During this time, the mount of ice will rise to 60% ~ 70% of the total sea area and there is more fast ice and hummocked ice. The overlapping phenomenon is serious. Because the structure of ice is tight and the ice is hard, the compressive strength is strong.

Table 1-13 The Classification of Ice in the China Sea

Speed	Class			Form	
According to the motion	According to the development process and consideration of the ice thickness	According to the type of appearance		According to the geometric dimensions	According to the shape features
Floating ice	New ice (C) Pancake ice (I) Ice rind (P) Plate ice (B) Grey-white ice (H) Thick ice (U)			Flat ice (N) Rafted ice (Z) Hummocked ice (D)	
Fast ice		Coastal ice Stranded ice Anchor ice Ice foot off the coast			Flat ice Rafted ice Hummocked ice Ice mound An iceberg

3. Melting period (disappearance)

   From the ice beginning to melt to the end of the melting, there are usually only 15 ~ 20 days. During this period, the ice grows thin due to increasing temperature. Fast ice and hummocked ice begin to melt into floating ice and there ismore crushed ice and thin ice. Because its structure is loose and pressure resistance is poor, the ice is crisp and fragile.

1.6.1.3 The Condition of Sea Ice

Ice condition refers to the severity and the extent of damage. Usually it is expressed in the size of the ice thickness. Here we introduce the ice condition of the China Bohai Sea and the north part of the China Yellow Sea.

1. Ice conditions in an average year

   The ice conditions of the China Bohai Sea and the north of the China Yellow Sea are not very serious in average year, as shown in Table 1-14.

   As seen from Table 1-14, the ice conditions among the three bays of the China Bohai Sea (Liaodong Bay, Bohai Bay, and Laizhou Bay) are different. The ice conditions are worst in northern Liaodong Bay, and the ice thickness is up to 100 cm; the ice conditions of Bohai Bay are better than that of Liaodong Bay, but worse than that of of Laizhou Bay. The ice conditions of the north of the China Yellow Sea are generally better than those along the coast of the China Bohai Sea, and the ice thickness is just 10 ~ 30 cm as shown in Figure 1-33.

2. Serious ice conditions in certain years

In history, there have been serious ice conditions in the China Bohai Sea. For example, in 1936, 1947, and 1969, the China Bohai Sea faced serious ice conditions , which are shown in Figure 1-34.

   As shown in Figure 1-34, in serious ice conditions years, the China Bohai Sea except for the center of the sea area and the channel is almost all covered by sea ice. In northern Liaodong Bay, the ice thickness is 30 ~ 150 cm and the thickest is up to 250 cm.The ice conditions of the Haihe

Table 1-14 The Ice Conditions of the China Bohai Sea and the North of the China Yellow Sea in an Average Year

Sea Area	Freezing Period	Breakup Period	Age of Ice	Scope of Ice Area, n mile	Ice Thickness, cm
Liao dong Bay — The north, The middle, The south	In the middle of November	At the end of March	Four months	10 ~ 30 away from shore	15~40, max 100 5 ~ 25 Less than 10
Bohai Bay	In late November to early December	In the middle of March	More than three months	5 ~ 20 away from shore	10 ~ 30 In thhe north 20 ~ 40
Laizhou Bay	In early December or in the middle of December	In the middle of March	About three months	3 ~ 15	10 ~ 30
North of the China Yellow Sea	In early December	In the middle of March	Three months	10	10 ~ 30

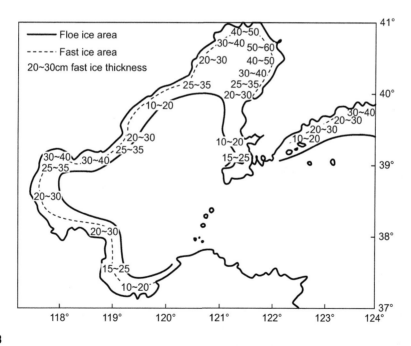

FIGURE 1-33

Distribution of ice conditions of the China Bohai Sea and the north of the China Yellow Sea in average years.

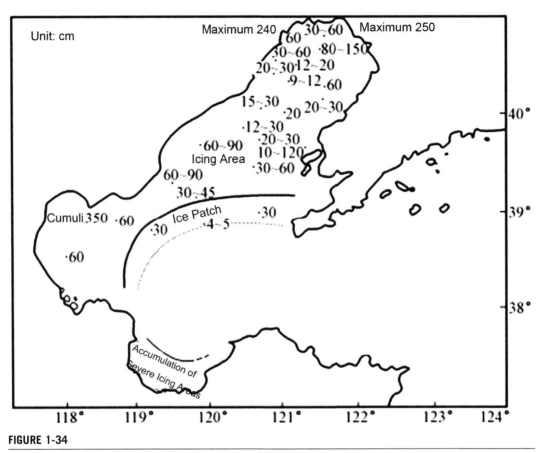

FIGURE 1-34

Distribution of ice conditions in the China Bohai Sea in serious ice condition years.

River estuary are the most serious in Bohai Bay, where ice width is up to 70 n miles, ice thickness is 30 ~ 45 cm and the thickest is up to 150 cm. The stack height of sea ice is up to 35 m in Dagu Kou. In Laizhou Bay the ice conditions are relatively better, but along the coast the ice width is up to 30 ~ 40 km and accumulation is serious.

1.6.1.4 Thickness of Sea Ice

The thickness of sea ice is an important index relating to the ice conditions and is a basic parameter to calculate ice load. Therefore, we need to know the ice thickness when researching ice. There are a lot of methods to determine ice thickness. Besides direct observation, there is also a method of heat balance. This method is based on heat conduction theory, with consideration of the heat conduction of this temperature difference according to the hypothesis. The temperature of the ice surface is equal to the atmospheric temperature and the temperature of the ice at the bottom is equal to the seawater temperature. This method uses the transmission area to calculate the thickness of sea ice. In addition, there are also semi-empirical or semi-theoretical equations to calculate the thickness of sea ice. Among them, the Chinese Marine Environmental Forecasting Center put forward a semi-empirical and semi theoretical equation that is close to reality by comparing with actual observation data. As shown in Figure 1-35, the simplified expression is as follows:

$$H = \alpha\sqrt{\left(\sum FDD - 3\sum TDD\right) - K}$$ (1-252)

In the equation, H is the calculated ice thickness; FDD is Freezing Degree Day (the average temperature is below $-20°C$); TDD is Thawing Degree Day (the average temperature is above $0°C$); K is the cumulative number of days for which the average daily temperature is lower than $-2°C$ when ice first generates; and α is the growth factor of sea ice thickness, namely the increasing size of sea ice thickness per effective FDD, in cm. The coefficient α is affected by many factors, such as water depth h_0, where the initial measured ice thickness is, H_0, sea area depth, h, the cumulative value $\sum(T_{ai} - W_j)$ of the difference of the temperature T_{ai} and water temperature W_j, the offshore distance, and the thickness of the snow on the ice, etc. It can be simplified as

$$\alpha = H_0\sqrt{\frac{h_0}{h}} \times \frac{1}{\sum_{j=1}^{n}(T_{ai} - W_j)}$$ (1-252')

These theory and the semi-empirical and semi-theoretical equation above are just a calculated equation under the condition of lacking of observed data of ice thickness. Direct observation is the best way to determine the ice thickness.

The data of ice thickness obtain by direct observation shows that both the ice thickness and the extreme value of ice thickness are random variables. That is to say, ice thickness is the distribution value. How do we transform the distribution values of ice thickness into a single ice thickness value to apply in actual offshore oil engineering? Here are two methods in common use.

1. Use the ice thickness that appeared once over many years as the single ice thickness value
   This method involves defining the maximum ice thickness $H_{max,n}$ that may appear in the stipulated recurrence interval as a single ice thickness value H, that is

$$H = H_{max,n}$$ (1-253)

By this method, according to the geographic location and the development prospects of oil and gas fields, etc., the China Bohai Sea is divided into nine zones by the China Bohai Petroleum Company, as shown in Figure 1-36.

According to the collection of 500 points of sea ice thickness data , the flat ice thickness (cm) for recurrence intervals of 10a, 25a, 50a and 100a are shown in Table 1-15. The table also lists the corresponding statistical parameters C_V and C_S.

2. Take the average of annual ice thickness extrema as a single ice thickness value
   This method takes the average (mathematical expectation) of annual ice thickness extrema over many years as a single ice thickness value. The annual ice thickness extremum represents the maximum ice thickness value $H_{max,1}$ that occurs once in a year. The commonly used distributions are
   a. Probability distribution of extremum type I
      If the annual extremum ice thickness $H_{max,1}$ obeys the extremum I distribution, the single ice thickness value H is

$$H = E(H_{max,1}) = \frac{V}{\alpha} + K$$ (1-254)

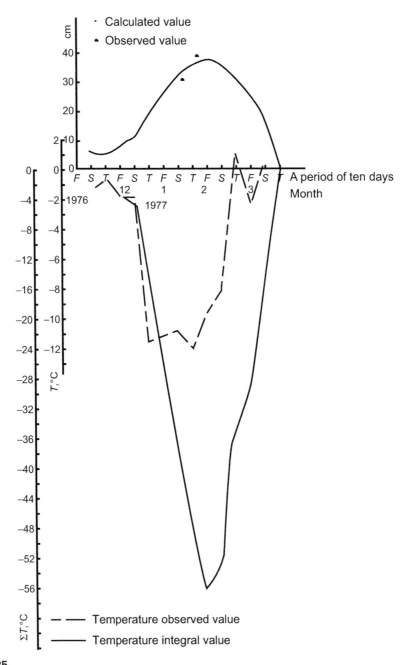

FIGURE 1-35

Comparison of actual measured data and the results which calculated by the ice thickness calculation equation proposed by China.

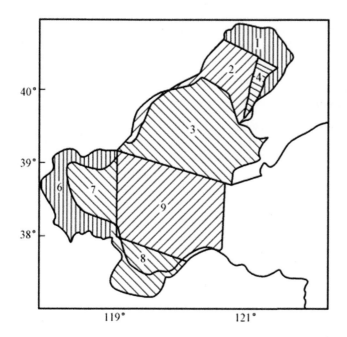

FIGURE 1-36

Sea ice thickness partitions in the China Bohai Sea.

In the equation, E represents mathematical expectation; V represents Euler's constant, generally $V=0.2775$; and α and K are the distribution parameters of extremum type (they can be obtained by finding the probability distribution of years of statistical data from the Bohai Petroleum Company shown in Table 1-16).

Table 1-15 Flat Ice Thickness (cm) for Different Recurrence Intervals in the Nine Sea Areas of the China Bohai Sea

| Sea Area | Statistical Parameters | | Recurrence Interval | | | |
	C_V	C_S	10a	25a	50a	100a
1	0.48	1.56	28	34	40	45
2	0.44	0.81	27	33	38	41
3	0.52	0.68	27	31	36	39
4	0.52	0.95	39	46	53	58
5	0.85	2.84	15	21	27	33
6	1.26	2.8	12	18	24	29
7	1.24	2.76	13	18	23	29
8	1.16	3.35	13	20	27	33
9	1.72	3.7	10	16	23	29

Table 1-16 The Distribution Parameters and Statistical Characteristic Values of the Annual Maximum Ice Thickness in the Nine Sea Areas of the China Bohai Sea

Sea Area	Average	Variance	Standard Deviation	Pearson III Type Curve		Distribution of Extreme Value Type I	
				α	β	α	K
1	17.16495	67.884	8.23918	1.643655	0.15560	0.155665	13.4570
2	17.75913	61.05887	7.81402	6.098663	0.31599	0.164134	14.2425
3	15.95731	68.85347	8.29780	8.650519	0.35445	0.154565	12.2230
4	22.87552	141.4974	11.8953	4.432133	0.17699	0.10782	17.5222
5	7.441344	40.00743	6.32514	0.495933	0.11134	0.20277	4.59477
6	4.821409	36.90533	6.07498	0.510204	0.11758	0.21112	2.08742
7	4.949126	37.66174	6.13692	0.5251	0.11808	0.208989	2.18726
8	5.678809	43.39409	6.58742	0.356427	0.09063	0.194697	2.71420
9	3.500151	36.24352	6.02026	0.292184	0.08979	0.213039	0.79079

b. The probability distribution of Pearson type III

If $H_{max,1}$ is a random variable and the data found by the Bohai Petroleum Company obeys the probability distribution of Pearson type III, the distribution parameters α and β have been given through the statistics gathered for many years. As shown in Table 1-16, the single ice thickness values H is

$$H = E(H_{max,1}) = \frac{\alpha}{\beta} \tag{1-255}$$

c. Lognormal distribution and Weibull distribution

According to the 20a flat ice statistical data for the annual extremum value $H_{max,1}$, the China University of Petroleum found out the probability distribution about $H_{max,1}$ in the three sea areas of the Liaodong Bay in China Bohai Sea. The hypothesis test showed that the lognormal distribution (LN) and Weibull distribution (WBL) were the best fits. So, the single ice thickness value H can be expressed as

$$H = E(H_{max,1}) = LN(1.39892, 2.6424) \tag{1-256}$$

$$H = E(H_{max,1}) = WBL(0, 1.6298, 1.0801) \tag{1-257}$$

In the equation, the numbers in brackets after the LN are the average μ and the variance σ^2 of the logarithmic normal distribution. And the numbers in brackets after the WBL are the position, scale, and shape parameters of Weibull distribution. According to the logarithmic normal distribution, the probability density curve (P_{df}) of the annual ice thickness extremum in the three sea areas of the Liaodong Bay is shown in Figure 1-37.

1.6.1.5 Sea Ice Strength
Sea ice strength is another basic parameter of sea ice and it is directly related to the force of sea ice acting on offshore oil engineering. The compression strength of sea ice is related to the force of the vertical structure, and the bending strength of sea ice is mainly related to the tilting structure.

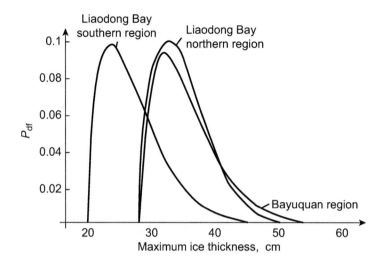

FIGURE 1-37

The probability density curve of the annual ice thickness extremum value in the three sea areas of the Liaodong Bay.

1. Determination of sea ice compression strength

   When sea ice interacts with a vertical structure, a crushing failure is happening, so the ability of sea ice to resist compression is called compression strength. It is usually expressed as the uniaxial unconfined compressive strength of sea ice, and the methods to determine it are as follows.

   a. Field test methods

   This method is directly measured at sea, so it is related to the different conditions such as sea area, coastal ice, or sea ice. Bohai Petroleum Company offers the compression strength of sea ice at different temperatures in Liaodong Bay according the field testing, as shown in Table 1-17 and Table 1-18.

   b. Indoor experimental methods

   Usually the sample ice is removed off about 5 cm, then the rest is divided into two layers: upper and lower. Finally 10 cm × 10 cm × 25 cm prism specimens are created for experimentation in the indoor test machine. Table 1-19 shows the compression strength of sea ice observed by indoor experiments under the different temperatures in Liaodong Bay.

   In addition, equations can be applied to calculate the compression strength of sea ice, but they are mostly empirical equations, so they are discussed no further here.

Table 1-17 Compression Strength of Sea Ice in Liaodong Bay under Different Temperatures

Ice Temperature,°C Compression Strength, Mpa	−2.5	−3.3	−6.4	−7.0
Maximum	2.53	2.64	3.38	3.69
Average	1.68	2.00	2.01	2.70

Table 1-18 Compression Strength of Coastal Ice in Liaodong Bay under Different Temperatures

Ice Temperature,°C Compression Strength, Mpa	−2.5	−3.3	−6.4	−7.0
Maximum	2.53	2.64	3.38	3.69
Average	1.68	2.00	2.01	2.70

2. Determination of bending strength of sea ice

   When sea ice experiences interactions with a tilting structure and a pyramidal structure, bending failure or surface bending failure (downward) of the root will occur. Thus the resistance capability of sea ice to bending and cracking is called bending strength. Generally the bending strength of sea ice is only 1/2 ~ 1/5 of the compression strength, therefor, it is important to take advantage of this characteristic to solve the design problem of anti-ice structures in offshore oil engineering. Generally the methods of determining sea ice's bending strength are as follows.

   a. Cantilever beam bending test at the scene

      Using this method, China Bohai Petroleum Company has done field tests many times over two consecutive years in Liaodong Bay. And under different ice temperatures and ice thicknesses it has obtained an average σ_{BM} of sea ice bending strength of (282 ± 80) kPa and (259 ± 52) kPa respectively.

   b. Indoor three-point bending or four-point bending test

      According to the indoor test used by the Bohai Petroleum Company, the bending strength of sea ice at different temperatures is shown in Figure 1-38.

      In addition, there are some empirical equations or theoretical equations to calculate the bending strength of sea ice. For example, according to the theory of an elastic beam, for a rectangular section of sea ice, if its length is L, width is b, and height is h, when the load acting on the midspan of the sea ice beam is P, bending failure occurs. And then the bending strength σ_{Bu} of sea ice can be calculated as follows:

$$\sigma_{Bu} = \frac{3PL}{2bh^2} \tag{1-258}$$

3. The probabilistic properties of sea ice's compression strength

In an ocean engineering structure, the most common problem is the extrusion failure of a vertical structure, so here we only introduce the probabilistic properties of sea ice compression strength. For sea ice compression strength, there are many influencing factors that cause that the test data of sea ice compression strength to have a great deal of dispersion, so sea ice compression strength becomes a random variable. And so we need to obtain the probability distribution.

Table 1-19 Compression Strength of Liaodong Bay Sea Ice during an Indoor Experiment for Different Temperatures

Ice Temperature,°C Compression Strength, Mpa	−15	−10	−7	−3
Maximum	4.40	3.43	2.65	2.05
Average	4.93	3.74	3.20	3.02

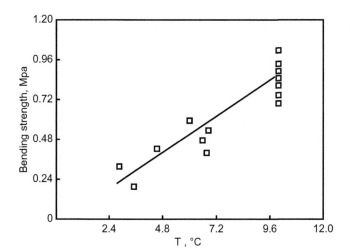

FIGURE 1-38

Bending strength of sea ice at different temperatures in China Bohai Sea.

According to the annual maximum data of sea ice compression strength measured at over 24 years at four ocean observatories (Huludao, Bayuquan, Qinhuangdao, Changxing Island) in Liaodong Bay, China Bohai Sea, China University Of Petroleum applied the least square to estimate the distribution parameters using the Weibull distribution, normal distribution, lognormal distribution, the extreme value type I distribution and so on, and used the K - S method to inspect and compare the error estimate. They found that the best conclusion was the Weibull distribution. Therefore, we can obtain a probability density (P_{df}) distribution curve of the annual maximum of flat ice compression strength in the four sea areas of the Liaodong Bay, as shown in Figure 1-39.

Meanwhile, after the probability distribution of sea ice compressive strength is known, according to the annual maximum probability density function $P(\sigma_C)$, we can calculate the average $\overline{\sigma_C}$ of sea ice annual maximum compression strength through the method of calculating mathematical expectation in the following equation:

$$\overline{\sigma_C} = \int_0^\infty \sigma_C P(\sigma_C) d\sigma_C \qquad (1\text{-}259)$$

In Table 1-20, $\sigma_C \sim WBL(\alpha, \eta, \beta)$ is given, that is to say that when σ_C obeys the Weibull distribution, the position (α), the shape (β), and the scale (η) in the probability density function are as listed in following table. In Table 1-20, the value of D_n and correlation coefficient r can be obtained by the K-S method of inspection and the estimated error s is also given by the least square method.

1.6.2 The Static Ice Force of Sea Ice

Broadly speaking, all the external forces acting on an ocean engineering structure can be seen as dynamic load. But when the dynamic amplification of structural response caused by dynamic load is not obvious, it can be treated as static load. So the ice force from sea ice that acts on the structure can be divided into static ice force and dynamic ice force. The former are the external forces that don't cause structural response dynamic amplification, and the latter are the external forces that

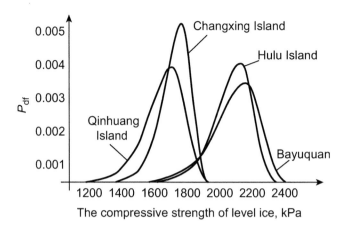

FIGURE 1-39

The probability density distribution curve of sea ice compression strength in the four sea areas of the China Bohai Sea, Liaodong Bay.

cause structural response dynamic amplification. Here we begin to introduce static ice force analysis.

1.6.2.1 Static Ice Force Acting on a Vertically Isolated Pile

This static ice force refers to the scenario when the structure is surrounded by a huge ice sheet and under the action of tide and wind; here an extrusion force on the structure is caused by large ice sheets moving integrally. Obviously, when ice is destroyed by extrusion (it achieves the ultimate compressive strength of sea ice), the extrusion force is at a maximum. Therefore, the extrusion force on a vertically isolated pile caused by sea ice mainly depends on the compressive strength σ_C of the sea ice. Of course, there are other influencing factors such as the shape of the structure (expressed in coefficient m), the contact conditions between the structure and ice (expressed in coefficient K), the local extrusion conditions between the structure and ice (expressed in coefficient I), the width B (or diameter D) of the sea ice contact with the structure, the thickness H of sea ice, etc. This means that the situation

Table 1-20 Parameters When Flat Ice Compression Strength obeys Weibull Distribution

Distribution Parameters	Huludao	Bayuquan	Qinhuangdao	Changxing Island
		Weibull Distribution		
α	0.483674	0.488155	0.367447	0.620483
η	2029.024536	2054.093018	1621.477417	1098.54161
β	20.330875	17.139069	15.760188	14.571115
s	0.221843	0.21624	0.207801	0.100686
r	0.980705	0.981676	0.98309	0.9959953
D_n	0.120821	0.125466	0.151879	0.096433

with static ice force F acting on a vertically isolated pile is a function $F(\sigma_C, m, K, I, B, H)$. The main influencing factors are σ_C, H, and B (or D), but the static ice force and the three main influencing factors have a direct first-power relationship or exponential relationship with H and D. This has become the key problem for sea ice researchers for many years.

According to the research over these years on the form of the function representing the static ice force that acts on a vertically isolated pile, the established mathematical model can be divided int: the plastic analysis model, the creep analysis model, and the fracture analysis model, among others. However, the creep analysis model regards sea ice as creep material and considers the time factor, so it's more difficult to analyze. Because the fracture analysis model analyzes sea ice force using the method of fracture mechanics, there are a lot of difficulties in accurately using quantitative analysis accurately. Therefore, nowadays sea ice researchers mostly focus on the plastic analysis model.

The plastic analysis model calculates the static ice force acting on a vertically isolated pile directly through plastic mechanics. This is because when ice comes into the fully plastic condition, the yield stress and ultimate strength remain in a steady value state. At present, there are more than 10 equations to calculate the static ice force by the plastic analysis model, and the equations are as follows:

1. First power function type represented by Korzhavin
   a. Korzhavinequation
      The following was proposed in 1962 by Korzhavin of the former Soviet Union:

$$F = I\left(\frac{v}{v_0}\right)^{-\frac{1}{3}} mKDH\sigma_c \qquad (1\text{-}260)$$

   In the equation, v represents ice velocity and v_0 represents ice reference velocity. The ratio of them reflects the strain rate, which has an effect on the static ice force. Generally, v_0 is 1.0m/s. Therefore, when $\frac{v}{v_0} = 1$, Equation (1-260) can be written as

$$F = ImKDH\sigma_c \qquad (1\text{-}261)$$

   Obviously, this model regards the static ice force as a function of first power. Based on the same consideration, the offshore fixed platform specification made by the China Classification Society also puts forward the following suggestion:

$$F = ImKDH\sigma_c \qquad (1\text{-}262)$$

   For a circular cross-sectional area, $m = 0.9$, the contact coefficient $K = 0.45$, and the local extrusion ratio $I = 2.5$. In Equation (1-262), D is the diameter of the cylindrical structure, H is the ice thickness, and σ_c is the ultimate compressive strength of sea ice.
   b. API equation
      When making the API specification, the American Petroleum Institute adopted the Korzhavin calculation model and put forward the following two equations:
      • Specified in the API RP 2A in 1984

$$F = CDH\sigma_c \qquad (1\text{-}263)$$

      • Specified in the API RP 2N in 1989

$$F = If_cDH\sigma_c \qquad (1\text{-}264)$$

In Equation (1-263) and Equation (1-264), API recommends that $\sigma_c = 1406 \sim 3516$ kN/m, coefficient $C = 0.3 \sim 0.7$, coefficient $I = 1.2 \sim 3.0$, and contact coefficient $f_c = 0.4 \sim 0.7$.

c. Afanasev equation

Afanasev gives another equation that refers to the Korzhavin model, namely

$$F = ImDH\sigma_c \tag{1-265}$$

The difference with the Korzhavin equation [Equation (1-261)] is that there is no consideration coefficient K, but the local extrusion coefficient I is decided by the ratio of pile diameter D and the ice thickness H, when $1 < \frac{D}{H} \le 6$, $I = \left(1 + 5\frac{H}{D}\right)^{\frac{1}{2}}$; when $0.1 < \frac{D}{H} \le 1.0$, $I = 4.0 \sim 2.5$.

d. The equation in Canadian specifications

In its relevant specification, Canada gives the following equations referring to Korzhavin model:

- Equation put forward in Canadian lighthouse specification

$$F = m'DH\sigma_c \tag{1-266}$$

- Equation put forward by the Canadian Standards Association

$$F = DHP_c \tag{1-267}$$

In Equation (1-266) and Equation (1-267), the comprehensive coefficient $m = 0.4 \sim 0.7$, the sea ice uniaxial compressive strength $\sigma_c = 1406 \sim 1753$ kN/m, and the ice floe effective stress $P_c = 703 \sim 2812 \text{kN}/m^3$.

e. Equation in the specification of the former Soviet Union countries

The equation in the CHИЛ 2.06.04 specification of the former Soviet Union is given as below in reference to the Korzhavin equation:

$$F = ImDH\sigma_c \tag{1-268}$$

In the equation, for a cylinder, as with Equation (1-261), the coefficient $m = 0.86 \sim 0.9$; for a square section, $m = 1$; coefficient I is also the same as Equation (1-261). When the width of the ice floe is 15 times as the cylinder diameter, $I = 2.5$; when the structure is relatively wide, $I = 1.0$. But Equation (1-261) considers the coefficient K and generally $K = 0.4 \sim 0.7$. Equation (1-268) doesn't consider the coefficient K.

2. Exponential function presented by Schwarz

a. Schwarz equation from Germany

This equation was put forward by the German scholar Mr. Schwartz in 1976 and again in 1991. It is different from the Korzhavin model above. It is not a power function but an exponential relation among the static ice force F, the ice thickness H, and the width (or diameter D). Static ice force functions are also proposed by China Bohai Petroleum Company, etc. They are also introduced in the following.

- Equation put forward by Schwarz in 1976

$$F = 3.56D^{0.5}H^{1.1}\sigma_c \tag{1-269}$$

Equation (1-269) is based on the test results of a model experiment in Iowa in the USA and the measured data from a 60 cm diameter bridge pier in the river.

- Equation put forward by Schwarz in 1991
  According to the discussion at the international sea ice meeting in 1990 in
  Hamburg, Germany, and an actual ice force measurement during a field test in China Bohai
  Liaodong Bay, a new static ice force calculation equation was put forward in 1991:

$$F = 0.5D^{0.5}H^{1.1}\sigma_c \qquad (1\text{-}270)$$

b. Equation put forward by the China Bohai Petroleum Company

$$F = 0.36D^{0.5}H^{1.1}\sigma_c \qquad (1\text{-}271)$$

In Equation (1-269) and Equation (1-270), σ_c is the sea ice uniaxial compressive strength, and
D and H are diameter and ice thickness, respectively. In the equation, indexes 0.5 and 1.1 and
the coefficient were obtained by field testing of the static ice force. In Figure 1-40, a comparison
between the measured data of the static ice force measurement from a field test by the Bohai

(a)

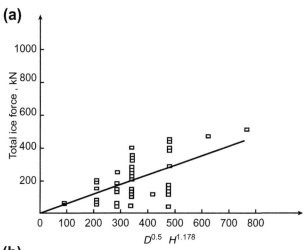

FIGURE 1-40

Comparison between the measured static
ice force and the calculation results.

(b)

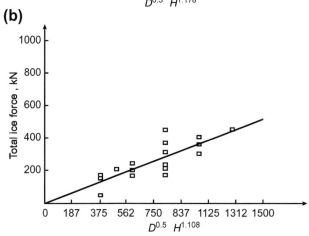

Petroleum Company and the computational result of Equation (1-270) is given. The comparison results show that both fit better. So, although in the Schwartz equation there is a mistake that dimension is not conform after considering index, it is fit for the reality of sea ice in China Bohai Sea.

c. Equation put forward by Hamayaka in Japan

This is also in the form of the Schwartz model. That is, in Equation (1-272), the coefficient C is respectively 5.0 and 6.8 for a circular section and rectangular section:

$$F = CD^{0.5}H\sigma_c \tag{1-272}$$

3. Comparison of equations of static ice force acting on a vertically isolated pile

The China Classification Society has compared the above 12 static ice force calculation equations [Equations (1-261) – (1-272)], and using the average value of each equation calculation result as the base number, has given the ratio of equation calculation results [the Equation (1-264) coefficient can be different, so it is listed twice] and the average, as shown in Figure 1-41.

Figure 1-41 shows that there are five calculation equations where the ratio is below 1.26, but calculation results with the equations are low and the risk is bigger. There are seven calculation equations when the ratio is above 1.51, and calculation results with these equations are high and too conservative; furthermore they all belong to the Korzhavin calculation model, which is a power

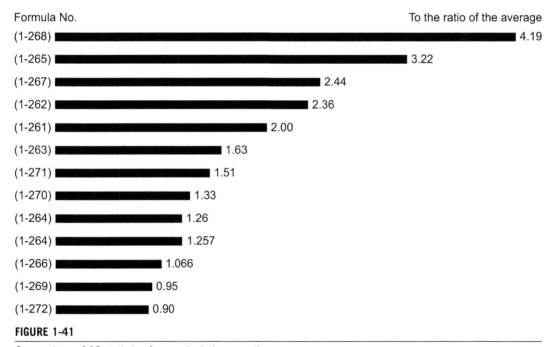

Formula No.   To the ratio of the average
(1-268) ————————————————————————————————— 4.19
(1-265) ———————————————————————————— 3.22
(1-267) ————————————————————— 2.44
(1-262) ———————————————————— 2.36
(1-261) ——————————————— 2.00
(1-263) ————————— 1.63
(1-271) ———————— 1.51
(1-270) ——————— 1.33
(1-264) ——————— 1.26
(1-264) ——————— 1.257
(1-266) —————— 1.066
(1-269) ————— 0.95
(1-272) ————— 0.90

FIGURE 1-41

Comparison of 12 static ice force calculation equations.

function type. Only the ratio of therevised Schwarz equation [Equation (1-271)] was tested by means of an on-site static ice force test by the Bohai Petroleum Company and its value is 1.51. It is not conservative and the risk is smaller, so this calculation equation is most suitable for the China Bohai Sea.

Therefore, we suggest that one should use Equation (1-271) to calculate the static ice force acting on a vertically isolated pile in an offshore oil engineering structure that constructed in the China Bohai Sea. Of course, if the structure is a clump of piles, we should also consider the shadowing effect and the blocking effect. When the pile spacing is equal to or below 1.5 times the pile diameter, the torque decreases by 30% on average; when the pile spacing is above 1.5 times the pile diameter, we should not consider the effect of the pile group.

1.6.2.2 Static Ice Force Acting on an Isolated Tilted Pile or a Vertical Pile with a Cone

Besides vertical piles, there are tilted piles and vertical piles with cones in the field of offshore oil engineering structure. Sea ice contact with a tilted pile or with a cone could cause sea ice bending failure. But as has been pointed out at the start of this section in general the bending strength of sea ice is 1/5 to 1/2 of the compressive strength. Therefore, a vertical pile structure with a cone can resisting ice well. We need to research the static ice force acting on a tilted pile and the cone of a vertical pile.

1. The basic principle of computing the static ice force acting on a slope
   There are many equations for a two-dimension model or three-dimensional model to calculate the static ice force on a tilted pile or the cone on a vertical pile. But their basic principles are the same:
   a. The total static ice force acting on the slope is composed of two parts: one is the horizontal force suffered by the structure when sea ice reaches bending failure; the second is the horizontal component of the thrust when the broken ice climbs along the ramp after bending failure.
   b. The horizontal force suffered by the structure mainly depends on the bending strength of sea ice when bending failure happens. We can regard sea ice floe as a semi-infinite beam on the basis of flexibility, and use beam-bending theory to calculate the bending stress by making it equal to the bending strength of sea ice.
   c. The thrust caused by the broken ice that climbs along the ramp after sea ice bending failure is relative to gravity itself and the friction of the ice climbing along the ramp. Therefore, according to the equilibrium relationship between the thrust and the two forces, the required thrust of the ice climbing along the ramp and the horizontal component acting on the structure can be calculated.
   d. The total static ice force acting on the cone of a vertical pile or a tilted pile is the superposition of two horizontal forces.
2. Two-dimension calculation model
   This model is suitable for use in the situation when sea ice is reacting on a wide slope, but does not work for a narrow slope. When the two-dimensional calculation model is established, the condition of the interaction force between the sea ice and the sloped structure is as shown in Figure 1-42.
   Here are the methods and steps of establishing a two-dimensional calculation model.

FIGURE 1-42

Force analysis when sea ice is reacting on the slope.

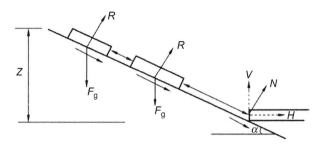

a. The vertical component N and the horizontal component H acting on the ice floe
As shown in Figure 1-42, the positive pressure N that is perpendicular to the slope can be divided to the horizontal component H and the vertical component V; they are

$$H = N \sin \alpha + \mu N \cos \alpha \tag{1-273}$$

$$V = N \cos \alpha - \mu N \sin \alpha \tag{1-274}$$

In these equations, μ is the coefficient of friction between the ice and slope; the angle α is as shown in Figure 1-42.

The horizontal component required when advancing ice floe is bending failure.

Suppose that the ice floe can be regarded as the beam of a rectangular section where the width is b and the thickness is t. The bending stress under the action of the bending torque M is σ_f, and σ_f has reached the bending strength σ_{Bu} of sea ice; then

$$\sigma_f = \frac{6M}{bt^2} = \sigma_{Bu} \tag{1-275}$$

But when the semi-infinite beam on the basis of flexibility is affected by the boundary load V, according to the bending theory of beams, the maximum bending torque M should be

$$M = \frac{IV}{e^{\frac{\pi}{4}}} \sin\left(\frac{\pi}{4}\right) \tag{1-276}$$

In the equation, l is the characteristic length of the structure; it is defined as

$$l = \left(\frac{D}{K}\right)^{\frac{1}{4}} \tag{1-277}$$

In Equation (1-277), K is the foundation coefficient. For an ice floe beam, it should be

$$K = \frac{1}{4}\rho_w gb \tag{1-278}$$

In Equation (1-278), ρ_w is the density of seawater, g is the acceleration of gravity. D in Equation (1-277) is the bending rigidity EI, where the ice floe's elastic modulus is E and the torque of inertia of a cross section is I.

Substitute Equation (1-278) into Equation (1-277) and after changing D into EI, substitute it into Equation (1-276), then substitute M into Equation (1-275). Then we can obtain V as

$$V = 0.68\sigma_{Bu}b\left(\frac{\rho_w gt^5}{E}\right)^{\frac{1}{4}}$$ (1-279)

After obtaining V, substitute it into Equation (1-274), calculate N, and then substitute N into Equation (1-273); we can obtain the horizontal force H as

$$H = 0.68\sigma_{Bu}b\left(\frac{\rho_w gt^5}{E}\right)^{\frac{1}{4}}\left(\frac{\sin\alpha + \mu\cos\alpha}{\cos\alpha - \mu\sin\alpha}\right)$$ (1-280)

b. The friction and gravity that trash ice needs to overcome when climbing a slope.
   After the ice floe reaches bending failure by reacting to the slope structure, it is stopped by trash ice on the slope. So when the subsequent ice floe climbs along the slope after bending failure, first it needs to have thrust that can cause the trash ice to climb up. Then it needs to overcome the friction F_f where ice has a positive pressure R on the slope and the self-gravity of the ice F_g; they should be

$$F_f = \mu R$$ (1-281)

$$F_g = \frac{Z}{\sin\alpha}bt\rho_i g$$ (1-282)

In Equation (1-282), as shown in Figure 1-42, Z is the climbing height, ρ_i is the mass density of sea ice, and $\frac{Z}{\sin\alpha}$ is the total length of the ice floe on the slope.
c. Determine the horizontal component of thrust when trash ice climbs along the slope
   Obviously this thrust P should be the sum of the friction F_f and component force $F_g\sin\alpha$ of gravity F_g along the slope, namely

$$P = F_f + F_g\sin\alpha = \mu R + \frac{Z}{\sin\alpha}bt\rho_i g(\sin\alpha) = \mu\left(\frac{Z}{\sin\alpha}bt\rho_i g\cos\alpha\right) + \frac{Z}{\sin\alpha}bt\rho_i g\sin\alpha$$

$$= \frac{Z}{\sin\alpha}bt\rho_i g(\mu\cos\alpha + \sin\alpha)$$

(1-283)

So, according to the relationship between Equation (1-279) and Equation (1-280), the horizontal component H' equals tthe vertical component $P\sin\alpha$ of thrust P multiplied by $\left(\frac{\sin\alpha + \mu\cos\alpha}{\cos\alpha - \mu\sin\alpha}\right)$, namely

$$H' = P\sin\alpha\left(\frac{\sin\alpha + \mu\cos\alpha}{\cos\alpha - \mu\sin\alpha}\right) = Zbt\rho_i g(\mu\cos\alpha + \sin\alpha)\left(\frac{\sin\alpha + \mu\cos\alpha}{\cos\alpha - \mu\sin\alpha}\right)$$ (1-284)

d. The horizontal component required when subsequent ice floe reaches bending failure
   Because the subsequent ice floe is stopped by trash ice at the front on the slope, when it reaches bending failure, the horizontal static ice force H'' acting on the slope structure should be the

sum of the horizontal force H given by Equation (1-280) and the horizontal component H' [Equation (1-284)] given by the thrust making trash ice climb along the slope; therefore

$$H'' = H + H'$$

$$= \left(\frac{\sin \alpha + \mu \cos \alpha}{\cos \alpha - \mu \sin \alpha}\right)\left[0.68\sigma_{Bu}b\left(\frac{\rho_w g t^5}{E}\right)^{\frac{1}{4}} + Zbt\rho_i g(\mu \cos \alpha + \sin \alpha)\right] \qquad (1\text{-}285)$$

e. The horizontal total static ice forces from the ice floe that act on the slope structure
If this total horizontal static ice force is H''', it should be the sum of the horizontal force H'' acting on slope structure and the horizontal component $P \cos \alpha$ of the thrust P climbing along the slope when subsequent ice floe reaches bending failure; so the total static ice force H''' can be given as follows:

$$H''' = H'' + P \cos \alpha =$$

$$0.68\sigma_{Bu}b\left(\frac{\rho_w g t^5}{E}\right)^{\frac{1}{4}}\left(\frac{\sin \alpha + \mu \cos \alpha}{\cos \alpha - \mu \sin \alpha}\right) + Zbt\rho_i g\left[\frac{(\sin \alpha + \mu \cos \alpha)^2}{\cos \alpha - \mu \sin \alpha}\right] +$$

$$Zbt\rho_i g\left(\frac{\mu \cos \alpha + \sin \alpha}{\tan \alpha}\right) = 0.68\sigma_{Bu}b\left(\frac{\rho_w g t^5}{E}\right)^{\frac{1}{4}}C_1 + Zbt\rho_i g C_2 \qquad (1\text{-}286)$$

The total horizontal static ice force on the unit width of a slope structure is as follows:

$$\frac{H'''}{b} = 0.68\sigma_{Bu}\left(\frac{\rho_w g t^5}{E}\right)^{\frac{1}{4}}C_1 + Zt\rho_i g C_2 \qquad (1\text{-}287)$$

3. Three-dimensional calculation model
In order to solve the problem of calculating the slope static ice force on a wide-bodied structure, a three-dimensional calculation model is put forward. The biggest difference between it and the two-dimensional calculation model is that after the ice floe reacts on the slope, it is regarded as several tapered beams. The circular cracks and radial cracks of the ice floe split into several tapered beams, as shown in Figure 1-43.
The basic principle of the three-dimensional calculation model is exactly the same as the two-dimensional model. The three-dimensional model just regards the bending failure of the ice floe as a frustum wedge (tapered beam) on the flexible support, and this is slightly different from the two-dimensional model.
At present, there are three categories and ten equations for the three-dimensional calculation model, but there are mainly five equations that are influential at the international level, namely Edwards-Croasdale, Croasdale, Ralston, Kato, and Hirayama-Obara. Here we do not introduce them one by one, but only compare them analyze their common ground and the comparison, and then put forward the best equation.

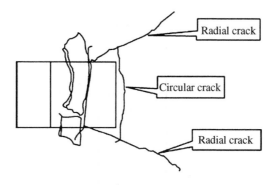

FIGURE 1-43

Ice floe damage in the three-dimensional calculation model.

The common ground of these five calculation equations is as follows:

a. The calculated static ice force is regarded as composed of two parts, namely the static ice force when the ice floe reaches bending failure and the static ice force from the ice climbing along the slope.

b. The first item of the static ice force is regarded as a function of the bending strength σ_{Bu} of sea ice and the square of sea ice thickness t. When there is a coefficient B, the static ice force when ice floe reaches bending failure is expressed by the function $B\sigma_{Bu}t^2$.

c. The second item of the static ice force when the ice climbs along the slope is regarded as a function of ice thickness t, the square of the structure diameter D at the waterline, and the acceleration of gravity g. When there is a coefficient A, it is expressed by the function $AgtD^2$.

d. The horizontal total static ice force H acting on the slope structure can be expressed by the superposition of the two horizontal static ice forces, and it can be written as

$$H = AgtD^2 + B\sigma_{Bu}t^2 \tag{1-288}$$

Then, according to the related equations and the corresponding experimental data, we can determine the two undetermined coefficients A and B through regression analysis.

To compare the calculation accuracy of these equations, two group contrast diagrams are given based on the Hirayama-Obara model and the Kato model, as shown in Figure 1-44(a) and (b). In Figure 1-44(a) and (b), the ordinate is the predicted force which is calculated by the equation, and the abscissa is the measured force in the laboratory test. Figure 1-44(a) is the contrast among the Hirayama-Obara calculation model and the other three models; Figure 1-44(b) is the contrast among the Kato calculation model and the other three models; Figure 1-44(c) is the contrast among four calculation models, all based on the Wessels model. The comparison results show that the calculation accuracy of Hirayama-Obara model is the highest [Figure 1-44(c)].

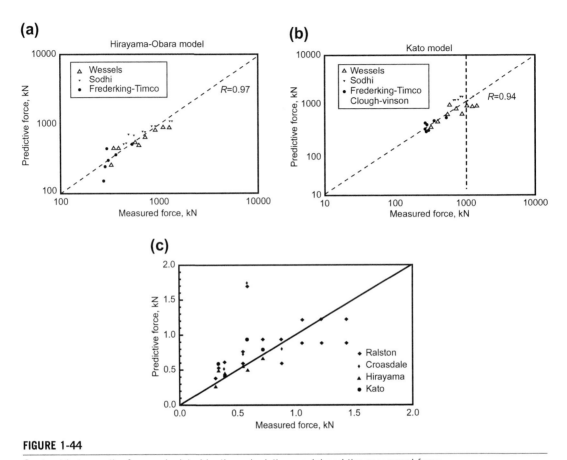

FIGURE 1-44

Contrast between the force calculated by the calculation model and the measured force.

The equation of the total horizontal static ice force H calculated by the Hirayama-Obara model is as follows:

$$H = \left[\frac{0.7\rho_i gtD^2 \xi \frac{Z}{D} \left(1 - \frac{Z}{D} \tan \alpha \right)}{\sin \alpha} \right] + 2.43\sigma_{Bu}t^2 \left(\frac{D}{l_c} \right)^{0.34} \qquad (1\text{-}289)$$

In the equation, α is the angle between the slope and horizontal plane; l_c is the characteristic length of sea ice, $l_c = \left[\frac{Et^3}{12\rho_w(1-v)^2} \right]^{0.25}$; v is the Poisson ratio of sea ice; ρ_w is the mass density of sea water; E is the modulus of elasticity of sea ice; Z is the height that the vertex of ice climbs; and ξ is the testing coefficient. The meanings of other symbols are same as in prior equations.

In addition, as can be seen from the Hirayama-Obara equation, if the slope structure is narrow, the first part to the right of the equals sign in Equation (1-289) can be left out because trash ice can be eliminated in time. It doesn't need to overcome the horizontal force acting on the structure when ice climbs up along the slope. Then Equation (1-289) can be rewritten as

$$H = 2.43\sigma_{Bu}t^2 \left(\frac{D}{l_c}\right)^{0.34} \tag{1-290}$$

But the coefficient $B = 2.43$ in Equation (1-288) was obtained from a laboratory test by Hirayama-Obara, and it is different from field measurements. In a field measurement by Dalian University of Technology in the China Bohai Sea, researches obtained the coefficient $B = 3.2$ for Liaodong Bay of the China Bohai Sea, so the equation of the total horizontal static ice force H that applies to a narrow slope (cone) structure on the Liaodong Bay sea area of the China Bohai Sea is

$$H = 3.2\sigma_{Bu}t^2 \left(\frac{D}{l_c}\right)^{0.34} \tag{1-291}$$

In the equation, l_c represents the fracture length of the ice floe.

1.6.2.3 Probability Distribution of Static Ice Force Acting on an Isolated Pile

For ocean engineering structure design in an ice zone, we adopt the design standards of extreme conditions, namely using the flat ice thickness and the extreme value of ice strength as design criteria. But this design is not only conservative, causing economic waste, but also affects the safety and reliability of the structure. So, it would be better to use the probability method for reliability design, namely regarding some or the entire design variable as a random variable, finding the probability distribution, and thus calculating the probability (reliability) that the strength is greater than the stress or that the forecast life is greater than the design life given the safety factor and its reliability. If we want to carry on reliability design of ocean engineering structures in an ice zone, we must know the probability distribution of the static ice force in the engineering structure; here we look at the probability distribution of static ice force on an isolated vertical, isolated tilted, or cone pile.

1. Probability distribution of static ice force acting on an isolated vertical pile
   According to the probability distribution of ice thickness that obeys the extreme value I distribution (as in Figure 1-37 above) and the probability distribution of the compressive strength of sea ice that obeys the Weibull distribution in the China Bohai Sea, the China University Of Petroleum adopted a comprehensive probability analysis method to do joint probability analysis and drew the conclusion that the static ice force acting on a vertical cylindrical structure obeys the extreme value I probability distribution, as shown in Figure 1-45.
   The analysis results from the joint probability analysis method show that when the extreme value of static ice force that China's current design criteria adopts for once every 50 years and once in a century values is applied to probabilistic analysis, the probability is 0.0061 and 0.0026 respectively. Namely the recurrence interval is 163 years and 384 years, respectively. This again proves that the existing design criteria are conservative.
2. The probability distribution of static ice force acting on an isolated incline or pile with cone.

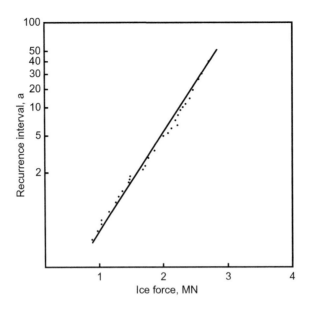

FIGURE 1-45

Probability distribution of the maximum static ice force of flat ice in the China Bohai Sea.

According to the five types of calculation models introduced earlier, the Dalian University of Technology, cooperating with the Bohai Oil Corporation, compared the predicted values of static ice force with the measured values for the static ice force acting on a pile with a cone, creating the frequency histograms and probability distributions shown in Figure 1-46. These frequency histograms show that the static ice force obeys a normal distribution, especially in Figure 1-46(e). Earlier we suggested that we should adopt the Hirayama-Obara model [as shown in Equation (1-289)] to calculate the total horizontal static ice forces on the inclined or cone structure, so it is very obvious that the probability distribution of the static ice force is a normal distribution.

1.6.3 Ice-Induced Vibration of Sea Ice and Dynamic Ice Force

Due to the action of sea ice, an offshore engineering structure generally has different levels of vibration, and this is known as ice-induced vibration. When ice-induced vibration happens, the ice force causing structural vibration is called the dynamic ice force. Only through solving the issue of calculating dynamic ice force can a response to ice-induced vibration be achieved, so these two problems are introduced together.

1.6.3.1 Analysis of Ice-Induced Vibration

1. Ice-induced vibration analysis model
   The forced vibration model and self-excited vibration model are commonly used to analyze ice-induced vibration.

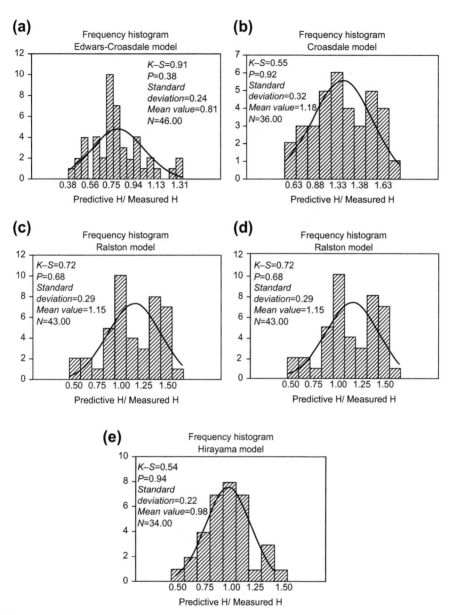

FIGURE 1-46

Probability distribution of static ice force using the predicted values calculated by the five types of calculation model compared with the measured values.

a. Forced vibration model

For this model ice-induced vibration is a forced vibration, and the dynamic ice force of sea ice is the disturbing force $F(t)$ of this forced vibration. Suppose that $[M]$, $[C]$, and $[K]$ represent respectively the mass matrix, damping matrix, and stiffness matrix of the structure, and that $\ddot{X}, \dot{X}$, and X represent respectively the acceleration, velocity, and displacement of the structure. Then the differential equation of this forced vibration is

$$[M]\{\ddot{X}\} + [C]\{\dot{X}\} + [K]\{X\} = [F(t)] \tag{1-292}$$

According to the differences in the description of interference force $F(t)$ in this model, it is divided into the Motlock model and the Engelbrektson model. In the former, $F(t)$ is the function of the deformation δ after sea ice acts on the structure; in the latter, the disturbing force $F(t)$ is a function of the "zigzag" dynamic ice force as shown in Figure 1-47. From Figure 1-47, the time that the maximum dynamic ice force F_{max} reduces to the minimum dynamic ice force F_{min} is 1/4 of period T, where $T = \frac{\Delta l}{v}$, which depends on the one-time destructive length of the ice floe and the ice velocity $v\Delta l = Ch$, which is the function of the ice thickness h; the coefficient C is generally $0.1 \sim 0.3$.

b. Self-excited vibration model

In this model, ice-induced vibration is categorized as self-excited vibration, and the energy of sea ice movement is the energy source of the consistent self-excited vibration of the ocean engineering structure. Due to the sea ice providing the required energy, the structure offsets the damping in each period of self-excited vibration, and the self-excited vibration of the structure is controlled, with the equiamplitude vibration in accordance with its natural frequency.

The self-excited vibration model includes the Maattanen model and ice force oscillator model. In the former, the $F(t)$ in Equation (1-292) is regarded as the dynamic ice force vector $F(u_r)$ related to the relative velocity u_r between the ice floe and structure, while $F(u_r) = F_0 - [\phi]\{\dot{X}\}$: F_0 is the static ice force vector when the ice velocity is u_0; $[\phi]$ is the

FIGURE 1-47

The function of the "zigzag" dynamic ice force.

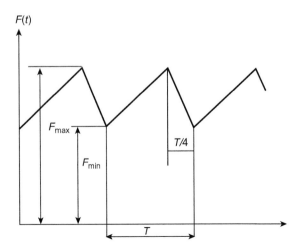

ice force coefficient matrix. In the ice force oscillator model, ice-induced vibration is a coupling vibration, and it is the result of mutual coupling of two dynamic systems in ice and structure. So the ice force oscillator equation and the dynamic equilibrium equation of a linear structure can be combined into the following simultaneous coupling vibration equation:

$$\ddot{C} + \omega_0\left(-\alpha - \gamma C + \beta C^2\right)\dot{C} + \omega_0^2 C = b\left(v_0 - \dot{X}\right) \tag{1-293}$$

$$\ddot{X} + \alpha\xi\omega_n\dot{X} + \omega_n^2 X = BC \tag{1-294}$$

where $C = \frac{\sigma}{\sigma_b}$ is dimensionless ice force coefficient; σ, σ_b are respectively the instantaneous ice stress and the ultimate stress of a brittle zone; $\dot{C}, \ddot{C}$ are respectively the velocity and the acceleration; ω_0 is the natural frequency of the ice force power system; v_0 is the ice velocity; $X, \dot{X}, \ddot{X}$ are respectively the displacement, velocity, and acceleration of the structure's power system; ζ is the critical damping ratio; α, β, γ, b depend on the specific coefficient of ice features; ω_n is the natural frequency of the structure; and B is a parameter.
B can be obtained as

$$B = \frac{ImKDH}{A_0 M} \tag{1-295}$$

where, I, m, and K are respectively the embedment coefficient, shape coefficient, and contact coefficient calculating the static ice force; D is the width (or diameter) of the structure; H is the ice thickness; A_0 is a certain reference displacement; and M is the structure mass.

The ice force oscillator model can be used to predict the frequency lock. The phenomenon of frequency lock is that the vibration of the structure will compel the vibration frequency of sea ice to be fixed close to the natural frequency of the structure and no longer vibrate with the vibration frequency of the ice force power system when the natural frequency of the structure dynamic system is close to the vibration frequency of the ice force power system. The phenomenon of frequency lock has been confirmed by the Bohai Oil Corporation according to the vibration on the measured platform and vibration of ice around the measured platform. Therefore, the forced vibration and self-excited vibration can be combined, that is, within the frequency lock region the ice-induced vibration is described by the theory of self-excited vibration, and outside the frequency locking region the ice-induced vibration is described by the theory of forced vibration. The wide frequency-locking band of the ice-induced vibration is the self-excited vibration characteristics; the narrow frequency-locking band is the forced vibration characteristics.

2. Analysis of the response to the ice-induced vibration
The response to the ice-induced vibration is displacement, velocity, acceleration, and strain and stress on the structure caused by the ice-induced vibration. Therefore, the response analysis of ice-induced vibration involves solving the differential equation established by the vibration model. Usually for complex structures, such as offshore oil platforms, the finite element discretization process should be used to establish the finite element model of structure, and then calculations should be done by the SAP5 calculation program. For something such as the fixed production platform commonly used in offshore oil engineering, using finite element analysis and

actual measurement, we can find three features in the response of ice-induced vibration in the China Bohai Sea:

a. Effects of ice force direction

Usually the vibration has a direction from north to south in January each year, and the frequency is above 1 Hz. Torsional vibration also occurs when sea ice turns.

b. Effects of sea ice breaking modes

Usually ice-induced vibration is most serious when caused by single-layer flat ice and layered ice, Hard single-layer flat ice where the ice thickness is 15 cm will cause a large enough vibration, and at this time the sea ice breaking mode is compression; with a bending or buckling breaking mode, the structure won't vibrate.

c. Effects of sea ice broken frequency

Figure 1-48 gives the displacement response spectrum of the SW fixed platform measured by the Bohai Oil Corporation, under three kinds of breaking modes, including compression, bending, and buckling. The measurement shows that the frequency at the maximum peak of the displacement response spectrum is consistent with the breaking frequency of sea ice at the maximum amplitude of ice force. From Figure 1-48, the different breaking modes of sea ice have different displacement response spectrum peaks, related to the frequency when sea ice is broken. In the China Bohai Sea, the ice-induced vibration frequency of the structure is usually between $0.8 \sim 2.5 \text{Hz}$, which is consistent with the frequency range of the maximum peak in the measured displacement response spectrum, as shown in Figure 1-48.

1.6.3.2 Dynamic Ice Force Acting on an Isolated Vertical Pile

The dynamic ice force acting on an isolated vertical pile can be analyzed by the time domain or frequency domain analytical method.

1. Time domain analysis

The time domain analytical method first needs to obtain the time history curve of sea ice forces on structures by actual measurement, as shown in Figure 1-49, then analyze it. At present, the time domain analysis model includes the time history model, Motlock model, Kivisild model, and the improved Engelbreak model. There are only two models that are commonly used.

a. The time history model

The analytical method of this model is used for the compression breaking mode of sea ice. By actual measurement, we take a time history curve (Figure 1-49) of the dynamic ice force acting on the structure when the sea ice is in the compression mode, and then select the maximum peak $[F(t)]_{\max}$ as the dynamic ice force $F(t)$ in the curve; that is

$$F(t) = [F(t)]_{\max} \tag{1-296}$$

b. The improved Engelbreak model

This calculation model is used for the buckling breaking of sea ice. Through actual measurements by the Bohai Oil Corporation in the China Bohai Sea, the time history curve of ice force is as shown in Figure 1-50 when the ice buckles. Simplifying the curves in Figure 1-50, we can obtain the "zigzag" model as shown in Figure 1-51. From Figure 1-51, when sea ice reacts with structure, the ice force will soon reach the maximum $F_{\max}$, and after a

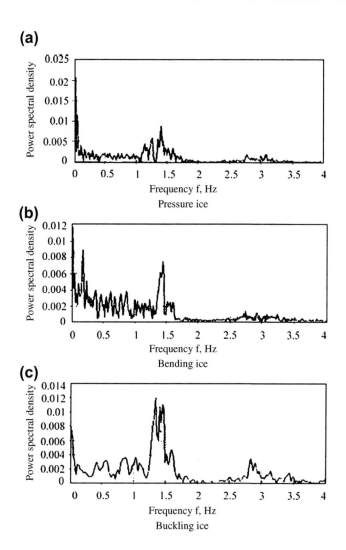

FIGURE 1-48

The displacement response spectrum of the SW fixed platform in the China Bohai Sea under three kinds of ice breakage.

short timet_1, the ice force will reach the minimum F_{min} when the ice buckles, and then after a period of time $T - t_1$, the next period T of the subsequent ice reaction will repeat.

If we suppose the dynamic ice force is $F(t)$ at a certain time and the angle between the oblique line of ice force rising and the horizontal line is θ, we can obtain

$$F(t) = F_{min} + t(tan\ \theta) \tag{1-297}$$

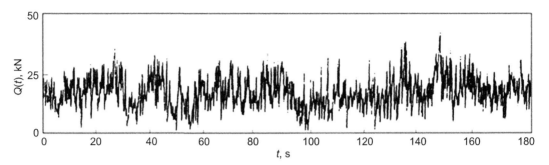

FIGURE 1-49

The time history curve of dynamic ice force when the ice is being crushed.

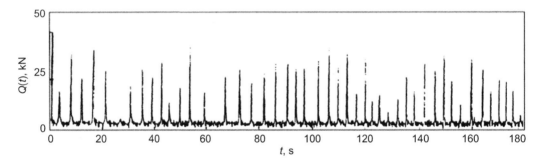

FIGURE 1-50

The time history curve of the ice force when the sea ice is buckling in the China Bohai Sea.

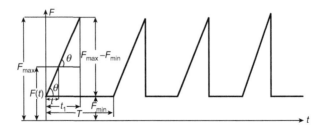

FIGURE 1-51

The simplified model of the dynamic ice force time history.

But from Figure 1-51

$$\tan\theta = \frac{F_{\max} - F_{\min}}{t_1} = m \qquad (1\text{-}298)$$

where m is the slope of the oblique line that the ice force is raising.
Substituting Equation (1-298) into Equation (1-297), we get

$$F(t) = F_{\min} + t\left(\frac{F_{\max} - F_{\min}}{t_1}\right) = F_{\min} + tm \qquad (1\text{-}299)$$

In this equation, t_1 is the time parameter associated with the crushing and clearing of ice floe. According to actual observations, it can be taken as 1/3 of the period T, and the period T should be the ice velocity u divided by the ice breaking length l_b, namely $T = \frac{l_b}{u}$. But considering that the ice breaking length l_b is usually regarded as seven times than ice thickness H, the period T is obtained as

$$T = \frac{7H}{u} \qquad (1\text{-}300)$$

Substituting $t_1 = \frac{t}{3}$ and Equation (1-300) into Equation (1-299), we get

$$F(t) = F_{\min} + t\left(\frac{F_{\max} - F_{\min}}{\frac{1}{3}\frac{7H}{u}}\right) = F_{\min} + 3tu\left(\frac{F_{\max} - F_{\min}}{7H}\right) \qquad (1\text{-}301)$$

2. Frequency domain analysis
   Frequency domain analysis is a method that adopts the spectrum of dynamic ice force to describe that force. According to actual measurements on a sea platform, the Bohai Oil Corporation finds that only compression and buckling breakage happens, and there isn't bending breakage when a vertical cylindrical structure (jacket leg on the platform) reacts with sea ice. Therefore, here we only introduce the spectrum in those two cases.
   a. The dynamic ice force spectrum when sea ice breaks through compression
      The Bohai Oil Corporation selected 37 ice force time history curves according to actual measurements, made the ice force spectrum, applied the transfer function $2rhH_T$, and used the power spectrum density $S(\omega)$ of a single point (head-on freezing point) pressure on the cylindrical structure to express the power spectra density function $S_{F(t)}(\omega)$ of total dynamic ice force acting on the right circular cylinder:

$$S_{F(t)}(\omega) = (2rhH_T)^2 S(\omega) = 4r^2 h^2 H_T^2 S(\omega) \qquad (1\text{-}302)$$

where r is the radius of the cylinder, h is the thickness of the ice, and H_T is the function coefficient of ice thickness, which can be selected from Table 1-21.

Table 1-21 Value of Coefficient H_T for Different Ice Thicknesses

Ice Thickness h, cm	10	20	30	40	≥ 50
H_T	0.8030	0.8047	0.8072	0.8110	0.8172

In Equation (1-302), the power spectrum density $S(\omega)$ is obtained by

$$S(\omega) = A(\omega)^P \exp(-B\omega^q) \tag{1-303}$$

In the equation, the coefficients A, B, p and q are respectively approximately 135.23(A), 6.70(B), 2.20(p), and 0.30(q) according to the actual situation in the China Bohai Sea.

b. The dynamic ice force spectrum when sea ice breaks through buckling

As above, according to the time history curve of actual measurement, the Bohai Oil Corporation makes $S(\omega)$, and gives $S_{F(t)}(\omega)$:

$$S_{F(t)}(\omega) = 2.91r^2h^2S(\omega) \tag{1-304}$$

But the coefficients in the expression of the power spectrum density $S(\omega)$ of head-on freezing point pressure on the cylinder are, respectively, $A = 840.63$, $B = 8.52$, $p = 3.1$, and $q = 0.3$ when sea ice breaks through buckling. These are different than the coefficients when sea ice is broken through compression.

1.6.3.3 The Dynamic Ice Force Acting on the Cone of an Isolated Tilted or Vertical Pile

For an isolated tilted pile or a cone slope on a vertical cylindrical structure, ice bending breakage will happen when it reacts with sea ice. At this time the dynamic ice force can be analyzed by time domain analysis and determined by the maximum peak in the time history curve of actual measurements. We also can use frequency domain analysis, which uses the ice force time history curve of actual measurements to express the spectrum and the power spectral density function of the dynamic ice force. Here we show the simplified function expression for dynamic ice force when sea ice is broken through bending, given by Bohai Oil Corporation and the Dalian University of Technology.

1. The ice force time history curve from field measurement
   Figure 1-52 gives the dynamic ice force time history curve when sea ice is breaking through being with actual measurements on the Liaodong Bay of the China Bohai Sea.
2. The simplified dynamic computational model of sea ice
   Considering that the ice force will decline from the peak to zero after the second crushing, the curve in Figure 1-52 can be simplified for an approximate model as shown in Figure 1-53.
3. The dynamic ice force function according to the simplified model
   Supposing a random time t ($0 \leq t \leq T$), if the dynamic ice force is $F(t)$ at time t, Figure 1-53 is obtained:

$$F_0 - F(t) = t(\tan \theta) = t\left(\frac{F_0}{\tau}\right) \tag{1-305}$$

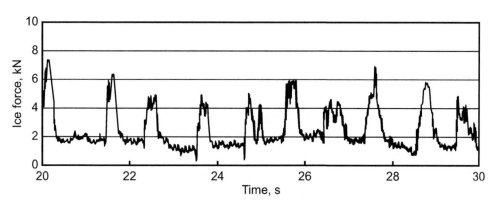

FIGURE 1-52

The dynamic ice force time history curve acting on a cone structure using actual measurements in the China Bohai Sea.

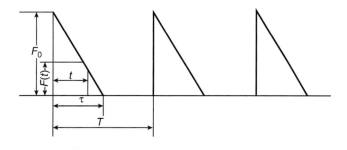

FIGURE 1-53

A simplified model of the dynamic ice force acting on the cone structure.

Namely,

$$F(t) = F_0 - t\left(\frac{F_0}{\tau}\right) = F_0\left(1 - \frac{t}{\tau}\right) \tag{1-306}$$

If we consider the period of $\tau \leq t \leq T$, it should be

$$F(t) = \begin{cases} F_0\left(1 - \frac{t}{\tau}\right) & 0 \leq t < \tau \\ 0 & l \leq t < t \end{cases} \tag{1-307}$$

4. Determining the related parameters in the function for dynamic ice force
  a. The peak F_0 of the cyclical dynamic ice force
    F_0 is the amplitude of dynamic ice force acting on the cone structure, and it should be the total horizontal static ice force acting on the cone structure. According to the Hirayama-Obara model described above, F_0 should be

$$F_0 = 3.2\sigma_{Bu}t^2 \left(\frac{D}{l_b}\right)^{0.34} \tag{1-308}$$

  b. The period T of dynamic ice force
    This depends on the ice breaking length l_b and ice velocity u when sea ice is broking through bending, according to the actual measured data. And when the l_b is seven times than ice thickness t_i, T is

$$T = \frac{l_b}{u} = \frac{7t_i}{u} \tag{1-309}$$

  c. The time that the sea ice climbs along the cone slope
    It is difficult to analyze the time τ by theory. But according to the ice force time history curve (Figure 1-54) measured in the China Bohai Sea, the time that ice climbs along the cone slope is about 1/3 of the dynamic ice force period.
  d. Comparison between the simplified dynamic ice force function and actual measurements

Looking at the contrast between the displacement response of the cone structure calculated by the dynamic ice force function and the dynamic response (displacement) from the actual measurements on the sea, as shown in Figure 1-55, the response peaks are consistent. Therefore, the simplified dynamic ice force function given by Equation (1-307) can be used in the China Bohai Sea.

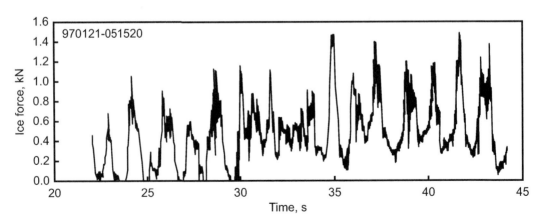

FIGURE 1-54

The dynamic ice force time history curve when sea ice is breaking through bending.

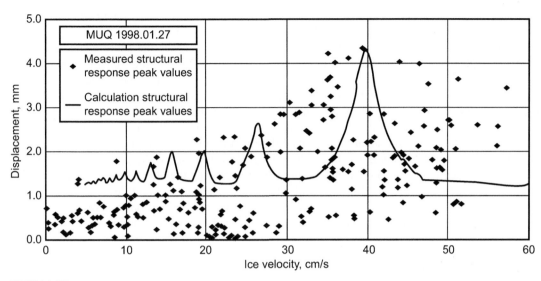

FIGURE 1-55

The contrast between the result calculated by the simplified dynamic ice force function and the displacement response from field measurement on the ice platform JZ20-1MUQ.

1.7 Tsunami and Earthquake Force

In most cases, tsunamis are caused by undersea earthquakes, and both of these will cause serious disasters, causing great harm to offshore oil engineering. In offshore oil engineering, these two factors should be taken under consideration for things such as planning shore base and oil terminals, prefabrication and placement of offshore oil engineering structures, design of engineering structures, and ensuring both the smooth offshore construction of engineering structures and safe operation during the service period. Therefore, we will introduce both tsunamis and earthquakes in a single section here.

1.7.1 Tsunami

A tsunami is the phenomenon where the sea level rises anomalously due to various causes. These different causes mainly include undersea earthquakes, undersea volcano eruptions, submarine landslides, slope collapses, coastal landslides, etc.

1.7.1.1 Tsunami Formation

At the beginning of a tsunami, the center forms with short height (generally only tens of centimeters) and large wavelength (up to 300 km or more), spreading with a great velocity (up to

800 km/h) to the shore. When tsunami waves reach the shore, with the water depth decreasing, the velocity slows down, but the wave height increases; sometimes it forms a "steep wall" of 10 meters, causing a sudden increase of the water level, and sometimes the shore water decreases suddenly, with the water level increasing after $10 \sim 15$ min, resulting in the tsunami. Whether the shore water level suddenly increases or decreases at the beginning, the appearance of these two cases is inseparable with the seabed deformation at the center of a tsunami. The reason why the seabed suddenly rises at the center of a tsunami is that the sudden change of water volume, caused by the tsunami, forms a pressure wave spreading to the shore in the form of solitary waves. Similarly, if at the center of tsunami the seabed deformation involves sinking and collapsing, and not lifting of the seabed, this is because a low-lying place forming at the center makes the surrounding seawater converge to the center of the tsunami, causing the phenomenon of the decrease of water level first at the coast. However, with the gradual recovery of water level at the center of tsunami, under the force of inertia, the water level there is lower than the normal level again, and at the same time, a water wave is produced that spreads around until transmitted to the shore, where the shore water level will increase.

1.7.1.2 Tsunami Propagation

1. Propagation velocity of tsunami
   If C represents the velocity, the Lagrange equation is available, namely,

$$C = \sqrt{gd} \qquad (1\text{-}310)$$

where g represents the acceleration of gravity, m/s^2, and d represents depth, m.
If the change of water depth is complex, it can be divided into several areas (such as n) where the depth is approximately equal. Take the average depth as d_i, and then calculate the average water depth d_m according to the following equation:

$$\frac{1}{\sqrt{d_m}} = \frac{1}{n}\sum_{i=1}^{n} d_i^{-\frac{1}{2}} \qquad (1\text{-}311)$$

So the propagation velocity C of a tsunami wave can be written as

$$C = \sqrt{gd_m} \qquad (1\text{-}312)$$

2. The required time for a tsunami to spread from the place of occurrence to the calculation point

Suppose the distance from the center of a tsunami to the calculation point and the time required are respectively R and T, and the propagation distance of the crest line along each direction is different; then we can obtain

$$t = \int_0^r \frac{dr}{\sqrt{gd}} = g^{-\frac{1}{2}}\int_0^r d^{\frac{1}{2}}dr \qquad (1\text{-}313)$$

If the depth from the center of tsunami to the calculation point is variable, its distribution is expressed by $f(r) = d^{-\frac{1}{2}}$; then Equation (1-313) can also be written as

$$t = g^{-\frac{1}{2}} \int_0^r f(r)dr = 0.32f(r)dr \tag{1-314}$$

In order to simplify the calculation of the time t that represents the time for the tsunami to spread from its origin to the shore, we can also divide the depth of water, which is roughly equal, into several segments with the help of the submarine topography of the propagation direction. For each segment, the water depth of each segment is d_i and the length of each section is l_i, so we can calculate the time required t_i for each segment. That is,

$$t_i = \frac{l_i}{c_i} = \frac{l_i}{\sqrt{gd_i}} \tag{1-315}$$

1.7.1.3 Tsunami Damage

1. Dynamic destruction of a tsunami
   When tsunami waves are near the coast, the water depth is decreasing, the wave height is increasing greatly, and the front slope steepening, forming finally into a relatively smooth surface, a crest known as the "water wall." It comes toward the shore with huge energy and high velocity, bringing great dynamic destruction.
2. Static destruction of tsunami

This damage is caused by erosion when the water is in recession. In a shallow water area with a wide sea surface, due to friction at the bottom, the tsunami waves lose a lot of kinetic energy and the water level rises. As the water level rises, this will flood large tracts of coastal areas. The water in the flooded area gathers a large amount of potential energy, forming a strong reverse current that continuously runs from the coast to the sea. This will cause huge damage to the engineering structures on the sea.

1.7.2 Calculation of Earthquake Force

Usually the earthquake force refers to the inertial force and dynamic water pressure when an earthquake occurs on the surface (layer). In accordance with the relevant specifications of China Classification Society, when the basic earthquake intensities (standards of the Seismological Bureau) are at the 7, 8, 9 level, the earthquake force can be calculated in accordance with the recommend calculation of the specification. But if the basic earthquake intensity is over 9 or the offshore engineering structure is very important, there is a need for additional specialized calculation. The recommended calculation method is discussed in the following.

1.7.2.1 Calculation Method with a Single Point System

This method regards the structure as a single point system to calculate the horizontal total seismic inertial force P_H, i.e.,

$$P_H = CK_H\beta mg \tag{1-316}$$

1. Quality m

This is the single point quality of engineering structures (kg). Taking an offshore oil platform as an example, it is the quality located in the deck of platform, which includes:

   a. The structure quality of the upper deck, the quality of equipment on the deck, and the quality of 75% of the spare parts and supplies

   b. The structure quality under the deck

   c. Reduced quality of other quality related to platform deck

2. The horizontal seismic coefficient K_H

The value of K_H should be selected from Table 1-22 in accordance with the seismic intensity.

3. Dynamic amplification coefficient β

The value of β should be selected from Figure 1-56 based on the natural vibration period T. The three curves in the figure respectively are three kinds of site soil, in which site soil I is slightly weathered and moderately weathered bedrock; site soil II is generally stable soil except for site soil I and site soil III; and site soil III refers to saturated loose sand soil, soft plastic, and soft state clay and silt.

In Equation (1-316), g is the acceleration of gravity (m/s²), and C is the comprehensive effect coefficient. Because the earthquake recurrence period is longer, standards recommend taking C as 0.9.

1.7.2.2 Calculation Method for a Many Point System

Suppose there are n mass points on a structure, and at each mass point i the lumped mass is m_i; then for the mass point i of a structure, the seismic inertia force P_{ij} in the horizontal direction of the j order modal vibration mode can be calculated as follows:

$$P_{ij} = CK_H\gamma_j\phi_{ij}\beta_jm_ig \qquad (1\text{-}317)$$

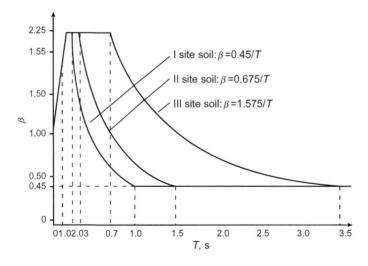

FIGURE 1-56

Dynamic amplification coefficient β when calculating the earthquake force.

$$\gamma_j = \frac{\sum_{i=1}^{n} \phi_{ij} m_i}{\sum_{i=1}^{n} \phi_{ij}^n m_i} \qquad (1\text{-}318)$$

where γ_j represents the participation coefficient in the j mode of the structure; ϕ_{ij} represents the horizontal displacement of particle i in the structure in the j mode, which is calculated according to the dynamic response analysis of the earthquake; and β_j represents the corresponding dynamic magnification factor in the j order vibration mode, which can be obtained from Figure 1-56.

The means of C, K_H, and g in Equation (1-317) are the same as in Equation (1-316).

1.7.2.3 Calculation Method of Dynamic Water Pressure p Acting on the Horizontal Part of an Elongated Member in an Arbitrary Direction during an Earthquake

Suppose the angle between the vibration direction of earthquake i and component l is $\phi(i, l)$; if the volume of the part which is immersed in the water is V, the dynamic water pressure p on the horizontal part of the elongated member can be calculated as follows:

$$p = CK_H\beta(C_M - 1)V\gamma \sin^2 \phi(i,l)\text{KN} \qquad (1\text{-}319)$$

In Equation (1-319), C_M is the inertial force coefficient, selected according to the inertial force coefficient C_M in the Morison equation given by Equation (1 - 214); γ is the volume-weight of seawater (kN/m^3); and coefficients C, K_H, and β are the same as in Equation (1-316).

1.7.3 Dynamic Response Analysis for an Earthquake

1.7.3.1 The Vibration Differential Equation

When an earthquake occurs, the ocean engineering structure will produce vibration, which causes a dynamic response. When analyzing the dynamic response of an earthquake, we should first establish the vibration differential equation. The vibration differential equation is usually composed of inertial force, damping force, elastic recovery force, and disturbing force (exciting force) which is balance with the elastic recovery force. Suppose the disturbing force on the unit length of structure is $f(z,t)$, the structure quality is M, C and k respectively represent damping coefficient and stiffness coefficient, and the vibration displacement and velocity are respectively u and $\dot{u}$. We can produce the vibration differential equation of the structure in the earthquake as

$$M\ddot{u}_t + C\dot{u} + ku = f(z,t) \qquad (1\text{-}320)$$

In the equation, $\ddot{u}_t$ represents the resultant acceleration. Because the term of inertial force includes both the inertial force caused by the vibration acceleration $\ddot{u}$ of structure (velocity is $\dot{u}$, displacement is u) and the inertial force caused by the ground (stratum) acceleration $\ddot{u}_g$ (velocity is $\dot{u}$, displacement is u_g) during the earthquake, the acceleration is the resultant acceleration $\ddot{u}_t$ of both accelerations, namely

$$\ddot{u}_t = \ddot{u} + \ddot{u}_g \qquad (1\text{-}321)$$

1.7.3.2 Disturbing Force of Vibration (Excited Force)

During an earthquake, the disturbing force of vibration caused by a structure is composed of the drag force f_D and inertial force f_M caused by the earthquake force and hydrodynamic force acting on the structure.

1. The drag force (resistance) f_D acting on the unit length of structure

$$f_D = C_D \rho \frac{1}{2} |u_r| u_r \approx \phi_D \sigma_{ur} \sqrt{\frac{8}{\pi}} u_r \qquad (1\text{-}322)$$

In Equation (1-322), u_r is the relative velocity between the water particles and the structure, and if the horizontal velocity of water is u', u_r should be

$$u_r = u' - \dot{u}_t = u' = (\dot{u} - \dot{u}_g) \qquad (1\text{-}323)$$

2. The inertial force f_M acting on the unit length of a structure

$$f_M = \frac{1}{4} D^2 \rho (\dot{u}_r) = \phi_M \dot{u}_r \qquad (1\text{-}324)$$

In the equation, $\dot{u}_r$ is the relative acceleration:

$$\dot{u}_r = u'' - \dot{u}_t \qquad (1\text{-}323)$$

where u'' is the horizontal acceleration of a seawater particle.

3. During the earthquake, the total wave forces $f(z, t)$ acting on a unit length of the structure

Taking Equation (1-322) and Equation (1-324) into superposition, we can derive that $f(z, t)$ is

$$f(z,t) = \phi_D \sigma_{ur} \sqrt{\frac{8}{\pi}} u_r + \phi_M \ddot{u}_r$$

$$= \phi_D \sigma_{ur} \sqrt{\frac{8}{\pi}} [u' - (\dot{u} + \dot{u}_g)] + \phi_M [u'' - (\ddot{u} + \ddot{u}_g)] \qquad (1\text{-}326)$$

In Equation (1-326), u' and u'' are a function of z and t, so the total wave force $f(z, t)$ is also a function of z and t.

1.7.3.3 Velocity and Acceleration of the Ground (Stratum) during an Earthquake

As shown in Equation (1-326), during an earthquake, the disturbing force of structure vibration is relative to the velocity $\dot{u}_g$ and acceleration $\ddot{u}_g$ of the stratum. Usually there are two methods to get $\dot{u}_g$ and $\ddot{u}_g$: time history analysis and frequency domain analysis.

1. Time history analysis method

This method first needs to obtain the time history curve of surface acceleration $\ddot{u}_g$ during the earthquake. Figure 1-57, for example, is two time history curves of instantaneous surface acceleration during the 7.8 earthquake of Tangshan City near the China Bohai Sea, China, at

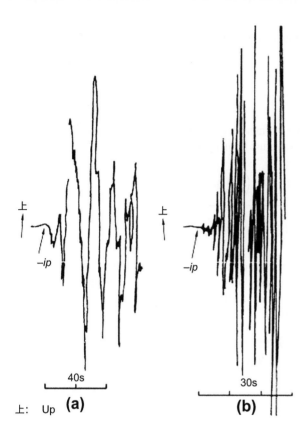

FIGURE 1-57

The time history curve of surface acceleration during the Tangshan earthquake near the China Bohai Sea.

40s

30s

上: Up **(a)**   **(b)**

上: Up

3 o'clock in the morning, July 28, 1976. So, if we can get a certain amount of time history curve samples about surface acceleration in given sea areas that belong to the seismic belt, a frequency histogram can be drawn and based on the statistical data of the curve peak, the probability density curve can be obtained. For example, Shtanko once obtained the result that the acceleration $\ddot{u}_g$ obeys the exponential distribution according to this method. If $f(\ddot{u}_g)$ represents this probability density function, the expression $f(\ddot{u}_g)$ given by Shtanko is

$$f\left(\ddot{u}_g\right) = A\left(\ddot{u}_g\right)^{-3} \tag{1-327}$$

In the equation, A is a coefficient that is obtained from the actual measurement. Obviously, after obtaining the probability density distribution $\ddot{u}_g$ in accordance with the requirements of the given recurrence interval, we can obtain $\ddot{u}_g$ and calculate $\dot{u}_g$.

2. Frequency domain analysis method

This method is the response spectrum analysis method; that is, it is a frequency domain analysis method of the dynamic response u_g, $\dot{u}_g$, and $\ddot{u}_g$ of the structure during the earthquake which can be made into a spectrum, from which the statistical characteristics of the response, such as variance $\sigma_{\ddot{u}_g}^2$ and mean value $E(\ddot{u}_g)$ (expectation) can be determined. For example, the Earthquake

Table 1-22 The Value of K_H for Different Earthquake Intensities

Earthquake Intensity	7	8	9
The K_H value	0.1	0.3	0.4

Research Institute of the University of Tokyo in Japan has given the power spectral density function $S_{\ddot{u}_g \ddot{u}_g}(\omega)$ of surface acceleration $\ddot{u}_g$ (stratum) during an earthquake:

$$S_{\ddot{u}_g \ddot{u}_g}(\omega) = \frac{\left| 1 + 4 r_g^2 \left(\frac{\omega}{\omega_g} \right)^2 \right| S_o}{\left| 1 - \left(\frac{\omega}{\omega_g} \right)^2 \right|^2 + 4 r_g^2 \left(\frac{\omega}{\omega_g} \right)^2} \tag{1-328}$$

where r_g represents the damping ratio of the stratum (take $r_g = 0.6 \sim 0.7$ for hard stratum); ω, ω_g represent respectively the natural circular frequency of structure vibration and surface vibration (usually we take ω_g as $5 \sim 6\pi$ rad/s); and S_o represents the earthquake intensity parameter (it can be determined from Table 1-22 by referring to K_H).

1.7.3.4 The displacement Response of the Structure during an Earthquake

Substitute $f(z,t)$ in Equation (1-326) into the vibration differential equation in Equation (1-320), and we can obtain

$$M\left(\ddot{u} + \ddot{u}_g\right) + C\dot{u} + ku = \phi_D \sigma_{ur} \sqrt{\frac{8}{\pi}} \left[u' - \left(\dot{u} + \dot{u}_g \right) \right] + \phi_M \left[u'' - \left(\ddot{u} + \ddot{u}_g \right) \right] \tag{1-329}$$

Then merge the terms of inertial force and damping force, sort, and write

$$(M + \phi_M)\ddot{u} + \left(C + \phi_D \sigma_{ur} \sqrt{\frac{8}{\pi}} \right) \dot{u} + ku$$

$$= -(M + \phi_M)\ddot{u}_g - \phi_D \sigma_{ur} \sqrt{\frac{8}{\pi}} \dot{u}_g + (\phi_M) u'' + \phi_D \sigma_{ur} \sqrt{\frac{8}{\pi}} u' \tag{1-330}$$

Obviously, solving this teo-order constant coefficient linear nohomogeneous differential equations can get the dynamic response of the structure under the action of waves during the earthquake, namely the displacement u.

1.7.4 Combinatorial Computing of Seismic Loads and Other Loads

In Equation (1-316) and Equation (1-317) given above, we can only calculate the earthquake force simply, that is, the horizontal to total earthquake inertia force P_H because of the earthquake action on the structure. However, during an earthquake, the structure is also under the action of wind, waves, current, and sea ice in the ice zone; thus the structure bears loads including wind load, wave force, and ice load in winter along with other loads. Therefore, we need to consider the problem of combining earthquake loads and other loads. Of course, this combination is not simply adding up their extremes. Considering their different recurrence intervals, we can combine them according to the relevant specifications.

1.7.4.1 Load Combination Calculation Methods

The so-called load combination, which is different from the load effect combination, is the combination of the external forces of the structure. The load effect usually refers to internal force generated by components, such as shear, bending torque, torque, and axial force after a variety of external forces acting on the structure. For a load combination, some calculation expressions are given in certain specifications at home and abroad, among which the expression given in the Chinese TJ9 -74 specifications is:

$$P = K\left[G_K + \phi_0\left(W_K + \sum_{i=1}^{n} Q_{ik}\right)\right] \tag{1-331}$$

1. Constant load G_k

   This is the fixed load, not changing with time, e.g.,
   a. The weight of the structure itself
   b. The weight of equipment and other objects on the structure
2. The wind load W_k

   The wind load should be calculated as the F_w in Equation (1-9) given in this chapter.
3. Movable load Q_{ik}

   The movable load Q_{ik} usually consists of two categories, i.e.,
   a. The persistent movable load

      The sea ice load belongs to this category, and it should be calculated in accordance with the probability distribution of the maximum value (extreme) of this load in the design reference period.
   b. The temporary movable load

      This is the load that appears with high frequency and short duration in the design reference period. For example, the earthquake load belongs to this category, and such movable load should also be calculated in accordance with the probability distribution of the maximum load (extreme) of this load in the design reference period.
4. Load combination coefficient ϕ_0

   This is the coefficient of appropriate reduction, because the maximum (extreme) that could be adopted is too conservative during the superposition of a number of load combinations. Because it changes with many factors such as the type of structure component and the type of load combination involved, it is very complex and not convenient. The calculation equation of this coefficient is given in some specifications, but at present in engineering, we take $\phi_0 = 0.85$, which should only be used in combinations of two or more than two ($n \geq 2$) loads.
5. The safety coefficient K

   This is determined according to the methods of general engineering.

1.7.4.2 Internal Force (Load Effect) Calculation Method for Combinations

The petroleum and natural gas industry standards SY/T10009 1996 gives the equation for the calculation of earthquake force and other load combinations according to two cases.

1. When the direction of the inertial force caused by the gravity load of structure is the same as the direction of internal force caused by the earthquake load
   The total internal force after combination is Q, which is calculated as follows:

$$Q = 1.1D_1 + 1.1D_2 + 1.1L_1 + 0.9E \qquad (1\text{-}332)$$

   a. The first kind of fixed load D_1
      This refers to the weight of the structure itself, and should be calculated in accordance with the design size of the structure multiplied by the standard quality density.
   b. The second kind of fixed load D_2
      This refers to the fixed load of the weight of the equipment and other objects on the structure, and it should also be calculated in accordance with the design size of structure multiplied by the standard quality density. Because it belongs to the permanent load, the probability in the design reference period is equal to 1, which generally can be seen as obeying the normal distribution.
   c. Movable load L_1
      This can be divided into the persistent movable load and the temporary movable load. Loads such as ice load, wind load, and the combined load of wave and tide belongs to the former class; the latter is a movable load with short duration, such as the changing weight. These two types of movable load should be calculated in accordance with the probability distribution of the maximum (extreme) in the design reference period, but the probability distribution of maximum value of the temporary movable load generally can be taken as obeying the extreme value I distribution.
   d. The earthquake load E
      It should also be calculated according to the probability distribution of the maximum value in the design reference period.
2. When the direction of the inertial force caused by the gravity load of a structure is opposite to the direction of the internal force caused by the earthquake load

This situation is that the direction of inertia force caused by the weight load in D_1, D_2, and L_1 is opposite to the direction of inertia force caused by the earthquake load E; in this case, Q should be

$$Q = 0.9D_1 + 0.9D_2 + 0.8L_1 + 0.9E \qquad (1\text{-}333)$$

The meaning of the symbols in Equation (1-333) are the same as in Equation (1-332).

1.7.4.3 The Probability Distribution of a Movable Load When Calculating the Load Combination

When applying Equation (1-332) or Equation (1-333) to calculate the load combination, we need to calculate in accordance with the probability distribution of the maximum value in the design reference period.

1. Probability distribution of sea ice load F_{LI}

   F_{LI} obeys the log-normal distribution, and its probability density function $P(F_{LI})$ is

   $$P(F_{LI}) = \frac{1}{S_{\lg F_{LI}}\sqrt{2\pi}}\exp\left[-\frac{1}{2}\left(\frac{\lg F_{LI} - u_{\lg F_{LI}}}{S_{\lg F_{LI}}}\right)^2\right] \tag{1-334}$$

   In Equation (1-334), $u_{\lg F_{LI}}$ and $S_{\lg F_{LI}}$ respectively are the mean value and standard deviation of the logarithm values of ice load F_{LI}.

2. Probability distribution of earthquake load F_{LE}

   F_{LE} obeys the exponential distribution; its probability density function $P(F_{LE})$ is

   $$P(F_{LE}) = \lambda e^{-\lambda F_{LE}} \tag{1-335}$$

   For coefficient λ in Equation (1-335), when acceleration of the surface is greater than 0.1 m/s^2 during the earthquake, Shtanko recommends that $\lambda = 3$.

3. Probability distributions of wave and tidal force

   Because the wave force and tidal force are directly related to the velocity of water particles, they directly depend on the wave height. We can use the probability distribution of the wave height, namely the Rayleigh distribution, to express the probability density function $P(F_{LW})$ of the wave force and tidal force:

   $$P(F_{LW}) = \frac{F_{LW}}{\sigma_{F_{LW}}^2}\exp\left(-\frac{1}{2}\frac{F_{LW}}{\sigma_{F_{LW}}^2}\right) \tag{1-336}$$

   In Equation (1-336), F_{LW} is the wave and tidal force; $\sigma_{F_{LW}}^2$ is its variance.

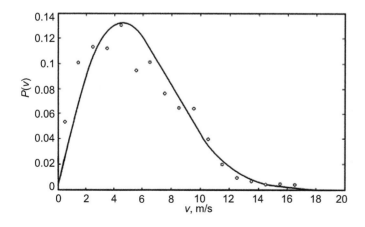

FIGURE 1-58

Probability distribution function curve of wind velocity.

4. The probability distribution of gravity changing with weight

Generally, according to the extreme value I distribution, the probability distribution function $P(F_{LV})$ should be

$$P(F_{LV}) = \exp\{ - \exp[- a(F_{LV} - K)]\} \tag{1-337}$$

In Equation (1-337), a and K are the distribution parameters for the extreme value I distribution, determined through observation and statistics.

5. The probability distribution of the wind load

From Equation (1-9), the probability distribution of the wind load depends on the probability distribution of wind velocity, but the long-term observations and statistics of the Bohai Oil Company and the First Institute of Oceanography of State Oceanic Administration gives the probability distribution function curve for winds of the China Bohai Sea, shown in Figure 1-58. Therefore, the probability distribution function $P(F_W)$ of wind load F_W can be expressed as

$$P(F_W) = \frac{F_W}{20.72} \exp\left(-\frac{F_W^2}{41.44} \right) \tag{1-338}$$

Offshore Oil and Gas Drilling Engineering and Equipment

Drilling is one of the most important methods in the development of oil and gas fields. The technology of drilling engineering is directly related to the results and output of oil and gas development; it is also related to the economic benefits of oil and gas fiweld development. The cost of well drilling and completion accounts for approximately half of all offshore oil and gas development cost. Obviously the biggest way we can reduce the cost of offshore oil and gas field development is by improving drilling speed and quality, which would cut the days drilling ships have to work and further optimize drilling technology. This chapter mainly introduces the key aspects of offshore drilling engineering and equipment (elements that are the same as the inland will not be covered).

2.1 Offshore Oil and Gas Drilling Platform (Ship)

The so-called "platform" is a workplace. Thus offshore drilling platform (or offshore drilling workplace) is the same as the drilling location on land. According to the operation characteristics, an offshore drilling platform can be divided into two categories: fixed offshore drilling platforms and mobile offshore drilling platforms. The former is fixed to the seafloor and can't be moved, but the latter is mobile and can be reused. The following reviews the two categories, introducing their structure and characteristics.

2.1.1 Fixed Drilling Platform

We first review the characteristics of fixed drilling platforms.

2.1.1.1 The Types of Fixed Drilling Platform

A fixed drilling platform is an offshore structure (for the purpose of offshore oil and gas drilling) fixed in position on the seafloor using a pile foundation, mat foundation, or other methods that produce supporting pressure. A fixed drilling platform is generally divided into two types: rigid fixed drilling platforms and flexible drilling platforms.

1. Rigid drilling platforms
   A rigid drilling platform is a permanent fixed drilling platform that doesn't shift under the influence of marine environment loads. It can be divided into two types: pile foundation platforms and gravity platforms.
   a. Pile foundation platforms
      This type consists of a pile inserted into the seafloor that undertakes the vertical loads and

Offshore Operation Facilities: Equipment and Procedures. http://dx.doi.org/10.1016/B978-0-12-396977-4.00002-0

resists horizontal loads. The most widely used platform is a jacket platform; the monopod and the tripod tower-type platform are also widely used.

b. Gravity platforms

This type is directly and steadily located on the seafloor on the basis of its own gravity, rather than a pile. The most widely used platform is a concrete gravity platform, beyond that the steel gravity platform and mixed gravity platform are widely used.

2. Flexible drilling platforms

A flexible drilling platform is a deep water drilling platform that can sway a certain angle around the fulcrum within a permissive range under the influence of marine environment loads. In order to meet deep water needs, this type of platform usually uses a slender steel jacket as a pile that inserts into the seafloor through tubes; cement is injected between the pile and tubes, creating one assembly of the pile and tubes that is attached to the steel jacket structure. Some flexible drilling platforms may use guy line or add camels to generate a restoring force. This type can be divided into string tower-type flexible platforms, camel tower-type flexible platforms, and others.

At present among these types, the most widely used fixed drilling platform is the pile foundation platform. Following are details about this platform.

2.1.1.2 Pile Foundation Jacket Fixed Platform

1. Structure

a. Jacket

As shown in Figure 2-1, one jacket is a steel frame structure, which is composed of some vertical tubes, certain slope tubes, and many horizontal and oblique connecting rods. The number of tubes and their position are the same as that of the piles. The inner diameter of the tubes is slightly larger than the outside diameter of the piles. So the piles can be easily inserted to the seafloor through the tubes. A jacket can be used as a positioning and orientation tool for drilling piles during construction. The loads of the platform can be delivered to piles across jackets. A jacket can also install a marine fender for the convenience of supporting a ship's mooring platform when drilling a well.

b. Pile

A pile is a steel stake with a sharp point. Knocked into the seafloor with a pile hammer, the piles undertake all the gravity of the platform and deliver the loads to the foundation soil layer through the friction near the piles and the resistance of the piles' sharp points.

- The axial carrying capacity

  The axial carrying capacity refers to the capacity of piles to undertake vertical loads. It depends on the piles' length, cross-section shape, dimensions, material, and the properties of the foundation soil layer. If a hard formation is located in the shallow seafloor mud, we just need to knock the pile sharp point until we hit the hard formation. Thus the pile is also called a supporting pile. If there isn't a hard formation, the axial carrying capacity mainly depends on the friction of the pile and the soil all around it, as shown in Figure 2-2 (a). This pile is called a friction pile. If the axial carrying capacity of friction piles is not sufficient, an explosion can be used to enlarge the pile sharp point. After the explosive under the pile sharp point detonates, soil near the pile point will explode and become a large hole. The

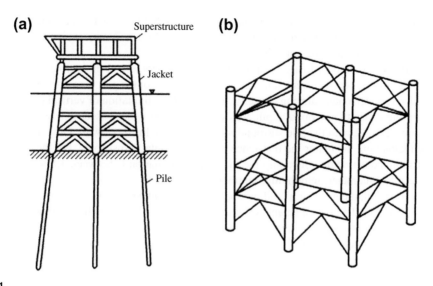

FIGURE 2-1

Structure of a pile foundation jacket fixed platform.

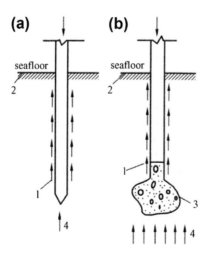

FIGURE 2-2

Diagram of two types of friction pile: 1–friction surrounding the pile; 2–ground; 3–concrete; 4-resistance of pile point.

nonsolidified concrete in the tubular pile falls into that hole. After the concrete solidifies, the enlarged pile point will be formed as shown in Figure 2-2(b).

- The horizontal carrying capacity
  In addition to vertical loads, the pile undertakes the horizontal thrust force produced by winds, waves, ocean current, and sea ice. The capacity of the pile to resist this thrust force is called the horizontal carrying capacity. Because of the limitation of vertical pile's horizontal carrying capacity, the oblique pile was generally most widely used widely [shown in Figure 2-1(a)], with the pile surrounding the jacket inclined outward. Obviously, the bigger of inclination, the better the resistance of the horizontal thrust force, but the harder the offshore construction. Generally the maximum inclination is between 1:8 and 1:11.
- The piles' diameter, length, and number
  The piles' diameter, length, and number should be determined by the working conditions. For example, the piles of an anti-ice platform should be long, small in number, and of large diameter, so as to increase the carrying capacity of the pile foundation. Incline piles should be used as much as possible. But for a platform that will be removed in the future, if we neglect the requirements of anti-ice, the piles should be short, large in number and of small diameter; for the convenience of pile extraction, straight piles should be selected. It is rare to remove a fixed platform.

c. Superstructure
   The superstructure of a jacket fixed platform usually consists of all kinds of trusses or beams (main beams, secondary beams, small beams) undertaking mechanical and other loads and decks.
- Main beams
  As shown in Figure 2-3, the first parts above the piles are the main beams. The main beams usually use a large amount of joist steel as their material. The number of main beams should depend on the span of the piles. Generally, if the span of the piles is large, the number of main beams should be small to resist the bending moment and allow full play of the properties of the material. The position of the main beams should be determined so that each main beam takes on an equal load, which requires equal strength, and has an equal section area so as to make manufacturing convenient.
- Secondary beams and small beams
  If the interval between two main beams is too large, a secondary beam should be arranged between them. Equally, if the interval between two secondary beams is too large, small beams should be arranged between the two secondary beams. In a word, the span of a panel, the bending moment, and the thickness of a panel are accordingly decreased. The arrangement of the secondary beams is perpendicular to the main beams, and the arrangement of the small beams is perpendicular to secondary beams, as shown in Figure 2-3.
- Upper and lower deck
  The superstructure usually is divided into two layers: the upper platform and the lower platform. As shown in Figure 2-4, the method of connecting the upper deck and lower deck is either through columns or trusses. In this way, the transmission of platform loads is as follows: upper loads → panels → small beams → secondary beams → main beams → piles → subsea foundation.

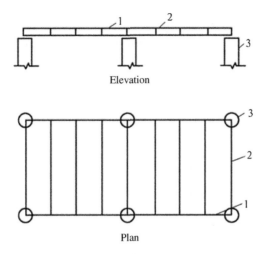

Elevation

Plan

FIGURE 2-3

Beam arrangement in a jacket fixed platform: 1–main beams; 2–secondary beams; 3–piles.

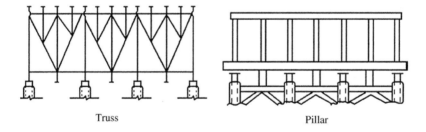

Truss   Pillar

FIGURE 2-4

Method of connecting upper decks and lower decks.

2. Elevation of the jacket platform

Figure 2-5 outlines the elevation of the components of the jacket platform.

a. Elevation of the top pile

The main beams and main trusses of the lower deck are usually installed on the panels of the pile. So the elevation of the top pile is the same as the elevation of the lower deck. At the design high tide level, the appropriate gap above the wave crest should be taken into account.

b. Elevation of the top jacket

The elevation of the top jacket is usually lower than the elevation of the top pile. According to construction experiences in China, the value is lower about 0.5 m.

c. Elevation of the bottom jacket

According to practical engineering experience in the Bohai Sea, China, the jacket is inserted into the seafloor at about 2 m. The jacket can be dragged by mud and kept steady during construction.

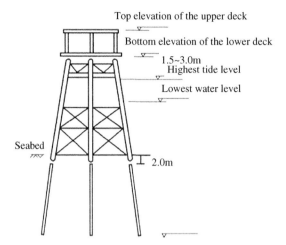

FIGURE 2-5

Elevation of jacket platform components.

   d. Elevation of horizontal tie bars at all layers

The jacket uses a support component that separates the upper trusses and lower trusses, mainly to prevent ice accumulation and decrease the area that must resist ice, which is necessary for offshore fixed platforms in ice regions. So the elevation of the upper trusses should consider the position above high tide level that is not impacted by ice in winter. The horizontal positions of the jacket should use tie bars, except the positions near the wellhead.

   e. Elevation of the bottom pile

The elevation of the bottom pile is determined by the buried depth, diameter, properties, and gravity of the platform. All points of the piles should be in the same soil stratum. The elevation of the bottom piles should not vary widely.

   f. Elevation of the deck

The elevation of the bottom deck should be the same as the elevation of the top pile's panels. The height between the upper deck and lower deck should be determined by the equipment arranged on the lower deck and its requirements. Saving steel and making equipment convenient for transportation and installation should be taken into account.

3. Layout of the jacket platform

The layout of a jacket platform includes the arrangement of equipment on deck and the arrangement of piles. Figure 2-6 shows the arrangement of equipment on the upper and lower decks of China's Bohai No. 12 fixed platform. The upper decks are as shown in Figure 2-6(a) and the lower decks are as shown in Figure 2-6(b).

Piles are arranged after the layout of equipment on decks has been determined. All piles should bear equal loads in the pile foundation, but in reality it will be only close to equal. As shown in Figure 2-7, a double oblique piles layout is usually used. The two basic pile arrangement types are the 8 piles style and 16 piles style.

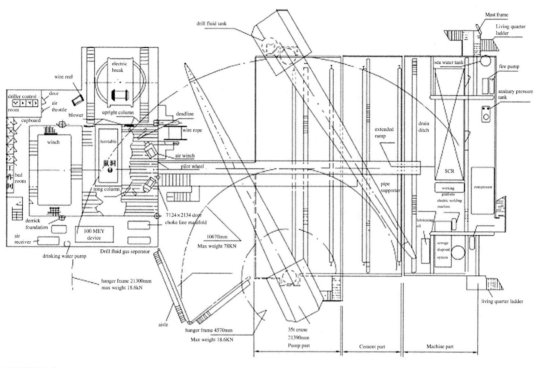

FIGURE 2-6

Floor plan of deck jacket fixed drilling platform.

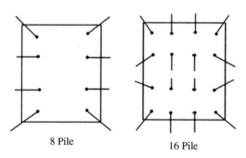

8 Pile   16 Pile

FIGURE 2-7

Pile foundation layout of jacket fixed drilling platform.

4. Construction and towage of the jacket fixed platform

   The structural steel and modules of the jacket fixed platform on the deck are all welded, constructed, installed, and assembled on shore. Respectively, jackets and upper modules are transported to the well position by barging or integral transport. Then they are installed on the seafloor with a floating crane.

a. Barging

As shown in Figure 2-8 (a), the jackets for a fixed platform are constructed on shore. After they are finished, jackets can be slid along the guide to a barge for hauling or be hung up and put on the barge for hauling with a crane, as shown in Figure 2-8(b). Figure 2-8(c) shows a new technology for transporting two jackets in one barge, which was created in the development of offshore oil and gas construction in the Bohai Sea, China.

b. Integral transport

By this method, after each end of the jacket is sealed, the whole jacket is hauled to the well position through its own buoyancy. Figure 2-9 (a) shows a whole jacket floating on the surface; Figure 2-9(b) shows the installation of the upper modules after the whole jacket arrives in good position. Figure 2-9(c) shows the final jacket fixed drilling platform.

2.1.2 Mobile Drilling Platform

A mobile drilling platform refers to a platform that can be moved from one drilling site to another after finishing the drilling work. The mobility of a mobile offshore drilling rig could

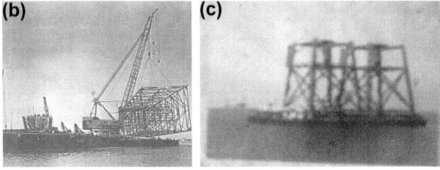

FIGURE 2-8

Fixed platform jackets on barge.

FIGURE 2-9

Integral transport of jackets and installation of fixed platform.

depend on a self-propelled rig or towing by other power-driven vessels. For example, mobile drilling platforms include submersible drilling rigs, jack-up drilling rigs, semi-submersible drilling rigs, floating drilling barges, and so on. The submersible drilling rig and jack-up drilling rig contact the seafloor through mats or legs when they are on the move. They are also called seafloor drilling rigs.

2.1.2.1 *Bottom-Supported Drilling Platform*

A bottom-supported drilling platform with a pontoon can not only float in the sea but also lower down to the seafloor for offshore drilling activities. When they are active, they are first towed to the well location by towboat (some types are self-propelled), then flooded so they sink. When they are on the seafloor, the anti-slide piles can be installed, and at last the drilling activities can proceed. After finishing these activities, the water is drained off and they are inflated until they are floating on the surface of the water, where they can be moved to other well positions. No. 2 and No. 3 of Shengli in the China Bohai Sea are bottom-supported drilling platforms that were designed and built independently. No. 4 of Shengli is a bottom-supported drilling platform whose technology was imported from foreign sources and modified as necessary domestically.

1. Structure

 a. Buoyancy tank

As shown in Figure 2-10 (a), a buoyancy tank can be a type of hull separated into some ballast tanks or be a type of pontoon. The buoyancy tank can sink or float through the method of flooding water and draining air or the method of draining water and flooding air.

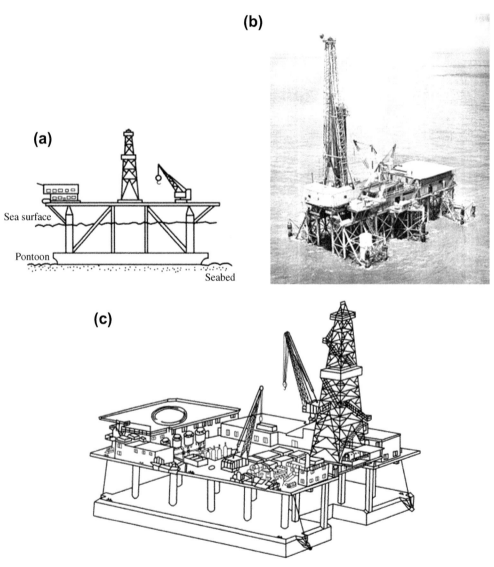

FIGURE 2-10

Submersible drilling rig.

b. Pillar

As shown in Figure 2-10(b), the pillars usually use steel trusses to connect the platform deck and buoyancy tanks. The height of pillars depends on the depth of water, and generally the value is about 20 m to 30 m. If we add large diameter steel bottles or pontoons at the four corner pillars, as shown in Figure 2-10(b), the depth can be increased and stability can be improved as well as the vertical speed. In addition, anti-slip piles can be added to improve the stability of the submersible drilling rig.

c. Platform decks

Mechanical equipment in the form of squares, rectangles, triangles, and so on can be installed on platform decks that are welded to the pillars. One side of the deck should be open for the convenience of moving the rig. There is hoisting equipment for the convenience of transporting equipment from attending ships. The layout of the platform designed and built by China for No. 1 of Shengli, a submersible drilling rig, are shown in Figure 2-10(c).

2. Merits and faults

The merits of a submersible drilling rig are as follows. It is fixed firmly during drilling. It is also mobile and flexible after well completion, and easily turned into a production platform, storage platform, activities platform, etc.. In a word, it is adaptable. The faults of a submersible drilling rig are as follows. The height of the upper deck is fixed and the depth of work is limited. It has a high resistance to the process of towing and there is less area on the platform. It is easy to move when the seafloor is softer and seriously scoured. So anti-slide, anti-erosion, and anti-hollowed measures must be taken, such as adding anti-slide piles.

2.1.2.2 Jack-up Drilling Platform

A jack-up rig is a type of mobile platform which can be fixed on the seafloor for operation and float on the sea for mobility, consisting of a barge-shaped hull and a number of legs that can go up and down on their own and be attached to the ground.

1. The types of jack-up rig

According to the structure of the leg, the jack-up rig can be divided into two types.

a. Jack-up rig with cylindrical legs

As shown in Figure 2-11 (a), the legs for this type are a cylindrical steel structure.

b. Jack-up rig with truss legs

As shown in Figure 2-11(b), the legs for this type are a truss steel structure.

2. Construction

Figure 2-11(c) shows the construction diagram of a jack-up rig.

a. The legs

The legs inserted into the seafloor are used to support the platform above the seawater. There are two types, cylindrical and truss. There are usually three or four cylindrical legs with a diameter from 2 m to 10 m. They go up and down with the hydraulic or gas power. There are usually three or four truss legs. Each of the legs is a truss triangle or square shape. If we look at No. 1 of Chinese South Sea and No. 2 of the Exploration jack-up drilling rig, the legs have a triangle shape with each edge 11.5 meters in length. The three points of the triangle have three cylindrical pipes with each pipe 0.76 meters in diameter. The legs are inserted 8 meters into the seafloor. The truss legs all use a series of gear racks to climb up and down mechanically.

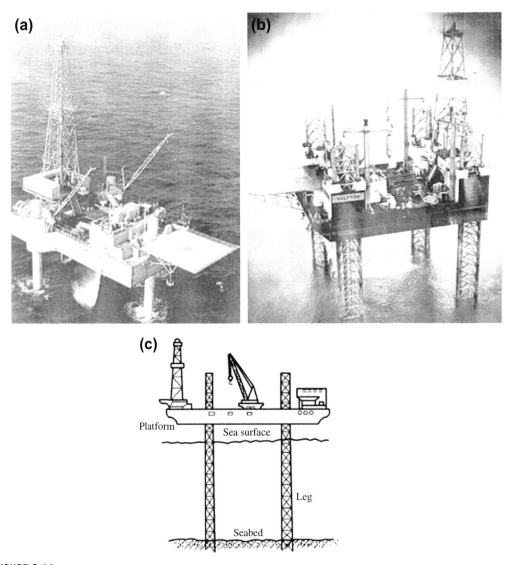

FIGURE 2-11

Construction diagram of a jack-up rig.

When the working depth of a jack-up drilling rig is more than 150 meters, boxes need to be added at the bottom of the legs with a diameter of around 10 to 14 meters; the must have an adequate support area (such as 70 to 150 square meters). Some of the legs are made in inverse conical style with a 60 to 120 degree slope. If the work depth of a jack-up drilling rig is smaller than 150 meters, rectangular pads need to be added on the bottom of the legs with a length of 50 to 60 meters and thickness of 3 to 4 meters to support 1000 to 2000 square meters. The pad can be either a single or a double type . The single pad has one A-shaped pad that is open in the center.

b. Working platform

The working platform is the deck of a barge; it is used to install all kinds of equipment. When it is ready, the working platform is towed to the well position by towboat, as shown in Figure 2-12 (b). After the drop anchor is fixed, the legs are inserted into the seafloor by a leg lifter mechanism. After the legs are preloaded, the leg lifter mechanism lifts the barge to a certain height above the water surface (leaving some room in consideration of the high tide level). Then the drilling of the well can start, as shown in Figure 2-12(a). After the completion of a well, the barge is lowered to the water surface, the legs are pulled to the towing position, and the working platform can be towed to a new well position, as shown in Figure 2-12(c). In order to adapt to different depths, the lifter mechanism is needed to accomplish the work of

(a)

(b)

(c)

FIGURE 2-12

Structure of jack-up drilling platform.

lifting and lowering the barge and piles. These components will be detailed in a later section. When the jack-up rig is at work, we should keep the position of the platform fixed. When the jack-up rig is being towed, we should keep the legs fixed.

2. Merits and faults

a. The merits

The jack-up rig has good stability that is reliant on the legs supporting the work platform by insertion into the seafloor. With good adaptability, the jack-up rig can be used for different seafloor soil conditions and different depths up to 100 meters. With good mobility, the jack-up drilling rig is convenient for repeated use. It can be economical; the jack-up rig has a lower cost without the pad, it needs little steel, and is easy to construct.

b. The faults

The jack-up rig is hard to tow, because the legs are long and the jack-up drilling rig does not resist storms well. The operations in placing it into and out of position can be complex, and the jack-up drilling rig is sensitive to waves. When the pads at the seafloor are suffering erosion, the foundation of the seafloor is damaged and pads sliding can occur; it is easy to make the platform slide and change the well position. The depth of work is limited to 0 to 100 meters, otherwise the length of the legs, size of sections, and gravity all will increase with the increasing water depth. In addition the work state and work stability all will become worse. Large legs have issues of vibration.

At present the jack-up drilling rigs in service in the world make up to 60 percent of all mobile offshore drilling rigs. So the spread of jack-up drilling rigs is very extensive. At present the jack-up drilling rig is being used in offshore drilling work in China, such as No. 4 of China's Bohai, No. 5 of China's Bohai, No. 7 of China's Bohai, No. 8 of China's Bohai, No. 10 of China's Bohai, No. 1 of South China, No. 3 of South China, No. 4 of South China, No. 2 of Exploration, No. 1 of Shengli, No. 5 of Shengli, and No.6 of Shengli. No. 5 of China's Bohai and No. 7 of China's Bohai were designed and built by China. Furthermore, more are being built in shipyards in China.

2.1.2.3 Semi-submersible Drilling Rig

The semi-submersible drilling rig is another offshore mobile drilling rig that is similar to the submersible drilling rig and composed of a work platform at the top, columns in the middle, and a buoyancy tank at the bottom. After using self-propulsion or being towed to the well position, firstly the rig casts its anchors and is set in position, then the lower buoyancy tank and ballast tank for the columns are flooded to exhaust the gas, so the entire platform starts to submerge. The rig is shown in Figure 2-13 (a); when it reaches the predetermined depth, the work of drilling can be proceed. During the drilling process, mooring system or dynamic positioning system is needed to keep the well in position; a compensation device is also needed to keep the proper drilling pressure on the bit and to avoid leaving the well bottom. After the completion of the well, the ballast tank is filled with gas and the water is drained. After the lower buoyancy tank emerges from the water, the rig may be moved to other well positions, as shown in Figure 2-13(b).

1. Types

According to the number of columns, semi-submersible drilling rigs can be divided into many configurations, such as four columns, six columns, eight columns, or more, as shown in Figure 2-13.

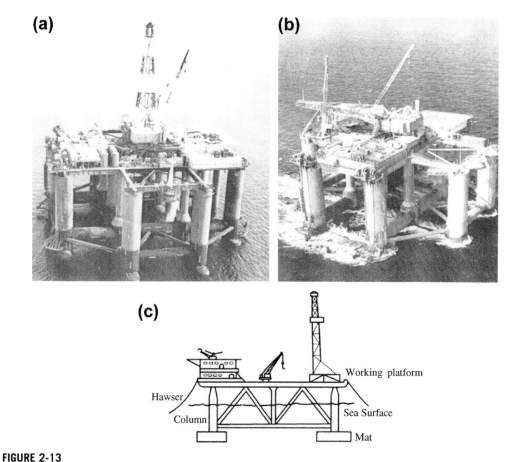

FIGURE 2-13

Semi-submersible drilling rig photos and diagram.

According to the method of movement, we can also divide them into two types: self-propelling and towing. Other powered ships should tow the latter.

2. Construction

The main components of a semi-submersible drilling rig are as follows.

a. Lower hull (including buoyancy tanks and pads)

As shown in Figure 2-13(c), the lower hull is usually in the shape of a ship. The lower hull is usually rectangular or trapezoidal. The inside of the hull is divided into some ballast tanks. Rectangular ballast tanks are symmetrical and accessible to control the draining and flooding of water. Trapezoidal ballast tanks can adapt to the nonuniform loads of the upper platform; furthermore, they have lower resistance to mobility and have a good head wave properties. Looking at No. 3 of Exploration, a semi-submersible drilling rig designed and built in China (as shown in Figure 2-14), its size is as follows: length × wide × height =90m × 14m × 4m, the draft while working is 20 meters, and the draft while being towing is 5.6 meters.

FIGURE 2-14

Semi-submersible drilling rig (No. 3 of Exploration).

- **b.** Upper hull (i.e., platform)
  As shown in Figure 2-13(c), the upper hull is usually made of steel or concrete with an open hole or "V" shape in order not to be obstructed by the well when moving the rig after the well has been completed. The upper hull mainly includes the workplace, production devices, and living facilities. It is usually divided into the main deck and upper deck. Taking No. 3 of Exploration as an example, the main deck has a height of 30 meters, a length of 91 meters and a width of 71 meters, while the length of the upper deck is 35.2 meters.
- **c.** Columns
  As shown in Figure 2-13(c), the columns are usually cylindrical steel structures. The ballast tank, storeroom, and elevator are inside the cylindrical steel. The columns mainly connect the lower hull and the working platform to transfer the load of the platform to the lower hull. The columns of the semi-submersible drilling rig have many styles; No. 3 of Exploration has six columns 9-meter diameters.
- **3.** Merits and faults
  - **a.** Merits
    The semi-submersible drilling rig has many merits. It has a large range of work depths, with depths that can reach 200 to 300 meters using the mooring method. With good stability, when it is working at draft depth, the hull submerging depth can reach 20 to 30 meters, and it can adapt to rough sea conditions; the mooring system or dynamic positioning system can improve the stability. It has good mobility, and it can move with a draft depth of 7 to 8 meters. If a self-propelling rig is used, this merit is even more of a benefit. The large deck areas are convenient for drilling or able to be changed to a production platform. In all, the semi-submersible drilling rig has both the submersible drilling rig's merits and the drilling ship's merits.

b. Faults

The semi-submersible drilling rig also has some faults. It has a lower self-propulsion speed (most of them have a loaded speed of less than 8 knots), while the drilling ship can reach 8 to 14 knots. The cost of construction is high (compared to a jack-up drilling rig); generally the cost can be above $100 million. The carrying capacity is limited and it is more sensitive to load. In sum, the semi-submersible drilling rig can not only meet the requirements of variable water depth but also better solve the problems of stability and mobility. Compared to other options, it has more potential for future development. Especially from the view of exploration and development in the conditions of deeper and rougher seas in future, the semi-submersible drilling rig will represent the main direction of development. At present, semi-submersible drilling rigs serving are No. 3 of Exploration designed and built in China, as well as No. 2 of South China, No. 5 of South China, and No. 6 of South China, which were imported from foreign countries.

2.1.2.4 Drillship

A drillship is a mobile offshore drilling rig where a mono-hull ship, catamaran, triple-hull ship, or barge is adapted or built for use as a platform. After reaching the well position, the drillship first gets into position with the mooring system or dynamic positioning system, and then begins to drill a well. The ship is floating when working. From the action of wind and waves, the drillship will encounter a heaving motion, sway on all sides, float on the surface, and so on. Hence, it needs all kinds of measures such as a drilling string heave compensation device and swing angle reduction device as well as dynamic positioning and so on to guarantee displacement of drillship within the allowable limits, so as to properly drill a well.

The drillship usually is made of steel or concrete; the latter can reduce the need for steel and prevent corrosion. The drillship can be a mono-ship, as shown in Figure 2-15 (a), (b). But this mono-ship must finish the well under water. Otherwise, the drillship can damage the wellhead equipment when it moves. If the drillship is a catamaran, as shown in Figure 2-15(c) (like drillship No. 1 of Exploration designed and built in China), it won't collide with the wellhead device because of the open hole between the two ships. So, it can finish the well on the water surface. No. 1 of Exploration is 99.23 meters long and 38 meters wide; each piece is 14.3 meters wide. Its main power is 2×2000 hp, while it has a 30 to 100 meter operating depth and a 3200 meter drilling depth, and it self-propelled speed is less than 12 knots.

1. Merits

The drillship has many merits, such good mobility, a high self-propulsion speed, and the ability to handle a wide range of variable loads. Owing to the large waterline area, the influence of variable loads (such as drill pipes, casting pipes, drilling fluids, raw materials, cement, fuel oil, water, and so on) on draft is very little. It has a large storage capacity, strong self-sustaining ability, and large working depth. If moored with the anchor, the working depth can reach to 200 to 300 meters. If it uses a dynamic positioning system with a thruster controlled by a computer, the work depth is unlimited and the value can reach 6000 meters.

2. Faults

The drillship also has some faults, such as having poor stability especially when mooring with the anchor; being heavily influenced and sensitive to wind and waves, which can affect the efficiency

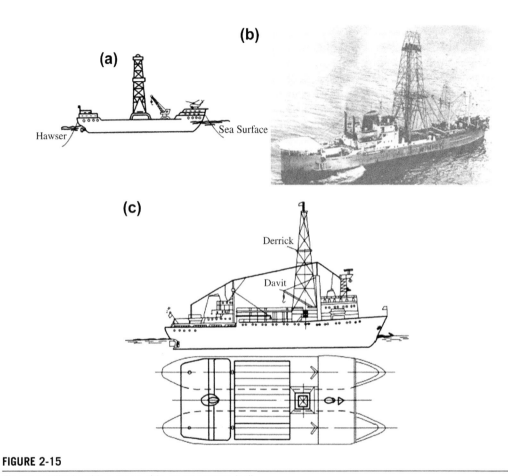

FIGURE 2-15

Floating drillship with a mono-ship or catamaran hull.

of the drilling operation; having a limited deck area for decks; and being expensive (more so if the drillship has a dynamic positioning system).

Altogether, because of the drillship's outstanding advantages, it is suited for drilling in deep water fields. It is becoming a more and more important tool in deep water oil detection. After finishing the development of No.1 of Exploration in 1974, China hasn't developed it further. It is time to catch up to the new devices.

2.1.2.5 Technical Characteristics of a Mobile Offshore Drilling Rig

Submersible and jack-up drilling rigs can be merged into a floor contact drilling rig category because they need pads or legs to contact the seafloor to support the work platform. Likewise the semi-submersible drilling rig and drillship can be merged into the floating drilling rig category because they float on the surface when at work. Their technical characteristics are showed in Table 2-1.

Table 2-1 Technical Characteristics of Mobile Offshore Drilling Rig

Category Characteristic		Floor Contact		Floating	
		Submersible	**Jack-up**	**Drillship**	**Semi-submersible**
Work depth (m)		5–30	10–160	30–500 (mooring) 6000 (dynamic positioning)	30–500 (mooring) 3000 (dynamic positioning)
Drilling depth (m)		4500–9000	4500–12000	6000–12000	6000–12000
Seabed conditions		Clay, sand clay, small inclination seafloor	According to soil properties	All kinds of seabed, with attention paid to anchor hold power	Same as drillship
Safety margin	Wind speed (m/s)	70	70	60	70
	Current (kn)	4	4	4	4
	Wave height(m)	About 10–15 (according to different depths)	About 16 (except special designs)	18–30	18–30
Dynamic properties in waves	Insert piles	No problem	No problem	A little poor	Good
	Floating toward	Huge problem	Huge problem		
Mobility	Towing resistance	Very large	Very large	Small	Big
	Wave intensity	Problem	Foot vibration problems with large devices	No problem	No problem if draft is increased
	Movement speed	Very slow	Very slow	Fast (10–15kn)	Slow (fast if auxiliary propeller installed)
Stability of drilling operation		Good	Good	Poor (easy for wind and waves to influence)	Better
Positioning capability		Good (easy for soil properties of seabed to influence)	Good	Easy for wind and waves to influence, better with dynamic positioning	Better than drillship
Loading capacity (variable loads), t		About 3000	800–2500	Above 4000	About 2000
Cost (maximum), 1E4 dollars		-	>3200	>6000	>7000

2.1.3 Selection of Offshore Drilling Rig

In the section, we review the process of selecting the appropriate type of offshore drilling rig.

2.1.3.1 Main Factors

We must review the requirements of the drilling operation: whether the drilling well is an exploratory well, an appraisal well, or a production well; and the methods used to complete the well, such as whether they are above the water surface, under the water surface, and so on. Other factors include:

1. Marine environmental conditions: this includes water depth, wind, waves, current, sea ice, and so on as well as the distance from the land.
2. Technical characteristics: whether the technology provided by the drilling rig at present meets the requirements or not.
3. Economic factors: this includes cost of construction, project investment, or rental and operating cost.

2.1.3.2 Basic Selection Principles

For exploratory wells, appraisal wells, and some production wells, we should select a mobile offshore drilling rig. According to different water depths we should select different types of drilling rigs, as shown in Figure 2-16.

- If the water depth is less than 10 to 15 meters, we should select a submersible drilling rig.
- If the water depth is between 15 and 75 meters, we should select a jack-up drilling rig.
- If the water depth is between 75 and 200 meters, we should select a semi-submersible drilling rig with anchored moored positioning or adrillship.
- If the water depth is more than 200 meters, we should select a semi-submersible drilling rig with dynamic positioning or a drillship with dynamic positioning.

  If the production wells are more concentrated, we should mainly consider selecting a fixed drilling rig. According to different water depths, we should select different types. We can refer to the following principals.

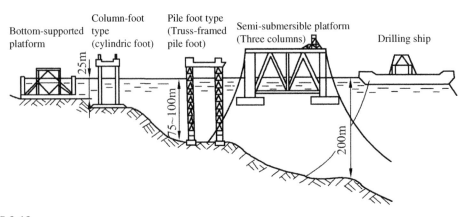

FIGURE 2-16

Comparison diagram of different types of mobile offshore drilling rig for different water depths.

- If the water depth is less than 300 meters, we should select a pile foundation jacket fixed drilling rig. After finishing the drilling operation, the rig can be used as a production platform.
- If the water depth is between 300 and 600 meters, we should select a guyed-tower flexible fixed platform or tension leg flexible fixed platform. The platform can expand into a production platform when necessary.
- If the water depth is less than 160 meters, the seabed is flat, and we have the advantage of building the channel and harbor of a concrete-gravity platform, we should select the concrete-gravity fixed platform to drill a well and produce oil.
- If oilfield development uses a floating production system or early production system, we should select a suitable mobile drilling rig according to water depth. When we need to produce oil, the rig should cooperate with the production system.

  For some special offshore drilling operations, such as installing a platform before drilling a well, hoisting equipment, and so on, we should select a jack-up drilling rig, because other mobile drilling rigs are easily influenced by the state of the sea.

  The selection of a drilling rig is just one part of total oilfield development, so we should select a suitable rig on the basis of a comprehensive analysis of the technological and economic aspects. Table 2-2 shows a comparison of all types of rigs and production systems for reference.

2.1.4 **Going Up and Down an Offshore Jack-up Drilling Rig**

At present, we can divide the equipment used into two categories: electric gear driven equipment and hydraulic cylinder equipment.

2.1.4.1 *Moving Up and Down Using the Electric Gear-Driven Machine*

A jack-up drilling rig such as No. 1 of South China or No. 2 of Exploration is composed of truss legs with each leg section a triangle. There are three cylinder legs in three triangle vortexes with outside diameters of 0.76 meters. The movement of the platform up and down and depends on the installation of an electric gear-driven machine on each leg. The lifting methods of the device are as follows.

1. Electric gear-driven device

   Because these jack-up drilling rigs all have three legs, the entire platform has nine electric gear-driven devices, and each one is composed of the four same units, as shown in Figure 2-17. Each unit is composed of an alternating-current motor, gear reducer, as well as a small gear for output. The components of the entire device are as follows.

   a. Motor

   The motor can be alternating current or direct current. The three-phase alternating current motor with a power of 14.72 kW, voltage of 600 V, and frequency of 600 Hz is generally used. The power is transferred from motor to reducer.

   b. Gear reducer

   The gear reducer is a two-stage reduction. The first reduction is from a 12-tooth gear to a 76-tooth gear. The second reduction is from a 17-tooth gear to a 54-tooth gear. Except for the driving gear and driven gear, other gears are all sealed in oil.

   c. Rack

   As shown in Figure 2-17, racks are welded on two sides of the cylinder legs. The racks are used to mesh with the output gear of the reducer so as to make the legs move up and down.

Table 2-2 Comparison of All Kinds of Offshore Drilling Rigs and Production Systems

Number	Item	Pile Foundational Fixed Rig	Gravity Fixed Rig	Submersible Rig	Jack-up Rig	Drillship	Semi-submersible Rig	Tension Leg Rig	Floating Production System	Subsea Production System
1	Suitable depth (m)	<300	15–160	<30	<150	30–6000	30–2000	150–600	80–600	30–600
2	Maximum depth (m)	416	220	45.7	183	2386	-	-	250	250
3	Maximum number of wells	96	60	-	30–40	-	15	30	15	-
4	First use	1899 (English)	1973 (Norway)	1889	1953	1956	1962	1984	1975	1962
5	Materials	Steel, concrete, wood	Steel, concrete	Steel	Steel	Steel	Steel (concrete)	Commonly steel	Steel, concrete	Steel, concrete
6	Construction	Pile foundation, jacket, deck	Basic container (oil storage), legs, deck	Lower hull, columns, upper hull (including deck)	Ships (deck), legs	Ships, mooring system	Lower hull, columns, deck	Deck, shell (columns + beams), mooring system (tension + foundation)	Floating rig, subsea production, manifolds and risers, SPM	Subsea wellhead, subsea Christmas tree, submarine pipelines
7	Main function	Permanent foundation, drilling, production, oil storage, mooring	Drilling, production, mooring, large volume for storage of oil, better fixed structure	Drilling exploratory wells in shallow sea region or tidal range	Drilling, production, offshore construction, changing to production platform	Drilling exploratory wells in deep water, self-propelling or dynamic positioning, as FPS	Drilling, mooring, storing oil in heaving conditions, as FPS, large work areas	Drilling a well and production in deep water, as FPS in heaving sea area	Drilling, production, mooring, oil processing, off-loading	Supporting subsea wellhead, connecting to production pipelines
8	Features	1. Mature technology 2. Provides a stable and good work place over sea	1. Huge oil storage capacity, small impact to environment 2. Big deck area, good work conditions 3. Small offshore construction workload, could be moved 4. Safety and reliability, low maintenance expenses, long work life	Good mobile property	1. Mature technology, the most widely used 2. Good mobile properties, could be used as early production system 3. Relatively low cost	1. Good mobile properties 2. No water depth restrictions 3. Could be used as FPSO	1. Deep draft, good stability 2. Big deck area 3. Good mobile properties 4. Could be used as FPSO	Small environmental impact, stable work area, better than semi-submersible platform, a promising supporting structure	1. Short construction cycle, quickly from oil field into production 2. Reuse, adaptable, flexible 3. Small impact due to water depth, good for underground soil, earthquake; 4. Suitable for early development, the development of marginal fields and deep water development	1. Wellhead at the bottom of the sea, reducing investment or stay ahead of schedule 2. Suitable for deep water floating production system development

	Problems	Abroad	Domestic
9	1. Costs increase with depth 2. Huge offshore construction workload, should not be moved 3. Long period of construction 4. Poor capacity for dealing with platform changes.		
10	Development Overview (up to 1990)	Almost 10,000 platforms. The most widely used support structure.	34 platforms
	1. High soil requirements at coast 2. Need deep water construction site 3. Long construction period.	23 platforms put into use	Being researched
	Large use limitations, so increasingly less used	36 platforms (1990)	3 platforms
	1. Applicable only at depths less than 200 meters 2. Small usable platform area	415 platforms (1990)	11 platforms
	1. Big cost 2. Large environmental impact 3. Small hull area	58 platforms (1990)	1 platform (made in 1974)
	1. Sensitive to the platform load 2. Big cost (compared with jack-up platform) 3. Water depth restrictions	171 platforms (1990)	4 platforms
	1. Sensitive to the platform load 2. Current technology is not mature	The first installation was on July 15, 1984	None
	1. Technology matures gradually 2. Production affected by climate 3. The wellhead number generally less than 10	Being developed	6 platforms put into operation
	1. Being developed, technology maturing gradually 2. The nature of the crude oil is limited and is not suitable for the development of high pour point crude oil	More than 600 platforms	None

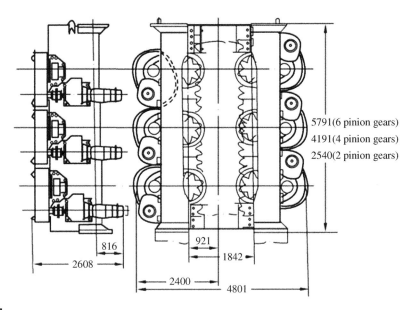

5791(6 pinion gears)
4191(4 pinion gears)
2540(2 pinion gears)

FIGURE 2-17

Electrically driven gear lifting device.

Generally the norm length of a rack is 10 meters, the distance of the circular pitch is 254 millimeters, and the thickness is 127 millimeters. The gears and racks as well as the axis are all made of alloy steel treated by heating.

d. The spring plate friction disc safety brake

This is a shaft safety brake mounted on the motor. A brake plate friction disc is mounted on the motor shaft. The friction plate is pressed by a fork, and fork movement functions through the suction of an electromagnet; the energized electromagnet is connected to the motor switch. Thus, when the motor is turned off, the electromagnet is energized to produce suction, promoting the fork while pressing the friction plate. The motor shaft is completely broken to ensure safety. When the electromagnet is de-energized, the friction plate is supported by spring force.

e. Shock absorber

There are spring steel pad shock absorbers, and each pile leg is connected to the platform with them, with a setup generally composed of seven elastic steel pads. The purpose is to reduce the impact of legs on the platform deck when legs initially contact the seabed.

f. Solid pile wedge

This aims at fixing the platform with the pile legs. There are eight wedges on each leg of one thigh, with four on the upper side and four on the lower side. Consequently, there are 72 wedges on the entire platform with three legs and nine thighs.

2. Lifting operation of electric gear-driven device

Pinions from the gearbox are symmetrically arranged in pairs on both sides of the cylindrical thigh and are meshed with the rack on both sides of the thigh at the same height, to offset the horizontal component force. Each thigh has four pinions meshing with the rack simultaneously from top to bottom. There are 36 pairs of pinions on the entire platform with three legs and nine thighs meshing with the rack on the thigh, and thus when the motor synchronizes, the three pile legs of the platform are synchronically transmitted through the gears and racks, gradually lifting up. The entire platform lifting operation of the motor can be mainly controlled by the platform control room, and also by each pile leg. Legs can drop down due to their gravity, but they should control the rate of decline by the brakes to ensure safety. The platform itself rises to a certain height on the surface of the water and down to the water in the conditions of pile legs fixed to the seabed. Because the motor and gearbox unit of the lifting device are installed on the platform deck, it still uses an electric gear transmission method, which is the same principle as the pile lifting its leg to operate.

2.1.4.2 Hydraulic Cylinder Drive and Lifting Device

1. Hydraulic lifting of the truss legs
   The truss pile legs of a jack-up drilling platform adopt a hydraulic cylinder drive, and the lifting device is shown in Figure 2-18. The device basically consists of the following:
   a. Hydraulic cylinder
   Each pile leg of the platform is made up of six hydraulic cylinders. The lifting capacity of the hydraulic cylinder is determined by the total weight of platform. Generally when the platform's total weight ranges from 1500 to 4000 t, the lifting capacity of the hydraulic cylinder should be about 200 to 1000t. When lifting, the various legs of the hydraulic cylinder should generate synchronized action.
   b. Pin rack rail plate and pinhole in pile legs
   As shown in Figure 2-18, each thigh of each pile leg is welded through a pin rack rail plate, and there are pinholes located a certain distance apart (in accordance with the stroke of the hydraulic cylinder); when the pin-driving hydraulic cylinder is operating, it can push the pins into the pinholes by driving the piston rod.
   c. Pin driving hydraulic cylinder
   As shown in Figure 2-18, there are upper and lower components in all three pairs, and six hydraulic cylinders on each pile of three legs driving pins to enter or exit the pinholes of the pile leg.
   In addition, like the lifting operation of electric gear-driven devices, there is a solid pile wedge shock absorber and brakes.
   The operational methods of the hydraulic cylinder driving truss leg movements are shown in Figure 2-19. Figure 2-19(a) shows the pile lifting operation of the leg from the jack-up platform in a floating state; this is comprised of the following steps:
      i. Insert the upper pin: As shown on the left in Figure 2-19(a). Insert upper pin A into the hole of the leg's lift ridge and connect the hydraulic cylinder to the leg.
     ii. Hydraulic cylinder piston up stroke: As shown on the right in Figure 2-19(a), the hydraulic cylinder shall push the piston up.
   iii. Insert into the lower part of the pin and pull out the upper pin: As shown on the right in Figure 2-19(a). First, insert lower pin B into the lower pinhole in the pile leg's lifting ridge. The leg is caught by the lower part of the platform connected with the pins. Then the upper pin A is pulled out from the pinhole, so the piston shall return.

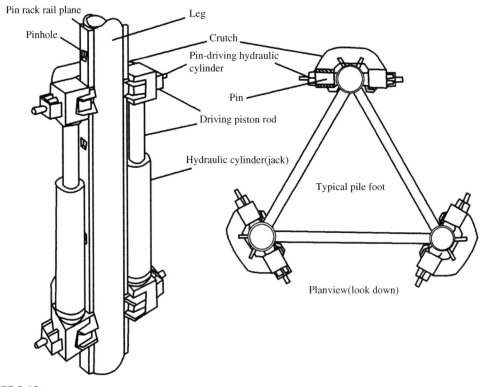

FIGURE 2-18

Truss leg hydraulic lifting device components.

    iv. The upper dead point returns to the lower dead point: When the upper pin A is pulled out, the hydraulic cylinder unloads, so the cylinder piston can be retracted to lower dead point for the next up trip.

    v. Lifting up stroke again: First insert the upper pin into the leg's pinholes, and then, pull out lower pin B from the pinhole, and push the piston up from the hydraulic cylinder, which in turn shall lift up. Generally speaking, each stroke is approximately 1.5 ~ 1.8 m (5 ~ 6 ft), so when the cycle continues, it can lift about 15.24 to 18.28 m (50 to 60 ft) each hour. The same approach for the lifting operation of the platform deck is used with fixed pile legs, with the procedure being the same as the pile leg lift operation, as shown in Figure 2-19(b).

2. Hydraulic lift methods of a cylindrical pile leg

Hydraulic lift methods are generally used in cylindrical pile leg jack-up platforms. China's "Bohai One" jack-up drilling platform has four cylindrical pile legs; each is 73 m long and the outer diameter of the cylinder is 2.5 m, and its lifting device is composed as follows:

    a. Pile leg guides

Figure 2-20 (a) shows the block diagram of the "Bohai One" leg guides. Guides of 75 mm on both sides of the cylindrical pile legs are welded, and there is a square hole on the guide plate

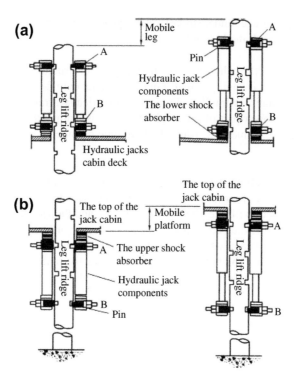

FIGURE 2-19

Truss leg movements and platform hydraulic lift operation.

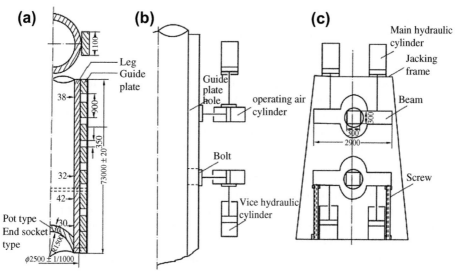

FIGURE 2-20

Cylindrical pile leg hydraulic lifting device.

every 900 mm (hydraulic cylinder stroke). Each square hole is 350 mm long for a bolt, in order to link the pile foot with the hydraulic cylinder.

b. The fixed pile frame

As shown in Figure 2-20(c), the frame is a truss type steel structure, welded with the platform deck, and its upper part is equipped with a main hydraulic cylinder. Each cylindrical pile leg is equipped with four main hydraulic cylinders, symmetrically arranged along the cylinder circumference. The lower part of the solid pile frame is equipped with four sub-cylinders through the fixed pile frame legs, while the hydraulic cylinder is connected to the platform deck.

c. Hydraulic cylinder

As shown in Figure 2-20(c), the upper and lower part of each leg in a fixed pile frame is installed with one of the four master cylinders and the secondary hydraulic cylinders. The master cylinder is used to lift the pile foot or platform, and the vice hydraulic cylinder is used to support the platform when changing "hands." Each pair of hydraulic cylinders lifts a horizontal ring square beam, which can make the pneumatic insert into the guide plate openings or exit from it. The bolt into the piston rod of the hydraulic cylinder can cross the ring beam and make the pile legs fuse. Otherwise, it is disengaged. Thus, by using of bolt insertion and withdrawal, the leg or platform lift and changing of hands can be carried out. The stroke of the "Bohai 1" hydraulic cylinder is 900.

d. Pump

The entire hydraulic system is equipped with a main fuel pump and low pressure oil pump. The main oil pump is a radial piston pump driven by a three-phase AC motor, which is used to provide each hydraulic cylinder with oil that goes through the upper and lower hydraulic cylinder into the oil distribution valve to control the fuel quantity; when going back to the oil, the slide valve needs to stop at an intermediate position. The low-pressure oil pump usually adopts a double gear pump to improve the flow fuel into the cylinder of air, which usually does not use this pump.

e. Accumulator

This is a compressed air energy storage device; the aim is to maintain a stable oil hydraulic cylinder, and the air pressure can be adjusted automatically.

In addition, a fulcrum wedge for solid contact of the legs and platform, shock absorbers, and brakes are provided; they are not explained in detail here.

Take the cylindrical leg hydraulic lift operation of jack-up rig as an example. It is ducked into seaside at the end the platform deck rising from the water conditions, as is shown in Figure 2-21:

i. Pull out the upper ring beam latch: As shown in Figure 2-21(b), the legs of the original pile are into the seabed conditions [Figure 2-21(a)], and the upper ring beam is pulled to lock the pins from the pile out of the pinholes in the legs of the guide plate, which is shown in Figure 2-21(a), (b).

ii. The oil-filled hydraulic cylinder pushes the piston down: As shown in Figure 2-21(c), after the oil fills the hydraulic cylinder upper chamber, the piston pushes down a stroke, thus the platform deck lifts a stroke, but the piston has reached the lower dead point.

iii. Pull out the lower ring beam lock lift for "hands": As shown in Figure 2-21(d), the locking pin is pulled out of the leg pin hole in the lower ring beam, and the entire platform deck is hooked temporarily by the locking pin of the upper ring beam in the pin hole of the insert legs caught for "change hands".

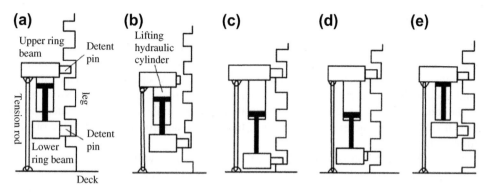

FIGURE 2-21

The cylindrical pile leg jack-up hydraulic lifting operation.

    iv. T oil-filled hydraulic cylinder chamber pushes the piston back to the upper dead point, as is shown in Figure 2-21(e). The piston in the hydraulic cylinder should return to the lower dead point and complete the "change hands" process.

    v. The upper ring beam lock pin inserts into the upper pinholes, then goes to another itinerary, as is shown in Figure 2-21(e), when the upper ring beam lock pin is inserted into the upper pinholes the hydraulic cylinder chamber into the oil, the platform deck lifts a stroke. Obviously, the platform deck can be lifted as required with a repeated cycle. Similarly, if the platform floats on the water, the above steps can lift the platform and move the pile legs down.

2.1.5 Stable Positioning of Offshore Floating Drilling Platform (Ship)

In floating conditions under the environmental loads of wind, waves, current, and ice, the platform moves, and all of these factors will affect the operation of the job. To solve this problem, it is necessary to keep it at a predetermined position, so that it can work in stable conditions. Now we will describe the methods to maintain stable positioning.

2.1.5.1 The Effect of Offshore Floating Drilling Platform (Ship) Movement

1. Heaving motion

A platform (ship) rises and falls with the heaving motion of the waves. It will make the drill bit and drill string rise and fall. This could make the pressure of the drill bit unstable at the bottom of the well, and affect the efficiency of drilling. What's worse, it could take the drill bit out of the bottom of the well, which makes drilling function abnormally. At the same time, drill strings being subject to bending can also exacerbate fatigue. So, it is necessary to add a drill string heave compensation device for the platform (ship) and telescopic drill pipe in the drill string; these problems will be explained in detail in 2.8 in this chapter.

2. Rocking motion

Namely, this is the swing of the hull. It will make the drill string bend. The shaking of the traveling block derrick swimming system will intensify, and the drilling fluid returning from the wellhead will shake too. When the swing angle is wide, it may make parties prolapsed from the heart prolapsed from kelly holes, all of which will affect the normal drilling operations. So, a shake reduction tank can be installed on the platform (ship) to reduce the swing by changing the natural frequency of the platform (ship). In addition, the method can also be adapted to reduce hull swing by using the countermovement of the liquid in the water tanks. A separate anti-shake tube having nothing to do with the hull can be installed, as shown in Figure 2-22, which can also be useful to resist the torque of the sea caused by wave instability on the hull, and thereby reduce the hull swing. Of course, if we increase the breadth of the ship, an anti-rolling effect can be obtained.

3. Drifting motion
Namely, this refers to the horizontal and vertical movement of the hull along the horizontal plane. This displacement will change the relative position of the hull to the underwater borehole, which increases the bending of the borehole drill string. What's worse, when sticking occurs, and the drill string cannot even reenter the original wellbore, it will affect normal drilling operations. Therefore, there must be limits to the drift of the drilling platform (ship), and these provisions are demonstrated as follows.

a. Operating state
In normal drilling, the drift should not exceed 5% to 6% of the working depth.

b. Nonoperating state
Namely, this should stop drilling, but if the wellhead and blowout preventer (BOP) are still connected, the drift can be increased to no more than 10% of the working depth.

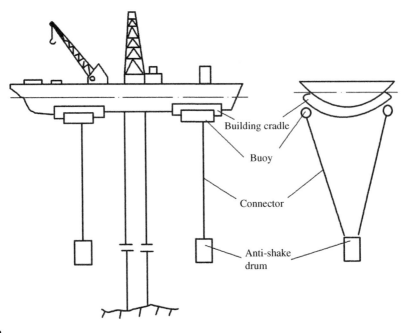

FIGURE 2-22

Reduce the rocking motion of the hull with the anti-shake tube.

c. Relief state

The riser is disengaged from the wellhead and BOP, so the drilling platforms should bow facing the wind and waves.

In order to ensure the drilling platform (ship) drifts within the allowable range, the anchor positioning method and dynamic positioning method are commonly used. They are described separately below.

2.1.5.2 Mooring Positioning of Offshore Drilling Platform (Ship)

Mooring positioning is commonly used as a positioning method. Generally, when at a depth of less than 300 m, and with moderate winds, this method is able to meet the positioning requirements of drilling operations.

1. Composition of the mooring positioning system

    a. Anchor

A floating drilling platform (ship) commonly uses a drogue (momentum anchor), and Figure 2-23 shows the structure and composition of the anchor and the name of each part. Figure 2-24 gives the geometry of the two anchors that are commonly used in a floating drilling platform (ship), and commonly used anchor geometry and technical specifications are listed in Table 2-3.

In order to maintain the positioning of the floating drilling platform (ship), high reliability requirements for the tension of the mooring system are needed; therefore, an anchor is generally chosen that can withstand a load 50% larger than a predetermined maximum positioning load (design load), as is shown in Table 2-3.

The fluke angle on the anchor grip has a great impact. It should be determined according to the seabed conditions. Usually it is chosen to be 50° for a soft seabed and from 30° to 50° for a sandy seabed.

The working vessel's anchoring program is conducted as follows. First of all, the work boat should drag the lanyard (short ropes connected to the anchor crown) to the anchor position, then the drilling platform (ship) releases the anchor cable or chain, as is shown in Figure 2-25 (a). When the work boat meets the design requirements of the correct position relative to the platform (ship), the anchor is slowly dropped to the bottom of the sea, letting the point of anchor contact the seabed, as shown in Figure 2-25(b). Finally, the anchor cable (or chain) in

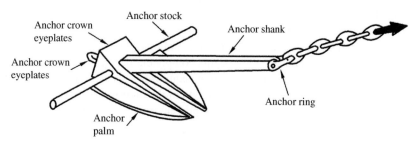

FIGURE 2-23

Structure and composition of anchor.

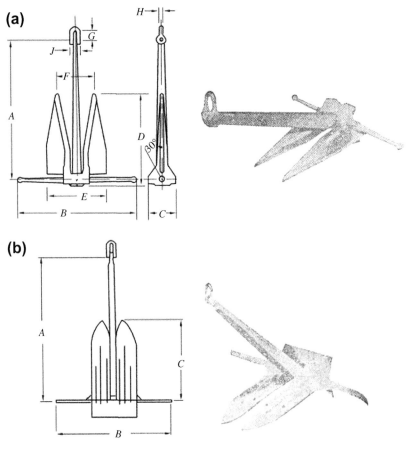

FIGURE 2-24

Two commonly used geometries for the anchor.

the drilling platform (ship) is pulled slightly taut, so that the anchor is caught in the seabed, and the lanyard is relaxed, as is shown in Figure 2-25(c).

b. Anchor cable

There is an anchor chain and wire rope in the mooring floating drilling platform (ship), sometimes in conjunction with a rope and chain, and mooring anchor cable is a general term used for this component. The selection of this component depends on many factors, such as mooring load, water depth, mooring operations equipment, anchor cable storage platform facilities, and so on. The size, strength, and length of the mooring anchor cable are decided by the size and shape of the platform, mooring depth, environmental conditions, and the allowed offset of the platform (ship).

i. Wire rope

At the same breaking strength, wire rope is much lighter than the chain, so gentle catenary wire rope can be formed, which is one of its advantages.

Table 2-3 Commonly Used Geometries for the Anchor

Anchor Weight, lbf	Geometries, in			Anchor Buckle Size, in	Design Load, lbf	Guaranteed Load, lbf
	A	B	C			
200	42	59	26	3/4	6000	9000
3000	129	109	72	1 3/4	60000	90000
6000	144	143	92	2 1/4	120000	180000
9000	160	170	100	2 3/4	180000	270000
12000	186	197	113	3	210000	315000
15000	204	224	126	3 1/2	240000	360000

Floating drilling platform (ship) mooring wire rope has two sizes, (6 × 19 and 6 × 36) as shown in Figure 2-26. The former has 6 stocks, and each has 15 to 26 strands; the latter has 6 stocks too, and each has 27 to 49 strands, all of which includes padding wire. Both use single strand wire rope of core refining plow steel, and the minimum breaking strength is up to 361 t (6 × 19) and 555 t (6 × 36). Mooring wire rope is usually right-hand twisted.

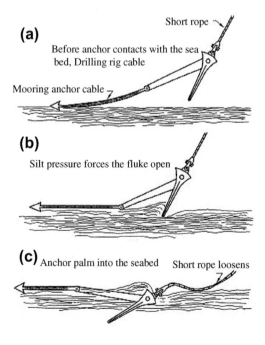

(a) Short rope
Before anchor contacts with the sea bed, Drilling rig cable
Mooring anchor cable

(b)
Silt pressure forces the fluke open

(c) Anchor palm into the seabed   Short rope loosens

FIGURE 2-25

Work vessel anchoring program.

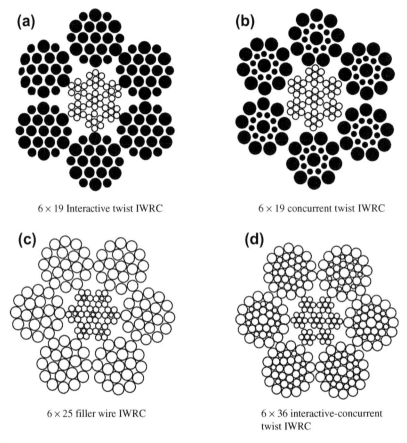

FIGURE 2-26

Two types of mooring rope.

The mooring rope should be galvanized on the surface to improve its corrosion resistance, but the strength will decrease by 10%. In order to eliminate the rebound straightening tendency of steel wire and strand, a "preformed" technique is adopted, that is, each wire and each strand have been preformed before being twisted into wire rope. A zinc rope cap is the main wire rope attachment for connection with the anchors, whose breaking strength is only 95% of the wire rope. The position closer to the connecting portion and the anchor rope is easier to fatigue, and a predetermined section of mooring rope 5 m (15ft) is often to remove re-watering rope cap.

The main failure mode of the mooring rope is fatigue damage; as the mooring load increases, fatigue accelerates. Take of mooring wire rope of diameter 3/4 in (1905mm) and a 6 × 25 single-strand wire rope core as an example. When the load is equivalent to 20% of the breaking strength, fatigue life is 10^6 cycles, while when the load reaches 30% of the breaking strength, the fatigue life is reduced to 1×10^5 cycles.

ii. Anchor chain

The mooring anchor chain of a floating drilling platform (ship) is like the Chinese word which links with the chain, as is shown in Figure 2-27, the flash welding chain and the writing off chain are commonly used. The size of the anchor chain ring is in multiples of the round nominal diameter. The pitch is four times the diameter, the external length is six times of the diameter, the gauge length of the test for the chain ring is six chain pins or 26 times the bar diameter, and the specific diameter of the chain ring should be decided by the mooring load requirements of the breaking strength of the anchor chain. Because the main failure mode of the anchor chain is fatigue damage and fatigue life decreases as the anchor chain load increases, it is generally understood that the anchor chain tension load does not exceed 1/3 of the breaking strength of the anchor chain or anchor chain load.

The end anchor ring (U-shaped anchor ring) is used to connect the anchor chain with the anchor, as is shown in Figure 2-28 (a); when the mooring anchor cable uses wire rope and a chain cable hybrid, the connection is shown in Figure 2-28(b). The wire rope is connected to the anchor chain through the end of the anchor ring, then the end of the wire rope or anchor chain is connected to the anchor.

c. Deck machinery

This refers to the machinery and equipment of the mooring system on the deck. The wire rope and cables are composed as follows.

i. Wire rope mooring

The major equipment needs are a winch and outboard guide, as is shown in Figure 2-29 (a). Generally, a winch drum uses grooved rollers, and the distance S from the drum to the guide pulley [Figure 2-29(b)] can be calculated as follows:

$$S = 14.5W_d \tag{2-1}$$

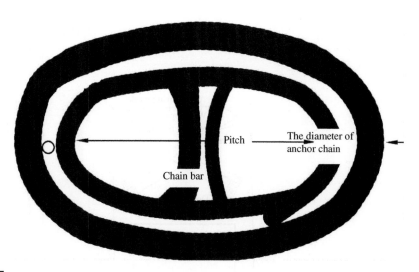

Pitch   The diameter of anchor chain

Chain bar

FIGURE 2-27

Mooring anchor chain link.

FIGURE 2-28

Anchor chain, anchor, and
mooring wire rope connection.

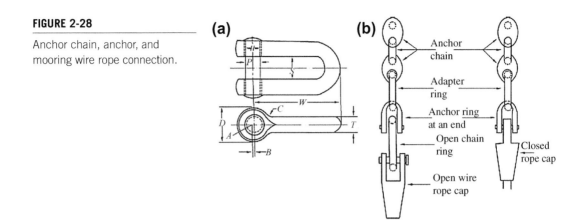

The θ angle, as is shown in Figure 2-29(b), is the rope angle that normally should not be greater than 2°, and the wheel groove radius used as a guide should be no less than 200 times the diameter of the strand's external wire. The wire rope tension meter is identical to the drilling weight indicator.

ii. Anchor chain mooring

There is an anchor chain manipulation device, anchor chain limiter, and guide on the deck as shown in Figure 2-30 (a). Figure 2-30(b) shows the semi-submersible platform arrangement between the anchor chain wheel and guide. Tension measurement is determined by the anchor chain limiter element, but it can only be measured when the anchor chain is stopped.

FIGURE 2-29

The mechanical layout of the wire
rope mooring on the deck.

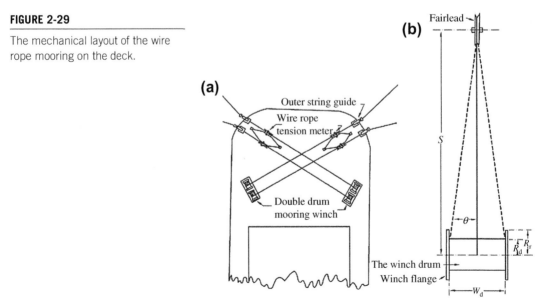

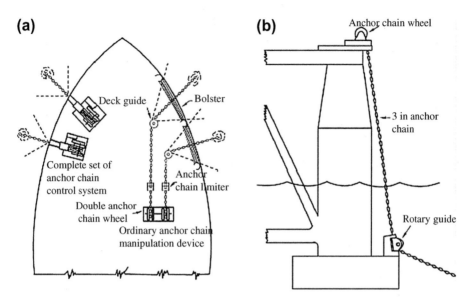

FIGURE 2-30

The machinery layout of the anchor chain mooring on the deck.

2. Requirements for the floating drilling platform (ship) mooring anchor cable
   a. Shortest length of a chain
   The length should be greater than the length of the catenary of the ultimate bearing tension of the anchor chain (1/2 of the anchor chain breaking tensile strength), to ensure the anchor always precedes and is horizontal to a section of anchor chain on the seabed, and ensure that the anchor is subject to a horizontal tensile condition, as is shown in Figure 2-31 (a). The chain length of a floating drilling platform (ship) is usually 8 to 11 times the water depth, with shallow water preferable to a large value. When the water depth is less than 100 m (328 ft), such as in the South China Sea drilling operations, the shortest chain length is often around 823 m (2700 ft).
   b. Initial chain length
   Generally, it is required to be 30.48 ~ 45.72 m (100 ~ 1500 ft) longer than the shortest length of a chain, which can be determined by the nature of the soil. The anchor chain will be back in a certain period grip test. If the initial length is too short, it can't guarantee that the chain length is the shortest.
   c. Anchor chain pretension
   This is set based on a platform displacement of up to 5% of water depth, and the anchor chain tension should be within 1/3 of its breaking tensile strength.
   d. The tension of the anchor chain when the anchor grip is tested
   The maximum anchor chain tension test is generally no more than 1/3 of the strength of the anchor chain breaking force.

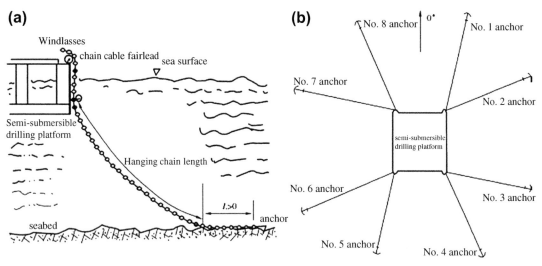

FIGURE 2-31

Mooring system of floating drilling platforms and semi-submersible drilling platform mooring mode.

3. The mooring modes of offshore drilling platforms (ships)
   Different drilling platforms require different types of mooring, but the design criteria is unique, i.e., the type of mooring should ensure the restoring force of the mooring system to have the capacity to resist substantial wind and waves in each direction.
   In addition to the small coring drilling platform using three to four anchor cables, others generally use eight anchor cables.
   a. Floating drilling vessels and semi-submersible platform mooring mode
      A semi-submersible drilling platform uses eight anchor cables. In the seas around China, the mooring type used is shown in Figure 2-31(b). Drilling pontoons can use this mode too, but the angle between the anchors should vary according to the platform or vessel's structure.
   b. Jack-up drilling platform mooring mode
      In order to adjust the position of the platform to reach the accurate place, jack-up drilling platforms in the platform should reach a predetermined well positioning. A drop anchor with three anchors and an anchoring angle of 120° is usually used to reach an accurate position. In the seas around China, a jack-up drilling platform usually uses such an anchoring mode.
4. Stress analysis calculation for anchor cable
   Floating drilling platform mooring positioning must take into account the marine environment loads of waves, wind, and current. If the platform happens to displace horizontally, the force of the mooring system will change, which is specifically explained in the following.
   a. Environmental load and gravity acting on the anchor cable
      i. Gravity
         If the weight of the anchor cable unit length is W, and the unit length of anchor cable buoyancy is B, then the unit length of anchor cable gravity P can be written as

$$P = W - B \tag{2-2}$$

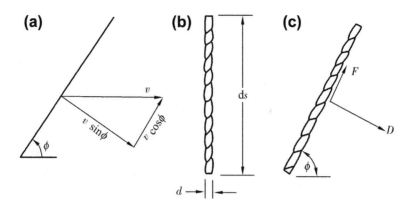

FIGURE 2-32

Analysis of anchor cable withstanding a load.

ii. Resistance (drag)

If the anchor cable is in an upright state, as is shown in Figure 2-32 (b), the diameter is d, the length is ds, and the current velocity is v in accordance to the resistance calculation method described in Chapter 1, the resistance R per unit length can be written as

$$Rds = \frac{1}{2}\rho C_D d v^2 ds \tag{2-3}$$

where ρ is seawater specific gravity and C_D is a drag coefficient, all of which can be selected in accordance to data given in Chapter 1.

In fact, we have an angle between the mooring systems and ocean current direction, as is shown in Figure 2-32(c). The resistance Rds shall be divided into the anchor cable law component of Dds and tangential component Fds. If we set the current velocity of the normal component to $v_N = v \sin \phi$, and that of the cut component to $v_T = v \cos \phi$, Dds and Fds can be expressed as

$$Dds = \frac{1}{2}\rho C_D d v_N{}^2 ds = \frac{1}{2}\rho C_{DN} d v^2 \sin^2 \phi ds \tag{2-4}$$

$$Fds = \frac{1}{2}\rho C_{DT}\pi d v_T^2 \cos^2 \phi \tag{2-5}$$

So by Equations (2-4) and (2-5), we can obtain

$$D = \frac{1}{2}\rho C_{DN} d v^2 \sin^2 \phi = R \sin^2 \phi \tag{2-6}$$

$$F = \frac{1}{2}\rho r C_{DT}(\pi d)v^2 \cos^2 \phi = \pi r R \cos^2 \phi \tag{2-7}$$

In equation (2-7), $rC_{CN} = C_{DT}$; r is the ratio of C_{CT}/C_{DN} and it is generally desired that it is from 0.01 to 0.03. We generally want C_{DN} to be from 0.9 to 1.2. $(\pi d)ds$ is the anchor cable surface area of the length ds.

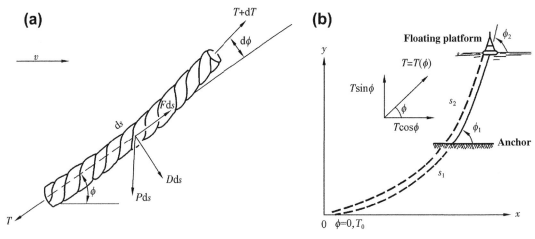

FIGURE 2-33

Anchor chain tension analysis.

b. Cable tension acting on the anchor

As shown in Figure 2-33 (a), if the tension acting on the anchor is T, after the anchor length changes ds, the change of the tension is dT. Then the static equilibrium equation along the normal direction of anchor cable is

$$(T + dT)\sin d\phi - Dds - P\cos \phi ds = 0 \tag{2-8}$$

In Equation (2-8), $\sin d\phi \approx d\phi, dTd\phi \approx 0$, and we can obtain

$$Td\phi = (D + P\cos \phi)ds \tag{2-9}$$

Similarly, considering the anchor cable along the tangential direction, the static force balance equation can be written

$$-T + (T + dT)\cos d\phi - P\sin \phi ds + Fds = 0 \tag{2-10}$$

or

$$dT = (P\sin \phi - F)ds \tag{2-11}$$

Using Equations (2-9) and (2-11), we can write the ratio $\frac{dT}{T}$:

$$\frac{dT}{T} = \frac{\frac{P}{R}\sin \phi - \frac{F}{R}}{\sin^2 \phi + \frac{P}{R}\cos \phi} \tag{2-12}$$

We integrate from point $0(\phi_0)$ to point $0(\phi_0)$ in the formula , and we make $\phi = 0$ at point 0 (the level of the anchor cable at the bottom of the sea). The initial tension is T_0, as shown in Figure 2-33(b), and the anchor cable tension at $\phi = 0$ is T. By Equation (2-12) we obtain

$$\frac{T}{T_0} = \exp \int_{\phi_0}^{\phi} \frac{\frac{P}{R}\sin\phi - \frac{F}{R}}{\sin^2\phi + \frac{P}{R}\cos\phi} d\phi \qquad (2\text{-}13)$$

In this way, if the environmental load R and gravity P have been calculated and the initial anchor tension was given, then the tension at any point of the anchor cable can be calculated by Equation (2-13) [for example, in Figure 2-33(b), $0(\phi_1)$]. Obviously, the tension T varies with different positions.

c. Anchor cable level resilience

Figure 2-33(b) shows the horizontal component $T(\phi)$ of the anchor cable tension T should be

$$T_H = T\cos\phi \qquad (2\text{-}14)$$

The anchor cable horizontal restoring force should be T_H. If we substitute Equation (2-13) we can get

$$T_H = T\cos\phi = T_0\exp\int_{\phi_0}^{\phi} \frac{\frac{P}{R}\sin\phi - \frac{F}{R}}{\sin^2\phi + \frac{P}{R}\cos\phi}\cos\phi\, d\phi \qquad (2\text{-}15)$$

But as we know from Figure 2-33,

$$dx = ds\cos\phi \qquad (2\text{-}16)$$

If we substitute Equation (2-9) into Equation (2-16), we can derive

$$dx = \frac{T d\phi}{(D + P\cos\phi)\cos\phi} \qquad (2\text{-}17)$$

However, dx is the horizontal displacement when the anchor cable is located in corner ϕ. This indicates that the anchor cable tension T, the level of resilience T_H (it should be $T\cos\phi$) and the anchor cable angle ϕ are a function of horizontal offset of the floating drilling platform (ship) dx. As long as dx changes, angle ϕ, tension T, and the level of resilience T_H will change. Therefore, if the horizontal offset of the drilling platform (ship) can be calculated according to the extreme conditions of a maximum storm, taking into account the drilling offset, and in accordance with the calculation method for an anchor cable restoring force vector, we obtain the maximum exposure of each anchor cable tension to breaking strength, according to the allowable design of the anchor cable anchor rope mooring system.

d. Catenary equation

At the boundary conditions of the mooring system, we use a mathematical method to describe the anchor cables sagging under their own weight, which is called the anchor cable catenary equation. It can be expressed by an anchor cable at any point coordinates x and y.

If Equation (2-6), $D = R \sin^2 \phi$, is substituted into Equation (2-9), we can write

$$T d\phi = \left(R \sin^2 \phi + P \cos \phi\right) ds \tag{2-18}$$

Namely,

$$ds = \frac{T d\phi}{R \sin^2 \phi} + P \cos \phi \tag{2-19}$$

Then by substituting Equation (2-13) into Equation (2-19) we get

$$ds = \frac{T_0 \exp \int_{\phi_0}^{\phi} \frac{\frac{P}{R} \sin \phi - \frac{E}{R}}{\sin^2 \phi + \frac{P}{R} \cos \phi}}{R\left(\sin^2 \phi + \frac{P}{R} \cos \phi\right)} \tag{2-20}$$

Using the quadrature formula with Equation (2-20), we have

$$\frac{Rs}{T_0} = \int_{\phi_0}^{\phi} \frac{\exp \int_{\phi_0}^{\phi} \frac{\frac{P}{R} \sin \phi - \frac{E}{R}}{\sin^2 \phi + \frac{P}{R} \cos \phi}}{\sin^2 \phi + \frac{P}{R} \cos \phi} \tag{2-21}$$

If we are provided the anchor cable arbitrary point (x, y) at the longitudinal abscissa x and y respectively, and the relationship of its slight changes dx and dy, then the anchor cable arc length ds is

$$dx = ds \cos \phi \tag{2-22}$$

$$dy = ds \sin \phi \tag{2-23}$$

Thus, s in Equation (2-21) is replaced with x and y:

$$\frac{Rs}{T_0} = \int_{\phi_0}^{\phi} \frac{\exp \int_{\phi_0}^{\phi} \frac{\frac{P}{R} \sin \phi - \frac{E}{R}}{\sin^2 \phi + \frac{P}{R} \cos \phi}}{\sin^2 \phi + \frac{P}{R} \cos \phi} \cos \phi d\phi \tag{2-24}$$

$$\frac{Ry}{T_0} = \int_{\phi_0}^{\phi} \frac{\exp \int_{\phi_0}^{\phi} \frac{\frac{P}{R} \sin \phi - \frac{E}{R}}{\sin^2 \phi + \frac{P}{R} \cos \phi}}{\sin^2 \phi + \frac{P}{R} \cos \phi} \sin \phi d\phi \tag{2-25}$$

x, y in Equation (2-24) and (2-25), respectively represents the abscissa and ordinate of any point (x, y) between two anchor cable points, $0(\phi_0)$ and $0(\phi)$, and the application coordinates x, y can depict the shape of the curve of the anchor cable sagging under its own weight. Obviously, this should be solved with a special program using a computer.

2.1.5.3 Dynamic Positioning of Offshore Floating Drilling Platforms (Ships)

Another positioning method for an offshore floating drilling platform (ship) is called dynamic positioning, which is a new technology that doesn't drop an anchor to keep the hull in position. Thrusters and a propulsion propeller are used to maintain position, so it is called dynamic positioning.

Dynamic positioning is achieved by acoustic or other advanced measurement techniques to measure vessel position, then a computer is used to calculate the original vessel's position to make a necessary adjustment in each direction with the required force and torque value. A signal is sent to the thrusters and also the propeller, which resist the forces and moments from the load of the marine environment, and adjusts their positions in real time, to maintain the position of the hull.

Dynamic positioning technology for floating drilling platform began in the late 1970s with the "Sedco 445" and "Discoverer 534" semi-submersible and floating drilling platforms, whose working depth reached 800 to 1100 m. Through the years, propeller power is continuously improving, and floating drilling platforms (ships) have been able to maintain their positions under adverse sea conditions in ultra-deep waters, with positioning accuracy up to 1% (hull horizontal displacement divided by water depth). Therefore, especially in the conditions of deep water, dynamic positioning technology will be widely used. A dynamic positioning system usually consists of three components: the measurement system, control system, and implementing agents.

1. Dynamic positioning measurement system

An acoustic measurement system is usually used to measure the relative position of the platform (ship) to the center of a turntable with a subsea wellbore, and this measurement system is composed as follows.

a. Hydrophone

This receives the measurement of the acoustic pulse signal. It is arranged according to a certain geometry layout called an array. In accordance with the triangles, the three vertices of the triangle install a hydrophone centroid of a triangle in kelly making up the core. It can also be arranged in a rectangular installation hydrophone in the four vertices of the rectangle, and the array is shown in Figure 2-34. The hydrophone array is arranged on the bottom of the hull, and the hull size limits the size of the group array, which is called a short baseline acoustic measurement system.

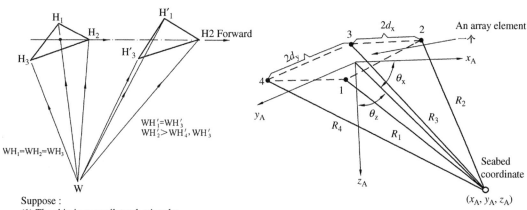

Suppose :
(1) The ship is an equilateral triangle;
(2) Reference point (kelly bushing) is at the geometric center of the triangle
(3) Sound pulse generator is at the center of the well
(4) No pitch, roll, and heave

FIGURE 2-34

Hydrophone array of acoustic measurement system.

b. Sonar (acoustic generator)

This is a device creating acoustic signals in accordance with the specified time interval. For a short baseline group array, sonar is located on the bottom of the sea, and is generally placed at the center of the seafloor borehole; if a long baseline group array is adopted, sonar should be located on the bottom of the hull in a quasi-roundabout, and the hydrophone array is arranged at the bottom of the sea.

With displacement of the platform (ship), the time sound waves take to travel from the sonar to each hydrophone will be different. For example, in Figure 2-34(a), when the hull is not moving, the single pulse sound waves arrive at the three hydrophones simultaneously. When the hull is forward and offset, single acoustic pulse arrives simultaneously at hydrophone H1 and H3, H2 hysteresis reach hydrophone; as hull offset moves forward and backward, and hydrophone H2 and H3 are further from the sonar than H1, the single pulse sound will lag reaching hydrophones H2 and H3. Thus, according to the different times of the single pulse reaching each hydrophone, the distance between each hydrophone, the sonic speed, and the depth of the water, the hull offset position can be calculated.

Take the hydrophone array in Figure 2-34(b) as an example; if the distance of the sonar to hydrophones 1, 2, 3, and 4 respectively are R1, R2, R3, R4, we have

$$\left.\begin{array}{l} R_1^2 = (x_A - d_x)^2 + (y_A - d_y)^2 + z_A^2 \\ R_2^2 = (x_A - d_x)^2 + (y_A - d_y)^2 + z_A^2 \\ R_3^2 = (x_A - d_x)^2 + (y_A - d_y)^2 + z_A^2 \\ R_4^2 = (x_A - d_x)^2 + (y_A - d_y)^2 + z_A^2 \end{array}\right\} \tag{2-26}$$

where x_A, y_A, z_A are submarine sonar location coordinates; and d_x, d_y are the longitudinal direction and the transverse distance between hydrophones.

The propagation velocity of sound in seawater is U_L, and generally $U_L = 1500$ m/s; from this the time of the sonar sound pulse transmission to each hydrophone t should be

$$t_1 = \frac{R_1}{U_L}, t_2 = \frac{R_2}{U_L}, t_3 = \frac{R_3}{U_L}, t_4 = \frac{R_4}{U_L} \tag{2-27}$$

Namely,

$$t_n = \frac{R_n}{U_L} (n = 1, 2, 3, 4) \tag{2-28}$$

Thus, when the hull is offset, following the above-described method, the time of the acoustic pulse arriving at each hydrophone is t_n, and the sonar distance from each hydrophone R_n is calculated, the offset of the hull position can be obtained. Then when compared with hulls in situ, the amount of displacement can be obtained.

The above calculation is based on the borehole at the center located on the seabed sonar, but the actual sonar is not exactly right in the wellbore center; it is offset. For this, you need geodetic coordinates (North/South, East/West) to indicate the positioning reference system. In addition, the inclination also needs to be corrected where the plane of the pitch and roll is caused by the hydrophone pendulum or gyroscope with a vertical reference device measured.

In order to solve the case of an interruption in the acoustic signal, very often an alternate positioning reference system is needed, and the most simple and commonly used method is the rope tensioning system. This system draws a wire from a floating drilling platform (ship) to the seabed, tenses using a weight, and determines the offset angle through the rope inclination. However, the deflection of the wire rope caused by ocean currents will affect the accuracy of measurement; therefore, it should be regularly calibrated with an acoustic measurement system.

2. Dynamic positioning control system

   A dynamic positioning control system is mainly composed of computer input data which are produced by a variety of sensors from which it is given the force required to correct the floating drilling platform (ship) position and torque.

   a. Flow diagrams

   Figure 2-35 shows the dynamic positioning control system flow chart; a brief description follows:

   i. Computer gets floating drilling platform (ship) position data from the tracking ship systems, and it gets the hull direction from the gyro compass.

   ii. Computer compares the hull (platform) position with coordinates (x_0, y_0, θ_0), which is the orientation of the platform and the original location of the hull directly above the borehole, to maintain a normal direction to the location.

   iii. Calculation of the force and torque to push the hull back to the wellbore just above the original position, using data including the wind, the position, and the orientation of the hull.

   iv. The controller issues a directive to the propeller. First the right thrusters are chosen, with generally the center of the most distant propeller used to generate torque, while the other boosters produce thrust. To determine the lateral and forward propeller we need the desired

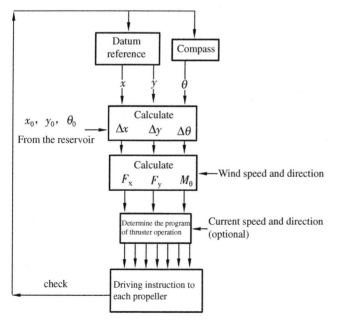

FIGURE 2-35

Dynamic positioning control system procedures.

net thrust and power available. Then the propeller thrust selected is provided to calculate the thrust and torque. Finally, we check the response from the use of the propeller.

b. Two methods to calculate the thrust and moment of thrust

i. Based on feed-forward data

This is commonly used to calculate the required thrust to resist gusts. Feed-forward means the data before the construction of the platform (ship); it is obtained by a wind tunnel test applied to the hull that generates force and moment data. Thus, we can store the role of the different wind speeds obtained from the test, the effect of different angular directions of wind on the hull, and torque data in the computer. When using dynamic positioning, as long as the wind speed and direction in measured in accordance with the vessel (platform), the thrust and moment of thrust can be directly gotten from the database.

ii. Based on the feedback data

This is a method based on using the change of the position and orientation of the measuring system on the floating drilling platform (ship), in accordance with the response of the hull to the marine environment loads, to calculate the thrust and moment of thrust to maintain the required position of the hull. Taking the hull wave forces and other forces for an example, we'll describe the details of the calculation. Assuming the hull is offset on the x-axis direction by the wave force, this offset by can be expressed by Δx. If the original and the present position of the hull in the x-axis coordinates are x and x_0, then

$$\Delta x = x - x_0 \tag{2-29}$$

Consider the thrust and moment of thrust as F_x, F_y, and M_θ. The thrust along the x-axis direction F_x for example is calculated as follows:

$$F_x = A_x \Delta x + B_x \frac{d\Delta x}{dt} + C_x \int \Delta x dt \tag{2-30}$$

In Equation (2-30), the coefficients A, B, C are dynamic response coefficients, which change with the platform (ship) quality and the depth of water, and are obtained by the wave force dynamic response test.

The first term on the right-hand side of Equation (2-30) $A_x \Delta x$ is a proportional term, when the offset Δx increases in direct proportion, the thrust needed to reset the hull increases. It is similar to the spring restoring force, where the coefficient A_x is similar to the stiffness of the spring coefficient.

The second term on the right-hand side of Equation (2-30) $B_x \frac{d\Delta x}{dt}$ is a differential term, which represents that the hull speed deviated from the wellhead location $\frac{d\Delta x}{dt}$ and a corresponding increase in the thrust force is required to reset the hull. It is similar to the damping force, proportional to speed, where the proportional coefficient B_x is similar to the damping coefficient.

The third term on the right-hand side of Equation (2-30) $C_x \int \Delta x dt$ is an integral term, which represents hull stability in the steady-state force around the wellhead position drifting back and forth to overcome the hull. For example, when the floating drilling platform (ship) is in the initially place, it will have such reciprocating drift. After a certain period of time t, it can make the hull in stable position at the wellhead through the third thrust.

3. The executive body of the dynamic positioning system
The dynamic positioning executive body is mainly composed of the thrust and moment of thrust generated by propeller and drive unit, and the propeller is the core. There are two commonly used thrusters for dynamic positioning: one is a fixed section moment, variable-turning DC motor drive pusher; the other is a constant speed, variable-pitch pusher, driven by an AC motor. The former has been rarely applied due to the limited ability to push for rapid reactions; the latter adopts hydraulic control to change the section moment, which can be carried out more quickly, and has up to 3,000 shaft horsepower, so it's commonly used in straight-wing systems. We will introduce it in the following chapter.

a. The work principles of a straight-wing propeller
A straight-wing propeller is a vertical impeller propeller with variable pitch. As shown in Figure 2-36, provided the forward speed is v_0, and it is equipped with a straight wing disc around the vertical in the forward direction of the center axis of rotation, the angular velocity is ω_R. At any point M on the straight wing disc, the instantaneous rate of synthesis speed v_0 is calculated with the circumferential velocity ω_R vector and the tangent to the cycloid, as shown in Figure 2-36(b). If the v_0 direction is taken as the x-coordinate direction, perpendicular to the x-axis and lateral to the disc coordinate y, the movement starts at $t = 0$, and the center of rotation at the origin position is O, as shown in Figure 2-36(a), the cycloid equation of said disc track at any point M is

$$\left.\begin{array}{l} x = v_0 t + r\cos(\omega t + \phi_0) \\ y = r\sin(\omega t + \phi_0) \end{array}\right\} \tag{2-31}$$

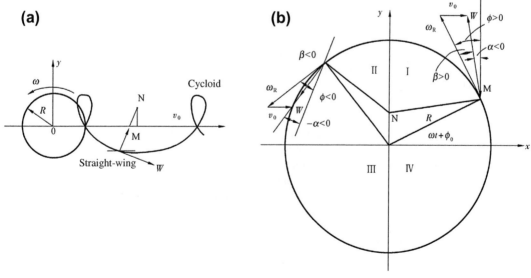

FIGURE 2-36

The working principle of the straight-wing propeller.

The magnitude and direction of aggregate velocity W the express pusher thrust given size, direction, and directly related to the size of W and the α angle. The α angle is called the angle of attack, which is the angle between the chord line of the straight wing and the disc aggregate speed W, and is positive when the rotation of the disc is counterclockwise, and negative with the disc rotates clockwise, as shown in Figure 2-36(b), $a = \phi - \beta$. ϕ is the angle between the straight chord line and the rotational speed ω_R, and β is the angle between the aggregate speed W and the rotational speed ω_R.

α reflects the speed of W, but it is not the same as shown in Figure 2-36(b) in the II, III, and IV quadrants. For example, in I and IV quadrant, $\alpha > 0$, in II and III quadrant, $-\alpha < 0$, which means the following: the straight wing rotation, if it can't around its own point, then the thrust in Section I and iv quadrant is positive ($\alpha > 0$) to offset most negative thrust ($-\alpha < 0$) produced in II, III quadrant, so that the thrust will be very small. Therefore, a steering mechanism is required in the propulsion system to make a straight-wing swing itself while rotating, so that different positions may be straight wing to the state through the ω_R into an angle ϕ, so the angle α always is in a good position, greatly reducing the scope of the negative thrust.

The principles of this control mechanism are shown in Figure 2-37.

i. The center of actuating mechanism N is located at O of the pivot center [Figure 2-37 (a)]: In this case, the blade is tangent to the state of operation, and the resultant speed W is coincident with a straight-wing chord line, as shown in Figure 2-36(b) for attack angle $\alpha = 0$, that is to say, there is no thrust.

ii. The center of actuating mechanism N moves left [Figure 2-37 (b)]: In this case, the blade is controlled; the blade chord line perpendicular intersects the center of the steering mechanism at the N point, as shown in Figure 2-37(b). Each of the blades does an oscillating movement around its own axis, the direction of the leading edge of the blade is first outwardly semicircular, thus the water flows into the passage of the blade, but when the direction of the guide section in the rear is inwardly semicircular, the flow then leaves the blade path in the same direction, resulting in a spray of water column back down, as shown in Figure 2-37(b) by the arrows. The back spray water column reaction force is S, which pushes the boat forward. The direction of the thrust S has a right angle with the ON line, which is proportional to the distance of ON; the longer the ON line, the greater the thrust S is.

iii. The center of actuating mechanism N point is moving left while moving forward [Figure 2-37 (c)]: The pusher jet of water and thrust S remains vertical to the ON line, but the thrust S with the longitudinal axis of the hull is inclined at a certain angle, thus, the thrust S will be generated by the longitudinal axis of the hull and the transverse axis direction of the two components, so the hull turns left.

iv. The center of the actuating mechanism N point moves forward as shown in Figure 2-37(d): In this case, as described above, the thrust generated by S is transverse to the hull, which will promote the aft lateral movement.

v. The center of actuating mechanism N point moves to right [Figure 2-37(e)]: At this point, the opposite of the case of (b) occurs as the thrust force produced by the pusher is backwards, and the hull moves backwards.

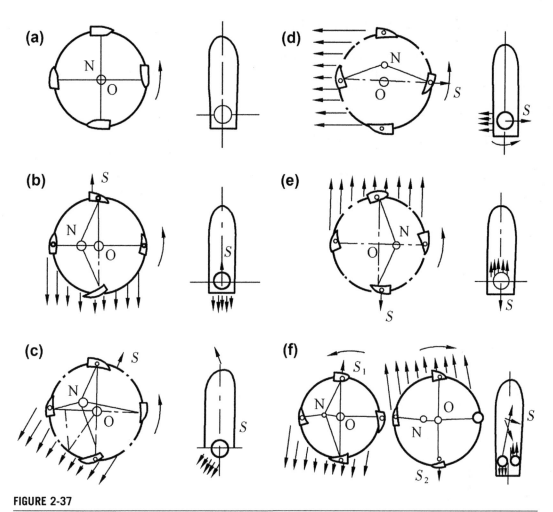

FIGURE 2-37

Principles of the operating mechanism.

  vi. The rear of the hull is equipped with two straight wing pushers [Figure 2-37(f)]: One
      propulsion generates the forward oblique thrust S_1, and the other generates the backwards
      oblique thrust S_2, and the two thrusts are in opposite directions, and are on two different
      sides; therefore, the thrust S_1 and S_2 constitute a couple, so that the hull rotates. If the head
      and the end of the hull are equipped with one pusher, it not only makes the hull rotates, but
      also allows the hull to move vertically, horizontally, or diagonally, so that the hull can
      move in all directions and recover from rotations, which is the working principle of the
      pusher.

b. The workings of a straight-wing propeller
   i. Thrust

The straight-wing propeller variable pitch of the propeller thrust T can be given with the diameter D of the propeller and the speed n. If the quality of sea water density is ρ, then it can be calculated as follows:

$$T = \rho n^2 D^4 K_T \tag{2-32}$$

The K_T of Equation (2-32) is the thrust coefficient, which can be checked in Figure 2-38 according to the pitch of the propeller P to propeller diameter ratio $\frac{P}{D}$ (pitch P is measured at 0.7R, where R is the radius of the propeller).

   ii. Push torque

The push torque Q of variable-pitch straight-wing propeller thrust can be calculated as

$$Q = \rho n^2 D^5 K_Q \tag{2-33}$$

The meaning of ρ, n, D, in Equation (2-33) is the same as Equation (2-32); K_Q is the push moment coefficient, which may be checked by the curve in Figure 2-38, and the pitch P, as shown in Figure 2-39, can be calculated by the following equation:

$$P = 2\pi r \tan \phi \tag{2-34}$$

where R is the radius of the measuring pitch and ϕ is the pitch angle.

   iii. Propeller efficiency

The efficiency η_a of the propeller can be produced by the coefficients K_T and K_Q of thrust T and moment of thrust Q and the propeller seawater relative linear velocity v_A with the following formula:

$$\eta_a = \frac{v_A K_T}{n D K_Q} \tag{2-35}$$

FIGURE 2-38

Calculate the propeller thrust and moment of thrust with the K_T and K_Q coefficients.

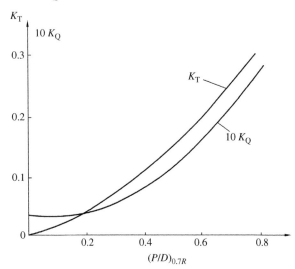

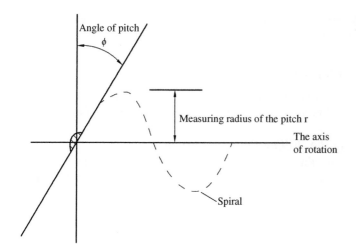

FIGURE 2-39

Propeller pitch and pitch angle.

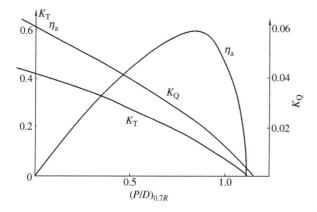

FIGURE 2-40

Propeller efficiency curve.

It can also be checked directly by the curve in Figure 2-40.

From the above analysis, according to the marine environmental load limit conditions and the allowed offset in drilling requirements, the required thrust T and push torque Q can be obtained, according to Equations (2-32) and (2-33), in accordance with the number of revolutions of the motor n, which can decide the diameter D of the propeller.

2.2 Offshore Oil and Gas Drilling Wellhead and Wellhead Devices

The wellhead which is the top opening of the drilling borehole and the casing head that is installed at the wellhead, they are collectively referred to casing head and the devices installed above the wellhead which are called wellhead device. We will introduce surface wellheads, underwater wellheads, subsea wellheads, and wellhead devices according to different locations of the wellhead in this section. This section will highlight and give a detailed introduction to subsea wellheads and wellhead devices because subsea wellheads and wellhead devices are specific to offshore oil and gas drilling.

2.2.1 Surface Wellheads and Wellhead Devices

The surface wellhead is often used in the drilling operations of offshore fixed platforms and jack-up type platforms. Sometimes the surface wellhead is in the semi-submersible drilling platform.

2.2.1.1 The Wellhead Device during the Drilling Operation Process

It is required to install wellhead devices in a timely fashion after well cementation In order to ensure the safety of the drilling and to fix the casing, we should install a BOP stack, drilling fluid discharge pipeline and so on during the drilling operation process.

1. Before the first spud in [Figure 2-41 (a)]: Generally, riser 1 is directly connected to the platform and the opening is connected to the drilling fluid discharge pipeline 12 to deflect drilling fluid.
2. Before the second spud in [Figure 2-41(b)]: Now, we need to shed the riser section on the platform and install surface casing head 6 to seal the annular gap between the two casings and let it withstand the weight of inner casing 2 after well cementation with the surface casing. Install the BOP group on casing head 6. The BOP group includes a cross joint (9); two ram preventers (10),

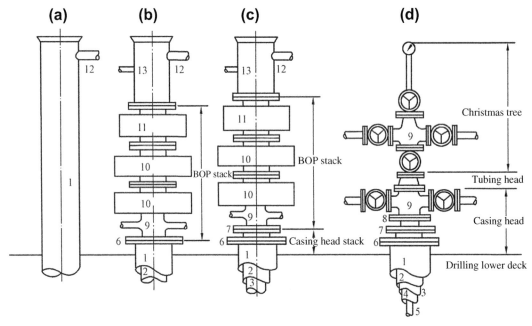

FIGURE 2-41

Surface wellhead and wellhead device: (a) before the first spud in; (b) before the second spud in; (c) before the third spud in; (d) well completion. 1 - Riser; 2 - Surface casing; 3 - Intermediate casing; 4 - Oil casing; 5 - Tubing; 6 - Surface casing head; 7 - Intermediate casing head; 8 - Oil layer casing head; 9 – Cross joint; 10 - Ram preventer; 11 - Annular preventer; 12 - Drilling fluid discharge pipe line; 13 – Short piece.

an annular blowout preventer (11) and a short piece (13). Take the openings in the deflector short piece 13 and then connect drilling fluid discharge pipeline 12 to deflect drilling fluid.

3. Before the third spud in [Figure 2-41(c)]: Firstly, we should unload the links of surface casing head 6 and the BOP group. Then, install the intermediate casing head 7 on the surfing casing head 6. Lastly, install the original BOP group and deflector short piece 13 to deflect drilling fluid after well cementation with the intermediate casing.

2.2.1.2 The Wellhead Device after Well Completion [Figure 2-41(d)]

We need to push down the well tubing 5, then knock down the BOP group and install oil string head 8 and cross joint 9 and the tubing head and Christmas tree after cementing with the oil string 4 during the well completion operation. The tubing head is used to hang the tubing and close the annulus between the tubing and the casing. It is directly mounted on the oil string head and then installs the Christmas tree on the tubing head. Figure 2-41(d) is the surface wellhead and the wellhead device used in the South China Sea.

2.2.1.3 The Wellhead Safety Device (BOP)

The BOP group is an important safety device for offshore drilling. It is used to prevent blowouts. The size of the water surface wellhead BOP group directly decides the height from the rig floor to the upper deck of the platform. If any blowout signs or blowout happens we should turn off the BOP, which will close the well annular space, as shown in Figure 2-42; if necessary, it can also cut off the drill pipe to close all wellbores drastically during drilling. The BOP group generally contains three to five BOP. Generally, each BOP has an entrance through the fill up line and an export through the choke line.

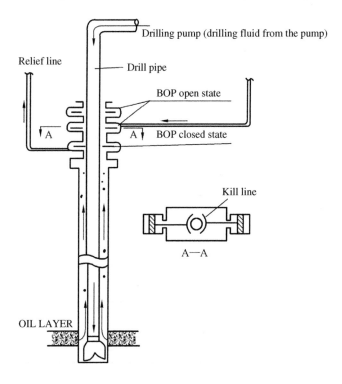

FIGURE 2-42

The action principles of the BOP during drilling.

a. Shear blind preventer

Generally, it is a ram type as the shown in Figure 2-43 (a). It can cling to the drill pipe via a hydraulic ram to close the annular space between the drill pipe and casing. During an emergency situation the drill pipe can cut off the drill pipe to close all wellbores drastically. The double ram type BOP is shown in Figure 2-43(b) below.

b. Two-part blind preventer

The core of this BOP is divided into two halves, which can be used to cling to the drill pipe and close the annulus.

c. Universal BOP

Its annular core has both full closure of the drill pipes of various sizes and variety functions of Kelly bar, as shown in Figure 2-43(c). The core shape of the two-partsblind preventer and universal BOP is

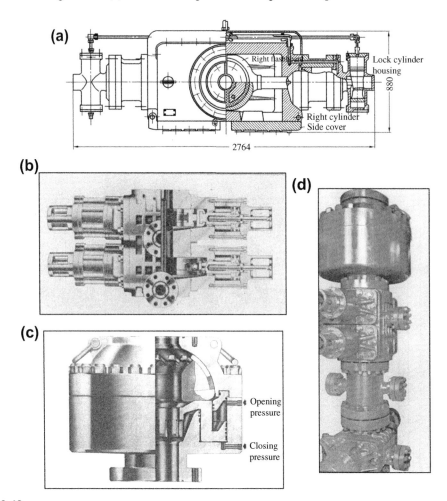

FIGURE 2-43

Several forms of the BOP and the BOP group.

the same, but it cannot be arbitrarily vitiated through drill pipes with different sizes and shapes. Figure 2-43(d) gives the full picture of the entire BOP group.

2.2.1.4 The Wellhead Hook-up Device (Casing Hanger, Tubing Hanger)

1. Casing head and casing hanger

   The casing head is connected to the upper end of the casing string, which consists of acasing hanger and its cone seat. As shown in Figure 2-44 (a), the upper end of the casing hanger connects to an upper flange; the lower end connects to a cross joint, and the lower part of the cross joint is welded by a lower flange. Therefore, there are upper and lower flanges and two annular exports, which constitute a casing head short piece.

   The wells of an offshore oil field generally have a multilayer casing and an annular space, so there are multiple casing heads. The lowest casing head is mounted at the top of the riser; its upper flange is respectively connected to the lower flange of the intermediate casing head; the lower flange is a thread or welded sleeve. The upper and the lower flange of the intermediate casing head is connected to the upper and lower casing head, respectively. The upper and the lower flange of the uppermost casing head are respectively connected to the lower flange of the tubing head and the upper flange of the bellowing sleeve. A casing head of an oilfield in the South China Sea is shown in Figure 2-44(b): a 339.7 mm (13 3/8 in) casing head with 508 mm (20 in) casing head synthesis of a whole; the lowest part of the casing head with a size ranging from 179 .4 ~ 539.8 mm (7 1/16 ~ 21 1/4 in) used to support a 114.3 ~ 406.4 mm (4 1/2~ 16in) casing; the

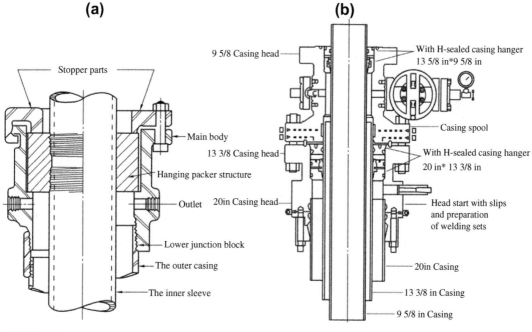

FIGURE 2-44

Casing head and the casing hanger.

intermediate casing head; the lower flange size ranging from 179.4 ~ 539.8 mm (7 1/16 ~ 21 1/4 in), which supports a 114.3 ~ 406.4 mm (4 1/2 ~ 13 3/8 in) casing.

The casing hanger is placed in the cone seat of the lowest part of casing head or intermediate casing head to hang the following smaller casing string and make a seal for the hanging casing and casing head cone seat. It should be able to withstand the weight of the casing hanging. Therefore, the tensile load, the weight load from the pressure in the well casing, and radial load of the conical shoulder are the main loads of the hanger.

2. Tubing head and tubing hanger

   a. Tubing head

   As shown in Figure 2-45 (a), it is installed on the end of the uppermost casing head and consists of tubing hanger and its cone seat. Its upper end usually links to an upper flange and the lower connects to a cross joint. What's more, the lower part of the cross joint is welded to a lower flange. In this way, it has upper and lower flanges and two annulus exports, which constitute a short section of tubing head. Some lower parts of the tubing heads are not flanged, but use a threaded connector, so it can be directly connected to the upper end of the production casing. The upper generally uses a flange and lock screw for clamping the tubing hanger. An offshore oilfield generally uses an upper and lower tubing head with a flange, as shown in Figure 2-45(b) for an oil field in the South China Sea with the tubing head.

   b. Tubing hanger

   It is sited on the cone seat of a tubing head to hang the tubing string and to provide a seal for the cone seat and hanging tubing. Because offshore oil and gas wells generally have subsurface safety valves, the tubing hanger must have an aisle for a hydraulic control line; in addition, in order to use ESP to recover oil, it should have a an aisle for a subsurface oil release valve and a seal for the electric cable.

   Figure 2-46 (a) is a diameter of 4 7/8 outer tubing hanger of FBB; Figure 2-46 (b) is a tubing hanger with a ESP cable aisle.

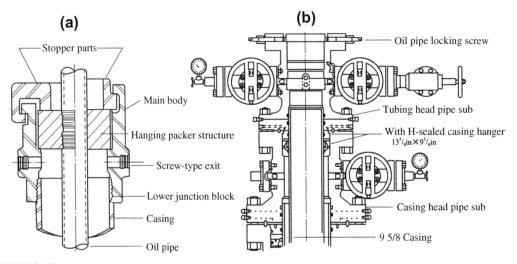

(a)

Stopper parts
Main body
Hanging packer structure
Screw-type exit
Lower junction block
Casing
Oil pipe

(b)

Oil pipe locking screw
Tubing head pipe sub
With H-sealed casing hanger 13 5/8 in × 9 5/8 in
Casing head pipe sub
9 5/8 Casing

FIGURE 2-45

Tubing head structure.

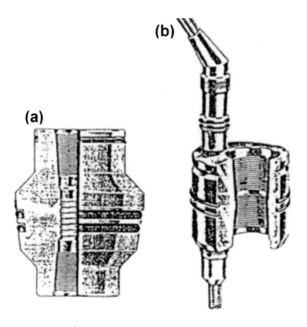

(b)

(a)

FIGURE 2-46

Types of tubing hanger.

2.2.2 Underwater Wellheads and Wellhead Devices

The underwater wellhead is used to sink a wellhead that is on a floating body into the water at a certain depth; the wellhead device is installed on a floating body that withstands the weight of the wellhead device. The floating body refers to the ABS (Atlantis Buoyant System), which is a deep water artificial buoyancy seabed, developed by the Norwegian Atlantis company. This underwater wellhead and wellhead device is only a preliminary success: Norway Atlantis conducted a test in March 2003 in the North Sea for full-size ABS sea towage, but it is still not formally put into production yet. CNOOC Oilfield Services Limited reached an agreement with Atlantis in 2005 that jointly set up a company in the South China Sea and tried to use the ABS to test of deep-sea offshore drilling. The biggest advantage of the wellhead and wellhead device is that it can be used in deep-sea drilling, generally at a working depth up to 400 ~ 4000 m, and the floating body of the ABS on which the water wellhead and the wellhead device sits can be submerged in water depth of 200 to 600 m.

2.2.2.1 The Overall Composition of the Atlantis Buoyant System (ABS)

1. Between the ABS and floating drilling platform
   Standard riser connections are used for the mooring or dynamic positioning floating drilling platform and also standard gas-liquid pipelines are used to control the system, as shown in Figure 2-47.
   b. On the ABS
   The main installation on the ABS is a standard blowout preventer (BOP) group and its control system, the connector that connects to the above riser, and the tension connector (telescopic joint) that connects to the lower casing string. The ABS is a cylindrical floating body with an

FIGURE 2-47

Underwater wellhead and wellhead device.

approximate diameter of 15 to 20 m and an approximate height of 6 ∼ 10 m. Flexible cable is used to connect the floating body and the surface floating drilling platform so that after the well completion operation and in cases of emergency the surface platform can be evacuated in time. The weight of the entire wellhead device is taken by the floating body, which does not increase the burden of the water surface platform.

c. Between the ABS and seabed

A conventional sleeve through a back joint links to the conventional casing after the well completion operation, as shown in Figure 2-47.

2.2.2.2 The Installation and Production of the Atlantis Buoyant System (ABS)

As shown in Figure 2-48, the installation and production steps of ABS are as follows:

Step 1 [Figure 2-48(a)]: Conduct the first drilling at the surface drilling platform, drill out the borehole and put down surface casing, and do the well cementation.

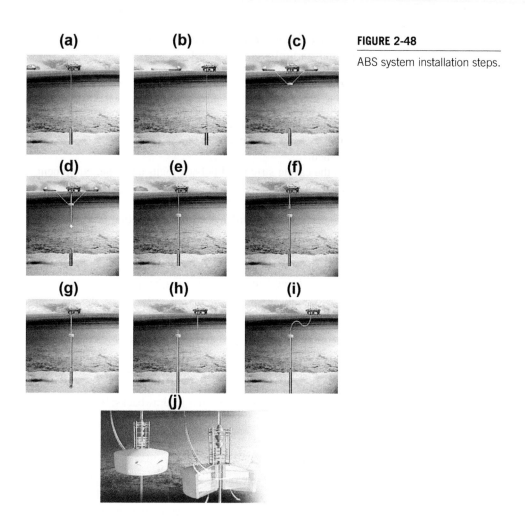

FIGURE 2-48

ABS system installation steps.

Step 2 [Figure 2-48(b)]: Drag the floating body to the well location.

Step3 [Figure 2-48(c)]: Charge the floating body water ballast and sink it into the water until it reaches the exact location, the given depth on the well.

Step 4 [Figure 2-48(d)]: Move the tieback casing from the floating body central hole, and then gradually land the casing at the subsea.

Step 5 [Figure 2-48(e)]: Connect the landing tieback casing and cementation surface casing through the tieback joints at the subsea. Then, inject the gas to the floating body to drain the water.

Step 6 [Figure 2-48(f)]: Land the BOP group and riser from the platform on the floating body, and connect it to the wellhead on the floating body.

Step 7 [Figure 2-48(g)]: Start the second drilling process and continue.

Step 8 [Figure 2-48(h)]: Shut down the BOP group on the floating body depending on security needs; the floating body departs from the offshore platform, but it is easy to return to the well location.

Step 9 [Figure 2-48(i)]: Take an oil test and conduct early production, transform the original drilling platform into an oil platform, and start oil production early.

2.2.2.3 The Advantages of an Underwater Wellhead

1. Applies to deep water and different water depths
   Due to the floating body (ABS) it can be used in deep sea, and it can also adapt to different water depths. The working depth currently ranges from 400 up to 4000 m.
2. Full use of existing conventional equipment
   Like the riser system, BOP group, etc., the casing string and tieback joints and other equipment can apply to all existing conventional equipment.
3. Can still use the operating procedures of shallow water
   Drilling and well completion operations can stay in accordance with the operating procedures of the shallow water.
4. Reduce the risk of deep-water drilling operations
   The risk of well control in drilling operations, riser relief emergencies and drilling fluid discharge into the sea after emergency relief can be reduced because the floating body can be easily integrated with the platform at the surface of the water.
5. Reduction of the operating costs of deep-water drilling

   The floating body withstands the weight of the entire wellhead to reduce the platform itself negative tam and save the transformation of the platform on the surface water that is required for investment. This means that one does not need to design or build new platform, as the existing second- or third-generation platform will be able to handle the deep-sea drilling.
       Of course, the underwater wellhead and the ABS system also has its drawbacks; it is sensitive to factors in the marine environment such as wind, waves, current, and other factors.

2.2.3 Subsea Wellheads and Wellhead Devices

A subsea wellhead locates the wellhead and wellhead device on the seabed, which bears the weight of the wellhead. This category is generally divided into the subsea wellhead of a floating drilling platform (ship) and the subsea wellhead of the sjack-up platform: the former often refers to a subsea wellhead system and the latter often refers to a mud line casing suspension system.

2.2.3.1 The Subsea Wellhead of a Floating Drilling Platform (ship)

The subsea wellhead of a floating drilling platform (ship) is used in deep water and the wellhead device is installed on the subsea wellhead with riser connections to the platform. We can use the guide rope on the base of the permanent guide frame to guide the wellhead device to the wellhead before the installation of the riser. By using this subsea wellhead we can conduct well completion operations in deep water and achieve the installation of a Christmas tree in the subsea and we can also use tieback equipment and tools to perform completion operations on the surface production platform.

1. Structure of the wellhead
   As shown in Figure 2-49, the floating drilling platform (ship) in deep water is used in the standard subsea wellhead. A high press casing (wellhead head) of 508.0 mm (20 in) goes into the 762 mm (30 in) conductor head and locks them with a locking block. Similarly, the 476.3 mm (18 3/4 in)

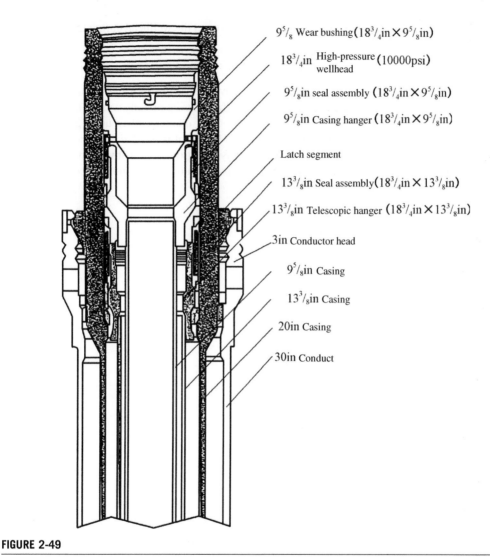

$9^5/_8$ Wear bushing$(18^3/_4\text{in}\times9^5/_8\text{in})$

$18^3/_4\text{in}$ $\genfrac{}{}{0pt}{}{\text{High-pressure}}{\text{wellhead}}$ (10000psi)

$9^5/_8\text{in}$ seal assembly $(18^3/_4\text{in}\times9^5/_8\text{in})$

$9^5/_8\text{in}$ Casing hanger $(18^3/_4\text{in}\times9^5/_8\text{in})$

Latch segment

$13^3/_8\text{in}$ Seal assembly$(18^3/_4\text{in}\times13^3/_8\text{in})$

$13^3/_8\text{in}$ Telescopic hanger $(18^3/_4\text{in}\times13^3/_8\text{in})$

3in Conductor head

$9^5/_8\text{in}$ Casing

$13^3/_8\text{in}$ Casing

20in Casing

30in Conduct

FIGURE 2-49

The wellhead device of the floating drilling platform (ship) in the subsea.

casing head and 244.48 mm (9 5/8in) casing head are set into casings of 508.0 mm (20 in) and 476.3 mm (18 ¾ in). Their work pressure standards are 68.9 MPa and 103.4 MPa. The lower end of the casing hanger layers is connected with the corresponding sleeve, which the internal thread of the upper portion connects to the landing tool. After the cementing, a sealing assembly is used to seal the annulus packer between the casings.

The high pressure casing head connects to the hydraulic connector of the lower part of the subsea BOP group during drilling operations. We can relieve the BOP group, connect the wellhead anticorrosion cap, and squeeze in anticorrosion fluid to protect the wellhead during the temporary abandonment.

2. Subsea equipment

    a. Wellhead disk

        As shown in Figure 2-50, this is placed in the subsea with the drill string and seats at the subsea to determine the well location and fix the subsea wellhead. It is generally welded by steel plate and a reinforcing bar into an octagonal box with an outer diameter often up to 3 m and an inside diameter of about 3/4m; the height is about 3/4m and it is is filled with concrete in the middle. The inner wall of the wellhead disk has a bayonet groove, so that we can use the drill string and a special connector and send it to the subsea, and then spin out the connector,

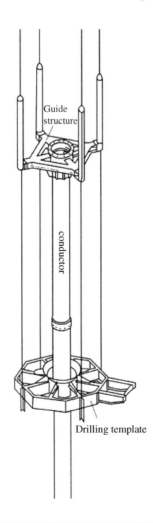

FIGURE 2-50

The wellhead disk and guide frame of the floating drilling platform (ship) in the subsea.

leaving the wellhead disk in the subsea. The upper part of the wellhead plate hole is tapered so that the conductor head can be aligned by its own weight.

b. Guide frame and conductor

As shown in Figure 2-50, the conductor head that is used in the first spud in is hung at the wellhead disk: the conductor diameter is about ¾ m. The length at the subsea location is about 30 to 60 m and each is 15 ~ 20 m. Each conductor is welded, uses a threaded joint link, or uses latch connectors. The guide frame is welded onto the conductor head. The guide frame has four pillars, and each pillar is tied by guide rope to guide the above BOP group and connectors to slide into the four pillars of the guide frame along the guide rope. Besides the guide frame there are two pillars for the installation of electrical imaging apparatus, which is placed via the guide rope. The center of the guide frame has a center string that connects with conductor head to locate the conductor head and hang the casing.

A kind of retrievable semi-fraction guide frame has been applied in the offshore oil recently. It connects with the conductor head and lands together with conductor and sits on the wellhead disk. It is semi-divided, as shown in Figure 2-51; we can use a ROV (Remote Operated Vehicle) underwater robot to open it and bring it back for recycling.

c. Casing head group

This was mentioned above, and shown in Figure 2-49.

d. BOP group

The BOP group usually has three to five BOPs; we should install two universal BOPs in order to land drill strings in case the BOP is shut down when conducting snubbing operation. The other BOP can be a blind ram; semi-blind ram, or upper body full-blind, lower body semi-blind ram double-use BOP. The entire BOP group has an exterior framework, and the four casing strings on the frame can be set with the guide rope of the bottom guide frame, in order to send the rope down the four uprights and slide the bottom guide frame in place, as shown in Figure 2-52. The framework is equipped with a hydraulic valve box or an electrical switch box.

The BOP group uses electric and pneumatic remote control. The pneumatic remote control system is used in the event of a blowout. The electric remote control is a three-throw plunger oil pump that uses a motor-driven triplex plunger pump on the platform deck for control. The

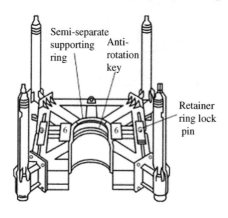

FIGURE 2-51

Retrievable semi-fraction guide frames.

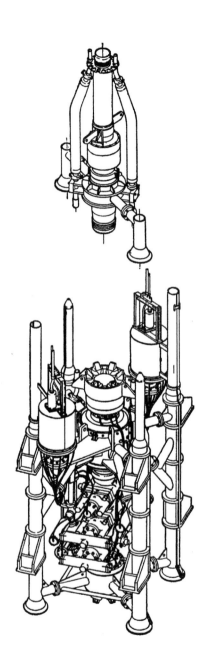

FIGURE 2-52

Subsea wellhead BOP group.

oil pressure is up to 210 Pa and the high-pressure working fluid moves through the rotary joint. The hose connects with the BOP to drive BOP fully closed clings drill pipe and puts spray and killing well operation. The pneumatic remote control system is a compressor and piston pneumatic pump on the platform to achieve hydraulic control.

e. Connector

There are in total two hydraulic connectors: one of them is on the lower end of the BOP group that connects the BOP group with the casing head group; the other is on the top of the BOP group for the connection of the BOP and impermeable flexible couplings of string, as shown in Figure 2-53. The hydraulic connector is a card block that consists of connector 1, connector 2, card block 3, hydraulic cylinder 4, card action ring 5, piston rod 6 as shown in Figure 2-54. The connector's up and down joint connect together, relying on the tightness of the card blocks. The card block is composed of two parts, each of which contacts the tapered surface. A part of the card block called action block ring 5 connects piston rod 6 of hydraulic cylinder 4; thus, the expansion of the piston rod will take the action ring up or down the link, which is to say another part of the card block can be pressed or locked. When the card block loses, which can make the upper joint and lower joint disengage with an angle of 30° or larger, the drilling platform (boat) can quickly evacuate the well location to avoid major loosening. The card block action ring is driven by six cylinders used to promote the action ring upward to loosen the card block and four cylinders used to promote the action ring down to press the card block pressed; the thrust ratio of both is 1:1.8. This connector is suitable for deep water, and the allowable load is up to 560 t.

f. Flexible joint

As shown in Figure 2-53, this is located between the hydraulic connector and riser. It can adapt to the angular displacement tmade by the offshore environment loads because it allows for a rotation angle (typically 10°). At the same time the equipment under the hydraulic connector including the BOP group, etc. also can avoid affecting the angular displacement. The structure and composition of the flexible connector is shown in Figure 2-55 (a). The upper and lower section each has a upper bearing 2, a lower bearing 4, an upper rotating ring 1 and lower rotating ring 3; the upper rotating ring 1 is connected with the riser, thus the riser string tension passes through rotating ring 1. Standed by upper bearing 2, so the tension passes through rotating ring 3 to lower bearing 4 is withstood by lower bearing. Sphere 5 is in the middle; the drilling fluid pressure through the ball passed to the swivel sphere itself is not subjected to axial load. The permissible loads of such flexible couplings are up to 560 t and can be used in deep water up to 1800 m.

Figure 2-55(b), (c) presents another pressure balanced flexible joint. It introduces sea water or drilling fluid into the fluid cylinder so that it pushes the piston to pressure the hydraulic oil to let the high-pressure fluid into the sphere contact surface, so that it can be used to balance the downward pressure or upward tension that the sphere contact surface withstands, keeping the sphere from bearing the axial load so it can adapt to the rotation of impermeable string. Generally, the allowable rotation angle is 10°, but the allowable load is lighter, only 370 t. There is also a multi-ball type flexible joint, which also uses the pressure balance principle, and just takes a sphere into three spheres. The whole connector is divided into three segments where each segment has a sphere that each allows a rotation of 3°, and due to the use of multiple spheres thereby reduces the abrasion of each sphere.

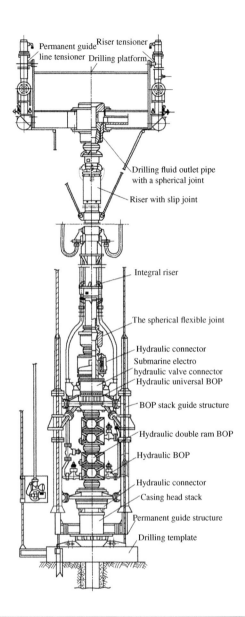

FIGURE 2-53

Chinese homemade II floating drilling subsea equipment.

g. Riser

This forms a channel from the subsea to the wellhead platform deck. When drilling you can import the drill string and various kinds of drilling tools, circulate the drilling fluid, and cut off the water at the same time; therefore the channel is called a riser or riser string. In addition, because of the riser can be scaled up and down within a certain scope, and can rotate at an

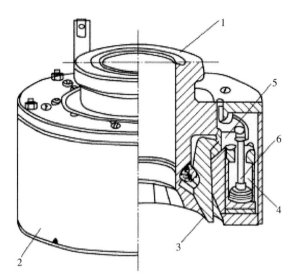

FIGURE 2-54

Hydraulic fixture block connector: 1 - up joint; 2 - down joint; 3 - card block; 4 - cylinder; 5 - card block movement ring; 6 - piston rod.

angle, it can adapt to the heaving and swinging movement of the floating platform (ship). The length of a single riser is generally 15 to 16 m with quick disassembling type coupling connecting each riser. Generally, there should be a tight rope and tension machine to keep the riser in a state of tension in order to avoid bending under axial pressure when the working water depth is more than 31 m; sometimes a float is added outside the string in order to increase buoyancy. If the working depth is over 250 m, the riser shall be specially designed. The calculation of buckling length L_{er} of the riser is according to Equation (2-36):

$$L_{er} = 3.44 \left(\frac{EI}{Q}\right)^{\frac{1}{3}}$$
(2-36)

where E is the elastic modulus of the riser, kg f /cm^2, I is the rotary inertia of the riser cm^4, Q is the weight of ariser unit length in the water, kg f /cm, and L_{er} is the critical buckling length, cm.h.

(a)   **(b)**   **(c)**

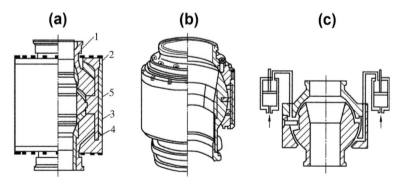

FIGURE 2-55

Flexible joint: 1 - upper swivel; 2 - upper bearing; 3 – lower turn ring; 4 - lower bearing; 5 - the ball.

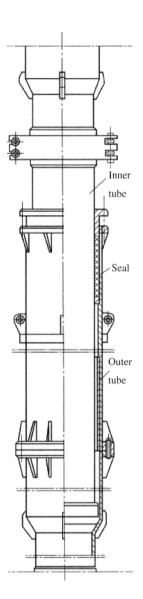

FIGURE 2-56

Extension risers.

Extension riser

Generally installed on the upper of the riser, as shown in Figure 2-53, an extension riser has an inner tube and outer tube, as shown in Figure 2-56. The inner tube and outer tube can do relative up and down movement, so the riser can adapt to the heaving movement of the floating platform (ship). Generally, the length of a scale riser is about $15 \sim 16$ m, and depends on the wave height and tidal range of the sea, generally the Chinese version is 14 m.

Figure 2-53 is the diagram of subsea equipment used in the "Exploration 1" floating drilling ship that is designed and made by China.

3. Surface equipment

   When using a subsea wellhead, some special equipment should be installed on the surface platform deck in order to cooperate with the subsea equipment. In addition to the ball joint fluid discharge pipeline (Figure 2-53) on the top of the riser and the control system of the BOP group on the platform deck, the main equipment includes the tension system that is used for the guide rope of the riser and to put the guide frame under tension.

   a. The tension system of the guide rope of the subsea guide frame

   The tension machine is used to tension the lead rope through the pulley system, and keep constant tension, as shown in Figure 2-57 (a).

   b. The tension system of the riser string

   This takes the tension rope of the tension machine pulley, connects it to the top of the riser, and then puts it through the tension machine to make a tight rope to maintain constant tension, as shown in Figure 2-57(b).

   c. The tension machine of the riser and the guide rope

The structure of the tension machine is shown in Figure 2-58. The pulley on the top of tension machine is supported by the piston rod and the piston in the cylinder, so through adjusting the air pressure in the tank we can put hydraulic pressure of the piston in the storage accumulator equal to the tension phase of the rope. In this way, when the rope tension reduces (when the platform is sinking) the piston bar and piston will not extend until reaching a balance; on the other hand, when the rope tension increases (when the platform is rising) the piston bar and the piston will go back again until it is balance. Therefore, we can deal with heave compensation for the platform and keep the guide rope of the riser string and the guide frame in a constant tension through the extension and retraction of the hydraulic cylinder piston and piston bar under a given constant tension.

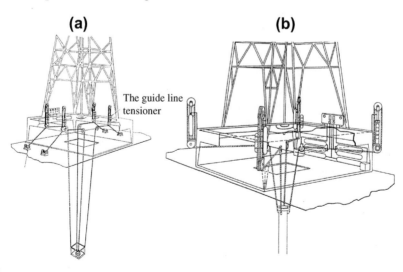

(a)   **(b)**

The guide line tensioner

FIGURE 2-57

Riser and guide rope tension system.

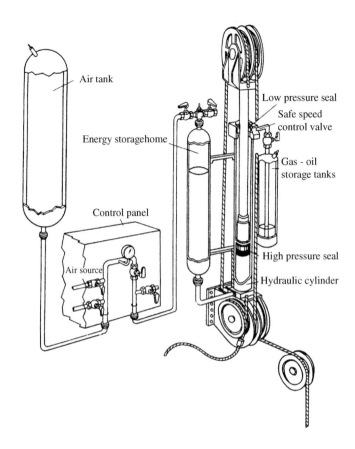

FIGURE 2-58

Risers and the guide rope extension machine.

2.2.3.2 The Wellhead of a Jack-up Platform

When using the jack-up platform to drill and develop a well, we need to install a mud line suspension system in the subsea in order to hang the layers of casing and the tieback production string. Usually, the mud line suspension system is called the subsea wellhead of the jack-up drilling platform. Using this subsea wellhead can relieve the load of the platform deck because the weight of each layer casing is hung on the mud line suspension system and supported on the subsea.

The bottom of the mud line suspension system connects to the corresponding casing. The top has the threading of the landing tool and tieback tool. The highest force is up to 103.4 Mpa. The structure of a mud line suspension system is shown in Figure 2-59.

1. Φ762 mm (30 in) conductor: As shown in Figure 2-59, 13 is a Φ762 mm (30 in) conductor and 6 is "ALT2" with outside and internal screw threads.
2. Φ508 mm (20 in) string: As shown in Figure 2-59, 14 is Φ508 mm (20 in) string and 10 indicates an "LS" type connector joint; 2 is the landing and tieback tool which has function of sent and tieback thread; 3 is the mud line hanger and the 508 mm casing; 14 is hanging below it.

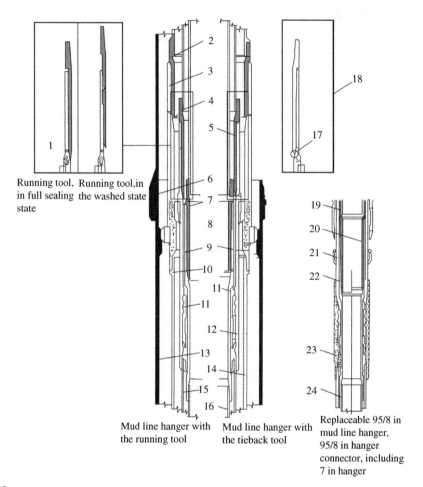

FIGURE 2-59

Mud line suspension of jack-up platform (wellheads): 1 - flushing hole; 2 – 20 in down and tieback tool (with LS type connector); 3 - 20in mud line hanger; 4 - 13 3/8 in down tools; 5 - 13 3/8 in tieback tool; 6 – 30 in "ALT2" male and female; 7 - 9 5/8 in down tool; 8 - 9 5/8 in tieback tool; 9 - 9 5/8 in mud line hanger; 10 - 20 in "LS" type connector; 11 - 9 5/8 in slotting clamp setting ring; 12 - 13 3/8 in mud line hanger; 13 - 30 in catheter; 14 - 20 in casing; 15- 13 3/8 in casing; 16 - 9 5/8 in casing; 17 - metal to metal seals; 18 - tieback tools, in full gear, and sealing state; 19 - 7 in down tool; 20- 7 in tieback tool; 21 - $9\frac{3}{8}$ in mud line hanger; 22 - 7 in mud line hanger; 23 - 9 5/8 in hanger joint; 24 - 7in casing.

3. Φ244.5 mm (9 5/8 in) casing: As shown in Figure 2-59, 16 is Φ244.5 mm (9 5/8 in) casing; 11 is the slotting clamp ring setting; 9 and 21 are the mud line hanger; 8 is the tieback tool; 7 is the landing tool.

4. Φ177.8 mm (7in) casing: As shown in Figure 2-59, 24 is Φ177.8 mm (7 in) casing; 22 is the mud line hanger; 19 is the landing tool; 20 is the tieback tool.

In Figure 2-59, the 1 is the flushing hole; the left picture shows the flushing and sealing of tools. The picture on the right side gives the full meshing state of sealing tool 18; the 17 represents the metal to metal seal.

2.2.4 Subsea Drilling Template

The subsea drilling template also belongs to the wellhead. It is a wellhead that can make many wells in one well location when drilling and developing wells in the subsea. The subsea drilling template can drill more than one production well at one well location when the oil field is a producing platform at the development stage. After finishing the production of the platform, we can withdraw the offshore drilling platform and tieback each casing to the production platform and produce the oil. In this way, we shorten the period of oilfield development and construction; this enhances the economic benefit. Therefore, it has a wide range of application from the North Sea to Brazil.

2.2.4.1 The Type and Structural Characteristics of the Subsea Drilling Template

1. The certain distance type template

    As the name suggests, the certain distance type template is a slot chassis that has a given distance among the hole drilling wells. It is one of the most simple subsea drilling templates structures. Figure 2-60 presents a type of the certain distance type template with four wells. The frame of the template is welded by pipe and can be used as a permanent guide base. A recyclable direct component (pilot) is installed on base so as to use the standard guide and BOP group. There is a well hole sleeve on each well seat to install the casing head and wellhead device. The base also has the pile pipe sleeve (guidance) so that it can be piled into the subsea through the pipe to fix the template. At the top of each well slot all wells have a funnel structure to put the permanent guide frame.

    Because there is no built-in leveling device in the certain distance type template, it can only be used with a seabed slope less than 5°. Generally, it cannot be applied for a water depth that is

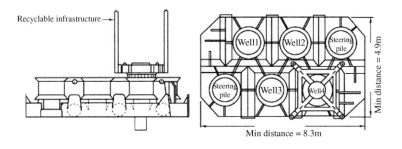

FIGURE 2-60

Unitized subsea templates of four wells.

deeper than 60 m or where the well slot number is more than 6. The certain distance type template can also be used in the subsea mud line suspension system of a jack-up platform system. The minimum distance should be in 1.8 ~ 2.1 m or more to put the permanent guide frame between wellhead seats on the template. The whole template size should be able to pass through the deck opening or the moon pool.

2. Unitized template

The unitized template is mainly used in a drilling operation that has known reservoir characteristics and a certain development pattern for wells; it is especially suitable for an area that has many wells to drill. The number of well slots on the template depends on the number of development wells plus some unforeseen wells. Commonly used well number are 9, 12, 15, 18, 20, and 24. Now the number is up to 32 wells. At present, the unitized template's working depth is up to to 150-200 m in the North Sea and Brazil.

The unitized template is a sheet structure that uses large diameter (generally 62 mm or larger) pipes to be welded together. The template frame has several parallel well arrays and each row has a layout of several well holes each a certain distance. Figure 2-61 is a unitized template that has four rows each consisting of six wells.

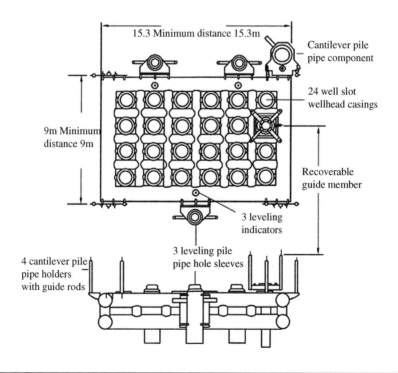

FIGURE 2-61

Subsea unitized template of 24 wells.

The structural features of the unitized template are the following:

a. The template frame is welded by different specifications of pipe.

b. The cantilever pile pipe component is used for positioning and guiding the pile tube chassis.

c. The hole sleeve is set for a 762 mm (30 in) wellhead in the subsea.

cd The permanent guide frame, which can be recycled without divers' help, is mounted inside the socket of the template guide frame.

e. The heavy three or four leveling device is landed by hoist on the floating crane or platform to the subsea leveling seat. It can find the flat hole sleeve with the help of the pile and through the pile hole import the hole sleeve which has slips inside. Therefore, the template can be constructed outside the roughness of the seabed.

3. Packaged template

The packaged template is a kind of subsea drilling template that is assembled by many frames with wellheads locked together, so it has great flexibility. Usually it is made up of two parts: one is the base of the initial frame and another is a hanging (bracket type) multiple or single well frame. The initial frame is generally a single well frame as shown in Figure 2-62, while another frame of a single well is hung on the initial infrastructure by interlock. The initial frame can also be a multiple well (and thus there are three wells). In this way, the number of drilling wells on the whole template can increase by coupling with the hanging single well frame.

Because this template can increase the number of wells, it can be used not only for exploration wells, but also for development wells. Also we often choose this kind of drilling template when the number of drilling wells is not certain. Usually we can take the initial basic frame of the packaged template as a permanent guide frame when drilling an exploration well in the tectonic area that has high possibility of oil production. In this way, we can recover the basic frame if the exploration well has no oil; if the exploration well has oil we can hang from the basic frame and assemble many single well frames to add the number of the wells, and finally form a subsea packaged template system that has at the center the discovery well and also contains several development wells just as shown in Figure 2-63 for a subsea packaged template system of five wells.

The packaged chassis can connect cantilevered flow line components to the outside of the initial basic frame and use the subsea Christmas tree to produce oil so as to shorten the time for a well to be put into production. In addition, we can also hang positioning components for the production platform outside to locate the oil production platform accurately to the template, and tieback the wellhead after drilling the required number of wells.

The packaged template installs a level indicator and two screw jacks and is equipped with surveillance cameras, so we can pass through the screw jack to create a precise leveling.

4. HOST (Hanger Over Subsea Template)

HOST is a new product developed in the mid-1990s that has been used in Norway's North Sea oil field since 1996. A five-well hanging combination chassis has also been used in the South China Sea in the 20th century as shown in Figure 2-64.

This kind of HOST has the following main advantages:

1. Decomposition of the whole for convenient manufacturing and transportation

This decomposes the unitized template into a center module and a number of single well guide

Connection to the tieback string

The initial infrastructure

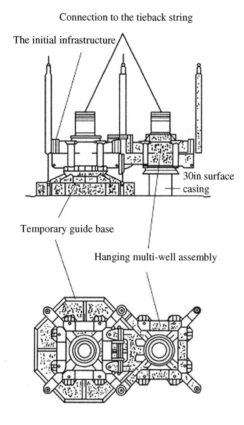

30in surface casing

Temporary guide base

Hanging multi-well assembly

FIGURE 2-62

Subsea packaged template system of two wells.

modules. The number of guide modules depends on the needs of drilling during oil field development. In this way, the small module are more convenient than the integral type. Also, the small module size is only 4.5 m × 5.8 m × 2.0 m (in the Chinese South Sea), so transportation is also very convenient. For example, the semi-submersible platform's moon pool size is 5.5 m × 6.0 m. It is convenient to use the crane to take the module to the moon pool of the semi-submersible platform and then tow it down to the subsea through the moon pool.

2. The structure is simple and easy to install and operate

During installation we should fix the center module first, and then the other modules are installed respectively around the center module. Then the wellhead and the Christmas tree that are used in well completion are installed on each guide module through the guide lines and guide string in accordance with the procedures used in drilling. All these assignments are in accord with the normal operation procedure using a conventional drilling platform.

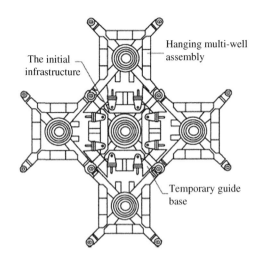

FIGURE 2-63

Subsea packaged template system of five wells.

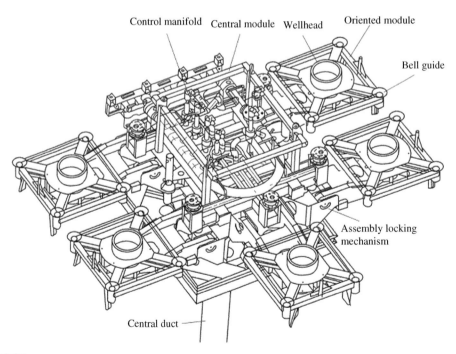

FIGURE 2-64

HOST of five wells in the South China Sea.

3. Save steel and reduce the operation cost
   In general, an equal number of wells can save about 25% in steel, and the installation saves about 40%.
4. Strong adaptability to different oil field scales and depths

The template can meet the needs of different scales of oilfields and well numbers, and can also adapt to small- and medium-sized oilfield development or different water depths.

2.2.4.2 Installation of a Drilling Subsea Template

1. Installation of the certain distance template
   Because the size of the certain distance template is bigger than the moon pool of a platform (ship), it cannot cross through the moon pool. We must use a floating crane or barge to sink the chassis to the subsea through steel wire rope.
   a. Installation of subsea template of a jack-up drilling platform
      Usually, the method is divided into landing the template first or landing on the jack-up platform.
      Land the drilling subsea template first: This method can be chosen when the oil field's reserve is proven and the jack-up drilling platform with a subsea drilling template are used to drill the development well. When using this method we should use a mud line suspension system to drill the first well, and install the packaged drilling template at the last 762 mm (30 in) riser of the first well. In this way, we can use the riser as a guidepost and use the drill pipe to land the drilling template and riser on the subsea. According to the design requirements, generally, we should land the lower template frame above the mud line to set aside 1 m of space for the diver to operate; the upper frame reaches above the mud line about $3 \sim 4$ m, which equals the maximum height from the mud line hanger catheter to its upper joint. Once the template attaches to the bottom of the subsea we can use cement to fix it. The drill string can also be unlocked and moved to another wellhead to drill.
      Alternate landing for the subsea drilling template: The alternate landing refers to reinstalling the new drilling template on the temporary abandonment wellhead. This installation process is usually divided into two steps.
      Step 1: Land the center riser. First of all, we weld a support bearing ring on the center riser to support the drilling template and land it with the riser during drilling. Second, we have procedure of back-off from above the center riser when abandoning the well. After the recovery of the riser the center riser will remain in the subsea.
      Step 2: Return to the well location to also land the template: First, reset operations on the jack-up platform, and then with the maritime transport of drilling template use the hoist hook on the platform hoist the template from the transport to the platform deck orbit. Then we should assemble the locking devices such as an attachment on the platform deck and use the drill string to land the template in the subsea with the assistance of a diver to make an exact landing that lets the template land on the bearing ring of the center riser in the subsea as shown in Figure 2-65. After it is stable, take two fixed piles (with a 914.4 mm diameter pile hole and 762 mm diameter pile pipe on both sides of the template respectively with the middle annular fixed with cement), and connect the fixed pile with the template to complete the installation work.

FIGURE 2-65

The installation of the subsea drilling template of jack-up platform.

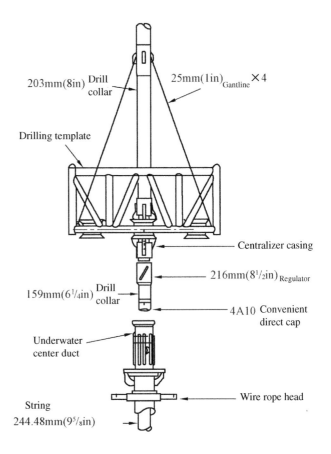

b. Installation of subsea drilling template of a floating drilling platform (ship)

First the preparation work must be carried out: drag the semi-submersible drilling platform to the selected road, clean up the subsea, and get the installation equipment such as the pile pipe [with 762 mm (30) in casing], solid pile of cement, winch wire rope, etc., ready. Then, we transport the drilling template to the sea and begin the important part of the preparation work. The template is lifted to a flat barge first, then installed to the semi-submersible platform. The template on the flat barge is rigged by a working ship to the road of the semi-submersible platform. Then ship works with another working ship to first drag the template and flat barge to the space below platform as shown in Figure 2-66 (a). When the drilling template aligns the turntable of the platform, we can connect the sling of the template to the sling of the system of rings of the drilling rig hoist on the platform; at this time, we can use a drill winch to lift the template from the barge first and after the flat barge is dragged by the two working ships out of the lower drilling platform space fix the drilling template on the platform, as shown in Figure 2-66(b). Now we are ready to land the template on the subsea. Figure 2-66(c) shows the condition of the chassis when installed.

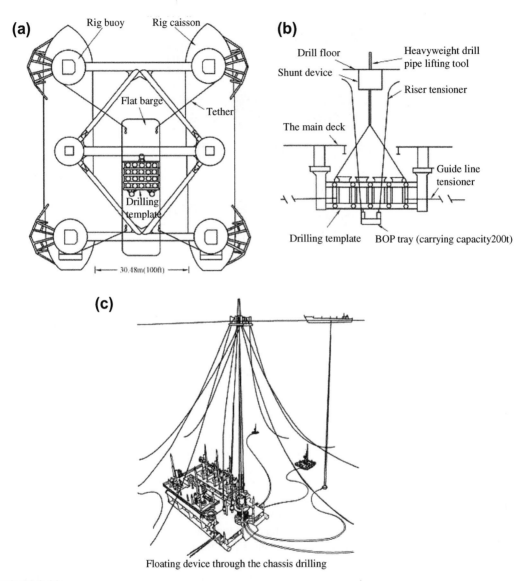

(a)
Rig buoy   Rig caisson
Flat barge
Tether
Drilling template
30.48m(100ft)

(b)
Drill floor   Heavyweight drill pipe lifting tool
Shunt device
Riser tensioner
The main deck
Guide line tensioner
Drilling template   BOP tray (carrying capacity200t)

(c)
Floating device through the chassis drilling

FIGURE 2-66

Drawing the subsea drilling template to the bottom of the platform and fixing it.

The steps of landing a subsea drilling template on the subsea (Figure 2-67) are as follows:
- **i.** Lift and land [Figure 2-67(a)]: Use a drill winch to lift and land on the subsea.
- **ii.** Fill and fix [Figure 2-67(b)]: Usually we land the $\Phi762$ mm tube through the pile hole and fix the annulus space with cement.
- **iii.** Install the wellhead [Figure 2-67(c)]: Install the guide frame and BOP groups, etc.

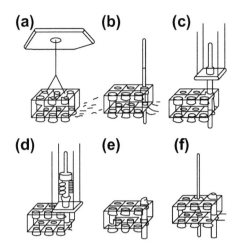

FIGURE 2-67

Installation of a subsea drilling template for a floating platform.

    iv. Conventional drilling [Figure 2-67(d)]: Drilling according to the conventional process.
    v. Unlock the wellhead [Figure 2-67(e)]: Set out the BOP groups and with the guide frame seal the drilling well.
    vi. Drill new well [Figure 2-67(f)]: Move the drilling rig on deck and drill a new well on the subsea drilling template.

2. Installation of unitized template

The installation steps of the unitized template are as follows:

    a. Drag with the crane barge: Use crane barge to ship it to the location.
    b. Unlock it and float it on the water: The template can be unlocked and made to float on the surface of the well.
    c. Hang to eliminate buoyancy: Connect the special wire of the template to rig of a ship (vessel) or working ship and open the valve and pour water into the ballast to eliminate the buoyancy.
    d. Land directly on the wellhead: Use the hoisting system of the platform (vessel) or the ship's hoisting equipment on the drilling rig to land the template slowly on the subsea.
    e. Use a pile and level and solidify the template: Drive the steel pile through the template pile holes and after leveling the template cement the grommet to solidify it.
    f. Drill using the process: Begin the drilling operation in accordance with the designed process one by one.
    g. Finish the well completion operation: After each drilling for the tubing, install the tubing hanger and subsea Christmas tree on the wellhead.

3. Installation of a packaged template

The installation steps for a packaged template are as follows:

    a. Land the guidance device: The main part of the guidance device is the temporary location of packaged template above it with a guide frame and four guide ropes. Send this guide device to the subsea first.

b. Land the template body: Using the 910 mm pile of the pile driver or the hoisting system of a pile drilling rig, land the main part of the drilling template with guide lines and locate it on the guide base.

c. Support cantilever parts: Use a screw jack to support the cantilever parts of the drilling template.

d. Level the template body: In order to guarantee the placement of the template body on the subsea in a horizontal position we should level the body after the driving the 760 mm pile through the pile hole, and pour the grommet into the empty cement to solidify it.

e. Drilling on the template wellhead: Drill the template well first and according to the order finish drilling the two wells on the template.

f. Land the cantilever base: After finishing the two wells on the template, land the cantilever base; make sure it is accurately landed on the guide pile and drill in sequence until the drilling is finished.

g. Finish the well completion operation: Install the subsea Christmas tree and get ready to put it into production.

2.2.4.3 Leveling of the Subsea Drilling Template

Generally, the horizontality requirement should be maintained at 0.5° so that it can keep the temple level. In order to prevent the abrasion of the drill string during drilling and reduce the bending stress on the template which can make it easy to buckle under the sea.

1. Leveling of the packaged temple

A packaged chassis generally uses a screw jack leveling system as shown in Figure 2-68. In the figure, 1 and 2 are the upper and lower shelves of the template structure; sleeve 3 is welded on the template structure, 4 is the screw jack; at the bottom of sleeve 3 is a female element 5; 6 is the centering guide; the upper part of the screw has a short section 7 and the bottom has a joint 8 which is borne by plate 9 and there is a guide bell mouth 10 on the upper portion of sleeve 3.

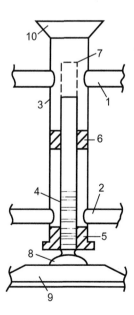

FIGURE 2-68

The leveling system of a packaged subsea drilling template.

When the drilling template is landed on the subsea, the drilling machine on the platform (ship) can be used to land the drill collar on the subsea. Let the drill collar pass through bell mouth 10, and it connects to short section 7 above screw 4 and creates a left or right rotating drill string after the connection. The drill collar will rotate the screw left or right and can make the template structure welded on sleeve 3 move up or down to level it.

Usually the whole subsea drilling template includes the installation of 3 ~ 4 screw jacks to constitute a three point or four point leveling device.

2. Leveling of unitized and certain distance templates

The two templates both use a hydraulic leveling system and the leveling steps are shown in Figure 2-69.

a. Drill a pile hole of 762 mm (30 in) (pile hole well)

First, weld the guide socket (guide pile seat) of string onto the template; then the chassis is landed on the subsea and the drill stem centralizer is used so the string can guide the jack through the pipe using Φ974.4 mm (36 in) to drill the hole.

b. Land the Φ762 mm (30 in) pile legs (foot) and the guide pile cap

Connect the pile legs running tool on a section of the drill string and put the tool, the pile, and its guide frame into the pile hole; install the cement shoes to prepare for cementing.

c. Land the Φ762 mm (30 in) pile leg (foot) and cement

After the Φ762 mm pile leg is landed in the pile hole, inject the cement through the cement shoes.

d. Land the template leveling tool and level the template

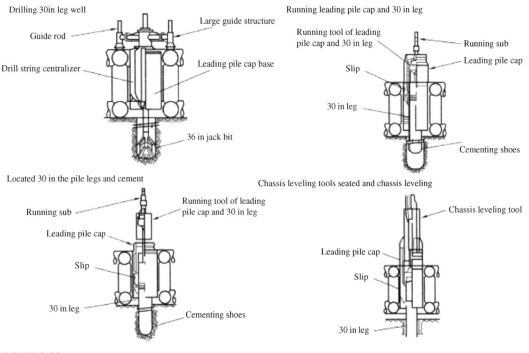

FIGURE 2-69

Leveling of unitized and certain distance templates.

First, land the template leveling tool and let it lock onto the guide pile cap and calculate the hydraulic fluid that is pumped into the leveling tool; let the reaction force of the fluid apply to the top of the pile and when the template is lifted to the level position use the heavy slip to fix it.

In this way, we need to level other piles until all piles are leveled to finish the leveling work after taking out the leveling tool.

2.3 Early Preparatory Work of Offshore Oil and Gas Drilling

Before we start offshore oil and gas drilling, we need to conduct a series of preparations, which mainly involve drilling design. This includes the design of platforms (ships) in place, wellbore structure, wellhead devices, casing, cementing, drilling fluid, drills, hydraulic parameters, drill assembly, and the design and control of wellbore trajectory. It also includes formation evaluation, BOP stack design, drilling program design, special operating procedures design, the schedule of the projects and materials planning, cost estimates, towing to the location of the drilling platform (ship), dropping anchor mooring, preparation work before spud, and so on. This section will introduce the distinctive elements thatare different from onshore drilling in early preparatory work.

2.3.1 Design of Offshore Drilling Platforms (Ships) in Place

The term "in place" means that the platforms (ships) stay at a predetermined drilling location. The issue of offshore drilling platforms (ships) staying in place is very important in ensuring that the subsequent drilling operation is carried out safely in the harsh marine environment. So we must conduct the design of it in place. For the so-called platforms (ships) in place is we design a floating drilling platform's (ship's) heading, the shortest chain length and orientation, the anchor tension test requirements and pretension value, the self-jack-up platform's heading, the plug-in depth of a pile, quantity of ballast, and air gap requirements, according to the seabed survey, water depth, weather, and sea conditions.

2.3.1.1 Offshore Positioning of the Drilling Platform (Ship)

1. Accuracy requirements of positioning
   With the improvement of positioning technology, we have become increasingly strict on the accuracy requirements of in place design of the platform (ship) well location. At present, the permissible error in China is a circle with a radius of 30 m when taking the designed well location as center.
2. Method of platform (ship) positioning
   Currently, the GPS global positioning system is commonly used in offshore positioning. It's a kind of high-precision positioning global coverage system. It can provide the exact location of the drilling platforms (ships) at all times, and display the information on the screen in the form of data and icons using the satellites in space and standard stations on Earth. When towing, we input designed routes, and we can know the location, speed, and course of the actual routes of the platform (ship) position and the deviation from the designed routes. By entering the well location coordinates and platform heading, we can display at any time the difference of the actual and designed location of the platform (ship). When dropping anchor, if the boat installs a trailer tracking device, and the orientation and

the distance of the anchor is entered, the accuracy of the anchor can be shown to guide the platform (ship) in place precisely and drop the anchor. While the work boat doesn't install trailer tracking equipment, the anchor boat can only use the radar equipment on board to determine the orientation of the anchor. It also can confirm whether the distance meets the design requirements, according to the chain scope of the platform (ship) or by reference to radar ranging.

2.3.1.2 Confirmation of the Platform's (Ship's) Heading

In order to confirm the heading of the platform (ship), first we should consider the various forces from the marine environment, mainly wind, followed by the force of surges, waves, flow, and so on. Next we should consider docking of the supply ship, taking off and landing of helicopters, blowing out of the burner and explosion zoning of the platform, and the escape route in the event of combustible gas. Usually, the dominant wind and flow direction of the operating season (against the wind, the upper stream) is chosen as the heading.

2.3.1.3 Anchoring Technical Requirements for a Floating Drilling Platform (Ship)

The semi-submersible drilling platform or drilling floating boat are required to stabilize the position with a mooring system, which was discussed in the first section of this chapter. The technical requirements of the mooring system is one of the important pieces in the design of offshore drilling platforms (ships); the main elements are as follows.

1. The chain scope
   Usually the chain scope is the length of the catenary anchor chain and anchor in front of the level of lying on the seabed and coupled with a length of anchor chain from the windlass to the next guide sprocket, to the total length, as shown in Figure 2-70. The chain scope depends on the water depth, and is usually 8 to 11 times the water depth. When working in shallow water, the chain scope should be a high value. When the water depth is less than 100 m, according to the experience of the South China Sea, generally, the length of the shortest chain scope is not less than

FIGURE 2-70

Semi-submersible drilling platform of cable system.

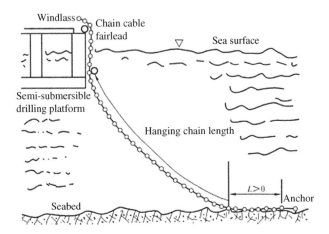

823 m (2700 ft), and should be greater than the anchor chain to withstand extreme catenary tension Generally, of course, the longer the chain scope is, the better the mooring force will be. The initial chain scope, which is generally longer than the shortest chain scope, can be determined according to the nature of the seabed soil, and this chain scope is usually about 30 ~ 45 m (100 ~ 150 ft).

2. Chain tension (anchor holding power) test

   When a floating drilling platform (ship) uses an anchor to position, the anchor chain tension test should be carried out, namely the expected maximum anchor tension test. Usually it doesn't exceed 1/3 of the breaking anchor chain tension. If the tension test can't meet this request, we should make the fluke angle larger, or increase the tandem anchor.

3. Anchor chain pretension

   Usually the anchor chain pretension is determined in accordance with a platform (ship) displacement of up to 5% of water depth and the anchor chain reaching 1/3 of the tensile strength of the anchor chain breaking point.

2.3.1.4 Jack-up Drilling Platform in Place

For jack-up drilling platforms, we propose the following technical requirements for design in place.

1. Pitching pile

   We should first confirm whether a normal pile can be used based on the findings of the submarine. If a normal pile can be used, we should take the appropriate measures to develop a viable option, and predict the insertion depth of the pile.

2. Ballast program

   For design in place, the principle of formulating a jack-up hull ballast program is as follows: a lifting mechanism is gradually loaded to the maximum supportive static load, and the platform is raised smoothly until the platform becomes even and not inclined. Suppose the maximum static supportive load of the lifting mechanism is G (kN), the platform's own weight is T (kN), and the variable load before ballast on the platform is W (kN); the ballast amount should be

$$Q = G - T - W \ (kN) \tag{2-37}$$

3. The height of the platform

   After the jack-up platform pitching pile is in place, the platform deck should be increased to a certain height H above the water. The height should be given in the design in place, and is commonly calculated by the following formula:

$$H = \frac{2}{3}h_1 + h_2 + 1.2 \ (m) \tag{2-38}$$

   In the equation, h_1 is maximum wave height, m; h_2 is maximum tide level, m; and values are rounded after calculation.

4. Prevention measures for ocean currents scouring the platform pile foot

   From seabed surveys, if it is known that ocean currents seriously scour the platform pile foot, we should work out the solutions in the design in place.

Of course, if there are other aspects that affect the safety of jack-up platform drilling operations, we must work out the solutions and emergency plans.

2.3.1.5 Mooring Mode

When drilling, the floating drilling platform (ship) relies on the positioning of the mooring system. Before the jack-up platform drill, whether it ballasts platform to stub, or uninstall platform to lift the ship and operations process also needs anchor positioning. Therefore, we should give the mooring mode in the design in place. This issue has been addressed in the first section of this chapter, and it can be found in Figure 2-31 and related content.

2.3.2 Marine Engineering Survey of the Well Location (Seabed Survey)

A seabed survey is when, before the drilling platform (ship) reaches the well location, we conduct a survey on undersea rock formations, water depth, seabed topography, shallow gas, and soil. It's an important element of the preparatory work of offshore drilling operations.

2.3.2.1 Purposes of the Seabed Survey

1. Give the submarine topography, and provide the basis for the arrangement of the anchor and selection of the wellhead position.
2. Find out the relevant data related to the seabed anchoring or plugging ability.
3. Measure the relevant data of the water depth, currents, and waves, and provide a basis to estimate the platform working conditions and forces on the mooring and riser.
4. Ascertain obstructions on the well location, surrounding areas, and the route, and provide a basis for the laying of submarine cables and submarine oil and gas pipelines.
5. Confirm the possibility of the existence of shallow gas, and identify its depth and thickness.

2.3.2.2 Subsea Research Content and Method

1. Echo sounding
   This is a method of measuring the depth of water which uses a high-frequency recovery device to get continuous depth recording data. The general requirements for a floating drilling platform (ship) should be a side length of not less than 2000 m square within the measured vertical and horizontal spacing of a 200 m measuring line. The measuring range for jack-up drilling platform is to take a side length of 1000 m square. In short, we must make the anchor placement platform (ship) fall within the measuring range, so that the measured water depth data can provide a reliable basis for the design of the platform (ship) mooring system.
2. Aim of lateral sonar measurements
   The goal is to provide a well location in the vicinity of any obstacles as well as uneven seabed. The measurement method uses side-scan sonar along the survey line with dual scan, and at least one cross-line should be through the center of the designed well location. The sonar measurements can reach 150 m to each side of the survey line. When the water depth is greater than 210 m, this lateral sonar measurement method will encounter difficulties.

3. Shallow strata graphic section measurement
   This can give an answer to the possibility of the existence of shallow gas, and also provide data for the possible biting depth of a jack-up drilling platform leg. Usually we use the method of acoustic measurements.
4. Undersea field measurements
   - On a jack-up drilling platform: We send divers to find out the bottom topography. The focus is to investigate the currents scouring the seabed.
   - On floating drilling platform (ship): Generally by test, the soil strength of submarine upper surface is given after the test for the purpose of determining the ability to withstand the subsea wellhead.
5. The collection of the main parameters of currents and waves
   This can be obtained from the meteorological department. If this is lacking, field measurements should be taken.

For designed well locations which have been identified using a jack-up drilling platform, 40 m deep cored wells should be drilled, with $1 \sim 15$ meter coring once per meter, $15 \sim 30$ meter coring once per $1 \sim 5$ meters, and below 30 meter coring once per 3 meters, to obtain soil samples, water content, bulk density, clay liquid limit, plastic limit, and compression, and for triaxial shear tests, sand particle size determination, and direct shear tests. Of course, if there is adjacent well information that can be obtained, we should not drill core wells.

2.3.3 Well Bore Structure Design

The key to drilling design is well profile design, which is an important part of pre-drilling. It is used to design the casing depth and level. For offshore drilling, we should design the well profile according to the characteristics of the subsea drilling and usual practice.

Due to the factors of the marine environment, casing string used in offshore drilling should not only be able to guarantee the drilling borehole up to the design depth safely and successfully, but also be able to support an underwater wellhead weight of about 100 t in the seabed surface, or support a wellhead weight about 30 t on the surface of the water to under the water about 20m. It should also be able to resist the currents and waves caused by the bending moment on the entire casing string and wellhead system. After years of practice, a set of standards has formed for offshore oil and gas drilling well structures in China.

2.3.3.1 Standard Casing String Structure

The structure of offshore oil and gas drilling casing string generally consists of five layers, namely catheters, surface casing, two middle casings, and an oil and gas layer tail pipe, as shown in Figure 2-71.

1. Catheter
   This is also called the surface catheter, and its role is to establish the wellhead, support the wellhead and BOP stack, and so on. The depth that it sets below the seabed is generally 30 to 60 m. In China, an offshore floating and drilling platform (ship) uses casing whose outside diameter is $\varphi762$ mm (30 in), inside diameter is $\varphi711.2$ mm (28 in), and steel grade is X52 or B. In jack-up

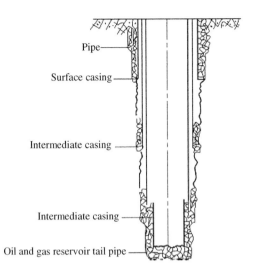

Pipe

Surface casing

Intermediate casing

Intermediate casing

Oil and gas reservoir tail pipe

FIGURE 2-71

The standard casing string structure for offshore oil and gas drilling.

or fixed drilling platforms, the tube is named into the riser, and it is generally made of φ762 mm (30 in) casing. But in shallow sea φ508 mm (20 in) casing production is available.

2. Surface casing

This is mainly used to seal the surface of the unconsolidated formation, and to establish the wellhead and wellhead equipment. In China the set depth for offshore drilling is usually about 500 m, and we commonly use φ508 mm (20 in) casing; sometimes we also uses φ339.73 mm (13³⁄₈ in) casing.

3. Intermediate casing

This is used to seal the complex formation to ensure drilling up to the design depth safely and smoothly. The casings commonly used in China offshore drilling are φ339.73 mm (13³⁄₈ in) and φ244.48 mm (9⁵⁄₈ in) casing.

4. Oil and gas casing or tail pipe

Oil and gas casings used in offshore drilling usually are φ244.48 mm (9⁵⁄₈ in) and φ177.8 mm (7 in) casing; oil and gas tail pipe usually use φ177.8 mm and φ114.3 mm (4¹⁄₂ in) casing.

2.3.3.2 Commonly Used Casing Structures

There are several types of well structure commonly used in China offshore oil and gas drilling.

1. The standard well structure type

This type of well structure is generally used for deep and ultra-deep wells, and can also be used in complex, high-pressure gas wells. The well structure is shown in Table 2-4 and the use of the φ215.9 mm borehole and the φ117.8 mm tail pipe, or the φ152.4 mm borehole and the φ114.3 mm tail pipe should be based on the stratigraphy and well depth.

Table 2-4 Offshore Drilling with Standard Well Structure Type

Borehole size	mm	914.4	660.4	444.5	311.15	215.9	152.4
	in	36	26	17½	12¼	8½	6
Casing size	mm	762.0	508.0	339.73	244.48	177.8	114.3
	in	30	20	13⅜	9⅝	7	4½
Casing name		Surface conduit	Surface casing	Intermediate casing		Tail pipe	

2. **The strengthening well structure type**
   This type of well structure is generally used for a maritime complex high-temperature and high-pressure deep well, and the well structure is shown in Table 2-5.
   In Table 2-5, the segment of the φ355.6 mm borehole and the φ298.45 mm tail pipe uses a 222.25 mm × 355.6 mm eccentric drill for drilling, and then is trapped into the φ298.45 mm flush casing. The segment of the φ311.15 mm borehole and φ244.48 mm casing pipe use a 200.03 mm × 269.88 mm × 311.15 mm eccentric drill for drilling. The requirements of the trapped φ244.48 mm casing string are that the casing trapped into the φ298.45 mm tail pipe or under is flushing casing, and the upper part should be common coupling casing.

3. **The simplified type I well structure**
   This type of well structure uses the φ339.73 mm casing instead of the φ508 mm surface casing, thus simplifying the well structure, as shown in Table 2-6.
   a. Floating drilling platform (ship)
      The φ476.25 mm standard subsea wellhead and the φ339.73 mm casing are connected by the φ508 mm × 339.73 mm which has a dedicated higher nipple to support and to ensure the integrity of the subsea wellhead system.
   b. Jack-up platform and fixed platform
      For a mud line suspension, we send up a hanging ring outside the 339.73 mm casing mud line hanger and hang a mud line support ring in the φ762 mm casing. Other casing hangers still use

Table 2-5 Offshore Drilling with Strengthening Well Structure Type

Borehole size	mm	914.4	660.4	444.5	355.6	311.15	215.9	152.4
	in	36	26	17½	14	12¼	8½	6
Casing size	mm	762.0	508.0	339.73	298.45	244.48	177.8	114.3
	in	30	20	13⅜	11 5/4	9⅝	7	4½
Casing name		Surface conduit	Surface casing	Intermediate casing	Intermediate liner	Intermediate casing	Tail pipe	

Table 2-6 Offshore Drilling with the Simplified Type I Well Structure

Borehole size, mm	914.4	444.5	311.15	215.9
Casing size, mm	762.0	339.73	244.48	177.8
Casing name	Surface conduit	Surface casing	Intermediate liner	Tail pipe

the standard type, and the casing head also doesn't change, so we still use the standard casing head.

4. The simplified type II well structure
   This type of well structure is simpler. It has only three layers of casing, and it uses φ339.73 mm casing as surface casing and also as intermediate casing. This type can only be used for shallow wells. This type of well structure is as shown in Table 2-7. This type uses a standard wellhead, and for the mud line hanger or the transition between the wellhead and the casing string refer to simplified type I.

5. Other types of well structure

When drilling cluster wells on a fixed production platform, some use φ762 mm casing as a riser, or φ609.6 mm (24 in) casing, and some use φ508 mm casing as a riser, and adopt the method of piling to construct.

As to how to use the given data, such as the formation pore pressure, fracture pressure equivalent density curve, swabbing pressure equivalent density values, well kick condition allowed values, formation fracture pressure safety value added, pressure allowable value, and well design depth, it should be according to the principles of the borehole pressure balance system to calculate and determine the depth of casing for the first time, the middle casing depth, the surface casing depth and ducts, etc. Because the well bore structures are the same as the design on land, they will not be repeated here.

2.3.4 Towage and Anchoring of Offshore Drilling Platform

Towage and anchoring of the offshore drilling platform is an important preparation before drilling, and they are introduced in the following.

2.3.4.1 Towage of Offshore Drilling Platforms (Ships)

With towage, the key is the connection between the towing ship and drilling platform (ship). Due to the fact that the structures of platforms (ships) are different, the towing methods are different too. We introduce several methods below.

1. Towage of jack-up drilling platform
   In general, a jack-up platform can be towed away in accordance with the towing speed requirements of the towing vessel (ship). The connection between the platform and the towing

Table 2-7 Offshore Drilling with the Simplified Type II Well Structure

Borehole size, mm	914.4	444.5	311.15
Casing size, mm	762.0	339.73	244.48
Casing name	Surface conduit	Surface casing	Oil-string casing

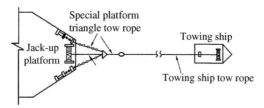

FIGURE 2-72

Towage of jack-up drilling platform.

ship is shown in Figure 2-72. Put the platform towing triangle cable on the work boats, and connect it with streamers on the towing boat. After connecting, you can drag the platform away by the towing boat.

2. Towage of semi-submersible drilling platform

   a. Self-propelled

   Some semi-submersible drilling platforms have the ability to self-propel, and thus they can move through this method. However, because the self-propelling speed is generally slow, they are usually self-propelled only a short distance, while for long distance we use towage to improve the speed and the purpose of steering.

   b. Conventional towing

   As shown in Figure 2-73, towing uses the same method of the jack-up platform. The front of the semi-submersible drilling platform connects to a drag towage, and at the same time it starts the propeller of the platform itself; thus it can be towed at the requested speed.

   c. Towing using anchors and cables

   Sometimes when movement is for short distances, in order to reduce the operating time, we do not use the dedicated triangle streamer. Instead, we use the anchors and cables of the semi-submersible platform to tow. This towing method is as shown in Figure 2-74, and it uses two anchors (anchor No. 1 and No. 8) and cables in the front of the platform. Connect the anchor and cables with the streamer to tow. Sometimes two towages can also drag the main anchors No. 1 and No. 8. But this method should not be used for long distance towing or in rough sea conditions, because the coordination between two towing ships and the platform is not easy. Anchor and chain towing can still cooperate with self-propelled movement.

3. Displacement of a floating drilling ship

Because the floating drilling ship has a strong self-propelling ability and the speed is higher, its self-propelled movement can achieve the necessary movement and speed requirements.

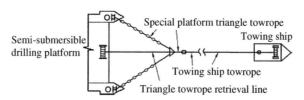

FIGURE 2-73

Conventional towing of semi-submersible drilling platform.

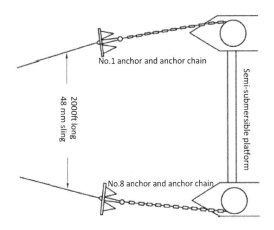

FIGURE 2-74

Semi-submersible drilling platform towage with anchor and anchor chain.

2.3.4.2 Approaching and Mooring in Place of the Offshore Drilling Platform (Ship)

Usually when offshore drilling platforms (ships) are close to the well location during towing, sailing, anchoring, and approaching the well location (placing) are carried out at same time, so this entire process is also called platform (ship) approach. In order to make the platform (ship) approach successful in one attempt as much as possible, generally go against the wind and current, and if the wind is inconsistent with the current, take the direction of the leading role. When we say against the wind and current this means making the platform (ship) heading just contrary to the direction wind or flow. Obviously, this will help control platform as well as help anchoring and piling. However, as already introduced, the platform heading is predesigned, and in the wells the direction of wind, and the flow are randomly changing, thus bringing difficulties to keeping drilling platforms (ship) in place. Generally under conditions that given the weather, sea conditions, and platform (ship) approach, anchoring, taking place conditions are under monitor, approaching often uses the following methods for different types of drilling platform (ship).

1. Jack-up drilling platform approach and mooring in place
   a. The wind blows from stem to stern on the right.
      As shown in Figure 2-75, in order to set up the platform approach against the wind from the No. 3 anchor position, when it reaches the No. 3 anchor position, the platform will throw the No. 3 anchor itself, continue moving forward along the approach route's edge, and releasing the No. 3 anchor chain. When the platform reaches the well, and the towage stops towing, stabilizing the platform at the well location, the platform stops releasing the No. 3 anchor chain. Then, by means of working boats, as the previously described anchoring methods and procedures, throw the No.2 anchor. When the No. 2 anchor stops the well, we can do locating and piling.
   b. The wind blows from stem to stern on the left
      At this time, in order to go against the wind, the platform should be approaching from the direction of the No. 2 anchor. The platform throws the No. 2 anchor until the platform reaches the well location. Then, after throwing the No. 3 anchor well and assisted by the working boat anchor, we can do the locating and piling to complete the platform approach and mooring in place.

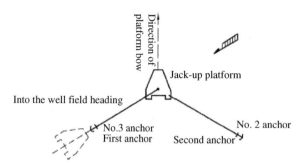

FIGURE 2-75

Jack-up drilling platform approach and mooring place.

2. Semi-submersible drilling platform approach and mooring in place

The example platform discussed below is equipped with eight anchors, and we use it to introduce the following methods.

  a. First case: Wind is level 2~4, and the wind blows from stem to stern

   i. Wind blows from the platform on the right: As shown in Figure 2-76 (a), the platform approaches from the No. 4 anchor position, throws the No. 4 anchor, and stops throwing the No. 4 anchor when it reaches the well location. Then the towage stabilizes the platform at the well location. Thereafter, the platform can, with help from the work boats, in the order No. 5→No. 1→No. 3→No. 6→No. 8→No. 2, throw these anchors; then the platform mooring is in place.

   ii. When the wind is on the left of the well: At this point, the platform approaches from the No. 5 anchor position, and throws the No. 5 anchor, then throws the rest of the anchors according to the order No. 4→No. 8→No. 6→No. 3→No. 1→No. 7→No. 2.

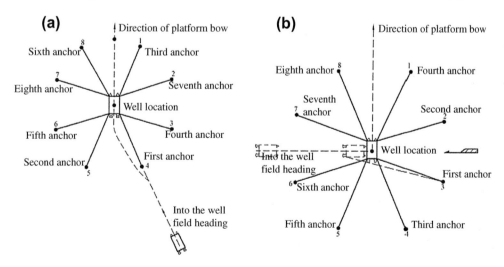

FIGURE 2-76

Semi-submersible platform approach and mooring place.

b. Second case: Wind is level 6, and wind direction to platform wells makes a 90° angle.

    i. Wind blows from the platform on the right: At this point, the platform approach is against the wind, assisted by the working boats, and first the No. 3 anchor well is thrown, then the No. 2 anchor; only then can you adjust the ship heading at the well location. Finally, we throw the rest of the anchors according to the order No. 4→No. 1→No. 5→No. 6→No. 7→No. 8; then the moor is in place, as shown in Figure 2-76(b).

    ii. When the wind is on the left of the well: At this point, the platform approaches against the wind. First, it throws the No. 6 anchor well with the help of working boats, then throws the No.7 anchor, adjusts the heading of the platform at the well location, then throws the rest of the anchors according to the order No. 5→No. 8→No. 4→No. 3→No.1.

c. Third case: Wind is level 6; wind direction is 180° or 0° to platform wells.

    i. The wind blows from the end of the platform: At this point, the platform approaches from the No. 4 anchor location following the wind, and throws the No. 4 anchor; when the No. 4 anchor chain releases long enough and the anchor has grasped the seabed very well, the platform is controlled. After that, the platform throws the No. 5 anchor with the help of working boats, then the platform moves to the well location, and throws the rest of the anchors according to the order No. 6→No. 3→No. 1→No. 8→No. 7→No.2; the process is as shown in Figure 2-77.

FIGURE 2-77

Semi-submersible drilling platform approach and mooring place for the third situation.

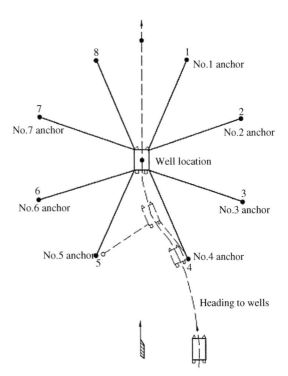

ii. The wind and platform heading into 0° on the well location: At this point, the platform approaches from the No. 4 (or No. 5) anchor, and throws the No. 4 (or No. 5) anchor across the anchor location; when the platform is towed to the well location, it throws the No. 8 anchor with the help of work boats, and at last throws the rest of the anchors according to the order No. 7→No. 2→No. 6→No. 3.

3. Floating drilling ship approach and mooring in place

The floating drilling ship uses the same approaches and methods as the semi-submersible drilling platform, so it will not be repeated.

2.4 Operating Procedures of Offshore Oil and Gas Drilling

The drilling operation stage will begin after the drilling platform (ship) is towed to the well site and the anchors dropped in position. This section will introduce all the operating procedures of normal offshore drilling.

2.4.1 Before Drilling Operations

There are some preparations before offshore drilling.

2.4.1.1 Checking the Seafloor

Before offshore drilling, we should check the seafloor around 70 m of the wellhead and find out whether there are some barriers obstructing drilling. For floating drilling platforms or ships, we could use a ROV (Remote Operation Vessel) to dive into the seafloor and check by camera or sonar system. For jack-up platforms, we only need to analyze the information from the seafloor survey rather than ROV.

2.4.1.2 Determining the Well Depth

Before calculating the well depth H, we should detect the distance L between the rotary table surface and the seafloor according to seafloor survey data. Design the conductor driving depth C and the pocket length R under the conductor shoe. Then, the well depth H could be determined by the following:

$$H = L + C + R \tag{2-39}$$

All units in Equation 2-39 are "m" and usually the value of R for floating drilling platforms or ships is 1 to 6 m and for for jack-up drilling platforms is 2 to 8 m. The larger the sediment sand content, the longer the pocket and the larger the value of R. Otherwise, the value of R should be smaller.

If the water depth measured by the seafloor survey is L(m) for which the influence of tidal range is not excluded and the altitude of a rotary table surface is M (m), then we have this equation:

$$W = L - M \tag{2-40}$$

The value of M in Equation 2-40 is constant for floating drilling platforms and ships as long as the platforms are loaded to the drilling ship draught. For instance, M = 25 m for the semi-submersible platform of "South China Sea II" and M = 23.2 m for "South China Sea V." For jack-up drilling platforms, M = the length of air + the distance between the bottom of the platform and the rotary table surface.

This can measure the water depth. Weld a depth measure plate onto the drill string while drilling and there are score marks according to the weight indicator slightly slacking off while drilling. For floating drilling platforms and ships, the distance between the rotary table surface and seafloor can be measured by score marks on drill string when lowering the temporary oriented base.

2.4.1.3 Lowering Temporary Oriented Base Plate

This operation before drilling is only implemented for floating drilling platforms and ships. A temporary oriented base plate is used for leading the drill bit and conductor to the site of well hole on the seabed. This device is welded with steel plate and rebar, and filled with concrete in the center. It also has a bayonet groove on the internal wall, which is convenient for lowering to the seafloor by drill string, as well as a special joint and a hole in the center. The external diameter of the whole temporary oriented base plate is 3 m and internal diameter is about 3/4 m with the height of 3/4 m. The base plate can be retained on the seafloor after it is sent to the bottom of the sea and the joint rotated out. The inclination angle of the base plate should be less than 5 °. The base plate should not be buried.

The oriented base plate will sink into the sludge if the mud of the seabed is soft. Once inclination happens, it will be hard for the drill tools to lower into the well hole. So usually a temporary oriented base plate will not be used. However, the temporary oriented base plate is needed in either of the following situations:

1. If the ocean current is large or the water is deep, the drill bit and conductor will be hard to seed into the well hole without conducting an installation.
2. If the seabed is hard and the drill bit rotates on the seabed, a vertical and tidy well hole will not be generated.

2.4.1.4 Performing a Seabed Penetration Experiment

This experiment is implemented only for the floating drilling platform. And it is better to conduct during the mooring operation of the platform or ship because time might be saved. The aim of the seabed penetration experiment is to confirm whether the seabed could support the weight of the temporary oriented base plate. A depth measuring board is often used as the experiment tool and will be welded onto the drill string. If the penetration depth is less than 0.5 m and the penetration pressure is up to 20.19 kN/m^2, the lowering operation is qualified and the seabed can support the oriented base plate. Otherwise, we need to expand the supporting area of the base plate.

2.4.1.5 Preparing to Lower the Permanent Oriented Conduct Shelf and Conductor

This is the final piece of preparatory work before drilling and is only conducted for operations with a subsea wellhead. The purpose of this before the drilling operation is to prepare for installation of the subsea wellhead and lowering of the permanent conduct shelf and conductor. This preparatory work should be performed after the temporary oriented base plate is lowered in place if that plate is used. It could be performed during the mooring operation if there is no need to use a temporary oriented base plate. This preparatory work should be conducted in the following steps:

1. Lifting and lowering of the permanent conduct shelf to the moon pool door of the platform or ship and checking it. The conductors that have been assembled and connected to the central hole of the permanent conduct shelf are lowered. They are then hanged on the moon pool door of the platform.

2. Lowering of the conductors through the permanent conduct shelf and connecting the conductors according to the design requirements.
3. Connecting the conductor head joints onto the conductors and connecting the lowering tools onto the conductor heads.
4. Leading the conductor into the base plate of the permanent conduct shelf with a heavyweight drill pipe of ϕ 127 mm (5 in) and locking with a locking pin bolt.
5. Pulling out of the lowering tools and covering the conductor head with a cover plate.

2.4.2 Drilling Operations for the Conductor Section

A well hole of ϕ 914.4 mm (36 in) should be drilled first according to the designed well bore structure for lowering the ϕ 762 mm (30 in) conductor. The operation during this stage will be called the drilling operations for the conductor section and the following are its major methods and steps.

2.4.2.1 Drilling Operations for the Conductor Section of Floating Drilling Platforms or Ships

1. Lowering drill assembly
   For a well bore of ϕ 914.4 mm, the following assembly is common:
   Drill bit (ϕ 660.4 mm) + joint with float valve and inclination measuring base + fixed arm expanding reamer (ϕ 914.4 mm) + 2 drill collars (ϕ 228.6 mm) + 7 to 10 drill collars (ϕ 203.2 mm) + joint + 15 heavyweight drill pipes (ϕ 127 mm) or drill collars (ϕ 228.6 mm) with float valve joint + drill bit (ϕ 914.4 mm). If the seabed is hard, float valve joint + drill bit (ϕ 444.5 mm) + drill collars (ϕ 228.6 mm) can be connected directly. The well hole could be expanded to ϕ 914.4 mm after it's been drilled initially. A conduct arm or four conduct soft ropes should be installed up to the expanding reamer (ϕ 914.4 mm) if the temporary oriented base plate was lowered. Tie the drill string in the center so that the drill bit can be conducted into the well hole of the temporary oriented base plate.
2. Running the drill bit to the seabed
   Once the drill bit sent by drill pipe is near the drill base plate on the seafloor or seabed, heave compensation equipment should be launched to reduce the lowering speed and the status of the drill bit in the well hole of the base plate should be supervised by underwater television. If we have confirmed the drill bit has been sent into the well hole of the base plate, the height between the seafloor and rotary table surface should be measured and the well bore depth that has been drilled practically should be calculated. The conduct arm can be recovered after the drill bit runs into the well hole and the real drill operation can be started.
3. Starting normal drilling
   The weight-on-bit at the beginning of drilling should be lower and should be less than 50 KN usually. Moreover, pump displacement should be lower, for example, 1500~1890 L/min can be used. The RPM should be 20~30 r/min. Then the RPM can be accelerated to 70~90 r/min (drill bit with ϕ 660.4 mm) or 100 r/min (drill bit with ϕ 914.4 mm) after the expanding reamer runs into the well bore. Pump displacement can be increased to 3800~4100 L/min normally after the first joint has been drilled. Generally, seawater can be used for drilling with the rate of penetration of 10~20 m/h. Clear the well hole by pumping 3974~4770 L of high viscosity drill mud and

ream the hole one time after completing each joint drilling. In order to avoid the drill bit being pulled up to the well hole when connecting the next joint and creating difficulty in sending the drill bit back into the well hole, a draught of water should be adjusted by loading or unloading the platform after completing the first joint drilling. The duration of normal drilling to the designed depth should not be long to prevent the well depth from being affected by tidal range variation. Timely circulate one cycle with seawater after drilling to the designed depth and pump 5560 L of high viscosity drill mud into the well hole. Next, clear the well hole with seawater again. After that, lower the inclination equipment and measure the well inclination. Generally, the well inclination should be less than 1 degree and the drilling platform should be migrated 10 m away from the previous well site and then should drill again if the inclination can't meet the requirement. Run the drill string again to detect sand settling if necessary. The next operation can be started if sand settling is minimal and conductors can be sent in place. Inversely, if there is too much sand settling, and reaming and clearing the well hole would have no good effect; conductors cannot be sent in place. The depth of conductors should be adjusted or the well depth should be properly increased to extend the pocket length.

4. Tripping out drill bit and lowering conductors

   After completing formal drilling, the drill bit can be tripped out. However, it is not permitted to pull up the drill string by rotating the rotary table because the drill bit would be entwined on the conduct ropes. After tripping out the drill bit, conductors can be lowered in order. When lowering conductors, install the conduct arm or conduct ropes 2 m up to the float shoe to lead the conductors into the well hole if the drill base plate was lowered onto the seafloor. When lowering conductors, the conductor string should pass through the central hole of the permanent conduct shelf, which is placed on the moon pool door of platforms. The conductor joint should be connected to the top of the conductor string and then connected with sending tools. Lead the conductor into the base plate of permanent conduct shelf with heavyweight drill pipe of ϕ 127 mm (5 in) and lock with a locking pin bolt. Launch heave compensation equipment to slow the speed of lowering and compensate for heave movement when the conductor string is near the temporary oriented base plate or seafloor. Recover the conduct arm after the conductor runs into the well hole. Then lower the conductor continually until the distance between the top of the conductor joint and the bottom of the sea meets the designed requirement, and place a part of the weight load on the temporary oriented base plate. If the conductors can't be sent into the well hole, the platform should be migrated and drilling should be restarted.

5. Implementing cementing operations

   Before cementing, control the pump pressure at 5.52 MPa, circulate with seawater, and observe the inclination status of the seafloor permanent conduct shelf by ROV or underwater television. The degree of inclination should not exceed 1 degree and the bow orientation of the permanent conduct shelf and the platform should be the same. Once the inspection is passed, normal cementing procedures can be implemented. During cementing, prevent conductors from migrating upwards and release the weight of the whole conductor string gradually and load it on the temporary conduct base plate. After completing cementing, wait for concretion, separate the sending tools and pull them up 2 m, and then launch the pump to clear the pipes with seawater. Heave compensation equipment should be adjusted well to ensure the conductor is static when waiting for concretion. Recover the sending tools after the cement hardens.

6. Installing risers

There is no need to lower risers and install a splitter system if the next well section does not to be appraised. Otherwise, risers should be lowered and a splitter system should be installed after cementing the conductors. Lower the hydraulic connector of the conductor first and connect it to the risers. Then, lower the risers and telescopic joints on the top of the risers in order according to the designed depth. Next, install the taut line and tensioner system of the risers. Lastly, install the splitter system and do a functional test with seawater. The test sealing pressure should be up to 0.272 MPa. Heave compensation equipment should be launched to eliminate the effect of heaving movement when the hydraulic connector has been laid on the conductor head along the conduct rope of the permanent conduct shelf. Check whether the height of the hydraulic connector on the wellhead base plate meets the design requirement after it has been laid. Lock it after it has qualified.

Figure 2-78 illustrates the main drilling procedures of the conductor section for floating drilling platforms or ships. Part (a) in this figure shows the situation where the drill bit and drill pipe run through the central hole of the conduct shelf to drill a well hole for conductors of ϕ 762.2 mm (30 in). Part (b) in this figure is the situation of preparing for sending a conductor of ϕ 762.2 mm into water after installation. Part (c) in this figure is the situation of lowering a conductor head of ϕ 762.2 mm (30 in) and the permanent conduct shelf onto the seafloor. Part (d) in this figure is the situation where the whole conductor string, conductor head, and the permanent conduct shelf lay on the seabed.

2.4.2.2 Drilling Operations for the Conductor Section of Jack-up Drilling Platforms

When it comes to the issue of drilling operation for the conduction section of jack-up drilling platforms, the drill assembly, technological measures, and parameters of drilling, as well as the methods of clearing, inclination measuring, sand settling detection, and cementing are all the same as those for floating drilling platforms or ships described above. Therefore, there is no need to repeat those and here only some special issues are detailed.

1. Special issues related to the conductor lowering operation

   a. Locate the mud line support ring in the conductor about 2 m under the seabed when lowering the conductor so that the conductor head can be supported by the seabed.

   b. Locate the tieback joint connection about 1 m above the seabed for utilization in the future. The first conductor joint on the seabed should be connected if there is no need to tie back. This connection should also be located 1 m above the seabed.

   c. When lowering the conductor, the bottom joint of the last conductor should be kept away from the conductor incision location for installation of the casing head in the future. This is also convenient for cutting conductors.

   d. The conduit should be loaded on the rotary table and cushioned with sleeper after lowering. Then lower the internal pipe with an insert joint and connect it with 1 to 2 stabilizers. After the insert joint inserts into the conductor float shoe, pour seawater to check on the sealing effect. After it has qualified, pull out the insert joint and discharge the seawater. Then insert the float shoe again and prepare for cementing.

2. Special issues related to installing the wellhead

   a. Remove the sleeper cushioned on the rotary table first and free the conductor. Then incise the conductor at the designed location.

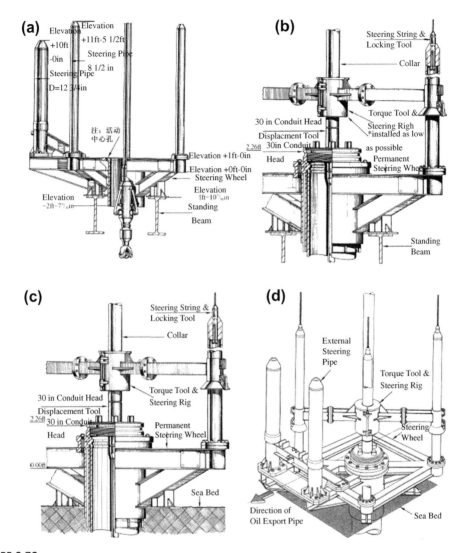

FIGURE 2-78

Preparing, sending, lowering, and laying of conductors.

b. Install the wellhead and wellhead equipment using ϕ 914.4 mm (30 in) with a blowout preventer system, as shown in Figure 2-79. Lengths A and B in the figure can be determined according to the height of the platform structure.

c. A simple wellhead without blowout preventer can be used if there is no shallow gas. In this case, the simple wellhead has two types according to whether it needs to bail sand or not. They are shown in Figure 2-80 (a) (bailing of sand is required along with drilling with

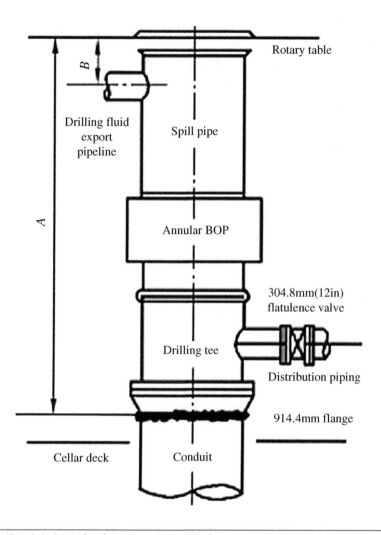

FIGURE 2-79

Wellhead and wellhead devices of self-elevating drilling platform.

drilling mud) and Figure 2-80(b) (bailing of sand is not required and we can drill with seawater). Size A and B in the figure should be determined according to the height of the platforms.

2.4.3 Drilling Operations for Surface Casing Sections

The aim of the drilling operation at this stage is lowering a surface casing of ϕ 508 mm (20 in) and drilling a wellhole of ϕ 660.4 mm (26 in). The main elements and steps of surface casing portion of the drilling operation follow.

FIGURE 2-80

Two simple wellhead types for jack-up platforms with different requirements.

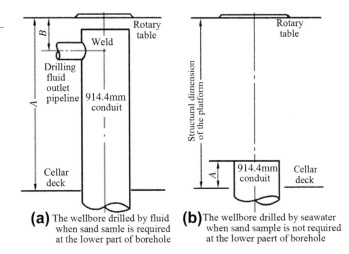

(a) The wellbore drilled by fluid when sand samle is required at the lower part of borehole

(b) The wellbore drilled by seawater when sand sample is not required at the lower paert of borehole

2.4.3.1 φ 444.5 mm (17 in) Pilot Hole Drilling Operation

The purpose of this operation is drilling a well hole of φ 660.4 mm (26 in). We drill a well hole of φ 444.5 mm diameter first and then expand it to φ 660.4 mm. This is why it is called a pilot hole drilling operation.

1. Drill assembly for drilling pilot hole

   The normal offshore drill assembly structure for a φ 444.5 mm pilot hold drilling operation is: Drill bit (φ 444.5 mm) + joint with float valve and inclination measuring base + 2 drill collars (φ 228.6 mm) +joint + 10~13 drill collars (φ 203.2 mm) + joint + 15 heavyweight drill pipes(φ 127 mm)

2. Technological parameters of drilling pilot hole

   a. Drilling a cement plug

   Seawater is often used to drill and the common technological parameters are weight-on-bit is 65~110 kN, RPM is 60~70 r/min, and pump displacement is 4100 L/min.

   b. Drilling a pilot hole

   Pilot hole drilling starts after cement plug drilling is finished; this is usually drilled with drilling mud, which should be replaced before continuing drilling. The technological parameters in this stage are weight-on-bit 20~65 kN, RPM is 80~90 r/min, and pump displacement is 3600~4100 L/min.

3. Technical processing of shallow gas

   a. Preparation

   If shallow gas is predicted to exist, implement a splitter system functional test before drilling, check the cement pump to make sure it is in good condition, and store high-density drilling mud.

   b. Processing method

   Pump in high-density drilling mud as soon as possible. Inject a cement plug to seal the gas layer in time after the well hole is stable. Circulate with high displacement drilling mud if the well kick still exists. Evacuate personnel, disjoint the wellhead, and remove floating drilling platforms or ships if the well kick is serious.

Well hole clearing, well inclination measuring, sand settling processing are all the same as we detailed for conductor operations so here the narration is abbreviated. However, it should be noted that there are two drilling operation methods for a surface casing section. The first one is drilling a well hole of ϕ 444.5 mm first and then expanding it to ϕ 660.4 mm; the second one is drilling a well hole of ϕ 660.4 mm directly. The latter is often used in the situation of a floating drilling platform or ship without lowering risers or a jack-up drilling platform with a simple wellhead.

2.4.3.2 Expand Well Hole of ϕ 444.5 mm (17 in) to Well Hole of ϕ 660.4 mm (26 in)

We can directly expand the ϕ 444.5 mm well hole if electrical logging is not needed. Otherwise, expansion of the well hole should be implemented after electrical logging.

1. Drill assembly for well hole expansion
   We use drill bit (ϕ 444.5 mm) + hydraulic expanding reamer (ϕ 660.4 mm) + other drill tools as with the pilot hole drilling operation described above.
   The drill assembly above is mainly used for floating drilling platforms (ships). If we have a jack-up drilling platform, a fixed arm expanding reamer of ϕ 660.4 mm is needed to replace the hydraulic expanding reamer. The other components are the same as described above.
2. Drilling parameters for well hole expansion
   When expanding the well hole, the RPM and pump displacement parameters for both drilling a cement plug for sealing shallow gas and normal section drilling are the same as for pilot hole drilling. But the weight-on-bit should be enhanced to $90 \sim 110$ kN when drilling a cement plug and $45 \sim 89$ kN when drilling for a normal section.
   Clear the well hole, measure well inclination, and process sand settling after drilling at the designed depth and then pull up the drill tools. Trip out should be implemented after the shallow gas is controlled. Trip out of the subsea wellhead at the bottom of the sea for a floating drilling platform or ship and then replace the drilling mud in the risers with seawater. Pull out the risers after trip out is completed.

2.4.3.3 Drill Well Hole of ϕ 660.4 mm (26 in)

1. Drill assembly
   Commonly a drill bit of ϕ 660.4 mm is used or drill bit (ϕ 444.5 mm) + fixed arm expand reamer (ϕ 660.4 mm), connected with drill tools with the same structure as for drilling the pilot hole.
2. Drilling parameters
   When expanding the well hole, the RPM and pump displacement parameters for both drilling a cement plug and normal section drilling are the same as those for pilot hole drilling. But the weight-on-bit should be enhanced to $110 \sim 134$ kN when drilling a cement plug and $66 \sim 110$ kN for normal section drilling.
   In addition, drill tool lowering and tripping-out, well hole clearing, well inclination measuring, sand settling processing are all the same as for the conductor operations introduced previously.

2.4.3.4 Lower φ 508 mm (20 in) Surface Casing

We implement the procedures for lowering the casing after installing the float shoe and completing the coupling of the casing string. The casing lowering operation can be distinguished into two situations.

1. Casing lowering operation for floating drilling platform or ship

   Commonly we install conduct ropes and a conduct shelf 5 m up to the float shoe and add tieback ropes. This makes it convenient to recover the conduct shelf when the casing is lowered about 10 m into the conductor. Heave compensation equipment should be launched to prevent knocking before the casing reaches the base plate. We lower drill pipe of φ 127 mm after the casing is lowered, matching the design requirements; the length should be 20 ~ 25 m up to the float shoe for pumping cement. The casing head is lowered to the seabed by heavyweight drill pipe of φ 127 mm and sending tools, then it sits on the conductor head of φ 672 mm and locks to it. When the casing head sits, heave compensation equipment should also be launched and the casing head should be pulled up with 90 kN pressure to test that whether it sits in place and is locked well. Next, pump cement into the drill string; it flows back to the casing head through the annulus and then flows to the surface through the conductor head. After the cement hardens, clear the well hole and recover the sending tools for the casing head and drill pipe.

2. Casing lowering operation for jack-up drilling platform

   The casing lowering operation procedures for a jack-up drilling platform are basically the same as the operations for a floating drilling platform/ship. The main differences are:

   a. Sitting of mud line hanger

      The wellhead of a jack-up drilling platform is a mud line hanger rather than the casing head for a floating drilling platform/ship. Therefore, when the mud line hanger is near the supporting ring of the conductor, the casing lowering speed should be slower. Because there is no heave compensation equipment, sitting can only be controlled by observing the weight indicator. After sitting on the supporting ring, add 90 kN pressure to ensure it is in place.

   b. Drill string with insert joint

      We lower a drill string with an insert joint into the casing for cementing. The insert joint should be inserted into the float collar or float shoe by 10 kN of pressure. Pump cement after cementing pipes and lines are connected to the top of the drill string. An elastic centralizer of 127 mm × 508 mm should be installed onto the first and second drill string joints to ensure that the insert joint was inserted into the float collar or float shoe correctly.

   c. Wash cement in the annular space of the casing

During the cementing process of a jack-up drilling platform, cement paste will be returned to the annular space of the casing on the mud line hanger and it should be washed with water. During washing, lower the casing weight on the mud line hanger and rotate the casing until the washing hole of the hanger comes out. Next, connect the casing circulating head and open the pump for washing. Then fasten the casing with the torque required.

2.4.3.5 Lower φ 476.25 mm (18 in) Blowout Prevents for Floating Drilling Platform/Ship

This procedure is specially implemented for a subsea wellhead used by a floating drilling platform/ship. The specific steps are as follows.

1. Preparations before operation

   These preparations are all completed on the deck of the platform. First, test the blowout preventer. Then, install the control system of the blowout preventer and control box and test the pressure. Next, install the hydraulic connector and flexible joint, which connects the risers and casing head.

Let the oriented pillar pass through the oriented rope. Last, connect two section risers and install the well killing and relief pipeline onto the risers. After the pressure test passes, the lowering operation can be started.

2. Lower blowout preventer with risers

   Connect the two section risers onto the blowout preventer and lower the blowout preventer under the sea surface directly along the oriented rope. Then continue to lower the blowout preventer with risers. Test the pressure of the kill and choke lines for each $2 \sim 3$ sections of risers; the pressure should be up to 80% of the rated pressure. Once the blowout preventer reaches to the seabed base plate, heave compensation equipment should be launched to prevent impact. Keep the height between the blowout preventer and the oriented pillar of base plate to at least 2 m to prevent the oriented rope from rubbing off.

3. Lock casing head and connector

   Lock the hydraulic connector under the blowout preventer and wellhead pipe end. Pull up with 130 kN pressure with heave compensation equipment to check whether they are locked.

4. Connect expansion joint of riser pillar

   Commonly, we connect the expansion joint for heaving compensation on the top of riser string and pull out the internal cylinder of the expansion joint to install the splitter (diverter system) after the lowering of the blowout preventer is completed.

5. Run wear bushing to casing hanger

   The function of wear bushing is to protect the casing head and casing hanger and prevent them from wearing down due to the drill bit, tool joints, and tools during drilling operation.

Wear bushing has four locking pins and can mesh with the J-shape groove on the nut of the casing hanger. So that it could let wear bushing in place, maintain location during drilling in, tripping, and anti-wearing. The lowering operation of wear bushing is implemented by sending tools. The wear bushing for a casing of ϕ 508 mm (20 in) is ϕ 476.25 mm (18 in). The wellhead and wellhead equipment of ϕ 476.25 mm (18 in) which is installed for a floating drilling platform/ship is shown in Figure 2-81. Figure 2-82 shows the major drilling operation procedures of the surface section of the casing for a floating drilling platform/ship: part (a) is the drill hole for lowering a casing of ϕ 508 mm (20 in); part (b) is the lowering of the casing; part (c) is the casing going into the casing head; and part (d) is the connection between the blowout preventer and casing head.

2.4.3.6 ϕ 539.75 mm (21 in) Wellhead Installation for Jack-up Platform

The wellhead and wellhead equipment of a jack-up platform are all located on the sea surface and they are different from those of a floating drilling platform/ship. Thus, after completing the lowering of the surface casing and cementing, the wellhead should be installed according to the following steps.

1. Cut conductor and casing

   At first, trip out casing slips to let casing be in a free condition, and lift the diverter system to cut the conductor at the welding line of the base flange of ϕ 672.0 mm. Next, lift the diverter and cut the casing of ϕ 508 mm neatly at a proper height on the conductor incision. Last, pull out the casing that has been cut and take off the diverter system of ϕ 762 mm.

2. Weld standard casing head

   The standard casing head is ϕ 539.75 mm with a pressure of 79 MPa, and the inside and outside welding method should be used. Let a circular iron plate sit on the casing head on the ϕ 762 mm conductor and weld it to the well according to the diagram in Figure 2-83. A pressure test should

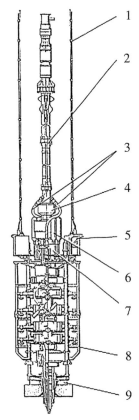

1: BOP has control tube.There are two tubes among which one is spare.each one has many control tubes.

2: Marine riser.It not only guides drill string into the hole,but also is the channel of drilling fluid and rock debris returning to the platform.

3: Choke line and kill line of BOP

4: Lower riser assembly: the flexible joint makes up for the horizontal motion,while the higher marine rise makes up for the vertical motion.

5: There are two BOP control box.They are seperately connected with the BOP control tube and can provide hydraulic function.Moreover,they can be pulled out for maintenancing and repairing.

6: Hydraulic accumulator is providing power for the BOP quick switch.

7: The upper connection device is used to unlock and trip out the marine tube system.

8: Guiding frame of BOP

9: Drilling system :temporary guide chassis、permanent guide chassis and wellhead.

FIGURE 2-81

Φ476.25 mm wellhead and wellhead devices of floating drilling platform.

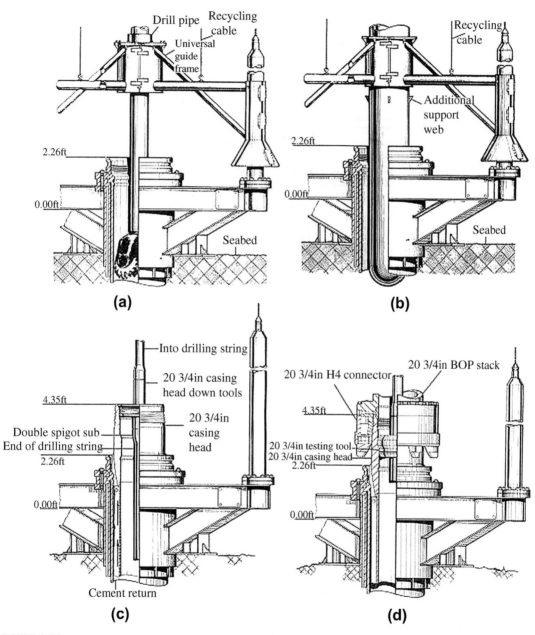

FIGURE 2-82

Drilling process in surface casing of floating drilling platform.

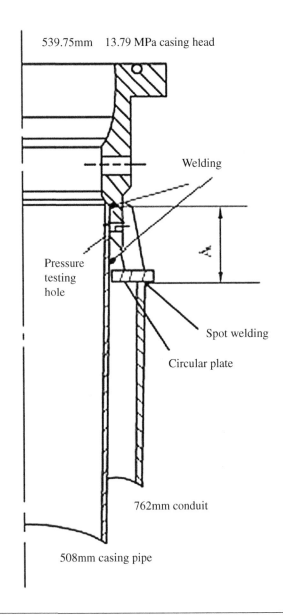

539.75mm   13.79 MPa casing head

Welding

Pressure testing hole

Spot welding

Circular plate

762mm conduit

508mm casing pipe

FIGURE 2-83

Casing head installment and welding of self-elevating drilling platform.

be done for the welding line; the pressure should be up to 60% of the casing collapsing strength and last 10 min.

3. Install blowout preventer system

There are three types of seawater drilling wellhead and wellhead equipment on the sea water for jack-up platforms in China. They are shown in Figure 2-84 and part (a) is the most commonly used type. After determining the type of wellhead equipment, the blowout preventer system,

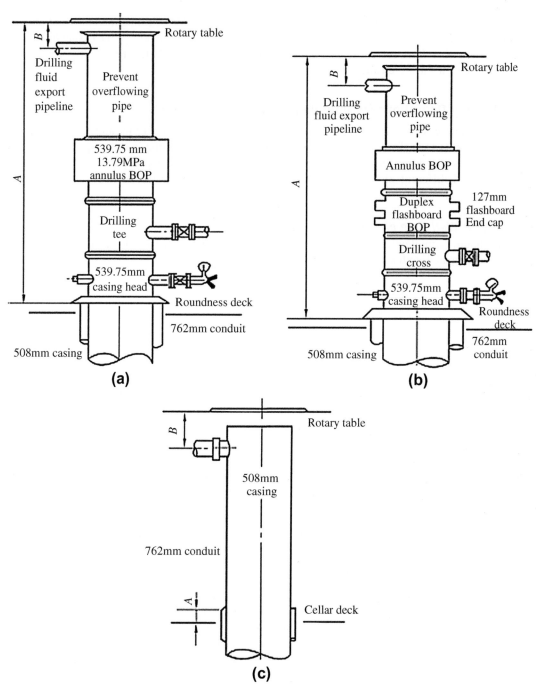

FIGURE 2-84

Wellhead and wellhead devices of self-elevating drilling platform used in China.

which contains blowout preventer groups, overflow prevention pipes, drilling mud flow-out pipe, and kill and choke lines, can be installed. We must do a pressure test to 60% of the casing internal pressure strength. Last, lower the wear bushing to the casing head. At this point, all drilling operations for a surface casing section will be completed.

2.4.4 Drilling Operations for Intermediate Casing Sections

The intermediate casing section refers to a well hole of ϕ 444.5 mm (17 in) and casing sections of ϕ 339.73 mm (13 in). The elements and steps of the drilling operation in this stage is generally the same as the procedures for the surface casing section. The following paragraphs only introduce the special and different elements of major procedures as compared with the surface casing section.

2.4.4.1 Drill Well Hole of ϕ 444.5 mm (17 in)

This step should be implemented after full preparation (the same as in the surface casing section); the differences compared with surface casing section are as follows.

1. Drill assembly structure
   The commonly used structure is the following:
   Drill bit (ϕ 444.5 mm) + joint with float valve and inclination measuring base + 2 drill collars (ϕ 228.6 mm) + stabilizer (ϕ 444.5 mm) + joint + 13 drill collars (ϕ 203.2 mm) + drill jar + 2 drill collars (ϕ 203.2 mm) + joint + 14 heavyweight drill pipes (ϕ 127 mm) or according to the designed requirements
2. Drill technological parameters
   a. Cement plug drilling stage: commonly used weight-on-bit of $110 \sim 130$ kN, RPM of $50 \sim 60$ r/min, and pump displacement of 3400 L/min
   b. Regular drilling stage: commonly used weight-on-bit of $40 \sim 130$ kN, RPM of $70 \sim 100$ r/min, and pump displacement of $3200 \sim 3400$ L/min

2.4.4.2 Lower ϕ 339.73 mm (13 in) Casing and Cementing

The operations in this step are different for a floating drilling platform/ship and jack-up platform.

1. Floating drilling platform/ship
   Nowadays, the method of sending casing by drill pipe to sit on the subsea wellhead is commonly used in Chinese offshore drilling. This method should implement the following assembly procedures.
   a. Assemble casing hanger joint
   First, connect a converting joint to sending tools on a heavyweight drill pipe of ϕ 127 mm; then, connect the sending tools onto the drill pipe converting joint. Next, connect the pressure balance valve and casing plug assembly in sequence; finally, lift and lay the casing hanger and two pin end joints onto the wellhead, which is connected to the deck, lower the sending tools into the hanger, which has been connected to the well, and lock and fasten the assembly. After assembling is complete, lift and lay it on the deck and prepare to lower it to the seafloor. This is shown in Figure 2-85.

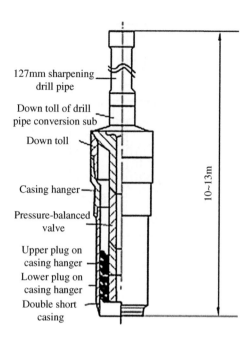

127mm sharpening drill pipe

Down toll of drill pipe conversion sub

Down toll

Casing hanger

Pressure-balanced valve

Upper plug on casing hanger
Lower plug on casing hanger
Double short casing

10~13m

FIGURE 2-85

Assembly of casing hanger when down the technical casing in the floating drilling platform.

b. Assemble drill pipe cement head joint

First, put the drill pipe plug and release ball into the drill pipe cement head on deck and fasten it well by releasing the rod. Then connect a heavyweight drill pipe of ϕ 127 mm to the bottom end of the cement head on the drill block and connect the drill pipe lifting sub to the top end. Fasten two left hand threads on the cement head and check the sealing ring. After this is complete, lift and lay it on deck and prepare for cementing. The whole assembling structure is shown in Figure 2-86.

2. Jack-up drilling platform

Because a mud line hanger is used to suspend the intermediate casing for a jack-up drilling platform, the casing hanger joint of a floating drilling platform/ship is replaced by a casing mud line hanger joint during assembly, and all other assembly steps are the same as for a floating drilling platform/ship.

The procedures of lowering the casing and cementing for an intermediate casing of ϕ 339.73 mm are the same for either a floating drilling platform/ship or jack-up drilling platform and need not be repeated here.

2.4.4.3 Install Wellhead and Wellhead Equipment

1. Floating drilling platform/ship

When it comes to a floating drilling platform/ship, the installation methods for a wellhead equipment like a blowout preventer group for an intermediate casing of ϕ 339.73 mm (13 in) are

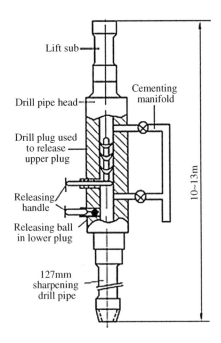

Lift sub

Drill pipe head

Cementing manifold

Drill plug used to release upper plug

Releasing handle

Releasing ball in lower plug

127mm sharpening drill pipe

10~13m

FIGURE 2-86

Assembly of drill pipe cement head in the floating drilling platform when down casing.

the same as for a surface casing and there is no need to repeat it here. The operation procedures are shown in Figure 2-87 and illustrated in the following.

a. Figure 2-87(a): In order to install the casing head (well head), connect the intermediate casing of ϕ 339.73 mm (13 in) as well as its sending tools between two short joints of casing of ϕ 339.73 mm.

b. Figure 2-87(b): Lower the blowout preventer group of the casing section of ϕ 339.73 mm together with a hydraulic connector of ϕ 339.73 mm into a casing head of ϕ 339.73 mm with combined risers of ϕ 406.4 mm (16 in) and ϕ 609.6 mm (20 in); connect them well and test the pressure up to 10.346 MPa.

c. Figure 2-87(c): Test pressure for the blowout preventer group with a connected drill pipe and assemble the tools that have been sent.

d. Figure 2-87(d): Lower wear bushing through the drill pipe and sending tools. At this time, installation of the wellhead equipment is complete.

2. Jack-up drilling platform

The wellhead of a jack-up drilling platform is on the sea surface and its installation method is the same as for the surface casing illustrated above need not be repeated. Figure 2-88 is the wellhead equipment after installation is complete for the intermediate casing of a jack-up drilling platform.

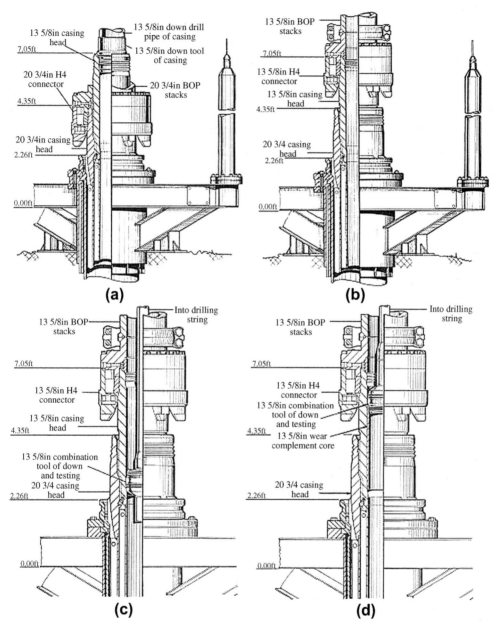

FIGURE 2-87

Installment of Φ339.73 mm technical casing in the wellhead.

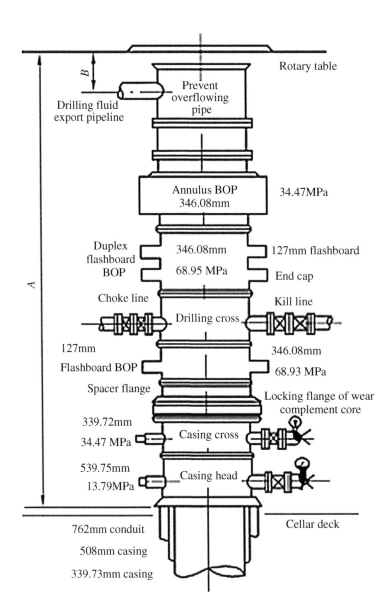

FIGURE 2-88

Wellhead devices of Φ339.73 mm technical casing in self-elevating drilling platform.

2.4.5 Drilling Operations for Production Casing Sections

2.4.5.1 Drill φ 311.15 mm (12 in) Well Hole

As with the operations for surface casing sections and intermediate casing sections illustrated above, after full preparation, a well hole of ϕ 311.15 mm for lowering the production casing can be implemented. The differences are as follows.

1. Drill assembly structure
   Drill bit (ϕ 311.15 mm) + fishing ring + joint with float valve and inclination measuring base + 12 drill collars (ϕ 203.2 mm) + joint + 1 heavyweight drill pipe (ϕ 127 mm) + joint of input check valve base + 14 heavyweight drill pipes (ϕ 127 mm), or according to the designed requirements
2. Technological parameters of drilling
   a. Drilling cement plug: Weight-on-bit is $110 \sim 140$ kN; RPM is $40 \sim 60$ r/min; and pump displacement is $2270 \sim 2600$ L/min.
   b. Regular drilling: We start drilling a new formation for 30 min: weight-on-bit is $45 \sim 100$ kN; RPM is $70 \sim 100$ r/min; and pump displacement is $2270 \sim 2600$ L/min. Then, we continue drilling to the design depth: weight-on-bit is $90 \sim 155$ kN; RPM is $80 \sim 100$ r/min; pump displacement is $2270 \sim 2600$ L/min.

2.4.5.2 Lower φ 244.48 mm (9 in) Casing, and Cement and Install Wellhead Equipment

The operation methods and steps of lowering the casing, pumping cement and cementing, as well as installing wellhead equipment for production casing (9 in) are the same as for surface and intermediate casing introduced above. The major procedures for a floating drilling platform/ship are briefly illustrated in the following and shown in Figure 2-89.

a. Figure 2-89(a): Assemble casing hanger and its sending tools onto short casing joint of casing string, and prepare for lowering.
b. Figure 2-89(b): Lower oil-string casing (9 in) along with its hanger and sending tools, and let them sit on the casing head (13 in).
c. Figure 2-89(c): Lower the sealing assembly to seal the annular space between the intermediate casing (13 in) and production casing (9 in). The sealing assembly and its sending tools are connected to a drill pipe joint and lowered by drill string. Then they sit on casing hanger and are fastened by threads. Their torque should be up to 16.269 kN·m.
d. Figure 2-89(d): Downing wear bushing onto the sealing assembly (13 in × 9 in). This is implemented by drill pipe. Test tools and sending assemble tools.

The pumping of concrete and cementing start after lowering the casing; the installation of the wellhead equipment and blowout preventer group start after lowering the sealing assembly. Because these operations are the same as for an intermediate casing, there is no need to repeat them here.

The operations of lowering the casing, cementing, and installing the wellhead for production casing of a jack-up drilling platform are all the same as the methods and steps for intermediate casing. So it is not introduced here again. Only the commonly used wellhead equipment schematics for a jack-up drilling platform after lowering a production casing of ϕ 244.48 mm and cementing are shown in Figure 2-90.

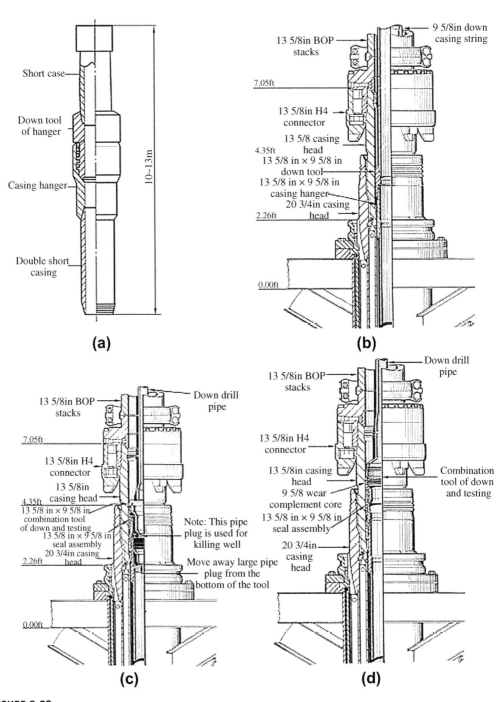

FIGURE 2-89

Operation process of oil-string casing in floating drilling platform.

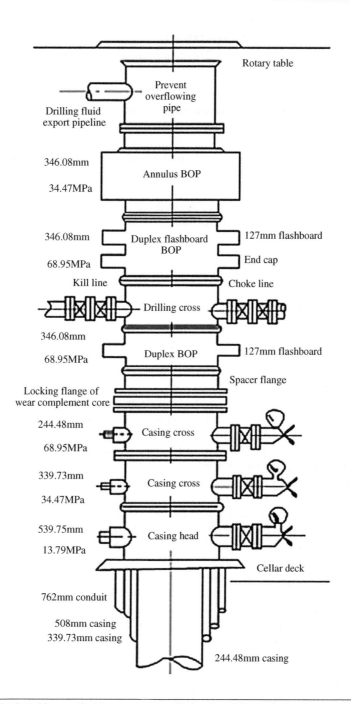

FIGURE 2-90

Wellhead devices of Φ244.48 mm oil-string casing in self-elevating drilling platform.

2.4.5.3 Implement Oil and Gas Test Operations

After cementing an oil-string casing of ϕ 244.48 mm as designed, oil and gas testing should be implemented through the following steps.

1. Drill cement plug until it reaches the required artificial bottom hole
   a. Drill assembly: The commonly used structure is drill bit (ϕ 215.9 mm) + fishing ring + joint with float valve + 15 drill collars (ϕ 165.1 mm) + joint + 15 heavyweight drill pipes (ϕ 127 mm).
   b. Drilling technological parameters: weight-on-bit is 85~130 kN; RPM is 50~60 r/min; and pump displacement is 1500 L/min.
2. Test casing pressure
   Testing of the casing pressure occurs after the well hole circulation is clear. The test pressure should be up to the design value or 80% of the casing internal pressure strength.
3. Measure cementing quality
   Lower electric measuring instruments after pulling out the drill string.
4. Implement oil and gas test operations until we meet the design requirements.

2.4.6 Drilling Operations for Tail Pipe Section

2.4.6.1 Tail Pipe Section of ϕ 177 mm (7 in)

The aim of this drilling operation is to drill a well hole of ϕ 215.9 mm (8 in) for lowering a tail pipe of ϕ 177 mm (7 in).

1. Drill well hole of ϕ 215.9 mm
   a. Drill cement plug: Drill a cement plug first before drilling a new formation.
      i. Drill assembly: The commonly used assembly is drill bit (ϕ 215.9 mm) + fishing ring + joint with float valve and inclination measuring base + 15~18 drill collars (ϕ 165.1 mm) + joint + 1 heavyweight drill pipe (ϕ 127 mm) + joint of input check valve base + 14 heavyweight drill pipes (ϕ 127 mm).
      ii. Drilling technological parameters: Generally, weight-on-bit is 80~120 kN; RPM is 50~60 r/min; and pump displacement is 1500 L/min.
   b. Drill new formation: Start drilling a new formation after drilling cement plug is complete.
      i. Drilling technological parameters: Generally in the beginning weight-on-bit is 65~85 kN; RPM is 50~70 r/min; and pump displacement is 1200~1500 L/min. After drilling to the designed depth, the weight-on-bit is 80~120 kN; RPM is 80~100 r/min; and pump displacement is 1140~1500 L/min.
      ii. Drill assembly: The commonly used assembly is drill bit (ϕ 215.9 mm) + joint with float valve and inclination measuring base + 2 drill collars (ϕ 165.1 mm) + stabilizer (ϕ 215.9 mm) + 12~15 drill collars (ϕ 165.1 mm) + drill jar (ϕ 165.1 mm) + 2 drill collars (ϕ 165.1 mm) + joint + 1 heavyweight drill pipe (ϕ 127 mm) + joint of input check valve base + 14 heavyweight drill pipes (ϕ 127 mm), or according to the design requirements.
2. Lower tail pipe of ϕ 177.8 mm (7 in) and cement
   After electrical logging, assemble the drill pipe cement head for cementing on the drill block, assemble the tail pipe hanger, ball seat coupling joint, and float shoe on deck, and then lower the tail pipe; send the tail pipe by drill pipe of ϕ 127 mm until it reaches the bottom hole. If it is

slacking off, launch a pump to wash. Next, connect the cement head and cement in the regular way. After solidification, drill a cement plug to the top of the tail pipe. The drill technological parameters are as follows: weight-on-bit is $85 \sim 130$ kN; RPM is $50 \sim 60$ r/min.

3. Test pressure seal between overlaps of tail pipe
   After the well bore circulation is clear, a pressure test should be implemented to check the seal conditions between overlaps (well bore) of tail pipe. Generally, a positive pressure test method is used for oil and gas wells. But a negative pressure test method should also be implemented for a gas well.
   a. Positive pressure test method
      This method lifts the drill bit away 8 m from the top of tail pipe. It tests the pressure between the overlapping well bore at the blowout preventer with a cementing pump. We test the pressure for 10 min and check that the pressure is up to the design requirement.
   b. Negative pressure test method
      This method lifts the drill bit, lowers a drill pipe with a packer to 8 m from the top of tail pipe, connects a high pressure pipeline, injects seawater with a cementing pump to maintain pressure, sets the packer to observe reflux, and confirms whether the packer sets well or not. We check if the overlaps of the tail pipe seal well, testing the pressure for 10 min to the design pressure. Pull out the packer after the test passes, and carry out subsequent operations.

4. Implement oil and gas test operations
   If an oil and gas test is needed, we should implement the following steps for well completion.
   a. Drill assembly: The commonly used structure is drill bit (ϕ 152 mm) + fishing ring + joint with float valve + 15 drill collars (ϕ 120.65 mm) + joint + 9 heavyweight drill pipes (ϕ 88.9 mm) + drill pipe (ϕ 88.9 mm) with the length of drilling to the designed depth + joint + drill pipe (ϕ 127 mm).
   b. Drilling of cement plug: The commonly used drill technological parameters are weight-on-bit of $60 \sim 80$ kN; RPM of 50 r/min; and pump displacement of $1300 \sim 1750$ L/min. Drill until reaching the artificial bottom hole required.
   c. Tail pipe pressure test: After the well bore circulation is clear, test the pressure of the tail pipe to the design requirements, and check that it does not exceed the test pressure value of tail pipe overlaps.
   d. Electrical logging of cementing: If the cementing quality does not meet the requirements, corresponding measures should be implemented; if the cementing passes after examination by electrical logging, oil and gas test operations can be implemented.

2.4.6.2 ϕ 114.3 mm (4 in) Tail Pipe Section

In order to lower a tail pipe of ϕ 114.3 mm (4 in), a well bore of ϕ 152.4 mm (6 in) should be drilled in this drill operation stage. The elements, methods, and procedures are all the same as for the drilling operation for a tail pipe section of ϕ 177 mm (7 in). We drill the well bore, lower the tail pipe, pump cement, examine the sealing pressure between the tail pipe overlaps, and conduct an oil and gas test. The following only briefly introduces relevant differences.

1. Drill assembly for drilling well hole of ϕ 152.4 mm
   a. Commonly used drill assembly for drilling cement plug: Drill bit (ϕ 152.4 mm) + fishing ring + joint with float valve + 15 drill collars (ϕ 120.65 mm) + joint + 12 heavyweight drill pipes

(ϕ 88.9 mm) + drill pipe (ϕ 88.9 mm) (which should meet the requirements of tail pipe (ϕ 177.8 mm) and well hole below) + joint + joint of input check valve base + drill pipe (ϕ 127 mm), or according to the designed requirements.

b. Commonly used drill assembly for drilling new formation: Drill bit (ϕ 152.4 mm) + joint with float valve + 2 drill collars (ϕ 120.65 mm) + stabilizer (ϕ 152.4 mm) + 15 drill collars (ϕ 120.65 mm) + joint + 12 heavyweight drill pipes (ϕ 88.9 mm) + drill pipe (ϕ 88.9 mm) + joint + joint of input check valve base + drill pipe (ϕ 127 mm).

2. Drill technological parameters for drilling well hole of ϕ 152.4 mm

a. Commonly used drill technological parameters for drilling cement plug: Weight-on-bit is 60~80 kN; RPM is 40~50 r/min; and pump displacement is 1300~1750 L/min.

b. Commonly used drill technological parameters before drilling first 30 m of new formation: Weight-on-bit is 45~65 kN; RPM is 50~60 r/min; and pump displacement is 950~1200 L/min.

c. Commonly used drill technological parameters after drilling first 30 m of new formation: Weight-on-bit is 45~80 kN; RPM is 50~60 r/min; and pump displacement is 950~1200 L/min.

3. Drill technological parameters of well completion operation

a. Commonly used parameters for drilling a cement plug to the top of tail pipe (ϕ 114.3 mm): Weight-on-bit is 60~85 kN; RPM is 40~50 r/min; and pump displacement is 1300~1950 L/min.

b. Commonly used parameters for drilling a cement plug to the artificial well bottom: Weight-on-bit is 22~45 kN; RPM is 40~50 r/min; and pump displacement is 1000 L/min.

4. Commonly used drill assembly for well completion operation

Drill bit (ϕ 95.25 mm) + joint + 15 drill collars (ϕ 88.9 mm) + joint + drill pipe (ϕ 60.33 mm) + joint + drill pipe (ϕ 88.9 mm) + joint + drill pipe (ϕ 127 mm), or according to the designed requirements.

5. Prevent well kick and well leakage

The speed of drill string trip in and out should be controlled to prevent well kick caused by pressure swabbing and well leakage caused by pressure surging. This matter should always be paid attention to when the drill bit is tripping in or out of the tail pipe and well hole. In addition, when pulling out sending tools to the top of the tail pipe (ϕ 177.8 mm) and into the casing (ϕ 244.48 mm), the lifting speed should be slow to prevent pressure swabbing.

2.5 Completion of Offshore Oil and Gas Drilling

Completion is the last link in oil and gas drilling engineering, and plays an important role in the development of offshore oil and gas fields, directly affecting the age and production capacity of the wells. In addition, offshore completion is too expensive, and the reasonable design and careful construction of completion can reduce the investment cost of offshore oil field development and improve the development benefits, so completion of drilling engineering of offshore oil and gas fields is of special significance. Therefore, completion design, completion procedures, completion tests, and so on are introduced in this section.

2.5.1 Completion Design

Completion design is the basis for completion; optimizing completion design scientifically and reasonably is the key to completion. Completion is not a single operation technology, but a comprehensive collection of technology and system engineering. Thus, to achieve good completion design, offshore completion work should look at the characteristics of offshore oil and gas engineering, and closely cooperate with other related engineering disciplines. First, before the completion design, the relevant data from reservoir geology, reservoir protection, oil production technology, underwater environment, engineering geology, drilling engineering, and so on, as well as the standards and specifications on completion design should be fully collected.

2.5.1.1 Basic Principles That Must Be Followed in Completion Design

1. Meeting the requirements of development and production engineering plans
   We must meet the requirements of the production position, mining methods, and dynamic monitoring of all kinds of wells in the offshore oil and gas field development plan.
2. Meeting the demands of drilling and offshore engineering plans
   We must meet the requests made by drilling engineering for production casings, cementing quality, and reservoir protection, and coordinate with the subsea wellhead equipment and jacket platforms, and the production needs.
3. Meeting the requirements of safety, high efficiency, and environmental protection
4. Meeting the requirements of reservoir protection and reformation measures

We must try to minimize the damage to reservoir completion and meet the needs of the necessary reformation measures in the future.

2.5.1.2 Selection of Completion Methods

1. Classification of existing completion methods

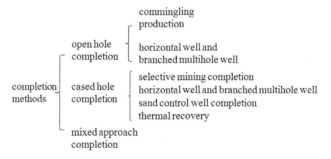

2. Selection principles
   a. Open hole completion is suitable for situations where the reservoir and interlayer rock are hard and not easily broken, such as limestone, dolomite, and granite hole type reservoirs or agglutinate and solid sandstone.
   b. In addition to the hard rock and rugged sandstone, all should be selected by the method of cased-hole completion, and production layer of sand should be sand control completion, such

as the application of lining tube completions or gravel pack completion methods which are filling certain size gravel between the liner and the borehole wall.

c. In the production process, production horizon adjustment, production logging, and stimulation process measures should be used the completion that commingling became points, points replace commingling as well as the card blocking layer process requirements.

d. As for special drilling technology, such as horizontal wells, multi-well, and large displacement wells, you should study the relevant completion.

2.5.1.3 Perforating Process Design

1. Selection of perforating modes

a. Existing perforating modes

i. Cable perforation

This is a method where the casing perforating gun is run down by the cable, the casing pup joint on the top of the reservoir is used to position the perforating depth, and the perforating gun is fired by the electric detonators to shoot through the oil and gas production layer, which can meet the requirements of high-density, deep penetration, large aperture perforation, and multi-layer perforation.

ii. Tubing conveyed perforation

This is a new perforating method where the perforating gun is conveyed to the perforating horizon by the tubing. This method in which large diameter perforating guns are used is suitable for high-pressure oil and gas wells, and highly deviated and horizontal wells.

iii. Through-tubing perforation

This method after completing the next production string, installing wellhead Christmas tree and production systems, plugging device, is to install the lubricator on the wellhead Christmas tree and cable sealing device, and cable down into the perforating gun from the tube perforating reservoir; the structure of the perforating gun is shown in Figure 2-91. This method is safe and suitable for high-pressure oil and gas wells, and a well can be put into production immediately after perforation.

iv. Through-tubing extending perforation

The difference between the through-tubing perforation and through-tubing extending perforation is that the perforating gun is in a vertical state when it is run down, but it is in a horizontal state after reaching the perforating horizon and turns into a large diameter perforating gun.

You can select the perforation modes with the reference to Table 2-8.

2. Selection of perforating technology

a. Styles of perforating technology

i. Overbalanced perforating technology

The formation is shot through when the pressure of the completion fluid in the wellbore is higher than that in the formation; this is called an overbalanced perforating technology.

ii. Underbalanced perforating technology

The formation is shot through when the pressure of the completion fluid in the wellbore is lower than that in the formation; this is called underbalanced perforating technology. Underbalanced perforation can eliminate the perforating fluid in the

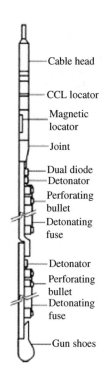

Cable head

CCL locator

Magnetic locator

Joint

Dual diode
Detonator
Perforating bullet
Detonating fuse

Detonator
Perforating bullet
Detonating fuse

Gun shoes

FIGURE 2-91

Structure of perforating gun for the method of through-tubing perforation.

formation, and the perforation tunnel can be made clean again to build a clean smooth flow channel of oil and gas, which is the reason this technology has been widely used. Usually it is used through the tubing conveyed perforation and through-tubing perforating method.

3. Determination of the perforating negative pressure
There are many sandstone reservoirs in the offshore oil and gas fields in China, and they are often divided into nondense formations and dense formations. A larger negative pressure can be selected in the situation as the former will not have sand after perforation; the latter has sand production easily after perforation, a reason the negative pressure should be selected reasonably for that type of formation.

a. Determination of the minimum perforating negative pressure
The minimum perforating negative pressure can be determined by the following equations for the nondense formation and the dense formation:
Oil reservoirs:

$$P_{\min} = 6.89 \times \frac{3.5}{K^{0.37}} \tag{2-41}$$

Table 2-8 Comparison between Several Perforating Methods

Perforating Methods	Cable Perforation	Tubing Conveyed Perforation	Through-Tubing Perforation	Through-Tubing Extending Perforation
Diameter of guns (mm)	73~177.8	35~54	73~177.8	42.9
Styles of bullet	Deep penetration and large diameter	Deep penetration	Deep penetration and large diameter	Deep penetration
Weight of bullet (g)	15~66	1.8~17	15~66	22
Density of holes (hole/m)	13~39	13~19	13~46	13
Depth of holes (mm)	400~800	146~615	400~800	678
Diameter of holes (mm)	7.1~31.3	5.4~14.5	7.1~31.3	
Phase (°)	120,90,60,45,30	80,90,60,0	120,90,72,60, 51.4,45,30,20	180
Range of negative pressure	Can not be negative pressure	Can be controlled or isobaric	Meets the requirements of negative pressure	Can be controlled or isobaric
Adaptive depth of wells	Diameters of casing pipes range from 114.3~245mm, vertical wells, deviated wells within.	Diameters of pipelines ≥ 60.3mm and diameters of casing pipes ≤ 245mm, vertical wells, deviated wells within.	Diameters of cased tubing range from 114.3~245mm, vertical, deviated, and horizontal wells.	Diameters of pipelines range from 60.3~114.3 mm, vertical wells, deviated wells within.
Range of application	Ordinary wells	Production wells, recharging wells	Ordinary, high-pressure oil and gas wells; sand, low permeability, and difficult wells	Production wells, recharging wells
Perforating effect	Perforating injury will affect the production capacity.	The small aperture and lower depth will perhaps partly affect the production capacity.	Flushing of the holes and high production capacity.	The aperture is small will perhaps partly affect the producing capacity.

Gas reservoirs:

$$P_{\min} = \frac{2.5}{K^{0.17}} \qquad (2\text{-}42)$$

where K formation permeability and $P_{\min}$ is the minimum perforating negative pressure, MPa.

b. Determination of the maximum perforating negative pressure

The 80% compressive strength of the casing pipe (50% compressive strength of the old casing pipe), or 80% working pressure is selected as the maximum perforating negative pressure for the dense formation. The maximum perforating negative pressure can be determined by the following equations for the nondense formation.

2.5.1.4 Completion String Design

Completion string design is an important part of completion engineering, and the ultimate goal of completion engineering is achieved through completion string design, so it is necessary to conduct completion string design before completion operations.

1. Basic principles that must be followed in completion string design
   a. The tubing strings must meet the requirements of the oilfield development program.
   b. The tubing strings must be able to be tested safely.
   c. The raising and running down of tubing strings must be reduced.
   d. The tubing strings must adapt to the working conditions on the deck.
2. Classification of completion strings
   Completion strings are usually classified into the following categories:
3. Structure of offshore flowing completion strings
   a. Single tube commingling completion strings

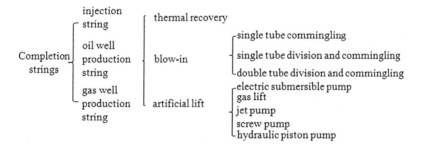

This structure of a completion string is shown in Figure 2-92. 1 is the hydraulic control line of the safety valve, 2 is the flowing short joint, 3 is the ground control subsurface safety valve, 4 is the eccentric operating drum, 5 is the slipping cover, 6 is the positioning joint, 7 is the permanent plugging device, 8 is the sealing assembly, 9 is the sealing drum, 10 is the operating drum, 11 is the pipe with a hole, 12 is the locking operating drum of the hanging production test instrument (NO- GO), and 13 is the guiding pipe of the coiled tubings.

The completion string structure is suitable for a multilayer system for commingling, while perforating production; it is suitable for the spray production of preproduction, but also for jet pump production in later periods, particularly for offshore platforms without production capabilities.

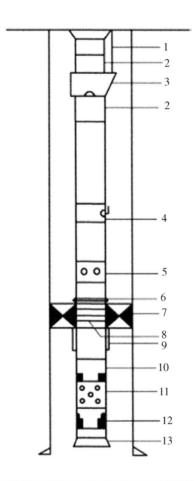

FIGURE 2-92

Structure of single tube commingling completion strings.

b. Single tube division and commingling completion strings
This structure of the completion string is shown in Figure 2-93. 1 is the flexible joint, 2 is the plugging device of a single hydraulic base, 3 is the thickening tube, and the names of other tools are not labeled as they are the same as those in Figure 2-92.
This structure of the completion string is suitable for a two-layer oil reservoir where separate recovery, uncontrolled commingling recovery, and controlled commingling recovery are needed, which is operated with the help of a wire rope.

c. Double tube division completion strings
The scope of application is same as the above. The structure of this completion string is shown in Figure 2-94. The names of the tools are same as those shown in Figure 2-92 and Figure 2-93.
This completion string structure is suitable for two-layer and separate recovery oil reservoirs where the difference among pressure systems, the physical properties of crude oil, and the formation parameters is large. The strings can be produced in the early days of development,

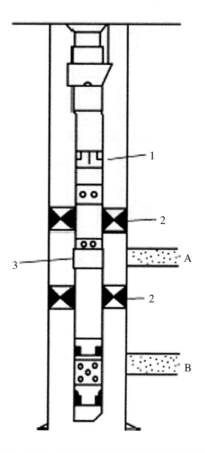

FIGURE 2-93

Structure of single tube division and commingling completion strings.

then the jet pump can be manufactured; also the A and B layers that drive each other are commingled. However, this string structure is only applied to production casing where the size is equal to or greater than 244.5 mm.

d. The completion strings for flowing production of sediment control wells

This refers to the completion strings used in an oil well with flowing production capacity where the gravel packing and sand control of the casing and open well has been finished. Because gravel packing and sand control have monolayer and multilayer types, the structure of the completion string is divided into two types. The monolayer completion string with sand control and flowing production is shown in Figure 2-95 (a), and the three-layer completion string with sand control and flowing production is shown in Figure 2-95(b). In Figure 2-95(a), 1 is the subsurface safety valve, 2 is the eccentric operating drum, 3 is the slipping cover, 4 is the sand control blocker (SC -1L), 5 is the sand control screen, 6 is the NO–GO operating drum, and 7 is the sump packer.

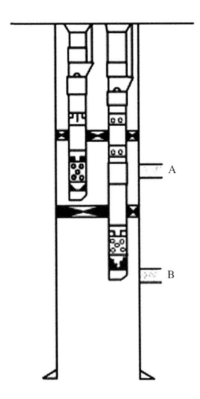

FIGURE 2-94

Structure of double tube division completion strings.

4. Selection of pipelines

Pipeline is not only the channel from the bottom to the ground for oil flow, but also the carrier of the down-hole tools needed for artificial lift oil production. Design and selection of pipelines is a key part of the reasonable production of oil and gas wells and enhanced oil recovery. Therefore, a reasonable choice of pipeline as well as tubing completion is an extremely important step.

a. Styles of tubing

Tubing is made up of external thread, a body, and internal thread (coupling). Nowadays, the tubing commonly used in oilfields in China includes API (American Petroleum Institute) tubing, non-API tubing, nonmetal tubing, and so on. But the tubing popularly used in offshore oilfields is API tubing, which is manufactured based on the API standard. The length of this tubing is usually 8.53 ~ 9.75m (28 ~ 32 ft); its strength, stress, hardness, and other specifications can be obtained in the API standard.

b. Selection of diameter of tubing

There are two kinds of methods to choose tubing diameter size: the experience method and node analysis method. The latter is more accurate, but the process is complicated, so it is not used as much. The current method is mainly used in offshore drilling engineering. The

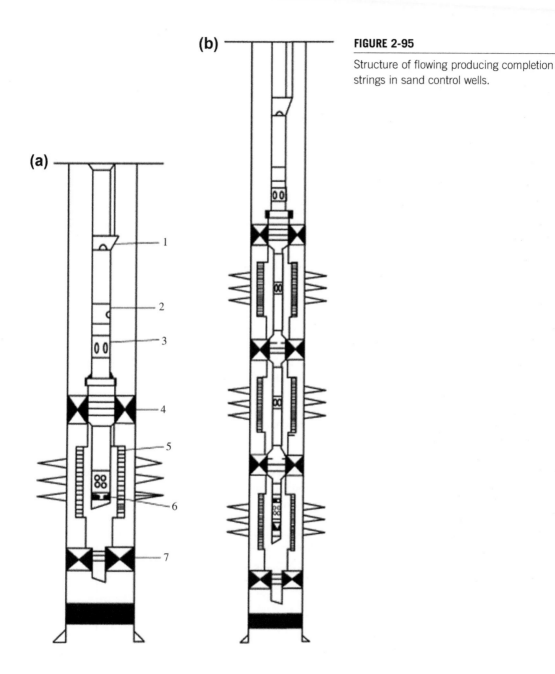

(a)

(b)

FIGURE 2-95

Structure of flowing producing completion strings in sand control wells.

empirical method mainly depends on years of field experience, using the flow rate to determine the diameter of the tubing. Of course, if you have other similar oil fields or oil well data, they also can be used as a reference. Below, according to the empirical method, and in view of the different methods of production, the recommended values for the diameter of tubing are given.

Table 2-9 Selection of Tubing Size in Flowing and Gas Lift Wells

Size of the Casing (in)	Maximum Size of Tubing (mm)	Nozzle Size (in)	Nozzle Size (mm)	Maximum Gas Production in Theory[1] bbl/d			
4	102		60	2000	200	15	400
4	114		73	5000	800	25	700
5	140		89	7500	1200	40	1100
6	168		114	15000	2400	80	2300
7	194		140	20000	3200	120	3400
8	245	7	178	60000	9550	$\gg 100$	$\gg 2800$

[1]Maximum gas production is usually reduced by the wellhead pressure, production capacity, ratio of gas and fluid, and nozzle size.

    i. Flowing and gas lift wells: The selection of tubing size can depend on the maximum production of oilfields as shown in Table 2-9.

   ii. Two-string completion: The selection of tubing size can depend on the size of the casing as shown in Table 2-10.

  iii. Pumping well: The tubing should be selected based on different pumping styles, maximum oilfield production, and the pump size, as shown in Table 2-11. The data comes from the assumption that the stroke of a pump is 3657.6 mm (144 in), the stroke frequeny is 15 times/minute, the efficiency of the pump is 100%, and the head of the electric submersible pump is 1000 m (3000 ft).

  c. Selection of tubing material

The tubing material is selected according to composition, density, humidity, pressure of oil and gas flow, and production requirements.

    i. Preventing carbon dioxide corrosion

Plain carbon steel tubing should be selected when the partial pressure of carbon dioxide is less than 0.2 MPa in the oil and gas flow; tubing of different grades of steel should be selected based on different partial pressures of carbon dioxide, or ordinary carbon steel

Table 2-10 Selection of Tubing Size for Two-String Completion Wells

Size of the Casing in	mm	Maximum Size of Two-string in	mm
$9\frac{5}{8}$	245	$3\frac{1}{2} \times 3\frac{1}{2}$	89×89
$8\frac{5}{8}$	219	$3\frac{1}{2} \times 2\frac{7}{8}$	89×73
$7\frac{5}{8}$	194	$2\frac{7}{8} \times 2\frac{7}{8}$	73×73
7	178	$2\frac{7}{8} \times 2\frac{3}{8}$	73×60
		$2\frac{7}{8} \times 5$ (concentric tube)	73×127 (concentric tube)
$5\frac{1}{2}$	140	$2\frac{1}{6} \times 1.9$	52×48

Table 2-11 Selection of Tubing Diameter in Pumping Wells

Size of the Casing		Size of the Tubing		Size of the Pump		Max. Production[1]	
in	mm	in	mm	in	mm	bbl/d	
Sucker-rod pump							
	89	1.9	48	1.5	38	550	100
4	102		60	1.75	44	800	150
	114		73	2.25	57	1300	200
	140		89	2.75	70	1900	300
Electric submersible pump[2]							
	114		73			1750	300
	140		89			4000	650
7	178	5	127			10000	1600
	245	7	178			35000	5550

[1]Assuming that the stroke of the pump is 3657.6 mm (144 in), the stroke frequency is 15 times/minute, and the pump efficiency 100%.
[2]Assuming that the head of the electric submersible pump is 1000 m (3000 ft).

tubing should be sprayed with paint, or injected with chemicals to form a layer of protective film on its surface when the partial pressure is 0.2 ~ 0.21MPa.
ii. Preventing against sulfureted hydrogen corrosion
Tubing corrosion will cause the rupture of the tubing when there is sulfureted hydrogen in the accompanying gas in wells. A corrosion speed test and design optimization design should be used on the carbon steel, 9 Cr steel, 13 Cr steel, 22 Cr steel, and so on. Then the most economical and practical material should be chosen from among them.
d. Selection of tubing coupling
External upset coupling strength is higher than the strength of the pipe body, so completion in the sea in general uses external upset tubing collars; integral coupling is rarely used.
e. Selection of tubing thread
Tubing thread is mainly selected based on the requirements of temperature, pressure, and safety. The API thread is usually selected when the temperature is lower than 138°C in a well. But metal-to-metal sealing thread must be selected to ensure the sealing of tubing when there are and in well, or the temperature is higher than 138°C in the well. It is required to used metal-to-metal sealing thread in offshore gas fields.
5. Design of wellhead equipment
The design of wellhead equipment presented here only covers the design of the tubing head and Christmas tree.
a. Design of tubing head
The structure of a tubing head is shown in Figure 2-45(a) and (b), and the following should be considered in its design.

i. Pressure

In general, a tubing head is selected according to the highest pressure at the mouth of the well during the injection or operation period, and the pressure grade should match with that in the down-hole tubing. The grades of working pressure are 7.0MPa, 13.8MPa, 20.7MPa, 34.5MPa, 69MPa, 103.4MPa, and 138MPa (1000psi, 2000psi, 3000psi, 5000psi, 10000psi, 15000psi, 20000psi). The minimum working pressure of the tubing head equals the closing pressure at the wellhead. Generally during completion, the tubing head should be selected for the situation where the rated working pressure is equal to the formation fracture pressure. In addition, the rated working pressure of a tubing head should match with the rated working pressure of the tubing hanger.

ii. Internal size

Generally the internal size should be same as the internal diameter; the external diameter can be selected based on the allowable size in offshore completion.

iii. Amounts of tubing

The tubing head of hanging single tubing or multitube tubing is selected based on the amount of tubing required in production.

iv. Channels for cable in pipelines

The well pipe string in the cable and all kinds of injection pipeline and hydraulic control lines are chosen to match with the multiple channel tubing string, sealing joint, and adapter flange as well as the cable penetrator.

v. Other parts

The secondary sealing types for the casing and size of the lower the pipelines should be selected according to the casing sizes, and the adapter flanges should be selected based on the flange sizes of the upper main valves and tubing spools.

b. Design of Christmas tree

i. Styles

Christmas trees used in offshore oil and gas fields include overwater Christmas trees and subsea Christmas trees, and the latter will be introduced in the third chapter. Christmas tree is divided into X-mas tree and subsea tree. But the overall style includes the main valve, safety valve, wax valves, and valves made of whole wing parts. The distance between valves should be small, and the valves should be space-saving and high pressure, so oil and gas wells offshore platforms generally use a unitary Christmas tree. The structure of the unitary Christmas tree popularly that used in offshore oil and gas wells is shown in Figure 2-96. In addition, in accordance with the category of production wells and completion production methods to divide, Christmas tree can be divided into self-flowing wells, electric submersible pump wells, gas lift wells, screw pump wells and gas wells Christmas tree, they have a single tubing with dual tubing of the points. Christmas tree is installed single pipe tubing completions in a single wellhead device, in addition to its split with the overall style, there monoplane with wings of the points. It uses Stone which is wings, whose tee is monoplane. A Christmas tree with double tubings is shown in Figure 2-96. The wellhead equipment and the Christmas tree for flowing and gas lift wells are shown in Figure 2-97.

ii. Pressure

The rated working pressure of a Christmas tree is equal to the minimum rated working pressure of all the parts, namely the part with the minimum rated pressure determines the

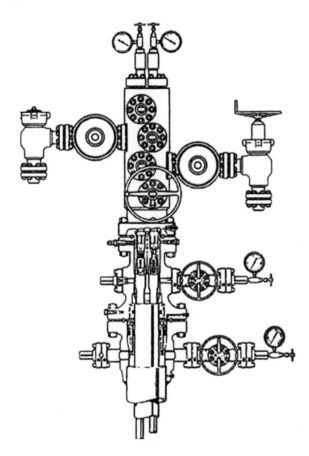

FIGURE 2-96

Unitary Christmas tree with double tubings for an offshore platform.

rated working pressure of the whole tree. The pressure rating of the whole tree should be checked to ensure it can meet the requirements of the production.

iii. Drift diameter

The vertical drift diameter of the Christmas tree assembly is selected based on the API standard. The drift diameter of a Christmas tree and the size of the inside gauge are shown in Table 2-12.

iv. Other parts

Selection of the hydraulic or pneumatic valves and the safety valves on the offshore platform or in the well is determined by the control panel on the platform and the safety requirements. Stop valves with two-way sealing, convenient operation, and simple maintenance are the popular product for the offshore platform. The connection between the Christmas tree hat and stop valve end should meet the requirements of hydraulic pressure observation, production tests, circulation killing, and squeezing and injecting operations. The control range and adjustment scale of the throttle nozzle should be

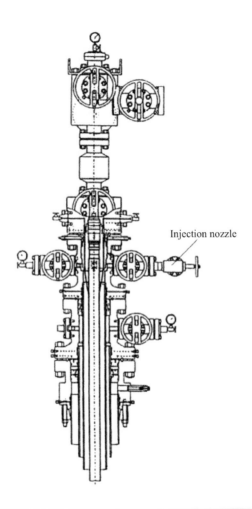

Injection nozzle

FIGURE 2-97

Equipment at wellhead and Christmas tree for flowing and gas lift wells.

selected based on the scope of production. The size and pressure rating of inlet and outlet flanges should be around that of the front and back flanges. There are a flange type and a flange-stud type for tubing spools and tees, and their size, rated working pressure, and material should match with the connecting parts according to the API standard.

The elements of the entire completion design, in addition to the above mentioned aspects, should also include the design of sand control, completion fluid design, related technical measures design such as acidification and plugging during the completion process, construction design and safety, a pollution prevention emergency plan, design of tools and equipment, etc., but the content is basically the same as for onshore completion design, and will not be described here.

Table 2-12 Drift Diameter of Christmas Tree and Size of Inside Gauge

Nominal Size and Diameter in (mm)	Flange or Clamp		Min. Vertical Drift Diameter in (mm)	Specifications of Tubing		
	Old Nominal Size (m)	Grades of Nominal Pressure (psi)		External Diameter in (mm)	Mass lb/ft	Internal Gauge Diameter[1] in (mm)
1 13/16 (46.0)	—	—	—	1.660 (42.2)	2.0	1.286 (32.52)
	1 1/2	2000, 3000, 5000	1 11/16 (42.9)	1.909 (48.3)	2.9	1.516 (38.52)
2 1/16 (52.4)	2	2000, 3000, 5000	2 9/16 (65.1)	2 3/8 (80.3)	4.7	1.901 (48.22)
2 9/16 (65.1)	2 1/4	2000, 3000, 5000	2 1/16 (52.6)	2 7/8 (73.0)	4.5	2.34 (69.62)
3 1/8 (79.4)	3	2000, 3000, 5000	3 1/8 (79.1)	3 1/2 (88.9)	9.5	2.867 (72.82)
				4 (101.6)	11.5	3.351 (85.12)
4 1/16 (103.2)	4	2000, 3000, 5000	4 1/16 (103.2)	4 1/2 (114.3)	12.73	3.837 (97.32)
1 1/16 (42.9)	1 11/16	10000, 15000	1 11/16 (42.9)	1.900 (48.3)	2.9	1.516 (38.52)
1 13/16 (46.0)	2	10000, 15000	1 13/16 (46.0)	2.603 (52.4)	3.25	1.657 (42.12)
2 11/16 (52.4)	2 1/16	10000, 15000	2 1/16 (52.4)	2 3/8 (60.3)	4.9	1.901 (18.22)
2 9/16 (65.1)	2 9/16	10000, 15000	2 9/16 (65.1)	2 3/8 (73.0)	6.5	2.347 (59.62)
3 1/16 (77.8)	3 1/16	10000, 15000	3 1/16 (77.8)	3 1/2 (88.9)	9.3	2.867 (72.82)
4 1/16 (103.2)	4 1/16	10000	4 1/16 (103.2)	4 1/2 (114.3)	12.75	3.833 (97.32)

[1]Internal gauge diameter should meet the requirements of internal gauge of tubing with external upset ends in API Spec 5A2, except for the tubing of in (52.4 mm) with integral joints, the tubing of in with internal upset ends, and the thread tubing of external upset ends.

2.5.2 Completion Procedure

2.5.2.1 Drilling Reservoir

Drilling of the reservoir is the first step of completion; the drilling assemblies and drilling parameters are introduced when the drilling operation for tail pipes of mm and mm was introduced before.

2.5.2.2 Installation of Wellhead

The subsea wellhead with dual tubing completion is taken as an example to show the procedure for installation of a wellhead.

1. Running one of the dual tubings
   a. Running the directional bushing
      The directional bushing is run to guide the first running tubing, which is run to the sealing assembly of mm by the running tool that is connected at the bottom of the drilling pipe; the procedure is shown in Figure 2-98 (a).
   b. Running the hanger of the first tubing
      First of all, the installation and pressure test of the hanger and testing joint should take place on the platform deck; second the running tool (right thread) above the hanger (left thread) should be connected with the thread into the string; third the running tubing and the running string are connected together; fourth the hanger is run slowly into the sealing assembly of mm under the guidance of the directional bushing; and finally the running tool and string are retrieved. The procedure is shown in Figure 2-98(b).
2. Running the other dual tubing
   The procedure and method of the running operation is the same as shown in Figure 2-98(a): first the directional bushing is run down, and then the tubing hanger is run. The running procedure of the other dual tubing and its hanger is shown in Figure 2-98(c).
3. Running the sealing assembly of tubing hanger
   A tubing hanging sealing assembly is used for the tubing hanger sealing; it is connected to its running tool with the left thread, and run with a drilling pipe into the tubing hanger. The running tool is locked with the tubing hanger by a locking key, then the running tool and the tubing hanger are unlocked and brought up after the sealing pressure test at 10.341MPa (1500lbf/ in²). The running procedure of the sealing assembly of a tubing hanger is shown in Figure 2-99 (a).
4. Running the tieback tool when a rope is used
   A tieback tool is run down to test operation with the rope. It is first connected to a drilling pipe and run to the sealing area of the tubing under the guidance of a directional bushing, then locked to the left thread of the well. After the test, the tieback tool and running tool are brought up together. The installation of the other dual tubing is exactly same as the method mentioned above. Finally, the directional bushing will be raised. The operation procedure for the running tieback tool is shown in Figure 2-99(b).
5. Installing the wet Christmas tree at the subsea wellhead
   a. The controlling equipment at the wellhead are moved from the seafloor [shown in Figure 2-100 (a)]
      The killing well and choke line are first released on the deck, and the connector at the underwater wellhead is released; finally the BOP assembly and riser are all brought up. The

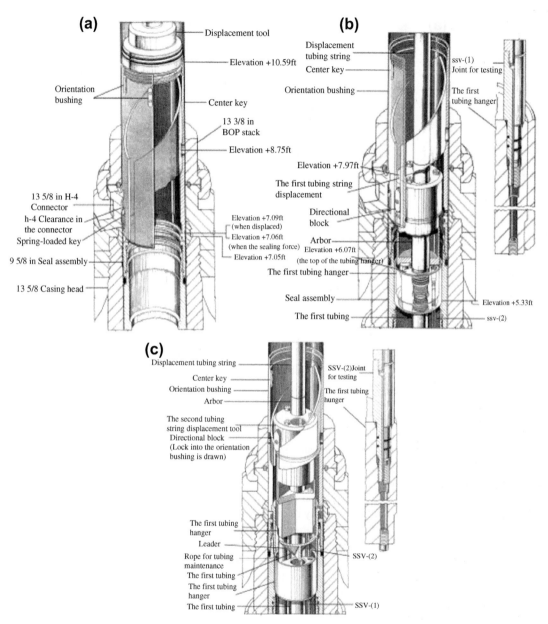

(a)
- Displacement tool
- Elevation +10.59ft
- Orientation bushing
- Center key
- 13 3/8 in BOP stack
- Elevation +8.75ft
- 13 5/8 in H-4 Connector
- h-4 Clearance in the connector
- Spring-loaded key
- 9 5/8 in Seal assembly
- 13 5/8 Casing head

(b)
- Displacement tubing string
- Center key
- Orientation bushing
- ssv-(1) Joint for testing
- The first tubing hanger
- Elevation +7.97ft
- The first tubing string displacement
- Elevation +7.09ft (when displaced)
- Elevation +7.06ft (when the sealing force)
- Elevation +7.05ft
- Directional block
- Arbor
- Elevation +6.07ft (the top of the tubing hanger)
- The first tubing hanger
- Seal assembly
- The first tubing
- Elevation +5.33ft
- ssv-(2)

(c)
- Displacement tubing string
- Center key
- Orientation bushing
- Arbor
- The second tubing string displacement tool
- Directional block (Lock into the orientation bushing is drawn)
- SSV-(2)Joint for testing
- The first tubing hunger
- The first tubing hanger
- Leader
- Rope for tubing maintenance
- The first tubing
- The first tubing hanger
- The first tubing
- SSV-(2)
- SSV-(1)

FIGURE 2-98

Running procedure of dual tubing and its hanger.

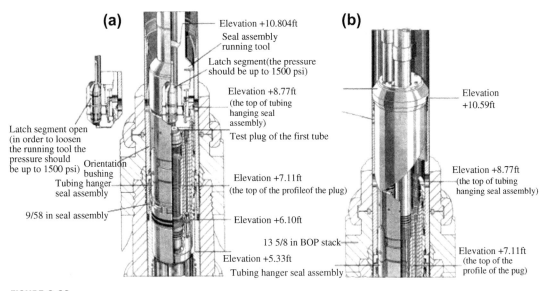

(a)

Elevation +10.804ft

Seal assembly running tool

Latch segment(the pressure should be up to 1500 psi)

Elevation +8.77ft (the top of tubing hanging seal assembly)

Test plug of the first tube

Latch segment open (in order to loosen the running tool the pressure should be up to 1500 psi)

Orientation bushing

Tubing hanger seal assembly

9/58 in seal assembly

Elevation +7.11ft (the top of the profileof the plug)

Elevation +6.10ft

13 5/8 in BOP stack

Elevation +5.33ft

Tubing hanger seal assembly

(b)

Elevation +10.59ft

Elevation +8.77ft (the top of tubing hanging seal assembly)

Elevation +7.11ft (the top of the profile of the pug)

FIGURE 2-99

Running procedure of sealing assembly of tubing hanger and tieback tool.

situation where the controlling equipment at the wellhead are moved from the seafloor is shown in Figure 2-100(a).

b. The Christmas trees are prepared on the platform [shown in Figure 2-100(b)]

The end of a Christmas tree should be connected to the wellhead connector well on the deck; the upper part of the tree is connected to the completion riser after a pressure test. Then in order to prepare to run the Christmas tree, the control panels with hydraulic control and hydraulic lines are tested and assembled, and the guide ropes pass through guide strings.

c. The Christmas tree assemblies are run to the wellhead [shown in Figure 2-100(c)]

The Christmas tree assembly is slowly run to the wellhead by a riser along the guide rope and the heave compensation system is started to avoid collision when it it close to the wellhead. When it is located at the permanent guide frame at the wellhead, the Christmas tree assembly is connected to the hydraulic connector; finally the pressure test for the Christmas tree, riser, and hydraulic control system will be conducted until they are qualified.

d. The manifold blocks are run and installed [shown in Figure 2-100(d)]

The manifold blocks and the cage frame are run by a drilling pipe. First the running tool is connected to the drilling pipe and the manifold block with screw thread. Then the manifold block that combines with its cage frame is run to the Christmas tree manifold by the drilling pipe along the guide rope. Finally, the running tool and drilling pipe are released and recycled.

6. Running and installing the cable for oil pipelines and hydraulic pipelines

The connectors for oil pipelines, cables, and hydraulic control pipes are placed in the external frame of the guide bar located to the side of the Christmas tree. First the connectors are connected

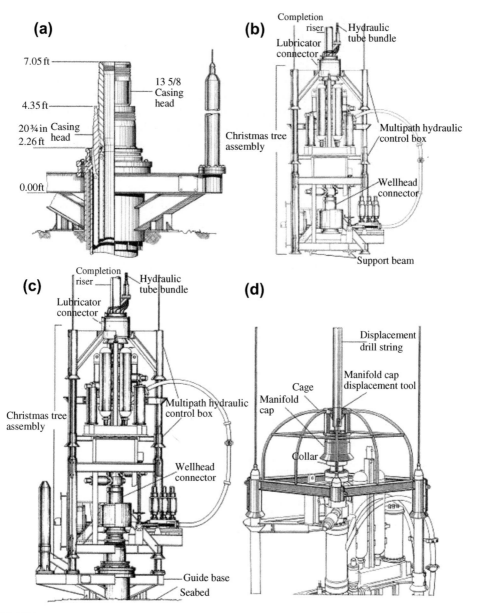

(a)

7.05 ft

13 5/8 Casing head

4.35 ft

20¾ in Casing head
2.26 ft

0.00 ft

Christmas tree assembly

(b)

Completion riser

Hydraulic tube bundle

Lubricator connector

Multipath hydraulic control box

Wellhead connector

Support beam

(c)

Completion riser

Hydraulic tube bundle

Lubricator connector

Multipath hydraulic control box

Christmas tree assembly

Wellhead connector

Guide base

Seabed

(d)

Displacement drill string

Manifold cap displacement tool

Cage

Manifold cap

Collar

FIGURE 2-100

Installation procedure of wet Christmas tree.

to the external frame on the barge deck, and the pipelines and cables are linked with the connectors. Then the guide rope is passed through the guide cylinder of the external frame and the external frame is slowly run to the two guide cylinders of the Christmas tree permanent guide along the guide rope, with the help of the tensioner on the ship. The cables and hydraulic controlling soft pipelines are bundled on the tubing to avoid intertwining with the Christmas tree. After the external frame is located and positioned well, the hydraulic control system is started to make all of the hydraulic connectors on the external frame connect with the Christmas tree. Finally, the guidelines are cut off and retrieved, and the cables, hydraulic pipelines, and other pipelines are all hung with light tension. All of these are shown in Figure 2-101.

2.5.3 Oil Test Operations

Oil test operations are the connecting link between drilling and oil recovery, so they are included in the completion of drilling engineering.

2.5.3.1 Purpose and Task of Oil Test Operation

The purpose of oil testing is that the oil and gas in oil and gas reservoirs are transported to offshore platforms (ships), and then the well drilling data and technical parameters can be obtained by oil testing, so the exploitation value of the oil well and oil field and production mode of oil wells can be determined reasonably. The detailed tasks are as follows:

1. Confirming the thickness, lithology, and exploitation values of the oil and gas reservoirs.
2. Finding out the physical properties and chemical constituents of oil, gas, and water.

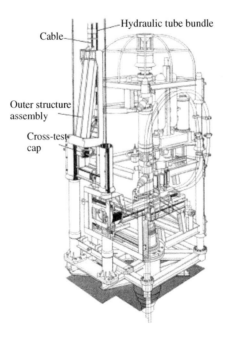

FIGURE 2-101

Installation and running procedure of cable for oil pipeline and hydraulic pipeline.

3. Finding out the oil and gas area of oil fields, the boundary between oil and water, and the types and productivity (exploration wells) of oil and gas reservoirs
4. Determining the parameters for production and figuring out the oil reserves, so the basis can be provided for a development plan for the oil field.

2.5.3.2 Oil Testing Methods

An oil well is drilled through several oil and gas stratum, and the depth, lithology, oil content, gas content, and reservoir pressure of them are often not the same, so oil testing must involve hierarchical testing and slicing. According to the method of hierarchical oil testing, generally the well is drilled to the design depth, and the casing is run for well cementation, so all reservoirs are sealed with cement and casing. Then down-up perforation and testing take place, that is to say after a test, the tested reservoir is sealed with injected cement, and then the layer above it is shot for testing. The cycle is repeated until the entire stratum is tested.

2.5.3.3 Oil Testing Procedures

The oil well testing completion process includes the following aspects.

1. Well washing and circulation
   The drilling fluid that meets the requirements for performance and density is used to suppress the wells to prevent blowout from the shot reservoirs.
2. Perforation
   The layer is shot in accordance with the requirements and methods introduced in previous completion design.
3. Influent
   Because there is drilling fluid in the well, and its pressure is generally greater than the pressure of the reservoir, oil can't flow into the well when the layer is shot. At this time, we need to take technical measures to induce out oil and gas, which is called an influent. The commonly used methods in the offshore oil field are the following.
   a. Spray method
      Use a very small amount of liquid to replace the well killing fluid. Water is commonly used to replace the drilling liquid and oil is used instead of water if that does not work, or compressed gas is injected to the wells to replace the well fluid for induction.
   b. Suction method
      The oil flow is sucked into the well approach. It's usually used in the application when the previous method will not work.

2.5.3.4 Oil Testing Process on the Deck

Offshore drilling platforms (ships) are generally equipped with a set of oil testing equipment, including furnaces, separators, burners, air compressors, manifolds, metering tanks, and related measuring instruments. The oil testing process flow diagram for the No. 5 drilling platform on the South China Sea is shown in Figure 2-102. As shown, oil from the wellhead first flows into oil manifold and choke manifold, then into the heating furnace to be heated via a ground testing tree (testing Christmas tree) and safety valve. After heating, the crude oil flows into a three-phase separator for separation of oil, gas, and water. The separated water is let into the sea after being processed; the separated gas is

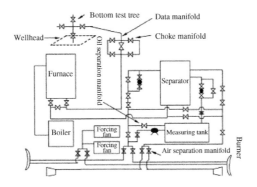

FIGURE 2-102

Oil testing process on the No. 5 drilling platform on the South China Sea.

transferred through the blowing fan into the two burners for burning; the separated oil is measured in the metering tank. If it can't be output, it will be pumped into the two burners to be burned.

2.6 Abandoned Wells and Tieback of Offshore Oil and Gas Drilling

After offshore drilling is completed or if the well is not immediately put into production, withdrawing the drilling platform or boat is called abandonment. A well can be for a short time so that a future platform or ship go back to the well mouth again and also can be abandoned for a long period. At a certain stage of the drilling operation, an emergency withdrawal of a drilling platform or boat is called a temporary abandonment. After abandonment, the drilling platform or boat will find the original well again, return to the complete or incomplete the well mouth and put it into operation or continue drilling operations. We need to restore the original well device and go back to oil production or continue drilling operations, so this operation is called tieback. In this section, we introduce abandoned wells and tieback in the context of offshore drilling operation.

2.6.1 Abandonment and Protection

2.6.1.1 Abandonment Category

1. Permanent abandonment
   Some wells such as appraisal wells will lead an offshore drilling platform (ship) to withdraw; the well may be rejected, and we will no longer return to a wellhead after exploration and evaluation if there is no exploitation value. Barren or dry well will be abandoned as well; all of these scenarios are called permanent abandonment.
2. Temporary abandonment
   There are two cases of temporary abandonment. One case is when an appraisal well finds the oil and gas layer with value, but the area is not yet ready for a production platform, or we have finished drilling production wells but the production platform has not been built. Drilling platform (ship) can be temporarily removed, after production platform to build well, and then again to return to the wellhead, using the production platform to get into oil recovery operation. Another

condition of abandonment is during emergency, where we must remove the platform (ship) temporarily, as with typhoon strikes or severe winter ice and other emergencies such as earthquakes and tsunamis; these scenarios are all examples of temporary abandonment.

2.6.1.2 Operating Procedures for Abandonment after Drilling Is Complete

1. Operating procedures for permanent abandonment

If you have decided to permanently abandon the well after drilling is complete, your operating procedures are as follows

a. Cement plug

For packing the hydrocarbon reservoir, oil and gas layers of an open hole should be packed by a cement plug during abandonment and the length of every cement plug length should be less than 200 m. Some parts with special requirements are as follows:

   i. Inject an 80 ~ 120 m cement plug between an open hole and the casing shoe, which shall be not less than 50 m on the casing

   ii. General it should down FZSV bridge plug between the atmospheres, top of atmospheres, and squeeze cementing plug is at least 30 m in the perforation below the bridge plug

   iii. Seal the reservoir perforated interval with a 50 m cement plug;

   iv. If the top of the stern tube is atmosphere, generally it should be down to a small bridge plug; or it should be a cementing plug, and we should squeeze the cementing plug 30 ~ 50 m into bridge plug.

b. Casing cutting

The casing cutting depth should be outside the casing cement segment, and coupling should be avoided. The casing above the cutting position should not be an open hole. We cut casing in two ways: one style is a mechanical cutting method with a cutter; another method is to use a chemical explosion cutting method. For an incision without cement, we should inject a cement plug.

c. Pipe cutting

The method is same as for cutting casing except we cut at the seabed under 4 m.

d. Dismantling BOP

We demolish it when the hole is safe and stable.

2. Operating procedures for temporary abandonment

If we temporarily decide to abandon a well after drilling is complete, but will find it again in the future and tieback, the operating procedures are as follows.

a. Maintain the density of the drilling fluid

The density of casing drilling fluid must be not less than that of the drilling oil and gas layer.

b. Inject cement plug

In addition to completely sealing the oil and gas layer, we should also inject 2 ~ 3 cement plugs in the oil and gas layer above different parts the well; each cement plugs should be 50 ~ 80 m in length, and the final cement plug should be around 250 m below the seabed.

c. Wellhead corrosion cap

After the upper part of the wellhead is backed off, the subsea wellhead should be infused with preservative liquid, and we should put on a wellhead cap.

2.6.1.3 Procedures for Emergency Abandonment and Recovery in Drilling

1. Abandonment and recovery of floating drilling platform (ship)
   a. Assignment of hanging tool installation
      The hanging tool refers to hanging the drill string when drilling. This tool will pick up at the appropriate position of drill string, and move inside the casing head or blowout preventer, and then you can hang the drilling string below it. If there is back off on the running tools, you can lift the drill string above the running tools. After closing the one-way back pressure valve inside the drill string hanging nearby the tools, and clinging the ram preventer of the hanging tool and the blowout preventer of suspension tools, you can lift the drilling riser and the annular blowout preventer to the deck safely after removing the hydraulic connector. With only part of the BOP wellhead and most of the drill string hung by the hanging tool left at the bottom of the sea, we are ready for temporary abandonment and evacuation of the floating drilling platform.
      Moreover, if in the future we return to the wellhead, we can still use the original suspension tools to pick up the hanging tool and continue to use the down-hole drill string.
      When installing a hanging tool, you should first lift the drill above the casing shoe around 10 m, plus the distance of the seabed to the wheel, or minus the distance of the seabed to the rotary table when drill is less than 10 m above casing shoe. Figure 2-103 shows the installation of running tools on the drill string and their location within the casing head.
   b. Evacuation assignments of floating platform
      Both a pontoon or semi-submersible drilling platform should relax guidelines, connect the buoy after marking, and place the buoy on the surface, after lifting the riser and the blowout preventer, in order to find the wellhead again. Then you should carry the anchor according to the method in the section in this chapter discussing a floating drilling platform (ship) mooring system. Set sail and move to other places when all the activity equipment on the platform (ship) is fixed. But one more program is needed for the semi-submersible platform, which is to lift the semi-submerged platform to the survival waterline, and if it will not meet the living water line, we need to send off part of the drilling fluid.
      If the platform (ship) has a large displacement by dropping anchor suddenly, breaking of the anchor chain, or running into big storms, we should conduct emergency relief measures to make the floating platform (ship) withdraw. To carry out the emergency relief procedure, you should first to unlock the hydraulic connector on the blowout preventer quickly, to release the whole water column on the top of the hydraulic connector; meanwhile, the down-hole drilling string should be seated on the ram blowout preventer to lock the flashboard and we can snip the drill string by cutting the flashboard, to separate the drill string below the blowout preventer and above it; this should close the blind flange BOP in time. Once the cut drill string is lifted off the blowout preventer, we can use the riser tension compensation device, the BOP tripping out the blowout preventer, the upper part of the blowout preventer to lift the deck floor and the cut upper part of the drilling string. At this time, because the water pipe string or the upper part of the drill string have been tripping out of the wellhead device on the bottom of the sea, it will not be affected by the large distance of the floating platform to work well-off. The wellhead device is at the bottom of the sea and the part of the drill string under the cut locked in the blowout preventer can be tied back when the platform comes back to the wellhead.

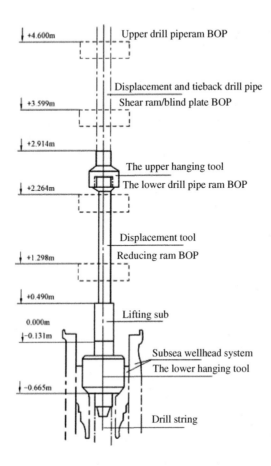

FIGURE 2-103

Installation and location of suspension tools in abandonment.

Of course the floating platform (ship) can be made stable in the well position by adjusting other anchor tensions outside of the broken anchor or relying on the DPS, then evacuating the platform with hanging tools in the way described before.

c. Resuming operation of the floating platform

This procedure generally include the following contents.

i. Adjust the mooring system or mooring again

A semi-submersible platform should pull a chain to the original position after finding the guide rope buoy, then tieback the guideline, and adjust the mooring system according to the condition of guide rope tilt so that the drilling operation is the ballast water line of platform. The drilling pontoon should be in accordance with the original mode of mooring again after reaching the location.

ii. Tripping the upper BOP and riser

Test the connector function before tripping, test the choke flow and pressure of kill lines in the tripping process, start the heave compensation device when it is located at the connector, and appropriately adjust the cables and connectors to verify whether they are locked.

iii. Tieback and lift emergency suspension tools

Open the upper shear ram (BOP), pass down the running tool of the hanging tool, screw on the clasp to tie back to the emergency suspension tools, and then open the lower pipe ram (BOP), lift up the emergency suspension tools, and unload the suspension tools and one-way back pressure valve.

iv. Changing the drilling fluid and draining gas pressure

Replace the seawater with drilling fluid (before the shear ram opens), release the ring gas pressure in the air with a choke manifold, and release the gas pressure with the pipe manifold.

2. Abandonment and recovery of jack-up drilling platform

a. Suspension of down-hole drilling tools

A jack-up drilling platform is different from a floating drilling platform (ship). It hangs drilling string, not emergency suspension tool but pressure test plug of BOP as a hanging tool (usually a $\Phi 311.15$ mm hole is with a Φ 339.73 mm pressure test plug, Φ 444.5 mm hole is with Φ 508 mm pressure test plug, and Φ 215.9 mm is hole with 273.05 mm Φ pressure test plug). Make the one-way back pressure valve connect the drill string and pressure test plug, then lower the drill string, and make the test plug sit on casing head and hook the drill inside the well more than 15 m above casing shoe on the drill string. We can abandon and withdraw the platform when the pressure test plug is seated well and the BOP is closed. Generally we should connect the recovery tool to the drill string before the start of the suspension application, and then recycle the casing head wear bushing down into the well.

b. Evacuate the drilling platform

The first step is to fasten all activity equipment on the platform; then we adjust the load distribution of the platform, and evacuate.

c. Resuming platform operation

When the jack-up drilling platform returns to its original location, first we must put out the gas pressure in the annulus and drill pipe from the casing spool; then, open the blind flange blowout preventer, trip in the tieback tools by drill pipe, and pull out the pressure test plug used to hang the down-hole drilling tools. When the drilling fluid circulates through one cycle, we can continue the drilling operation again after removing the test plug and one-way back pressure valve.

2.6.1.4 Protection of the Drilling Hole during Abandonment

According to the different stages of the drilling operation, the protection measures are as follows:

1. Well hole protection during abandonment when in the catheter phase

As far as possible pull out the catheter and then fix it on the deck. If the catheter can't be pulled out, we can deal with it in the following two ways:

a. For a jack-up platform: Pass down the catheter in place according to the requirements of the original design; then, inject cement paste.

  b. For a floating drilling platform (ship): If a temporary guide plate is not down, the catheter will reach the place, and back off the running tools; if a guide plate is down, we should lift the running tools after cementing.

2. Borehole protection in the stage of drilling a Φ 660.4 mm hole
Immediately stop drilling, circulate drilling fluid inside the borehole, as well as 1.5 times the borehole volume of high-viscosity fluid, and then start working.

3. Borehole protection in the stage of drilling Φ508 mm or Φ339 mm holes and other stage
Borehole protection for this phase is conducted according to the following two scenarios.
When the casing has not been down into the casing shoes, we should immediately lift out the casing or connect the casing hanger (mud line hanger), hanging the casing temporarily in the borehole.
When the casing is out of the casing shoes, we should quickly trip the casing, and cement the well quickly. The floating platform (boat) should lift out the running tool in a timely manner.

4. Wellbore protection when expected formation drilling of oil and gas or other abnormal situation occurs has not been seen.
At this point, if we are electric logging, we should immediately carry out the electric logging tools, and then drill down with a hanging tool within the shoe; if we are boring, then we should immediately stop drilling, circulate drilling fluid, and the drill to the shoe to around 10 m, hanging the tool inside the casing shoe.

5. Borehole protection of expected drilling formation own oil and gas or abnormal high-pressure layers.
When we are electric logging, we should immediately carry up the electric logging tools, and then at the casing shoe upper around 10 m down to the bridge plug, and drill pipe above the bridge plug that cementing plug is 30 ~ 50 m, hanging the drill string in a safe place.
When we are drilling, we should immediately stop drilling, trip out, trip down the slick pipe, inject a 80 ~ 120 m cement plug at the top of the oil and gas layer, then at the casing shoe inject a 70 ~ 100 m cement plug, and hang the drill pipe in a safe place.

2.6.1.5 Wellhead Protection When Abandoning at Sea

1. Wellhead protection at the surface
The method generally used in China is to infiltrate above the surface a large diameter welding pipe string or concrete in the platform out of the well's riser before evacuating the drilling rig (ship). Then we seal it and mount a beacon, such as Bohai ever used the method to infiltrate into Φ around 10 m 1.2 m above the surface of wellhead protection tube for 9 well, but only 1 well protected to platform back, other well protecting tube suffered the impact of the wind and waves and are damaged, obvious it cannot be long-term effective measures to protect the wellhead. Therefore, catheter, only pillar and mat jacket frame have been proposed later, which are combined single well protection and with test scheme of the platform structure, for further exploration.

2. Protection of underwater submarine wellhead
This is generally done by drilling down to the bottom chassis when you make a test well or drilling production wells by drilling horizontal wells as described in 2.2 of this chapter. Figure 2-104.

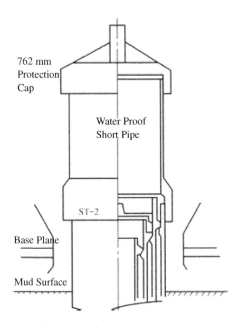

762 mm
Protection
Cap

Water Proof
Short Pipe

ST–2

Base Plate

Mud Surface

FIGURE 2-104

Underwater submarine wellhead protection.

2.6.2 Platform (Ship) Returning to the Location

In the case of temporary abandonment described before, the offshore drilling well platforms all need to return to the location and continue drilling operations. A drilling platform (ship) may withdraw when the exploratory well or evaluating well have value after evaluation, and we need to transfer the completed production platform to the original well position for installation and production. Therefore, looking for the original wellbore demands putting the platform accurately in place, aimed at the wellhead, so it becomes a very important link in offshore drilling operations.

2.6.2.1 Look for the Wellhead

The first step for the platform (ship) in returning to the location is looking for the wellhead, and usually there are two methods as follows.

1. Sonic location system

   As shown in Figure 2-105, when the towing platform (ship) is in the location above well position, adjust the platform (ship) so that is within 100 m of the subsea wellhead that uses an acoustic positioning system. An acoustic generator is installed on the bottom of the wellhead, and an acoustic wave receiver is installed at the bottom of the platform (ship), with the change of the sound waves transmitted to the receiver. According to the first section in this chapter you should use the principles of the dynamic positioning system of the displacement measurement platform (ship) principle and use a computer system to calculate the gap between the current position and

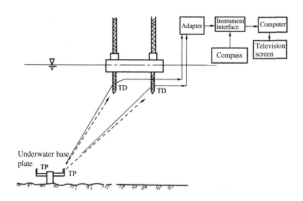

FIGURE 2-105

Platforms returning to the location using the acoustic positioning and GPS. measurements.

the original wellhead position; with the help of a tug or a dynamic positioning system, we can gradually adjust position, make the platform (ship) return to the location, and keep the difference within 2% of the depth of the water.

2. Underwater laser TV technology

As shown in Figure 2-106, for a subsea wellhead, water quality often becomes mud as we get close to the sea floor. It is not a good idea to use a water camera monitoring platform for wellhead operation, so we commonly use laser TV technology underwater. It installs a laser probe in the subsea wellhead, and then transmits to the platform at the surface by cable, and the condition of the platform is display with a laser TV. We can obtain good results using this technique in conditions with a surface layer transparency of $1 \sim 1.3$ m.

2.6.2.2 Platform Gets into Position

The platform must do some work in addition to looking for the wellhead when it returns to the location to get in position. Platform location work should be done as introduced in 2.3 in place of the design

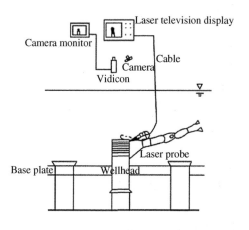

FIGURE 2-106

Platforms returning to the wellhead position using underwater laser TV.

requirement. Anchor work needs to be done according to the first section of this chapter where the floating drilling platform (ship) mooring mode and the anchorage of the jack-up platform were detailed.

2.6.3 Tieback

The yieback operation is the whole process of tying back each layer of casing of predrilled wells to the surface platform and install wellhead equipment with tieback pipe and other tools.

The tieback operation generally ties back using a marine riser such as Φ 508 mm (20), Φ 339.73 mm (13 3/8 in), and Φ 244.48 mm (9 5/8 in). A jack-up platform also ties back using a marine riser like Φ 762 mm (30). As for the order of tieback, usually it is from big to small, that is tie back the riser first, and then tie back the Φ 339.73 mm (13 3/8 in) and Φ 244.48 mm (9 5/8 in) pipes.

A riser can prevent the two layers of casing from being corroded by seawater within it, and resist the impact of the wind and waves. So in particular we should conduct a stress analysis and do a strength calculation to ensure its service life. As for the two layers within the riser casing, we just require it is tested only in order to meet the completion requirements. It often makes riser and the wellheads not aimed because of the seabed wellhead installation error and to return to the location of the new production platform built in the process of construction and installation error, and brings difficulties to the tie-back operation, therefore tieback is a complicated operation on the sea.

2.6.3.1 Tieback Tools

The primary tieback tool in the tieback operation is a joint, and it is a major connector between the tieback column and the wellhead. Its main function is to lock and seal. It can be divided according to size as follows.

1. **Φ 508 mm (20) tieback joint**
   Φ 508 mm (20) joints are divided into two types.
   a. Internal locking tieback joint
      As shown in Figure 2-107 (a), the upper part of the Φ508 mm (20 in) internal lock joint with is connected with the Φ 508 mm (20 in) riser with a guide bell, and a lock is set internally. When the tieback joint is seated in the casing head of the seabed wellhead, we can lock it with the casing head of the wellhead by thread with the torque tool spin lock set. The seal between the tieback joint and wellhead is a rubber ring, and inside the lock set is a zigzag tieback internal thread of Φ339.73 mm (13 3/8 in). Figure 2-107(b) shows the connection between the Φ 107 mm (20) tieback joint with internal thread and Φ 476.25 mm (18 3/4) wellhead equipment.
   b. Outside locking tieback joint
      As shown in Figure 2-108 (a), the difference between the outside locking tieback joint and internal locking tieback joint is that it is locked with a lock block. When spinning the lock, in order to make it move down, we can push the lock block into the mouth of the well outside the casing head locking groove to lock it. Figure 2-108(b) shows that it is locked into the Φ 476.25 mm (18 3/4 in) wellhead.
2. **Φ 339.73 mm (13 3/8 in) tieback joint**
   This joint has only one style, an internal lock, and it makes the casing tieback of zigzag thread into Φ 508 mm (20) in joint lock back close set of internal saw tooth thread by the weight of tieback

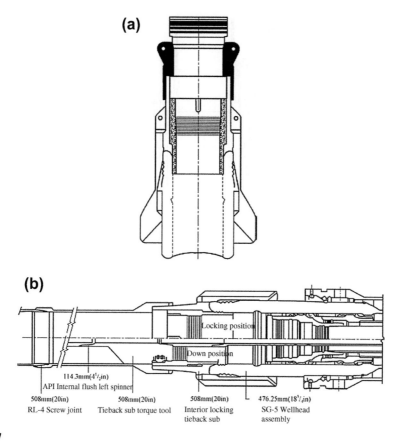

FIGURE 2-107

Lock type joint and tieback lock connection.

casing, and it is with the help of thread engagement to realize locking. Then it uses "O" shape circle seal the seal in Φ 244.48 mm (9 5/8 in) inside the casing hanger. Figure 2-109 shows the assembly conditions of the Φ 339.73 mm (13 3/8 in) tieback joint.

3. Φ244.48 mm (9 5/8 in) tieback joint
   This tieback joint only has one internal lock type; it matches with the standard tieback casing hanger. Under the back joint there is a department for back or move with the external thread lock close short section, when the joint with Φ 244.48 mm (9 5/8 in) tripping below casing hanger, then spin lock with the twisting force under tools, to make it mesh the casing with Φ 244.48 mm (9 5/8 in), then tieback can be realized. It realizes a metal surface seal depending on the torque. Figure 2-110 shows the assembly conditions of the with Φ 244.48 mm (9 5/8 in) tieback. Table 2-13 presents the technical characteristics of three kinds of commonly used tieback tools in China.

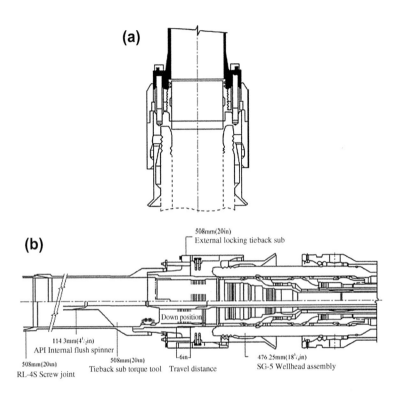

FIGURE 2-108

Outside tieback joint locking into the wellhead.

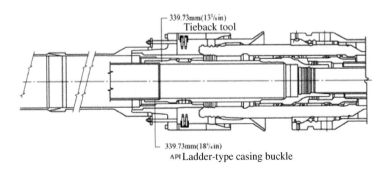

FIGURE 2-109

Φ339.73 mm (13 3/8in) tieback joint assembly.

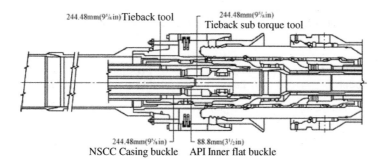

FIGURE 2-110

Φ244.48 mm (9 5/8 in) tieback joint assembly.

2.6.3.2 Tieback of Subsea Wellhead

The tieback refers to use floating drilling platform (ship) drilling, the well at the bottom of the sea, and use the method of well completion in water surface and installation tree when the well is completed to product, so it needs to use the casing head to stay at the bottom of the sea after drilling, tieback the casing to the platform.

1. Inspection before tieback operation

   First we must check each joint thread, seal and thoroughly clean them, then coat with butter before the tieback operation. Second, we should check the function of the tieback joint and torque tools, and verify the condition of the locked tieback joint and torque tool spin position.

2. Tieback operation order

   For seabed wellhead tieback, first tie back the Φ 508 mm (20) riser, then tie back the Φ 339.73 mm (13 3/8 in) casing, and last tie back the Φ 244.48 mm (9 5/8 in) casing.

3. Tieback operation procedure

   The two styles of casing tieback operation procedure are basically the same, and they are in accordance with the following steps.

   a. Taking up the anticorrosive hat

   When the production platform is in position, it should trip down the special tool with lifting system from the platform, and the casing head rot proof cap casing head taking up, and into the interval sets.

   b. Cleaning the wellhead

   Trip down tools with injection of joints and steel wire brush, with the help of the ROV; clean the wellhead tieback and threads thoroughly.

   c. Connecting tieback pipe

   Connect the tieback joint and the buckle drill column ion well, and connect the joint by fast joint, and install the centralizer according the design requirements of the tieback pipe.

   d. Locking tieback connection

   Trip the tieback pipe and joint down, make it seat in the casing head, then reverse torque tools to tighten them down, forcing the pawl and casing head of H-4 of section mesh (lock back

Table 2-13 Technical Characteristics of Three Kinds of Commonly Used Tieback Tools

Company Name	Tieback Tool Model	Lock Turning Laps	Test Pressure MPa	Torque $N \cdot m$	Super Lift Pressure kN
FMC	508 mm (20 in) interlocking	5 (right)	6.8	2712	89.04
	339.73 mm (13$\frac{3}{8}$ in) interlocking	1 (right)	20.7	2712	178.09
	244.48 mm (9$\frac{5}{8}$ in) interlocking	1 (right)	34.5	2712	178.09
ABB	508 mm (20 in) external lock	12 (left)	Watering test	20337	133.57
	508 mm (20 in) interlocking	6 (left)	Watering test	2033	133.57
	508 mm (20 in) nut interlocking	3.5 (left)	Watering test	20337	133.57
	339.73 (13$\frac{3}{8}$ in) Gravity setting type	String down distance 127–140 mm	17.2	4067	445.22
	244.48 mm (9$\frac{5}{8}$ in) Gravity setting type	String down distance 177.8 mm	24.1	13558	543.17
CAMERON	508 mm (20 in) Gravity setting type	String down distance 48–60 mm	3.5	678	89.04
	508 mm (20 in) rotation interlocking	4 (left)	3.5	6779	89.04
	339.73 (13$\frac{3}{8}$ in) Gravity setting type	String down distance 127–140 mm	10.3	5423	44.52
	244.48 mm (9$\frac{5}{8}$ in) Gravity setting type	String down distance 177.8 mm	24.1	6779	44.52

joint). Apply about 20337 N·m torque, then complete the connection between the tieback column and the casing head, and make sure it is locked.

e. Inspection of tieback seal

Through the rotation of the torque tool from the platform, make the combination seal compression (s.a.), within the tieback string injection water, and then inspect the sealing through the presence of tieback leakage.

f. Carrying out the torque tool

We can carry out the torque tool when we have ensured that the seal is sufficient through inspection.

g. Installation of platform wellhead

Cut the pipe column and tie back to the surface according to design requirements, and then install the casing head and the oil tree to the platform wellhead.

2.6.3.3 Tieback of Mud Line Suspension System

The tie-back operation refers to tie back to the water surface on receiving the tieback operation of well production when completed by flatten jack-up drilling bench drill production. Because the jack-up drilling platform uses a seabed mud line suspension system during drilling, we should tie back all casing of the mudline suspension system at the bottom of the sea when complete.

In order to tie back the mud-line suspension system, the first step is the Φ 762 mm (30) riser, the second is each Φ339.73 mm (13 3/8 in) casing. The inspection before tieback and the steps of the tieback operation are basically the same as for the seabed wellhead tieback operation class, so no more explation is needed here. Figure 2-111 presents typical tieback conditions for the mud line suspension system used in the jack-up platform. In the figure the left side is the drilling status, and the right side is tieback state, explained in the following

1. Φ508 mm (20) casing drilling and tieback uses the same coarse thread on the mud line suspension. From the figure you can see the Φ508 mm mud line casing string and tripping in and the Φ508 mm tieback tools (joint).
2. The Φ340 mm (13 3/8 in) casing uses the upper mud line suspension thread and protects its own tieback threads while drilling [Figure 2-111 (left)]. The Φ340 mm casing has a dashed mud flow diverter in its lower part, and it can clean the Φ340 mm and Φ508 mm pipe thread sufficiently. They use the lower part of the thread of the mud line suspension when tying back.
3. Φ244.5 mm (9 5/8 in) casing uses the thread of the upper mud line suspension while drilling [Figure 2-111 (left)] in order to protect its own tieback thread.
4. The Φ177.8 mm (7 in) casing during drilling (the left side of Figure 2-111) and during tieback after completion (the right side of Figure 2-111) uses the same coarse thread of the mud line suspension.

2.6.3.4 Installation of Wellhead Platform

Here, we are discussing installation after the wellhead and riser of the mudline suspension system and each layer of casing has been tied back to the platform. the general installation steps for a surface wellhead platform are as follows.

1. Cut the riser with Φ762 mm (30 in) and the surface casing with Φ508 mm (20 in)

The cutting position should be decided according to the set of the inside step length of the tube head.
2. Install the Φ508 mm (20 in) casing head

The casing head is installed on the casing by welding. The installation direction of the casing head is the same as the tubing spool, and in accordance with the requirements for completion, you should choose a good angle with the axis. We should do a pressure test to 60% of the casing internal pressure strength voltage after welding that lasts 5 minutes.

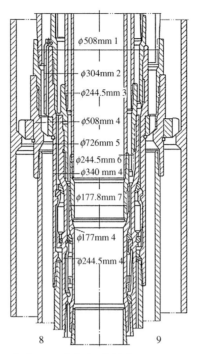

1: down and tie-back tool
2: flush diverter
3: down tool
4: mudline casing leg
5: support ring
6: tieback tool
7: down and tieback tool
8: drilling regime
9: tieback production status

FIGURE 2-111

Tie-back of mudline hanger system in self-elevating drilling platform.

3. Cut Φ339.73 mm (13 3/8 in) technical casing and install Φ339.73 mm (13 3/8 in) casing head
   First, you should install the slips and seal assembly of the Φ339.73 mm (13 3/8 in) casing in the Φ508 mm (20in) casing head, and test the pressure of the seal (same as for the Φ508 mm casing head); then cut and weld and test the pressure in the same manner as step 2.

4. Cut oil layer Φ244.48 mm (9 5/8 in) casing and install the Φ279.4 mm (11 in) casing head
   After tying back Φ244.48 mm (9 5/8 in) casing, we lift 80% ~ 90% of the weight of the casing above the running tools, and set the Φ244.48 mm (9 5/8 in) casing slip and seal assembly inside the Φ339.73 mm (13 3/8 in) casing head; we cut the Φ244.48 mm (9 5/8 in) casing according to the height of the sealing bushing. We use the welding method, and install the Φ279.4 mm (11 in) casing head, then test the pressure until it is up to the standard, that is 60% of the voltage regulator for 10 min.

2.7 New Technologies for Offshore Drilling in China

2.7.1 Optimized Drilling Technology in the Bohai Sea in China

There are marginal oil fields in the Bohai Sea of China, and it is predicted that the proven reserves of marginal oil fields is about 5×108 t. Among the factors constraining the costs of a marginal oil field, drilling cost is the most important factor besides the reasons related to development programs. The cost of drilling and completion takes up about half of oilfield development funds. Therefore, in order to develop a marginal oilfield effectively, work in the Bohai Sea has been looking to increase drilling speed, reduce drilling cost, and optimizing drilling technology. Due to significant efforts in recent years, there have been a lot of innovations to optimize drilling technology, which have greatly shortened the well building cycle to just a third of the average well. At the same time, drilling cost has been greatly reduced, with savings of 27% over the standard budget. Here we present a brief introduction to the optimized drilling technology and sea drilling developments in the Bohai Sea.

2.7.1.1 Dense Cluster Well Group with Focus on Rapid Drilling and Cementing Surface Technology

1. Dense cluster well group

   A drilling well group has been adopted that is a dense cluster wellhead platform designed for 35 well slots. Consider 16 wells with a wellhead spacing of 2 m * 2 m and there are 4 rows * 4 mouths = 16 wells/platform. If we decrease the spacing to 1.5 m * 1.8 m the well number can increase to 7 rows * 5 mouths = 35 wells/platform. The platform control area is expanded from 11 km^2 to 22 km^2. In this way, each wellhead platform control area expands, which greatly saves investment, and makes the relative average operation time increase by 50%. Thus the drilling cost is reduced greatly.

2. Fast drilling and cementing surface

   This refers to the drilling and cementing of the surface casing section. Some water surfaces of the Bohai Sea contain an unconsolidated sandstone and mudstone layer, which leak easily. If a riser drilled into the mud becomes shallow, there could be losses. In order to solve the formation problems of the surface casing section of drilling, we always use a pendulum drill assembly with a PDC bit, which can optimize drilling parameters, and has a large displacement, high revolving speed, light penetration pressure drilling, and a hole deviation controlled within 0.5 degree. Adopting bevel and sloping field character could jump between drilling surface, which makes drilling wells and the crossover operation of cementing well be separated by distance and time. This could prevent leakage.

2.7.1.2 High-Density Cluster Directional Well Technology

High density refers to increasing the lot number of a wellhead platform. As a result, many wells populate the wellhead making platform area finite, the well spacing decreases and bottom hole displacement and angle increases, and therefore directional well deflection, anti-collision, and magnetic suspension are difficult. For this reason, a scheme of the drilling platform center location and platform structure will be selected and the location and layout of cluster wells along with the drilling sequence will be optimized. According to the principles of reducing the largest hole deviation and achieving the shortest interval to the target displacement, the design of the hole structure should be also optimized. The optimization of the string assembly and drilling parameters may result in the fact that the average deviation reaches 0.1 degree and the displacement reaches only 10 m per meter, which realizes the optimum fast high-density drilling of a directional well.

2.7.1.3 High Density Cluster Well Anti-collision Technology

Because of the high density of 35 wells focused on a wellhead platform, whose wellhead center distance is only 1.8 m × 1.5 m, the borehole anti-collision problem is very serious. In order to solve this problem, COMPASS software for anti-collision calculation and borehore trajectory design and WELLPAN software for drill bit hydraulic parameter calculation and stress analysis have been developed. The straight section of the inclined surface well technology (deviation not more than 1°) is adopted. A reasonable choice is PDC drill + rate gyroscope + MWD + the adjustable motor drilling tool structure. The kickoff point is chosen as shallow outer and deep inside. It uses

2.7.1.4 Shielding Temporary Plugging Technology to Protect the Reservoir

There may be different degrees of damage of the surface-to-surface layer in the progress of drilling. In order to control formation damage in the process of completion work, CNOOC has been developed a shielding temporary plugging agent with emulsified paraffin. It will join the rest of the drilling fluid, act as a temporary plugging agent, and reasonably control the positive pressure difference after drilling into the reservoir, which makes the annular return velocity of drilling fluid not exceed the maximum of 2 m/s. Combined with timely monitoring of drilling fluid particle size distribution after drilling into the reservoir and of drilling fluid before and after conversion, we can protect the reservoir and achieve remarkable effects and benefits. The productivity index goes from 0.81 m^3/ MPa·d·m to 1.5 m^3/ MPa·d·m, while the drilling damage skin factor reduces from 24.43 to 2.16.

2.7.1.5 Long Open Hole Single-Stage Double Seal without Waiting for Setting Time Cementing Technology

An open hole section with a length of 2000 m originally needed at least a two-stage cementing. Thus drilling a cement plug and setting will cost a lot of rig time, and the economic benefit is poor. A single-stage double sealing cementing technique can be used to cement two intervals through accurate control of casing shoe coagulation time, which breaks general cementing and affects setting time and realizes cementing without waiting. This creates a single-stage double seal without waiting cementing technology. The key technology is optimization of the slurry formula and determination of the single-stage double sealing fluid volume. The slurry formula is to reduce water mud filtrate as much as possible, making fluid lose control within 30 ml, free water within 0.8%, and thickening time within 180-220 min. Accurately determining fluid volume controls the annular space quantity density and plays a decisive role in the two periods before and after the cementing. Fluid volume is determined through controlling the total fluid volume. First according to the different hole caliper log volume values, we determine the rear slurry and liquid volume. Then according to the oil cap depth we determine the length of the tail slurry cementing. The rear pulp empty volume sets a theoretical amount of additional annulus volume according to the experience value log volume. The total amount in the front minus rears plasma quantity is named for the required amount of fluid in the middle.

2.7.1.6 Grand Slam Open Hole Logging Technology

This is a set of advanced applied optimal fast logging methods in order to meet the need of optimized drilling and maximizes the logging technology in accordance with the requirements of the logging services on the premises. It mainly includes the following three aspects.

1. Grand slam open hole logging technology
   Usually in the development of well completion logging needs to go up and down the cable three times to complete the data acquisition task. Using the grand slam open hole logging technology, it can be admitted to a trip to the well completion task of all geological information data, accord with the requirement of optimized drilling. This technique is carried out on the instruments under the reasonable combination that the resistivity instrument, acoustic instruments and radioactive instrument adopt the concatenated composition mode, at the same time, ball joint universal short section is set between the instruments, which improves the flexibility of instrument string. This technique also adopts a down-hole signal digital transmission method, which can improve the cable transmission ability and reliability of information.

2. Cementing quality measurement technology that doesn't occupy drilling rig time
   Generally after drilling, casing, and cementing, a well needs to wait for about 24 h of setting time; this does not comply with optimized drilling. Therefore, taking a cantilever beam of the steel structure at the top of wellhead as the logging framework, will not occupy the hoisting system of drilling rig. This mix of cementing quality logging and drilling greatly reduces the idle time of the rig.

3. Engineering services new logging technology
   The new technology of all-round cementing quality detectors (SBT) logging has solved the cementing quality problem for difficult wells. It will eliminate unnecessary squeezing cementing operations, and provide a clear filling cement decision which is a basis to the well really needed to remedy.

2.7.2 High Temperature and Overpressure of Natural Gas Drilling Technology in the South China Sea

The high temperature here is referring to overpressure wells that have an expected or actual bottom hole temperature greater than 150 °C and bottom hole pressure greater than 68 MPa, or a formation pore pressure greater than 1.80 g/cm^3 equivalent density of the drilling fluid. The Yinggehai and Qiongdongnan basins of the South China Sea are a prime location of high temperature and overpressure wells. In recent years, through the joint efforts of CNOOC, scientific research institutes, and institutions of higher learning, we have developed scientific research work related to this issue. And we have obtained satisfactory results that the efficiency of drilling relating to the high temperature and overpressure gas drilling of the southeast Yinggehai basin can be greatly improved.

2.7.2.1 Prediction and Detection Technology for High Temperature and Overpressure Formation Pressure

The main research achievements in this respect are the following:

1. Pressure prediction methods before drilling into the formation.
   Based on the idea that the formation of abnormal and high pressure is mainly caused by the mudstone

under compaction in the Yinggehai and Qiongdongnan basins of the South China Sea, we can divide the methods into two classes, and the high pressure aspect uses different forecasting methods.

a. Primary overpressure

This type of overpressure is related to stable rock formations and the phenomenon of corresponding overpressure under compaction. The seismic velocity prediction method can be used to forecast, and this uses acoustic logging, VSP, and surface seismic measures for seismic waves. Then using seismic p-wave velocity we can predict the formation pressure before drilling, after improving the resolution, signal-to-noise ratio, and eliminating of the influence of structure suppressing noise and multiple waves.

b. Source overpressure

This is caused by the fact that diaper structure, fault, sand bodies, previous drilling and other factors make collusion between the deep layer and shallow. Such overpressure compaction does not comply with the law, and it is difficult to use the seismic velocity method to predict it. Instead, with the method of simulation analysis, we can set up two models. ModelIrefers to source overpressure caused by fracture and corresponds to the previous drilling model; ModelII mainly analyzes the stratum lateral extension of the corresponding source overpressure, thus it can make up for the lack of speed in predicting formation pressure.

2. Method of monitoring super pressure in the drilling process

We usually adopt the method of formation pressure detection to monitor during the process of drilling. But because of the special geological condition and stress distribution characteristics, it is very difficult to adopt traditional methods in the Yinggehai basin. Rock strength is proposed for this method. Through the indoor unit simulation analysis of a large amount of real drilling data and field inversion, we establish a model of the relationship between the rock strength and the formation pressures. Thus according to the actual sample measurement results of rock mechanics, we can test and monitor of formation pressure. Practice has proved that the rock strength method has high precision and high applicability.

3. Forecasting method for formation overpressure in undrilled areas during the process of drilling

According to overpressure forecasting for the drilled formation in the stratum that has not yet been drilled, we have developed a new method according to the Yinggehai basin in the South Sea. Besides using the vertical seismic logging (VSP) data and seismic logging while drilling (SWD) bit data to predict the lithology, this method also uses conventional logging data combined with surface seismic data at the same time. We can predict formation pressure under the bit through a fitting extrapolation method.

4. Pressure test after drilling into the formation

Pressure detection after drilling into the formation refers to the use of all information of the drilling period to inspect and verify the results of these sections before application of drilling pressure. The main purpose is to fix the prediction of drilling formation pressure. Here are some common measures.

a. Logging data test

First through correction of well logging data, automatic segmentation of logging curves, it chooses a log characteristic value, and establish a formation pressure detection model. We can get pure mudstone well logging data by using different depths for the pick up point as the basis of improving the accuracy of the logging data. Then we can compute the depth point of pore pressure and fracture pressure. We calculate the pore pressure by the equivalent depth method and the fracture pressure by the Eaton method.

b. Effective stress test

This method first establishes an experience model of a longitudinal wave propagation velocity in porous rock, and then the function relationship between the porous rock p-wave velocity v_p and the influencing factors (the vertical effective stress p_e, rock porosity Φ, and speed of rock skeleton v_0) are given as follows:

$$v_p = v_0 - A_\phi \phi + A_p p_e - B_p e^{\frac{-P_e}{C_p}} \tag{2-45}$$

Then by using the actual data, we can compute the nonlinear regression calculation to acquire the coefficients for $A\Phi$, A_p, B_p and C_p in (2-45). Then we can directly calculate effective stress p_e by (2-45), according to the upper normal compaction of acoustic velocity and validate it for use for effective stress. Practice shows that this method for mudstone gives priority of applicability to the continuous deposition of sand shale profile.

Additionally, because the above effective stress test does not apply to defects for mudstone compaction and the causes of the abnormal pressure, we propose an experience model which increases consideration of the effect of the shale content V_{sh} and rock density ρ that is suitable for a complex sand shale profile. We would combine it with the actual data regression of the Yinggehai basin. After determining the undetermined coefficients, the calculation formula of rock p-wave velocity v_P would be given as below:

$$v_p = 5.543 + 0.093\rho - 6.885\phi - 1.182\sqrt{V_{sh}} + 0.397\left(p_e - e^{-14.7P_e}\right) \tag{2-46}$$

This method is suitable for the sand shale profile and would not be restricted by compaction causes.

2.7.2.2 Stratum Drilling Process for High Temperature and Overpressure

With the high temperature and overpressure drilling practices of the Yinggehai basin in the South China Sea, we have created a productive high overpressure stratum drilling process. It includes a shallow gas drilling program, BOP testing, and a process of high-temperature, high-pressure drilling and well control. High temperature and overpressure drilling optimization design and drilling technology works before the material preparation. This drilling process has strong flexibility and compatibility with well bore structure design. It has many advantages including good thermal stability, anti-sloughing, anti-pollution, and high-density drilling fluid design with good rheology under high temperature. It adopts equilibrium pressure cementing technology. Additionally, it conforms to the requirements of the cementing equipment working pressure and has large capacity processing abilities for high-temperature, high-density drilling fluid with high-density cement slurry mixed with rubber parts to work with the high-temperature drilling rig equipment and advanced management modes, etc.

2.7.2.3 High Temperature and Overpressure Formation Testing Technology

Formation test engineering is usually divided into a halfway test and well completion test. High-temperature and overpressure wells commonly have used the well completion test. We can achieve various results, such as the nature of fluid properties and reservoir capacity through completion tests.

High-temperature and overpressure well tests are difficult and require complex techniques, so we often need to use some special equipment and advanced technology to complete them. High-temperature and overpressure well pressure test technology not only requires new technology domestically, but also there is a good chance to draw lessons from abroad. CNOOC's scientific research has earned great achievements in the following several aspects in recent years.

1. Safety systems under the shaft of high-temperature and overpressure wells
   In order to guarantee the safety of a high-temperature, overpressure well under the shaft, a new calculation model has been established to solve the problem of the deformation calculation of the test string. The new model has introduced tubing space deformation after friction occurs between the tubing and casing into the tubing string stress and deformation calculation. The optimal design method considers tool parameters, well conditions, and operation factors, such as intensity of the testing string to determine the safety coefficient as well as platform and weather factors. And analysis software has been compiled that is suitable for high temperature and overpressure tubing string in a sea well.

2. High-temperature and pressure well testing technology.
   Because of the different rules of gas-liquid two-phase flow distribution between high temperature and overpressure wells and general wells and the hydrocarbon fluid phase state problem, we have established a new unitary two-phase flow equation, which determines the discretionary and the step length of partial differential equation. This not only ensures convenient calculation, but also ensures that the important characteristics of two-phase flow and presents accurately the high temperature and overpressure in well testing string water multiphase flow pattern of oil and gas distribution. In addition, we have already set up a test model when testing tube temperature that can calculate the wellhead flow temperature, pressure differential, and bottom hole pressure. At the same time, we also established a math model for high-temperature and pressure in a homogeneous medium or for composite nonlinear unsteady seepage. The model will give the analysis method of curve fitting to transient well test, high temperature and pressure gas back pressure deliverability test.

3. Test ground data monitoring and safety control system for high-temperature and overpressure wells
   We integrate the above high-temperature and overpressure well test technology to develop a method with measurement accuracy of 1% on the surface temperature and pressure flow parameters, such as oil, gas, gas composition, and measurement of the content of poisonous gas. The compiled software can also help the mathematically established model, with real-time program evaluation and calculation of safe production, and real-time adjustment and control, which can help avoid major accidents on the ground during the test.

2.8 Offshore Drilling Rigs

The rig is the mechanical equipment for drilling oil and gas wells with a power-driven, lifting, rotating, drilling fluid circulation system and other auxiliary operations. For offshore drilling mechanical equipment, the structure, design calculations, and use of the equipment is essentially the same as for land. Therefore, there's no need to elaborate on these factors here. Here, we only do a brief introduction to the special problems of the offshore drilling rig.

2.8.1 Special Requirements of the Offshore Drilling Rig

When designing or selecting an offshore drilling rig and taking into account ease of use, the special requirements of the offshore drilling rig are summarized in the following

2.8.1.1 Maximizing Drilling Depth

The first special requirement is requiring the rig to provide greater drilling capacity. The main indicator of drilling capacity is drilling depth for rigs. Especially with offshore oil drilling and production getting to deep water development, this requirement is especially pronounced for the following reasons:

1. Working depth is gradually deepening.

From a global perspective, working depth in offshore drilling in 1998 was 2300 m, and at that time Oil and Gas magazine published that the American Noble Drilling Company used floating drilling rigs to get to a working depth of 2305.2 m, which was the highest in the world. In 2006, the working depth of the world's offshore drilling reached a record 3051.35 m. It can be seen that the working depth of offshore oil drilling is deepening.

2. Deepening of the reservoir away from the bottom of the sea.

In 1998, "Discoverer Enterprise" was the strongest drill in the world, and its drilling depth capacity was 10700m (35000 ft). "Sai Deke quick" and three other semi-submersible drilling platforms were built in 2000, and their working depth is 2590 m (8500 ft), but their drilling depth capacity is 35000 ft (10700m). In 2006, the deepest ocean depth of wells has reached 8071 m. Obviously, the reservoir is deeper than the seabed, which requires offshore drilling rig to further enhance drilling depth.

2.8.1.2 Maximizing Lifting Capacity

As mentioned above, with the increasingly development of offshore oil into the deep sea, as well as the depth of the reservoir moving away from the seabed, offshore drilling rigs requires as much as possible to improve lifting capacity. Improving rig drilling depth (drilling capacity) which requires solving the following two issues.

1. Improving hoisting winch power

In order to save on the cost of construction and installation of offshore platforms, now the radius of platform-centered horizontal wells has been expanded from 3 km to 4 ~ 5 km, and thus this requires the hoisting winch of the offshore drilling rig to provide greater lifting capacity. Of course, as working depth increases and the distance of the reservoir moves away from the seabed, it also requires the lifting winch to increase the lifting capacity to meet the need for a hoisting that can lift the large-diameter drill string (6 in), deep-water riser, and large deep-hole casing. To this end, the winch must further improve in power. For example, in 1998, the world's first maximum power winch was manufactured by Ames Scott and is known as the fourth-generation electric drive winch; it has power of 5000 hp, and hook load capacity of 907.4 t. However, in 2000, the "Sai Deke Ekspressr" semi-submersible drilling platform had a winch manufactured by Ames Scott, and its power was 6600 hp (HitecoAHD type). The "Nevis • prospectors I" floating drilling winch has power up to 7000 hp. Offshore drilling rig winch power is definitely increasing.

2. Improving lifting system strength

The main equipment for increasing lifting capacity is the rig winch, but the various unit of the entire lifting device system, with the increase of the weight and the load, must also be increased accordingly in strength. As mentioned above, if the hook load capacity is up to 9074 t (200×104 lb), then the lifting system hook, rings, elevator, traveling block, wire rope, crane and derrick equipment, and other units are required to have a corresponding increase in strength.

2.8.1.3 Trying to Enlarge the Turntable Diameter

In order to make larger diameter risers and other underwater apparatus go through the turntable during offshore drilling, the turntable maximum initial hole diameter must be bigger than turntables used with the same load onshore rigs, which is a special requirement compared with a land rig. Moreover, with deep sea development, the turntable drift diameter also shows a constant increasing trend. For example, originally, the maximum initial hole diameter of offshore rigs was mostly 1257 mm (49.5 in), and the minimum was not less than 953 mm (37.5 in). However, the "STENADON" semi-submersible drilling platform built in 2001 was equipped with a turntable manufactured by Varco, and its drift diameter was up to 1536.7 mm (60.5 in), which is currently the maximum size drift diameter for a turntable. But note that the turntable drift diameter made in China is only 952.5 mm.

2.8.1.4 Minimizing the Volume and Weight

Offshore drilling is different from land drilling, and it uses a fixed drilling platform or mobile platform (ship). Therefore, the well field area and space are extremely valuable. For each equipment unit of the drilling rig system, we are required to reduce the volume and weight as much as possible; this is due to the following reasons

1. Need for carrying capacity
   Starting with the platform and the strength of the hull itself, we need the drilling system to minimize weight, to decrease the platform or hull load. For a floating platform (a ship), due to the need for stability, we require the drilling system to minimize the height. For the platform legs and the floor plate of a bottom-supported platform, the soil-bearing capacity requires the drilling system to minimize weight and reduce pressure.
2. Economic efficiency
   One of the important economic indicators of offshore drilling platform (ship) is the cost per square meter. Therefore, this requires the drilling system to take up as little deck area as possible, and to reduce the volume and the height as much as possible, in order to minimize the use of the deck.
3. Need to work safely
   In order to ensure drilling safety, we require the various mechanical equipment volumes to be as small as possible, using a compact structure, to provide comfortable working conditions to the staff.

2.8.1.5 Integration of Machine, Electricity, and Liquid as Much as Possible

The integration of machine, electricity, and liquid refers to using electricity and fluid to control drilling rig winch speed and drilling pump flow rate and to regulate the drilling pressure, rotating torque and speed, downhole drilling technical parameters, and so on. The drilling system is gradually realizing full automation under the command of computers; because offshore drilling cost is far more expensive than land drilling, how to combine favorable working conditions, saving time, and rapid and high quality drilling has become the main conflict. Therefore, the requirements for mechanization and automation are much higher than for onshore drilling.

There are several forms of the integration of machine, electricity, and liquid of the current offshore drilling rig which have been put into use.

1. A full hydraulic drive drilling rig

   This form adopts a hydrostatic transmission for the main winch, rotary table, and drilling pump; the hydrostatic drive drilling rig replaces the conventional rig. For example, the Marine Hydraulic Company was established by the cooperation of the American library beauty company and Norway in 1990s, produced a full hydrostatic drive drilling rig that used a hydraulic motor to drive the winch and the turntable and a hydraulic cylinder to drive the drilling pump, and was applied to the semi-submersible drilling platform of Norway's national oil company. This rig winch power is up to 4000 hp, the drift diameter of the turntable is from 952.5 mm (37 in) to 1257 mm (49in), the maximum capacity of the drilling pump is 2500 L/min, and the maximum pressure is 34.5 MPa. The main advantages the full hydraulic drive drilling rig are:

   a. It is easy to connect with a computer, amplified by the microcomputer - electric - fluid, realizing the drilling rig rotary, lifting movement and stepless adjustment of the drilling pump flow, automatic control of the computer program.

   b. The pressure-torque indication and overload protection make the drilling process safe and reliable.

   c. You can optimize drilling process parameters through a computer, which can make full use of power, improving work efficiency.

   d. It reduces the weight of equipment and saves on space. Take a rig winch as an example. This form has power of 4000 hp, with a total weight of 37 t; the power of a 1625DE winch of a conventional drilling rig is only 3000 hp, but the weight has reached 39.53 t.

   e. Save wear and noise caused by mechanical transmission and the replacement time of worn-out parts.

   f. It is easy to combine the hydraulic top drive with a automatic discharge robotic drilling system, and with a drill string heave compensation device, you can realize full automation of the offshore drilling operation.

2. Hydraulic cylinder lifting drilling rig

   This form of the drilling machine is composed of a large cylinder supported by a vertical steel frame structure and liquid cylinder stroke amplification system consisting of a compound pulley and top drive system combined with a tap. For example, in 2001, a hydraulic cylinder lifting rig installed at the "Future Western I" semi-submersible drilling platform manufactured by a marine hydraulic company had a lifting height of 36 m, lifting capacity of 750 t, and drilling capacity of 10060 m (33000ft). The advantages of this form of drilling rig are:

   a. Reduction of the cost of equipment and drilling. The cumbersome and expensive conventional winch and turntable are eliminated, and the floor area decreases, with an estimated cost reduction of about 30%.

   b. It can achieve drill string heave motion compensation and automatic drilling, so that to replace Heave Compensation Equipment and automatic drilling equipment, a significant reduction in investment.

   c. Eliminates the need for the derrick, crane, and traveling block system.

   d. The combination of the lifting hydraulic cylinder and top drive can realize drilling pressure regulation, automatic drilling, and optimization of drilling process parameters, so as to improve drilling efficiency by about 30%.

3. Automatic drilling rig

   The so-called fully automated rig is an automatic control rig that integrates the rotational movement of the top drive, the lifting movement of the winch, and the remote operation of the drilling pump, the joint, and removal of the drilling pipes together. It is not only able to achieve automatic control of each unit, but also uses a computer to optimize the drilling process parameters, creating a drilling rig with a high degree of automation. This fully automated rig has been produced by the U.S POOL Company, and industrial tests are being carried out in the Gulf of Mexico for offshore drilling operations.

2.8.1.6 Trying to Make the Equipment Modular

In order to facilitate the rig's installation and movement and to save construction time at sea, an offshore drilling rig system tries to make equipment modular. These modules can include the winch, the turntable, the base of the rig (mobile derrick base), drilling pump, drilling fluid, motor control center (MCC), the power module, etc. The reason that the equipment is required to achieve modularity is mainly based on the following two aspects.

1. From the view of offshore and deep-water drilling

   At present, Chinese offshore drilling is promoting optimal and fast drilling technology. Abroad, the drilling speed of the United States Unocal in the Gulf of Thailand is the world's fastest. For a 3500 m well, the drilling time is only 6 to 7 d. After the Chinese learned the foreign advanced drilling technology, optimal and fast drilling techniques were applied, such as the 18 - 1 oilfield in the Gulf of Bohai Qikou; it has already drilled three production wells, with an average well depth of 3561 m; the average well construction cycle is only 18.82 d. For one of the fastest wells drilling only took 13.82 d (the average cycle of similar wells is 57 d in the same locale), and the time of well construction cycle was shortened by 2/3, with drilling costs reduced by 27%. Obviously, with the well construction period shortened to such extent, saving time during the transportation of equipment and installation of the drilling system, will gradually become the main issues. Therefore, the equipment modularity of the drilling rig system is increasingly prominent.
2. Analysis from the perspective of very shallow sea and offshore drilling

   In these waters, the rig equipment can "get in," "stand," and "come out," which is a prominent issue. Rig equipment modularity is one of the important ways to solve the transportation problem in these waters, which will be discussed in detail in Chapter 5. Among the topics are how to decompose equipment into modules and how to transport and install such modules.

2.8.1.7 AC Variable Frequency Drive

The AC variable frequency drive is a power drill system driven by generator AC diesel power, with a silicon-controlled rectifier drive DC motor to drive each equipment unit on the rig; usually it is called AC-VFD-DC technology. It is used to equip the semi-submersible drilling platform. United States of America TDS Varco-BJ company production of roof driving device as an example, it is used in the AC-VFD-DC 746kW (1000HP) DC motor (GE752). This AC-VFD-DC driving technology is a new technology developed in recent years. However, for drilling system on offshore drilling platforms, it has completely replaced the traditional DC generator drive DC motor (DC-DC driver) technology. AC-VFD-DC has the following advantages over DC-DC technology.

1. Improvement of the performance of speed adjustments
   The technology of AC-VFD-DC drive power overcomes the small range and poor speed adjustment performance. The winch, rotary table, and drilling pump can be driven using stepless speed adjustment range.
2. Saving space
   AC-VFD-DC technology eliminates the need for mechanical transmission gears, torque converters, and chain gears, and has a smaller alternator volume than DC generators. In short, all of these features save space.
3. Improvement of transmission efficiency
   Due to the greatly reduced mechanical wear compared with mechanical transmission, and reduced hydraulic torque of the drill rig, the transmission efficiency of the drilling system is improved by about 11% with AC-VFD-DC
4. Improvement of service life
   Because it avoids the abrasion of the DC generator commentator and brush, the service life is improved.
5. Improvement of the stability of the platform
   Because the AC-VFD-DC diesel engine generator has flexibility in the system set up, the platform deck equipment arrangement is easier, thereby reducing the center of buoyancy of the floating drilling platform (ship), improving the stability of the entire platform (ship).

2.8.1.8 Using a Top Drive Device

The top drive device, referred to as a TDS (Top Drive System), is a device where a power swivel at the top of the derrick directly drives drill string rotation. Soon after 1982 when the United States Varco-BJ company first produced the TDS, it was applied to offshore drilling. In 1988 offshore drilling platforms (ships) using TDS reached 50% (floating drilling vessel), 66.7% (bottom-supported platform), 69.7% (jack-up platform), 71.7% (semi-submersible platform), and 75% (drilling barge), and all new manufacturing offshore drilling platforms (ships) are equipped with top drive devices. Therefore, the top drive device has become a basic requirement in the offshore drilling rig; the TDS can save drilling time, which is conducive to the implementation of optimal and fast drilling.

2.8.1.9 Using a Heave Compensation Device

A heave compensation device is a drill string heave motion compensator, referred to as a DSC (Drilling String Compensator), which is one of the necessary pieces of equipment in the drilling system of a floating drilling platform (ship). An onshore drilling system does not require DSC, which is a special requirement of the offshore drilling rig. The offshore floating platform (ship) heaves with the ocean waves. The heave motion located in the bottom of the bit in the drill string driven, up and down the heave motion, to put it mildly is not stable, more than a mild inflation then bit from the bottom, can't drilling. Therefore, it must be equipped with a DSC device, and can be installed together with a large hook, or installed in a crane or buckle rope. There are several types, which will be introduced in later.

2.8.1.10 Using the Automatic Discharge Drilling Tool

Offshore drilling rig system is used in the automatic discharge requirements which have the drilling device, mainly based on the following two points.

1. Matching with optimal and fast drilling technology

   As already mentioned, at present, offshore drilling is promoting optimal and fast drilling technology. However, to improve the speed of the drilling operation and shorten the drilling cycle, one of the important elements is saving time in pulling and running the drill pipe; automatic arrangement of the drilling device can satisfy this requirement. For example, the automatic arranging drilling device manufactured by the Japanese MES Engineering Shipping Company and Japan KHI Heavy Industries Company include different kinds of control arms, a drill pipe clamp, a turntable, a turntable mobile frame, and a computer. It needs just 6 min to transmit a vertical root, 3 min to transmit a pipeline, and only 2.5 min to transmit a sleeve, which saves on operation time.

2. Device matched with a top drive

   Because offshore drilling platforms (ships) in general have been equipped with TDS, in order to match well with the TDS, it is necessary to equip an automatic arrangment drilling device. One of the main advantages of TDS is that it can save a lot of time during the drilling operation. For example, it takes 2/3 of the time necessary to drill a root using conventional rotary drilling. And if it is used in conjunction with the automatic discharge drilling device, then the vertical root connection time can be further shortened, further improving the drilling efficiency.

2.8.1.11 Using a Special Derrick and Basement

Offshore drilling process is in the drilling platform (a ship), from the characterizes that platform deck area is limited and the requirements on a platform (a ship) drilling several wells start, it requires the use of special design and derrick base, the different points with the land derrick and substructure are as follow.

1. Change the size and structure of the derrick

   Because of restrictions of the platform (ship) deck area, a drill derrick can't be fixed with an anchor line. So steps should be taken to increase derrick base size (or substructure size), increase the horizontal size of the derrick mast, and strengthen the thigh and fixed the connection structure measures to make up for the shortcomings of using the guy rope.

2. Adding a movable type of basement

   Because offshore drilling with a platform (a ship) often involves drilling multiple wells, the mobility will often make the derrick and substructure move within a certain range vertically and horizontally in all directions, In addition, in the platform for ship (ship) end or side of the platform (a ship), also needs to set sail, the derrick and the base away from the wellhead. Therefore, this requires that the basement be designed into a double deck, the upper moving transversely along the lower layer, and the lower moving along the track of the deck, which makes it convenient to move the derrick and basement to another well head after completing a well.

3. Addition of a mobile guide rail for the traveling block

   For offshore drilling, under the actions of winds and waves, the derrick swings with the platform, so that the traveling block system swings followed by the derrick, all of which affect normal drilling. So we need to add operation guide rail-travel car in the derrick space to limit the movement of the runner wagon.

2.8.1.12 Equipment for Cementation and Well Testing

Cementing offshore drilling and oil testing operation are already provided in a platform (ship) complete cementing equipment. The test for oil equipment, this point is different with land drilling. The special requirements of manufacturing for offshore drilling platform (a ship) include the use skid-mounted equipment; its composition is as follows.

1. Complete sets of cementation equipment
   This includes a diesel engine, three cylinder pumps, stirrer group jet mixer, metering box, and center console; they are respectively arranged in several skids.
2. Complete sets of well testing equipment
   This mainly including the heaters, separators, burners, manifolds, measuring instruments, and so on; they are respectively arranged in several skids. In general, they are arranged on the main decks of the platform near the drilling floor. So, crude oil comes out of the wellhead tubing head, goes through the testing of the Christmas tree and safety valve, moves to the flow manifold, then is dragged by the tubing, and enters the heater. The heated crude oil goes to the seeparator, seperated water goes into the sea, separated gas goes to the burner to burn, and crude oil goes to the metering tank for measurement.

2.8.2 Types and Technical Performance of Offshore Drilling Rigs

2.8.2.1 Types and Technical Performance of Foreign Offshore Drilling Rigs
Table 2-14 gives the type of the main unit rig for the offshore drilling platforms for several foreign companies, and also details their technical performance.

2.8.2.2 Major Technical Performance Aspect of Chinese Production Drills
In order to facilitate the selection of Chinese offshore drilling rigs, the main technical performance aspects of production rigs are presented in Table 2-15.

In addition to the rigs shown in Table 2-15, the China Baoji petroleum machinery factory in recent years has developed an AC variable frequency electric drilling rig (AC-SCR-DC) which drills welss to a depth of 12000 m. The rig provides favorable conditions for Chinese offshore drilling.

2.8.3 Special Equipment in Offshore Drilling Rig Systems
The special equipment in the offshore drilling rig system, such as the top drive device, drill pipe discharge device, and drill string heave compensation device are summarized in the following.

2.8.3.1 Top Drive Device
As mentioned before, current offshore drilling practice has been to use the top drive device. Therefore, the following presents the top drive system.

1. Structure
   Figure 2-112 shows the assembly diagram of the TDS-9S type top drive device powered by a 350 t dual AC motor drive and produced by Varco in the United States. From Figure 2-112, we see that its core equipment consists of a $2\times(350\sim700$ hp) AC drilling motor drive system and a $360°$ power swivel head. Above they are connected with the existing hook-traveling block system, and with the pipe operation system below. The drive system can be seen in Figure 2-112(b): it is integrally connected with the faucet assembly.

Table 2-14 Foreign Offshore Drilling Rig Type and Technical Performance

Countries, Companies, and Model / The Main Parameters	National Oilwell Company, United States			Emsco, Inc, United States		Dreco, United States			Norwegian Maritime		Denver, Gordon, United States		
Rated power HP (KW)	2000 (1492)	3000 (2238)	4000 (2983)	2000 (1492)	3000 (2238)	2000 (1492)	3000 (2238)	4000 (2983)	1400 (1044)	2500 (1865)	2000 (1492)	2500 (1865)	3000 (2238)
Drilling depth, m (ft)	3962~6096 (1300~20000)	9144 (30000)	6096~12192 (20000~40000)	3962~7620 (13000~25000)	4877~9144 (16000~30000)	6700 (22000)	9144 (30000)	12192 (40000)	-	-	4600~6400 (15000~21000)	6400 (21000)	9150 (30000)
Winch rope diameter, in	—	—	—	—	1	—	—	2	1	1	—	—	—
The main roll size mm (in)	762~1429 (30~56)	914~1556 (36~61)	914~1575 (36~62)	1066~1829 (42~72)	762~1473 (30~58)	893~1575 (35~62)	—	—	—	—	—	—	—
Winch block number	4	4	4	4	4	—	—	—	—	—	4	4	4
Dial diameter, mm (in)	375 or 495953 or 125	495 1257 (49)	495 1257 (49)	—	—	375或495或953或1257	495, 1257 (49)	495, 1257 (49)	—	—	953 (37)	—	—
Rig dimensions L × W × H, mm	7772× 4267× 289	7696× 3899× 2996	9728× 3419	7874× 4879× 2978	7747× 4267× 2978	—	—	—	—	—	7870× 3200	7920× 3220	8530× 3220
The winch NW, kg	34346	3538	—	35982	37289	—	—	—	3306	5016	2630	31700	3540
Drilling pump model	12-P-160×2	A-1700 PT×3	14-P-220×	FB-1600×2	FC-2200×2	12-T-1600×2	12-T-1600×3	12-T-2000×2~3	—	—	PZ-10	PZ-11	PZ-11

Table 2-15 Rig Main Technical Performance Aspects

Countries, companies and Model. The main parameters	National Oilwell Company in the United States				Emsco, Inc, United States		Dreco, United States			Norwegian Maritime		Denver, Gordon, United States		
	1320—UE 1320—UDBE	1625—DE 1625—UDBE	E3000 E—3000—UDB	2040—UDBE	C—2—11 Electric drive	C-3-11 Electric drive	D2000E D2000AC	D3000E D3000AC	D4000E	U—914 —EC	—EC U—1220 —EB	1500E	2100E	3000E
The pumps rated power, kW (Hp)	1193 (1600)×2	1193 (1600)×3	1268 (1700)×3	1642 (2200)×3	1193 (1600)×2	1642 (2200)×2	1193 (1600)×2	1193 (1600)×3	1492 (2000)×(2~3)	—	—	1007 (1350)×3	1193 (1600)×2 ×3	1193 (1600)×3
Pump piston sizemm (in)	101.6~184.2 (4,5,6,7,7)	101.6~184.2 (4,5,6,7,7)	127, 165~191 (5,6,6~7)	127, 152.4~203.2 (5,6,7,7,8)	140,152~178 (5,6~7)	140,152~203 (5,6~8)	139.7,152.4~203.2 (5,6~8)	139.7,152.4~203.2 (5,6~8)	139.7,152.4~203.2 (5,6~8)	—	—	(5,6,6,7)	(5,6,6,7)	(5,6,6,7)
Rated working pressure of the pump MPa (lbf/in2)	34.5 (5000)	34.5 (5000)	34.5 (5000)	41.4 (6000)	34.5 (5000)	50.5 (7332)	34.5 (5000)	34.5 (5000)	34.5 (5000)	—	—	34.5 (5000)	34.5 (5000)	34.5 (5000)
Pump piston strokemm (l)	304.8(12)	304.8(12)	304.8(12)	356(14)	305(12)	381(15)	304.8 (12)	304.8 (12)	304.8 (12)	—	—	254(10)	279(11)	279 (11)
Pump pumping speed, min-1	120	120	—	105	120	100	120	120	120	—	—	115	115	115
The pump total mass, kg	24810×2	24810×3	18756×3	39000×3	23559×2	37576×2	31080×2	31080×3	34619×(2~3)	—	—	16788×3	18784×2	18784×3
Pump Dimensions (L × W × H) mm	5309×1997×1950	5309×1997×1905	—	5544×2311×2139	4877×2413×2083	6019×2895×2070	5347×2831×1905	5347×2831×1905	5347×2831×1956	—	—	—	—	—
Rig Model	ZJ15J		XJ100K	XJ135KQ	ZJ20	ZJ32J	ZJ45L	ZJ45J	ZJ45D	ZJ60L	ZJ60S	ZJ60D	ZJ60S	ZJ80
Nominal drilling depth, m	1500		2000	2500	2000	3200	4500	4500	4500	6000	6000	6000	6000	8000
Device total power, kW	1056		303	500/68	662	1886	1886	1886	2100	2940	2800	2800	—	3500

Continued

Table 2-15 Rig Main Technical Performance Aspects—Cont'd

	900	1350	1350	1350	2250	3000	3000	3150	4500	4500	4500	5850
Maximum hook load, kN	50	—	70	70	115	160	160	160	220	220	220	220
Maximum weight of the drill string, t												
Enhance the system maximum tethered	4×5	4×5	4×5	4×5	5×6	6×7	6×7	6×7	6×7	6×7	12	7×8
Hoisting rope diametermm (in)	26	28.5	28.5	29(1)	32(1)	35(1)	35(1)	35(1)	38(1)	35(1)	38(1)	42(1)
Enhance the system pulley diametermm	915	915	915	915	1270	1270	1270	1270	1524	1524	1524	1524
Swivel stemThrough-hole diameter, mm	64	64	64	64	75	75	75	75	75	75	75	75
The winch rated power, kW	330	510	478/650	510	740	1100	1100	1100	1470	1470	1470	2210
The winch block number (up /down)	3/1	4/1	3/1	6/2	6/2	6/2	6/2	4	6/2	4	4	4
The dial opening diameter, mm	445	445	444.5	444.5	520.	698.5	520	698.5	698.5	952.5	952.5	1257.3
The dial block number (up /down)	3	3/1	3/1	3/2	3/2	3/2	3/2	2	6/2	2	2/2	2
Quantity and power of drilling pumphp	1×735	1×350	1×368 (500)	1×370 (500)	2×960 (1300)	2×740 (1000)	2×740 (1000)	2×960 (1300)	2×1180 (1600)	2×960 (1300)	2×960 (1300)	2×1180 (1600)
The derrick effective height , m	34.5	31~33	31.5	31.5	43	43	43	43	44.8	44.8	45	45.8
The drill height , m	3.5	4	4	4	4.5	4.5	4.5	7.5	9	7.5	9	9
The effective height of the rig floor, m	2.88	3	3.74	3.2	3.44	3.5	3.5	6.45	7.42	5.6	7.42	7.5
explanation												

J: tape drive; L : chain transmission; D : AC - SCR - DC power transmission. There are three types of ZJ32J rig.

(a)   **(b)**

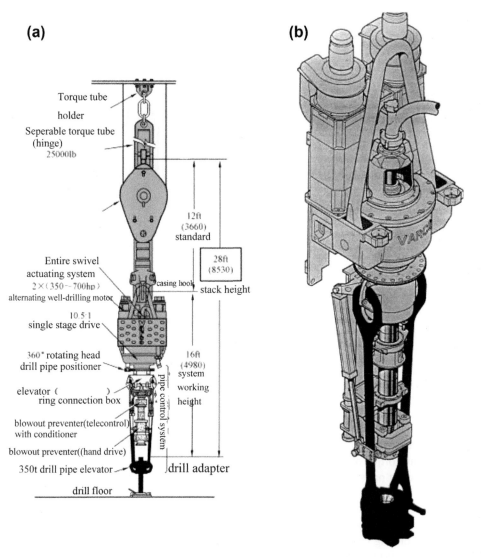

Torque tube

holder

Seperable torque tube
(hinge)
250001b

12ft
(3660)
standard

Entire swivel
actuating system
2×(350--700hp)
alternating well-drilling motor

casing hook

28ft
(8530)
stack height

10.5:1
single stage drive

360° rotating head
drill pipe positioner

16ft
(4980)
system
working
height

elevator (　　　　)
ring connection box

pipe control system

blowout preventer(telecontrol)
with conditioner

blowout preventer((hand drive)

350t drill pipe elevator

drill adapter

drill floor

FIGURE 2-112

TDS-9S type top drive.

Figure 2-113 shows the assembly diagram of the top drive device's drive system. From the figure, the top drive is with the motor shaft pinion gear meshing with the gear of the gear box with single-stage transmission 10.5:1, driving the rotating head to and thereby driving the drill string. The following discussion of the TDS – 3 type is used as an example to introduce its composition.

  a. Motor assembly: Includes the motor, gearbox, motor cooling and lubrication system of the gearbox, air brake equipment, rotating head, sliding system, and so on.

FIGURE 2-113

Drive apparatus of the top drive system.

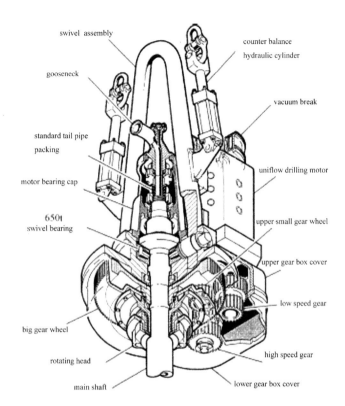

swivel assembly

gooseneck

standard tail pipe
packing

motor bearing cap

650t
swivel bearing

big gear wheel

rotating head

main shaft

counter balance
hydraulic cylinder

vacuum break

uniflow drilling motor

upper small gear wheel

upper gear box cover

low speed gear

high speed gear

lower gear box cover

  b. Guide wheel assembly: This is a welded structure, to make the TDS move vertically, and to bear the anti-torque of the TDS rotating drill string on the guide wheel.
  c. The supporting parts of the motor: For example, when the motor starts, this includes the resistance and temperature indicator and so on.
  d. The faucet assembly and extended faucet U lift plug.
  e. Tube operating system: This includes the torque wrench, pipe string within the BOP and the regulator, rotating head and rings, elevator and ring tilt regulator, etc.
  f. TDS control system: This includes the hydraulic circuit, power switch control panel and instrumentation, etc.
2. Technical performance
  The TDS by Varco has the following technical performance parameters:
  a. Drilling motor system: The continuous motion power is 746 kw (1000 hp), intermittent operating power is 732.9 kw (1250 hp), gear speed ratio is 5.33:1, torque transmission is 37400 N·m (183 r/min) and 49800 N·m (intermittent motion), and the maximum RPM is 250r/min.
  b. Guide wheel assembly: The opening degree of the guide wheel is 1575 mm, and the wheel diameter is Φ152 mm.

Table 2-16 Technical Performance of the Various Types of TDS by Varco

Project	TDS–3S	TDS–4S H	TDS–4S LO	TDS–6S
Maximum continuous drilling torque, N · m	42000	39400	61700	76900
Maximum make-up torque, N · m	46200	43400	67900	81400
Maximum continuous drilling torque, speed, r/min	165	175	110	195
Maximum speed, r/min	250	250	—	240
Gear ratio	5.33:1	5.08:1	7.95:1	5.33:1
TDS moving parts some of the weight,	16.04	18.04	—	21.35
Exploded torque, N · m	47400	46100	—	94900
Lifting capacity, t	650	650	—	650/750
moving part of the working height, m	6.35	6.35	—	18.44
The lower part of the faucet, The upper part of the elevator drill pipe robot				
Project	PH–60		PH–85	
Exploded torque, N · m	81400		115300	
Adaptable drill size, in	3~5		3~6	

c. Drill arranging device: The lifting capacity is 500 t rushing torque is 81400 N·m, the hydraulic power is for a 14MPa working pressure and a 113.6L/min flow amount, and the drill pipe diameter is 127 mm (5 in) (w/41/2in within flat connector).

The main technical performance parameters for other models of TDS by Varco can be obtained from Table 2-16. Furthermore, the technical performance of the DQ series hydraulic drive TDS and DC-driven TDS can be seen from instructions in China.

3. Major advantages

a. Saving drilling auxiliary time and improve drilling efficiency
Because we can save time in making a connection, putting the bar into the mouse hole and moving back, we can save drilling auxiliary time, and generally can improve drilling efficiency about 15% ~ 20%.

b. Reduction of drilling accidents and improvement of the safety and reliability of drilling
Because TDS can fully circulate the drilling fluid, it reduces the time of stopping and circulating drilling fluid when making a trip. It is convenient and fast to connect the power water swivel on the drill column to circulate drilling fluid, so as to avoid or eliminate easily accidents in the borehole, such as the bit freezing and covering up drilling. And it also can rapidly circulate the drilling fluid to pressure wells, in order to prevent well kick and blowout. This improves the safety and reliability of offshore drilling.

c. Suitable for drilling directional wells and horizontal wells, while reducing drilling cost
Because TDS can lift the drill string while rotating the drill string (back reaming), the friction resistance is greatly reduced when drilling high angle wells and horizontal wells, which increases drilling ability and saves drilling costs. It generally can save 1/3 of the investment cost.

d. Get a longer and more continuous sampled core
Because we use TDS to obtain a continuous core, it isn't restricted by the kelly length and making a connection. So after using the TDS, the continuous core length is usually 2 to 3 times that of the coring length of ordinary rotary drilling.

e. Reduction of the space occupied by the turntable on the drill floor and the drive component, which makes drilling more safe

f. It is easy to create and provide a drill-over ability of the ideal length
When common rotary drilling finishes each kelly length needs to lift above the hole bottom about 9.14 m (30 ft), making a connection; but TDS can easily connect the short joint of the drill-over of the ideal length under the dynamic water tap [usually around 2.13 m (7 ft), floating offshore drilling needs to increase the heave distance of the ship (platform)]. When connected with the triple rod, it generally only needs to lift the drilling string away from the hole bottom about 3.05 m (10 ft). In this way, it can further save 2/3 of the time on lifting every single joint. Also, because the time of lifting the drilling string from the bottom of the hole is less, it reduces the possibility of well hole collapse, and improves the safety and reliability of offshore drilling.

2.8.3.2 Drilling Rod Laying Device

As mentioned before, in order to cooperate with the top drive and improve the drilling speed, mechanized and automated drilling rod laying devices are generally provided in the drilling platform (ship) to decrease offshore drilling operation time.

1. Types of drilling rod laying device
   At present, there are two main types of drilling rod laying device on the drilling platform.
   a. Types of upright drilling rod laying device
   This type of drilling rod laying device lays the drill pipe so it stands on a vertical box on the drill floor, leaning on the derrick. This model has the following advantages:
   i. Lifting a (lower) empty elevator can be done at the same time as screwing (removing) on a buckle, increasing the drilling speed.
   ii. Because of the vertical drilling rod, the area occupied on the deck is small.
   Disadvantages: The derrick is too heavy. With the heavy waves, the derrick bears not only a big wind load, but also the dynamic load caused by waves, which coupled with the platform (ship) swing and drill pipe leaning on the derrick make it easy to cause anaccident. Because there is still a second layer platform operation, the roughneck is unsafe.
   b. Horizontal drilling rod laying device
   This type of drilling rod laying device lays the drill pipe in horizontal shelves outside the drill floor. During drilling operations, it transports drill pipe to the drill floor to lift. This model has the following advantages:
   i. Equipment with a low center of gravity is conducive to floating platform (ship) stability.
   ii. Because the laying is horizontal, the loading condition of the derrick is greatly improved.

   iii. Strong adaptability. It can be used for tripping, making a connection, casing and other operations, and applies to a variety of drilling rigs.

   iv. It doesn't need a second layer platform operation, which is safe.

   However, there are some disadvantages:

   i. Occupies a larger deck areas

   ii. Bending deformation with throwing stand

It's easy from the analysis of the advantages and disadvantages to see the uses of the two types: for the pelagic floating drilling platform (ship), the horizontal type generally is suitable. For example, the floating drilling platform "Exploration 1" uses a horizontal drilling rod laying device designed in China. In the China South Sea, the maximum wind speed can reach 50m/s, therefore we suggests that platforms there use the horizontal type. For the drilling platform (ship), if the carrying capacity of the derrick can meet the requirements, the vertical drilling rod laying device can be used, such as with the "Found No. 2" floating drilling platform in America. Because the derrick is specially designed, it can work in a maximum wind speed of 16.5 m/s, maximum wave height of 7.8 m, and maximum heave of 2.1 m.

2. Vertical drilling rod laying device used by offshore drilling platform (ship)

We will look at the vertical drilling rod laying device used by the offshore drilling platform "Found No. 2," shown in Figure 2-114, in detail.

  a. Main features

   i. The use of a hydraulic telescopic mechanism can make the traveling block give a pathway

As shown in Figure 2-114, when the stand moving mechanism extends out to catch the stand, the contractive hydraulic telescopic mechanism shrinks and takes the traveling block and hook back, deviating from the wellhead center along the guide rail of the derrick (installed opposite the stand moving mechanism), descending to the roundabout.

   ii. The use of multiple movement laying devices for various lifting operations.

As shown in Figure 2-114, there are three laying devices (above, middle, under) along the derrick. Among them, the above and the middle laying devices are used for cementing to bring the casing down; the under laying device is equivalent to a crane, moving power tongs on the drill floor, lifting the kelly bar into the mouse hole, and doing other auxiliary operations.

The laying device structure consists of the driver, laying the arm, and moving the manipulator. The driver can slide on track in the derrick, the laying arm is contractive, and the laying device can rotate at a certain angle. So the stand can be moved to any position of the laying frame by the driver and laying arm.

   iii. The use of the hydraulic cylinder lifting mechanism is to lift the stand for a short distance

The stand needs to be lifted for a short distance, and moved by a mechanism which can bring it into the laying frame. But the elevator is opened, hunting and hook has contracted sliding down a short distance, so hydraulic lifting mechanism can only be used to achieve the stand lifted for a short distance. As shown in Figure 2-114, the lifting mechanical hand gets stuck under the lower single connector shoulder, using the wire rope of the lifting hydraulic cylinder to drive the manipulator upward along the vertical rail of the intermediate laying device.

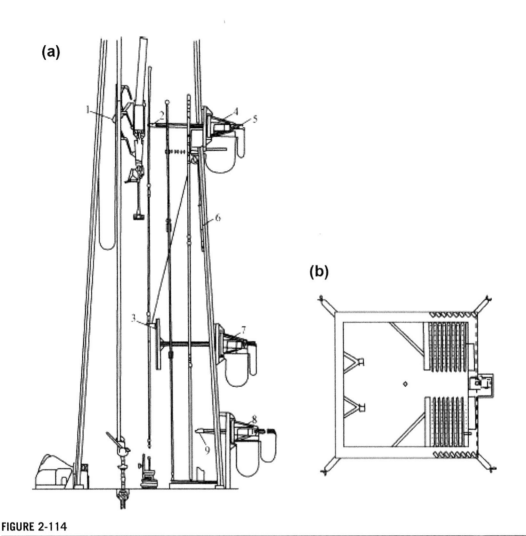

FIGURE 2-114

Vertical pipe racking device on the "Founder II."

b. Operating procedures

The operating procedures of the vertical pipe laying device in tripping are shown in the following.

i. Lifting drilling rod stand

Use the elevator to stick the drill string, release the power slip, and the traveling block runs up and will lift the drill string. At the same time, the upper and middle laying devices are near the borehole centerline position. When the elevator is just over the upper laying device, the upper and the middle laying arm promptly move to the center line of hole and grasp the stand by the manipulator, but the drill string is free to move up and down with small lateral movement. At this time, the manipulator of the middle laying should hold the bottom of the lowest single joint.

ii. Release of connector screw thread

When the stand is lifted above the power tongs, the traveling block stops rising, the slip sticks the drill string, and the power tongs move close to loose the buckle.

iii. Laying device holds the stand

After the connector looses the buckle, the stand is off. At this time, the drill runner in the two-layer platform contracts the piston rod of cylinders and tensions the steel wire rope; the manipulator of the intermediate laying device declines along the stand, to the bottom single connector shoulder, which can promote and suspend the whole stand. At the same time, the elevator opens, and the traveling block and hook contract, sliding down along the track, then back to the wheel.

iv. Laying device drives the stand

When operating the laying device, we can put the stand into the stand box and arrange it in order. As shown in the top view of Figure 2-114, each side can lay 87 stands, and the drill collar is at the last virgin stock. A total of 10 drill collar stands, and is fixed by bolt.

v. Elevator sticks the next stand.

When using the laying device to put the stand into box, the power tongs move away from the wellhead, and the traveling block and hook extend to the wellhead position. The elevators stick the drill column top joint and are ready to lift the next stand.

The above is the hoisting operation procedure. The tripping operation is completely the same, just in reverse order.

The capacity of the vertical pipe laying device "Found No. 2" is a 4800 m drilling rod (ϕ 5 in) plus 10 drill collars (ϕ 7 3/4in). It can also lay combined drill string (ϕ 3 1/2in and ϕ 4 1/2 in) and lay casing of (ϕ 13 3/8 in). The tripping and hoisting speed is one stand per minute.

3. Horizontal drilling rod laying device used in offshore drilling platform (ship)

For a horizontal drilling pipe rod laying device, the stand should be transferred to the drill floor in the trip in hole, so we need to solve the problems of horizontal transfer and sending on the drill floor.

a. Scheme of horizontal transfer stand

i. Chain transfer scheme

As shown in Figure 2-115, we put all the stands on three chains driven by electric devices installed along the stand. Then, along with the movement of the chain, the stand can be

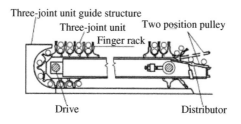

FIGURE 2-115

The chain horizontal transfer root device.

FIGURE 2-116

Horizontal transfer root lifting device.

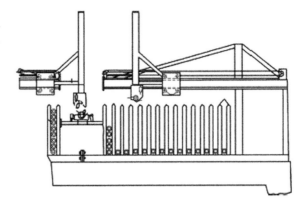

transmitted to the horizontal center line below the drill floor. It can also be driven by a pneumatic or hydraulic motor.

ii. Cranes and mechanical hand transfer scheme

This scheme uses cranes and a mechanical hand to achieve the stand horizontal transmission, as shown in Figure 2-116. The mechanical hand has three sets along the whole length of the stand, which can be extended or retracted by the hydraulic cylinder, to adjust the position of the manipulator in the vertical direction. The manipulator jaw can be opened or locked by the operating hydraulic cylinder. The mechanical hand and crane can be moved horizontally or vertically by the automatic control.

b. Scheme of transferring stand to the drilling floor

After the stand is transmitted to the horizontal center line below the drill floor, we must further solve the problem of sending the stand to the drill floor. The solution to this problem is as follows.

c. Scheme of the car sliding on the track

This scheme is to put stand that is pre-lifted in drilling or is proposed sent to laying frame after breaking out a joint in the car, as 1 and 4 shown in Figure 2-117 .The car slides along the rail with the traction of the steel wire rope derived from the winch, with a roller bracket supporting the stand (as seen in 2 in Figure 2-117). The roller frame is driven by a cylinder or hydraulic cylinder, and the car has a side plate and groove to stick the noumenon of stand. In addition,

FIGURE 2-117

Horizontal drill pipe discharge device on the drilling platform (ship).

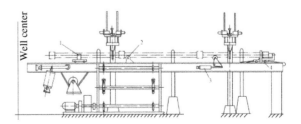

there are drill pipe transfer and positioning devices, so that in tripping the upper end of stand can aim at the borehole, while in hoisting the end of stand moves from the borehole to the car.

d. Truss motion scheme of supporting stand

This scheme will put the drill pipe stand on the roller bracket of a truss structure, and the roller bracket supports the stand with a hydraulic cylinder or cylinder, or catches the stand to send the stand to the floor by the movement of the truss. The upper and lower ends of the truss have rollers, sliding respectively along the two horizontal and vertical tracks. With the roller sliding, the truss transfers from the horizontal position gradually to the vertical position; this will send the stand to the floor. As ith the above scheme, this scheme also requires installation of a drill pipe transfer and positioning device. The advantage of this program is that the stand is on the truss, so it can avoid bending deformation in the process of lifting and landing.

Figure 2-118 (a) shows the horizontal drill pipe laying devices with a truss supporting stand on the rig floor of a floating drilling ship; Figure 2-118(b) gives a schematic diagram of the truss supporting stand on the rig floor.

Of course, if both the top drive device and automatic laying drill pipe device are used at the same time, the operation procedure is different from the rotary table, and that is not detailed here.

2.8.3.3 Heave Compensation Device

The heave compensation device is the drilling string DSC, which is a device to compensate for the raising and sinking motion of the floating drilling platform (ship) with ocean waves. It is special equipment for a floating offshore drilling rig that is not used in an onshore drilling rig. Here we will specifically introduce it.

The way to use this equipment is to add a set of drill string heave compensation devices in the floating drilling platform (a ship) for drilling rig components in order to maintain the drill string without raising and sinking the platform (ship).

The heave compensation system generally adopts a hydraulic transmission. For example, there is double hydraulic cylinder between the traveling block and hook, and the cylinder is connected with the traveling block, as shown in Figure 2-119. When the platform raises and sinks, the traveling block drives the cylinder hydraulic cylinder for periodic movement up and down, but the piston and the big hook remain immobile. At this time, the whole drill string weight is supported by a hydraulic piston below. Liquid pressure can be kept constant and can be adjusted according to the requirements of controlling the drilling string tension and adjusting the pressure of drilling.

1. Structure types and working principles of the heave compensation device

a. Heave compensation device installed between the traveling block and hook

As shown in Figure 2-119, the device consists of the following parts.

i. Tank

Two hydraulic cylinders are connected with the traveling block by the frame and move up and down with the platform.

ii. Piston

The piston for two hydraulic cylinders goes through the piston rod and is connected with the lower frame, which is fixed on the big hook, and the hook load is supported by the hydraulic below piston.

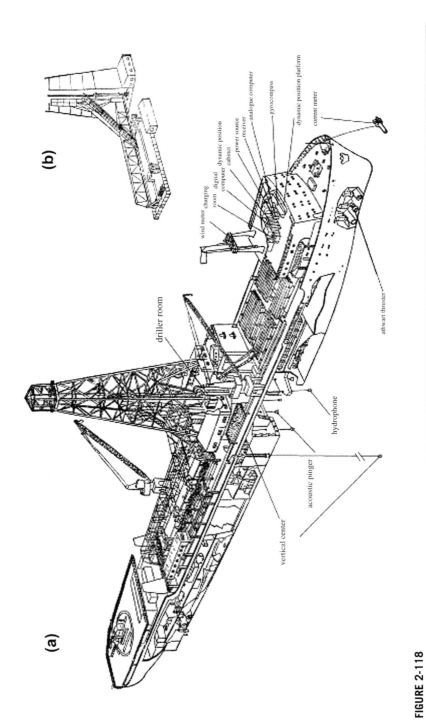

(b)

(a)

wind meter charging
room digital
computer dynamic position
cabinet
power source
receiver
analogue computer
gyrocompass
dynamic position platform
current meter

driller room

athwart thruster

hydrophone

acoustic pinger

vertical center

FIGURE 2-118

Diagram of horizontal drill pipe laying devices with truss supporting stand on the rig floor of a floating drilling ship.

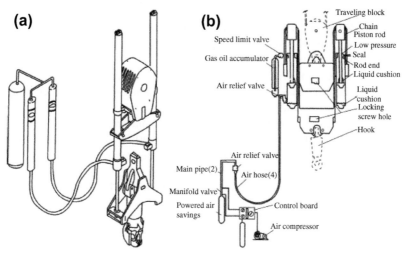

(a)

(b)

Speed limit valve

Gas oil accumulator

Air relief valve

Traveling block
Chain
Piston rod
Low pressure
Seal
Rod end
Liquid cushion
Liquid
cushion
Locking
screw hole
Hook

Air relief valve

Main pipe(2)
Air hose(4)

Manifold valve

Powered air
savings

Control board

Air compressor

FIGURE 2-119

Heave compensation equipment installed between the traveling block and hook.

iii. Storage

The storage device is connected with the liquid cylinder. There is a piston in energy storage. The lower end liquid through a tube is connected with the liquid cylinder, and the upper end gas through the pipeline is connected with a storage tank. In this way, the liquid pressure in the hydraulic cylinder is determined by the gas pressure in energy storage. Adjusting the gas pressure can change the liquid pressure.

iv. Locking device

This makes the two frames lock into a whole so that the traveling block and hook are connected together for tripping work. Figure 2-119(b) is a new type of compensation device used in the "China South Sea No. 2" semi-submersible drilling platform.

The working principles of the heave compensation device are as follows:

i. Normal drilling well

The drill string weight hanging on the hook is Q, the drilling pressure at the bottom of a well is W, and hydraulic pressure in the hydraulic cylinder of the compensation device is P_L. The balance relationship between them is as follows:

$$2p_L A_p + W + (-Q) = 0 \tag{2-47}$$

Put "Q= qL" into (2-47), and then the result is

$$p_L = \frac{qL - W}{2A_p} \tag{2-48}$$

where q is drill string weight per meter, N/m; L is the length of the drill string, m; and A_P is the hydraulic cylinder piston area of compensation, m².

By Equation (2-47) we can get the following:

- In order to keep the drilling pressure, the liquid pressure of P_L in the hydraulic cylinder is a certain value. In order to adjust the drilling pressure, we only need to adjust the gas pressure in energy storage.
- In order to realize automatic launching, the liquid pressure of P_L in the hydraulic cylinder is adjusted for that. The $2P_L A_P$ is slightly less than the drill string hanging weight and the displacement of the piston in the hydraulic cylinder is larger than the raising and sinking distance.

The force of the hook of normal drilling is shown in Figure 2-120.

ii. Rope operation

When wire-line operation such as electronic logging and well testing is carried out, because the appliance going down into the well is light, the heave compensation device can't play a role. Another sensor rope is fixed on the top end of the riser. And then through the hanging pulley on the hook, the rope is fixed on the derrick substructure. Thus, T_R, that is the tensor on the hook with the sensor cable, and is equivalent to the drill string hanging weight. Therefore, the heave compensation device also can play the role for motion compensation when the rope operates.

When the rope operates, working rope sending instruments are put into the hole from the winch through another pulley on the hook. There is a certain distance between the pulley and the pulley that the sensor rope goes through, which are fixed on the same bar. Because the drilling floor has heave motion, the fixed end of the sensor rope, the working rope, and the rope winch move up and down with the drill floor. So, two ropes on the

FIGURE 2-120

Current design of heave compensation device after the big hook.

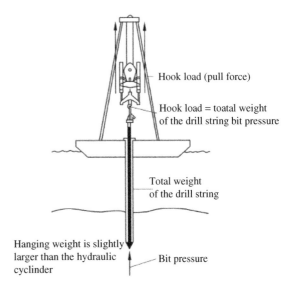

Hook load (pull force)

Hook load = toatal weight of the drill string bit pressure

Total weight of the drill string

Hanging weight is slightly larger than the hydraulic cyclinder

Bit pressure

pulley hook are loose or tight, which will cause the force on the hook from the two ropes to be large or small. But when the hydraulic pressure in the hydraulic cylinder of the heave compensation device is fixed and the sensor rope is loose and tension decreases, the constant hydraulic pressure will push the piston up and lead the hook to lift, and the sensor rope recovers tension. If the sensor rope force increases, because the constant pressure is less than tension of the sensor, the piston and the hook is pulled down and sensor rope is loosened. In this way, not only do the sensor rope and working rope keep the tension for the hook, but also they can compensate for the heave motion and keep rope operations normal.

The force on the hook from the rope operation is shown in Figure 2-120. Figure 2-121

b. Heave compensation device at the crane.

The structure of this device is shown in Figure 2-122, and the crane heave compensation device consists of the following components.

i. Floating crane

This moves in the vertical track through the rollers. In addition to the common crane pulley, there are two auxiliary pulleys. The shaft of the auxiliary pulley is connected with the axes of the pulley crane by a link. Fast rope and dead rope are started through two auxiliary pulleys. Thus, when the crane moves along a vertical track, only an auxiliary pulley shaft moves. There is no relative motion between the wire rope auxiliary pulley and the pulleys, which extends the life of the wire rope.

ii. Master cylinder

This is used to support a floating crane which is the equivalent of a total of four large springs that are slanted; gas is supplied by the gas compressor on deck.

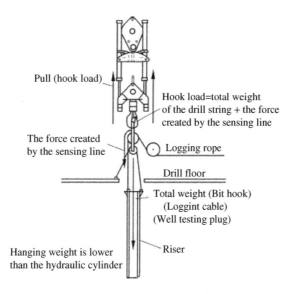

FIGURE 2-121

Analysis diagram of load on the hook for rope operation.

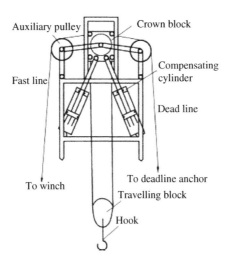

FIGURE 2-122

Heave compensation on the overhead traveling crane.

   iii. Cylinder

A total of two are placed vertically and oil supply is from the deck. It is only used as a safety cushion hydraulic cylinder for a hydraulic buffer to overcome the inertia effect of the load hook.

   iv. Energy storage device

This is installed on the derrick and has a pipe connecting with the four main cylinders for adjusting the gas pressure in the main cylinder.

This device works as follows:

   i. Heave compensation

Compensation is achieved by a floating crane. When the floating platform is up or down, the derrick moves along the track up and down. Gas in the master cylinder is in compression or expansion, which is equivalent to a big spring. The crane and hook remain immobile basically and then heave motion can be compensated.

   ii. Control of drilling pressure

The driller uses the deck valve to control air pressure from the air tank to the master cylinder system, so that the well bottom drilling pressure is adjusted to the appropriate value.

   iii. Deliver automatically

During normal drilling, the pressure in the cylinder is slightly adjusted below the load on the hook. Thus, the floating crown block goes down along the track to realize automatic delivery under the drive of the hook load. When the floating crown block goes down to the lowest point, the driller looses the wire rope on the hoisting barrel to make the floating crown go up to the highest point, for continued automatic delivery.

   iv. Accident prevention

When the hook load decreases suddenly or the main cylinder leaks seriously, the hydraulic cylinder can be used to support the weight of drill string and slow it down to prevent accidents for security.

v. Rope operation

During rope operation, a sensor rope is added, where one end is fixed on a riser and the other end is led to the floating crown block from the outside of the derrick, and connected to the barrel of the rig after going through a pulley. In this way, the sensor rope gets loosened or tightened in accordance with rig movement. The floating crown block does the corresponding compensation movement under constant air pressure so that heave compensation in the wire line operation can be realized.

vi. Tripping operation

In tripping operations, the locking device is used to lock the floating crown block to prevent the floating crown block from sliding up and down with tripping drill string.

c. Heave compensation device on dead line

As shown in Figure 2-123, this is mainly composed of the following parts:

i. The fixed pulley group

The dead line is led from the crown block: first, it goes pasts a sensor pulley, transferring the pulling force into an electrical signal; secondly, it travels to the weight indicator; and finally goes through the fixed pulley group and the movable pulley, and the end of the dead line from the fixed pulley group is finally fixed on the deadline anchor.

ii. The movable pulley group

This can move within the framework and its stroke size is relevant to the pulling force of the dead line. The bearing block of the movable pulley group is installed on the crane, which has rollers sliding along the I-girder track on both sides. Ahead of the movable pulley group, there is a half-moon-shaped fork fixed on the crane, and the other end of the fork is linked to the piston rod of the hydraulic cylinder.

iii. Hydraulic cylinder

The fluid of the side of the piston in the hydraulic cylinder is connected with a low pressure accumulator, while the other side is connected with a high pressure accumulator. When the pulling force on the dead line decreases, after the signal from the sensor pulley, the command valve will be open. Pressure on the right hand of the piston increases, which pushes the piston to move left. The rope on the movable pulley group strains, and fluid in the left hand of the piston reflows into the low pressure accumulator. When the pulling force on the dead line increases, the command valve opens, the fluid pressure on the right hand of the piston decreases, and the piston moves right, loosening the dead line until reaching constant pulling force. The fluid reflows into the accumulator from the right hand of the piston.

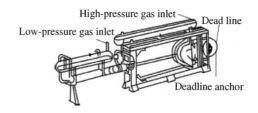

FIGURE 2-123

Heave compensation device installed on dead line.

iv. High pressure accumulator

The gas of the accumulator is supplied by compressed air. There is a safety valve on the top and a vent valve at the bottom.

v. Low pressure accumulator

The air flows into the low pressure accumulator through a filter, regulator along with the pipeline. There is still a vent valve on the top.

vi. Console

There are pressure gauges, weight indicators, indicator lights for the movable pulley group, pressure controllers and compressor start and stop mechanisms, etc., on the console.

As shown in Figure 2-123, the pulling force in the dead line can be changed by adjusting the air pressure in the regulator. Then the pressure in the bottom of the well can be adjusted and maintained by changing the dead line pulling force. In addition, the effective length of the rope can be adjusted by pushing the piston through the hydraulic pressure.

In conclusion, the structure of the heave compensation device installed on the dead line is complex due to the installation of an electrode control system, such as a sensor signal and telegraph, so the application has little been used. There are many advantages of the heave compensation device installed on the crown block, including a small covered broad area, short pipeline, less sealing, and no need for a high-pressure hose. However, its application has not been used widely because of the special large derrick and crown block.

The heave compensation device installed between the traveling block and the hook has been widely used recently. It does not require a special derrick and crown block, and the traveling block and the hook are common. But its disadvantages are frequent sealing of the hydraulic cylinder, a serious problem of hydraulic pressure leakage, a long pipeline, and big friction loss, which still needs further improvement.

2. Equipment selection for the heave compensation device

a. Selection of the drilling string for the heave compensation device

Based on the production requirements, the selection is decided by the drilling capacity of the drilling rig equipped on the platform. The hook load capacity of the compensation device is selected in accordance with the hook load of the drilling rig. In addition, we still need to choose the stroke of the hydraulic cylinder in accordance with the compensatory heave displacement required by the sea conditions.

The fundamental parameters that can reflect the technical performance of drilling string compensation device are

- The hook load produced in the heave compensation
- The hook load produced in the lock status
- Hydraulic cylinder stroke

b. Selection of accumulator and gas tank

i. Spring stiffness (K_1) in hydraulic cylinder and accumulator

If the fluid from the hydraulic cylinder through to the accumulator and the gas in the accumulator are regarded as mass less spring, then since in the accumulator, the pressure of the piston on the fluid is equal to the restoring force of the compressed gas, it can be concluded that the spring stiffness is

$$K_1 = \frac{A_p^2 p_0}{V_0} \tag{2-49}$$

where V_0, p_0 are respectively the volume and pressure of the accumulator and gas tank; and A_p is the piston area in the accumulator.

ii. The volume of the accumulator and gas tank

When selecting, first determine the average pressure p_0 and piston area A_p based on the work requirement, then according to the calculated spring stiffness, the volume V_0 can be brought out from Equation (2-49). Then, the number and the volume of the gas tanks and the volume of the accumulator can be determined.

The gas tank is usually placed horizontally or vertically in the appreciate place near the derrick and on the skid. An isolating valve should be put on the top of the gas tank to reduce gas volume quickly.

The piston structure is commonly adopted by the accumulator, which is placed horizontally or vertically. The diameter of the accumulator is usually about 0.5 mm. The length is determined by the calculation above. A boat with high heave frequency requires that the accumulator is placed as close as possible to the hydraulic cylinder.

The layout of the whole heave compensation device on the drilling pontoon is shown in Figure 2-124.

3. Vibration analysis of the heave compensation device

When an offshore floating drilling device is equipped with a heave compensation device, vibration still occurs in the hook. Thus, a vibration system of the heave compensation device is needed to analyze the vibration parameters.

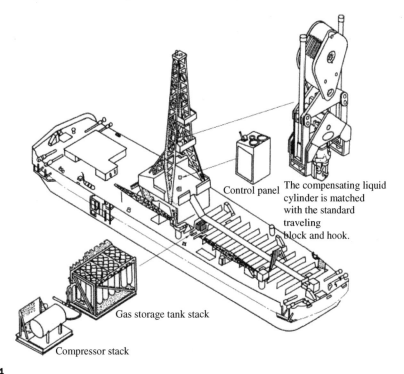

Control panel

The compensating liquid cylinder is matched with the standard traveling block and hook.

Gas storage tank stack

Compressor stack

FIGURE 2-124

Arrangement of heave compensation on drilling barge.

a. The dynamics model

Regard the drilling string heave compensation device used in offshore floating drilling as a dynamics model shown in Figure 2-125. The following are the assumptions:

i. Lumped mass

This part that bears the pulling force of the drilling string is regarded as an elastomeric center of the mass and the weight (W_F) of fixed equipment such as hook on the hook, and regard the concentrated equivalent weight of the drilling string as 1/3 by weight of all drilling strings with the help of Rayleigh's method in mechanical vibration theory. Then the concentrated mass on the hook M is

$$M = \frac{1}{3}\frac{W_{DP}}{g} + \frac{W_F}{g} \qquad (2\text{-}50)$$

ii. Stiffness concentration

Stiffness is an external force that is produced by the unit deformation of an elastomer. Concentrate the stiffness of the whole drilling string elastomer and the stiffness coefficient is K_2; the fluid in the hydraulic cylinder of the heave compensation device and the gas in the accumulator are regarded as a massless spring with stiffness concentrated and the stiffness coefficient is K_1.

iii. Damping concentration

Damping is a coefficient of damping force which is described in the form of mathematics. Damping coefficient C_2 is used to represent the friction caused by the drilling string in well bore, and C_1 is used to represent the friction damping coefficient of the liquid in the compensation hydraulic cylinder and the accumulator.

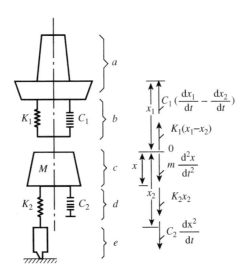

FIGURE 2-125

Dynamics model of the drilling string heave compensation device.

iv. Hypothesis of boundary condition

The drill collar and drill bit at the bottom of the drill string are connected with the downhole, which can be seen as a fixed end. The cylinder on the top of the heave compensation device is regarded as a support under the effect of vibration displacement, which is assumed to comply approximately with simple harmonic motion, ignoring the influence of the rotary line, which is the same as the heave displacement of the floating drilling device:

$$x_1 = \frac{H_c}{2} \sin \omega t \qquad (2-51)$$

where H_c is the reduced wave height after considering the draft of a platform and ship, and ω is the circular frequency of the sea wave.

According to the above assumptions, if the heave vibration is regarded as the main influencing factor and the influence of the drilling rig vibration in the downhole is ignored, the drilling string heave compensation system can be regarded as a forced vibration system under the effect of vibration displacement that is single degree of freedom with two damping parallel springs.

b. Force analysis

According to the dynamics model shown in Figure 2-125, a small unit is taken out from the hook. If the upward displacement of the hook is x_2, and the upward heave displacement of the platform is x_1, the main forces are as follows.

i. Spring restoring force

The spring restoring force is the force that wants to restore the deformation after the deformation of the elastomer caused by the external force. It should be equal to the deformation times the stiffness coefficient. Here, we look at the spring restoring force of the drilling string: since the drilling string is stretched when the hook moves upward, this restoring force is compressive, and the value is negative, i.e., $-K_2X_2$. With regard to the spring restoring force of the fluid and gas of the compensation device: since the fluid in the cylinder is compressed when the hydraulic cylinder body moves upward, its restoring force is tension, and the value is positive. The actual displacement should be the contrast between the heave displacement x_1 and the hook displacement x_2, i.e., $K_1 (x_1 - x_2)$.

ii. Damping force

Generally the damping force is regarded as viscous damping, which is equal to the product of the movement speed in the liquid and the viscous damping coefficient. Here we have: Damping force of drilling string: Because the drilling string moves upward along with the hook, the direction of the damping force should be opposite its movement direction. It is negative, i.e., $-C_2 \dot{x}_2$.

Damping force of the fluid in the hydraulic cylinder: Since the hydraulic cylinder body along with the platform moves upward and the fluid in the cylinder moves downward against the hydraulic cylinder, the direction of the damping force of the fluid in the hydraulic cylinder is opposed to the direction of movement. It should be positive, i.e., $+ C_1(\dot{x}_1 - \dot{x}_2)$.

Gravity: The gravity of the lumped mass (M) of the drilling string system is negative $-Mg$ downward. But because this gravity is equal and opposite to the restoring force $+K_2\delta st = +M g$ of the static elongation δst of the elastomer of this drilling string, offsetting each other, it can be ignored when calculating resultant force.

c. List the differential equation of movement

According to the Darren Bell principle, we can get

$$-Mx_2 - k_2x_2 + k_1(x_1 - x_2) - C_2x_2 + C_1(x_1 - x_2) = 0$$

After summing up, it also can be written as

$$Mx_2 + (C_1 + C_2)x_2 + (K_1 + K_2)x_2 = C_1x_1 + K_1x_1 \qquad (2\text{-}52)$$

Equation (2-52) is a linear second-order nonhomogeneous differential equation with constant coefficients. It is the differential equation of movement of the drilling string heave compensation system. From Equation (2-52), we have a single degree of freedom damped forced vibration system under the effect of the vibration displacement.

d. Determine the vibration displacement of the hook

From Equation (2-52), if the heave displacement x_1 of the floating drilling device, M, C_1, C_2, K_1, and K_2 are given, the vibration displacement x_2 can be figured out by solving this differential equation. Here we introduce the Euler solution to solve linear n-order nonhomogeneous differential equations. First apply the derivative mark of Cauchy:

$$D = \frac{d}{dt}, D^2 = \frac{d^2}{dt^2}$$

Thus Equation (2-52) can be written as

$$MD^2x_2 + (C_1 + C_2)Dx_2 + (K_1 + K_2)x_2 = C_1Dx_1 + K_1x_1 \qquad (2\text{-}53)$$

If $T_1 = x_2/x_1$, according to Equation (2-53),

$$T_1 = \frac{x_2}{x_1} = \frac{K_1 + C_1D}{MD_2 + (C_1 + C_2) + (K_1 + K_2)} \qquad (2\text{-}54)$$

In advanced mathematics, the Euler formula gives the relationship between the exponential function $e^{i\omega t}$ and a complex number field and trigonometric function:

$$e^{i\omega t} = \cos \omega t + i \sin \omega t \qquad (2\text{-}55)$$

Equation (2-55) indicates that any mathematical function of the simple harmonic oscillation in the natural world $\cos \omega t$ or $\sin \omega t$ can be represented by an exponential function. Thus, both the heave vibration displacement x_1 of the offshore floating drilling device and the vibration displacement x_2 of the hook can be represented. But $e^{i\omega t}$ is from the differential of $e^{i\omega t}$, it can be known that:

Therefore the derivative mark of Cauchy D can be written as

$$\frac{de^{i\omega t}}{dt} = i\omega e^{i\omega t} \qquad (2\text{-}56)$$

Plug Equation (2-56) into Equation (2-54), and we can get

$$T_1 = \frac{x_2}{x_1} = \frac{K_1 + C_1 i\omega}{M\omega^2 + (C_1 + C_2)i\omega + (K_1 + K_2)}$$

$$= \frac{\dfrac{K_1}{K_1 + K_2} + \dfrac{C_1 i\omega}{K_1 + K_2}}{1 + \dfrac{(C_1 + C_2)i\omega}{K_1 + K_2} - \dfrac{M\omega^2}{K_1 + K_2}} \tag{2-57}$$

By mechanical vibration theory, the fixed frequency ω_n of two series-wound springs and the critical damping ratio ξ of the critical damping vibration respectively are

$$\omega_n = \frac{K_1 + K_2}{M} \tag{2-58}$$

$$\xi = \frac{r}{r_c} \frac{C_1 + C_2}{2\sqrt{M(K_1 + K_2)}} \tag{2-59}$$

where r is the damping coefficient, and r_c is the critical damping coefficient.
Submit Equation (2-58) and Equation (2-59) into Equation (2-57), and it can be also written as

$$T_1 = \frac{\dfrac{K_1}{K_1 + K_2} + i\left(\dfrac{C_1\omega}{K_1 + K_2}\right)}{\left(1 - \dfrac{\omega^2}{\omega_n^2}\right) + i\left(\dfrac{2\xi\omega}{\omega_n}\right)} \tag{2-60}$$

From Equation (2-60), both its numerator and denominator are complex numbers. By advanced mathematics, modulus R is used to represent the size of a complex number, and argument θ is used to represent its direction. Thus, the complex number of the numerator can use the evolution of the sum of the square of both its real part and imaginary part to indicate the size of the complex number. If the numerator and denominator of the complex number in Equation (2-60) are represented by moduli, T_1 can be written in the form of the modulus ratio:

$$T_1 = \frac{x_2}{x_1} = \sqrt{\frac{\left(\dfrac{K_1}{K_1 + K_2}\right)^2 + \left(\dfrac{C_1\omega}{K_1 + K_2}\right)^2}{\left(1 - \dfrac{\omega^2}{\omega_n^2}\right) + \left(\dfrac{2\xi\omega}{\omega_n}\right)}} \tag{2-61}$$

Submit Equation (2-51) into Equation (2-61):

$$x_2 = x_1 \cdot T_1 = \frac{H_c}{2} \sin \omega t \sqrt{\frac{\left(\dfrac{K_1}{K_1 + K_2}\right)^2 + \left(\dfrac{C_1\omega}{K_1 + K_2}\right)^2}{\left(1 - \dfrac{\omega^2}{\omega_n^2}\right) + \left(\dfrac{2\xi\omega}{\omega_n}\right)}} \tag{2-62}$$

From Equation (2-62), if the parameters of the heave compensation device K_1, K_2 and C_1 are given, the parameter of the vibration system is ω_n, and the circular frequency of the sea wave ω and the heave displacement of the floating drilling device H_c are given, the vibration displacement of the hook x_2 can be calculated from Equation (2-62).

e. Determine the variation of the drilling pressure in the down hole

After the offshore floating drilling device is equipped with a drilling string heave compensation device and the vibration displacement x_2 still remains on the hook, there is variation ΔWb for drilling pressure in the down holes:

$$\Delta W_b = K_2 x_2 \qquad (2-63)$$

Figure 2-126

Submit Equation (2-62) into Equation (2-63), and a variation of drilling string ΔWb can be calculated, and we can check whether it is within the permissible range in accordance with the technological requirements.

4. Design calculations for the heave compensation device

a. Gas volume variation in the accumulator

As shown in Figure 2-127, the high-pressure fluid at the bottom of the piston is pressed into and discharged from the hydraulic cylinder, which has brought in the changes of gas volume at

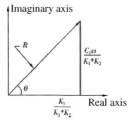

FIGURE 2-126

Diagram of complex representation.

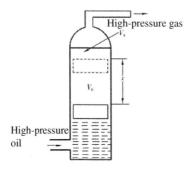

FIGURE 2-127

Gas volume variation in an accumulator.

the top of the piston. If the average volume of the original gas is V_0, the gas volume will reduce to V_x due to the x upward displacement of the piston, and then

$$V_x = V_0 - A_p x \tag{2-64}$$

where A_p is the effective area of the piston.

b. Gas pressure variation in the accumulator

According to the gas isothermal principle,

$$p_0 V_0 = p_x V_x \tag{2-65}$$

where p_0 and p_x are the numerical value before and after the change of pressure above the piston in the accumulator.

c. Determination of the stiffness K_1 of the fluid spring

According to the principle that the restoring force of the fluid spring should be equal to the force of pressure changing in the accumulator, you can write

$$(p_x - p_0)A_p = k_1 x \tag{2-66}$$

Submit p_x in Equation (2-65) and V_x in Equation (2-64) into Equation (2-66):

$$K_1 = \frac{(p_x - p_0)A_p}{x} = \frac{\left(\dfrac{p_0 V_0}{V_x} - p_0\right)A_p}{x}$$

$$= \frac{\left[\dfrac{p_0 V_0}{V_0 - A_p x} p_0\right]A_p}{x} = \frac{p_0\left[\dfrac{V_0}{(V_0 - A_p x)} - 1\right]A_p}{x} \tag{2-67}$$

$$= \frac{p_0 A_p}{x}\left[\frac{V_0}{(V_0 - A_p x)} - 1\right]$$

That is,

$$K_1 = \frac{p_0 A_p}{x}\left(\frac{V_0 - V_0 + A_p x}{V_0 - A_p x}\right)$$

$$= \frac{p_0 A_p}{x}\left(\frac{A_p x}{V_0 - A_p x}\right) \tag{2-68}$$

$$= \frac{p_0 A_p^2}{V_0 - A_p x} \approx \frac{p_0 A_p^2}{V_0}$$

Equation (2-67) indicates the following:

i. According to the given data V_0, A_p and p_0 of the drilling string heave compensation device, the stiffness K_1 of the fluid spring can be calculated.

    ii. With the help of Equation (2-63), x_2 can be determined in accordance with the permissive ΔW_b of the drilling process, then K_1 can be figured out based on Equation (2-62) and compared to K $_1$ from Equation (2-67), in order to reasonably select or design a drilling string heave compensation device.

    iii. Based on the required ΔW_b of the drilling process, K_1 can be calculated by Equation (2-63) and Equation (2-62), and the structure sizes of the drilling string heave compensation device such as A_p and V_0 can be determined from Equation (2-67) for structure design.

5. New heave compensation system implemented by the drawworks itself

With the progress of modern information technology and the integration of machines, electricity, hydraulics, and pneumatics, a new type of heave compensator has been invented abroad. With this heave compensator, there is no necessity to install a hydraulic cylinder-type heave compensator between the traveling block and hook, as shown in Figure 2-119. Instead, the drawworks itself is able to realize compensation of heave motion for offshore drilling. Therefore, the new heave compensation system used abroad is and implemented by the drawworks itself is called AHD (Active Heave Drawworks). The composition and working principles of this new heave compensator are shown in Figure 2-128 below.

a. Sensor-based system

This includes a heave acceleration sensor and dead line tension sensor, etc. as shown in Figure 2-128. The former can measure the heave motion data of the platform vessel while the

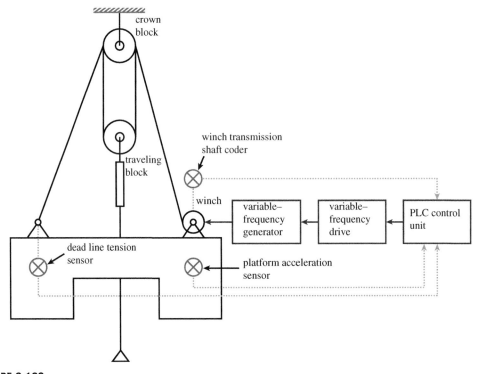

FIGURE 2-128

New foreign heave compensation system implemented by winch.

latter can measure the load change of the information on the hook through the changes of the dead line tension. All the information is transmitted to the programmable controller.

b. Coding system

This is mainly composed of a drawworks transmission shaft coder, and its function is to get the relative position information of the traveling block through the movement of the drawworks on the transmission shaft, and then send it to the programmable controller in a timely fashion.

c. Control system

This is mainly composed of programmable controller, as shown in Figure (2-128): a PLC control unit. Its function is to process the information which the sensor system and drawworks transmission shaft coder sent to the programmable controller, and then send the output signal to the drive winch motor after processing, so as to change the speed and direction of the motor output shaft and the drawworks drum shaft, and implement the active heave compensation. Obviously, through adjusting the rotational speed of the motor and transforming its steering, we can make an adjustment and transform the speed and direction on the transmission shaft and drum shaft of the drawworks, which actively compensates the heave motion on the hook. Therefore, the heave compensator can also be referred to as an active heave compensator.

6. Improved domestic heave compensation system implemented by the drawworks itself

In order to further reduce the energy consumption of the system and improve compensation precision, Zhejiang University, China University of Petroleum (East China), and other scientific research institutions put forward an improvement of the foreign heave compensation system implemented by winch itself in 2012; the improved composition is shown in Figure (2-129). The structural composition and principles of the improved heave compensation system implemented by the drawworks itself are as follows

a. Power input system

Active compensation: Driven by electricity, power is input through the outside gear ring of the transmission mechanism, as shown in Figure 2-129.

Passive compensation: Using a hydraulic drive motor, power is also input through the outside gear ring of the transmission mechanism, as shown in Figure 2-129.

Obviously, the way that they combine using the hydro-pneumatic spring in a hydraulic motor to achieve passive compensation and using an electric drive to achieve active compensation is a true innovation. This method gives full play to the advantage of both active and passive heave compensation and thus can be also called semi-active heave compensation.

Auto-bit feed: Driven by electricity, power is input through the sun gear of the transmission mechanism, as shown as Figure 2-129 and Figure 2-130.

b. Drawworks driving gear

The differential planet gear train, as shown in Figure 2-130, is adopted as the transmission mechanism of heave compensation drawworks. The power from the active compensation motor and the passive compensation hydraulic motor is input by the outside gear ring; the sun gear gets power from the bid feed motor; the planet carrier outputs power (shown in Figure 2-130 and Figure 2-129) thereby driving the drawworks drum and realizing composite motion.

c. Drawworks control system

Forward and reverse rotation of the drum: As shown in Figure 2-129, the PLC control unit, according to the detected unit heaving information, controls the active compensation motor,

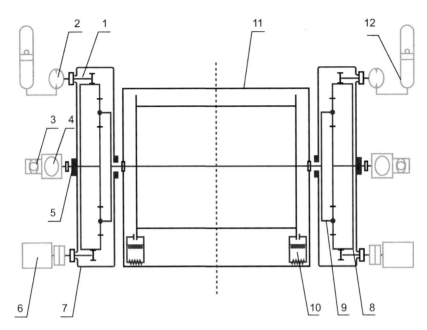

1- Outer ring gear power input shaft; 2- Passive compensation hydraulic motor; 3- Send drill motor; 4- Cycloidal reducer; 5- sun gear power input shaft; 6- Active compensation motor and variable –frequency drive system; 7- differential planet gear reducer; 8- outside ring gear; 9- planet carrier; 10- hydraulic disc

FIGURE 2-129

Improved heave compensation system implemented by winch itself in China.

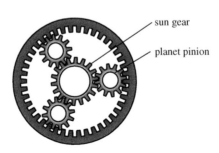

Figure of planet gear construction

FIGURE 2-130

Differential planet gear of drawworks driving gear.

driving the outside gear ring of the differential planet gear so that it compensates for heave motion by the forward and reverse rotation of the drum.

Auto-bit feed: The PLC controls the units, controls the auto-bit feed motor, and drives the sun gear thereby enabling the auto-bit feed based on the detected information about bit pressure change, as shown in Figure 2-129 and Figure 2-130.

d. Reducing energy consumption

It is equipped with an accumulator (gas-liquid converter). In the accumulator, before and after the plunger is liquid and gas, respectively. Then it is possible to carry a partial static load of drill string through the passive compensation hydraulic motor, and to overcome the remaining load in the process through the active compensation motor. Therefore, the active and passive compensation combine to give full play to their strengths, greatly reducing the energy consumption of the system and improving the accuracy of the compensation.

Compared with the new foreign heave compensation system implemented by thedrawworks itself and the traditional heave compensation system, the improved heave compensation system implemented by the drawworks itself proposed by our country in 2012 has the following main advantages.

i. Saving energy

Due to the use of the characteristics of hydraulic pressure and an accumulator, tgravitational potential energy can be released and stored periodically, thereby reducing the installed power and energy consumption of the compensation drawworks. In addition, the accumulator can also be used to reduce part of the energy consumption during the trip.

ii. Saving investment

Due to the combination of the hoisting and heaving function of the drawworks for offshore drilling, the improved set can spare the space for another special set for compensating the heaving of the drilling string, therefore reducing a major part of investment compared with the traditional set.

iii. Saving space

The space of the offshore platform is very precious, as a result of which, saving space for the offshore platform is of significant importance. The Chinese-improved drawworks with auto-set for heaving compensation has spared space compared with the traditional one. The spared space can be used to increase the live load of the platform, enhance the self-sustaining ability of mobile platform, and reduce the equipment and platform load.

iv. Improving performance

According to the simulating calculation results of the Chinese-improved drawworks with auto-set for heaving compensation, the displacement of post-compensatory hook S is between -0.1–0.1 m. This range can satisfy the needs of oceanic drilling and enhance the compensatory accuracy of relatively traditional and new types of foreign drawworks with auto-set for heaving compensation. In addition, it has such advantages as rapid response, lower center of gravity of the drilling system, and compensatory range off limits of the length of hydraulic cylinder. All these advantages are unparalleled compared with the traditional and new type of foreign set.

v. Joint movement

The Chinese set adopts an initial differential planet gear as a transitional mechanism to perform heaving compensation by driving the outer gear ring and auto-bit feeding by driving the sun gear. In this way, they are two different movements of the planet carrier and drum shaft of the drawworks, subtly performing the decoupling control and joint movement for these two movements. With the PLC control unit, the Chinese set can complete the tasks of compensatory heaving and auto drilling, which is one of distinctive advantages of the Chinese-improved winch with auto-set for heaving compensation.

Marine Petroleum (Gas) Engineering and Equipment

Offshore oil (gas) field development refers to oil and gas recovery, whereas offshore oil (gas) production engineering refers to a defined system consisting of specific technologies and management approaches that are able to efficiently and effectively extract undersea oil and gas. In this chapter, we will look at the main aspects, key equipment, and facilities of such as system. In order to highlight the unique characteristics of marine oil and gas fields, some technologies and equipment that are the same as those for onshore oil and gas fields are not repeated here.

PART I OFFSHORE OIL AND GAS GATHERING AND PROCESSING

Offshore oil and gas field production refers to the whole process of exploration, separation (preliminarily into oil, gas, and water), processing, short-term storage, and transportation (either by sea or by seabed pipeline) of crude oil buried beneath the sea. In the process, the steps of preliminary separation and processing can be called production processing, while the rest of the process is called oil-gas gathering and transfer. In this section, oil gathering and processing are described.

3.1 An Offshore Oil-Gas Gathering and Delivery System

The oil and gas from the wellhead is delivered to the test manifold and production manifold. The test manifold delivers the oil and gas to the separator on the wellhead platform for separation and measurement of the individual well's oil, gas, and water. The production manifold collects the oil and gas from each well and transports it to the production processing system, in which the liquid first goes through the process of multistage separation, including demulsification, dehydration, and desalting, etc., and then, if up to quality standards, is stored and transported by appropriate methods to meet the needs of users. This system of offshore oilfield process is called the offshore oil and gas gathering and transportation system.

3.1.1 The Task Undertaken by the Offshore Oil and Gas Gathering and Transportation System

1. Oil and gas gathering

   This refers to how the crude oil and natural gas (sometimes including water) are taken from the well on the sea floor, transmitted to the processing device through the Christmas tree and the production manifold, and tested through the test manifold.

Offshore Operation Facilities. http://dx.doi.org/10.1016/B978-0-12-396977-4.00003-2

2. Production processing

The mixture of oil, gas, and free water is separated on the offshore platforms (or floating boats) or the shore. If separation is completed at sea, the separated crude oil, after going through dehydration, stability, and measurement processes, is transported to the platform tank or oil tanker storage tank for temporary storage, and then sent to the shore by shuttle tankers regularly. The part of the separated gas that goes through the dehydration is used as fuel, and the rest is transmitted or sent to the torch for burning. The separated wastewater is released after it is processed. If this progress is completed on shore, each process that follows can also be finished on shore.

3. Storage and transport

The stored oil-gas is carried to users by carrying vessels (when temporarily stored at sea) or pipelines (including when it is stored on shore or temporarily stored at sea) after it is measured and pressurized.

4. Secondary oil recovery

In order to improve oil recovery, secondary oil recovery techniques have been adopted, including gas and water injection. So processes like seawater treatment, water injection, and gas injection are usually put into the oil gathering and transfer system.

3.1.2 The Pattern of Offshore Oil and Gas Gathering and Transfer

Typically, offshore oil gathering and transfer methods, according to the environment and position where the entire or part of the oil gathering and transferring system is located, can be divided into the all-sea method, the sea and land method, and the all-land method.

1. The all-land oil-gas gathering and transfer method

As shown at the bottom of Figure 3-1, after the crude oil is extracted from the well, the three-phase mixture of oil, gas, and water is transported to the bank directly by the submarine pipeline, and then the separation and other processing are conducted on the land. After reaching the necessary quality standard, it is stored and transported to the users. As the production, processing, and storage are all conducted on land and there is little offshore operation, the investment is lower and construction is quicker. But because the frictional resistance in the pipeline is bigger for three-phase mixing transportation, it is not suitable for offshore oil fields far away from shore.

2. The sea-land oil-gas gathering and transfer method

As shown in the middle of Figure 3- part of the oil gathering and transfer system facilities are on the sea, and the other part are on land. Generally the collection, separation, metering, dehydration, and other production processings are finished at sea, and the crude oil is transmitted by pipelines (gas uses other pipelines) for stabilization, storage, and transportation on land. This method has good adaptability to both the open sea and offshore. Because the method must use the submarine pipelines (crude oil and natural gas), it is not suited for complex seabed topography and unsteady crude oil.

3. The all-sea oil gathering and transfer method

As shown at the top of Figure 3-1, all processes of gathering, transfer, and production are on the sea for this method. After processing, the qualified crude oil is temporarily stored in a floating production storage and offloading (FPSO) unit, and waits for shuttle tankers at regular intervals. This method is especially suited for the open sea and deep sea oil fields (and can be used with all

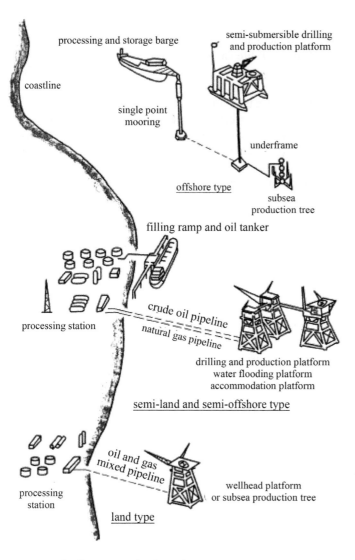

FIGURE 3-1

Three modes of the offshore oil gathering and transfer.

types of offshore production of course). Because lots of floating facilities are being used, the cost is relatively low. So some farther offshore stripper fields and marginal oil fields often use this method.

3.1.3 Choosing a Method of Offshore Oil Gathering and Transfer

The offshore oil/-gas gathering and transferring modes methods are important components of the offshore oil field development plan, so making a reasoned choice is necessary. There are many factors that influence the choice, so selecting a reasonable solution must be based on a comprehensive analysis of data and guidelines about the technology and economics involved.

1. Influential factors
   a. Oil deposit situation: This includes the oilfield area, recoverable reserves, oil and gas well production capacity, production methods, well life and oil-gas properties, etc.
   b. Oilfield geographic location: This includes the offshore distance, presence of islands near the oilfield, conditions of the onshore dock and the port, etc.
   c. Ocean engineering environment: This includes the oilfield water depth, submarine topography, seawater and seabed soil properties, weather, sea conditions, seismic data, etc.
   d. Construction conditions at sea: This include the technology level and equipment conditions related not only to offshore engineering construction, transportation, and pipeline laying, but also to the factories responsible for the construction of the offshore structure.
   e. Oil and gas marketing: This includes the domestic sale or export of the crude oil, the distance away from the center of consumption, the oil-gas transportation routes (sea or land transport), etc.
   f. Economic factors: This includes the crude oil or natural gas prices, purchasing materials price, investment estimates, operation cost estimates, repeated utilization ratio of facilities and earnings after the economic evaluation, etc.
2. Guiding principles for selection
   The guiding principles for the rational selection of the oil-gas gathering and transfer include:
   a. Technical feasibility
   This means that the chosen method of the oil-gas gathering and transfer should be feasible, no matter whether it is the technology adopted during the construction process or the technique adopted for the equipment in its design, manufacture, use, maintenance, etc. Once advanced technology is adopted, it should pass the test. For example, once the all-land oil gathering and transferring method is adopted, whether the oil deposit pressure and wellhead pressure meet the technology needs that will ensure the oil is transferred from the oilfield to the shore with a long-distance pipeline or not must be considered.
   b. Relatively high economic benefits
   We must look at the economic benefit compared with other methods. The economic benefit is the focus of the oil field development plan and choosing the oil-gas gathering and transfer pattern. Economic benefit is a comprehensive result affected by many factors, and from an objective view of the chosen method of the oil-gas gathering and transfer, the two most important factors are production and oil and gas prices. From the view of the influencing factors directly related to the oil-gas gathering and transfer method, there are six factors mentioned above, including the recoverable reserves, oil-gas properties, well production capacity, production method, etc. At the same time those factors are also associated with the evaluation of the economic benefit, for example the oilfield offshore distance is related to the oil field location; the marine engineering environment and construction conditions are related to the project investment and construction cycle, etc. (operations, repeated utilization ratio of equipment and so on).
   c. The low degree of risk on the implementation
   The risk is a comprehensive description of the probability of occurrence of a specific risk event and the extent of the damage. The degree of the risk refers to the size of the probability of risk occurrence. There are a lot of implementation risks that need to be forecasted and evaluated for the oil-gas gathering and transfer system. When a method of the oil-gas gathering and

transferring is chosen, the degree of risk is considered mainly from the three perspectives: the uncertainty of predicting the basic oil deposit situation, the uncertainty of predicting the project construction cycle, and the uncertainty of economic evaluation. For example, there is uncertainty in predicting oilfield recoverable reserves and oil and gas well production, forecasting sea conditions and weather during construction, and predicting oil and gas prices. When the method of oil-gas gathering and transfer is chosen, it is necessary to fully use the probability analysis method to analyze every major uncertainty and optimize the scheme of the lower the degree of risk.

3. Selection examples

   Take the Qinhuangdao QHD32-6 oilfield built in China's Bohai Sea in recent years as an example of illustrating how to choose an oil-gas gathering and transfer pattern.

   a. Analyze the basic oil deposit situation

   The QHD32-6 oilfield is a heavy oil field. Its crude oil properties are that viscosity and relative density are high, up to 616 mPa·s at 50 °C and 0.9546 respectively, meaning we have a heavy, thickened oil. Moreover, the oil deposit pressure is low: the initial pressure is 12.0Mpa and the minimum pressure is 5.5MPa. So oil production needs an electric submersible pump and relies on water injection. Therefore, a submarine pipeline is not suitable. Choosing the land-sea oil gathering method means that not only is the technology complex, but the the relative economic benefit is lower.

   b. Analyze the relative economic benefit

   For the QHD32-6 oilfield, which belongs to the category of heavy oil fields, production is very difficult and the cost is high. Higher capital costs are also needed in order to achieve the qualified crude indicator, which refers that the water cut of purified crude oil is less than 0.5% after processing (BS and W). So if the method of oil gathering and transfer mode is chosen, economy is an important indicator. However, amongst the three kinds of the oil-gas gathering and transfer, since the all-sea method generally adopts more floating facilities and the main facilities can be reused, it has relative superiority for QHD32-6. According to the above analysis, the all-sea method of oil-gas gathering and transfer mode was adopted for QHD32-6 oilfield development and construction. The scheme for the offshore oil-gas production facilities is (Figure 3-2):

   WHP (wellhead platform) + SPM (single point system) + FPSO (floating production storage unit)

3.1.4 Offshore Oil/Gas Gathering and Transfer Process

The oil/gas gathering and transfer process is a general illustration of the flow of the oil and gas in the internal field. The process begins in the offshore oil-gas wellhead and goes through the transmission and several processing processes, until the oil and gas become qualified products. Here we take the process of the all-sea oil-gas gathering and transfer mode for the QHD32-6 oilfield as as example. Figure 3-3 shows the flowchart.

As shown in Figure 3-3, the oil, gas, and water multiphase flow from offshore wells is sent respectively from three sets of the wellhead platforms (WHPB and WHPA are a set, WHPC and WHPD are a set, and WHPE and WHPF are a set) with two sets in series, to the single point mooring system (SPM). Then, the multiphase flow from the SPM first is sent into the free water separator

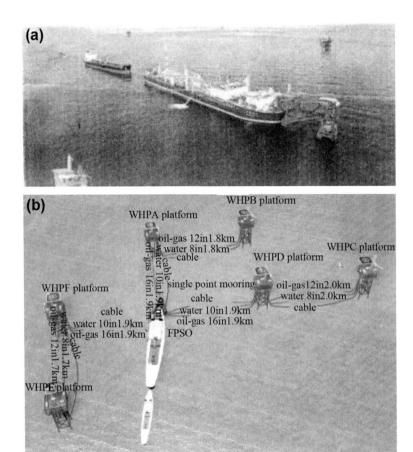

FIGURE 3-2

The scheme for the offshore oil and gas production facilities in QHD32-6.

(V-101) and heated treater (V-102) for processing on the FPSO by the hosepipe, which lowers the water cut in the crude oil below 30%. Then, it is sent to the electric dehydrator (V-103) for dehydration processing by the feed pump (P-101). After the electric dehydration process is complete and the oil is isolated from the free water, the crude oil is heated in the heat exchanger (HE - 101), and then is cooled down to about 80 °C by the water cooler (WC - 101).

If the crude oil is qualified after electric dehydration (BS and W less than 0.5%), it is directly sent into the cargo tank. Otherwise, it goes through the two levels of settling separation in the oil processing tank until qualified, and then is sent into the cargo tank. As for the gas separated from the free water separator (V-101), after separating liquid and dust, it is sent to the knockout tank as fuel to heat the medium boiler fuel. The gas separated from the heat processor (V-102), due to the low pressure, may go into the burning torch system. In addition, the water separated from the electric dehydrator (V-103) goes back into the free water separator (V-101), which can improve the operating temperature of the

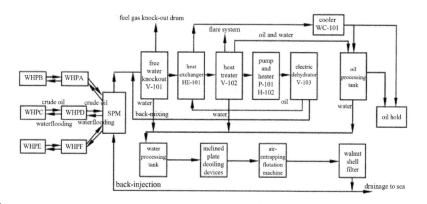

FIGURE 3-3

The oil-gas gathering and transfer method for the QHD32-6 oilfield.

separation, and also can help in reuse of the chemicals added into the electric dehydrator (V-103). Also, the production water from the free water separation (V-101), the oil processing tank, and heat treater (V-102) first goes through two levels of settling separation in the water processing tank, and then is filtered by the inclined plate oil removal device, gas flotation machine, and walnut shell until the oil in the water is not over 30 mg/L, with solids no more than 5 mg/L, and suspended particles not greater than 4 mm in diameter. At the condition that the processed water satisfies the requirements of injection water and discharging water into the sea (no back to the injection well), it is transferred to the wellhead platform by the SPM and seawater injection pipeline, and backed into injection wells. If the injection water from the processed water is lack, it also can be supplied from water source well of the wellhead platform. At the initial stage of oilfield development, the wells (also called advance drainage) don't need flood water for the first half-year and the produced water can be discharged into the sea.

It should be stated that in the oil-gas gathering and transfer process, in order to solve the distance issue of oil, gas, and water mixed transportation from the wellhead platform to the FPSO, the plan adopted is that the wellhead platforms are divided into three groups and each group has two platforms which are in series (to shorten the pipeline distance), and there are new technologies such as multiway valves and multiphase flow meters that can automatically measure the single well production are installed on the wellhead platform. The implemented technologies play an important role in choosing the method of the oil-gas gathering and transfer.

3.2 The Offshore Crude Oil Processing System

It is easy to see from the offshore oil-oilgas gathering and transfer process that the crude oil processing system includes the oil-gas separation, crude oil dehydration and desalination, associated gas processing and so on, which will be respectively introduced below.

3.2.1 Oil-Gas Separation

With the liquid's pressure decreasing during the process of the reservoir's liquid flowing from underground to the wellhead along the tubing, the gas dissolving in the oil under the conditions of the reservoir's high temperature and high pressure is continuously separated out and forms an oil-gas coexistence mixture together with the oil. Therefore, it is necessary to separate the gas from the oil to meet the production quality requirements, which is the oil-gas separation process. It is an important link in the oil-gas processing.

1. The basic methods of offshore oil-gas separation
   Offshore oil-gas separation commonly uses classification separation, namely, discharging the separated gas by reducing the pressure. Each of the gas discharges is called one stage separation, and each time the gas is discharged represents another level of separation. Because the end of oil-gas separation is always carried out in the normal pressure oil tank, it is common to take two oil-gas separators and an oil tank in series for oil-gas separation, which is three-stage separation. Also commonly used in offshore oil fields is a method of three-stage separation that takes high, medium, and low pressure oil-gas separators in series. The last separated liquid from the low pressure separator doesn't enter the oil tank, but is directly pumped to the cargo tank or outward. The gas after the liquid removal is used as fuel or for torch burning, and the wastewater containing a small amount of oil is discharged into the sea after processing (Figure 3-4).

2. The oil-gas separator used in offshore oil fields
   The oil-gas separator refers to the device separating the mixture of oil and gas. The separation effect of the separator can be reflected by the degree of clarification. The degree of clarification can generally be showed by the liquid ratio in the gas (how many droplets the gas contains) and the gas ratio in the liquid (how much gas the crude oil contains). A reasonable selection of oil-gas separators should take a high degree of clarification and the structure's economy into consideration.
   Separators used in offshore oil fields are divided according to their appearance into horizontal vertical types, and according to the function into an oil-gas two-phase separator and oil, gas, and

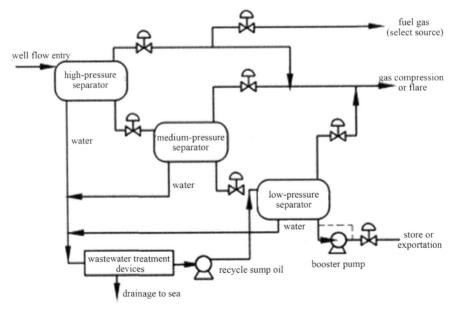

FIGURE 3-4

The three-stage oil and gas separation process for offshore oil fields.

water three-phase separator. The most commonly used separators in offshore oil fields are horizontal three-phase separators, as shown in Figure 3-5.

As can be seen from Figure 3-5, after the oil-water mixture enters into the separator from the inlet, it first enters into the preliminary separation section (inlet diverter, which is a kind of dish baffle). When the oil, gas, and water mixture with a certain velocity strikes the dish baffle, its speed and movement direction suddenly change, and the heavier liquid flows into the bottom of the container along the surface of dish baffle while the lighter gases escape upwards, so as to achieve the purpose of preliminarily separating the gas and liquid. There is a gas-liquid baffle beneath the dish baffle, in case the dropping water splashes the collecting liquid back into the gas again. This is the preliminary separation within the separator. After that, the fluid enters into the main separation section, and at the appropriate location the gas steam flow plates, which consist of a series of concentric parallel plates, are installed, and are fully distributed at the cross section of containers above the control level. After the gas passes the steam flow plate, the turbulent flow state after preliminary separation is greatly reduced, which promotes the droplets in the air flow to settle down under gravity's action. At the same time, oil-water separation is accomplished in the collecting fluid part. Leaving the main body of the fluid separation part, the fluid enters into the oil mist catcher part, which makes use of the separation principle of collision to capture the oil droplets of small size that were not removed in the gravity settling stage. To this end, there is a demister in the separator, which is a gauze pad consisting of lots of stainless steel gauzes whose have diameters between 0.12 ~ 0.25 mm and a gap rate of 97%. The gauze pad is similar to the strainer, but the effect is different, and the gauze pad generally can remove oil droplets of the size of 10 ~ 30 mm, relying on collision to capture oil mist. Under the anti-mist device, the intercepting part is equipped with oil spill board, and the oil-water interface is controlled beneath the board. In addition, the anti-eddy boards are set at the oil outlet and water outlet in case of producing the eddy which takes away the crude oil above the sewage and the gas above the crude oil. Moreover, a special injection pipe is installed at the bottom of the water layer in the separation device, which can regularly stir the sand and other solids that deposit at the tanker's bottom with the high-pressure seawater.

Table 3-1 takes the Wei 11-4 offshore oil field in the west of the South China Sea as an example, listing the sizes and the basic parameters of the separator, and comparing some parameters of the design value to the actual value.

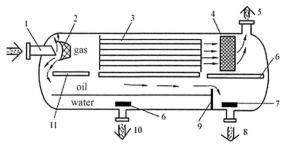

1--oil and gas inlet; 2--inlet diverter; 3--gas transportation plate; 4--demister; 5--gas outlet; 6--gas-liquid breakboard; 7--eddy-prevent plate; 8--oil outlet; 9--oil spill plate; 10--water outlet; 11--fluid baffle

FIGURE 3-5

Horizontal three-phase separator used in offshore oil fields.

Table 3-1 The Size and Basic Parameters of the Separator at the Wei 11-4 Offshore Oil Field

Type and Serial Number	Horizontal Three-Phase Separator (P - V - 102)		Horizontal Three-Phase Separator (P - V - 103)	
Size (Diameter × Length), mm	3000 × 10000		3000 × 9600	
Basic Parameters	Design	Actual Production	Design	Actual Production
Pressure, MPa	0.56	0.20	0.40	0.20
Temperature,°C	2.9 ~ 90	70	2.9 ~ 100	71
Minimum operating pressure, MPa	0.25	—	0.15	—
Maximum operating pressure, MPa	0.50	—	0.57 ~ 0.97	—
Liquid flow, m³/d	7800	9100	2000	4000
Crude oil flow, m³/d	1800	4000	1600	3200
Water flow, m³/d	6000	6800	1000	1800
Gas flow, m³/d	7800	10000	7000	9000

In recent years, a new kind of efficient three-phase separator has been applied to offshore oil fields. It integrates and applies different kinds of oil, gas, and water separation technology to the separator, such as mechanical, thermal, electrical, chemical methods, etc., and achieves high separation efficiency of the oil, gas, and water by careful selection and laying out of the internal components of the separator. Figure 3-6 is the structure diagram of an HNS efficient three-phase separator. Figure 3-7 is a sketch of a multifunctional efficient combined dehydrator. This efficient three-phase separator is made into a kind of skid-mounted device, which saves installation space, reduces the workload of installation, greatly simplifies the process, and is convenient for operation management as well as strongly adapted to the production situation. So this separator is more suitable for three-phase separation at offshore oil fields.

3. Special problems with oil, gas, and water separation for an offshore floating production system
   a. The effect of the movement of the floating production system on the equipment.
      The floating production platform of the FPSO or semi-submersible, pontoon type has six degrees of freedom of movement affected by the ocean environment (flow of wind, waves, etc.). Some of these movements have a great influence on the characteristics of the equipment for oil, gas, and water separation, such as the pitching (four directions), rolling, heaving, advancing, and retreating movements. Table 3-2 shows the effects of the movement forms on the equipment characteristics.
      As pointed out in Table 3-2, Figure 3-8(a) though (d) lists the impact of four kinds of movement on the fluid in the equipment in the table, while Figure 3-9 gives quantitatively the relative percentage that pitching movement increases the gas flow rate. It shows that when pitching motion occurs, the cross-sectional area of the container decreases, thus increasing gas velocity increasing, and that the degree of increase is associated with the increase of pitching motion angle.

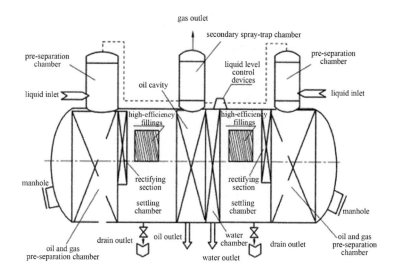

FIGURE 3-6

HNS high efficiency three-phase separator.

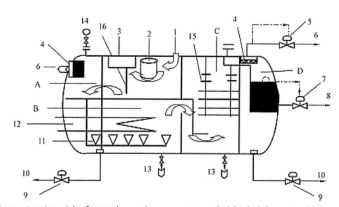

1--oil-gas-water mixture inlet; 2--rotated separation component near the inlet; 3--deflector; 4--oil mist extractor;
5--pressure control valve; 6--gas outlet; 7--control valve for oil outlet; 8--crude oil outlet;
9--interface control valve; 10--water outlet; 11--liquor distributor; 12--heating coil; 13--drain outlet;
14--pressure instrument; 15--suspending insulator; 16--spoiling flap;
A--predissociation section; B--heating and settling section; C--electric dehydration; D--buffering section

FIGURE 3-7

Multifunctional efficient combined dehydrator.

Table 3-2 Effect of Floating Production System Movement on the Equipment Characteristics

Movement Type	Effect on the Process Equipment Operational Performance	Diagram
Container level area lean	Vertical tilt caused by pitch will make the gas gallery in the container smaller and gas flow accelerates, increasing the amount of droplets carried amount and generating foam, thus worsening equipment operation performance.	Figure 3-8(a)
Primary liquid perturbation	The tilts of the container, resonance waves and other forces make the liquid in the container move. So these liquid absorbing the movement energy produce the primary liquid perturbation that slows down the separated speed of the bubbling from the crude oil and makes the oil and water mix with each other. The primary liquid perturbation decreases the separation efficiency.	Figure 3-8(b)
Resonance wave	The natural frequency of the liquid longitudinal wave in the container (period of 6 ~ 12s) is close to the pitch and floating hull surge frequency (period of 10 to 16 s), causing resonance waves, increasing the gas flow rate of certain areas, and leading to an oil/water two-phase mixture that reduces the operating performance of the process equipment.	Figure 3-8(c)
Secondary liquid disturbance	In order to eliminate resonance waves and reduce the secondary liquid disturbance, a weir plate part with a hole is usually installed in the container. A secondary disturbance occurs when liquid passes the weir plate hole and reduces the separation efficiency of oil and water.	Figure 3-8(d)

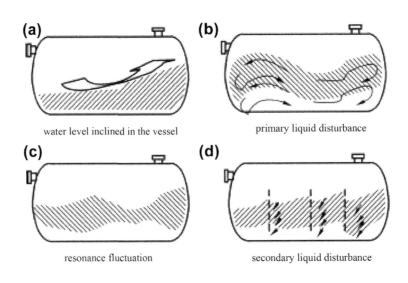

FIGURE 3-8

Floating production system motion's direct impact on equipment performance.

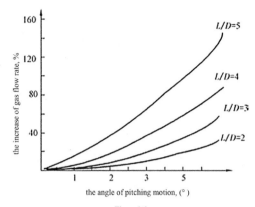

Figure 3-9
the relationship between gas flow rate and the angle of pitching motion
L--length of vessel; D--diameter of vessel

FIGURE 3-9

The relationship between pitching movement angle and gas flow rate increase.

b. Equipment failure caused by the floating production system's movement

   i. Because of the acceleration of the heave movement, the apparent weight of the float in the device changes.

   ii. The change of the liquid surface in the equipment caused by the movement of floating production system will result in alarm and shutdown from the hi-lo signal alarm.

   iii. The pitch and roll motion hangs the surface float on the buoy wall, which results the failure of the control reading.

c. Main measures to reduce the influence of movement on equipment

   i. Arranging containers reasonably

   The position and direction of the oil, gas, and water separation main facilities on the FPSO or floating platform are directly related to reducing the equipment movement amplitude. Because one of the most harmful movements is pitch, the axial of the separator should be placed along the smallest pitch direction, that is, as close as possible to the center of gravity of the ship, in order to minimize the heave movement.

   ii. Designing containers reasonably

   The structure of the oil, gas, and water separator, such as its size and geometric shape of internal components, etc., should be reasonably designed. To reduce the influence of floating production system movement, the structure of the offshore horizontal three-phase separator should be designed as much as possible to increase its diameter, and to decrease the length. From the perspective of internal components, if the number of weir plates is increased in the container, the primary disturbance will be reduced, but secondary disturbance will be increased. In order to minimized the overall disturbance, optimizing the design and using a reasonable number of weir plates should be considered. In addition, improving the geometry of the weir plate also can reduce the influence of rolling motions. Offshore field tests show that,when a well-designed horizontal three-phase separator on a vessel undergoes 6° rolling and 3° pitching, it still can maintain satisfactory operation. The patent for utility model No. 28626 that China University of Petroleum developed—a swinging resistance device reducing the liquid sloshing inside the container—is a measure designed to reduce the impact of floating production system movement on equipment through a reasonable design of the container's internals. In practice this device improves the separation efficiency for the separator of an offshore floating production system.

3.2.2 Dehydration and Desalination

With oilfield development, formation water may be extracted with crude oil, and water productivity especially grows faster for the water flood field, even reaching 90% of producing fluid at later stages of oilfield production. So solving the crude oil dehydration problem is very important. In addition, when the oil and water is extracted, large amounts of salts from the formation also are carried out, such as chloride, sulfate, carbonate, etc. And these salts tend to be dissolved in water, so the crude oil processing system needs both dehydration and desalination.

Crude oil with salt will corrode the equipment and pipeline and make them scale. Crude oil with water will increase the heat load in the process, and the cost and power consumption in the transportation process. Especially to protect the marine environment, treatment is still needed before the sewage with oil discharges into the sea, which takes additional effort. Thus the dehydration and desalination of crude oil becomes an important link in the crude oil processing system. In general, the moisture content of qualified crude oil for offshore oil field is 0.2% ~ 1.0%, and the salt content should be less than 50 mg/L.

1. Principles and method for dehydration of the emulsion

    Crude oil is not miscible with water (or is minimally miscible), and its physical and chemical properties are very different from water. Water separated from oil with simple sink sedimentation in a short period of time is called free water. As mentioned earlier, the water that is separated from the three-phase separator by the sedimentation principle is the free water. In production, crude oil and water are not simply mixed, but in a fairly stable emulsion state, and this kind of water is called water emulsion. A mixture of crude oil and water emulsion is called crude oil emulsion.

    a. Demulsification

    The dehydration process of the crude oil emulsion can be divided into two phases: demulsification and sedimentation. Demulsification refers to the process where the membrane, which forms due to the existence of a natural emulsifier in the emulsion's oil-water interface, is damaged under the action of the chemical, electrical, thermal, and other external conditions, and the water drops packaged in the membrane collide, merge, and coalesce. Due to the coalescence of the dispersed phase droplets, the droplets' diameters increase and the emulsion turns into a suspension, so that those droplets merge in further collisions and result in sedimentation separation, which is the mechanism of demulsification dehydration. Obviously, the method of using chemicals to rupture the membrane which forms for the oil-water interface is called a chemical demulsification method, such as using SP, AE, AP, AR demulsification, etc., which are used in China. For heavy and hyper-viscous oil, an electric demulsification method will be used, because chemical demulsification methods are not up to the quality standard for the water cut of crude oil. This is a physical method which utilizes the function of a high-pressure electric field on the droplets—weakening the membrane's strength at the oil-water interface, promoting the collision and merger of the droplets, and forming larger diameter droplets—and separates water from crude oil. Generally high-water-content crude oil first needs to go through thermal-chemical demulsification dehydration and then enters into the electric dehydrator after the reduction in crude oil moisture content, because the dehydrator operation is not stable when the electrical method processes high-water-content crude oil.

b. Sedimentation and separation

Water droplets in the crude oil emulsion collide and merge after the breakdown of their emulsion, which makes their particle sizes increase, and helps them settle down with the action of gravity and eventually separate from the crude oil. That is the second phase of emulsion dehydration. The speed of water droplets sinking in crude oil is slow, and often in a turbulent state, and the subsidence velocity, u, can be expressed by the Stokes formula, when the drop diameter is d_w:

$$u = \frac{d_w^2 g(\rho_w - \rho_o)}{18\mu_o} \tag{3-1}$$

In the equation ρ_w, ρ_o are the mass density of water and oil, kg/m^3; μ_o is the crude oil viscosity, Pa·s; and g is the acceleration of gravity, m/s^2.

As shown in Equation (3-1), the higher the viscosity of crude oil is, the smaller the oil water density is, and the slower the sedimentation velocity of water droplets in crude oil is. When processing a certain amount of the crude oil emulsion, for high viscosity heavy crude oil, the diameter of water droplets should be increased, which will lengthen the residence time of emulsion in the dehydrator. For example, the settling time of the high viscosity heavy crude oil in the formerly introduced QHD32-6 oilfield is up to 8 to 12 h. Offshore oil fields often follows the following empirical formula according to a residence time T = 15 ~ 30 min:

$$D = \sqrt{\frac{TQ}{60\pi L}} \tag{3-2}$$

The diameter D of the dehydrator can be designed in accordance to the above equation. L in this Formula is the effective length (m) of the dehydrator, and Q is for the dehydration processing fluid volume (m^3/h). When the dehydration proceeds on an FPSO, due to large volume of the oil tank, we can meet the requirements of high viscosity heavy crude oil for large sedimentation tanks. We should also point out that when the electrical method is used to process crude oil emulsion with high water cut, it is easy to produce an electric breakdown and make the dehydrator operation unstable. Therefore, demulsification and dehydration by the thermal-chemical method should be first carried out to reduce the moisture content.

2. Dehydration and desalting technology for crude oil

Due to space limitations, the offshore crude oil process combines the separation process of the oil and gas with the dehydration of the crude oil, that is to say, free water separation in the crude oil is completed in the three-phase separator, and the separation of the emulsion water is not only completed in the dehydrator containing the chemical and electric demulsification processes, but also in the secondary separator and dehydrator where a chemical demulsifier is injected in advance and makes the low pressure separator a thermal-chemical processor. Some of China's Bohai Sea and South China Sea offshore oilfields adopt a dehydration process like this. Some FPSOs in the Bohai Sea do not install an electric dehydrator, and only the adopt thermal-chemical dehydration process, which uses two separate tanks for sedimentation in rotation, and keeps the emulsion for only 8 h. The result is also very good. As far as the offshore fixed production process, due to the limitations of space, a large enough sedimentation tank can't be installed on it. In order to qualify oil products, a full dehydration test of the oil sample must be taken before oil field production to obtain the reliable equipment design data.

The desalting process of the crude oil is actually the dehydration process. Crude oil with high salt content can meet the moisture requirements after going through a series of dehydrations, and also meet the salt requirements (50 mg/L). Otherwise, fresh water should be injected into the crude oil, making the salt soluble in water. The crude oil with fresh water passes through another dehydrator in series and goes through dehydration again, in order to achieve the salt regulation requirements.

3. Electric dehydrator

Offshore oilfields generally adopt the horizontal electric dehydrator. This is because the horizontal dehydrator is not restricted by the offshore deck's height between layers, has a larger cross-sectional area to set the electrodes than the vertical electric dehydrator, and has a shorter settling distance of droplets than the vertical system, which is advantageous in the separation of water from the crude oil.

Figure 3-10 gives the structure diagram of the horizontal electric dehydrator. It can be seen from the figure that the dehydrator space is divided into two parts, the upper part for hanging electrodes, and the lower part for water separation. The upper electric space hangs the electrode grids (label 6 in Figure 3-10), and the distance between two electrode grids reduces from bottom to top in turn, so that the electric field intensity increases from bottom to top in accordance with the gradual reduction of the water content. Electrodes are connected by wires, which are extracted by the insulation rod with the shell. The crude oil with water enters into the multihole tanker liquid pipe (label 5 in Figure 3-10) through the enter valve, and goes from bottom to top through the electric field space, which makes the demulsification occur and the water droplets collide, merge, and settle down the bottom of the dehydrator. The water separated from the crude oil is ejected from the discharge pipe (1).

Liuhua oilfield in the South China Sea used the dynamic electric desalting device successfully in 1996. Liuhua oil field adopted the dynamic electric desalting technology that combines electrodesalting and electro-dehydrating. It is based on the double electric field technique. After the desalinated water enters into the device, it flows in reverse with the crude oil. Under the effect of the high-strength electric field, the desalinated water is broken into many tiny particles, and they mix with the reversed crude oil, which makes the desalinated water more fully contact with the salt water. When the electric field weakens, tiny particles will unite more easily, and the particle size increases and then settles, so as to realize both dewatering and desalting within a device. The technology and process consist of four stages: particle diffusion, particle mixing, particle combination, and particle sedimentation. Figure 3-11 shows the variation of the electric field in the dynamic electric desalting device and the cycle consisting of four stages. Figure 3-12 shows the structure diagram of the dynamic electric desalting device applied to the Liuhua oilfield. The process and the device, which combines electric desalting and dehydration, will be widely used in offshore oilfields.

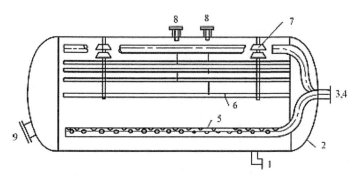

FIGURE 3-10

The horizontal electric dehydrator used in offshore oilfields.

1--releasing and emptying; 2--the shell of dehydrator; 3--cleansing oil outlet; 4--wet crude oil inlet; 5--distributing pipe for liquid flux; 6--electrode; 7--suspending insulator; 8--the hole for inlet wire installation for suspending insulator; 9--manhole

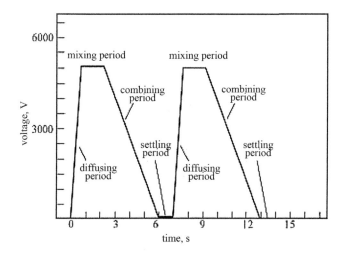

FIGURE 3-11

Electric field dynamic variation in the electric desalting device.

3.2.3 Associated Gas Processing

The gas extracted with crude oil is called the associated gas. These gases from the oil-gas separation system are carrying liquid (oil or water) to varying degrees, which is can easily make the pipe or equipment fail. Especially in winter, the water in these liquids freezes easily and can block the pipes. So a process is needed to deal with the associated gas. Processing the associated gas mainly cuts off the liquid (oil and water) in the gas, and the dry gas can be used locally, as fuel for the offshore platforms, sealing cylinder gas, purge gas, or output gas after compressing. The rest can be burned in the torch system.

As for the associated gas processing technology, here we take the process in the South China Sea Wei 11-4 center platform as an example to illustrate. As shown in Figure 3-13, the system is made up of the pigging receiver, slug flow trap (1), fuel gas scrubber (2), gas compressor (3), air cooler (4),

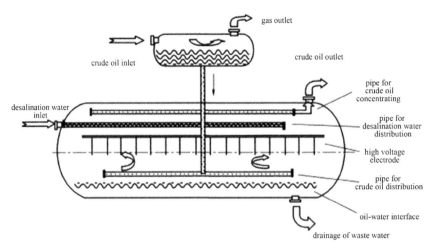

FIGURE 3-12

The dynamic electric desalting device applied to the Liuhua oilfield.

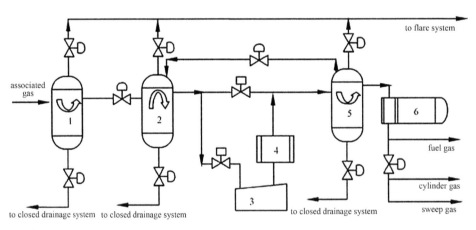

FIGURE 3-13

The associated gas processing system on the South China Sea Wei 11-4 center platform.

degassing chamber (5), gas heater (6), etc. The associated gas separated out from the center platform, through the pigging receiver, enters into the slug flow capture (1) and then into the scrubber (2). In the scrubber (2), the condensate oil that is separated from the gas because of the pressure drop in the submarine pipeline is washed away. Furthermore, the oil and water in the gas are separated by compression and cooling. To prevent the steam in the gas of the separator (5) from condensing again in the conveyance process, gas is needed to heat the gas heater (6). The gas out from the heater can be used as fuel, sealing, and purge gas. When the gas turbine breaks down and doesn't need the gas, it can be vented to the torch system forburning.

3.2.4 **The Closed Discharge**

Collecting and processing the sump oil and wastewater through the closed discharge in the crude oil processing and utility system, making the sump oil return to crude oil processing, and making the qualified wastewater discharge into the sea is called the closed discharge processing system.

1. The function of the closed discharge processing system

This is a secondary system for oil and gas production in an offshore oilfield, of which the main functions are:

a. Receiving the oil or the mixture discharged from the production line or the container equipment when in an abnormal state.

b. Receiving the oil and gas mixture discharged from the oil and gas production equipment after they shutdown.

c. Separating the received mixture gas-liquid, sending the gas to the torch for burning, and recycling the oil to the crude oil processing system for rehash.

2. The technological process and equipment of the closed discharge processing system

The closed discharge processing system is mainly composed of a closed pot and closed drain pump, etc. The height of the liquid level in the tank controls the pump's work conditions; therefore the closed drain pump is not continuously running.

The closed discharge processing system in most offshore oil fields is substantially similar. Take the South China Sea Wei 11-4 closed discharge processing system as an example to illustrate. Figure 3-14 presents the Wei 11-4 process chart for a closed discharge processing system. The closed discharge tank in Figure 3-14 is a horizontal two-phase separator, with a liquidometer, pressure gauge, liquid level switch, drain outlet, vent nozzle, sand outlet, etc. on the outside; and oil importer, separator, oil

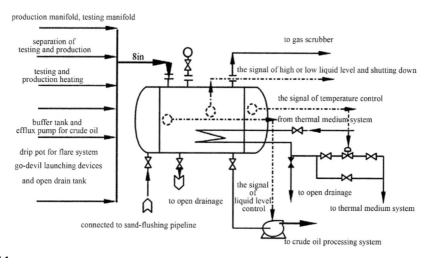

FIGURE 3-14

The closed drain system in the South China Sea Wei 11-4 oilfield.

export checkered plate, mist precipitator gas exports, sand washing machine, sand drain tank, coil heater, etc. on the inside.

The closed drain system is an intermittent operating system. As shown in Figure 3-14, the mixed liquor discharged by other processes and collected through 8 in manifolds enters into the jar from the jar's roof where there is a relief valve, The separated gas that comes from the roof tube is discharged to the torch system, and the separated mixed liquid is pumped into the process flow from the bottom discharge of the tank by a recycling pump. During normal operation, when the liquid level reaches a set high level, the pump will automatically open, and send the mixture outside the tank; when the liquid level reduces to the set low value, it will automatically stop.

The normal operating temperature in the tank is 50°C, and the water temperature is maintained by heating the coil pipes. When the temperature is lower than 50°C, the temperature control valve automatically opens, and the heat is transferred to the mixture liquor in the tank, heating the mixture. When the temperature rises to above 50°C, the temperature control valve automatically shuts down, so the flow rate of the heat reduces. The temperature control valves create an automatic control of the tank temperature, maintaining a constant temperature and controlling the tank temperature within the set temperature range.

3.3 Water Processing System for Offshore Oilfields

The water processing system for offshore oil fields mainly involves the water treatment system used for seawater injection and the oily water treatment system; in addition, there are shallow groundwater treatment systems for water injection and domestic water, and secondary sewage treatment systems for collecting the deck wash water and rainwater.

3.3.1 Water Treatment System for Water Injection

Offshore oilfield injection water mainly has three sources: seawater, which is the major source for offshore oil fields; reservoir water, which is the production associated with gas and oil; and the other shallow water at the bottom of shallow portions of the sea. Of course, the above three water sources could be injected together into the layers, but the technology is not yet mature enough.

1. Requirements for seawater injection quality
   An oilfield has two types of water flooding: edge water flooding and inner edge water flooding. Either way, its purpose is to maintain the reservoir pressure and achieve a high and stable yield; therefore, requirements for water quality are high. When seawater is injected, because the bacteria, algae, sea creatures, suspended solids and dissolved oxygen that are contained in the seawater will damage the oil layer, dealing with the water is necessary to make it reach a certain water quality standard. There is no across the board standard for offshore oil fields in China now. Here we give the standard for the Bohai Sea Suizhong 3-6 Field as a reference (Table 3-3).
2. Water treatment methods and equipment for injection water
   Figure 3-15 shows the flowchart for the seawater injection system in China's Bohai Suizhong 36-1 oilfield. The system is located in FPSO "Pearl." The seawater lift pump first raises the seawater to the FPSO from the seafloor, and then the coarse strainer and the filter successively filter the seawater in turn. Then the filtered seawater first is deoxidized in the deoxygenation tower

Table 3-3 Injection Water Quality Standards for Bohai Suizhong 36-1 Oilfield

Item	Water Quality Standard
Dissolved oxygen content	≤0.01
Hydrogen sulphide content	≤0.5
Total iron content	≤0.5
Ferrous iron content	≤0.5
Membrane filtration factor	>10
Suspended solid content	≤5
Suspended solid particle diameter	≥95
	≥80
SRB bacterium	≤100
TGB bacterium	≤10000
The average corrosion rate	≤0.076

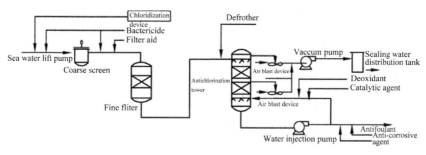

FIGURE 3-15

Injection water treatment system in the Bohai Suizhong 36-1 oilfield.

again, and then pumped by the water injection booster pump to the production wellhead platform through the seawater injection pipeline and assigned to the injection wells by the water injection manifolds on each of the wellhead platforms. The following takes this procedure as an example to illustrate the methods and equipment of water treatment.

a. The seawater lift pump

There are two kinds of seawater lift pump has two kinds: a submersible pump and vertical long-axis centrifugal pump. The lift pump used in the Suizhong 36-1 offshore oilfield, model F-P-1-2102-A/B, was designed with a displacement of 550m³/h and a discharge pressure for 1200kPa. In addition, the pumping water can also be used as cooling water for the tank oil.

b. Addition of chlorine

This refers to the sodium hypochlorite solution produced by the electrolysis that is pumped to the injection manifold to kill sea creatures and sterilize the liquid. Figure 3-16 gives a schematic diagram of the chlorine unit. In the unit, water is supplied by a water pump (1) (as shown in Figure 3-16), sodium hypochlorite and hydrogen is produced by electrolysis in a

FIGURE 3-16

The chlorine device in the water processing system.

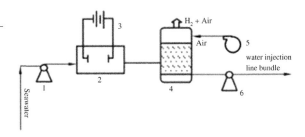

generator (2), the air is blown into the hydrogen removal tank (4) by the blower (5) to dilute the hydrogen, and then the diluted hydrogen is discharged into the atmosphere, and the sodium hypochlorite solution is pumped into water injection manifold by a pressure pump (6). Also, the 220V AC is rectified into a 24V DC for the generator by the rectifier (3).

c. Coarse filtration

This refers to filtering the suspended solids whose particle size is greater than 50 ~ 80mm in seawater. The F-X-731-A/B/C type coarse filter used in the Suizhong 36-1 offshore oilfield, which uses a metal mesh as the filter layer, can filter out 98% of the suspended solids whose particle size is greater than 100mm, and can execute automatic backwashing at a set time, qualitatively controlling the motor rotation.

d. Fine filtration

From the coarse filter the liquid moves into the fine filter, which involves three filter layers under pressure, intercepting 95% of suspended solids whose diameter is 5mm in water, further purifying the water. The three layers of the filter from top to bottom respectively are composed of anthracite, sand, and fine stone. This also can process the backwash operation.

e. Deoxidization

Seawater deoxidization encompasses three methods, namely gas stripping, vacuum deoxidization, and chemical deoxidization, among which vacuum deoxidization, characterized by a deoxygenation tower, is widely used in the offshore field. When the tower is in a vacuum, the oxygen pressure reduces, making more oxygen flash off the seawater into the gas phase, thus reducing the oxygen content in the water. The composition of the deoxygenation tower from top to bottom is as follows: the primary packing section, a grooved water dispenser, a secondary packing section, and a liquid storage section. The packing section is filled with polypropylene rings; when the water flows through it, it can disperse the seawater, significantly improving the superficial area of the water, which is good for oxygen molecules escaping into the gas phase. The water dispenser can maintain different degrees of vacuum between the two packing levels. The liquid storage section is used to carry out chemical deoxidization. Generally when using a vacuum pump, when the tower inside pressure reaches 1.8~5kPa, the oxygen content in water can be reduced to 0.05 ~ 0.1 mg/L. If the chemical agent is added in the liquid storage section and the liquid stays for 3 min, the oxygen content in water can be further reduced to 0.01 mg/L.

f. Pressurization of the seawater

As shown in Figure 3-15, the seawater at the bottom of the deoxidization tower with pressure from the water injection booster pump distributes to the injection wells through the water injection manifold.

g. The injection of chemicals

As shown in Figure 3-15, the deoxidizer and the catalyst are sent into the liquid storage section at the bottom of the deoxidization tower by the pump. The anticrustator and the corrosion inhibitor are sent into the water injection pipeline by the pump, preventing pipeline from scaling and slowing down pipeline corrosion.

3. The equipment and process for the water recycling system

When the sewage meets the water quality requirements with no more processing necessary, the wastewater from the water disposal system, first passing the buffer tank, and then going to the water injection manifolds through the pressurization of the injection pump, is distributed to the injection wells. Figure 3-17 shows the sewage reinjection process at the Chengbei oilfield in the China Bohai Sea. The sewage first enters into the purified water buffer tank (label 1 in Figure 3-17) with a breathing valve on its top, and gas can be injected through the tank's top which prevents oxygen from entering into it. The booster pump (2) pumps the sewage to the injection well, the power fluid system or the sea. The injected water is pressurized to 10MPa by the injection water pump, and distributed to the injection wells by the injection manifold.

4. The process and equipment of underground shallow water injection

Underground shallow water is one of the sources of injection water for offshore oilfield. For instance, there are large reserves of shallow water in the China Bohai Sea, which can be used as an injection water source. At present, most of China's offshore oilfields don't utilize shallow underground water treatment and injection work. Here we take the water injection process of the underground shallow water at the Qikou 18-1 platform in the west Bohai Sea as an example to explain the equipment and the process of using such equipment. Figure 3-18 presents the water injection process on Qikou 18-1 platform.

The salinity of underground shallow water and the amount of suspended solid particles water is higher, generally containing iron and manganese. So the procedure mainly carries out filtration step by step. The processing equipment is similar to the filtering of the seawater treatment process as mentioned above, composed of a coarse filter and fine filter, etc. As shown in Figure 3-18, the underground shallow water is raised from mining wells by the underground water pump, and gradually filters out the solid particles in the water from big to small through a triple filter—the grit catcher, coarse filter, and fine filter. Then the underground water, which is drained slowly, is sent to the water manifold by the injection pumps and distributed to the injection wells.

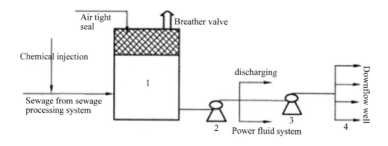

FIGURE 3-17

The sewage reinjection process at the Chengbei oilfield in the China Bohai Sea.

FIGURE 3-18

The shallow underground water injection process in a west Bohai sea oilfield.

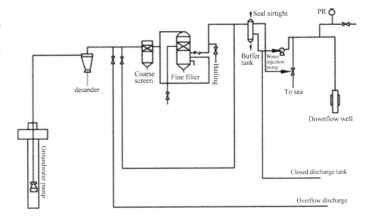

5. The method of mixed water injection

The water source for water injection in an offshore oilfield comes from three places—seawater, sewage and underground shallow water. The mixed water may have the following four combinations:

a. Combination of three water sources—seawater, sewage, and shallow underground water

b. Mixed injection of seawater and shallow water

c. Mixed injection of seawater and sewage

d. Mixed injection of shallow water and sewage

Because setting up three sets of water treatment systems at the same time takes up a lot of space, it is not advisable for the offshore platform (or floating ship). So mixed injection of two water sources is often used for an offshore oilfield. Under the condition of mixed injection, influenced by the water quality, temperature, pressure, etc., the scaling problem is crucial. Therefore, it has been a major focus of developing processing equipment and technology to find a reasonable proportion of two water sources by test and add equipment that injects antisludging agent and antiseptic, which makes the process have a certain amount of antiscaling and anticorrosion. At present there is still a lack of experience with mixed injection in China offshore oilfields, so we don't elaborate further here.

3.3.2 The Oily Sewage Treatment System

The main source of offshore oilfield wastewater is the associated water (the oil-reservoir water) extracted in the process of the oil production, including rain and the wash water of the deck and equipment. Oily sewage contains less iron, a higher degree of mineralization, generally more alkaline, lower hardness, disperse oil, creamy oil, dissolving oil, and some harmful substances that can easily lead to scaling and corrosion. Therefore, we must deal with it to meet the requirements for discharge into the sea or oilfield reinjection water. Table 3-3 gives the quality standards for reinjection water, and the quality standards for discharge into the sea. The oil content in China Bohai Sea is less than 30 mg/L, while in the South China Sea it is less than 50 mg/L.

Table 3-4 Major Oil-Bearing Wastewater Methods in China Offshore Oilfields at Present

Handling Method	Characteristics
Sedimentation method	Natural separation is achieved by the relative density difference between crude oil particles and suspended impurities and sewage; this method mainly used to remove oil and some of the larger diameter dispersed oil and impurities.
Coagulation method	Add coagulant in water, coalesce the small elaioplast to a big one, speed up oil-water separation, and remove small particles of dispersed oil.
Gas flotation method	Inject gas in water, making emulsified oil or tiny solid particles in water attach to the bubbles; they floa to the surface, creating water separation
Filtration method	Use quartz sand, anthracite, filter, and other material to filter sewage, and remove small particles and suspended solids in water.
Biological method	Use microorganisms to oxidize and decompose organisms in order to achieve degradating organisms and oil.
Cyclone method	Gravity differentiation by high-speed rotation; the oil emerges from the water.

1. Oily sewage processing method

   Oily sewage processing methods included a physical method and a chemical method, but a combination of these two methods is often used in the production process, as shown in Table 3-4.

2. Equipment used in oily sewage processing systems.

   Oily wastewater treatment systems adopt a wide variety of devices, such as the oil storage tank on an FPSO or fixed production platform, which can be used as the sediment dehydration unit. Filtering tanks can be divided into filter tanks with pressure and no pressure; we can also choose a vertical or horizontal filter. Table 3-5 presents the major oily sewage processing equipment corresponding to different processing methods.

   Below we selectively introduce the major equipment used in offshore oil fields.

   a. Impeller-type flotation device

   Figure 3-19 gives a sketch of the air impeller-type flotation device including its hydraulic characteristics. It is a sewage purification device that has two fluid channels, namely gas channel A and flow channel B. And it is divided into three areas, namely, the scum area, gas float area and two-phase mixing area, as shown in Figure 3-19(b). Gas from the top of the air float chamber enters into the liquid, which is gas channel A. At the same time, the liquid circulates upward from the bottom of the gas float chamber, which is fluid flow channel B. Liquid mixes with the gas at the two-phase mixing area where there must be enough gas injected in order to make the gas break into micro air bubbles under the action of sufficient shear force. Those bubbles adhere to the oil and solid particles in the liquid at the mixing area. After sufficient contact of the gas and liquid, the oil and solid flocculation constituent with air bubbles attached forms. In the gas floating area, the degree of turbulence must be reduced to a minimum to form a flow regime that is suitable for the rise of the flocculation constituent and removing it from the device. In order to achieve fast separation between bubbles and the oil beads in the gas floating area, coagulant, air floatation agent, and foaming agent are often added.

   b. Plate coalescer

   Figure 3-20 presents the horizontal plate coalescer used in the Wei 11-4 offshore oilfield in the South China Sea. Its interior mainly consists of a sewage distribution pipe, solid particle

Table 3-5 Main Equipment for Different Wastewater Treatment Methods in Offshore Oilfields

Number	Equipment	Main Corresponding Method	Number	Equipment	Main Corresponding Method
1	API rectangular multichannel separator	Sedimentation method	7	Fixed material coalescence filter	Coagulation method, filtering method
2	Settling tank	Sedimentation method	8	Active filter coalescence material	Coagulation method, filtering method
3	Pressurized container flotation device	Gas flotation method	9	Filtering tank	Filtering method, chemical method
4	Impeller type flotation device	Gas flotation method	10	Gravity filtering tank	Filtering method
5	Nozzle natural ventilation flotation cell	Gas flotation method	11	Single valve filtering tank	Filtering method
6	Coalescence plate	Coagulation method, sedimentation method, physical	12	Hydrocyclone	Cyclone method

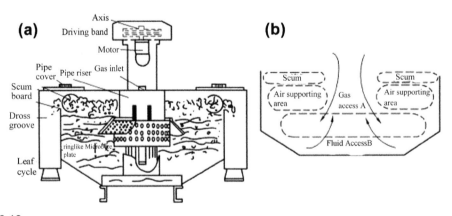

FIGURE 3-19

The impeller-type air flotation device and its hydraulic characteristics.

blocking plate, and eight groups of corrugated plates along the shell axial alignment with the horizontal plane at a 45° angle. When the oily sewage enters, larger solid particles first deposit in front of the solid particle blocking plate, and at the same time the sewage overflows it, and passes through the holes between the corrugated plates to the coalescer's exit. When the oily sewage flows through the holes between the corrugated plates, because the oil is lighter than

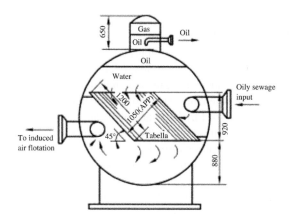

FIGURE 3-20

Plate coalescer for oily sewage treatment.

the water, the oil droplets adhere to the bottom of corrugated plate, forming smaller oil material at first; then the smaller oil material moves up along the corrugated plate, forming larger drops at the top of the plate floating on the sewage surface. Those drops are discharged into the oil trap in the open drain tank through a discharge outlet.

c. Process of oily sewage treatment system

Figure 3-21 presents the oily sewage treatment process for the Chengbei oilfield in the Bohai Sea. Oily sewage first goes through the coalescer, and the flocculants are added before the entrance. Through the flocculation and gravity separation in the coalescer, bigger particles of the crude oil and suspended solid float up, and are skimmed in the guide grooves. The processed sewage enters into the flotation machine through gravity, and by adding the gas for floatation, small oil carriers together with oil beads float to the top, and are skimmed out by the skimming device, while the sewage flows out from the lower export and is pumped into the filter. The filtered sewage enters into the buffer tank, and is used as injection water or power fluid if it reaches the qualified standard, while the rest is discharged into the sea.

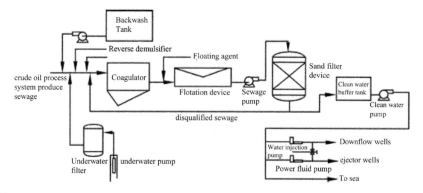

FIGURE 3-21

Process of oily sewage treatment for the Chengbei oilfield in the Bohai Sea.

3.4 Natural Gas Processing in Offshore Gas Wells

Natural gas produced from the marine gas wells usually contains the solid impurities, such as the sand, cuttings, liquid (the water, gas, condensate, etc.) and gas (the water vapor, hydrogen sulfide, carbon dioxide). The solid impurities may result in the attrition and jam of the pipeline and equipment. The water vapor may reduce the delivery capacity of the pipeline, when the water vapor in the natural gas is separated out in the form of liquid water, ice, or natural gas solid hydrate, and also can increase the tube pressure drop in the pipeline, or even block the pipe. In addition, the dissolving of the gas condensate in the gas in the water, as well as hydrogen sulfide, carbon dioxide, and other acidic gases, will also intensify the corrosion of the pipeline and equipment, and some gas impurities can even affect the quality of products and pollute the environment. Therefore, it is necessary to remove the solids in the gas through dust removal and water through the separation method, and to try our best to prevent the formation of natural gas hydrates under conditios of high pressure and low temperature by elevating the temperature of the gas or adding inhibitors. As for removing acidic gas such as H_2S and CO^2, because large devices and a lot of space are needed, this doesn't occur on the sea, and only the natural gas dehydration and reduction of the gas dew point is completed on the sea. Other processing is conducted after conveying the gas to the shore. From the viewpoint of the characteristics of offshore gas fields, the following focuses on the process of natural gas dehydration.

3.4.1 The Process of Natural Gas Dehydration

The dehydration process for natural gas, namely dry processing, reduces the dew point of the gas, and puts the gas at the highest delivery pressure and the lowest environmental temperature it can be at while not meeting the requirements of the dew point of residual moisture in the gas. Usually when the submarine pipeline delivers the gas, the dew point temperature should be $5 \sim 15\,°C$ lower than the minimum temperature along the pipeline, to ensure that liquid water does not appear in the gas delivery process.

Natural gas dehydration methods include the solvent absorption method, solid adsorption method, low temperature separation method, etc., but the solvent absorption method is most widely used in offshore natural gas dehydration. The most widely used solvent absorption is glycol compounds. Figure 3-22 presents the triethylene glycol dehydration process for offshore gas wells.

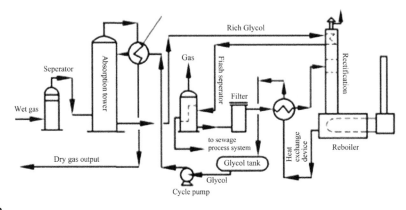

FIGURE 3-22

The triethylene glycol dehydration process for offshore gas wells.

Triethylene glycol's boiling point is higher than diethylene glycol's, and it also can achieve a larger dew point drop. It has low vapor pressure and good thermal stability and a small regeneration loss, so offshore natural gas dehydration uses it as the absorbent. The dehydration process of triethylene glycol is mainly composed of glycol absorption and regeneration. As shown in Figure 3-22, water-bearing gas (greenhouse gas) first enters into the imported separator, in order to remove the liquid and solid impurities carried in the gas, and then into the absorption tower, where the gas flows through each column tray from bottom to top and contacts with the countercurrent of the lean glycol liquid from the tower top; then the water vapor in the gas is absorbed by the glycol liquid. Dry gas after dehydration is discharged from the top of the tower and the rich glycol liquid absorbed water, which outflows from the bottom of the tower, enters into the flash separator after the heat exchange at the regeneration distillation column; the gas is either used as fuel or sent to the safety vent. The hydrocarbon fluid flowing out from the flash separator is collected by the oily sewage treatment system on the platform. The rich glycol solution, which comes out from the flash separator bottom, enters into the regeneration device after it passes the filter and poor or rich glycol heat exchanger. The rich glycol, after being concentrated in the regeneration device, overflows to the reboiler at the bottom, and enters into the regeneration device after cool down. Concentrated glycol is pumped into the absorption tower through an air or glycol heat exchanger, by the circulating pump. The absorption tower and heat exchanger equipment can be installed in the platform danger zone, while the reboiler must be installed on the platform safe zone leeward because it involves direct fire tube heating.

3.4.2 Main Equipment of Natural Gas Dehydration

Two kinds of equipment from Figure 3-22 are introduced in the following.

1. Absorption tower

   The absorption tower is also called the contact tower. This tower when it has a small diameter can use a packed tower style, and with large diameter should adopt the bubble column of the column tray type. The bubble column of the column tray type is commonly used on a sea platform, and usually the column trays number about $4 \sim 6$, with plate spacing about 0.6 m. The diameter of the bubble column D can be calculated according to the following equation:

$$D = \left[\frac{1.27G}{C}\right]^{0.5} \bigg/ \left[(\rho_1 - \rho_g)\rho_g\right]^{0.25} \tag{3-3}$$

   where G is mass flow rate of the treated gas, kg/h; ρ_1, ρ_g is the density of the liquid phase and gas phase in the absorption tower, kg/m3; and C is a constant (when the tray spacing is 0.61 m, take $C = 500$, and when the plate spacing is 0.76 m, $C = 550$).

   The circulation volume of the triethylene glycol solution in the absorption tower is large, which is good for dehydration, but the operating cost will increase. The circulation volume of the triethylene glycol commonly used on an offshore platform is $20 \sim 40$ L per one kilogram of abjection water.

2. Regeneration device

   In a triethylene glycol dehydration device, the regeneration system of the ethylene solution is composed of the regeneration distillation column and the reboiler. When the requirement for the gas dew point drop is not high, the regeneration distillation column operates under normal pressure; if the requirements are greater, the system should adopt vacuum recycling or stripping regeneration. The rich liquid of triethylene glycol generally enters from the middle part of the regeneration distillation column, and a packing or column tray can be used in the distillation column.

Vacuum regeneration is the process that connects the rectify pillars with the vacuum device and creates negative pressure in the distillation column to further evaporate water in glycol. A stripping regeneration setup installs a barren solution stripping column at the barren solution outlet of the reboiler, which can increase the contact surface of the barren solution of the triethylene glycol and the stripping gas from the reboiler, reduce the water vapor partial pressure on the surface of the solution, and improve the concentration of the triethylene glycol.

The heating modes of the reboiler of a regenerating triethylene glycol device includes direct burning heating, hot oil heating, electric heating, waste heat heating, etc., but in offshore platforms the waste heat of the power station is often used.

3.4.3 Gas Emptying Torch

The natural gas of the offshore platform and oil vapor from the process, except those gases used platform fuel, will endanger platform security and cause serious pollution if vented into the air. Therefore, effective measures must be taken to deal with this issue. The torch is a necessary safety facility to safely dispose of the vent gas; it is a open-air burning method for handling the exhaust.

1. Offshore platform torches

    Torches used on offshore platforms fall into four categores according to their structure:

    a. Cantilever tower torch

      This is a leaning tower structure, stretching out from the edge of the offshore platform, generally with the horizontal angle of 15° to 45°, and a torch arm length of about 30 ~ 60 m. On offshore fixed platforms that use this type, the specifications are generally a horizontal angle 15°, arm length 15~45m, and discharge capacity of about 110 x 104 ~ 150 x 104 m³/d.

    b. Vertical tower torch

      Due to the vertical structure, it is easy for the liquid in the emission gas to fall on the deck, causing fire, so this is rarely used on an offshore platform.

    c. Independent platform torch

      This is a torch where the burner is supported in a single platform that can connect with the main platform using a trestle of 100 m. Obviously, the investment on this type is higher.

    d. Ground torch

      This type burns gas completely in a cylindrical combustion chamber, without flame light, as shown in Figure 3-23. At the inner wall of the combustion chamber, there is a refractory layer with high-temperature-resistant material and a main burner or multiple small burners along the inner wall. This torch type is widely used in FPSOs.

2. Basic calculation parameters for a torch device

    a. Cylinder diameter d of the torch head

      According to the regulations of API RP-521, the cylinder diameter d should be determined by flow Mach number Ma:

$$Ma = (11.61)(10^{-2})\frac{w}{pd^2}\sqrt{\frac{T}{KM}} \tag{3-4}$$

where Ma is the Mach number, that is, the speed ratio of the gas flow in the cylinder body and transmission of sound in the torch; ω is gas discharge mass flow, kg/s; p is the gas pressure in the

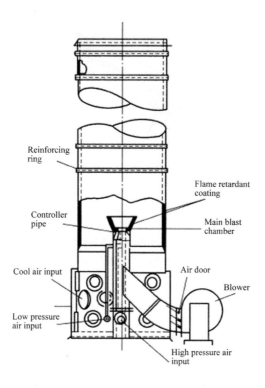

FIGURE 3-23

Structure of ground torch.

torch head, which is taken near the exit, kPa; M is the relative molecular mass of gas; and K is the gGas adiabatic index, $K = c_p / c_v$.

b. Torch arm length H

Figure 3-24 shows the relationship of related geometric parameters when calculating the torch arm length H along the axis. In the diagram, L is the flame length, θ is the offset angle of the flame axis with wind effects, and D is the distance efrom the flame center to the torch arm bottom point B. When the flame root is blown away from the cylinder top a certain distance a, we have

$$H = \sqrt{D^2 - \left(\frac{1}{3}L \sin \theta\right)^2} - \frac{1}{3}L \cos \theta - a \qquad (3\text{-}5)$$

3. Equipment of torch system

a. Torch head

This is the main piece of the torch system. And through it the gas vented is burned. At sea, the simple torch heads, smoke free jet or blast torch heads, and multi-tube torch heads are commonly used.

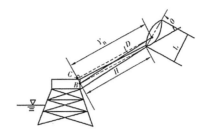

FIGURE 3-24

Geometric parameters related to torch arm length.

b. Ignition system

This provides endless flame at the end of the torch head, and usually consists of a system that manipulates the torch at the trachea (eternal lamp), a flame generator, and so on, forming a remote control ignition device.

c. Torch gas tank

In order to make sure that the gas going into the torch doesn't contain much liquid, and to avoid forming a "rain of fire" and endangering platform security, the torch gas tank should be installed at the front of the torch device.

PART II PRODUCTION FACILITIES FOR OFFSHORE OIL AND GAS FIELDS

The production facilities for offshore oil and gas fields are established for the oil and gas processing. This part will put forward some requirements for the production facilities and describe different facilities.

3.5 Requirements for Offshore Oil and Gas Production Facilities

Offshore oil and gas production facilities are different from land facilities, and the special requirements are discussed in the following

3.5.1 Requirements of Ocean Environment Loads

There are not only the loads of the wind, sea current, and waves at sea, but also the load of the sea ice in the ice zone and the seismic force in the seismic zone. Because these marine environmental loads are stochastic, the design and construction of offshore oil and gas production facilities shall meet the requirements of the intensity and stiffness under the random loads.

3.5.2 Requirements of Safe and Reliable Production

Because the space for offshore production facilities is small and the equipment and persons are relatively concentrated, the probability of both accidents and damage is large. So the requirements for the design of the deck layout of the production equipment and the safety and reliability of the structure are higher.

3.5.3 Marine Environmental Protection

Protection of the marine environment is clearly defined in international and Chinese laws. For example, there are strict requirements for crude oil leakage and the sewage pollution on the sea. So dealing with the sewage, emissions, and gas venting should be considered during the design.

3.5.4 Requirement of Efficiently Investing in Equipment

The size of the offshore production facilities determines the scale of the investment, but the size, efficiency, degree of automation, and general layout influence the scale of production facilities. It is necessary to act efficiently when selecting and arranging equipment. For example, a central control system, which can intensively control offshore oil gathering and transportation can reduce the scale of production facilities, and wellhead platforms without workers can meet requirements.

3.5.5 Self-Sustaining Capability

Because offshore production facilities are far from land, a complete supply system must be established to meet the requirements of production and life. At the same time, offshore production auxiliary facilities must also own the self-sustaining capacity which ensures normal production and people's safety for 7-10d. So in order to meet these requirements the ship's berthing and lifting equipment, the layout of the helicopter apron, and the arrangement of pumps and the storage warehouse should be appropriately designed.

3.5.6 Independent Power Supply

Offshore production facilities should own a perfect independent power supply system, which can generate the electricity by itself and continuously supply power. For example, a gas turbine driven generator can be established on the platform with electricity transferred to the power plant through the switchboard. In the platform group electric supply travels through submarine cable, and generally each of the platforms should have backup generators. In addition, there also are the emergency generators on the wellhead platform.

3.5.7 The Reliability of the Communication System

The communication system ensures that offshore production facilities can communicate with the outside world or contact each other, so establishing a reliable, excellent communication system must be considered an important task.

3.6 Classification of Offshore Production Facilities

There are many types of offshore production facilities, and they basically can be divided into the following three categories.

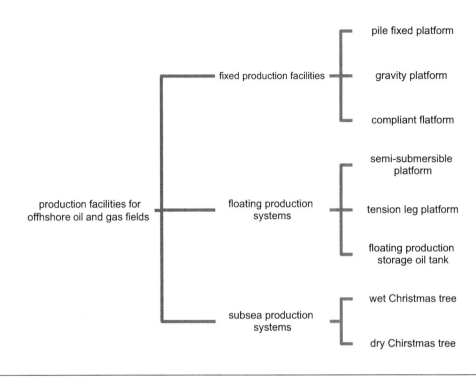

3.7 Fixed Production Facilities

The fixed production facilities are the structures that are fixed at the sea floor by a series of fixed methods such as a pile foundation and bottom supported foundation, and have certain characteristics of bearing capacity and stability. According to the purpose, the fixed production facility consists of the wellhead platform, production platform, storage platform, life power platform and integrated platform which integrates drilling, production processing, life. And according to the types of platform structures, the fixed production facilities can be divided into the pile foundation platform, gravity platform, artificial island and compliant platform. They will be introduced separately as follows by the structural types.

3.7.1 Pile Fixed Platform

The pile fixed platform is commonly a steel fixed platform, as shown in Figure 3-25. It is a popular form for the offshore oil production.

1. Structure of the steel fixed pile platform

   The jacket platform is the most popular one for the steel fixed platform. It mainly consists of four parts: the pile, jacket, jacket cap, and deck. Because the jacket cap and the deck module usually become one component, this becomes three components, as shown in Figure 3-26.

   a. The jacket

   This is a steel truss structure that is welded by low alloy steel pipe with a large diameter and thick wall. The main column of the steel truss structure is called the thigh. Because it is the

FIGURE 3-25

Fixed pile platform.

guide pipe during pile driving, this truss structure is called the jacket. There are 3, 4, 6, or 8 roots, according to the size of the upper part of the platform and the depth of the water. Between the thigh jacket, the horizontal transverse bracing and syncline are supported as reinforcement to transfer load and to add strength and stiffness.

b. Pile

The jacket is fixed at the seabed by piles. And the piles are divided into main piles and cluster piles. Main piles are driven into seabed through the inside of the thigh. The cluster piles are arranged around the bottom of the jacket.

c. Jacket cap

This is a part of the structure above the guiding pipe rack and below the deck module, belonging to the transition structure between the jacket and the module.

d. Module

This is also called the block, and the deck is composed of various modules. The platform can be a single deck structure or a multideck structure depending on the size of the platform. The drilling area is called the drilling module; the oil processing area is called the production module; the mechanical power region is called the power module; and the living area is called the life module.

2. Construction of the steel fixed platform

The construction of the jacket platform is a complex process, and it is often divided into land fabrication and offshore installation. Onshore various modules are precast during assembly of the

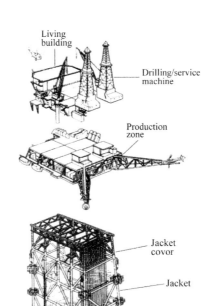

FIGURE 3-26

Composition of the steel jacket fixed platform.

steel structure and equipment and facilities, similar to a common metal structure. The construction method is different for offshore operations.

a. Lifting installation

The construction operation program is shown in Figure 3-27. A jacket is shipped to the location using the floating method, and then the jacket is dipped to the bottom using a floating crane; the

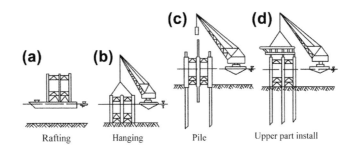

FIGURE 3-27

Lifting method procedure.

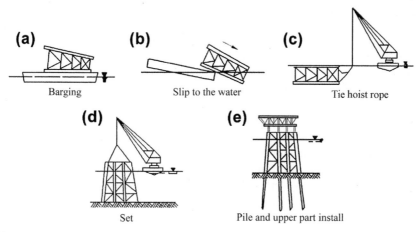

FIGURE 3-28

Operation procedures for slipping method.

ship pile frame legs are sent into the bottom of the sea with concrete and a steel pile leg between annulus consolidations. This method mainly uses a construction crane (crane), which limits the lifting capacity and lifting height; the equipment should be decomposed into several pieces.

b. Sliding installation

The sliding installation is shown in Figure 3-28. Firstly seal the top of the jacket well, and then use the barge with subaqueous slipway to carry the jacket to the site. At the scene, tilt the barge to slip the jacket into the water along subaqueous slipway, and then tie the hoist rope to the float jacket because of sealing the jacket. After that, eliminate the seal and pour water into the jacket, and then install the jacket at the seabed under the help of the derrick barge. Last, drive piles and install the upper blocks on the top of piles. Because the jacket bears a big bending force [Figure 3-28(b)] during the process of sliding the jacket into the water, the jacket structure should be strengthened appropriately.

c. Floating installation construction

The installation program of for the floating method is shown in Figure 3-29. Both ends of the jacket tube should be sealed in advance, and it floats in the water by its own buoyancy (when

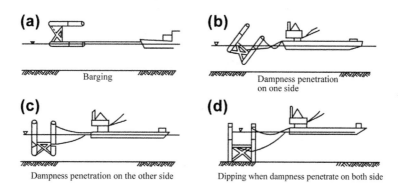

FIGURE 3-29

Operation procedures for floating method.

required we can add a buoy). The tug tows the jacket until the well, and then we irrigate into the pipes. One side of the jacket comes into the water and then the other side, so that it is evenly filled with water to sink to the bottom of the sea.

The use of these methods should be based on the specific conditions.

3. Use of the steel fixed platform in offshore oilfields

Because the oil and gas production facilities are different, the use of the platform is different and can be divided as follows:

a. The wellhead platform

This is the steel fixed platform for wellhead production and generally, a certain number of x-trees are installed on the platform, and fluid from wells through the x-tree is measured by the measurement system on the platform. So the wellhead platform does not handle the processing of the crude oil production system and power system, but system support and utility systems must be installed, and even a staff building; in addition the wellhead platform shall be provided with a rig. Figure 3-30 shows a typical wellhead platform in China.

It is worth mentioning that independent designed Chinese wellhead platform with no one guarding achieve the "five no", namely no living floor, no main power station, no production separator, no water treatment facilities, and no seawater lift pump. At present, this wellhead platform has been successfully applied in Bohai.

b. Production platform

This is also called the central platform. It is a large platform that can include the crude oil processing system, technology system, public system, power system, and staff area. It has the ability to process fluid, and also monitors the power, water injection and wellhead platform operations functions.

FIGURE 3-30

The first wellhead platform in China.

The production platforms are different because of different combinations of functions. For example, there are several combination schemes such as the platform with the functions of production, life and power, the platform with the functions of drilling, production, life and power, and the platform with the functions of life and power and so on.

The production platform brings together conveying fluid from the wellhead platform. After oil, gas, and water are separated from the crude oil processing system, qualified oil is put into the oil storage facility storage; gas can be used for power generation, gas lift, and in the furnace. Water after treatment can be qualified and can be injected into a reservoir or discharged into the sea.

A typical production platform is shown in Figure 3-31.

FIGURE 3-31

A typical production platform in China.

c. Storage platform

This is a crude oil storage tank platform; qualified crude oil store in the center platform is carried away by a shuttle tanker, but due to high investment for this platform, it is not commonly used. Figure 3-32 shows a maritime docks and storage platform in China.

4. Advantages and disadvantages of steel fixed platform

a. Advantages

The construction and installation technology are mature, and there is less investment and high reliability. The use of this platform is more economical in shallow and deep sea areas; it has good stability, with less influence from the sea, and is smooth and safe.

b. Disadvantages

With the increase of water depth, investment increases significantly. Because it's ability of suiting water depth is limited and can't move, the repeated utilization rate is low. And it takes a long period to construct and install it. However, offshore installation is very complex. In addition, its flexibility of matching the oil field production schemes is bad.

5. Development direction of steel fixed platform

Judging from the overall development trends, the quantity of this type of platform has increased year by year in the world. For example, from 1950 to 1970 the quantity increased from 40 to 1000, reaching 3800 in 1986; in the early twentieth-first century the use will double. The direction of its development is as follows.

a. Depth is increasing

The maximum water depth was 144 m in 1975, 259 m in 1976, in 1978 the maximum depth of the Cognal platform had reached 313 m and the maximum depth continues to increase year by

FIGURE 3-32

Maritime docks and storage platform.

year. At the beginning of the twentieth-first century, the maximum working depth of the Bull Winkel steel jacket fixed production platform had reached 412 m.

b. Production capacity is expanding

The number of well drilling platforms is expanding continuously as drilling capability increases; the quantity at the beginning was 8 to 10 wells, then 16 wells, 36 wells, and at present a platform can be drilled with 60 wells. The Gilda offshore oil field even has 96 wells in the USA. In the past platform areas were about 1000m² and now the maximum area of the Harmony fixed platform has reached 10730 m², an increase of 10 times. The production and processing capacity of crude oil for a large platform is 140000-280000 m³/d (Harmony and the Heritage platform in USA), and production capacity is continuing to expand now.

c. Structures are more and more diverse

In the recent years the structure of the steel fixed platform has developed very quickly. And another new structures come out in addition to the traditional structures, such as the jacket which has four thighs and two layer decks and can drill 6 ~ 12 wells with the depth shallow to deep, and the jacket which has 8 or 12 thighs and consists thousands of conductor casings and two drilling rigs.

i. The light fixed platform

This is mainly used to meet the needs of marginal oilfield development; lightweight platforms have a single leg, two legs, or three legs with different structures.

The jacket platform using a single leg is supplemented with water diagonal support, thereby reducing the amount of steel and offshore construction and also the cost. There are two basic types of single leg light platform.

The first type is shown in Figure 3-33, using a riser for the legs and underwater adding a diagonal brace for the lower structure; the bracing is fixed on the seafloor by the pile, which can have two, three, or four diagonal bracings and so on. There are one to three layers of the upper deck of this platform and it can be used as a single well or with four to six wellhead platforms and a central platform.

Instead of a riser for legs, the second type uses a larger diameter (3 m) single leg or leg column ring set on the riser as the support platform. The lower leg is located above the mud line, using three or four diagonal bracing and cross bracing on the seafloor, and Figure 3-34 is the first application of the single wellhead platform with three pile legs in China.

ii. The assembly type fixed platform

A steel jacket fixed conventional platform is usually constructed together. In order to facilitate maritime transport and recycling an assembly platform is sometimes used, with the jacket divided into a number of segments and assembled. Generally it is divided into two sections, such as America's Hondo platform. Some are divided into three sections such as America's Cognac platform, which is connected together through the jacket leg of 1830 mm steel inner casing, with cement poured to connect and form a whole body. The whole assembly platform is composed of the jacket, processing equipment, and deck module, and generally it can drill 9 to 62 wells.

For the marginal oilfield and the area where reservoir reserves has not be proven, the assembled jacket fixed platform has great advantages. It does not need to be built using a large-scale platform factory. Using the modular structure can improve production and reduce cost. In addition, the jacket can adapt to different depths by increasing or

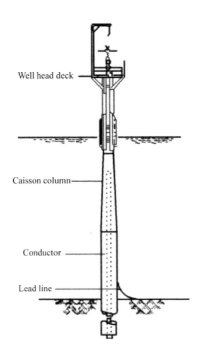

FIGURE 3-33

Single leg column of light platform.

FIGURE 3-34

China's first light wellhead platform with three pile legs.

decreasing the assembled segments of the jacket. Brazil has built three models (A type, B type, C type) of assembled platform and can adapt to the development of shallow and medium depth oilfields.

iii. The bucket foundation platform

This is named the bucket foundation platform because its shape is similar to an opening inverted bucket (Figure 3-35). In fact it is buried shallow, and has a greater length to diameter ratio (1 or higher) and short pile foundation compared with the traditional pile foundation. Compared with the traditional pile foundation, the main difference is that the sinking technology use pressure difference, instead of using the penetration method. There is also the very big difference in the foundation bearings.

The bucket foundation platform is developed using the base of the traditional pile foundation platform. On the beach, the steel for the pile foundation platform is half the cost of the platform structure, while piling and installation costs account for the other half of the total cost. With development in the deep sea, the cost not only increases sharply with the water depth, but also piling technique difficulty increases. Therefore, in order to solve the problem of the high cost of pile foundation platforms, as well as the long construction period and complex construction technology, the bucket foundation was developed. As early as 1996, Norway successfully established a wellhead platform of the bucket foundation in 2000 m water depth. And in 2000, China established four tethered platforms of multicylinder bucket foundation at the China Bohai sea JZ9-3 oil field. In 2002, the bucket foundation is applied to the FPSO system in 80 m water depth at the WenChang oil field in the South China sea.

The foundation of bucket foundation platform is a steel structure steel bucket, its roof and the legs of the jacket are fixed, connected steel pile, and it has a cylindrical skirt roof with an open bottom.

The key technology of the bucket foundation platform is the sinking technology. The methods of sinking the bucket foundation consist of the jetted method, diversion sinking method and negative pressure (suction) sinking method. And the negative pressure sinking method is the most common method at present. The pressure difference between the inside and outside of the bucket forms by discharging out the water in the bucket, and then it makes the bucket overcome the frictional resistance of the soil around the bucket to penetrate into the seabed. The exact method of calculating the soil

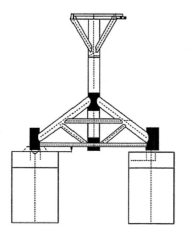
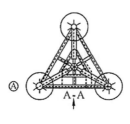

FIGURE 3-35

The bucket foundation platform in the Bohai SZ9 - 3 oilfield.

resistance outside the bucket has not been presented for years because of the complexity of the multi-phase medium interaction of solid, soil and liquid and the uncertainty that flow field influences the density and strength of the soil around it. In addition, the ways of bearing capacity between the bucket foundation and pile foundation is different. Rather than depend on the extrusion and friction between the outside bucket and soil, the large area at the top of the bucket can also provide a part of bearing force. And the adsorption force is also the advantage of improving the bearing capacity. However, it is known that both seabed soil scouring and liquefaction of sandy soil under certain conditions decrease the bearing capacity. In short, these uncertain factors make it very difficult to determine the bearing capacity of the bucket foundation. So it is highly necessary to solve these problems further in the use of the bucket foundation platform.

Of course, if the characteristic that the bucket foundation can be pulled out is used by changing the way of discharging water out the bucket into the way of pumping water into the bucket, the bucket foundation can be recycled and the economic benefit of the project can be improved further.

3.7.2 Gravity Platform

The gravity platform is different from other platforms; it does not need to insert a bottom pile to bear the vertical load, and completely relies on the stability of the weight at the bottom of the sea. According to different construction materials, they can be divided into three categories.

1. Concrete gravity platform
   Concrete gravity platforms on the shore and in the shallows have a long history, but it has only been used in the sea since the 1970s. The concrete gravity platform structure is shown in Figure 3-36, and it consists of a mat, deck, and column.
      i. The mat (base): This is the foundation of the platform. In order to resist the wind, waves, and flow, the mat has a large volume and can be used to store oil.
     ii. Deck: This provides working space for production and can be used for installing processing equipment, power, and various productions facilities.
    iii. Column: Columns are between the mat and the deck and is used to support the deck. The Kandipu concrete gravity platform in Figure 3-37 is a multifunctional platform that integrates drilling, production and storage. It consist of the base of cluster tanks(19) circular cylinder, columns and steel decks. It adapts to depths between 120 m and 140 m, and it can be used in areas with maximum wave height of 30 m. The overall weight is 350 t. The number of columns are three and they are slender at the top, and used to support platform's deck. Sixteen cylindrical foundations in the base's cylinders are used for storage, and one of three columns is used for the ballast tank to balance the platform. The circular cylinder column foundation uses a concrete shell structure, and the cylinder wall can bear great pressure and prevent oil leakage. A reinforced concrete apron and steel apron under the foundation are inserted into the sea, which can prevent horizontal slip of the platform; The sand is placed in three of nineteen cylindrical columns as the ballast to increase overall stability at floating and sit-on-bottom states. Two rigs, a helipad, and living rooms are installed on the deck. Two rigs run drilling tools through the two columns. This can help prevent the influence of waves and currents. Drilling pipe can be arranged according to the number of wells being drilled. Diesel generator sets, pumps, and oil storage and ballast system control devices are installed in the

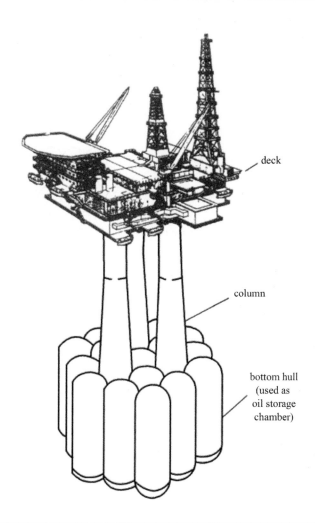

deck

column

bottom hull
(used as
oil storage
chamber)

FIGURE 3-36

Concrete gravity platform.

column of the ballast tank. 16 cylinders can store oil up to 140 to 200 t. The columns are divided into four groups, which can store the oil and discharge oil at the same. And the oil can flow into cylinders by gravity and be purified by the precipitation. After that, the oil is transferred to the shuttle tanker.

Figure 3-38 is the Selmer three-leg concrete gravity platform. This concrete platform has no mat structure, and a total of three columns gradually expand downward. Three columns are connected into integration at their bottom with horizontal transverse concrete box beams, forming three basic areas. Columns are connected at their top by the support structure of concrete slab. If it needs, one or three concrete columns can be installed on concrete box beams to support the deck or be as the wellbore. Three columns have a ballast tank, in order to ensure stability and installation of the air drag.

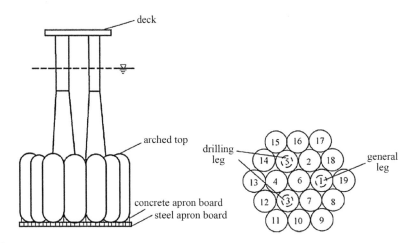

FIGURE 3-37

Kangdipu concrete gravity platform.

The spacing of the three columns can be 60 ~ 80 m; when it is 80 m, the deck area is up to 7000 m², with a towing deck equipment weight of up to 5000 T. As the column spacing becomes larger, the deck safe area and hazardous area can be completely separated. At the same time, three basic intervals also maintain a large spacing, easy to sit at the end of the water discharge under the foundation, drilling guide on the post, reducing the wave force; oil, gas, water treatment system and storage system and pump in the post, capacity of up to 200 ~ 270t; the oil outlet pipe and the J pipe are arranged in the column within, not only decreasing the wave force, and making pipeline connection convenient. The Selmer platform working depth is up to 300 m.

a. Construction

Construction of a concrete platform is quite complex, because it does not use the conventional construction method of the similar pile foundation platform. Figure 3-39 shows the construction process of the concrete platform; the steps are discussed in the following.

The first step: As shown in Figure 3-39(a) the lower half is constructed in dry dock, with the dry dock using a cofferdam construction prefabrication site. Because the heavy bottom area is too big to construct on a general dock or slide, there must be specifically a dry dock construction mat for the bottom half.

The second step: When construction reaches a predetermined height water is put into the construction site and becomes the same as sea level (it should be fixed to the mat inside by ballast water at this time), the dock gate is opened, the ballast water is excluded, and we tug the mat from the dry dock out as shown in Figure 3-39.

The third step: The mat will be towed to the construction area with a certain water depth. After it is anchored on the sea, the upper part is constructed using the slidingcmethod with the construction mat, as shown in Figure 3-39(c).

The fourth step: As shown in Figure 3-39(d) we continue and cast the heavy column pad with sliding construction.

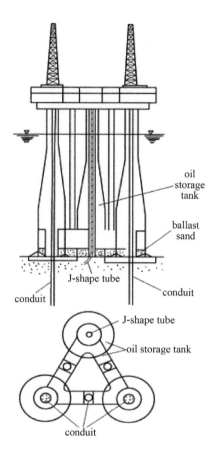

FIGURE 3-38

Selmer three-leg concrete gravity platform.

The fifth step: Using the tug, we drag th mat and the column into the deep sea to install the deck, as shown in Figure 3-39(e).

The sixth step: Send ballast water to the mat, pouring in the water until the water is at the upper part of the upright post, then install the deck, as shown in Figure 3-39(f).

The seventh step: Install various modules in the deck, as shown in Figure 3.39(g).

The eighth step: Discharge the ballast water, making the structure rise, and tow to a predetermined location, as shown in Figure 3-39(h).

The ninth step: After the platform is installed, inject water into the mat, s that the platform is installed on the seabed accurately, as shown in Figure 3-39(i).

c. Advantages of concrete gravity platform

　i. Save more steel: Under the same conditions, the Kangdipu platform uses about 1.2×10^4 ~ 1.5×10^4 t of steel. The steel jacket offshore platform is about 2.0×10^4 ~ 4.0×10^4 t.

　ii. Reduce cost: From the Beihai Norway oilfield, the cost can be reduced by about 20% with a jacket platform.

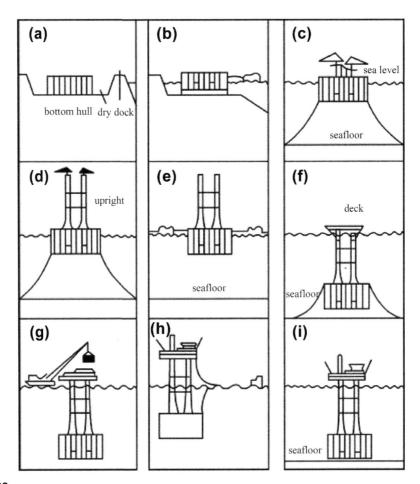

FIGURE 3-39

Process of concrete platform construction.

      iii. Offshore construction is simple: All construction work is performed in dry dock and at shore; we can installation requirements in deep water and we do not need seabed piling, saving a lot of offshore construction.

      iv. Anticorrosion and fire safety: Corrosion, fires, and explosions are handled well. Drilling can be worked through the post. The safety and reliability of production is improved.

      v. Longer service life: Compared with a steel jacket platform the life is longer and repair work is lowered.

    d. Disadvantages of concrete gravity platform

      i. The structure design requires a lot of analysis.

      ii. High requirements for the site and construction: The dry dock and the construction site with certain depth waters are needed. In addition, a large amount of wood template is also needed during the platform building.

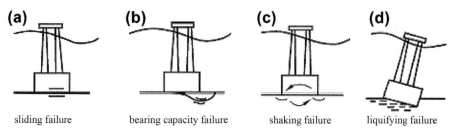

(a) sliding failure **(b)** bearing capacity failure **(c)** shaking failure **(d)** liquifying failure

FIGURE 3-40

Four forms of destruction of concrete platform foundation stability.

iii. The requirement of towing the whole platform is high: Compared with a steel jacket platform, the towing requires high safety and high reliability.

iv. High requirements for basic stability of the platform: Basic stability is an important factor in addition to the intensity of concrete platform. There are four main types of stability failure of a concrete platform as shown in Figure 3-40: Figure 3-40(a) shows foundation damage caused by the horizontal thrust of wind, waves, and flow, causing sliding; Figure 3-40(b) shows damage caused by horizontal and vertical loads which leads that the pressure of foundation's one end is bigger than the soil bearing capacity. Figure 3-40(c) shows shaking damage of foundation's two ends caused by variable loads-wave force, sea ice force etc. which cause foundation vibration. Figure 3-40(d) shows liquefaction damage caused by the disappearance of silt soil shear resistance under the action of wind, wave and earthquake.

v. The high performance requirements required in icy seas.

2. Steel gravity platform

The steel gravity platform as shown in Figure 3-41 is built by Italy. Its work water depth is 90 m. And it mainly consists of caissons supporting frames and the deck. And caissons can also be as

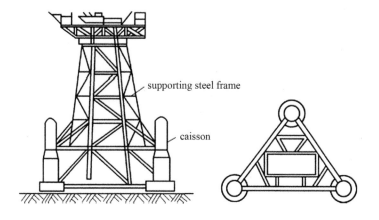

supporting steel frame

caisson

FIGURE 3-41

Steel gravity platforms built by Italy.

storage tank. During construction, the first three parts are prefabricated and combined into a whole on the shore, and then sent to the well.

a. Advantages: When being prefabricated it does not have to be in deep water; it requires a low amount of power during towing; and the bearing capacity requirements of the foundation are not high during use. Although the caisson storage is small, the overall economic benefit is higher than a concrete platform.

b. Disadvantages: Corrosion resistance, insulation, and oil storage capacity are lower than for a concrete platform, and steel usage is higher.

3. Mixed gravity platform

This refers to the mix of a concrete and steel jacket. This hybrid platform can be used in deep water. Compared with the concrete gravity platform, it has light weight, low cost, and convenient construction and installation; the mixed gravity platform can be divided into two types as follows.

a. Mixed type tower platform

Figure 3-42 shows a mixed platform in the Norway Beihai oilfield, where the depth of water is 310 m. The lower half of the concrete structure includes the base foundation and caisson on the foundation; while the upper part is the steel structure, namely a jacket which supports most of the deck, avoiding the use of a cantilever supporting structure. The caisson has seven cylindrical tanks, whose top is arched, and the tanks can be used for storage. Half of the upper and lower parts use temporary hinges that are connected to build the platform; the steel structure of the upper part of the first level is horizontally prefabricated, and then rotated through the hinge 90 degrees to an erect position.

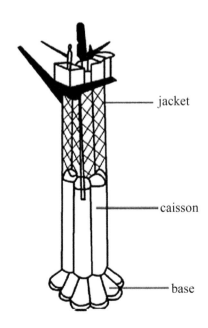

FIGURE 3-42

Mixed type tower platform in Beihai oilfield.

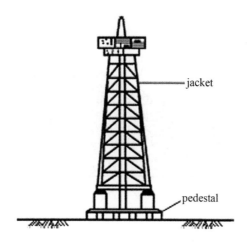

FIGURE 3-43

Gravity base jacket platform designed by TPG and LJA.

   b. Gravity base jacket platform

     Figure 3-43 is hybrid platform, composed of a jacket and concrete foundation above it, used in Beihai and on the east coast of the United States of America, and its working depth is up to 300-500 m. The bottom of the platform is a concrete base, which has concrete columns strengthened with steel connecting with a steel jacket above it. When the construction of concrete column reaches 2-3 m, first the foundation and concrete column should be installed, and then they are stood up and connect with the jacket. Compared to the hybrid platform and pile jacket platform, this can save on installation time. Also, because of its strong rigidity, it has good dynamic characteristics as well.

3.7.3 The Artificial Island

The artificial island is built in the sea, and the rig, oil and gas production equipment, public facilities, storage tank, and unloading terminal can be set on the island. The artificial island, categorized by wall forms, can be divided into a slope-type artificial island and caisson-type artificial island.

1. Revetment artificial island

   As shown in Figure 3-44, this is composed of gravel, with sandbags or stone pitching. When built, first the barge with a bottom opening sends gravel through the opening to the island, and then stacking sandbags makes it stand slightly out of the water, forming an underwater cofferdam, and finally we use gravel to fill the island body.

2. Caisson-type artificial island

   This is usually divided into a steel caisson enclosed-type artificial island, reinforced concrete caisson enclosed-type artificial island, and mobile polar caisson artificial island. The characteristics of the caisson-type artificial island is that a whole caisson or a plurality of steel or reinforced concrete caisson is in a closed loop, and the middle of the island is built with backfill sand. With regard to construction, the caisson in continental Shangxian was prefabricated and floated to the site for installation; the caisson can be adjusted for different underwater seabed heights, applicable to

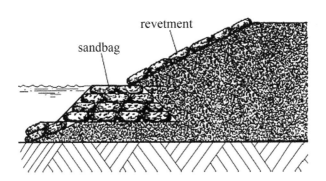

FIGURE 3-44

Revetment artificial island.

different water depths. When the artificial island is no longer in use, excluding the ballast water makes the caisson float up and then it is dragged to another place for repeated use.

The steel caisson enclosed artificial island has been used successfully at a water depth of 26 m, as shown in Figure 3-45.

3.7.4 Compliant (Flexible) Platforms

The compliant platform is also called the flexible platform, and is a kind of fixed platform under marine environmental loads. The platform can move a certain angle around the fulcrum to swing in the permitted range, which is why it is called a compliant platform. This platform is a slender steel jacket

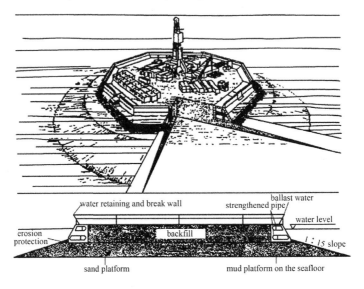

FIGURE 3-45

Steel caisson enclosed artificial island.

structure, and the cross section along the height direction does not change. Well slots are arranged in the middle of the platform, and some flexible platforms use a number of root piles for support at every angle. After piling through the conductor casing, the pile top shall be higher than the height of the mud line, and the top of the conductor casing is at a half of the platform height. Cement slurry is poured into the gap between piles and conductor casings, and after solidification piles and conductor casings are integrated into the assembly. The jacket is installed at top of the assembly. Such a large length provides enough elasticity to produce flexible resilience, and by adjusting the body length, different structural parameters of the system environment can be adapted, and there are some flexible platform that use guy rope (cable) or flotation to produce resilience.

1. Types of compliant platform
  a. Guy rope tower platform
     This kind of platform has a base on the seabed, but it is fixed by the lateral tension of the anchor guy rope which is close to the surface and connected with the platform. That is the reason why the platform is called the guy rope tower platform. Figure 3-46 shows the composition of the guy rope tower platform structure; the base is a closed truss reinforced case. In the shallow water, it is not necessary to pile for fixing the base. But if the water is deeper than 300 m, the base is fixed by piling and the depth of the piles' penetration is about 140 m.

     The guy line is composed of galvanized steel wire, with diameter of about 12.7 cm and generally there are 16 ~ 24 for different setups according to the water depth. The tower is a steel truss; the cross-section can be square, rectangular, or triangular, and the cross-sectional area generally is different. The deck is used in the drilling and production process, with the area and number required varying with the number of wells and workers. This platform is suitable for deep water, whose economic depth is 240 ~ 480 m; the maximum water depth is 600 m, and the Exxon Company has put it into use in Mexico.

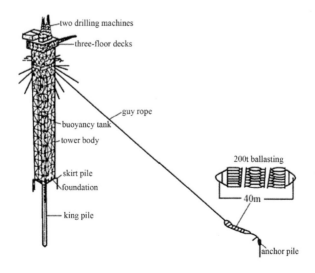

FIGURE 3-46

Guy line tower platform.

b. Bottom fixed strap tower platform

Its structure is characterized by a guy line, and it also will have a fixed base for the tower body on the seafloor, as shown in Figure 3-47. Main piles fixed on the base bear the vertical load, and the surrounding short pile and catheter are used to transfer torque and shear. Because the fixed base is adding a rigid torsion spring and it is short, the rolling period is less than the guy line tower platform and the use of steel is less when the water is very deep.

c. Floating-tower platform

This is composed of a flexible tower body and a floating box on the upper part of the tower body with the base fixed on the bottom of the pile as shown in Figure 3-48. Floating boxes of this platform can not only counteract the weight, but also generate restoring force. Compared with the guy line tower platform, we don't need guy lines and we reduce the axial pressure on the structure. Because there is no stretching tight rope, the base shear requirement is high and because of the increasing use of floating boxes, the use of steel is increased.

d. Bottom fixed floating-tower platform

The characteristics of the platform are the axial tube and a guide. The axial tube is positioned inside the body and there are three flexible tower bases in the four corners of platform for a

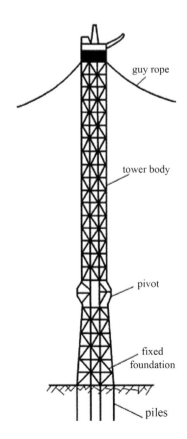

FIGURE 3-47

Bottom fixed strap tower platforms.

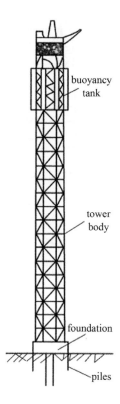

FIGURE 3-48

Floating-tower platforms.

total of 12; the diameter and wall thickness should be based on the sea and ocean environment loads. The upper axial tube is connected with the tower, and the lower is connected with the fixed base. It can bear the vertical load on the tower body and withstand wind, waves, and flow from a variety of forces, so that the tower body stays vertical. The axial tube is supported by the tower body and the guides connected with the base, in order to prevent bending. The guide allows relative motion for all parts, with the exception of the level of movement between the tower body and the base. Shear connector connecting the tower body with the base consists of several guides fixed on the tower bottom and the shear pins connecting with the base. When the shear force of the complaint platform is over the shear strength of the shear pins, they are cut off to ensure the safety of the tower body.

This platform uses less steel; the dynamic response characteristics are good; manufacture and installation are easy; it requires fewer offshore support vessels; and the cost of installation is low.

e. Flexible tower platform

As shown in Figure 3-50, its structure and its platform basically are similar to the floating-tower platform, but it does not have a floating box. Certain upper part of the tower bottom is closed and water can enter. So water volume in this part can be adjusted appropriately. Under the condition that the bending frequency is not affected by certain circumstances, it is

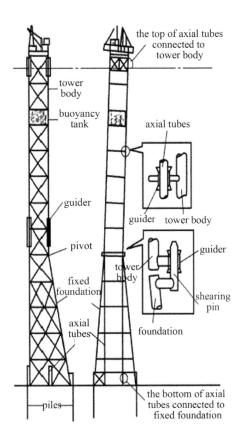

FIGURE 3-49

Bottom fixed floating-tower platform.

necessary to increase the swing frequency to remove the platform's intrincic frequency far away from the wave frequency.

f. Flexible pile tower platform

The main characteristic of this platform is that a steel pile will be distributed to four corners inserted into the seabed, and a guide has been extended upward to a flexible tower body and connected to the tower body as shown in Figure 3-51(a). Figure 3-51(b) shows the connection of the steel pile, a guide device, and a flexible tower body. The steel pile can produce enough overturning resistance, so the tower can swing and resist twisting and shearing.

g. Flexible pile tower platform with the base fixed

As shown in Figure 3-52, this is similar to floating tower platform with the base fixed, Due to the increase of a fixed base, equivalent to increase a rigid spring, its steel consumption save 30% to 40% steel than that of the guy line tower platform.

h. Composite leg flexible platform

This was designed by the McDermott Company, and is shown in Figure 3-53. This platform is a composite type, and the lower part has a fixed base jacket that goes through the legs into the

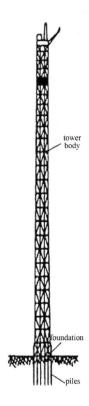

FIGURE 3-50

Flexible tower platform.

bottom of the sea; the upper part of the jacket is ordinary steel. In combination with a common pile grouting method with a rigid connection, the bottom of the leg and the fixed base are connected through the guide or steel pipe. Compared with the general deep-water platform, this platform presents the advantages of a simple structure, less steel consumption, segmented transport installation, installation without additional equipment, and less manufacturing and installation costs.

2. Main characteristics of adaptive platforms
   a. Sea adaptability. Because the flexible platform vibration cycle is long, and the rigidity is low, it can move with the wind, waves, and flow.
   b. Anti-fatigue properties. Because a flexible platform can be composed of piles, and the combination of the casing body and the jacket damps vibration, the amplitude of vibration reduces greatly to improve the anti-fatigue properties.
   c. Limited motion. Because it has articulated joints, large buoy and damper, which can restrict the deck with certain range, it isn't necessary to install a special device.
   d. Build simplicity. Because of the use of a number of repeated structure and standard components, design and manufacture become simple; the components are lightweight, lifting

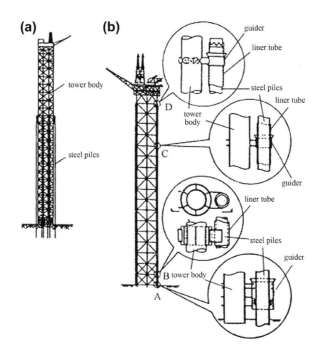

FIGURE 3-51

Flexible pile tower platform.

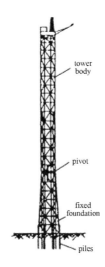

FIGURE 3-52

Bottom fixed flexible pile tower platform.

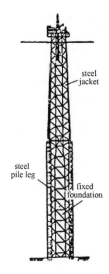

FIGURE 3-53

Composite leg flexible platform.

installation is easy, operating methods of conventional shipping and underwater construction can be used. Also, offshore construction is simple.

e. Economic cost. Because of its short construction period (generally two years), the cost of construction and steel, and reduced design complexity, the total cost is low. Take a water depth of 300 m as an example; the cost generally is about 3/4 of the jacket platform.

3.8 Floating Production Systems

Floating production systems refers to a full set of production facilities that integrate oil production, processing, storage and make use of the modified or special built semi-submersible platform, jack-up platform, tension platform and oil tank.

3.8.1 Floating Production System Types

1. Floating production system mainly for FPSOs
2. Floating production system mainly for semi-submersible platforms
3. Floating production system mainly for jack-up platforms
4. Floating production system mainly for tension leg platforms

3.8.2 FPSO Floating Production Systems

1. System composition

   This floating production system is a floating production storage and offloading (FPSO) system, which has production and processing equipment, water (gas) equipment, public facilities, and living facilities installed on a ship with storage and an offloading tanker. The oil and gas are sent to the single point mooring (SMS) device through pipelines at the seabed, and then sent to FPSO being connected with SMS, through a riser and soft pipe tied at SMS, eventually going to processing system. Qualified crude oil is stored by the FPSO, and after measurement calibration sent to a shuttle tanker. The composition of the whole system is shown in Figure 3-54(a). Figure 3-54(b) is a single point mooring device and FPSO.

The floating production system has another form, namely, the main body is only a oil tanker with the function of storing and offloading oil (FSO). Because the oil, gas, and water production and processing system is not installed on the oil tanker, the crude oil and gas just are processed on the other places. And qualified oil and gas is sent to the FSO to store through submarine pipelines. The oil tanker carry away periodically. A little old tanker can be modified into the FSO, and the construction period becomes short.

The floating production storage tanker can be built by two methods, namely constructing a new tanker or transforming an old tanker. The former is applicable to long-term oilfield development, while the latter is suitable for the early development of oil field, because of its short construction period.

2. Single mooring system (SMS)

The single mooring system is a part of the floating production storage and offloading (FPSO) system or shuttle tanker in the offshore oilfield; it is the terminal that completes the oil transportation and offloading oil operation. The single mooring system is an important part of the floating production system being FPSO or FSO as the main body and there are many types. The main types are introduced in the following.

a. Catenary anchor leg mooring system (CALM)

As shown in Figure 3-55, this is a mooring system based on catenaries and an anchor pull buoy, mooring and sending oil by float bowl. It floats a large diameter drum mooring buoy in the water, composed of six catenaries anchored to the seabed. A buoy is equipped with rotary joint, pipe, and mooring cable. A swivel hose and submarine pipeline are connected. The tanker (FPSO or FSO) moors to a buoy by rope, and a floating hose (floating pipe) is connected with floating tube by a single point: oil is loaded to the tanker by floating pipe. The following are the main components of the system.

i. Buoy

The buoy is a long cylinder; when the water depth is less than 200 m and SMS moores more than a $10 * 10^4$ t tanker, the diameter is $12 \sim 17$ m, and the height to diameter ratio

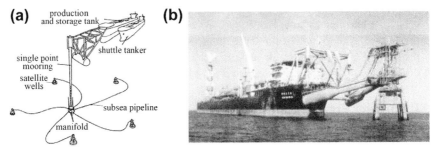

FIGURE 3-54

FPSO floating production system and single point mooring device.

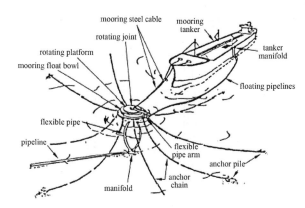

FIGURE 3-55

Catenary anchor leg mooring system.

is about 3/1 to 2/1. The inside is divided into four, six, or eight watertight compartments to be used as ballast. The buoy center is connected to the sea, where the hose is introduced and fastened to a rotary union; the float around the lower part has a skirt edge and a protective plate with rubber fender skirts and a cable rack. The protective plate skirt wall, buoy shell, and channel wall should be packed with anodic treatment for aluminum sheets as an antiseptic measure.

ii. Rotating platform

This is installed on the upper portion of float bearings. When the wind direction changes, it can rotate around the center of the float 360 degrees through the mooring force. The turntable is equipped with mooring arm and mooring cable, manifold and crane, and navigation lights.

iii. Rotary joint

This is located at the top of the buoy and generally made of a tubular sleeve, and it can rotate around the upper part of the fixed portion. The fluid entrance is arranged at the bottom of the fixed part of SMS, and the fluid outlet is set at the SMS rotating part. So SMS can continuously convey fluid to FPSO when rotating. The rotary joint is often a single channel (tubular sleeve) whose diameter is $10.2 \sim 61$cm. And there are also rotary joints with two or four channels.

iv. Anchor chain

This is commonly six chains, or sometimes there are four or eight, composed of many links. The upper of anchor chains is connected with the brake at the buoy bottom and uniformly arranged around the buoy, stretching out like the hanging chain. And they are anchored at the seabed.

v. The anchor pile

This is used for fixing the pile structure and is driven into the seabed, and it commonly has two kinds of forms, as shown in Figure 3-56. Figure 3-56(a) shows the form that directly puts cement steel piles into the mud, with part of the pile projected in the hole and the use of linked anchors. Figure 3-56(b) shows another form; it puts the pile into a portcullis type structure in the drilling hole, and anchoring force is greater than the force that Figure 3-56(a) shows.

FIGURE 3-56

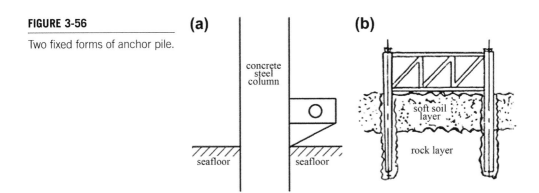

Two fixed forms of anchor pile.

vi. Hose

The hose from the float to the submarine pipeline often has a "lantern" shape (Figure 3-55) or "S," and the length is about two times of water depth. The former shape is used for shallow water, and the "S" shape is for deep water. The shape and length of hose ensures that when the buoy has the biggest displacement, it is still in a state of relaxation. Increasing oil amounts may also cause a change to bend pipe, but the upper and the lower ends shall be provided with a universal joint.

vii. Buoyancy tank

In order to make the hose between the pipelines and buoy keep a smooth curve, one or two buoyancy tanks are arranged on the proper position for suspension of the hose pipe.

If the mooring cable is replaced by a rigid arm (rigid soft arm), it becomes another form of CALM, as shown in Figure 3-57. The one end of the rigid arm are connected with the tanker with a hinge, and the other end is integrated with the swivel table on the buoy, and it is a "A" shape, with a closed-end box structure. It eliminates risks of breaking cables, collision between the tanker and the buoy, and frees the operation of change cables. In

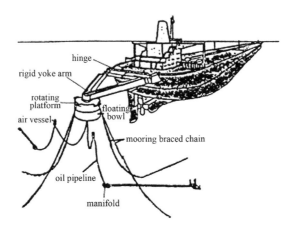

FIGURE 3-57

Rigid arm CALM single point mooring system.

addition, the rigid arm makes SMS be suitable that the tanker is fulled with oil at the starboard side, which simplifies the filling operation, it is used more and more widely.

The CALM system is often used at present, but the depth is shallow, generally less than 100 m, with a maximum of 150 m, and the mooring limit wave height is not more than 5.5 m.

b. Single anchor leg mooring system (SALM)

A SALM is characterized by a buoy and a single anchor leg mooring system; according to different structures of the single anchor leg, it can be divided into the following forms:

i. Single anchor leg mooring

As shown in Figure 3-58, with a heavy chain a long cylindrical buoy is fixed on the seabed. Because of the positive buoyancy of pontoons, the chain tenses, and the buoy on the sea can freely rotate. The base is fixed at the bottom of the sea with a pile; the rotating head is fixed on the base; and the water hose is connected and floating on the surface of the water. When the tanker arrives the tanker manifold is connected using a floating mooring hitch.

ii. The single leg post and cable type

As shown in Figure 3-59, a single anchor leg is composed of two sections, the upper chain and lower slender column. A universal joint is connected with the base, and the column can rotate with the buoys; this form can be used for deep water.

iii. The single leg column type

This is a column plus two universal joints or a plurality of columns and a universal joint connected to form a single leg. The former can be used at 100 m depth; the latter can go deeper than 150 m. Figure 3-60 is a schematic of a multiple column single leg mooring system.

If the SALM cable is replaced with rigid arm, it becomes a rigid arm single pile leg mooring system. Figure 3-61 shows a rigid arm single anchor oil mooring system (SALS) diagram; Figure 3-61(a) shows a single leg with a constant diameter leg, and Figure 3-61(b) is the single pile composed of an unequal diameter single column, both for the depth of 100 m. If the structure of single leg is transformed into multiple columns, the water depth can be deeper than 150 m. Also, if the structure of single leg is a column and

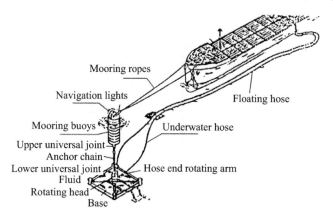

FIGURE 3-58

Chain type single anchor leg mooring system.

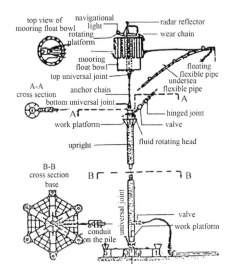

FIGURE 3-59

Mooring system with an upright post with a single pile anchor leg.

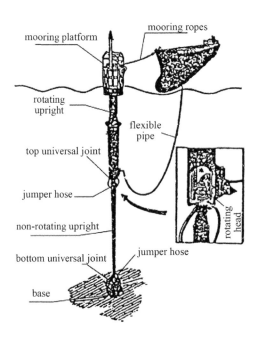

FIGURE 3-60

Multiple column pile leg mooring system.

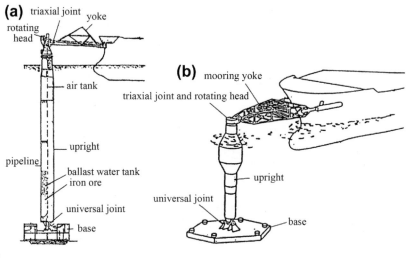

FIGURE 3-61

Arm type single anchor leg mooring system for oil storage.

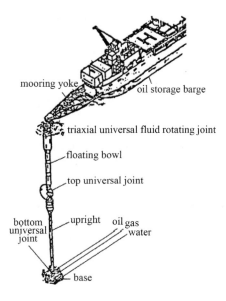

FIGURE 3-62

The rigid arm column and pontoon single anchor leg oil-storage mooring system.

pontoon type, as shown in Figure 3-62, the working depth can also be increased. The mooring system designed by the EMH company has been applied in a Libyan offshore oil field. And its work water depth is 167 m. It resists the wave with the height of 13~26.8 m and moors the tanker with the weight of 250,000 t.

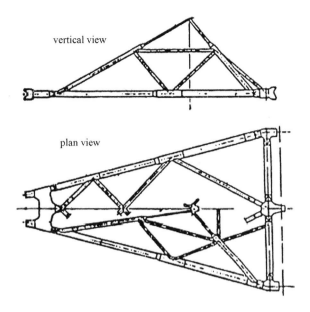

FIGURE 3-63

The rigid arm mooring system.

The key component of the single anchor leg mooring system for oil storage is the rigid arm and three axial joints. Figure 3-63 is the rigid arm's elevation and shape, and from the view of the picture it is the girder structure of the space truss made up of two parallel truss bars like "A". The rigid arm is often $40 \sim 60$ m, and its weight is up to $300 \sim 800$ t. Figure 3-64(a) is a three axis joint which connects the rigid arm and the top of the upright column; its function is to transmit the weight of the rotary part at the SMS upper, the weight of the rigid arm and mooring forces, and to reduce the dynamic effects of tanker's movement. The three axial joint allows the tanker to roll, pitch and rotate around the single

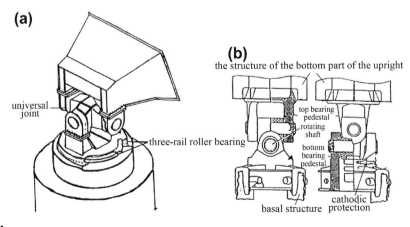

FIGURE 3-64

The universal joint and the three axial joint.

point with 360 degree. It consists of a common universal joint [Figure 3-64(b)] and a three track roller bearings. The universal joint allows the anchor leg to swing arbitrarily relative to the base. Each axle and bearing surface is coated with more than 5 mm of aluminum alloy to prevent wear, and the universal joint should be installed with cathodic protection system to prevent corrosion.

c. Articulated tower single point mooring system (ALT)

The characteristics of this single point mooring system is that it depends on a vertical tower which is hinged with the base at the seabed to moor. So it is called articulated tower mooring system. As shown in Figure 3-65(a), the floating tower is a upright steel structure, and at its bottom a universal joint is used to hinge with the base at the seabed. The lower part of the floating tower has a ballast tank and auxiliary floating tank; and its upper part below the water is installed with a main buoy. So the tower is called floating tower. Its top is a rotary table which has loading oil arms, the hose and the rotary joint. Oil from submarine pipelines is transferred to the rotary joint through the universal joint and the riser. And then it sent to the tanker through the horse.

A key component of the mooring system is a universal joint, as shown in Figure 3-65 (b). It can bear the load of the upper part of the tower equipment, but also can be used as the fluid channel. The number of channels with the diameter of $41 \sim 51$cm is one or two, and the fluid from the supporter under the universal joint enters into the channel, and then flows to the riser on the buoyant tower from the upper supporter. There are two tubular reverse sealed joints with thickness of about 10 cm separately at the entrance and outlet of the universal joint. The kind of the universal joint allows the buoyant tower to swing at the range of 30 degrees.

The buoyant tower is the main body of the mooring system, and it has three forms:

i. The combination of triangular steel truss and buoy, as shown in Figure 3-65(a), Because the main buoyant is set nearby the water surface, the float force concentrates at the upper of the mooring system, resulting in the larger restoring moment.

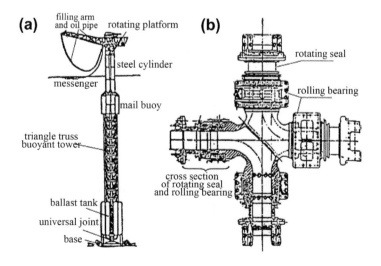

FIGURE 3-65

Articulated tower single point mooring system.

    ii. The combination of concrete structure and buoy. This type can bear larger loads because the weight of its tower body is large, its heart high, which leads it needs larger buoyant.

    iii. The all steel type; it is round steel tower structure. The steel column diameter is 6 to 8 m and the upper segment of the tower is the main body and provides the buoyancy. Its middle segment can appropriately control the tension of the universal joint, and space the distance between the buoyant centre and the gravity centre.

        An articulated tower mooring system is used to moor a shuttle tanker, and it is suitable for 100 to 300 m depth, with a wave height is not more than 5.5m; the moored tanker displaces 15×10^4 t.

d. The fixed tower mooring system

The feature of this mooring system is that the pedestal of the mooring tower is fixed on the seabed with steel pile anchors, so it is called the fixed tower mooring system. This system is used for permanent mooring tankers, and it also be used as a shallow sea oil and gas importer and exporter for a conventional offshore terminal, so the system is applied widely in floating production systems which use a production and storage tanker (FPSO) as the main structure in the Beibu Gulf Oilfield, South China Sea and the Bohai offshore oilfield. There are two types depending on the mooring tower structure.

    i. Column type

Figure 3-66 shows a column fixed tower mooring system applied in one oilfield in the South China Sea, Beibu Gulf and its mooring tower is a cylindrical tower with a diameter of 5.7 m; the pedestal of the tower is fixed at the bottom of the sea, with an an anti-collision ring on

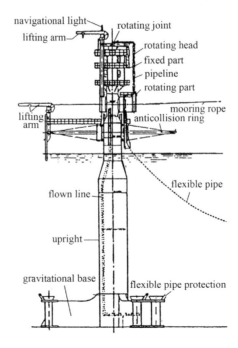

FIGURE 3-66

Column fixed tower mooring system.

the tower, and rotary joints on the top. One end is connected with the fixed riser, the other end is connected with the floating hose. It can resist typhoons in the South China Sea, and when a typhoon comes, the mooring tankers can be evacuated, and the free hose can sink to the bottom of the sea with its own weight.

iii. Jacket type

Figure 3-67 shows a jacket fixed tower mooring system applied in one oilfield in Bohai, China. The mooring tower applies a jacket structure, using a rigid arm instead of a cable. Therefore, it has both a rigid arm instead of a cable and the advantages of a jacket with its safety and high anti-wave capacity. But these two types are only applied in 50 ～ 60 m shallow sea.

The main disadvantage of the fixed tower system is that it does not have great compliance, so it should be used in combination with a soft system such as a cable, or a flexible yoke consisting of a tube frame.

A fixed tower mooring system shall install an anti-collision ring (Figure 3-66) to prevent collision from the oil tanker or other objects. The anti-collision ring, like a bicycle wheel, can use the large plastic deformation energy of the spokes and the rim to absorb the impact energy from collisions.

e. No-turntable single point mooring system (NRT)

The key characteristic of this single point mooring system is that it has no turntable, relying on a funnel buoy whose two outer layers can rotate relatively to realize mooring, as shown in Figure 3-68(a). The main components are as follows.

i. Funnel buoy

As shown in Figure 3-68(b), it has two layers. The outer layer goes through two up and down clamping devices connected the oil tanker hull, which can rotate with the tanker relative to the inner layer of the buoy; a floating box set in the inner layer provides the necessary buoyancy, and through the anchor ring it is connected with the mooring anchor chain. There is a channel in the center of the funnel buoy and a flexible pipe can be connected with the valve chassis at the top though this channel. The chassis is relatively

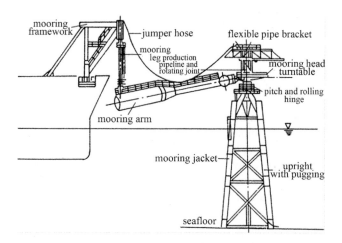

FIGURE 3-67

Jacket fixed tower mooring system.

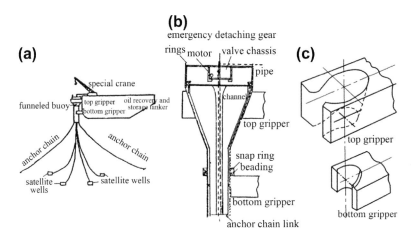

FIGURE 3-68

No-turntable single point mooring system.

fixed with the outer layer of the buoy in peaceful times and the motor can drive the relative motion when necessary.

ii. Mosaic clamping device

As shown in Figure 3-68(c), the funnel buoy is like a bolt inserted into the mosaic clamping device. It has in total two bows located in the oil tanker. By starting the crane of the bow to lift the buoy and insert the buoy, the buoy and the tanker will be connected together under the action of gravity; if necessary a locking device can be used.

This kind of no-turntable single point mooring system featuring low cost, a short construction period, easy repair, and the capability to test and repair wells on its own, is especially suitable for the early production of oilfields and marginal oilfield exploitation.

f. Advantages of single point mooring system

i. Adapting to any wind and flow changes: Because of an arbitrary rotation of 360° with the mooring point as the center, it can make the motion of the ship and mooring load decrease to a minimum.

ii. Elastic system can buffer the external shocks: It can reduce the stress of the cable greatly, reduce cable accidents, and prolong the life of the system.

iii. Mooring operation is simple and time-saving: The work of mooring, releasing cables, and retracting the hose is simple, and it needs fewer auxiliary ships and staff.

iv. Short construction period, low cost, quick recovery: Generally it can be completed in 6 ~ 12 months, and the investment is about 30% to 50% less than fixed mooring.

v. Enough depth to moor a large tanker: It can be arranged at a sufficient depth for large tanker mooring, with a high turnover rate.

vi. Multiple functions and wide range of application: It has the functions of storage, transportation, living and production, etc., and is not restricted by water depth and seafloor conditions, so it is used widely.

    vii. Applicable for early production and marginal oilfields: It can save on submarine pipeline cost for a marginal oilfield. For early production, it can bring a fast recovery of funds.

g. Installation and construction procedures of single point mooring system

Now we take the single anchor leg system (SALM) as an example (Figure 3-69) to give a brief description of the installation and construction procedures of a single point mooring system.

    i. Figure 3-69(a): Select the installation position and put three weights on the seafloor, indicate where they are, and take the pedestal and the risers to the marked locations with tugs.

    ii. Figure 3-69(b): Fill the pedestal ballast until we are at the horizontal position, and make the risers upright.

    iii. Figure 3-69(c): Connect three traction ropes with the pedestal via the seafloor fixed pulley bodies.

    iv. Figure 3-69(d): Use the traction rope to sink the pedestal to the seafloor slowly.

    v. Figure 3-69 (e): Pile with a piling barge through the pedestal to a certain depth into the seafloor and pour concrete.

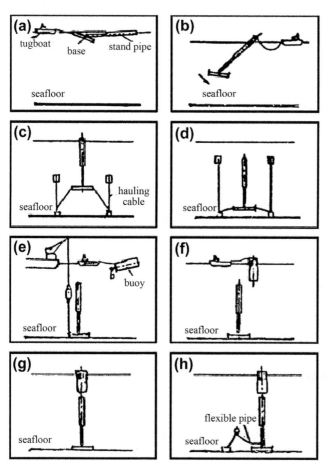

FIGURE 3-69

Installation procedures of the single anchor leg single point mooring system.

    vi. Figure 3-69(f): Place the buoy with the tug, and fill ballast so that it can stand up close to the main pipe.

    vii. Figure 3-69(g): Use the hydraulic jack to press the buoy, align the buoy to the riser connected to it, and tighten it with bolts.

    viii. Figure 3-69(h): Connect the subsea pipeline with the system using rubber hoses, and decompress.

3. Oil tanker

The oil tanker is the main body of the floating production system, and can be newly built or rebuilt. Whether we use newly built or rebuilt is determined by many related factors, such as oilfield life and duration of use. Generally if the long use duration is required, new construction is better than reconstruction. The oilfield development scheme is also one factor to be considered, for example, if we have an early development scheme, rebuilding the oil tanker takes less time, and is better than new built. Also, we should consider whether the existing oil tanker tonnage can be used; the general needs of a floating production system is a 10×104 t or larger oil tanker. If there is one available in a spot, rebuiling is proper. Economic conditions should be considered an important factor; in order to recover the investment as soon as possible for some small and medium oilfields and marginal oilfield, it is appropriate to use a rebuilt oil tanker.

Whether new or rebuilt, the oil tanker should be equipped with:

a. Production system: including water injection, gas injection, gas lift, artificial lift and so on

b. Oil, gas, and water processing system

c. Well-repair system

d. Public system: including the living system, sewage processing system, seawater processing system, and heating and oil transportation systems

e. Power system

f. Instrument system

g. Safety system

h. Rescue system

The overall layout of the ship is an important problem for building and rebuilding oil tankers. The basic principles of overall layout is effectiveness, economics and security. For example, mooring facilities should be set at one end of the ship or one fourth of the ship length. The living room should be at the upwind areas relative to the production equipment and close to the mooring facilities and the ship's end, which benefits the personal safety. The gas torch should keep away from the living house and helicopter deck. And the production facilities should be arranged near the ship's heart, on or below the main deck. If facilities are on the main deck, they should be set at the higher place to facilitate operation and repair. If facilities are below the main deck, they are protected from the environment influence and more safe. But this method increases costs and reduces the amount of oil storage.

3.8.3 Floating Production System with Semi-submersible Platform as the Main Body

1. Characteristics

The characteristics of this kind of floating production system str that the production equipment (such as the Christmas tree), water (gas) injection equipment, and oil, gas, and water processing

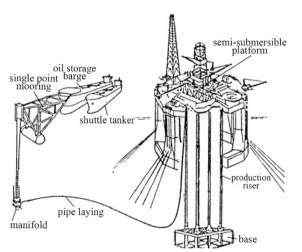

FIGURE 3-70

Floating production system with semi-submersible platform as the main body.

systems are installed in a newly built or rebuilt semi-submersible platform, which needs to be connected with another oil tanker to complete the oil loading and unloading functions (Figure 3-70). Oil and gas are delivered from a subsea wellhead via a production riser (rigid or flexible) to the production processing system on the platform, then go through the submarine pipeline and single point mooring system and enter into the oil storage tanker (FSO) after separation processing, and then are removed by a shuttle tanker.

2. Semi-submersible platform

This is the main body of this kind of floating production system, to distinguish it from the semi-submersible drilling platform, it is often referred to as the semi-submersible production platform; it is composed of the deck, column, and the hull (mat). Modifications can be installed for the eight main systems as with the oil tanker, and the principles and requirements of the overall layout of the platform should also be in accordance with the requirements of oil tanker modification.

3. Advantages and disadvantages

The advantages of a floating production system with a semi-submersible platform as the main body is good stability and adaptability to harsh conditions. Disadvantages are a small deck area, low carrying capacity, long modification time, and high cost.

3.8.4 Floating Production System with Tension Leg Platform as the Main Body

The main body for this kind of floating production system is the tension leg platform. A tension leg platform is a platform that utilizes the tension generated by a tight cable to balance the residual buoyancy force; therefore, it can be regarded as a vertical anchor line semi-submersible platform, as shown in Figure 3-71(a). The advantage of this floating production system is that it can be used for the development of deep-water oilfields with water depth up to 600 m.

1. Composition of the tension leg platform

The tension leg platform [Figure 3-71(b)] consists of four parts:

a. Anchor pile. This uses the piles driven into the seabed as an anchor, thus the name anchor pile. The bottom of each tension leg has an anchor connector and a lower flexible device, the

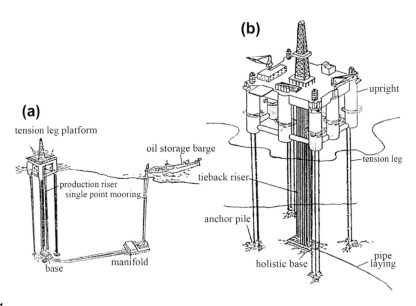

FIGURE 3-71

Floating production system with tension leg platform as the main body.

former providing automatic mechanical locking to fix the tension leg on the anchor pile and the latter allowing the tension leg to move relative to the locking device for oblique movement.

b. Anchor noose. This is a main component of the tension leg, and consists of wire rope or steel pipes. Steel pipe has rigidity, and is lightweight and easy to install; it can be handle a lot of usage. Cable diameter is often 17.8 cm with 7 pipes, with each pipe composed of several sections. The top of each tension leg has alternating load support and an upper flexible device connected with the platform, thus making the tension leg able to do oblique movements and bear lateral loads.

c. Column. Like the column of a semi-submersible platform this supports the weight of the whole platform; there can be four, five, six, or eight.

d. Deck. This is equivalent to the deck of a semi-submersible platform; in addition to the drilling equipment, there should also be oil extraction and production processing equipment.

2. Installation procedures of the tension leg platform

As shown in Figure 3-72, installation procedures for the tension leg platform are as follows:

a. Figure 3-72(a): First install the anchor pile in the predetermined submarine location, and then place the tension leg platform at a location about 40m away from the horizontal position of the anchor pile.

b. Figure 3-72(b): Lift the tension leg to the turnbuckle motion compensator that is the same as that on the drilling platform.

c. Figure 3-72(c): Move the tension leg platform, and adjust it to be accurately located above the anchor pile.

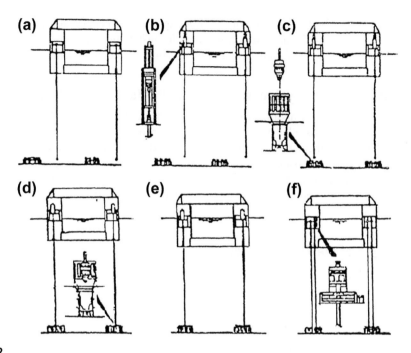

FIGURE 3-72

Installation procedures for the tension leg platform.

    d. Figure 3-72(d): Make the anchor joints (anchor connector) on the tension leg insert into the anchor pile.
    e. Figure 3-72(e): Lock the anchor joint.
    f. Figure 3-72(f): Install the remaining tension leg.
3. Advantages of the tension leg platform
    a. Simple and economical structure: Compared with the steel platform, the structure of a tension leg platform is simple, and when the water depth is more than 360 m, steel consumption is less than that of steel platform.
    b. Less offshore work: Construction is not influenced by climate, and thus the construction cost is less than for a semi-submersible platform.
    c. Strong mobility: It is convenient to transfer.
    d. Applicable to deep-water fields: It's not sensitive to the depth of water and earthquakes, so it applicable for deep-water scenarios.
4. Design essentials of the tension leg platform
    a. Tension of the mooring leg: The mooring leg should always be strained, that is the tension value is positive. In addition, the tension variation should be as small as possible.
    b. Platform motion cycle: In order to avoid resonance, we should make the cycle of the six degrees of freedom movement of the platform exceed the actual wave cycle.

c. Platform geometry: On the condition of not affecting the other performance, we should adopt a program that saves steel, and that is convenient for installation and reparation. We need to pay a lot of attention to the fatigue strength of tension leg, and consider corrosion.

3.8.5 Floating Production System with Jack-up Platform as the Main Body

This kind of production system is modified from the jack-up drilling platform, as shown in Figure 3-73. A modified deck should be installed with production and processing equipment. The oil and gas outflows from the subsea wellhead through the riser up to the deck for the separation process, and then through the vertical pipe returns back to the bottom. It is delivered via the submarine pipeline to the single point mooring, and this is transported to the tankers. This kind of production system is subject to the jack-up platform working depth, which is only in shallow water. Because it can move through floating, it can be reused.

3.8.6 Advantages of the Floating Production System

1. Can be quickly put into production: It takes 4 to 5 "a" represents "annum". for the general oilfield to go from discovery to formal development, but the floating production system takes only $1 \sim 2$ annum.
2. Lower investment: Take the Beihai oilfield as an example. The investment for a floating system is just 1/4 of that of the fixed platform. The lower economic limit of the required annual recoverable reserves for a floating system is $550 \times 10^4 \sim 700 \times 10^4$t, and for a fixed platform it is 2700×10^4t. For a single well oil production system, the lower economic limit of the annual recoverable reserves can be decreased to 40×10^4 to 50×10^4t.
3. Strong adaptability: It can replace larger or smaller production devices when there is a change of the oil production capacity, and it is easier to meet the requirements of later developments of the oilfield and the use of different technology.

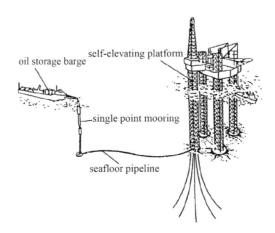

FIGURE 3-73

Floating production system with jack-up platform as the main body.

4. Flexible: With strong mobility, easy movement, and reusability the floating system can still be utilized after oilfield depletion. The drilling process can make the transition from discovery wells, exploration wells, or evaluation wells to production wells easy.
5. Applicable to deep water: Compared with a fixed platform, deeper waters require a much lower investment.

Because the floating production system has the above advantages, it is widely used in the world. Its working depth increases constantly (the maximum water depth is now more than 900 m) and its structures see continuous improvement (such as the 12-edge semi-submersible platform, semi-submersible platforms with articulated riser towers, etc.). The ability to avoid crises is a strength (it can resist against 100-year storms), and the production capacity improves constantly. Taking China as an example, the floating production system with a semi-submersible platform ("The South China Sea Challenge") as the main body has been successfully applied to the Liuhua 11-1 oilfield with a depth of 305 m. The modification cost of the semi-submersible production platform "The South China Sea Challenge," when modified with a semi-submersible drilling platform, was less than 2×10^8\$, while the jacket of a fixed platform costs 10×10^8\$, and a newly built tension leg platform needs 12×10^8\$. Moreover, consider the Lufeng 22 - 1 oilfield, which has the deepest working depth in China, at 333 m. It adopts a floating production system of multifunctional production and a storage oil tanker as the main body. It uses an FPSO with a dynamic positioning system, which keeps the ship at a pre-determined position without anchors. It should also be pointed out that China, as one of the world's few countries with large-scale use of FPSO, already has 11 FPSOs in service and 2 FPSOs under construction, and there are already 30×104 t ultra-large type FPSOs. The floating production system with an FPSOs as the main body in China is developing quickly and just getting started.

3.8.7 Research Directions for the Floating Production System

1. Adapting floating production systems for deeper waters in bad sea conditions
2. Making further efforts to improve stability, and improving the oil, gas, and water separation processing equipment and the separation efficiency while floating.
3. Use of all-weather production risers, so that outside of maintenance and emergency, they needn't break off and lift out.
4. There is the oil tank on the floating production platform. So it is necessary to shut in well when the oil tanker breaks off the mooring. At the calm sea area, oil can directly send to the oil tanker from the production platform, which will reduce investment.
5. Making use of a new generation of floating production system rigid vertical pipes for production, and installing the Christmas tree on the floating body, to adapt to deeper water.
6. Floating production systems suitable for large oil and gas field and condensate oilfields.

3.9 Subsea Production Systems

The subsea production system refers to the scenario where the Christmas tree is put into the seabed and a mixture of oil and gas from the subsea Christmas tree is delivered to the subsea manifold center through the subsea manifold. The subsea manifold center accomplishes well metering, fluid collection,

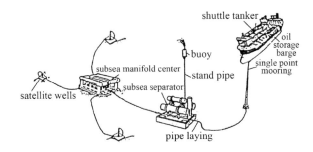

FIGURE 3-74

Subsea production system.

and pressurization, and then the well fluid is transported to the floating production system for processing and storage through a subsea pipeline; or metered at the subsea manifold center, then separated through the water separator. The separated gas going through the riser is delivered to the torch on the buoy for burning, while the separated oil through the subsea pipeline is delivered to the mooring tanker (FSO) for transportation to the shuttle tanker, as shown in Figure 3-74.

A subsea production system is composed of subsea facilities and a control system above the water. Subsea facilities include a subsea Christmas tree, subsea manifold center, and subsea separator. The subsea controlling system above the water is generally placed on the floating production system, and works through the subsea manifold center to control, shut off, and inject water, gas, and chemicals and manage maintenance and repair work for the subsea wellhead.

3.9.1 **Main Facilities of the Subsea Production System**

1. Christmas tree

   The types of the subsea Christmas trees can be classified according to the installation method (vertical, horizontal), workover method (through flow line TFL and non-TFL), completion method (single tube, multi-tube), whether the help of divers is needed and the well layout scheme (satellite wells, chassis wells) and so on. But the types of the subsea Christmas trees generally are divided into dry and wet types based on the structure types.

   a. Wet Christmas tree

   This is the most commonly used subsea Christmas tree currently, because the tree is completely exposed in seawater and directly contacts with seawater. Because this type is relatively simple, and because the anticorrosion performance of metal materials and the technology of the remote control from the surface to subsea have been continuously developed in recent years, the wet subsea Christmas tree is widely used. Here we give a brief introduction to its structure.

   Figure 3-75 shows a type of double-tube subsea Christmas tree with a satellite well through the oil pipeline (TFL), but no need for diver assistance in installation.

   i. Valves

   They are the main parts of this Christmas tree, including the main valve, oil suction valve, wing valve, slide valve, and program valve. The main valve through the oil pipe is

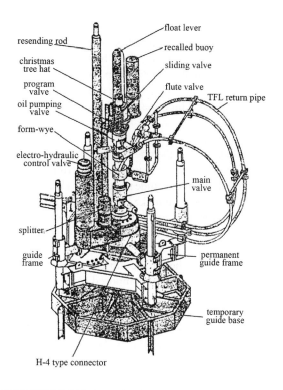

FIGURE 3-75

A type of wet Christmas tree.

connected to the tubing hanger with a flange. The wing valve located at the side of the T-cock opening above the master valve can be replaced without killing the well. All valves are installed in a solid valve body, with directly drilled holes in the solid body for valve cavities; the valves are equipped with a manual or hydraulic operating device.

ii. Connector

As shown in Figure 3-75, this piece is installed on the bottom and side of the Christmas tree, and used to connect with the oil pipeline and the wellhead respectively. Figure 3-75 shows the H-4 type hydraulic connector, produced by the Vetco Company in the United States, which is mainly composed of a wedge locking block, piston, eccentric ring, etc. Relying on the hydraulic pressure to drive the piston, and then driving the eccentric ring, it locks the block on the wellhead thread.

iii. Y-type tube

This is a kind of bent forging tube, providing a pathway for the oil tube in the well and the subsea pipeline. A connected is mounted on the Y type tube, which can fix and orient the steering gear. The pump down tool out of the oil tube through Y-type can turn into pipelines.

iv. Diverter

Diverters are installed and locked in the Y-type tube. It has a lock rod with a fishing neck, which can set orientation using the mutual occlusion method to lock the tool in the lock rod.

v. Tree cap

This is a cap that can provide a Christmas tree with a low pressure seal to prevent exoteric pollution, and protect the Christmas tree from being damaged from falling debris, so it is also known as the Christmas tree protection cap. In the tree cap we can install a single flow valve and discharge valve, so that the oil or the controlling liquid can replace the seawater in the installation and the working pipe, which can reduce the corrosion of the seawater after installation.

vi. Christmas tree guide structure

This can provide the guidance and the orientation for the Christmas tree, reducing the stress when installing the Christmas tree; it is used as the terminal support of the oil pipeline (TFL). The main composition of the guide structure is a leading truck, and the leading truck of the Christmas tree connector can make the tree align with the guide base accurately. In addition, there are truss structures connected to the top of the tree, and an anti-collision frame which protects the TFL loop line from being damaged during installation.

b. Dry Christmas tree

This is a kind of Christmas tree installed in the subsea wellhead section with an atmosphere; it does not contact the seawater. It is mainly composed of the following parts.

i. Subsea wellhead section (cabin)

This is a cylindrical container with sufficient strength to withstand water pressure, which can protect subsea trees and all kinds of valves and control equipment, and ensure they stay in normal working conditions, not affected by the surrounding high pressure and seawater corrosion.

The subsea wellhead section has two types: the vertical type and the horizontal type, as shown in Figure 3-76(a) and (b) respectively. Because the horizontal type can improve the ability of tools through the pipeline, as well as expand the indoor personnel operating space, the horizontal type subsea wellhead section is currently adopted more.

The horizontal type subsea wellhead section, as shown in Figure 3-76(b), has a diameter of 3.2 m, so that there is a 0.1 m gap with a 1.5 m radius ring pipe inside through the pipeline tools. It can contain dual tubing with a rated pressure of 344.5 Pa (ϕ5.1 cm) and a Christmas tree with a 5.1 cm circular channel. There is a bull nose shaped tube opening at the top left hand side at about 45° from the subsea wellhead section, for the incoming and outgoing pipelines. When installing the wellhead section, the holes can be sealed by a floating plug which can act as a rope buoy, relying on the differential pressure in and out of the section to compact the seal components on the outer surface of the wellhead section.

There is a flange connected with the reparation tanker on the cup-shaped junction gate at the top of the subsea wellhead section, in order to connect tightly with the reparation tanker. As shown in Figure 3-76(c), there is a closed tanker hole on the bottom of the junction gate, sealing the wellhead section, to maintain atmospheric pressure inside it under normal conditions; it opens only when the reparation tanker is connected above it. This junction gate opening should be large enough, in order to let the wellhead equipment and operating personnel pass through smoothly in case of well completion and work ending. It should allow 2 ~ 3 people to operate comfortably inside the wellhead section.

Installation of the wellhead section should be done after well completion, before the drilling platform has been evacuated. First, connect the wellhead section with the hydraulic

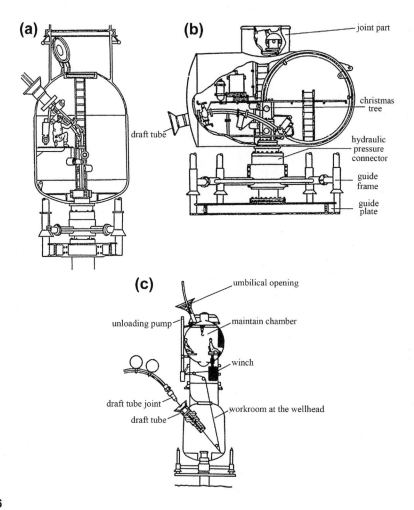

FIGURE 3-76

Subsea wellhead section.

connector together on the platform; then, lower the guide rope with the drill string into the guide frame, until the hydraulic connector below is located at the casing head of the wellhead. At present, foreign production of horizontal wellheads have already reached the 122 m, 244 m, 366 m depth and the largest diameter has reached 4.3 m.

ii. Subsea reparation (service) cabin

This is mainly for personnel and ropes or pumping tools and other equipment, to provide a channel from the surface supply ship to the subsea wellhead chamber. As shown in Figure 3-77, its shape is like a ball, on the top and bottom of which is a hatch, along with a sleeve on the bottom, and an apron. The spherical pressure shell of the subsea reparation tanker is attached to the propulsion module with four electric propulsions, so that it is provided enough thrust in the seawater to resist the current. High

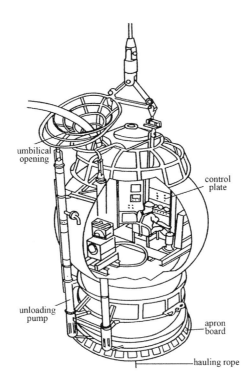

umbilical opening

control plate

unloading pump

apron board

hauling rope

FIGURE 3-77

Subsea reparation (service) cabin.

pressure is always maintained inside the spherical shell, which can supply fresh air, the temperature and humidity can be adjusted inside the regulating cabin, and exhaust gas is discharged. There is a umbilical cord hole at the top left side of the spherical shell, through which the exhaust gas can be discharged, the air and electric power can be supplied from the surface supply vessel, and the communications and data signal circuit can be transferred to the supply vessel. The spherical shell body is installed with rope lifting operation equipment and tools as well as a winch; and a subsea operation and construction monitoring system, including a camera, television, and cabin pressure, oxygen concentration, and other environmental factor monitoring system. The subsea reparation tank bottom cuff is provided with a ballast tank, an unloading pump for drainage is equipped in the cabin, and an apron plate is installed in the bottom cuff. For creating the apron seal (wet seal edge), the wet seal edge is installed in a wide flange with a thick rubber liner at the bottom of the sleeve. When the flange is set on the flange of the subsea wellhead section cup-shaped joint opening, a small amount of water drains out; with the pressure of the seawater the flange can squeeze the rubber gasket, creating a seal, so as to form a sealing channel between the tank hole of the lower flange and the wellhead cellar. On the top of the reparation module there is also a cabin hole for safety use, which is not in use usually.

c. The dry/wet Christmas tree

The characteristic of this is that it can change between dry and wet; the Christmas tree is wet during normal production; when under repair, the reparation (service) cabin is connected with the subsea Christmas tree, the water inside is emptied, making it a dry Christmas tree at atmospheric pressure.

A dry/wet subsea Christmas tree mainly consists of a low pressure casing, subsea oil extraction equipment, pipeline connector, and dry/wet conversion joint. The low pressure casing is a external pressure container designed according to a standard. We open a hole on it, in order to fit a ring and connect with the dry/wet conversion joint, forming a sealed container when creating a dry environment. The appearance of the dry/wet conversion joint is like a conical shell. When operating, the joint bottom is connected with the ring of the low pressure casing, and the joint top is connected with the reparation module. So, it converts into a dry Christmas tree for repair work.

d. Caisson (plug-in) type Christmas tree

This kind of Christmas tree is also called the lower mud line Christmas tree, characterized by a main valve, connector, and subsea wellhead arranged in the catheter below the seabed $9.1 \sim 15.2$ m. The wellhead device on the seabed is very short, generally just above the seabed $2.1 \sim 4.6$m (normally the subsea Christmas tree is above the seabed about 10.7 m). For developing an offshore oilfield, this can avoid ship, anchor, fishing, and other maritime operation damage and environmental destruction like the movement of an iceberg in shallow water, and allow the installation of protective cover devices economically and rationally.

Figure 3-78 shows a kind of caisson type Christmas tree produced by the Vetco Company in the United States. Its main part is divided into two sections, an upper part and a lower part. The lower part of the Christmas tree is connected with a sub-mud-line wellhead device through an H-4 type connector. The upper part of the Christmas tree is connected with the lower part of the Christmas tree by an H-4 type connector, and directly connected with the wellhead devices above the mud line. The wellhead devices above the mud line, including the protection cover, have a height of about 2 m, and are mainly composed of the leading truck of the upper part of the Christmas tree, guide line roller, no-diver pipeline connector, tree cap assembly, and protective cover. The Christmas tree cover can be welded by steel plate, hanging the earrings with steel wire rope down through the guide rope and positioning with the guide post, and then locking in a preinstalled groove.

The biggest advantage of the caisson type Christmas tree is that, after the upper part of the structure is removed, the Christmas tree can still use the wellhead below the mud line, and this kind of Christmas tree has simple structure, which is able to use TFL for the workover. But the biggest drawback is expense, about 40% higher than that of conventional wet tree, and its applicability has a certain limitations. At present, the maximum water depth has developed from 190 m to 300 m, and it seems that there is much potential for the applicable water depth.

e. The choice of subsea Christmas tree.

For the selection of subsea Christmas tree, the following factors should be considered:

    i. Operation water depth: For different water depths, the Christmas tree has different hydraulic pressures. We should select different Christmas tree pressures according to different water depths.

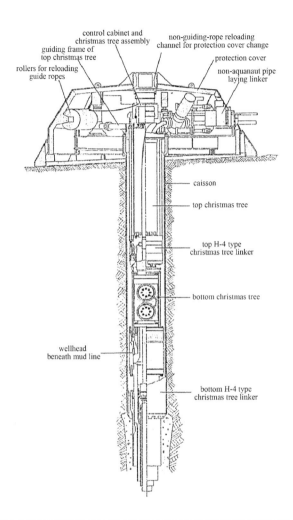

FIGURE 3-78

Caisson type Christmas tree.

    ii. Well reparation mode: If you choose the TFL well reparation mode, you need to select a TFL Christmas tree; otherwise select the non-TFL Christmas tree.

    iii. Oil reservoir characteristics: Reservoir pressure, temperature, and output should be all considered. High pressure requires the components to withstand high pressure, high temperature requires the seals to withstand high temperature, and high output requires large volume and diameter.

    iv. Well type: Christmas trees for oil wells and injection wells are basically the same, with slight differences; the structure of a Christmas tree for a gas well is quite different, as the throttles should use a needle valve, the components should be able to resist H_2S and corrosion, and valves should be configured in pairs.

v. Structure constitution: The structure should be as simple as possible. Importantly, there should be as few hydraulic should be as few as possible, and the component connections should minimize the use of flanges.

vi. Economic efficiency: Try to select a cheap Christmas tree with a long life and easy maintenance that is safe and reliable.

2. Subsea oil gathering manifold (oil gathering station)

This is used to gather the oil and gas of the nearby mouths of subsea oil and gas wells together, and then through an undersea pipeline transport the oil and gas mixed flow to a nearby floating production facility, fixed oil platform, or shore base for processing. Adopting an underwater oil gathering manifold can reduce the length of the submarine pipeline.

Figure 3-79 shows an underwater manifold for three wells to gather the oil. It is a horizontal cylindrical shell with seven pores, and three holes for three production wells; the bundles wearing into each hole include: two pipelines with a diameter of 5.1 cm, an annular space pipeline with a diameter of 5.1 cm (for gas lifting, preservatives or waterproofing agent injection, clearing wax, or other purposes), and a hydraulic line with a diameter of 3.2 cm. Of the remaining four draft pipe nozzles, one is for transporting the oil and gas mixture to the nearby production facilities through the pipelines; one is for the entrance of the pipeline from the nearby production facilities through the pipeline tools (TFL). Another is for introducing the power supply, communication cables, and hydraulic lines; there is also a spare hole. The top of the underwater manifold has a hatch, for connecting with the repair tanks for maintenance work.

3. Underwater manifold center (UMC)

The underwater manifold center is the main equipment of a new kind of single well bit multiwell subsea oil production system, as shown in Figure 3-80; the main equipment is an underwater manifold center (Figure 3-81).

The underwater manifold center has a variety of functions, such as drilling the submarine cluster wells through the chassis and linking to the satellite wells, gathering and controlling the oil and gas fluid produced from the chassis wells and satellite wells, and exporting oil and gas through the subsea pipeline to the nearby production facilities for processing. It can also allocate the injection water processed from the nearby production facilities to the injection wells; pour chemicals into the wells to prevent the waterproofing compounds from forming; can achieve no-diver

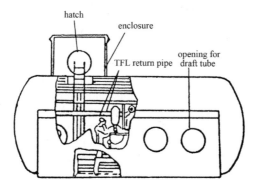

FIGURE 3-79

Subsea manifold for three wells.

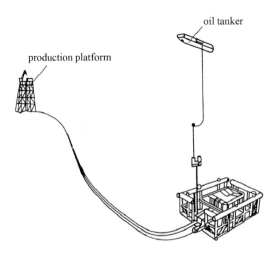

FIGURE 3-80

Single well bit multiwell subsea oil production system.

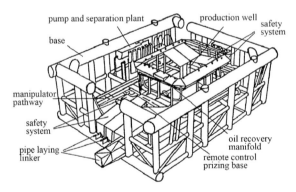

FIGURE 3-81

Underwater manifold center.

maintenance operations; and can support remote operation from the surface of the platform. The underwater manifold center consists of the following:

a. Large tubular chassis (seat)

The base as shown in Figure 3-81 has a 52 m length, 42 m width, 15 m height frame, made of large diameter steel pipe with a 2.7m upper tube and 1.4m lower tube, and it weighs 2120 t in the air, and 1785 t in the water, with a displacement of 2300 t. It can be divided into 44 watertight compartment tanks used as ballast tanks (pontoon), and the base is the main support structure with well-oriented columns and production facilities. The base is fixed to the bottom of the sea with four piles with a diameter of 760 mm, and the pile is infiltrated through the pile catheter on the base, and poured with concrete.

b. Wellhead device

This wellhead device drills multiwell chassis in a single well, including the ongoing underwater drilling wellhead equipment, and a Christmas tree for oil extraction after well completion, so that early production can be conducted. In order to use pumping tools to enter the workover, we usually equip the subsea Christmas tree with TFL, and adopt the 88.9mm ($3\frac{1}{2}$ in) tubing. Drilling wellhead devices and Christmas tree are all sent down to the seabed from surface working boats.

c. Oil production manifold

In addition to bringing the oil and gas fluid from the oil production wells on the chassis together, this can also collect the fluid from the satellite wells and the collected fluid is transferred via the pipeline connector to the nearby production facility for processing. The water after processing is assigned to injection wells via this manifold. The oil production manifold can act as a channel for the workover TFL tools and a channel for chemical injection, and as the pipelines for oil well testing, killing, and cleaning. In this way, the manifold is equipped with two 20 cm diameter main explorers (control oil and gas, distribute water injection) and two 20 cm diameter service explorers (oil well tests and TFL maintenance). In addition, there is also a 3.8 cm diameter chemical injection pipeline.

d. Pumps and separation station

The pump and separation station installed in the pedestal is shown in Figure 3-82(a). In order to avoid the difficulties of oil-gas-water mixed transportation, the oil and gas of the oil wells

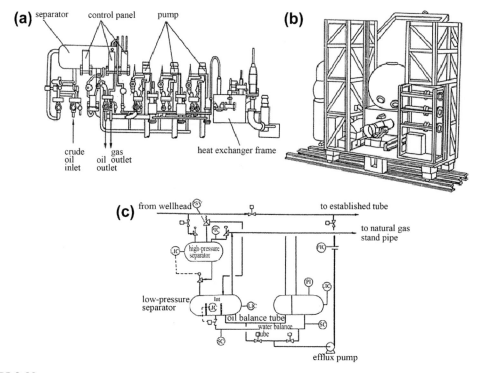

FIGURE 3-82

Pumps, separation stations, and service stations for underwater oil extraction.

can pass though the oil and gas separation to separate the gas, then reenter the multistage centrifugal pump, using the pumps to transport the oil through the pipeline connector to nearby production facilities for further processing. Figure 3-82(c) shows a typical underwater oil and gas separation process.

e. The underwater maintenance station

This can be sitting on the actuator track on the chassis, around the oil production manifolds, pumps and separating stations, and the control system, as shown in Figure 3-82(b). When signals are sent out from the water surface, the buoy is released from the subsea, and then the maintenance station floats up. When underwater maintenance operation is required, it is lowered to the actuator track on the chassis by ballasting with the help of its mats. When operations need personnels, the maintenance station carry men down into the subsea. Its total weights is 8.5 t, and is applicable to the water depth of 140 m.

f. Control system

The entire control system adopts hydraulic control, with pumps and ancillary equipment installed on the remote control skid base on the chassis. The oil pump adopts an electric drive, and the cable is driven. A gas accumulator is used to store enough gas for various valves, and the energy storage can be maintained 7 days in case of an accident.

The installation method of the underwater manifold center is as shown in Figure 3-83. It is prefabricated onshore, and placed in a launch barge for towing, then a tug and crane barge are utilized to leash the two sides of the manifold center after drowning, and it is towed to the well location; when it reaches the predetermined well location, it is suggested to use a 3000 t crane barge to lift the underwater manifold center to protect the rope from the tugs and the crane barge, and slowly tow it into the seabed.

FIGURE 3-83

The installation process of the underwater manifold center.

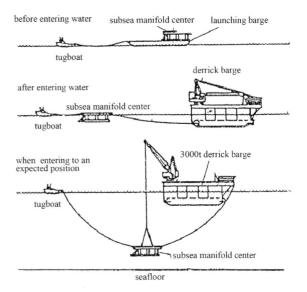

3.9.2 **Control System of the Subsea Production System**

According to classifications of the system's composition, there are eight typical structures of the subsea production and control system: direct hydraulic control, oriented hydraulic control, procedure hydraulic control, electro-hydraulic composite control, all-electric control, underwater autonomous control, and integrated buoy control.

Subsea control technology is a key technology that integrates the subsea production system and the production of the offshore oil field. In foreign countries, it is becoming a mainstream development model to rely on above-water devices and use the subsea system to conduct the production of the offshore oil field, whereas in China, the subsea production system and control technology are still at the initial stage. Every structural change for subsea production and control systems and every technological innovation reflects the urgent needs that the different periods of production of the subsea oil and gas source have for new functions of the control system. The main factors that lead to changes of structures and functions of the control system are the water depth of the development, the oil field scale, the tieback distance, and the economy. In engineering practice, the scheme of the subsea production system needs to integrate the production models of oil and gas field with the function and structure characteristics of every control system.

3.9.3 **Main Advantages of the Subsea Production System**

1. Conducive to the early development of oil fields: We can realize oil production within $1 \sim 2$ years after discovery, greatly reducing the time to put the system into operation.
2. Conducive to making full use of the drilled wells: Whether it is an exploratory well, appraisal well, or the well that has oil, you can carry on development and production first.
3. Conducive to developing remote sea fields of the well: Even if the individual well is located in remote marine waters outside the scope of the platform control, an underwater Christmas tree can be used to develop it, and then be tied back to the fixed oil platforms or floating production facilities.
4. Conducive to late stage infill development program: it is easy for subsea Christmas trees at the infill well heads to tie back to the old platform.
5. Conducive to avoiding adverse sea conditions: The underwater equipment is located in the seabed, avoiding the impact of surface wind storms and currents.
6. Conducive to recycling equipment for reuse: Subsea production equipment can be easily and economically recycled at the water surface for reuse.
7. Conducive to the development of the deep waters of oilfields: In some extreme conditions, such as the severe cold and heavy ice conditions in the Arctic Ocean, where the water depth is greater than 900 m, you can still use the subsea production system.

As it is convenient to recycle subsea facilitates and they can be reused, we can greatly reduce the cost of development of remote marine waters wells and post-encryption wells, etc. Early development of the oilfield and full use of the exploration wells and appraisal wells also greatly reduces the oilfield development cost.

3.9.4 The Subsea Production System of the Chinese Lufeng Oilfield

The Lufeng 22-1 oilfield in the South China Sea has a water depth of 333 m. It has adopted the subsea production system, creating a model to use one tanker to develop an offshore oilfield in China. Figure 3-84 shows the production system of the Lufeng 22-1 oilfield. It is mainly made up of:

1. Subsea Christmas tree
   Wellheads of five horizontal wells and a Christmas tree are located at the hinged combination chassis in the seabed and the underwater oil collecting manifold is on the chassis. The flow from five wells is gathered through two flexible production risers with a diameter of 203.2 mm (8 in), and tied back to a floating production storage vessel.
2. Floating production, storage, and offloading tanker (FPSO)
   The tanker "MuNing" has a length of 253 m, width of 42 m, and can store 10.2×10^4 t of crude oil. Processing equipment is installed in the rear of the main deck, and the daily processing capacity of crude oil is $0.9 \times 10^4 \text{m}^3$. It uses electric drive propulsion with diesel engine power and a dynamic positioning system on board, which keeps the ship at a predetermined position without using an anchor. When a typhoon comes, the mooring system can extricate and evacuate.
3. Turret production mooring system
   The mooring of the "MuNing" floating production is fixed at the submerged buoy by six suction anchors. The diameter of the suction anchor is 5 m and its height is 10 m. The weight of one suction anchor is 45 t. The submerged buoy is under the sea surface about 45 m. There are several groups of swivel joints installed at the mooring system. Electricity is sent to the underwater booster pump by one of them. Hydraulic power and chemicals are sent to the underwater equipment through one group of hydraulic swivel joint. The lines controlling signals cross

FIGURE 3-84

The offshore production system of the China Lufeng oilfield.

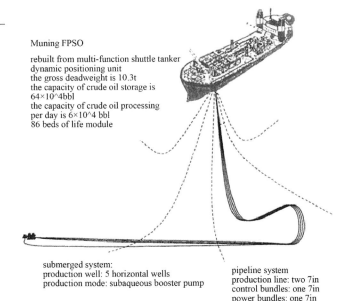

Muning FPSO

rebuilt from multi-function shuttle tanker
dynamic positioning unit
the gross deadweight is 10.3t
the capacity of crude oil storage is
64×10^4bbl
the capacity of crude oil processing
per day is 6×10^4 bbl
86 beds of life module

submerged system:
production well: 5 horizontal wells
production mode: subaqueous booster pump

pipeline system
production line: two 7in
control bundles: one 7in
power bundles: one 7in

through one group. There are another two crude oil production channels with a diameter of ϕ203.2 mm, through which the underwater oil and gas is sent to the "MuNing."

4. Hanger Over Subsea Template

This is called HOST for short. The overall chassis is divided into a center module and a number of wellhead-oriented modules. The size of the center module is 5.95 m × 5.45 m × 1.777 m, and it weighs 30 t. It is fixed first, and then the wellhead-oriented modules are assembled around the center module one by one. The central module connects with the conduit string of the wellbore and catheter wellhead, then down into the center of the borehole in the seabed. The wellhead-oriented module is a special mechanism to connect the central module to other wellheads, supporting and hanging the wellhead as well as putting the catheter into the wellbore. After well completion, the center manifold module is put down into the submarine center module, and the subsea Christmas tree will be installed. Compared with the conventional integral chassis for the same number of wells, this kind of suspension combination chassis can save 25% of the steel consumption, and 40% of the installation fee.

3.10 Offshore Oilfield Development and Production Facility Combinations

3.10.1 Combination Solutions

When developing offshore oilfields, given the different reservoir characteristics, geographical location, and marine environments, the production system always adopts different production facility combinations. The following are the usual combination solutions.

1. The first combination solution

WHP (wellhead platform) + FPSO (floating production storage and offloading) platform
This combination is made up of one or several wellhead platforms and floating tankers with the function of oil and gas processing, crude oil storage, and offloading. This production system is one where a simple single well measurement of the well fluid flowing from the wells is first conducted at the wellhead platform, and then it goes through the subsea pipeline system and is transported to the floating production, storage, and offloading (FPSO) tank by a single point mooring, for processing in the oil and gas processing system of the storage tanker. The qualified crude oil after processing is transported to a storage tanker for storage, and then carried away by the shuttle tankers.

   This combination solution is adapted by most of the offshore oilfields in China, and Figure 3-85 shows this combination solution of the Wenchang 13-1/13-2 oil fields in the South China Sea.

2. The second combination solution

WHP + CPP (central processing platform) + SPOD (storage platform and oil dock)
This combination solution combines one or several different functions of wellhead platforms and the center platform with the oil and gas processing system and an oil storage platform made of a number of crude oil storage tanks and the oil terminal. The first developed Bohai Chengbei oilfield in China belongs to this combination solution. As shown in Figure 3-86, the wellhead platform liquid of this solution is carried to separation processing in the drilling production platform, and the qualified crude oil after processing then is transported to the storage platform

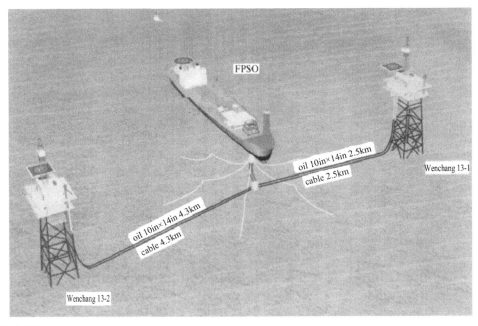

FIGURE 3-85

China Wenchang 13-1/13-2 offshore oil field production system combination solution.

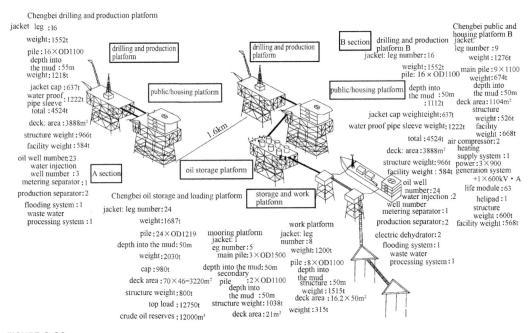

FIGURE 3-86

The production system of the Chenbei Oilfield in Bohai, China.

for storage. The shuttle tanker is moored in the oil terminal, and the crude oil of the tanks is transferred to it.

3. The third combination solution

SW (subsea wellhead) + FPSO

This combination solution is made up of a number of subsea wellheads and a production storage and offloading tanker with a oil-gas-water processing system, crude oil storage, and offloading. The South China Sea Liuhua 11-1 oilfield production system belongs to this combination solution. As shown in Figure 3-87, the subsea production system, composed of a subsea Christmas tree and underwater manifold, sends the crude oil to the FPSO for separation processing through a submarine pipeline, and the qualified crude oil after processing is stored in the oil storage tanks of the FPSO, and then transported to the shuttle tanker.

4. The fourth combination solution

OFP (offshore fixed platform) + LT (land terminal)

This combination solution is composed of a number of fixed offshore platforms (WHP or wellhead platform and central processing platform) and an onshore terminal with a certain

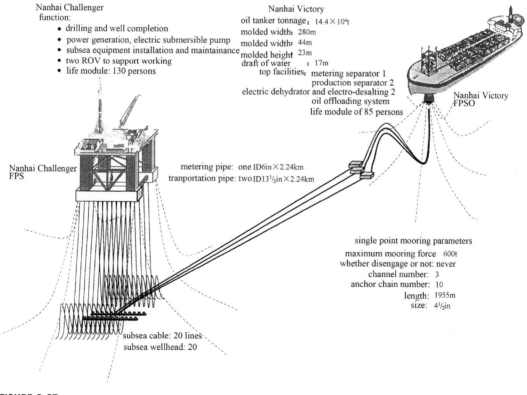

FIGURE 3-87

The South China Sea Liuhua 11-1 oilfield production system.

processing ability. This kind of production system is mainly applied in offshore gas fields or condensate oilfields and oil fields close to the shore, such as the South China Sea Ya 13-1 gas field (Figure 3-88), Bohai SZ36-1 condensate oilfield (Figure 3-89) and Pinghu gas field in the East China Sea, etc. All of them belong to this kind of combination solution. Gas fields of this kind of combination solution first dehydrate and pressurize the natural gas on a fixed platform, and then transport the natural gas via a subsea pipeline to the onshore terminal for processing to achieve the qualified products.

5. The fifth combination solution

OFP (offshore fixed platform) + AI (artificial island)

This kind of combination solution is composed of certain fixed platforms (WHP or wellhead platform and central platform) and production artificial islands with the function of production processing, crude oil storage, and offloading, and has not yet been adopted in China.

6. The sixth combination solution

CP (concrete platform)

The solution refers to conducting production processing, storing the crude oil, and outputting the oil on a concrete platform. The platform can be placed by a rig for drilling and well

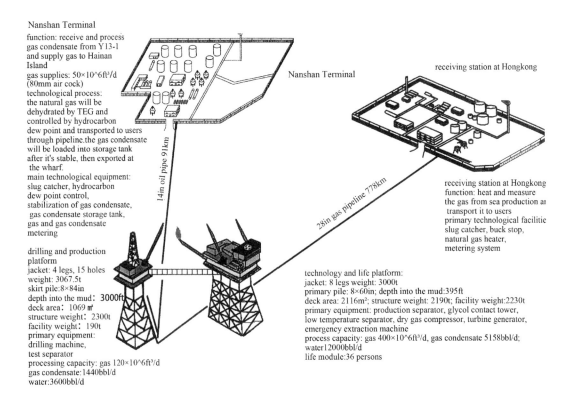

FIGURE 3-88

The South China Sea Ya 13-1 gas field production system.

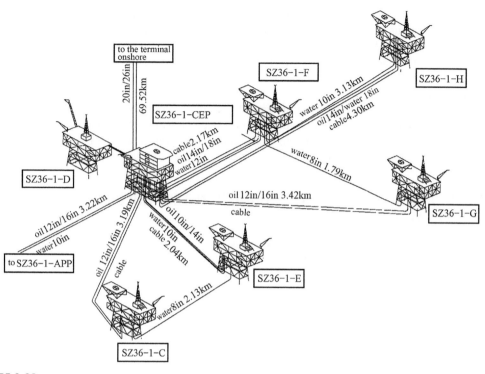

FIGURE 3-89

China Bohai SZ36-1 oilfield production system.

completion, and then conduct oil production, with a workover operation when needed. At present, there is no such offshore oilfield production system.

3.10.2 Selection Principles

When selecting between the above several combinations, you should consider the following factors.

1. Oilfield water depth
   Working depth has a significant impact on the selection of the combination solution. From an economic point of view, in shallow waters, when the water depth is less than 200 m, it is generally recommended to select a fixed platform production system (steel jacket platform or gravity platform). When the water depth is more than 200 m, due to a sharp increase in the cost of a fixed platform, the combination solution of subsea production systems and floating production facilities (such as the third solution) should be considered.
2. Oilfield location and scale
   If the oilfield is close to shore and has a certain scale, a solution that should be considered is to transport the pipeline ashore, and build a terminal with a oil and gas processing system and storage and transportation facilities onshore; you could also consider the solution of building an artificial island offshore.

If the oilfield is far away from the shore, or we have marginal oilfields with little yield, floating production systems should be selected, as the reusable features of floating production systems can be taken advantage of.

An artificial island solution will be adopted if the production is large and the depth of the sea is shallow (less than 10 m).

3. Seafloor topography

For a sea area with flat seafloor topography and hard soil, the concrete gravity platform (the sixth combination solution) should be considered. For a sea area with soft soil and uneven seafloor topography, a fixed platform (depending on the depth of the water) or other production facilities should be considered.

PART III OFFSHORE OIL (GAS) PRODUCTION TECHNOLOGY AND EQUIPMENT

Due to the difference of offshore oil and gas in reservoir structure, type of drive, depth, and fluid properties, mining methods also are not the same. This section will introduce the process and equipment for commonly applied production methods for offshore fields, such as natural flow and artificial lift, etc. Methods that are similar to land will not be covered here.

3.11 Mining Methods and the Selection of Offshore Oil and Gas Fields
3.11.1 Mining Methods

1. Natural flow

If the development totally depends on the natural energy of oil and gas layer, the fluid can flow through the wellbore to the wellhead from the oil and gas layer and enter into the ground processing system, maintaining the partial pressure, which is called natural flow. The general gas reservoirs are the main applications of natural flow, but parts of the gas reservoirs also use the drainage gas recovery in the production tail.

2. Artificial lift

This is also known as mechanical oil production, including sucker rod pumps, screw pumps, electric submersible pumps, hydraulic piston pumps, jet pumps, gas lifts, piston pumps, cavity gas lifts, electric pumps, submarine boosters, etc. This machinery and equipment can lift the fluid to the surface. Electric submersible pumps and hydraulic piston pumps are commonly used in offshore gas lifts.

3.11.2 Special Requirements of Offshore Oil and Gas Production for Production Facilities

The environmental conditions of offshore oil and gas fields are different from the land, thus the process of oil and gas production has special requirements for production facilities.

1. Small volume: The requirements for production facilities are that the structure should be as simple as possible in order to make them small, which minimizes the size and the area of the wellhead platform.

2. Lightweight: The production facilities must minimize weight to reduce the load of the platform and jacket and to simplify the structure of the wellhead platform.
3. Long operation period: Lengthen the production facility maintenance cycle and reduce maintenance time to improve the efficiency of offshore production so as to reduce the cost of operations at sea.
4. Wide range of applications: During development of oil and gas fields, when the reservoir pressure, physical properties of fluid and other aspects change, the mining method and ground production facilities remain unchanged.

In short, we want to solve the problem of improving the overall economic efficiency of oilfield development.

3.11.3 **Principles and Methods of Production**

1. Meet the requirements of the well production parameters and make it technically feasible. Well production parameters are the basis of selecting production methods and the main production parameters include fluid volume, fluid properties, formation properties, and production pressure and we should pay attention to their accuracy and reasonable range. According to the well production parameters and from the perspective of technical feasibility seen in Figure 3-90, we can select several production methods of artificial lift that are technically able to meet the requirements of oilfield development and provide a good mode of operation. Because there is not only one choice, in addition to technically feasibility, we need to evaluate the choices further from economic, reliability, and operability standpoints to make a choice.
2. Meet the requirements of offshore production facilities and reliability in production To adapt to the characteristics of offshore oil exploration, offshore production facilities must be small in volume, lightweight, have a long operation period and wide range of applications, and so

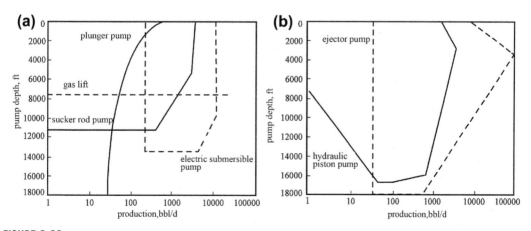

FIGURE 3-90

Selection of mining methods for offshore artificial lift.

on. So it is necessary to compare the production methods selected from Figure 3-90. Special attention also should be paid to the reliability and operability of electricity and gas meters and other auxiliary equipment, such as whether it is easy to control and reduce the extent of loss caused by human error.

3. Meet the requirements of the comprehensive evaluation index, and improve overall economic benefits

The comprehensive evaluation index includes technical, economic, reliability, operability and other aspects that can be refined as multiple indicators to compare with measures such as initial investment, working life, maintenance cycle, production and operating costs, the degree of localization, mechanical efficiency, power source, the degree of energy consumption, integrated economy, and so on. In short, a comprehensive economic method is selected from several mining methods by comprehensive evaluation. In accordance with the principles and methods of the selection, Figure 3-91 shows the logic diagram.

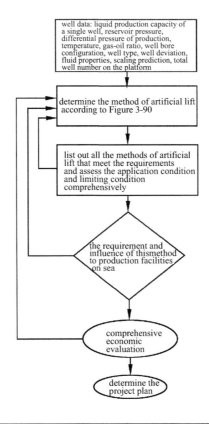

FIGURE 3-91

The logic diagram of the offshore oil field.

3.11.4 Select the Time When the Production Methods of the Offshore Field Transforms from Natural Flow to Artificial Lift

The main considerations, when selecting, are the following two factors:

1. Changes to the bottom hole flow pressure
   The bottom hole flow pressure constantly changes during oil exploration. When it changes to below a certain value so as not to raise the fluid column to the ground, we need to timely adopt the appropriate artificial lift methods.
2. Requirements of oil production
   Considering the overall efficiency of oilfield development, the oilfield should have appropriate production requirements. Therefore, when the well still has the natural flow ability but is reaching less than reasonable oilfield production requirements, we need a timely transfer to artificial lift from natural flow. It is necessary to transform to artificial lift to improve the yield of the oil well through the use of external energy to meet the requirements.

3.12 Offshore Production Technology and Equipment of Flowing Wells
3.12.1 Well Structure

An offshore well structure is shown in Figure 3-92.

1. Marine riser
   This is a type of structure that is different from those used on land, and it can isolate the drilling (drill string) columns or oil tube from strong corrosive seawater and can also be used to guide the drilling tool, production string, and other tools, and separate drilling fluid with seawater when circulating mud. They are generally made of 762 mm (30 in) diameter casing; the riser is at a depth of a 70~100 m mud line at sea generally.
2. Surface casing
   Basically this is the same as on land, generally using 508 mm (20 in) casing down into the mud line at 300 to 500 m depth. Individual circumstances use 340 mm (13/8 in) casing; accordingly the riser in this case is 508 mm (20 in).
3. Intermediate casing
   This is the same as the land, often using 340 mm (13 3/8 in) or 244 mm (9 5/8in) casing. If the well is deeper, we can choose the two layer of intermediate casings.
4. Oil casing
   Similar to the land scenarios, in order to ensure the normal production of oil wells, 177.8 mm (7 in) oil casing often is used. If the well is deeper, sometimes 177.8 mm (7 in) tail pipes hanging on the bottom of a casing of 244 mm (9 5/8 in) are selected, in order to save on the cost of drilling.
5. Oil tube
   This is a channel of oil and gas from the reservoir to the wellhead, and it commonly uses 1 11/16 in, 1 13/16 in, 2 1/16 in, 2 9/16 in, 3 1/16 in, and 4 1/16 in sizes and so on.

3.12.2 The Wellhead Device

The typical wellhead device for natural flow at sea is shown in Figure 3-93.

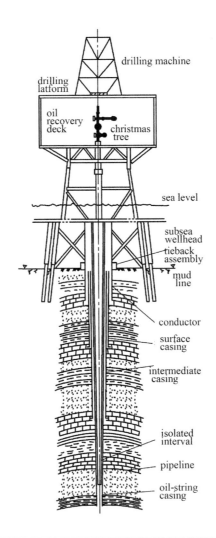

FIGURE 3-92

Well structure of an offshore oil well.

1. Wellhead

   A wellhead is actually composed of the casing head, casing spool, and tubing head, which is located in the lower part of the wellhead (Figure 3-93). The Chinese name a casing head by its joint name traditionally.

   a. Casing head and casing spool

   Its role is to connect layers of casing and to seal the gap between the layers of casing, to support the BOP stack when drilling, and to hold the tubing head and Christmas tree after completion.

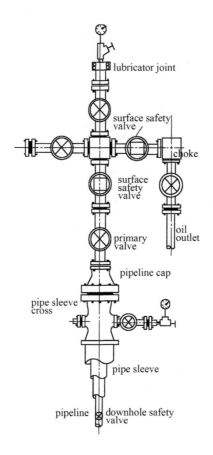

lubricator joint

surface safety valve

choke

surface safety valve

primary valve

oil outlet

pipeline cap

pipe sleeve cross

pipe sleeve

pipeline

downhole safety valve

FIGURE 3-93

The typical wellhead device for the offshore natural well.

   b. Casing tieback device

This is a special marine device installed in the mud line position (Figure 3-92) and its role is to detach or connect the casing above or below the mud line, which is applied during drilling breakdown or tieback. The mud line hanger and the subsea wellhead, which are respectively used on the self-elevating drilling platform and the semi-submersible drilling platform, are just different forms of the casing tieback device.

   c. Tubing head

This is called a big cross joint, and it is placed above the casing spool for hanging tubing, and sealing the tubing and annulus of the production casing.

   d. Tubing hanger

This is used to hang tubing in the tubing head. There are threads inside the tubing hanger which is used to connect with the back-pressure valve. After the perforation operation is completed, the back-pressure valve is loaded into the tubing hanger through the blowout preventer, tightened with threads inside the back-pressure valve. The back-pressure valve can seal the tubing hanger to prevent blowout after the blowout preventer is dismantled. After installing the Christmas tree, special running tools should be sent to dismantle the back-valve pressure from the top of Christmas tree, subsequently washing the well and inducing flow.

2. Christmas tree

It is installed at the tubing head. And its main function is to control the production rate of natural flow wells and measure required parameters of production wells (such as casing pressure, tubing pressure, and provide necessary conditions of operations of workover and replacing the parts).

The Christmas tree consists of the tubing cap, the main valve (gate), a tubing cross joint (small cross joint), wing valves (production gate), an overhead valve (workover gate), a Christmas tree cap, and a nozzle (restrictor). There are two main valves, one of which is manual-operated, the other is automatic. And so are the two swing valves.

An offshore Christmas tree can be installed in two places:
- Mounted on a platform on the water (Figure 3-92)
- Installed in the underwater wellhead on the seabed (see Part II of this chapter)

3. Cracked valve (cherry)

This is installed at the outlet of the Christmas tree, as shown in Figure 3-94. Its function is to use a small aperture throttle to control oil production. There are two common forms:
- Adjustable cracked valve
- Fixed cracked valve

The pressure before the cracked valve is called the oil pressure, namely the pressure of the bottom hole that lifts the liquid to the ground; the pressure after the cracked valve is called the wellhead back pressure, which includes piping resistance from the wellhead to the separator and the separator pressure. The casing pressure is the pressure between the tubing and the casing annular space, which reflects the changes in the conditions of well production. All of these pressure parameters should be measured to indicate well production performance.

3.12.3 Well Safety Systems

That the Christmas tree gets out of control or catches fires will cause serious damage for natural flow well because its tubing pressure is very high. In addition, because the arrangement of the offshore wellhead devices is crowd and traffic is inconvenient, the safety system is more important for saving lives. Figure 3-95 shows the offshore wellhead safety system.

FIGURE 3-94

The structure of the cracked valve of a natural well.

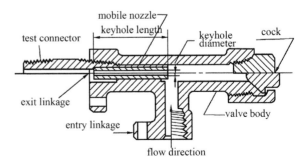

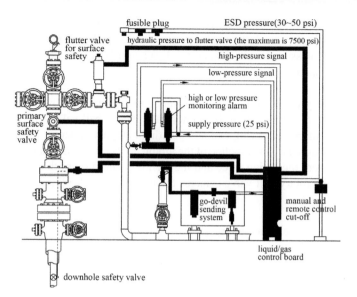

FIGURE 3-95

Wellhead safety system for an offshore oil well.

1. **Fusible plug**

   As shown in Figure 3-95, this is installed on the emergency shutoff pipeline near the wellhead. When the temperature is higher than a predetermined temperature (typically set to 123°C), fire breaks out, and the plug will melt so as to release a low control gas in the ESD pipeline, leading to shut-in and starting the fire extinguishing pump at the same time.

2. **High and low pressure alarm monitoring**

   As Figure 3-95 shows, this is mounted on the flow lines behind the cracked valve, with two monitors: high and low pressure. Signals will go to the control panel when the line pressure is too high or too low, which leads to shut-in and starting the alarm.

3. **Emergency shutoff station**

   This is also called the ESD (emergency shutdown) station, and can be installed on the helicopter deck, lifeboat station, and so on. The station, through both remote control and manual operation, can quickly cut off all wellhead oil flow from different parts of the platform manually in case of emergency.

4. **Liquid/gas control panel**

   A pneumatically powered hydraulic pump opens the surface and downhole safety valves by hydraulic pressure. And then the well can produce normally. When the temperature and the pressure in the well is abnormal, the gas inside the pneumatically powered hydraulic pump will vent and its hydraulic pressure drops, which make safety valves closed and the well shut-in. The wellhead control panel mounted in a central control room on the platform can also control all the wellheads.

5. **Ground safety valve**

   This is often called an SSV (surface safety valve), and it is installed at the wing valve position, or both at the wing valve and the main valve position (Figure 3-95). Generally, the normally closed valves with full open state and pressure startup are often chosen. It has two forms: pneumatic or hydraulic driving. The valve in a fully open state relies on the liquid (gas) pressure. In the case of an accident, with the driving pressure dropping, the valve will close automatically under the action of the spring. If the temperature is abnormal, the fusible plug is activated. When the pressure is abnormal, the high and low pressure monitor is activated. When it is needed, the control panel and ESD station can be manually cut.

6. The downhole safety valve

This is called the SSSV (subsurface safety valve) and is usually installed in the tubing below the initial point of wax at a downhole depth of 30 ~ 300 m. It is a normally closed ball valve with pressure startup, which consists of a compressed spring and an operating ball valve piston. Hydraulic pressure from the ground through the ½ in pipeline of the tubing hanger keeps the value open. When hydraulic pressure is released, the spring force and the well pressure prompt the upward movement of the piston, and the rotary ball valve changes to close position, leading to shut-in.

7. Gas source

The gas source for the entire well safety system can be provided by compressed air and natural gas, which is produced by wells. Because the use of natural gas requires a large gas-oil ratio, and safety is also poor, the existing compressed air systems are used as the gas source for offshore platforms.

3.12.4 **The Paraffin Removal Process for an Oil Well**

As regards natural flow, if the crude is high wax content oil, during the course of the movement to the wellhead, as the pressure and temperature is dropping, the wax separates out from crude oil and congeals in the tubing wall, which is known as wax deposition. Wax precipitation must be cleared out in a timely fashion, through a process called paraffin removal. There are many ways of knifing a well, such as mechanical paraffin removal, hot oil paraffin removal, electro-thermal paraffin removal, and chemical paraffin removal, and in an offshore oilfield we often use the chemical paraffin removal method.

Wax deposition is particularly serious in the pipe sections from the Christmas tree to the test manifolds, due to the lower temperature and pressure. If the pipe section is shorter, the steam or electric heating methods can be used to clear the wax. If the pipe segment is longer, usually a pigging transmission device is created at the outlet of the Christmas tree (Figure 3-95), and a receiving apparatus is established before the test manifolds. In this way, one to three ball pigs can be sent for wax removal in one day.

3.12.5 **Production Analysis of a Flowing Well**

Flowing well oil production uses the energy of the reservoir oil to spray the oil to the ground. The energy source is the reservoir pressure. The pressure that makes the crude oil from the reservoir flow to the bottom of the well is known as the flowing bottom hole pressure. When bottom hole flowing pressure is relatively high, the pressure can raise the oil to the wellhead, and even into the gathering manifolds and the separator. In addition, with the rising oil and gas flowing along the wellbore, the pressure will gradually reduce, and the gas will be separated out and expand its volume. The elastic expansion energy of gas is also one of the energy sources of the natural flow. Of course, as time goes on, the energy supplying the natural flow weakens, so the yield of the natural flow decreases.

Therefore, it is necessary to study the relationship between the pressure and the production and to analyze the flowing characteristics of natural flow.

1. The basic flow process of natural flow
   a. Seepage in the reservoir: flow from the reservoir to the bottom
   b. Flow in the wellbore: flow from the bottom to the wellhead
   c. Flow of crude oil through the choke after arriving at the wellhead
   d. Ground pipeline flow: flow from the crack to the separator
      When analyzing the production of natural flow, the first and second flow processes are mainly taken into consideration.
2. Flow characteristics of natural flow
   The IPR (inflow performance relationship) flow characteristic is the relationship between oil production and flowing bottom hole pressure, and the curve that shows the relationship is called the flow characteristic curve or the flow dynamic curve. It reflects the ability of the first flow process where the crude oil flows from the reservoir to the bottom hole, and also is the base for analyzing production performance. The IPR curve is related to the reservoir type, as shown in Figure 3-96.
3. Formula for single-phase flow yield
   According to the percolation theory, the yield formula of a well at the center of the infinite formation is listed as follow with the Darcy's law.

$$q_o = \frac{K_o h (p_r - p_{wf})}{141.2 \mu_o B_o \left(\ln \frac{r_e}{r_w} - \frac{3}{4} + S\right)} \tag{3-6}$$

where
   q_o represents oil production, bbl/d
   K_o represents effective permeability of the reservoir, mD
   h represents the effective thickness of the oil layer, ft
   p_r represents average reservoir pressure, psi
   p_{wf} represents bottom hole flowing pressure, psi

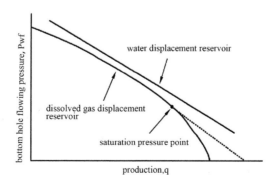

FIGURE 3-96

The typical IPR curve.

μ_o represents viscosity of crude oil, St

B_o represents the reservoir volume factor

r_e represents well drainage radius, ft

r_w represents wellbore radius, ft

S represents the skin factor

The parameter S is a coefficient relative to well completion, well contamination, stimulation treatment, and so on, and can be obtained from the analysis of actual production performance data. It is generally expressed by drill size, and other parameters may be obtained by the well testing, measuring wells, and logging data.

4. Formula for two-phase flow in production of oil and gas

Vogel provides a simple formula which is used to draw IPR (inflow performance relationship) curve of the dissolved gas drive reservoir, namely:

$$\frac{q_o}{q_{max}} = 1 - 0.2\left(\frac{p_{wf}}{p_r}\right) - 0.8\left(\frac{p_{wf}}{p_r}\right)^2 \tag{3-7}$$

where q_{max} represents the maximum theoretical yield of the wells when p_{wf} is zero, bbl/d. The symbols of the other parameters have the same meaning as in Equatoin (3-6).

Example 3-1: For the Z well in an oilfield in the South China Sea, the average reservoir pressure $p_r =$ 3600 psi, flowing bottom hole pressure $p_{wf} =$ 3400psi, and yield $q = 3600$ bbl/d. Try to use Vogel formula to draw the well IPR curves.

Solution:

a. Calculation of q_{max}:

$$q_{max} = \frac{q_o}{1 - 0.2\left(\frac{p_{wf}}{p_r}\right) - 0.8\left(\frac{p_{wf}}{p_r}\right)^2} = \frac{3600}{1 - 0.2\left(\frac{3400}{3600}\right) - 0.8\left(\frac{3400}{3600}\right)^2} = 36911 bbl/d$$

b. Calculation of the yield under different flow pressures:

$$q_o = \left[1 - 0.2\left(\frac{p_{wf}}{p_r}\right) - 0.8\left(\frac{p_{wf}}{p_r}\right)^2\right] q_{max}$$

q_o is calculated according to the above formula and is listed in Table 3-6.

c. Draw IPR curves

According to the data of Table 3-6, we can draw the IPR curve, as shown in Figure 3-97.

5. Analysis of the natural flow production of the multilayer offshore reservoir

Because of the higher costs of oil production in offshore fields, we are required to produce more oil in a relatively short period of time. So commingling, multilayered reservoir production is widely used.

Table 3-6 Values Calculated in Accordance with the Vogel Formula

Flow pressure, psi	3400	3000	2500	2000	1500	1000
Yield, bbl/d	3600	10253	17544	23695	28708	32581

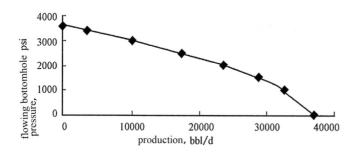

FIGURE 3-97

The IPR curve of the Z well in an oilfield in the South China Sea: 1 bbl = 0.1589 m^3; 1 psi = 6.894kPa

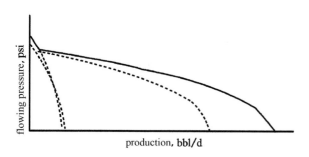

FIGURE 3-98

IPR curves for multilayer exploitation of wells.

a. The inflow characteristics of the commingling

In general, in the same strata or adjacent oil layers, if the oil layer's physical properties are little different, and the pressure system are the same, and the deliverability of each oil layer is little different, the commingling production technique is often used. The total IPR curve of multi-layer production well is the superposition of the IPR curves of each oil layer, as shown in Figure 3-98.

b. Commingling and single production options

The choice of both production methods is related to production performance and the different development stages of the offshore oilfield.

In general, in the early development stage, when multi-layers' features in one well meet the commingled conditions as the above, multi-layers commingling can be used. As the production of one oil well is going, the characteristics of different oil layers in one well will change at different levels, which will affect the efficiency of the commingling. For example, the high permeability layer is firstly flooded by the injection water which affect the producing rate of middle and low permeability layers. So the blocking measures is taken to the high permeability layer. And the timing of sealing the high permeability layer is determined according to two aspects that whether the ultimate recovery of the layer decreases and whether the production rate of one well meets production requirements. After economic evaluation, some wells can flexibly use multi-layers commingling, round mining or blocking.

3.13 Tools and Equipment of Marine Mechanical Production

Mechanical production is also called artificial lift. It is a production practice that uses different machinery and equipment to raise the liquid in the reservoir to the ground through artificial methods. Because a well at sea is a directional well, which limits the use of a rod, a rod pump is generally not used at sea. Therefore, offshore artificial lift generally uses methods such as gas lift, electric submersible pumps, hydraulic piston pumps, and so on.

3.13.1 Gas Lift

Gas lift is an artificial lift method that is the closest to natural flowing production. Firstly inject compressed gas into the bottom of the tubing, where it mixed with fluid that oil layers produce. So both the density of fluid that oil layers produce and the weight of fluid column in the oil tubing decrease, which lead the flow pressure gradient in the oil tubing to decrease. And then pressure difference between the bottom hole and the layer increases, which is the power that fluid in the oil tubing flows. So the method of the gas lift is the same as the natural flowing production. And the difference is that the gas lift comes out by the artificial method. Figure 3-99 compares the gas lift system with natural flowing production's.

1. Types of gas lift

    The key to the gas lift method is gas lift valves installed in the tubing with special design. These valves are installed along the tubing and provide channels for the injected gas from the casing to the tubing. They can be made into the shaft-like, and at each top end there is a wire rope. They are put down to the designed positions by the traction winch, as shown in Figure 3-100. In addition, they can also be pull out by the traction winch.

    Although there are several types of gas lift, the following are the main two.

    a. Continuous gas lift

    Gas with some pressure is injected into the tubing by opening the gas lift valve, and the liquid upon the gas lift valve mixes with the gas, becomes lighter as the gas mixes in, and is raised to the ground by the continuous expansion of the gas. This method is called continuous gas lift. Figure 3-101 shows the entire drainage process of a continuous gas lift.

    b. Intermittent gas lift

    If gas injection is done at a predesigned interval and crude oil is produced in a certain intermittent cycle, this is called intermittent gas lift. The intermittent gas lift device includes a ground control valve and a cycle-time controller on the gas injection pipeline to control the gas injection time and closing time. For intermittent gas lift, there are a few gas lift valves downhole. The higher valves are known as the drain valves and only the bottom valve is called the gas injection valve. The intermittent gas lift drainage process is similar to continuous gas lift. At the beginning, all gas lift valves are open, and then they gradually turn off at each level, until the bottom injection valve is open, as shown in Figure 3-102.

    Because intermittent gas lift uses the non-return valve, the bottomhole flowing pressure is very low. So it is suitable for low pressure and low yielding wells of which production rate are $0.16 \sim 80$ m^3/d. However, due to intermittent work of the gas lift, the bottom hole pressure is unstable, which is easy to lead that the reservoir flows sand. And the wave surfaces of the ground processing facilities fluctuate greatly, which is difficult to control and leads that the

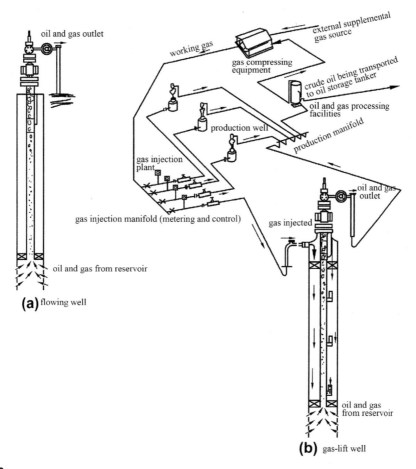

oil and gas outlet

working gas

external supplemental gas source

gas compressing equipment

crude oil being transported to oil storage tanker

oil and gas processing facilities

production well

production manifold

gas injection plant

oil and gas outlet

gas injection manifold (metering and control)

gas injected

oil and gas from reservoir

(a) flowing well

oil and gas from reservoir

(b) gas-lift well

FIGURE 3-99

Comparison of the gas lift production system and natural flowing production.

efficiency of facilities is very low. In contrast, continuous gas lift is suitable for higher production wells of which production rate are $16 \sim 11924$ m^3/d. It has high efficiency due to the full use of its energy. Furthermore, because of the stability of the bottom hole pressure flow, it is not easy for sand and wax to enter the oil well. Continuous gas lift does not apply to low-yield wells because of the high bottom hole pressure.

2. Gas lift valve

   a. Work principles

   The principles of the gas lift valve are similar to the general pressure regulating valve. As shown in Figure 3-103, when the spring of the gas lift valve pushes the valve stem that makes the valve block sit on a valve seat, and the production pressure multiplied by the valve seat area is greater than the spring force, the valve body will be out of the valve seat and the valve will open. Otherwise, the valve is closed. As can be seen from Figure 3-103, when considering

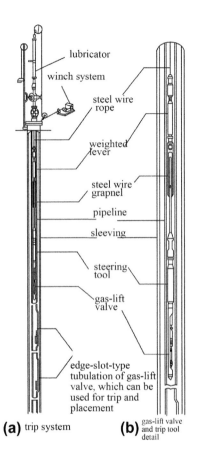

(a) trip system   (b) gas-lift valve and trip tool detail

FIGURE 3-100

The gas lift valve in the oil tube.

the injection gas pressure for the bellows area (cross-section), the inner pressure of the bellows force F_c that makes the gas lift valve close is as follows:

$$F_c = p_{bt} \times A_b \tag{3-8}$$

F_o is the force that makes the gas lift valve open:

$$F_o = p_1 \times (A_b - A_p) + p_2 \times A_p \tag{3-9}$$

Obviously, when $F_o > F_c$, the gas lift valve opens, from which we get

$$p_1 \geq \frac{p_{bt} - p_2 \left(A_p / A_b \right)}{1 - \left(A_p / A_b \right)} \tag{3-10}$$

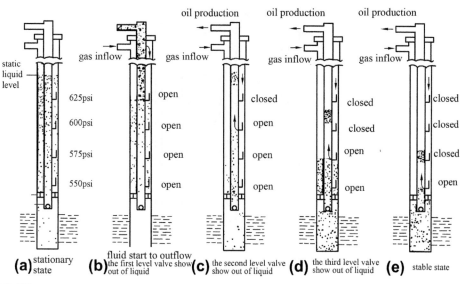

FIGURE 3-101

The process of continuous gas lift cleanup.

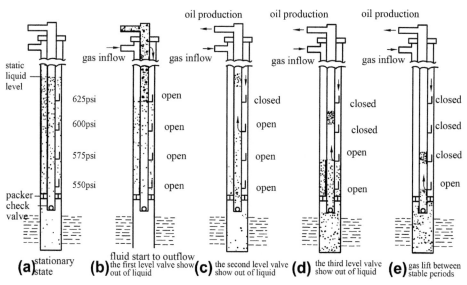

FIGURE 3-102

The process of intermittent gas lift cleanup.

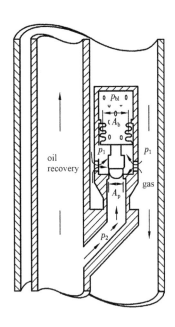

FIGURE 3-103

The work principles of a gas lift valve with annulus injection.

When the gas lift valve turns on, the gas from the tubing and the casing annulus enters into the tubing from the hole in the gas lift valve cylinder wall through the value.

b. Types

There are several methods to classify the types of gas lift valve, for example, according to continuous gas lift and intermittent gas lift, according to whether the main pressure that controls the opening and closing of the gas lift valve is the pressure of gas injection or the production pressure, according to whether the position of the gas injection is in the tubing or the tubing casing annulus, and according to whether the gas valve can be sent in by the wire line and is fixed or removable. The removable annulus injection spring gas lift valve is widely used at sea. The sizes of the gas lift valves used generally in offshore oilfields are 38.1 mm (1.5 in) and 25.4 mm (1.0 in). Figure 3-104 shows the structure diagram of the removable gas lift valve.

3. The process of the gas lift system

When the gas lift method is used, offshore platforms need to provide a set of separators and compressors for injection gas and so on, which is called a gas lift system.

a. The process for a gas lift system using ground equipment

Figure 3-105 is a diagram of the gas lift's system, which shows the arrangement of all equipment on the offshore platform.

This is the ground equipment process for gas lift production used in the South China Sea. The oil and gas mixture produced by the gas lift flows through the separator, and the separated gas goes through a multistage compression and is injected into the well to lift the oil, which recycles. Water precipitation occurs after each stage compression, and separation and desiccation can be carried out by molecular sieve. At the start of the system, because there is

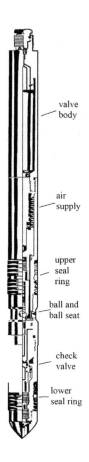

FIGURE 3-104

Removable gas lift valve structure.

no supply of natural gas, we need to produce nitrogen, which is separated from the air. Like natural gas, multistage compression is necessary. After the system begins to operate, we can use natural gas instead of nitrogen.

b. The process for gas lift production in an underwater wellhead

Figure 3-106 shows the diagram for gas lift production in a certain underwater wellhead in the South China Sea. The figure shows the distribution of underwater oil pipelines, gas pipelines, and the control line, as well as the control line junction except for the ground equipment on the platform.

4. The advantages and disadvantages of gas lift production

Gas lift production is a method of artificial lift that can make full use of the natural energy of the reservoir. This is because it has the following characteristics:

a. Flexibility

First of all, this can be adapted to a very large range of production. Take the continuous gas lift used in the South China Sea for an example; the same production string can be adapted to the

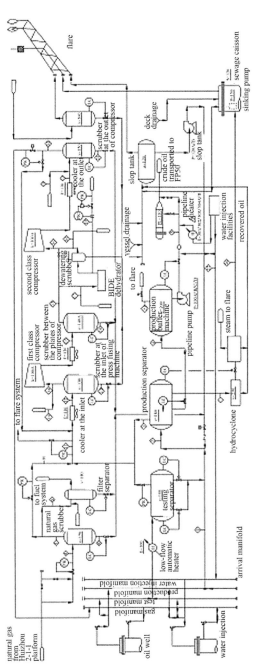

Figure 3-105 the flowpath of the ground equipment of gas lift system

FIGURE 3-105

Process flow diagram and facilities of the gas lifting system.

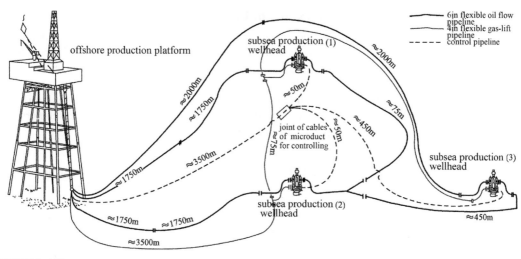

FIGURE 3-106

The process for the gas lift production system of the subsea wellhead at one of the fields in the South China Sea.

production range of 95 to 1590 m^3, which cannot be achieved in other artificial lift oil production. The gas lift system only needs to adjust the depth of the gas lift valve, the gas injection volume, and the gas injection pressure, which can be adapted to different production requirements. It doesn't need to carry tubing string or need to be redesigned, unlike other artificial lift methods. Second, it can be applied from the position of the initial static liquid level close to the ground where the static fluid level is near the bottom of the well when the well almost is dried up.

b. Economy

Gas lift does not require the carrying of tubing string and additional downhole equipment, and the major equipment is installed on the platform, which makes maintenance and allows the use of natural gas as fuel supply, with low operating costs for the entire system. When oil production decreases, we can use the wire rope to replace the gas lift valve to increase production. Because there are fewer moving parts in the gas lift system, the repair cycle is longer, and economic efficiency is higher.

c. Adaptability

Gas lift is suitable for large incline angle wells, larger dogleg wells, sand wells, high gas-oil ratio wells, and waxy and scaly wells, and it has high adaptability.

But it should be noted that there are also limitations in gas lift production. For example, the gas source must be able to supply enough gas, and this requires that the oilfield itself must have a minimum limit (10% for a normal gas lift) of the amount of dissolved gas. There must be a certain degree of the bottom hole flowing pressure, so it is not suitable for low-pressure wells. In addition, the gas lift method is not suitable for heavy oil and emulsion oil wells.

3.13.2 Electric Submersible Pumps for Oil Production

Electric submersible pumps for oil production are an effective, economic, maturing artificial lift method. Due to their great range for displacement and lift, high power, huge pressure of production, strong adaptability, simple technological process, long working life, convenient management, and significant economic benefits, it has been widely used. Offshore oilfield cluster wells and high-yielding wells are appropriate applications for this kind of pump. For example, in the SZ36-1 oil field in Bohai Bay, China and the W10-3N oilfield of the western part of South China Sea, electric submersible pumps for oil production are used after the end of the flow period.

1. Composition of the unit

    The electric submersible pump is a specially designed multistage high-pressure centrifugal pump. The submersible motor, protector, gas separator, multistage centrifugal pump, and armored cable of the entire unit are mounted at the end of the tubing down to the well and the armored cable is fixed to the tubing, as shown in Figure 3-107. Sometimes, we use the method of hanging the entire

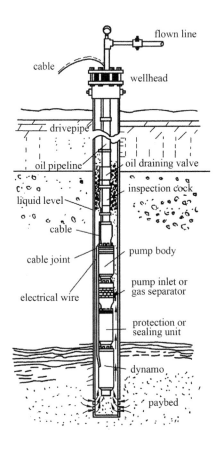

FIGURE 3-107

Electric submersible pump unit.

unit into the well with a cable suspension, which eliminates the need for tubing. Not only it this convenient for carrying but also the size and displacement of the pump can increase, and the friction of the liquid to the ground decreases. We introduce its composition in the following.

a. Submersible motor

The submersible motor structure is shown in Figure 3-108. Submersible motors are used in offshore oilfields, and are mostly closed-end three-phase induction motors installed in a steel tube that is filled with transformer oil, and whose diameter ranges from 1 to 0. 15 m. The transformer oil pressure is slightly higher than the fluid pressure of the depth where the motor is sent to prevent formation water from seeping into the interior of the motor. Motor power ranges from 10 to 500 kW and its maximum temperature is 95°C. Because the motor works in an underground high temperature environment for a long time, it has a special oil circulation system for lubricating and cooling the moving motor parts. The motor rotor and stator are

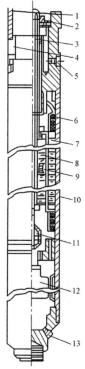

Figure 3-108 structural diagram of
submersible motor

1-ribbon cable 2-thrust bearing 3-axis 4-cable 5-fuel
feeding valve 6-lead wire 7-stator 8-rotor 9-alignment
bearing 10-shell 11-lubrication impeller 12-filter
screen 13-oil drainage valve

FIGURE 3-108

The chart of the submersible motor structure.

separately divided into a number of sections and each section has a separate winding that strings on the motor shaft, which forms a thin and long motor shape.

b. Electric submersible pumps

This is a multistage centrifugal pump including stationary and rotating parts and the structure is shown in Figure 3-109. The fixed part consists of the guide pulley, the pump housing, and the bearing jacket and the rotating portion includes the impeller, shaft keys, friction pad, ,bearing and retainer. Electric submersible pumps are divided into sections, and the sections are divided into levels again, and each level is a centrifugal pump. Submersible pumps can be divided into floating, semi-floating, and fixed categories according to whether the impeller is fixed. The number of electric submersible pumps used in an offshore oilfield range from 84 to 332; the highest is up to 400. The length ranges from 5 to 10 m, with the longest up to 20m. The outer diameters range from 85.5 to 102 mm, the displacement ranges from to 30 to 8000 m^3/d and the pump efficiency ranges from 44% to 52%.

c. Protection

The role of the protector is sealing and lubrication. It consists of two independent parts; the upper part is filled with grease to lubricate the thrust bearing for the centrifugal pump. The lower part is filled with transformer oil to replenish the transformer oil consumption in the motor steel pipe housing. The ends of the shaft of some protectors have a spline connecting the motor and pump with a special coupling. Some designs don't use the coupling in order to

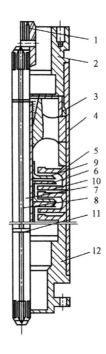

FIGURE 3-109

Structure diagram for an electric submersible pump.

make the motor directly connect with the pump shaft. Figure 3-110 shows a schematic structural view of a communicating protector.

d. Separation

This is located in the lower part of the multistage centrifugal pumps to separate the gas. If the borehole fluid contains a relatively high oil-gas ratio, we can pre-separate a portion of the gas with a separator, which can be released from the oil jacket annular first. If the gas-oil ratio is low, the gas separator is not required so this is positioned in the pump inlet. Figure 3-111 shows a configuration diagram of a rotary separator. This type of separator is suitable for offshore oil rig with a gas-liquid ratio higher than 10%.

e. Wellhead devices for an electric submersible pump

When we use electric submersible pump, the wellhead devices must allow the cable through, as shown in Figure 3-112. The wellhead has a socket connector above tubing cap and below the tubing hanger. The socket connector is connected with the ground cables and underground cable connector when installing the wellhead, which is very convenient, safe, and reliable. But they are costly and the cable occupies a large amount of space due to its thickness, so the tubing is forced into an eccentric position in the tubing hanger. Thus, another method of putting cable through the wellhead has emerged, which divides the cable into three strands that go through the tubing hanger and are connected with the ground cables and underground cable by a hinge joint. This method is low cost, and the tubing can remain in the center position. But the operation is complicated and the reliability is also slightly inferior.

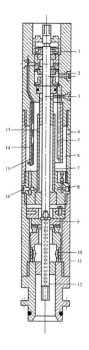

FIGURE 3-110

The schematic structural view of a communicating protector.

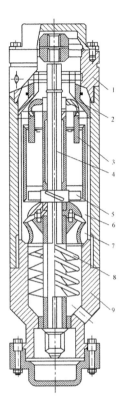

FIGURE 3-111

The configuration diagram for a rotary separator.

The top of a multistage centrifugal pump is generally also equipped with a check valve and drain valve so that we can discharge the oil of the tubing and centrifugal pump housing when we need to lift centrifugal pump unit to ground maintenance for service.

2. The advantages and disadvantages

The main advantages of an electric submersible pump are high flow rate, high lift, uniform flow, no effects from well deviation, good ability to carry sand and wax, simple ground equipment, convenient control, and better economy. However, due to the restrictions of the motor's heatproof degree, its applicable depth is not more than 3000 m. Changes in oil production and exploitation of high viscosity oil will affect the efficiency of the pump. It does not allow more than 2% volume fraction of gas into the pump. The equipment has a complex structure and the price is also expensive.

Example 3-2: Try to choose electric submersible pump oil production unit for an offshore heavy oilfield, given the following known relevant data of the offshore oil field:

Casing inner diameter: 177.8 mm (7 in)

Casing's masses per unit length: 23 lb/ft

Tubing outer diameter: 63.5 mm (2 ⅞ in)

Middle part's depth of the reservoir: 1615 m (5300 ft)

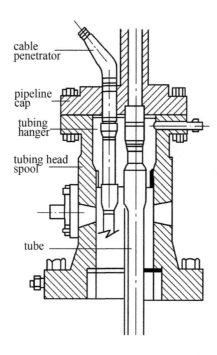

cable
penetrator

pipeline
cap

tubing
hanger

tubing head
spool

tube

FIGURE 3-112

The wellhead devices of an electric submersible pump.

Pump setting depth: 100 m above the perforated interval
Crude oil density: 0.959 g/cm³ (16.0°API)
Natural gas relative density: 0.68
Relative density of reservoir water: 1.02
Reservoir temperature: 54°C (130°F)
Reservoir pressure: 15 MPa (2133 psi)
Wellhead pressure: 0.35 MPa (50 psi)
Moisture content: 30%
Design fluid production rate: 270 m³/d (1700 bbl/d)
Production gas-oil ratio: 8.9 m³/d (50 ft³/bbl)
Flowing bottom hole pressure: 6 MPa (854 psi)

Solution:

a. Calculate the total lift when pumping clear water

  i. Relative density of liquid in the well: $0.959 \times 0.7 + 102 \times 0.3 = 0.977$

  ii. Reduced pressure head of oil pressure: $0.35 \times \frac{100}{0.977} = 36(m)$

  iii. Vertical lifting height: $1615 - 6 \times \frac{100}{0.977} = 1000(m)$

  iv. The loss pressure head of the tubing friction: according to the pressure head curve of the tubing in Figure 3-113, at the flow rate of 270 m³/d (1700 bbl/d) the loss pressure head of the tubing with inner diameter of 63.5 mm (2 ⅞ in) is 20 m/1000 m. Then the tubing friction loss pressure head for 1515 m should be $1515 \times 20/1000 = 30$ m.

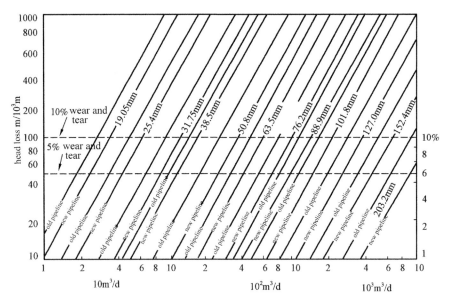

FIGURE 3-113

The curve of the oil tube pressure head.

    v. Total lift: 36 + 1000 + 30 = 1066 m

b. Identify the crude oil viscosity without the gas

It is known from the above that the crude oil relative density is 0.959 g/cm³ (16°API), when the pressure is one atmospheric pressure (1atm) and the temperature is 1.6°C. So the viscosity of the crude oil without gas is 140 mPa·s when the reservoir temperature is 54.40°C (°F), as shown in Figure 3-114.

c. Correct for the saturated oil viscosity

From Figure 3-115, for oil that has values of 140 MPa·s for oil viscosity without the gas and 50 for the gas-oil ratio, we amend to a saturated oil viscosity of approximately 65 mPa·s.

d. Convert viscosity units into SSU.

According to Figure 3-116, a viscosity of 65 Mpa·s can be converted to 400 SSU.

e. Correct the viscosity of crude oil with 30% water

From Figure 3-117, in accordance with a moisture content of 30%, we find a viscosity correction factor of approximately 2, so the oil viscosity after correction is 800 SSU.

f. Find the displacement, head, and power correction factors

If the pump efficiency is 70%, correction factors can be found according to Table 3-7 in the light of the corrected viscosity 800 SSU: the displacement factor is 85%, the head factor is 86%, and the power factor is 1147.0r.

g. Fix the pump performance parameters

Corrected lift: 1066/0.86 = 1240 m

Corrected displacement: 270/0.855 = 316 m³/d

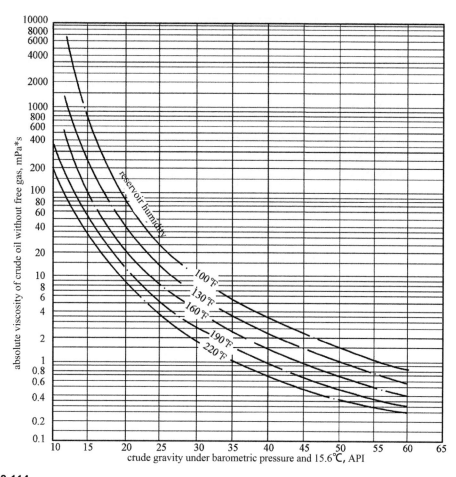

FIGURE 3-114

The crude oil viscosity without the gas under different reservoir temperatures.

h. Pump selection and performance parameters
  We can discern from a pump directory that the Richard CENTRILIFT400 Series N80 pump is the most appropriate selection. According to the pump characteristic curve shown in Figure 3-118, the pump performance parameters are a displacement of 320 m³/d, a single-stage lift of 16 m/level, and a pump efficiency of 64%.

i. Determine the total pump series.
  Total pump series: 1240/4.1=303 levels

j. Calculate the motor power P
  In accordance with the calculated motor power, the 450 Series 112 HP motor from the CENTRIIFT catalog is the best selection, with a voltage of 1270 V and a current of 56 A.

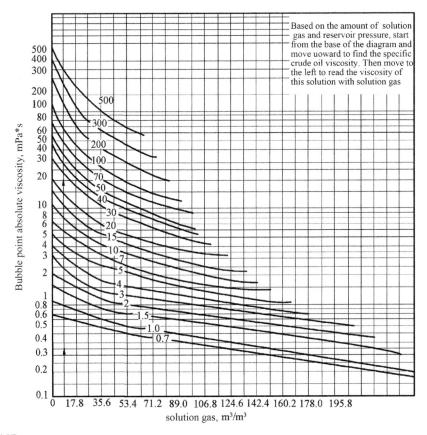

FIGURE 3-115

The saturated oil viscosity under the action of different reservoir temperature and pressure.

k. Select the cable

According to the motor current, as shown in Figure 3-119, the Richard #4 cable does not exceed a voltage drop of 30 V/305 m (1000ft), and the cable length is 1515 m. The total voltage drop is

$$\frac{25V}{305m} \times 1515m = 125V$$

l. Calculate the capacity of the transformer

$$\text{Transformer capacity} = \frac{\sqrt{3} \times 56 \times (1270 + 125)}{1000} = 136kV \cdot A$$

Because equipment replacement on an offshore oil field platform is relatively difficult, we should try to choose a transformer capacity that meets the production time span of the well

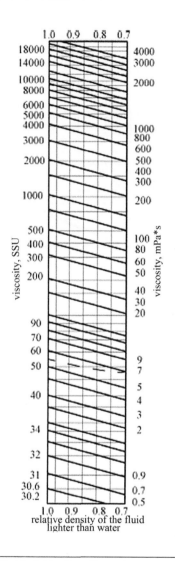

FIGURE 3-116

The viscosity conversion of the liquid that is lighter than the water.

during the life of the offshore platform. For example, the transformer used in SZ36-1oil field in Bohai Bay in China has 26 taps. And its voltage is from $700 \sim 2000$ V.

3.13.3 Hydraulic Piston Pump

Hydraulic piston pump production is a method that uses equipment on the surface to provide high-pressure power fluid that will be injected through underground pipes to manipulate a piston pump with reciprocating movement to lift crude oil.

FIGURE 3-117

The emulsification effect on the crude.

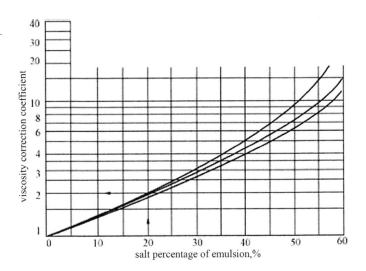

Table 3-7 The Effect of the Viscosity and Relative Density on the Pump's Performance (assuming the maximum pump efficiency is 70%)

Pumping Temperature Viscosity SSU	Maximum Efficiency Performance Parameters			
	Displacement,%	Head,%	New Efficiency,%	Power,%
50	100.00	100.00	69.5	108.0r
100	99.0	99.250	65.5	105.3r
300	94.0	94.5	55.0	113.1r
500	90.5	90.5	49.0	117.0r
800	85.5	86.0	43.0	117.0r
1000	83.0	84.0	37.0	116.0r
2000	74.0	75.5	32.0	112.6r
3000	67.0	69.5	27.0	108.8r
4000	61.0	64.0	25.0	101.2r
5000	55.0	60.0	10.0	92.4r

Note: In the table r is representative of the relative density of the crude oil at the pump temperature.

1. Working principles

   The oil production system of the hydraulic piston pump uses two parallel pipelines, one for the powered liquid pipeline, and another for the oil production pipeline. Sometimes we place small tubing in the larger tubing. Figure 3-120 shows the oil production system of the hydraulic piston pump.

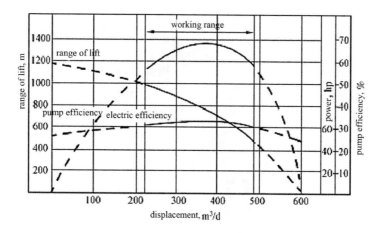

FIGURE 3-118

The pump characteristic curve.

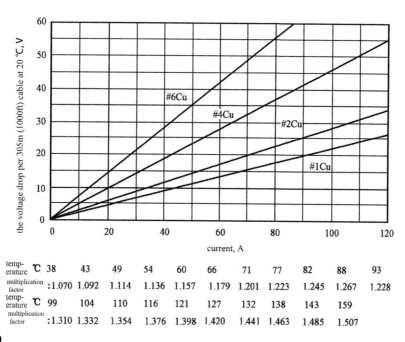

temp-erature ℃	38	43	49	54	60	66	71	77	82	88	93
multiplication factor :	1.070	1.092	1.114	1.136	1.157	1.179	1.201	1.223	1.245	1.267	1.228
temp-erature ℃	99	104	110	116	121	127	132	138	143	159	
multiplication factor :	1.310	1.332	1.354	1.376	1.398	1.420	1.441	1.463	1.485	1.507	

FIGURE 3-119

Voltage drop loss graph for the cable.

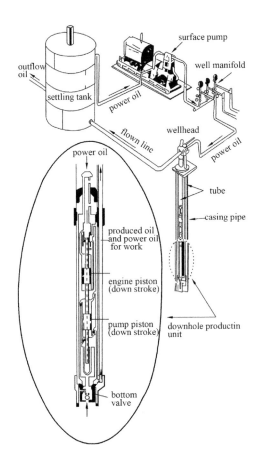

FIGURE 3-120

The production system of the hydraulic piston pump.

    The power fluid, which consists of clean crude oil, as shown in Figure 3-120, is sent to the hydraulic engine from the high-pressure pump on the platform through the large diameter tubing in the well, making the engine piston reciprocate and thus driving the alternating motion of the oil pump piston in the lower portion of the pump, which sucks the crude oil and sends it to the upper part. Mined crude mixes with the reactive power liquid, and then through the small capillary tube the mixed fluid is returned to the settling tank on the platform. Then the crude oil is extracted from the top of the settling tank and transferred to the high-pressure pump for cycling.

    The maximum displacement of the hydraulic piston pump depends on the well conditions, including tubing size, oil production capacity, power fluid characteristics, and so on. Offshore oilfields have successfully used this method to increase extraction of crude oil to 4500 m.

    On offshore platforms, we should consider the cost of power fluid settling and the cleaning process. Therefore, a closed power fluid system is commonly used. The system's ultimate power fluid is not mixed with the mined crude oil. Thus, it is necessary to add tubing in the well to make it return separately.

2. Advantages
   a. Suitable for lifting crude oil from deep wells.
   b. Flexibility: When the well yield declines, the hydraulic piston pump can make its displacement consistent with the well production and the bottom hole flowing pressure down to the lowest pressure. If the crude oil is very thick, the dilute work oil is used to dilute the thick oil to reduce the viscosity of the crude oil. It is suitable for multi-completion wells. Using a closed system can resist corrosion.
3. Disadvantages
   a. Safety complications: The working oil is flammable and its release may cause a fire disaster.
   b. Water treatment problems: The large amount of water used in the working fluid may result in high cost.
   c. Exhaust problems: Because of its sensitivity to gas, we needs to exhaust the gas, but if another pipeline is installed, the costs will be high.
   d. Cleaning problems: The crusting below the packer is difficult to clean.

3.13.4 The Choice of Offshore Oilfield Artificial Lift Methods

When the offshore fields need to use artificial lift oil production methods, we select the aforementioned methods. In addition to consideration of space limitations on the platform and suitability for inclined wells, we should consider the following questions:

1. Liquid production of the oil well
   In general, electric submersible pumps are more suitable for the work condition of large displacement. For those wells with high water content and low oil-gas ratio, the production of wells using electric submersible pump is larger. The lifting liquid volume has reached 17000 m^3/d for some foreign water source wells by using the electric submersible pumps. And the electric submersible pump has the largest capacity of lifting liquid. The next is the hydraulic piston pump and the last is the gas lift.
2. Gas content of the oil well
   A high gas-oil ratio is unfavorable for the electric submersible pump and hydraulic piston pump methods for oil production. Generally when the well bore gas volume is more than 10% of the pumping amount, the electric submersible centrifugal pump will stop working. Although the hydraulic piston pump can work in those conditions, its fluid production capacity will be greatly reduced. At that point, it would be necessary for us to increase the pump depth and install another exhaust to solve the problem. However, for a high gas-oil ratio, the use of gas lift method is very suitable.
3. Depth of the wells
   For electric submersible pumps and hydraulic piston pumps, we should make the pump sink to a sufficient depth below the producing fluid level in order to prevent cavitation and to overcome the throttling loss generated by a downhole separator, etc. For electric submersible pumps, a deeper submergence depth helps to reduce the influence of free gas. But as we increase the pump depth, the temperature also increases, creating a greater impact on the life of the motor and the cable. There have been reports that electric submersible pumps can work in the 230°C temperature. But usually electric submersible pumps are set up to work in the conditions not higher than 95°C,

which limits the application of electric submersible pumps. A hydraulic piston pump is generally more suited to a working depth for $3000 \sim 4000$ m and there are several individual reports of pump depth up to 5486 m abroad. The working depth of the gas lift method is subject to the restrictions of the injection pressure that the gas equipment can provide on the platform. The gas injection pressure for a gas lift is generally 2 to 20 MPa. But, when the pressure is higher than 10 MPa, the life of the gas lift valve will be greatly shortened; therefore, the maximum working depth of the gas lift method only has reached 3657 m.

4. Flexibility to adapt to different conditions
   The flexibility to adapt to changes in oil production is one of the indicators in evaluating the mechanical recovery methods. The gas lift method flexibly changes the maximum range of the daily fluid production rate through several artificial lift methods. It can change from a cubic meter to thousands of cubic meters. The hydraulic piston pump method follows next for flexibility, as it is relatively easy to change the stroke rate and stroke length per minute. So we can fairly easily change oil production within the rated current range. The electric submersible pump has the worst flexibility, but with the successful development of the current frequency converter, the flexibility of electric submersible pumps will likely improve.

5. Sand production rate
   Sand flow will bring erosion problems to the electric submersible pump and the hydraulic piston pump. In addition, if the top of the downhole pump is filled up with sand, it will be difficult for us to lift the pump. The gas lift method of oil production lift method, however, can deal well with fluid that has sand.

6. Economics of oil production
   When choosing the artificial lift oil production method there are four economic indicators:
   a. Initial investment
   b. Operating costs
   c. Well life and the useful life of the equipment
   d. The number of wells using artificial lift methods at the platform
      Each artificial lift method has an economic limit which depends on the circumstances of the specific analysis.

3.14 Offshore Gas Production Technology And Equipment
3.4.1 Mining Measures for Different Gas Reservoirs

There are different types of offshore gas reservoirs, and we should take appropriate measures for production according to the characteristics of different types of gas reservoirs.

1. Mining measures for an anhydrous gas reservoir
   Anhydrous gas reservoirs are also called gas reservoirs without edge/bottom water. We do not have to worry about flooding and water channeling in this case, so production measures with anappropriate large differential pressure can be used. The advantages are that we can:
   a. Increase the pressure difference between the large seam holes and a slight gap, so that the gas in the small gap can be easily discharged, thereby enhancing the gas reservoir gas production rate to meet production needs.

b. Give full play to the low pressure layers and the low permeability areas.

c. Purify the bottom of the well and improve the drainage condition at the bottom of the well. During later periods of production, a water-free gas reservoir may come across some problems such as lack of the energy used to lift the liquid, liquid loading (condensate water) at the bottom hole. At this conditions, the measures to solve these problems is to decrease the choke manifold back pressure on the platform and to blow off the gas production system regularly.

2. Mining measures for a gas (or oil) reservoir with edge/bottom water

   For this type of gas reservoir, the problem of premature water breakthrough that affects gas well production and the gas recovery ratio should first be solved. The measures commonly used for flood control are as follows.

   a. Controlling water breakthrough problem

   Gas wells should control the water breakthrough problem both before and after water breakthrough. This is generally achieved by controlling the critical flow and critical differential pressure.

   i. Anhydrous critical differential pressure is generally calculated as follows:

   $$p_r^2 - p_{wf}^2 = \{16(H - h_{dt})\rho_w[500p_r - (H - h_{dt})\rho_w]\}/\Psi \times 10^{-8} \tag{3-11}$$

   Where, p_r p_{wf} respectively are the formation pressure and flowing bottom-hole pressure, Mpa; H is the thickness of the gas layer, namely the distance from the top boundary to the gas-water interface, m; h_{dt} is the drilling thickness of the gas layer, m; ρ_w is the density of formation water, t/m^3; Ψ is a coefficient, which should be determined by drill ratios $h = \frac{h_{dt}}{H}$ and $\frac{H}{r}$; and r is the wellbore radius, m.

   ii. The anhydrous critical flow rate is usually based on the following equation:

   $$q_L = \left\{\left[a^2 + 4b\left(p_r^2 - p_{wf}^2\right)\right]^{\frac{1}{2}} - a\right\}\Big/2b \tag{3-12}$$

   where q_L is the critical water flow rate, 103 m^3/d; and a, b are coefficients, which are determined by the well test data. The other symbols are the same as Equation (3-11).

   b. Water shutoff

   Water shutoff should be based on different water conditions.

   c. Gas dewatering methods

   Two commonly used gas dewatering production methods commonly are:

   i. Gas dewatering for a single well

   ii. Gas dewatering through drilling a drainage well. Generally in the active water area of the gas reservoir, drilling drainage wells or changing flooded wells into drainage wells can slow the advancement of the water to the main gas wells.

   d. Black oil treatment

   For parts of gas reservoirs with edge/bottom black oil, you need to build the oil and sewage systems on the platform for the processing of the black oil. This is because the black oil generated in a gas well will affect separation and form a mixture with the water and the ethylene glycol that will also affect the normal operation of the glycol regeneration system. And the black oil may also hinder the delivery of subsea pipelines.

3. Production measures for a condensate gas reservoir

To prevent valuable heavy hydrocarbon components from separating out from the condensate gas in the production process in the formation and to improve the production rate of the condensate gas reservoir, the offshore production of condensate gas reservoir is generally divided into two phases.

   a. First stage: In this stage the condensed liquid is contained. The pressure should be kept above the critical pressure of production, in order to avoid the anti-condensate phenomenon and recover as much condensed liquid as possible.

   b. The second stage: This is the pure gas reservoir production phase. Measures should be taken in accordance with general gas reservoir production.

   Due to the economic limitations for offshore gas fields, we don't take the measure of recycling dry gas, which is used by onshore condensate gas reservoirs.

4. Production measures for gas reservoirs with corrosive fluids

For offshore gas reservoirs with corrosive fluid, in addition to the general gas reservoir production measures, we should also take corresponding rational production measures to deal with corrosive fluids.

Corrosive fluid in a gas reservoir are mainly two types of acid gas: the hydrogen sulfide with high corrosiveness and toxic and carbon dioxide. It generally is believed that the well with the partial pressure of hydrogen sulfide at the bottom hole greater than 0.345 kPa is called the smeller or sour gas well, and the reservoir with the hydrogen sulfide content more than 5% is called high sulfur gas reservoir. For these wells and reservoirs appropriate protective measures should be taken. If the partial pressure of the carbon dioxide in the well is more than 21 kPa, to take measures is necessary.

   a. Sour gas reservoir production measures

   As to sour gas reservoir production, the problem of corrosion from hydrogen sulfide should be addressed. Hydrogen sulfide's corrosive effect on metallic materials can be divided into three types: electrochemical weight loss corrosion, hydrogen embrittlement, and sulfide stress corrosion. Due to the fact that oil casings and submarine pipelines are in a closed state for long periods of time, electrochemical corrosion is generally light in these components. However, the electrochemical corrosion of sulfur-containing gas in high-temperature environments (above 80°C) is significant, with a speed of up to 4 to 6 mm/yr for carbon steel. The corrosive speed of sulfurous waste liquid at the bottom of a container or at the blind pipe is quite serious, up to 1 to 2 mm/annum for the carbon steel. Usually when the hydrogen sulfide partial pressure is greater than 0.345 kPa, designs should be in accordance with anti-sulfur specifications. Recently, anti-corrosion measures have had the following three aspects:

      i. Use of anti-sulfur materials

      Material has a reliable anti-sulfur capability when its hardness has an HRC $\leq$ 2. In addition, the critical value of sulfide stress cracks is 40% more than that of the allowable stress value. These parameters should be paid attention to when designing anti-sulfur measures and choosing anti-sulfur material. Usually in the offshore gas field, materials such as KQ-35, KQ-70 and KQ-100 (MPa) should be used for the wellhead device for gas production, and for valves one should choose KZ41Y-6.4(10.16) (for the flat valve), FJL44Y-16(32) (for the throttle valve), and FJ41 and FJ43 (for the vent valve). Oil casings should use materials such as J-55, C-75, DZ_1, and so on, and the downhole wire rope should use DL-659(depth of 3500m), and DL-600 (depth of 6000m).

   ii. Use of reasonable structures and manufacturing processs
     Machined pieces for cold work or welding should undergo high-temperature tempering treatment before use, and local welding equipment or pipelines should be slowly cooled to make their hardness lower than 22.

  iii. Selecting a suitable corrosion inhibitor
     For the oil casings and equipment of production and transportation, the corrosion inhibitor should be used to prolong their service lifespan. There are many common corrosion inhibitors, such as liquid nitrogen, fine pyridine, 1901 and 7251 and so on. The anticorrosive principle of these corrosion inhibitors is that inhibitors' molecules form a protective film on the metal surface which isolates the steel from hydrogen sulfide.

b. Production methods for gas reservoirs with carbon dioxide

Without water, carbon dioxide does not corrode steel, but when there is free water, the carbon dioxide dissolves in water to form carbonic acid:

$$CO_2 + H_2O \rightarrow H_2CO_3 \qquad \text{(3-13)}$$

Carbonate can make the pH value of the water decrease, and produces hydrogen depolarization corrosion on the steel:

$$Fe + H_2CO_3 \rightarrow FeCO_3 + H_2 \qquad \text{(3-14)}$$

The types of corrosion caused by carbon dioxide in the gas-condensate fluid are: pit corrosion, round moss-like corrosion and erosion, and so on. At section changes for pipeline and the contraction throttle position, as the flow rate increases, the corrosion intensifies. Generally when the air flow rate increases 3.7 times, the corrosion rate will increase by 5 times in the wellhead and tubing, which is very serious.

    The main factors affecting CO_2 corrosion are pressure, temperature, and composition of the water. At a certain temperature, with the increasing partial pressure of carbon dioxide, the pH value of the solution decreases. As the temperature increases, the solubility of carbon dioxide decreases, and the pH value of the solution increases. Some of the dissolved substances that have a buffering effect on the water can reduce the corrosion level of CO_2. But for the gas-condensate wells or condensate water, there are no dissolved substances, and at higher temperatures, the solubility of CO_2 is determined by the pressure. So the pressure is the main factor in controlling corrosion. That is, with increasing carbon dioxide partial pressure (the quantity of dissolved CO_2), corrosion gradually increases. In this way, we can use the CO_2 partial pressure to predict the extent of corrosion for gas-condensate wells, and usually carbon dioxide partial pressure can be calculated as

$$p_{CO_2} = p_T \times CO_2\% \qquad \text{(3-15)}$$

where, p_{CO_2}, p_T respectively are the carbon dioxide partial pressure and the total gas pressure (bottom hole pressure), MPa; and $CO_2\%$ is the volume fraction of CO_2 in the gas.

The criteria to determine corrosion by carbon dioxide partial pressure are as follows:

    i. When the partial pressure is less than 0.021 MPa: no corrosion
   ii. When the partial pressure is 0.21 ~ 0.021 MPa: corrosive
  iii. When the partial pressure is more than 0.21 MPa: severe corrosion

It should be noted that the corrosion rate of carbon dioxide is at a maximum when the temperature is 107°C.

For sulfur-free wells, although the corrosion is caused by the saline water with dissolved minerals, the judgement of its corrosion can cite the partial pressure standard of carbon dioxide. For sour gas wells, the above criterion is fully applicable.

As for gas or oil wells with carbon dioxide, the appropriate anti-corrosion measures for downhole tools should be taken. This includes the rational use of materials, the use of appropriate anti-corrosion coating, adding corrosion inhibitor, and other measures. One can refer to the anti-sulfur corrosion measures as well.

3.14.2 Production String for Offshore Gas Well

1. Basic requirements of the production string for offshore gas wells
   a. String structure and size must meet the requirements of the development program.
   b. The safety control device must be installed at the bottom of the well. For example, the downhole safety valve should be installed at the tubing and the packer should be installed in the casing annulus.
   c. The downhole tool structure is simple, safe, and of high reliability.
   d. As for gas wells with corrosive fluids, corrosion-resistant materials should be used for downhole tools and tubing.
   e. A special thread with good air tightness should be used for the tubing buckle type.
   f. For gas wells with high temperature and high pressure, we should minimize the rubber seals as much as possible
   g. For wells that contain corrosive gases such as H_2S or CO_2, the packer should be as close as possible to the gas layer to protect the casing.
2. Features of the production string commonly used by offshore gas fields
   Take the Pinghu gas fields in the East China Sea as an example to illustrate the above features as follows:
   a. The wellhead device and the tubing are made of corrosion-resistant material 13Cr.
   b. The tubing uses FOX threaded buckles with an outer diameter of $4\frac{1}{2}$ in, which has good sealing.
   c. The TME-6.5 downhole safety valve is used, which can be controlled on the ground and recycled by the tubing.
   d. The CAMCOHSP-hydraulic packer-type permanent packer is used.
   e. The G22-S seal assembly is used for the tubing sealing.
   f. Use "CMU" sleeve to come out the separate zone production.
   g. A TCP underbalanced perforating system is in use; it perforates first and then the production string is run.

Figure 3-121 shows the production string in the Pinghu oil and gas field.

For a gas condensate field, such as the JZ20-2 gas field in Bohai Bay, China, an eccentric barrel and an injection check valve should be installed on the production string, injecting the glycol from the platform to the column to prevent the formation of hydrates. Figure 3-122(a) shows the structure of the injection valve, and Figure 3-122(b) shows the structure of the eccentric cylinder.

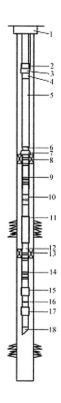

FIGURE 3-121

Production string used by the Pinghu oil-gas field: 1—tubing hanger; 2, 4—slipping collar; 3—safety valve; 5, 12—oil tubing; 6—cross over sub; 7—position indicator; 8, 13—packer; 9—locator seal assembly; 10—scale sleeve; 11—thick walled pipe; 14—inserted seal assembly; 15, 17—seating connector; 16—perforated tube; 18—wire guide shoe.

3.14.3 Gas Recovery by Liquid Drainage

Solving the problem of wellbore liquid loading in the gas well is a prominent issue in gas well production. When the well lacks energy and the wellbore liquid loading cannot be discharged, it not only affects gas well production, but also may cause the flow of the gas well to cease when there is too much liquid loading. Therefore, for a well with liquid loading, we must take effective measures for gas recovery by liquid drainage. Of course, for a water-gas reservoir, the problem of preventing water from entering the wellbore also needs to be addressed. And focusing on the reason for water breakthrough, we must create an effective water shutoff process. Obviously, the processes of water shutoff and draining are complementary.

At present, there are seven gas recovery methods by liquid drainage in use for water drainage/oil recovery, such as optimum tubing string, foam, a gas lift, a piston gas lift, a beam pumping unit, an electric submersible pump for drainage gas, and a jet pump. Table 3-8 shows a comparison of these seven methods, including applications, current achieved depths, economy, and flexibility, as a reference for selection of a gas recovery through liquid drainage methods.

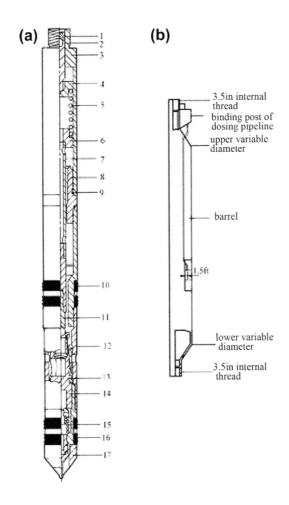

FIGURE 3-122

Fillup valve and side pocket mandrel at the production string: 1—concave tuner screw; 2—ball; 3—spring cavity; 4—spring upper connector; 5—spring; 6—spring down connector; 7—push rod; 8—connector; 9—ring with "O" shape; 10—packing with "V" shape; 11—the seat of regulating valve; 12—the container of the valve seat; 13—the cavity of the regulating valve; 14—the valve seat assembly; 15—spring; 16—packing with "V" shape; 17—jet nozzle.

3.15 New Technology and New Techniques for China's Offshore Oil Production

3.15.1 Extending Testing Techniques and the Use of Early Production Systems

Development evaluation for offshore oil fields, relies mainly on drill-stem testing (DST) for exploration and appraisal wells, but using this test's data as the basis of production evaluation often results in wrong decisions because the time for a DST is short (usually 6 to 12 hours). Therefore, in order to reduce the risk of development of offshore fields and avoid errors in design and economic losses caused by the lack of test data, before the formulation of development programs and design engineering programs, we need to extend tests (e.g., the Jinzhou Marine 9-3-D well had an extended test

Table 3-8 Comparison of Gas Recovery by Liquid Drainage Methods

Lifting Method (Right)/ Usage Comparison (down)	Preferred Column	Foam	Gas Lift	Piston Gas Lift	Beam Pumping Unit	Electric Submersible Pumps	Jet Pump
Largest amount of liquid discharge, m³/d	100 (tubing)	120	400	50	70	500	300
The deepest wells (pump), m	2700	3500	3000	2800	2700	2200	2800
Well type (inclined or curved wells)	Responsive	Suitable	Suitable	Limited	Limited	Limited	Suitable
Ground and environmental conditions	Suitable	Suitable for small devices	Suitable	Suitable for small devices	Generally suitable for large and heavy devices	Suitable for small devices with high-voltage power	Suitable for source of power away from the wellhead
High rate of natural gas liquefaction	Very suitable	Very suitable	Suitable	Very suitable	More appropriate for gas-liquid separator	More sensitive, suitable	More sensitive, suitable
Sand	Suitable	Suitable	Suitable	Limited	Poor	<5 %	Very suitable for no moving parts

Continued

Table 3-8 Comparison of Gas Recovery by Liquid Drainage Methods—cont'd

Lifting Method (Right)/ Usage Comparison (down)	Preferred Column	Foam	Gas Lift	Piston Gas Lift	Beam Pumping Unit	Electric Submersible Pumps	Jet Pump
Formation water fouling	Of defense, better	Very suitable for well washing	Of defense, better	Poor	Of anti-poor	Of defense, better	Of defense, better
Causticity (H_2S, CO_2)	Suitable corrosion	More appropriate for corrosion	Suitable	Suitable	High H^2S use is limited, poor	Poor	Suitable
Design difficulty	Simple	Simple	Easier	Easier	Easier	More complex	More complex
Maintenance management	Very convenient	Convenient	Convenient	Convenient	More convenient	Convenient	Convenient
Investment cost	Low	Low	Lower	Lower	Lower	Higher	Higher
Operation efficiency, %			Lower	Lower	<30	<65	the maximum operation efficiency only up to 34
Flexibility	Adjustable work system	Adjustable injection rate cycle	Adjustable	Good	Adjustable output	Adjustable frequency	Adjustable nozzle
Exemption period	Greater than 2 years		Greater than 1 years	For general scaling, 3 months		0.5~1.5 years	

time of 109 d). At the same time, offshore oil fields can also conduct early production and collect economic benefits, which kills two birds with one stone. This is why China has conducted independent studies related to extending testing techniques and early production systems.

1. Extending the testing process

   Platform facilities for extending the testing process and using early production system mainly depends on the pressure and the production of test wells, which generally use two-stage separation technology. The flowchart is shown in Figure 3-123. Because tests are conducted continuously, and the produced crude oil is sent to the tanker, the system should be equipped with a crude oil transport pump, fire safety systems, security alarm systems, and so on.

2. The combination of extending tests and early production facilities

   a. The first program

   (SEP/SSP)+ (SMS) + (FSO) + (ST)

   (jack-up or semi-submersible drilling platform) + (mooring system) + (oil storage tankers) + (shuttle tankers)

   This program transforms the jack-up or semi-submersible drilling platform into an integrated operating platform for drilling, extended tests, and early production. Crude oil produced during the extended testing period is sent to the oil storage ship through the flexible floating hose with a certain intensity. And then the crude oil is transferred to the shuttle tanker for the oil storage, and transported to the land by the shuttle tanker. This transfer of the crude oil is through the board or board method between the shuttle tanker and the oil storage ship. In order to facilitate the crude oil transfer by the floating hose and the shuttle tanker aboard the oil

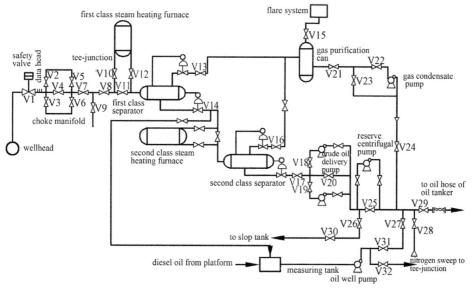

FIGURE 3-123

Flowchart for extending the testing process.

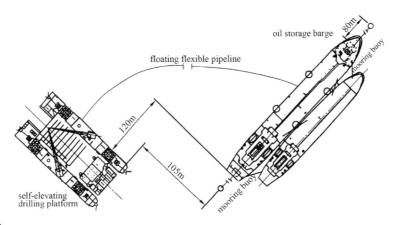

FIGURE 3-124

Layout of the first program system combining test extension and early production.

storage, the oil storage ship uses the two-point mooring system of the underwater buoy (negative pressure suction anchor two-point mooring). Figure 3-124 shows the layout of the first program system adopted in the Bohai Bay, China.

The applicable range of the first program is very wide. The jack-up drilling platform is applied when the water depth is 5 to 70 m, while the semi-submersible drilling platform can be used for water depths of 70 m or more.

b. The second program
(SEP) + (SMS) + (FSO) + (ST)
(semi-submersible drilling platform) + (mooring system) + (storage and offloading) + (shuttle tankers)
This program is almost the same as the former program except that it uses a jack-up drilling platform. So it is only applicable to the sea with better sea conditions and the water depth of only 5 ~ 10 m. When the wind is more than 5 levels, the platform should release at right. Its required total cost is low.

c. The third program.
(WHP)+ (FPSO) + (ST)
(wellhead platform) + (floating production storage and offloading) + (shuttle tankers)
This program applies to extended test wells where a simple small wellhead platform has been completed on test wells. The purpose of testing here is to measure the oilfield production capacity and do early production for recovery of the investment. The moorage of the oil storage ship and the shuttle tanker can use an anchor chain with eight-point mooring or two-point mooring.

3. Underwater buoy two-point mooring system and binocular suction anchor
An underwater buoy two-point mooring system anchors the storage and offloading platform on the sea for a long time, as shown in Figure 3-125. The system includes a friction chain and nylon mooring lines above the water and the buoy, mooring chain and cement anchor (negative pressure gravity anchor) under the sea.

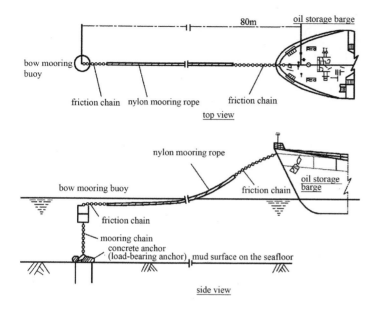

FIGURE 3-125

Underwater buoy two-point mooring system.

A binocular suction anchor is an inner tube suction anchor, as shown in Figure 3-126. It is installed by a remote-controlled pump valve system with the function of pumping and injecting water which monitors the status of the anchor chains in real time. This binocular suction anchor was designed for the Bohai Jinzhou 9-3 oilfield, where the surface soil at the subsea is upper soft and lower hard, and the fast shear strength of the soil under the mud line within 3 m is $4\sim8$ kPa and that with $3\sim6$ m is $6\sim20$ kPa.

In short, this mooring system structure is simple, safe, reliable and easy to handle, and its manufacture, installation, and utilization are convenient. At the same time, it is reusable, requires less investment, achieves good efficiency, and has good resistance to wind and waves.

3.15.2 Enhanced Oil Recovery (EOR) Technique Using Profile Control and Polymer Injection

In China's mined offshore oilfields, oil recovery is generally less than 30%. To improve oil recovery, we need to take further measures to enhance oil production. During the middle period of the water injection development or early phase, tertiary oil recovery technology can come out the purpose of enhancing oil recovery and makes the ultimate recovery increase by 10% to 20%. But offshore oilfields are so unique that many tertiary oil recovery techniques, proven by use in lots of land oilfields, have limited applicability in offshore oilfields. The new technologies of profile control and polymer injection, developed in offshore oilfields, are proposed as a development direction for offshore oilfield tertiary oil recovery technology.

1. Profile control technology

The technology of adjusting the injection profile of injection wells is called injection well profile control technology. In order to enhance well productivity, Chinese offshore oilfields widely use a

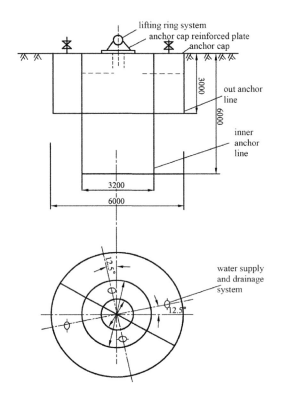

FIGURE 3-126

Structure diagram of the binocular suction anchor.

water injection process, but the problem is that the water breakthrough comes quickly and the moisture content of the liquid production rises rapidly. Water production has great influence on the economic benefits of an oilfield, as recovery of the investment for new oilfields will be seriously affected, while the economic exploitation period of the old oilfields will be shortened and even some of recoverable reserves will be lost. Therefore, we must shut off the water quickly and control the profile of wells with water breakthrough. The water shutoff process is similar for offshore oilfields and onshore oilfields, and thus it will not be restated here. The following is only a brief introduction to the offshore oilfield profile control technology.

a. Principles and methods of profile control

For water injection development of oilfields, the heterogeneity in the portrait and the landscape and the differences in oil-water viscosity result in the onrush or fingering of the injected water along the orientation of the high permeability zone and the fractures with least resistance between the water injection wells and the production wells; the injecting water bypasses the zones with low permeability and high resistance (as shown in Figure 3-127). Thereby, the swept volume of water and the water flooding effect is reduced. Even zones that can't be swept by the injecting fluid form a region of bypassed oil, which not only makes the degree of reserve recovery for low-to-moderate permeability layers decrease, but also makes

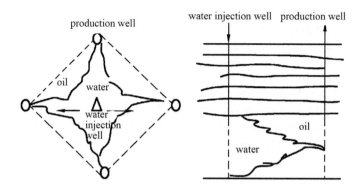

FIGURE 3-127

Profile control principles for an injection well.

the oil wells prematurely produce too much water, impacting the stable and high production of the oilfields and reducing the efficiency of water injection and the economic benefits.

In order to improve the effect of the water flooding, profile control technology needs to be employed to adjust the water injection profile. Profile control has two methods: mechanical profile control and chemical profile control. The mechanical profile control method is not suitable for serious heterogeneity, and cannot conduct deep profile control, so offshore oilfields are developing a chemical profile control method. The chemical profile control method injects a chemical agent in the injection wells, which reduces the water absorption of the high absorbent layers withchemical agents, and thus accordingly improves the water injection pressure, the water absorption of the low-to-moderate permeability layer, the water absorption profile of injection wells, and the water flooding situation.

b. Deep profile control agent

There are primarily two kinds of commonly used deep profile control agents—the gel profile control agent and the granular particle profile agent. The polymer delayed cross-linking deep profile control agent belongs to the gel profile control agent category. Among the possible options, the polyacrylamide-bichromated profile control agent system is the profile control agent for a high-permeability fractured layer, and the delayed cross-linker profile control system, consisting of a complex ion polymer, a cross-linking agent, and an auxiliary is the polymer delayed cross-linking deep profile control agent for a low-permeability reservoir. As to the granular particle profile agent method, it usually uses large doses of the injection, which makes it go deep into the oilfield. Through the function of its flocculation, expansion, and accumulation, etc., it narrows and plugs the channel of the reservoir to change the injection water flow. At present, the particle profile control agent often uses a two-fluid polyacrylamide-clay particles profile control agent system.

2. Polymer injection technology

At home and abroad, new developments have been achieved concerning the technology of polymer flooding to enhance oil recovery, so polymer injection technology for offshore oil production has seen new breakthroughs. For example, the water flood recovery factor of the Bohai heavy oil field is only 18.25%. Synthetically analyzing the permeability of the oil field itself, the underground oil viscosity, and the polymer flooding technology level, polymer flooding in the

Bohai oil field can enhance the oil recovery factor by 10% to 15%, but it should be implemented according to the characteristics of the offshore oil field. The special requirements of polymer flooding technology for the Bohai offshore oil field are as follows:

a. A polymer solution can only be made up with seawater of high salinity. At the same time, due to the requirements of environmental protection, the output sewage cannot be directly discharged, but must be reinjected, which requires that the polymer has good salt tolerance.

b. Due to the space constraints of operations at sea, polymers should have good solubility.

c. The cost of offshore oilfield polymer injection is high. The oil at the Bohai oilfield has high viscosity. Therefore, in order to achieve mobility control, it inevitably is required that the polymer solution has higher viscosity with the economic allowed. So the polymer must have the ability of increasing viscosity.

d. The wide well spacing in offshore oilfields requires that the polymer has good injection and shear resistance.

Currently, the new hydrophobically associating water-soluble polymer NAPS that is suitable for the harsh conditions of offshore oilfield has been successfully developed and applied by the Suizhong oilfield in the Bohai Sea.

3. Fixed string acidification technology

Acidification is used to relieve the formation damage that is generated during drilling and completion and the oil production, and is an important means to improve reservoir permeability and production per well. However, the cost of the acidification work is very high (sometimes 260×10^4 RMB per a well) and the operating time is long, because the offshore acidification operation needs to use the drilling rigs (or platforms) as the support vessels, and after the acidification operation the rapid discharge of residual acid requires the use of the electric submersible pumps or jet pumps and must lift and down the tubing four times. Therefore, it cannot meet the requirements of rapid production and cost-effective benefits for offshore oilfields. In order to change this situation, in part of the offshore oil field in Bohai Bay, China uses a fixed string acidification technology and achieves good results. This type of technology, under the premise of ensuring that the scale and effect of acidification doesn't change (compared with acidification with the use of drilling rigs), doesn't use the rig (or platform), doesn't occupy the "three" work boats in the long term, and doesn't move production strings. In order to avoid these "three don'ts," the new technologies used are as follows:

a. Use acid with very low dissolution rate for the anti-sand gravel layer.

b. Use the acid to release the core's contamination and to improve the core permeability.

c. Use the acid that hardly acidifies the cables, pumps and interfaces of the submersible pump.

d. The residual acid processing technology won't cause residual acid corrosion on the production process and the subsea pipeline.

e. The steering acidification technology of chemical particles temporary plugging is used to avoid the residual acid damaging the reservoir.

PART IV OFFSHORE WORKOVER OPERATIONS AND EQUIPMENT

During the flow process and artificial lift oil extraction for different oil wells, the downhole equipment may go wrong at any time, and there may also be reduction and even stoppage of the production because of production oil layer exhaustion. In order to recover normal production of oil wells, operations that take appropriate measures to clear trouble, replace downhole equipment, and adjust the parameters of oil well production are called workover operations. In this section, we will introduce workover operation and equipment.

3.16 Content and Features of Offshore Workover Operations
3.16.1 Reasons for Oil Well Failure

1. The well itself fails

   Failure of the well itself includes downhole sand plugging of the well, severe wax and salt precipitation in the wellbore, oil layer blockage, permeability reduction, oil-gas and water layer collusion, production oil layer exhaustion, etc. For instance, the crude oil in the China Bohai oilfield have generally high wax content, often up to 13 to 18 percent, and the freezing point is between 22 degrees and 35 degrees, so it has the characteristics of hypercoagulability and high wax content. In addition, the wellhead pressure and temperature are both very low, with the wellhead temperature of most oil wells between 20 degrees and 40 degrees. These two features combined to aggravate the wax precipitation problem. For instance, the freezing point of crude oil in the China Bohai BZ28-1 oilfield is 35 degrees, the wax content is 12 percent, and the wax precipitation point is 47 degrees, which results in the phenomenon of severe wax precipitation. Therefore, wax precipitation becomes the main failure point for oil wells of this offshore oilfield.

2. Failure of the well structure

   Oil well structure failures such as tubing fracture and leakage all can lead to the well failure. An article from OTC5313 introduces statistics relating to subsea wellhead equipment failure in America for a certain offshore oilfield (Table 3-9).

3.16.2 Main Elements of Workover Operations

Generally, in the initial stage of oilfield production, the main elements of workover operations include wax cleaning, sand washing and plugging, squeezing cement, perforating, replacing the packer, etc. In the later stages of production, because stratum energy consumption decreases capacity, in order to improve production of the oil well, we often need fracturing, acidizing, and other workover operations. In addition, downhole fishing is also one of the elements of workover operations. The above contents can be summarized as follows:

1. Lifting operations: Operations that lift tubing, downhole tools and equipment, etc., which are convenient to overhaul and replace, and then are again sent down to conduct the wellbore operations of pumping, bailing, mechanical wax cleaning, etc.
2. Fluid circulation operations: Operations such as sand washing, hot washing, squeezing cement, and drilling fluid circulation that need to use a pump to make the fluid circulate in the well.
3. Operations of rotating equipment: when running these operations such as drilling cement plugs, enlarging hole, re-drilling, deepening the wellhole and repairing the casing and so on, rotating the corresponding drilling tools is necessary by driving the rotary equipment.

Table 3-9 Failure Statistics of Foreign Oil Well Subsea Wellhead Equipment

Years / The number of failure wells / The total number of completion wells / The failure modes	1960–1969	1970–1978	1979–1984	Total
	66	66	177	309
Downhole well completion equipment	7	12	22	42
Connection and sealing device	6	6	3	15
Valve and regulator	1	8	0	9
Control system and hole extrusion	3	14	5	22
Lead line and pipeline	4	1	0	5
Catastrophic marine accident	1	4	0	5
Unknown factors	13	0	11	24
Total	35	46	41	122
Failure rate, %	53	70	23	39.5

3.16.3 Characteristics and Special Requirements of Offshore Workover Operations

1. Focus on operational safety

   The oil, gas, and water wells in the offshore oilfields are generally concentrated on one or several wellhead areas, often being arranged in cluster wells, and the measures and sidetracking operations should not affect the surrounding production wells. In addition, the distribution of equipment of an oil-gas processing system in the wellhead platform is often concentrated. Therefore, workover operations need very strict security requirements, especially with regard to safety measures for leakage, blowout prevention and explosion.

2. Fully functional devices

   Most of offshore wells are deviated or horizontal wells. Because of this feature, the workover equipment must have the rotary function. However, for these workover equipment, except that of normal workover functions, some of them must also have the sidetracking function. At the same time, there are sufficient power storage.

3. Ensure continuous operation

   Due to the high cost of offshore operations, we require continuous operation. Thus, we need to ensure that we have reliable equipment and a low fault rate, and that there is a variety of materials and auxiliary equipment on site, which are ready to work continuously.

4. Pay attention to environmental protection

   Offshore oilfields must observe marine environmental protection laws. For a variety of liquid emissions during workover operations there are strict requirements. If the liquid is unqualified, we are not allowed to discharge it, and you need to deal with it on the sea or carry it to land. Special attention should be taken to pollution during workover operations. Oil pollution on the sea should be eliminated.

5. Meet the dual-use requirements of drilling and repair

   We spend significant amounts on removing and replacing offshore equipment. Thus we need to require that the equipment on the platform meet the various operating requirements of the oil, gas, and water wells, and even require the equipment to have both drilling and repairing operating functions, for the purpose of satisfying the requirements of sidetracking operations, etc.

6. Service the subsea wellhead

   Because some offshore oilfields use subsea wellheads, the equipment should be able to conduct workover operations for oil and gas wells of the subsea wellhead, which requires special working ships and workover equipment.

7. Adapt to a variety of sea conditions

   The offshore engineering environment is complex. Severe sea conditions often occur, and under the conditions of salt spray, wind, waves, current and sea ice, etc., we require that workover equipment should meet the specification standards and meet the needs of the operation.

3.17 Offshore Workover Methods and Selection

To complete a workover task, we should choose different workover methods and processes based on the different operation contexts. The following briefly describes several common offshore oilfield workover methods, and when to choose these operatoins.

3.17.1 Workover Methods Commonly Used in Offshore Oilfields

1. Platform workover method for water wellheads

   If the offshore oilfield uses water wellhead equipment for well completion, the platform workover method can be adopted. There are several cases related to the platform workover method.

   a. Large wellhead platforms, artificial islands, or large-scale integrated production platforms with a large amount of wells

   The workover operation for this type of platform requires retaining a rig after completion. Besides undertaking workover operations, the rig also can be used for drilling future wells, infilling wells, or supplementing wells.

   b. Medium and small wellhead platforms

   The platform retains some drilling equipment after well completion, such as the turntable, drilling pump and drilling fluid purification system, derrick, and base, for the purpose of workover operations. Alternately we can only keep the base, preparing another workover derrick, drilling fluid circulation system, and turntable. The Chenbei oilfield in China Bohai Sea uses the latter method.

   c. Cantilever-type mobile drilling platform or special workover operation boat

   This method does not set up workover equipment in the original well completion platform; when workover operations are needed, the cantilever-type mobile drilling platform or special workover operation boat are temporarily deployed for the workover operation.

   d. Assembly platform workover equipment

   When repairing the well, this method does not use cantilevered mobile drilling platform or a workover vessel. The operators break down the workover equipment into several pieces, then transport them to large- and medium-sized platforms after well completion using a supply vessel or helicopter. Then the pieces are assembled together, and finally they are hanged on the wellhead universal base by cranes, to conduct various workover operations. Currently, the heaviest piece of assembly-type workover equipment weighs only 8 t, and the transportation is very convenient.

2. Workover method that turns a subsea wellhead into a water wellhead

If an offshore oilfield uses a subsea wellhead, when the well is complete, the subsea wellhead can be turned into water wellhead, and then we can use the several platform workover methods for a water wellhead listed just above. At this point, we need to use aspecial workover vessel (or platform) to achieve the following several steps:

a. Well kill: Before dismantling the subsea Christmas tree, we kill the well first.

b. Lift the subsea Christmas tree.

c. Tie back wellhead: Use the drill conductor to tie back the subsea wellhead to the platform, and make it change to a water wellhead.

d. Install control equipment for the water wellhead.

This method requires dismantling the subsea Christmas tree, so this can only apply to the overhaul operations, such as recompletions, replacing downhole equipment, casing repair operations, etc. As for the small repair jobs, we should use as much as possible the simple workover method.

3. Through flowline workover method (pump-down tool workover method)

The basic principle of this workover method is to use the pumps installed on the production platform to pump the power fluid to drive the propeller. Then the propeller drives the manipulating tools and workover tools into the well through the flowline to complete workover operations. Then, we take the workover tools and manipulating tools back to production platform by reverse circulation. Finally, we recycle the pump-down workover tools using equipment on the platform.

In order to achieve pump-down tool workover, you must use a specially designed TFL Christmas tree, as shown in Figure 3-128. This Christmas tree has dual tubing, two flowlines, and a flowline loop. Their purpose is to facilitate circulation in the workover operations. The use of the flowline loop (loop) is to pump down tools from the flowline down to the downhole tubing. In addition, there is a flowline bypass valve (switching valve) between the two loop pipes, to control the direction of circulation. Near the bottom well the two loop pipes are connected by an "H" shaped part (H-type connector) to adjust the bottom circulation.

The major equipment used for the pump-down tool workover method is as follows:

a. Water equipment

This includes the pumps (typically about 110 kW), manifold, instrumentation skid base, and BOP. The role of the BOP is to pump down tools into the oil wells under pressure (generally horizontal pressure), as shown in Figure 3-129.

b. Completion equipment

This includes the devices that must be installed on the platform (devices to go into the flowline channel and devices to take in or out workover tools), the equipment (dual flowline) between the wellhead and the platform, the wellhead equipment, and the downhole required equipment ("H"-shaped reversing parts, etc.). Among them, the wellhead equipment is shown in Figure 3-130. It is connected to the flowline by the valve's bent pipe, which is tangent to the center line of tubing

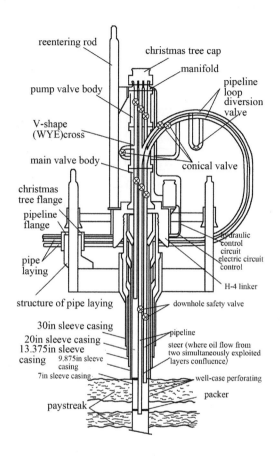

FIGURE 3-128

The Christmas tree used by the pump-down tool workover method.

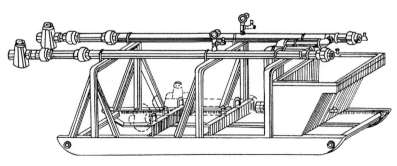

FIGURE 3-129

BOP used by the through flowline workover method.

FIGURE 3-130

The wellhead equipment used by the pump-down tool workover method.

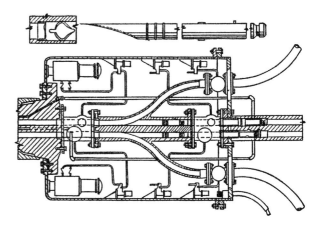

at the intersection. If this wellhead equipment fits the plug, you can apply it to pump-down tool workover operations; otherwise, removing the plug, this can also be applied to ordinary workover operations. When using pump-down tool workover methods, the tools can be pumped into the downhole or returned to the water surface through two flowlines with a bending radius of not less than 1.5 m on the wellhead equipment. These two pipelines communicate with each other near the bottom of the well, so the hydraulic medium can be recycled.

c. Pump-down workover tools

As shown in Figure 3-131, this mainly consists of a propeller [Figure 3-13(b)], accelerator [Figure 3-13(c)], toggle, jar, manipulation tools, workover tools (wax cleaning, perforation, sand washing, formation plugging tools, etc.), pressure gauges testing the bottom hole pressure, and so on. Manipulation tools can connect and release the pump-down parts with remote control.

d. Downhole control devices

These include safety valves, fixed valves, circulation control valves, etc. Circulation control valves are used for the pump-down tool system which has three strings and completes dual completion operation. Figure 3-132 shows the circulation control valves with two strings.

4. Rope workover method

The rope workover method is also a kind of through flowline workover technique, but it uses a steel wire rope to replace the pumping power fluid used by the pump-down tool workover method. The rope workover operation is mainly completed by a tool lever hanging on the wire. It uses the weight of the tool lever or the upper and lower shock of steel wire rope to complete the lifting and lowering operations in the tubing. Therefore, these operations can be completed by lifting and lowering the tool lever or controlling workover tools. These operations are commonly paraffin removal, cleaning up the clutter on the tubing wall, washing sand, cleaning up other debris in the bottom hole, squeezing cement, removing excessing consolidated cement, perforation and acid fracture and so on.

Rope workover operations require staff to work on the platform, and the required equipment can be divided into two categories: the water workover equipment and the downhole workover tools. The water equipment includes lifting systems (such as a lifting winch) and hydraulic

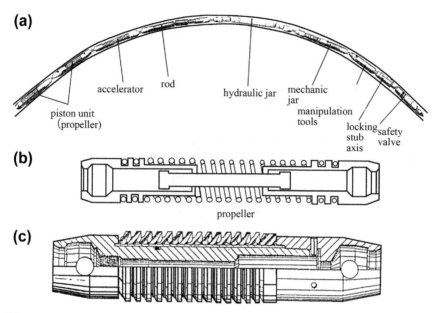

(a)

accelerator rod hydraulic jar mechanic jar manipulation tools locking stub axis safety valve

piston unit (propeller)

(b)

propeller

(c)

FIGURE 3-131

Composition of the pump-down workover tools.

control systems. The downhole workover tools include the pressure cutter and the suction tool, which are used to remove tube wax, iron rust, and other sundries; the impression tool used for lost downhole items; the running tool; the pulling tool; the tube perforating guns, etc.

For the wells of a subsea wellhead, when using the rope workover method to conduct workover operations, we have the following two methods to establish workover operation channels:

a. Use composite completion risers (riser) to establish workover operation channels

This method is the simplest method, namely, using composite completion risers to extend the downhole tubing, through the Christmas tree, to the platform, and conducting the workover operations with the tubing and the control line on the platform. Figure 3-133(a) shows a type of composite completion riser, and its main components are two tubes, a hydraulic workover controlling line, and the raising and lowering joints and seals. Two tubes are used to extend the oil well tubing to the platform; the control line is used to connect the hydraulic workover control system with the workover control manifold joints at the top of the Christmas tree, to achieve hydraulic remote control on the platform.

b. Using subsea equipment for the rope workover method

Figure 3-133(b) shows the subsea equipment system diagram for the rope workover method. It mainly includes two parts, the upper BOP assembly and the lower BOP assembly. We connect it with the subsea wellhead equipment using the subsea equipment. Then we can conduct rope workover operations on the platform.

5. The concentric tubing workover method

This method is based on the rope workover method, and also is a through flowline workover technology. The difference with the rope workover method is that a thinner flexible hose replaces

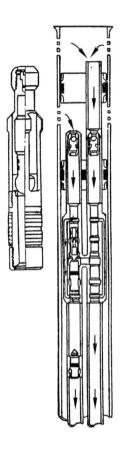

FIGURE 3-132

Circulating control valve in a pump-down tool system.

the extension tubing from the rope workover, and extends directly from the platform on the water to the bottom, which simplifies the workover program and makes it more convenient. The concentric tube workover method can complete the same workover content as rope workover method, and both processes are also exactly the same.

3.17.2 Choice of Offshore Oilfield Workover Methods

Generally when choosing between the above several workover methods, we should consider the following:

(1) For the workover scale, we should distinguish between a routine workover and a well overhaul. A so-called routine workover refers to smaller workover operations, such as wax cleaning, sand washing, squeezing cement, bottom fishing, perforation, etc. Obviously, these routine workover operations can be done using the simple and rapid pump-down tool workover method or the rope workover method. However, for well overhaul operations, such as recompletion, downhole

(a)

(b)

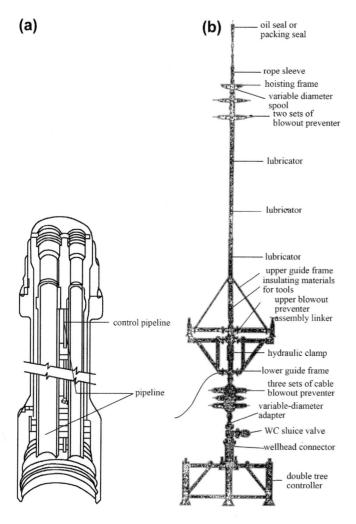

— oil seal or packing seal

— rope sleeve
— hoisting frame
— variable diameter spool
— two sets of blowout preventer

— lubricator

— lubricator

— lubricator
— upper guide frame
  insulating materials for tools
    upper blowout preventer
    assembly linker

— hydraulic clamp
— lower guide frame
  three sets of cable blowout preventer
  variable-diameter adapter
— WC sluice valve
— wellhead connector

— double tree controller

— control pipeline

— pipeline

FIGURE 3-133

Subsea equipment and composite completion risers for the rope workover method.

equipment replacement, and casing repair, and other large-scale workover operations, you should use the water wellhead platform workover method, or the workover method of turning a subsea wellhead into a water wellhead.

(2) As for the type of the wellhead, we should distinguish between the subsea wellhead and the above-water wellhead. Obviously, for the subsea wellhead, the only method of large scale workover is to turn the subsea wellhead to the above-water wellhead and the method of small scale workover is through flowline workover. For the above-water wellhead, both the method of the above-water wellhead platform workover for large scale workover and the method of the rope workover for small scale workover are used.

(3) As for the operating frequency, we should distinguish between frequent operations and scarce operations. For routine workover operations for subsea wellheads, in order to complete the operation quickly and economically, frequent operations should use the pump-down tool workover method; for infrequent operations for subsea wells, we can use the rope workover method or the concentric tubing workover method. Because both methods can use conventional completion methods and equipment for well completion, we can use the common wet Christmas tree without TEL. Thus, the cost can be greatly reduced.

(4) As for economic benefits, we should distinguish the simple equipment from complex equipment. For example, when using the through flowline workover technology, if choosing the method of pump-down tool workover, we need a specialized completion system and the completion equipment is extremely complex, which results in higher costs, reducing economic benefit. While using the method of concentric tube workover, the cost of concentric flexible hose is higher than the wire rope in the rope workover method. Thus, using different workover methods requires careful economic accounting and a review of the choices from the economic point of view.

3.18 Offshore Workover Rigs

In order to complete workover operations, workover rigs are needed. A workover rig mainly consists of lifting equipment (winches, derricks, etc.), loop flushing equipment (pumps and manifold systems, etc.), and rotating operation equipment (turntable, etc.). It is similar to an ordinary drilling rig. The operating load is relatively light, so the workover rig's power and other basic parameters, such as the volume, weight, and so on, are relatively smaller than the drilling rig. Though offshore workover rigs are similar to drilling rigs in composition, the installation is different at sea. They are often divided into several types. The following will introduce requirements for offshore workover rigs, and then introduce key features.

3.18.1 Particular Requirements for Offshore Workover Rigs

1. Take up as little deck area as possible: The offshore workover equipment and the layout model should occupy as little deck area as possible. For example, try to use a three-dimensional space layout program.
2. Minimize the weight of the workover rig: in order to make the operation more safe, the fixed loads and the variable loads of the workover rig must be limited during the design.
3. Make the offshore workover rig have as many production functions as possible: for the workover rig that is fixed on the platform, it is necessary to meet as variable kinds of functions in offshore oil production as possible. Except for the functions of lifting (upgrade) and rotation, it is best to make the rig have the functions of drilling and workover.
4. Try to meet the requirements of the marine environment: Workover equipment should try to adapt to the wet, salt-sprayed environmental conditions.
5. Try to resist severe sea conditions: The workover rig machine should withstand a variety of conditions, such as wind, earthquake, and overturning. Because the derrick requires no guy ropes, the derrick leg span is longer than a derrick with ordinary guy wire.
6. Try as much as possible to have the ability to move on the sea: In addition to workover rigs fixed on the platform, a variety of workover rigs with support systems can be installed on a movable platform or on board, and should have self-propelling ability.

3.18.2 **Workover Rigs That Share the Support System on the Fixed Platform**

The character of this type of the workover rig is that its electricity, gas, water, oil, etc. are provided by a utility system on fixed platforms. And living systems, lifting systems and communication systems are also provided by the utility system of the fixed platform. The deck and specialized region used for the workover operation are provided by the fixed platform. This category of offshore workover rigs currently used have the following types.

1. Mobile track workover rig

   This adopts a module structure. It is skid-mounted, and ordinarily includes three modules: the lower sled base, the upper base skid (rig skid), and the derrick. The workover rig can move on the track of the platform deck, which moves across the upper base (rig) on the lower base rail and across the lower base on the deck rail to move along the x-axis and y-axis in both directions. Thus, the workover rig can move on the deck to satisfy the workover needs of different wells on the platform. This workover rig has the following two types:

   a. Movable workover rig with workover equipment under the lower base

   For this type of workover rigs, the ancillary workover equipment such as mud pumps, drilling fluid tank system, and well control equipment all hang beneath the rig, and move with the workover rig together, while the duty room, generator room, temporary house, and material library are all grouped on the rig floor. This compact layout can save deck area, but the load of the wellhead deck rail is concentrated. Therefore some of the platform structures are relatively specific, as shown in Figure 3-134.

   b. The workover rig without ancillary equipment under the lower base

   As mentioned above, the workover ancillary equipment of the workover rig is installed around the tubing on the deck around the site, and the duty room, temporary house for workers, and

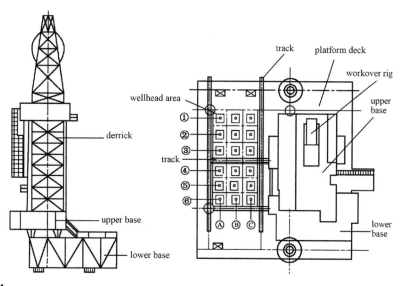

FIGURE 3-134

Layout of South China Sea SCS W12-1 platform workover rig.

material library are also installed on the deck, which takes a large area of the workover deck, but the advantage is that the load on the deck is distributed to the deck, and the load of the carrier rails of the deck and the tubing is relatively uniform. So there is no special requirement for the structure of the platform.

Figure 3-135 shows the layout of workover rigs on the China Bohai SZ36-1B zone platform. The workover technical performance of fixed platforms is shown in Table 3-10.

2. Simple workover rigs on a fixed platform

Currently, the simplest workover rig on a fixed platform in the Chinese offshore oilfields uses a tractor hoist as the workover rig. The structure of the tractor hoist is simple, the technique is mature, and its use is convenient. But it has high requirements for the workover deck, which restricts its use in the offshore oilfield. For example, the derrick of the tractor hoist needs to fix the guy wire, which requires a broad workover deck area. Obviously, some common platforms would face difficulty in meeting this requirement.

3. The drilling and workover dual-use machine

This also uses a mobile track, so its technical characteristics are the same as the mobile track workover rig. It can also be used to drill, so its power and transmission system are increased and matched to the drilling needs. Also the carrying capacity of the derrick should be strengthened. A drilling and workover dual-use machine not only does regular oil well jobs, such as the increasing production and augmented injection, but also conducts overhaul operations and well drilling and completion operations when the oilfields at a later stage drill the adjustment wells. Table 3-11 shows us some of the technical parameters of a drilling and workover dual-use machine being used in Chinese offshore oilfields.

4. Hydraulic pressure balance workover rig

There are two kinds of hydraulic workover rigs in offshore oilfields. The first one is the ordinary hydraulic workover rig that uses a hydraulic motor as direct power to drive the winches and the turnplates, but, currently, it is rarely used. The other is the hydraulic pressure balance workover rig with long and short stroke. The following will focus on such a workover rig.

FIGURE 3-135

Layout of Bohai SZ36-1B workover platform.

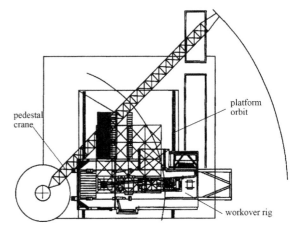

Table 3-10 Workover Technical Performance for Offshore Oilfields in China with Fixed Platforms

Workover rig model		2024	K-600	K-400	Willson30	HXJ60
Respective districts and platforms		South China Sea W12-1 oilfield	South China Sea W1-4 oilfield	Bohai Chengbei B platform	Bohai Chengbei A platform	SZ36-1 A platform
Workover rig manufacturer			Dreco	Dreco	LTV	Janghan Siji factory
Workover rig maximum load, t		150	90	60	90	90
Workover rig load rating, t		150	90	60	90	90
Diesel models, power and speed		CAT3412/740 hp/ 2100 r/min	CAT3406/400 hp/2100 r/min	CAT3306B/300 hp/2100 r/min	CAT3406/360 hp/2100 r/min	CAT3406/360 hp/ 2100 r/min
Diameter of wire rope, in		$1\tfrac{1}{8}$	1	1	1	1
Derrick	Derrick form	A	-	A	K	K
	Height	102 ft	96 ft	60 ft	102 ft	29 m
	Machine capacity on the second floor	14500 m($3\tfrac{1}{2}$ in tubing)/4420m ($4\tfrac{1}{2}$ in tubing)	2500m ($3\tfrac{1}{2}$ drill pipe)	3000 m ($2\tfrac{7}{8}$ in tubing)	3000 m ($3\tfrac{1}{2}$ in tubing)	3200m ($2\tfrac{7}{8}$ in tubing)
	Wind load rating	Full set of root 73 joint/ No set of root 93 joint	Full set of root 80 mph/ No set of root 108 mph	Full set of root 73 joint/ No set of root 93 joint	Full set of root 73 joint/ No set of root 93 joint	Full set of root 93 joint/ No set of root 108 joint
Winch	Input power	740 hp	-	402 hp	425 hp	250 kw
	Brake drum diameter × width	42 in×12 in	38 in×8 in	38 in×8 in	38 in×$8\tfrac{1}{2}$ in	10700 mm×360 mm
	Brake wrap angle	300°	300°	300°	300°	340°
Base	Drill box load, t	100	-	60	100	90
	Lower base dimensions	-	-	18 ft×24 ft×8 ft	19 ft×10 ft×8 ft	16 m×3 m×4.58 m

Continued

Table 3-10 Workover Technical Performance for Offshore Oilfields in China with Fixed Platforms—cont'd

Workover rig model		2024	K-600	K-400	Willson30	HXJ60
Drilling pump	Model	–	–	KT250	PAH	F500
	Diesel engine power	–	–	360 hp	310 hp	373 kw
	Maximum working pressure	–	–	340 MPa	8000 psi	26.3 MPa
	Maximum displacement	–	–	0.86 m³/min	1.1 m³/min	1.2 m³/min
Tethered lifting system		5×4	5×4	4×3	5×4	4×3
Transmission form		Diesel + converter + shaft	Diesel + converter + shaft	Diesel + converter + shaft	Diesel + converter + shaft	Diesel + converter + shaft
Workover rig model		K-80	IR160	HXJ80B	HXJ120	HXJ225A
Respective districts and platforms		SZ36-1B platform	SZ36-1J platform	QK17-3/Qk18-1	JZ9-3W and JZ9-3E platforms	QK17-2 platform
Workover rig manufacturer		Dreco	U.S. IRI	Nanyang Erji factory	Janghan Siji factory	Nanyang Erji factory
Workover maximum load, t		64	60	120	150	225
Workover rated load, t		64	60	80	120	150

Table 3-11 Technical Parameters of Drilling and Workover Dual-Use Machine

Workover rig type		HYX150	Remarks
Location		Bohai-QK17-2 platform	
Manufacturer		RG PETRO-MACHINERY	
Rated load, t		150	
Maximum load, t		180	
Derrick heigh, m		31	
Capacity of the monkey board, m		4 in DP or TB*3600	
Derrick wind resistance, knots	No load	107	
	Full load	93	
Winch diesel engine	Type	Detroit 8V-92TA	Pre-press stand-alone configuration, power at full load 80 t workloads
	Power, hp/r/min	480/21000	
Transmission case	Type	S5600H	Allison
	Parallel type	gear parallel	
Winch	Type	JC28A	
	Fast rope relay	280	
	Brake hub diameter*width, mm*mm	1270*267	
	Clutch type	ATD330H	
Auxiliary brake	Type	324WCB	EATN
	Braking torque, N·m	33870	
	Cooling water emissions, L/min	307	
Turnplate	Diameter, in	17½	
	Maximum operating torque	13729	
	Type of turnplate clutch	ATD218	
Upper base (L*H*W), mm*mm*mm		14000*6400*6000	H is the height from the deck
Lower base (L*H*W), mm*mm*mm		13000*10000*5000	H is the lower base height

a. Operating principles

The hydraulic snubbing unit is a kind of workover rig used on natural flow wells with high pressure to conduct workover operations directly without killing the well with circulating fluid.

As shown in Figure 3-136, the moving body of the short stroke hydraulic workover rig is fixed to the wellhead BOP by the connecting device. Its four cylinders are symmetric, and can

FIGURE 3-136

The composition of the hydraulic pressure balance workover rig.

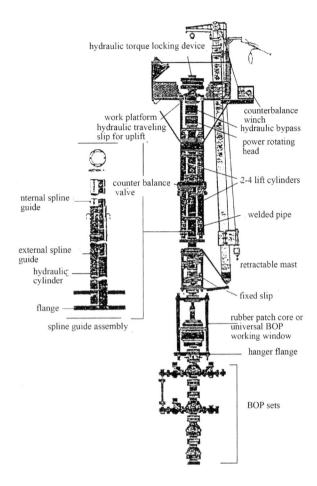

make the end of piston rod move up and down. And they are also connected with a traveling head that includes both a traveling slip and a power rotary head, which makes a trip and rotates the pipes; the power rotary head is hydraulically driven. When the bottom hole pressure is below 21 MPa, the rubber bushing is used to control the well pressure. If the pressure exceeds this value, you need to control well pressure with the BOP. At the time of operation, the BOP valve is open. During the upward trip, the lower slip (fixed slip) is open. The traveling slip goes up to the top with the master cylinder piston rod, and the tubing is lifted. During the downward stroke, the fixed slip is closed, holding the tubing, the upper traveling slip is open, and the hydraulic piston rod is down to the bottom position. Then the hydraulic piston rod returns back, and restarts the trip. Repeating this cycle, all up-down operations can be completed step by step. If there is a need for a job involving rotation, you can drive it through the power swivel head.

As for the long-stroke hydraulic snubbing unit, killing well equipment using the moving bodies is similar to the short stroke, but the stroke is lengthened. Therefore, it combines the cylinder with the derrick. The moving fluid cylinder is installed on the derrick, which makes the upper end of the hydraulic piston rod connect with the top of the derrick crane. So, when operating, the fluid cylinder moves up and down along the derrick, and is centralized by the guide wheel. The top of the liquid cylinder is equipped with the traveling pulleys of a pulley system, slipping up and down with the liquid cylinder. When the cylinder moves down, the slip head is led by the pulley system and moves up, which makes the stroke of the traveling head increases two times. The assembly pulley system is grouped symmetrically, with the middle assembly pulley used as the balance pulley to balance on the tension of both sides of the wire rope.

b. Technical performance

The hydraulic snubbing units are imported from abroad to China offshore oilfields; for example the Bohai offshore oilfield workover rig is a HRL-142 rig from the HydraRig Company in the United States. The technical performance parameters of the snubbing unit manufactured by the HydraRig company (including the length of the stroke) is shown in Table 3-12. Table 3-13 shows the long and short stroke hydraulic workover snubbing performance comparison.

c. Main features

A hydraulic workover rig doesn't kill the well with heavy kill fluid. Therefore it does not pollute and damage the reservoir, reducing the demand for follow-up measures to increase production; and it is suitable for installation at a higher wellhead location, allowing the wellhead position to go up to 18 m. The equipment is small in size, and it is lightweight, needs a small area for operation, and is easy to dismantle and move, which makes it particularly suitable for offshore workover operations. The simple operation and convenient maintenance,

Table 3-12 Performance Parameters of Hydraulic Workover Rigs

Type	HRS150	HRS340	HRS600	HRL120	HRL142
Maximum lifting capacity	150000	340000	60000	118000	142000
Maximum lowering capacity	66000	188000	26000	60000	71000
Power	230	305	380	230	305
Tubing size range	$1\text{-}2^{7/8}$	$1\text{-}7^{5/8}$	$1\text{-}8^{5/8}$	$1\text{-}2^{7/8}$	$1\text{-}5^{1/2}$
Rotational torque	2800	6600	6600	3000	3000
Stroke	10	10	14	36	36
Maximum hole size	8	11	15	8	8

Table 3-13 Performance Comparison between the Long and Short Stroke Hydraulic Workover Rigs

Stroke	Short stroke	Long stroke
Applicable well pressure range	≤ 107	<21
Single stroke	3.05	10.97
Lifting speed requirements	slow	Fast
Suitability for higher wellhead requirements	Higher	Normal

along with fewer staff requirements and saving costs used for the killing fluid and pumping equipment, make the economic return good.

d. Operational capability

It can be used for window sidetracking, drilling small holes, drilling cement plugs, setting down the packer, fishing, milling, lifting and lowering the column, and cement operations. It is now widely used, and also is considered to be an important supplement for conventional drilling and coiled tubing services.

3.18.3 Workover Platforms with a Self-Support System (Ship)

A workover system doesn't need to share a support system with the wellhead oil platforms; when a workover rig and all its support systems are fully available on the workover dedicated platform or onboard a ship, this is called a workover platform with a self-support system or a workover boat.

1. The self-elevating workover platform

Some of the workover platforms have the capacity to self-propel, but others without self-propelling capacity move by relying on a tugboat. In addition, some have no workover rigs, which means the workover rig is on the wellhead platform. They are only used as workover operations support platforms. For instance, for a lower power (230 hp) hydraulic workover rig the workover rig is placed on the wellhead oil production platform, and the diesel engine for the power equipment and the driven variable displacement pump are installed on the other self-elevating auxiliary platforms.

a. Structures

The jack-up workover platform in China is mainly the modified drilling platform. Both its left and right boards are equipped with hanging machines which can lift facilities of drilling and workover. At the front of the ship is the base of the workover rig with a derrick and workover auxiliary facilities. And the base can move forward and after, left and right in order to be adapt to the wellhead positions. The cantilever of the workover rig base stretches out and is aligned with the wellhead. And then the workover rig base is aligned with the wellhead by moving it. The drill floor can rise to the upper deck or helicopter parking apron of the production platform when running workover operations. The cabin and the upper deck store other workover facilities, such as the drilling pump, the tank of drilling fluid etc. The housing and the aircraft deck are often behind the deck. There are power generation equipment and oil-water storage facilities.

b. Structural parameters

The structural parameters for the "Zili" jack-up workover platform are given in Table 3-14.

c. Workover rig technical performance

The "Zili" workover platform is a XJQ450 workover rig (Figure 3-137); it has a single drum skid decoration hoist, mainly driven by the diesel engine HY500 TBD234V8 hydraulic mechanical transmission box; winch ski racks; derrick systems; water brake systems; hydraulic systems; and electrical systems. The workover machine technical performance parameters are shown in Table 3-15.

The jack-up workover platform is mainly applied on offshore platforms without satellite wells or small platform wells.

2. Workover barge

Some single wellheads are without platforms at sea, and the wellhead cannot support major workover equipment, so a specialized barge equipped with a workover rig is used, and you can

Table 3-14 Bohai "Zili" Jack-up Workover Platform Structure Parameters

Item	Zili jack-up platform
Existence of mat	Yes
Onboard size (length × width × depth). m × m × m	51.82 × 32.62 × 4.267
Mat dimensions (length × width × height), ft	170 × 132 × 13 (including the 2 in apron)
Leg (length × diameter), ft	173 × 6
Permitted variable load when operating, 10^3 lbf	3500

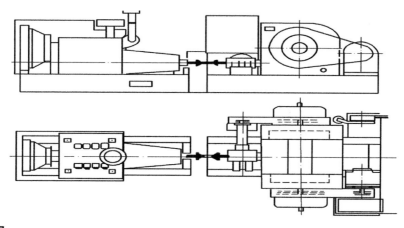

FIGURE 3-137

XJQ workover rig.

Table 3-15 XJQ Workover Technical Performance Parameters

Item	XJQ workover rig
Workover well depth, m	5500 ($2^{1/2}$ in tubing)
	4500 ($2^{7/8}$ in tubing)
	4000 ($3^{1/2}$ in tubing)
Hook load rating (wind speed is far less than 20m/s), kN	784
Hook load, kN	980
Hook maximum static load, kN	1195.6
Engine power, kW	331
Derrick height, m	44.5

take advantage of the inherent advantages of the workover barge. This workover barge is mostly used in the United States and most are self-propelled. For example, the workover barge of the American Dresser company is fitted with a double drum winch, diesel power generation facilities, diesel or electric drive, lightweight derrick standpipe box, turntable, power tongs, BOP stack (500 psi), pneumatic card tiles, drilling fluid circulation, and grout circulatory system. This workover barge is equipped with cantilevers that can stretch out the barge. And all of the derrick, the winch and the skid base of the derrick are mounted on the cantilever and can move vertically and horizontally. In order to ensure transportation and the personal security, there is a helicopter parking apron on the barge.

3. Use drilling pontoons or semi-submersible drilling platform to replace workover platform
    Sometimes, according to the special needs, we can use drilling pontoons or semi-submersible drilling platforms to do workover operations. Of course, these must be equipped with the appropriate workover well servicing tools and equipment.

4. Workover ship equipped with a hose workover rig
    This workover boat is a pontoon, on which are installed hose workover rigs, and its working depth is up to 75m. It is specially designed for underwater subsea wellhead completion workover operations. Figure 3-138 shows this workover boat structure; the workover equipment is installed on board, including the hose reel, heave compensation system, and rope operating winch. Workover pontoons use dynamic positioning, and can maintain stable hull through large waves, as Figure 3-138 shows. The workover vessel's heave compensation system keeps the tension of the hose by expanding and contracting the compensation wire rope, so that the hose remains fairly stable under the action of wind and waves. Due to the horizontal movement and the currents of the

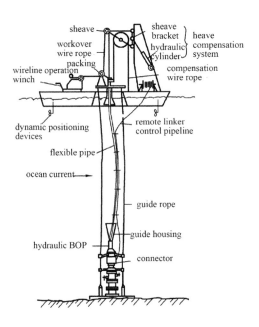

FIGURE 3-138

Workover ship equipped with a hose workover rig.

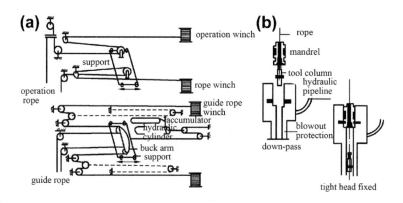

FIGURE 3-139

Rope workover on the workover ship.

floating boat, the hose produces a horizontal displacement, and tends to form a curve, in particular at its upper and lower ends. To this end, in order to control the gap between the rope tools and the inner surface of the hose at two points are equipped with a large radius guide cover.

5. Workover ship equipped with a rope workover rig

This kind of workover boat can be used for workover operation for at least 100 m depth subsea wellheads and is equipped with an onboard rope skid-mounted workover rig. The rope workover works like the rope workover rig operation described earlier; it is mainly composed of six parts, namely a rope winch, operating winches, mast frame, hydraulic console, horizontal heave motion compensation system, and working platform, as shown in Figure 3-139. In Figure 3-139(a), rope can be seen in the rig; it is drawn through the operating winch and rope winch through two horizontal pulleys. Because one of two horizontal pulleys is mounted the active bracket, the movement of the workover ship relative the subsea wellhead doesn't spread out by the rope. Similarly, a guide rope is drawn through the guide rope winch though the active pulley installed on the compensation lever bracket. The movement of the compensation lever bracket is controlled by the hydraulic cylinder of the accumulator. So the guide rope tension can be kept away from the influence of the heave motion of the workover ship.

The subsea workover equipment supporting the rope workover rig are the subsea BOP, hydraulic lines, active packing and rope tool string and so on, as shown in Figure 3-139(b). The lubricator used here has the same standards as a land lubricator, thus, the technology of the onshore rope workover operation can be a direct reference for the offshore subsea wellhead, but due to the wellhead being in the water, we need divers to assist in the subsea wellhead connection.

PART V COILED TUBING TECHNOLOGY AND EQUIPMENT

Coiled tubing equipment is hydraulically driven workover equipment, and there are two kinds: truck-mounted and the skid-mounted. This can replace common workover equipment and the wire rope workover rig to complete some of the workover operations, including completion and rope work, and are more and more widely used in offshore fields. Coiled tubing technology involves coiled tubing twisted on a drum that can be continuously withdrawn or lowered from the well without connecting thread. The coiled tubing is made of high-strength, low-alloy materials, which first is rolled straight and then welded together. Its length is designed according to specific needs, generally its inner and outside diameter respectively are 31.75 mm and 38.1 mm, the general total length is 4500 m, and the maximum of its outside diameter is up to 88.9 mm. Because coiled tubing technology does not use the shackle and pipe connection process, it can continuously conduct the cyclical operations of lowering the coiled tubing. Thus this saves the time of running and pulling tubing, and can effectively reduce the pollution and damage to the reservoir. The operation is safe and reliable, so it has been widely applied to drilling, completion, sand control, testing, logging and workover and other assignments. But due to the limitations of coiled tubing size and weight, it has no ability to operate when the depth is over 5000 m, the lifting weight of the platform's hanger is less than 18 tons, and the platform's area is small. This section will mainly introduce the overwater equipment, main downhole tools, and the application of coiled tubing technology to workover operations.

3.19 Overwater Equipment of Coiled Tubing Technology

As shown in Figure 3-140, the overwater equipment of coiled tubing technology is mainly composed of the drive guidance system (injection head), the drum, the coiled tubing, the operation room, the power source, and the BOP system.

FIGURE 3-140

Coiled tubing technology used on above-water equipment: 1 - injection head; 2 - drum; 3 - coiled tubing; 4 - the operating room; 5 - lubricator; 6 - BOP; 7 - wellhead; 8 - casing; 9 - production tubing; 10 - production packer; 11 - check valve; 12 - locator; 13 - hydraulic relief; 14 - through-tubing inflatable packer device.

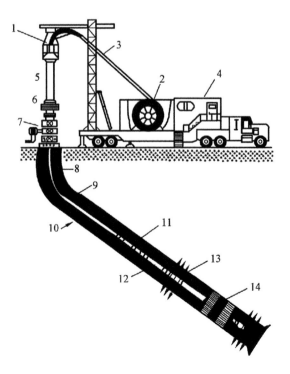

3.19.1 **The Driver Guidance Device (Injection Head)**

This device runs the coiled tubing into the well or pulled the coiled tubing from the borehole, and its structure is shown in Figure 3-141. Its technical performance and specifications are listed in Table 3-16. The following briefly introduce its composition.

1. The drive

The device is composed of two relative chain boxes, and in each of the chain boxes there are two inner chains and two outer chains, and two hydraulic power devices are installed on them, as shown in Figure 3-142. The hydraulic pump provides hydraulic transmission power to the chain box. Because the inner chains are two-way drive, they also can drive the outside chains to move bi-directionally. Because the outside chains are tightly pressed on the coiled tubing, producing an axial friction force that is greater than the weight of the coiled tubing itself, we can freely pull and run the coiled tubing and shock, etc. Currently, the Halliburton Company produces the product: the greatest pressure of the hydraulic pump is 17.2 MPa, the maximum bearing capacity is 36 t, and the maximum speed is 52 m/min.

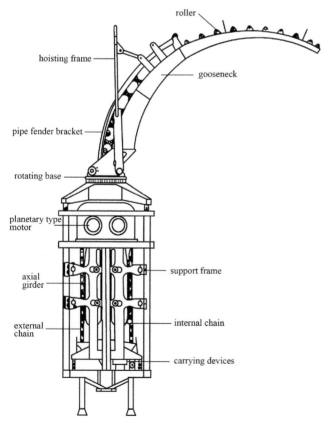

FIGURE 3-141

Coiled tubing injection head.

Table 3-16 Technical Performance of Coiled Tubing Injection Head

		15k	30k	80k Standard	80k DNV
Tubing size, in		1	1	1.252.25 1.502.75 2.003.5	1.252.25 1.502.75 2.003.5
Lift and lower pulling force	LBF (continuous)	19300	24000	71600	73200
	LBF (gap)		30260	85600	85365
Pressure coefficient, LBF/psi		7.72	12.104 (low speed) 30.260 (high speed)	14.32	24.4
Maximum speed, ft/min			112(low) 224(high)	188	93.5168
Speed rate, ft/min		192		1	0.56
Tubing length inside the injection head, in		20	30.42	32.25	32.25
Tubing length inside the injection head, in (seal to guide)		3.5	4.5	50	5.0
(seal to outside the chain)			22	27.2	27.2
(outside the chain to guide)			9.5	12.88	12.88
Weight, lb (guide)		4200	3500	16000	16000
Equipment size, in × in × in		42.5×42.75×136.5	50×50×196	76×181×262 75×76×6	75×181×262 75×75×145
Steering radius, in		48	72	96	96
Motor type		DOUBLE VANE	PISTON	DOUBLE VANE	DOUBLE VANE
Motor displacement, f		9.5	19	7.25	13
Transmission ratio		21.48	21.48	50.47	50.47
Chain wheel diameter, in		6.803	8.42	12.025	12.025

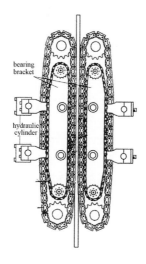

FIGURE 3-142

Chain drive device for injection head.

2. Guide piece

This is installed on the drive unit. The coiled tubing is located on a gooseneck track when it's output from the drum. The coiled tubing is also covered by a curved grille so it moves along the established track conveniently, which prevents the coiled tubing from derailing. Figure 3-143 shows the guide piece of the coiled tubing.

3. The load sensor

The sensor is located between the injection head bottom and the support frame, and connected with the wellhead device, as shown in Figure 3-144. The weight of the injection head and the

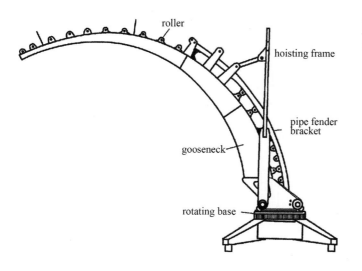

FIGURE 3-143

Guide means for injection head.

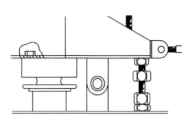

FIGURE 3-144

Coiled tubing load sensor.

coiled tubing in the well are measured separately because they are separated with the wellhead and the support frame by the load sensor. The load size of the coiled tubing is transferred to the display device in the operating room through the sensor and the hydraulic system.

4. The packing box

The packing box is connected to the bottom of the injection head using the flange. It is opened or closed by the spring load device. Whether coiled tubing is in the well or not, its seals can be conveniently disassembled, as shown in Figure 3-145.

3.19.2 **The BOP System**

1. The BOP stack

This consists of four ram-type blowout preventers, which can be controlled by hydraulic pressure and by a hand drive as well. From the top to the bottom, they are in turn for the full seal shear

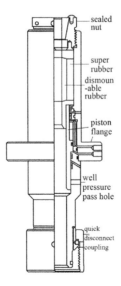

FIGURE 3-145

Coiled tubing packing box.

brake, the tubing hanging brake, the half seal brake, and the balance valve that balances the pressure. There is a 50.8 mm low torque valve between the shear brake and the tubing hanging brake, which can be used as a choke, and also can tie back the kill lines. In an emergency, by tying back the kill line through this valve, the well killing operation is conducted.

2. The accumulator

   The accumulators of the coiled tubing are respectively installed in the drive device and the BOP stack. The former provides the energy. Under the engine breakdown conditions, it can still make the outside chain press the coiled tubing, preventing the coiled tubing from falling. The latter is to provide the energy for the closing of the BOP when the power source operating the BOP stack failures. The liquid nitrogen is stored in the accumulator, and its oxygen content is less than 3%. The nitrogen pressure of the accumulator in the drive device is for $2.4 \sim 3.45$ MPa. The nitrogen pressure of the accumulator in the BOP stack should be less 1.4 MPa than the system pressure, and it is commonly for 6.9 MPa. The amount of the nitrogen should be enough to close the BOP stack valves for 1.5 times.

3. The lubricator

   This is installed between the packing box and the BOP stack, or between the BOP stack and the Christmas tree. When pulling and running downhole tools, the lubricators should be installed, and its structure is the same as in Figure 3-129.

3.19.3 **The Power Source**

This consists of the diesel engine, hydraulic pump, hydraulic tank, and hydraulic control system. The function of the power source is to provide hydraulic power for the drive device of the coiled tubing (injection), the tubing roller control system, the operations room, and the BOP stack control system.

3.19.4 **The Drum**

There is a motor at the bottom of the drum that can make the chains on the tubing drum rotate bi-directionally. The diameter of the drum depends on the length and the outside diameter of the coiled tubing. The major axis of the drum is hollow, and all kinds of liquid are pumped through it into the coiled tubing. The end of the coiled tubing is connected with the rotary head of the hollow drum through which the gas and the liquid are pumped into the well by the circulation pump. The design ensures the continuous circulation during the process of pulling and running the coiled tubing. In an emergency, closing the shutoff valve between the tubing and the axial beam can ensure safety. Figure 3-146 shows the structure of the drum.

3.19.5 **The Operations Room**

This is the workshop that can be lifted and lowered by hydraulic pressure, and in it the movement status of the coiled tubing and the condition of the wellhead can be directly observed. The instrument panel is installed in the indoor part of the workshop, and is equipped with controls for the drum and movement of the injection head, bi-directional switches and regulators that set speeds and control emissions from motor oil pipes, the diesel engine throttle, the engine emergency shutdown, the control switches for the

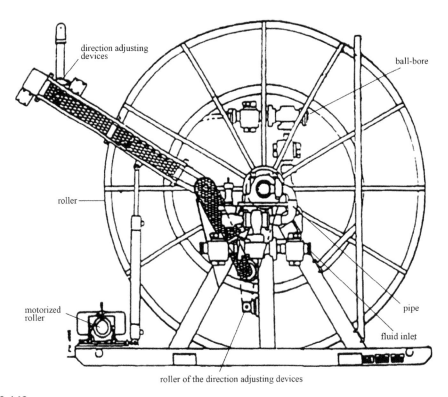

direction adjusting
devices

ball-bore

roller

motorized
roller

pipe

fluid inlet

roller of the direction adjusting devices

FIGURE 3-146

Coiled tubing drum.

blowout preventer, a wellhead pressure gauge, the hanging tubing heavy gauge, and other instruments. Also, there is a hand pump to be used for emergency shutdown under the condition of failure of the hydraulic system and accumulator.

3.20 The Coiled Tubing

3.20.1 Specifications and Technical Performance of the Coiled Tubing

At present, China is not able to produce coiled tubing, and the specifications and technical performance of CYMAX100 coiled tubing produced in foreign countries are listed in Table 3-17 for reference.

3.20.2 Coiled Tubing Operating Pressure

1. The coiled tubing operating pressure is the pressure in the tubing, which can be expressed by the wellhead pressure or the pump pressure.

Table 3-17 Specifications of the Type and Technical Performance of CYMAX100 Coiled Tubing

External Diameter in mm	Thickness in mm	Inner Diameter in mm	Mass lb/ft kg/m	Yield Load lb/ft kgf	Yield Stress psi kPa	Fracture Pressure psi kPa	Torque ft.lbf N.m	Volume gal/10³ft L/m³
1.000	0.087	0.826	0.848	24950	17300	18100	432	27.84
25.4	2.21	20.98	1.262	11310	119200	124700	592	0.345
1.250	0.087	1.076	1.081	31790	13900	14500	721	47.24
31.8	2.21	27.33	1.608	14410	95800	99900	977	0.586
1.250	0.109	1.032	1.328	39070	17300	18200	856	43.45
31.8	2.77	26.31	1.976	17710	119200	125400	1150	0.539
1.500	0.087	1.326	1.313	38620	11600	12100	1075	71.74
38.1	2.21	33.68	1.954	17510	79900	83400	1457	0.890
1.500	0.109	1.282	1.619	47630	14500	15200	1288	67.06
38.1	2.77	32.56	2.410	21600	99900	104700	1746	0.831
1.500	0.134	1.232	1.955	57500	17700	18600	1505	61.98
38.1	3.4	31.29	2.909	26070	122000	128200	2040	0.768
1.750	0.109	1.532	1.910	56190	12400	13000	1809	95.76
44.5	2.77	38.91	2.843	25480	85400	89600	2452	1.187
1.750	0.134	1.482	2.318	68080	15200	16000	2129	89.61
44.5	3.4	37.64	3.441	30840	104700	110200	2888	1.111
2.000	0.109	1.782	2.201	64750	10900	11400	2420	129.39
50.8	2.77	45.26	3.276	29480	75100	78500	3281	1.067
2.000	0.134	1.732	2.670	78550	13300	14000	2864	122.39
50.8	3.4	43.99	3.974	35810	91600	96500	3882	1.518

2. When the pressure in the coiled tubing is more than 34.5 MPa (wellhead pressure is more than 34.5 MPa, and pump pressure is more than 8 MPa), the life of the coiled tubing will reduce, so it should be operated according to specific standards and procedures, and should also be analyzed by the relevant departments.
3. The life of the coiled tubing is influenced by the drum's diameter, the radius of the tubing guidance device, the corrosion of the pump fluid and the tubing guide track, and other factors, and it also is associated with the different positions where the tubing is used. Generally, when the work pressure of the coiled tubing is below 34.5 MPa, the designed service life of the tubing can be guaranteed. When the service life of coiled tubing is intended to be more than a quarter of the design service life, the injection pressure of the high-pressure pump must be limited within the maximum working pressure of 34.5 MPa.
4. All equipment should be checked before the high pressure operation of coiled tubing.

3.20.3 Limitations of Using Coiled Tubing

1. No rotation
   Coiled tubing cannot be rotated, which is a usability limit. But this shortcoming can be overcome by mud motor driving where the coiled tubing can rotate by connecting with a dynamic crossover-type rotary joint. However, the remedy is limited to light loads.
2. Restrictions on intensity and size
   At present, the yield strength of the produced coiled tubing can only reach 45 t (100,000 lb), and its maximum diameter is 603 mm (2 in). As a result, due to the volume and weight of the drum and other factors, the applicable range of the strength and size of the coiled tubing is restricted.

3.21 Main Downhole Tools Used in Coiled Tubing Operations

The main downhole tools used in coiled tubing operations include the tubing connectors, the hydraulic release joint, the centralizer, the universal joint, the accelerator, the weighing rod, the jar, and the hydraulic pressure nozzle, among others. Here are a few of the special downhole tools.

3.21.1 The Tubing Connector

The tubing connector is the connector that connects the coiled tubing with different kinds of downhole tools, and its structure is shown in Figure 3-147. The architecture of the tubing connector, from top to bottom, is the end cap, the inner ring, the sealing ring, and the ontology. When the connector connects with the coiled tubing, first make a thread of the coiled tubing's end, then insert the tubing's end into the connector. We then tighten up the connector's ontologies with the coiled tubing, and then tighten the end cap to press the inner ring, which makes the coiled tubing firmly connect with the tubing connectors. The "O" shape packing is used to guarantee the sealing of the inside and outside of the tube.

3.21.2 Hydraulic Disconnection Joint

The hydraulic disconnection joint is the joint that disconnects the coiled tubing from the downhole tool under the connector, and its structure is shown in Figure 3-148. It is mainly composed of the joint, the

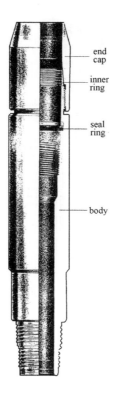

end
cap

inner
ring

seal
ring

body

FIGURE 3-147

Coiled tubing connector.

shear pin, the inner sleeve, the inner fishing head, the closing piece, and the bottom connector. When operating the disconnection joint, first throw the ball into the coiled tubing, suppressing and cutting off the pin inside the disconnection joint before pushing the inner sleeve down. Then the convex block is sent back so that the top connector is disconnected from the fishing head. Because a fishing head is on the top of the downhole tool, it can be disconnected from the fishing head, which can make the coiled tubing disconnect from the tool's assembly under the connector. If fishing operations are required in the future, they can again use the fishing head to complete the fishing. The applications of this hydraulic disconnection joint mainly divide into the following two aspects:

1. Disconnection of the tools that are designed to stay in the well, such as the bridge plug, etc.
2. If the downhole tool cannot be taken out if it happens to be blocked, it can be disconnected by the hydraulic disconnections joint, and stays in the well.

3.21.3 **Universal Joint**

The main use of the universal joint is to connect the coiled tubing with the tools for fishing operations or the tools that are lowered into a high-angle deviated hole; it allow the connected tool string to freely deflect in the well in order to reduce the possibility of sticking the tool.

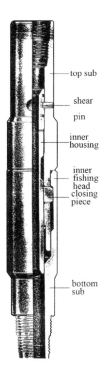

top sub

shear

pin

inner
housing

inner
fishing
head
closing
piece

bottom
sub

FIGURE 3-148

Coiled tubing hydraulic disconnection joint.

This universal joint, as shown in Figure 3-149, is mainly composed of the female nipple, locking cap, coat, spherical pin, nipple, etc. It is a ball and chain connection, with a deflection angle of 250°. The middle of this universal joint is hollow, so it can conduct circulation operations and withstand internal pressure as well.

3.21.4 Accelerator

The accelerator is on the sinker jar and the bumper jar, and beneath the universal joint. If the coiled tubing size is greater than 73 mm (2 7/8 in), a centering device should also be added between the universal joint and the accelerator. The accelerator is used to store energy, so that operations such as shearing the pin, opening and closing the scale sleeve, going fishing, and so on can be carried out. It should be used together with the sinker jar and bumper bar, equivalently adding energy to them. The accelerator is divided into upward and downward components, as shown in Figure 3-150. When assembling the tools, the appropriate downward connector is chosen to generate the appropriate shock direction (upward or downward). The accelerator energy is stored in the spring. If there is a downward shock, the spring is compressed; in contrast, if there is an upward shock, the spring is tensed. The force of compressing or tensing the spring is about 363 kg (800 lb), but the shock force can be up to 9 t (20000 lb).

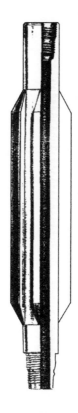

FIGURE 3-149

Coiled tubing universal joint.

3.21.5 Hydraulic Jetting Tool

The hydraulic jetting tool, shown in Figure 3-151, adopts the structure of the upper and lower ratchet wheels and centroidal axis on which the spring is installed. When the internal part of the hydraulic jetting tool is pressurized, the lower ratchet wheel makes the internal centroidal axis and the nozzle head turn 15°. When the pump pressure is released, the upper ratchet wheel can prevent the cenroidal axis and nozzle head from turning. When in operation, the nozzle head can rotate 15° if the pump pressure is released and then repressurized. Since each repeat causes a rotation of 15°, the tool can not only wash down all surroundings simultaneously, so that the tank of the tool and the perforation tubing are cleaned up effectively. Nozzle size selection is very important as the jet depends on the hydraulic pressure, so the selection of the nozzle jet should be in accordance with the pump pressure, tubing pressure, tubing size, and liquid properties being used. When using this tool, we can wash the sand with high-pressure liquid, remove a sand bridge, clean the perforation or open hole section, and clean the coiled tubing inner wall or sliding sleeve and scale in the working barrel.

There is the filter in the tubing matching the use of the jetting tool, which is usually installed on the jetting tool. Its main function is to finally filter the pumping liquid before the liquid enters into the nozzle jet to prevent the particles in the liquid from plugging the nozzle jet of the hydraulic jet.

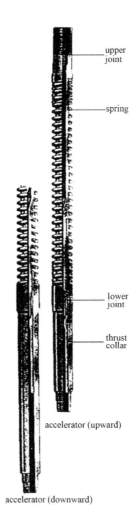

upper joint

spring

lower joint

thrust collar

accelerator (upward)

accelerator (downward)

FIGURE 3-150

Coiled tubing accelerator.

Apart from the above main tools, a through-tubing back pressure valve is also installed on the coiled tubing; it is under the connector of the coiled tubing. When the pump stops or the pipe leaks, it will automatically be shut down. It is often used as a security tool for acidizing operations.

In addition, when lowering the packer and the bridge plug, the coiled tubing is equipped with a circulating valve that can provide the cycling channel in the running tool. When the valve is opened, the layers above the packer can be processed, and the bridging agents can also add displacement to the upper packer congestion.

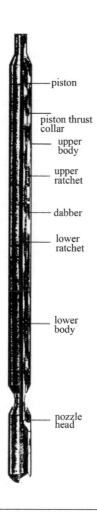

piston

piston thrust
collar

upper
body

upper
ratchet

dabber

lower
ratchet

lower
body

nozzle
head

FIGURE 3-151

Coiled tubing operations with a hydraulic jetting tool.

3.22 **Application of Coiled Tubing Technology in Workover Operations**
3.22.1 **Well Flow through Displacement**

Well flow through displacement can use liquid nitrogen or a liquid with a small density (such as diesel, etc.). Well flow through displacement with nitrogen gas is, in fact, gas lift. The nitrogen pumped into the wells by the nitrogen pump mixes with the liquid in the wells, which makes the hydrostatic pressure less than the formation pressure in the wellbore. So well flow can be achieved. In this way, when using this well flow through displacement technique, in addition to the original equipment of the coiled tubing technology, a nitrogen pump, nitrogen tank, high-pressure pipeline, etc., should be added. After the coiled tubing equipment is connected with the equipment for the nitrogen and the pressure in that equipment is tested, the coiled tubing is lowered into the wells about 35 m (100ft), and then the nitrogen can be injected, with the speed of the tubing being 15.25 ∼ 21.3 m/min, and the pump speed at 14 ∼ 5.6 cm³/min (500 ∼ 200 scf/min). In the process, the nitrogen is continuously pumped into the wells until reaching a certain depth, and then the well can flow naturally. Then, the coiled tubing is pulled. In the process of pulling the coiled tubing, the nitrogen is continuously pumped at 2.8 cm³/min, until the tubing is pulled out of the wellhead. And, of course, nitrogen injection can apply the intermittent injection method.

3.22.2 Sand Washing

Because the coiled tubing technique allow to wash the wellbore under the condition of no killing wells, it is easy to flush the sand at the bottomhole to the ground or wash off the scale on the well wall. But if using common tubing to wash sand and clean wells, the circulation work often is suspended due to the operation of connecting the tubing. As a result, we cannot continuously wash the sand and the wells. Accordingly, the use of the coiled tubing is superior for sand washing and well washing. And the sand washing and well washing can be easily performed if there is a hydraulic pressure jetting tool at the bottom of the coiled tubing.

1. Commonly used sand washing and flushing fluid
   a. Water
   This is commonly used as the circulation medium. Before operations, whether the formation can withstand the water pressure of the pipe string and the tubing's size for sand washing or not should be first considered. If the return rate of the liquid between the coiled tubing and the production tubing is less than 30.5 m/min (1000 ft/min), a drag reduction agent should be added into the water.
   b. Denitrifying water
   If the reservoir pressure is lower than the water column pressure, the pressure gradient of the circulating fluid can be decreased by mixing a small amount of nitrogen with water at the coiled tubing joints. The mixing of nitrogen and water is called nitration. Before the operation the proportion of nitrogen and water should be determined. If the nitrogen ratio is too high, the carrying capacity will be reduced, and the coiled tubing will be easily stuck.
   c. Bubble
   If the above two kinds of fluid are used and the sand still cannot be transported from the wellbore to the ground, then foam can be used as a kind of the medium. It can carry the sand in deep and large diameter wellbores to the ground with a lower pump speed and limited displacement.
2. The up-hole velocity of the annulus fluid
   The up-hole velocity of the annulus fluid is a critical parameter that determines whether or not some solids can be carried out of the well. The end velocity of the sand settling refers to the velocity at which the solid particles settle down in the liquid under their own gravity. The up-hole velocity of the annulus fluid of a vertical well or deviated hole with a deviation less than 30° is twice the end velocity of the sand settling. For a horizontal well, the up-hole velocity of the annulus fluid should be 10 times the end velocity of the sand settling.

a. Confirmation of the end velocity of the sand settling
Usually this is found out from Figure 3-152 according to the diameter of the sand grain size (horizontal).

$$v = 21.221Q/(D^2 - d^2) \tag{3-16}$$

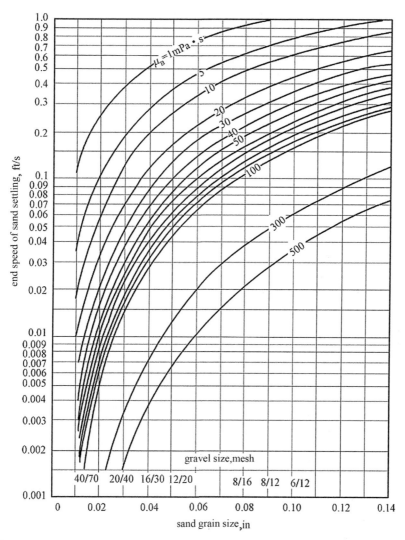

FIGURE 3-152

Sand size and sand settling velocity curve: 1 ft/s = 0.3048 m/s, 1 in = 0.0254 m.

   b. The up-hole velocity of the annulus fluid can be confirmed by the below equation:
   In the equation, v represents the annulus fluid return rate, m/s; D is the inner diameter of the
   tubing or casing, mm; d is the coiled tubing diameter, mm; and q is the pump displacement, l/s

3. The speed of lowering the coiled tubing
   The speed of lowering the coiled tubing has a relationship with the return speed of the annulus
   fluid. In the process of sand washing, the internal and external hydrostatic column pressure
   difference of the coiled tubing will be increased. If the speed of lowering the coiled tubing is too
   fast, the return speed of the annulus fluid may decrease to the limited speed of the sand settling,
   which may cause the solid particles to settle down and the coiled tubing to stick. Therefore, the
   appropriate speed of lowering the coiled tubing should be confirmed according to the pump
   speed, the speed of lowering the coiled tubing and the concentration of the solid particles in the
   return liquid, and it generally should not be more than 15 m/min when the distance between the
   coiled tubing and the sand surface is about 35 m.

4. Sand washing operation techniques
   After the coiled tubing, pumps, and mixing equipment are in place, first install a T-joint under the
   BOP and install the adjustable nozzle on the tubing. Then the coiled tubing can be put into the
   well. Wash the well in the lowering process. The coiled tubing should be pulled every 150 m to
   confirm the string load. If it is found that the return reduces or there is no return, the coiled tubing
   should be pulled continuously until there are returns. The lowering operation stops when the
   coiled tubing is lowered in front of the production string. Pull the coiled tubing slowly and
   circulate for a week. Circulating at least two weeks when it reaches the desired depth for well
   washing. During the circulation, the coiled tubing is pulled once every 15 min. After the
   completion of sand washing the coiled tubing can be pulled out, but the circulation should be
   continued in this process.

3.22.3 High-Pressure Cleaning

High-pressure cleaning is a kind of workover operations that makes use of high-pressure jet technology to clean the fouling and paraffin deposits inside the wellbore with coiled tubing. Because the
coiled tubing technology saves the time of running and pulling the production tubing and reduces the
operation's cost, it has been applied gradually in recent years.

   Offshore oil wells and injection wells have pipe scaling problems; the scaling is mainly barium
sulphate, silicate, calcium carbonate, calcium sulfate, etc. In order to make use of the coiled tubing
technology to remove the fouling, the Halliburton Company designs a high-pressure jet cleaning
system. The system mainly includes the shifting injection tool, the downhole filter, the above-water
filter, the circulating pump, and other equipment supporting the coiled tubing. In addition, it also
includes calculation software that optimally designs the nozzle size and quantity, the speed of running
and pulling the coiled tubing, and the flushing frequency.

   The high-pressure cleaning process and the sand washing process are basically the same. All of the
injection tools have been introduced in Figure 3-151. Here is a brief introduction to the filter equipment. All cleaning fluid shall be strictly filtered before it enters into the pump. Diatomaceous earth
filters or filter barrels are generally used, and the filtration precision is $2 \sim 10$ um. However, from the
entrance of pump to the coiled tubing entrance, the cleaning fluid should be filtered again, generally
with a sieve tube filter with 0.25 mm spacing. When the cleaning fluid reaches the well bottom, in order

to prevent the residual rust in the coiled tubing and the dirt in the drill fluid from blocking the eyehole or the jet nozzle (diameters of $0.76 \sim 2.29$ mm) on the jet tools, a sieve tube filter with gaps of 0.5 mm should be installed in the coiled tubing, which carries the fluid through to the final filter.

3.22.4 Acidification

Compared with ordinary tubing, coiled tubing technology for acidizing treatments can reduce the corrosion in the tubing, which is caused by contact of the acid and the tubing, and reduce the pollution on the reservoir, which is caused by the thread compound and other dirt. Also, the use of the coiled tubing can be applied at different depths and different positions, so more attention must be paid to the acidification operations of the coiled tubing.

Outside of the coiled tubing, the equipment used in coiled tubing acidizing treatment includes the mixing acid equipment and the acid tank. In addition, the downhole tool requires a tubing connector, back pressure valve and nozzle, etc. The acids that are generally used are hydrochloric acid, hydrofluoric acid, hydrochloric acid, and hydrofluoric acid among others, and their type and concentration should be reasonably selected according to the formation.

When conducting acidification, the first step is to put the coiled tubing and nitrogen device above the water in place and carry out pressure testing. Then put the coiled tubing into the hole with seawater or completion fluid circulating until the tube reaches the acidification position. After that, we can pump in the acid hierarchically, the volume of which should be calculated and designed by the pump pressure. During hierarchical pumping, it is feasible to pump in a certain amount of nitrogen for separation. Last, use the nitrogen or other fluids to displace the acid completely before pulling out the coiled tubes.

3.22.5 Cementing Plug

If the wells produce a certain amount of water, the cementing plug method often is adopted to block the layer. The operation of squeezing the cement using coiled tubing technology doesn't need to run and pull the production string, which can save on operational costs. This is very useful for offshore workover operations. Matching with the cementing plug, the through-tubing bridge and the packer can be lowered by the coiled tubing, which can also improve the sealing effect.

When using the cementing plug, except for the equipment associating with the coiled tubing technology, there should also be a cementing pump and ash cans, etc. When this equipment is installed, first we should carry out pressure testing. And then, we put the coiled tubing to the depth of well where the circulation will be killed. After preparing the cement milk, which consists of high quality cement, agent additives, and so on, the fixed component cement milk is pumped into the hole step by step. Then the coiled tubing should be lifted to the top depth of the cement plug, and the excess cement milk should be looped out. After that, the coiled tubing is lifted to a depth 500 m above the cement plug. After the cement solidifies, lower the coiled tubing continuously, explore the cement plug, and carry out pressure testing of the cement plug; if the completion well qualifies, the coiled tubing can be pulled.

3.22.6 Setting Down the Through-Tubing Bridge Plug and the Packer

Figure 3-153 shows a tool combination that uses coiled tubing to set down the through-tubing bridge plug or the packer (permanent or recoverable), and its application. Figure 3-153(a) shows the tool

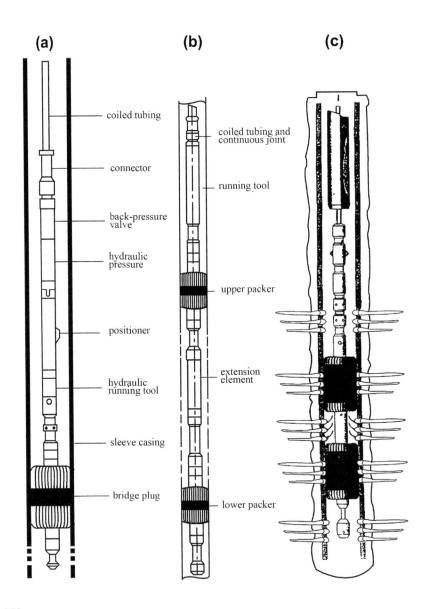

(a)   **(b)**   **(c)**

coiled tubing

connector

back-pressure valve

hydraulic pressure

positioner

hydraulic running tool

sleeve casing

bridge plug

coiled tubing and continuous joint

running tool

upper packer

extension element

lower packer

FIGURE 3-153

Coiled tubing for setting packer and through-tubing bridge plug.

string that sets down the through-tubing bridge plug. The string of tools from top to bottom are the connector, the back pressure valve, the hydraulic source, the locator, and the hydraulic running tool. Figure 3-153(b) gives the tool combination that uses coiled tubing to set down the packer. The coiled tubing connects the running tools through the connector and there is a bumper sub between the upper packer and the lower packer to adapt the heave motion of the coiled tubing as the water surface heaves. From Figure 3-153(c) it can be clearly seen that the packer that the coiled tubing sets down is selected for separate zone production.

3.22.7 **Salvage**

Coiled tubing technology, relying on its advantage that there is no need for running and pulling the production tubing, can be used for salvage operations, such as recycling all different kinds "fish," including the bottom tools that take off the coiled tubing, expansion bridge plugs, cabling falling into the well, and even the coiled tubing itself.

1. Using the coiled tubing to fish out the inside scoop neck tool
   This tool combination, as shown in Figure 3-154, uses hydraulic pressure to release the fishing string tool, but also uses pumping liquid to flush the cuttings covering or filling the scoop neck top.
2. Using the coiled tubing to fish the outside scoop neck
   Figure 3-154 shows the required tool combination, including the hydraulic release bell socket for the outside of the scoop neck. When the spear or the bell socket can't release the fishing string, the

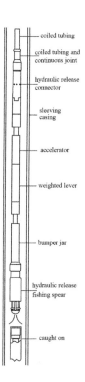

coiled tubing

coiled tubing and
continuous joint

hydraulic release
connector

sleeving
casing

accelerator

weighted lever

bumper jar

hydraulic release
fishing spear

caught on

FIGURE 3-154

and pulling neck coiled tubing salvage combination of tools.

hydraulic release tool can be installed below the jar, which takes the jar off the "fish." The fish is pulled out of the well by using the coiled tubing.

3. Using the coiled tubing to mill a stuck tool for flow control.

One of the fishing operations is to mill a tool that cannot be pulled out; this tool combination is shown in Figure 3-155. Use downhole power tools and the grind shoe to mill the flow control device; the grind shoe is installed with the hard alloy tooth. There is a binary-cycle valve installed on the downhole power tools, providing holes for cycle well washing. When the coiled tubing is pulled out of the well, the fluid for the well washing also circulates using the binary-cycle valve. There is a bootleg on the grind shoe or the downhole power drilling tool to carry the cuttings at the annular between the coiled tubing and the production tubing out of the well.

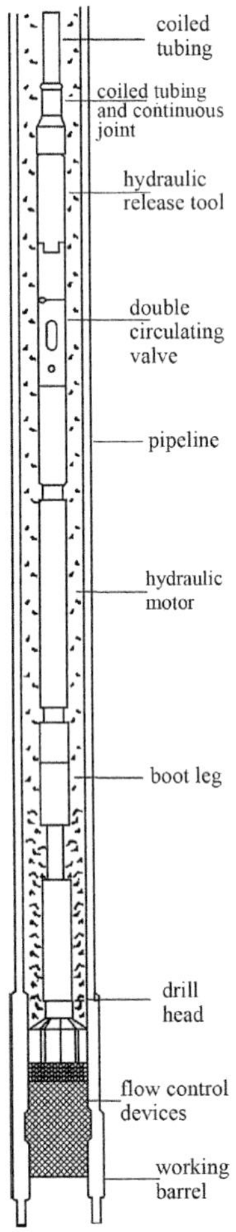

FIGURE 3-155

Combination of tools for milling operations with coiled tubing.

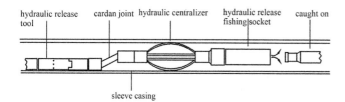

hydraulic release tool   cardan joint   hydraulic centralizer   hydraulic release fishing socket   caught on

sleeve casing

FIGURE 3-156

Combination of tools for salvage operation for highly deviated boreholes.

4. The salvage in a high angle well

In high-angle wells or horizontal wells salvage is very difficult, and the combination needed is shown in Figure 3-156. It uses a change in the coiled tubing pump speed to change the spear or socket extension angle of the hydraulic lifter above, so as to change the position of the end of fishing gear to put it in the right place on the links to "fish."

3.22.8 Incision

It has been used widely at the offshore fields that the hydraulic cutting tools installed at the coiled tubing cut off various downhole strings. For example, the hydraulic tubing cutter installed at the bottom of the downhole power drilling tools can cut off the production tubing. When operating, the motor and cutting knife are driven by the circulated fluid through the coiled tubing. The assemble is shown in Figure 3-157. The coiled tubing is connected with a set of the tool strings which includes the hydraulic release tool, a hydraulic centralizer, a double acting circulating valve, a hydraulic motor and a hydraulic tubing cutter and so on. The hydraulic motor consists of three pieces of the hardening steel hydraulic cutting knives. When completing the cutting operation, the hydraulic pressure make the knives return back and it can be pulled out. The hydraulic tubing cutter can cut the tubing with the diameter of about $48 \sim 88.9$ mm ($1.9 \sim 3 \frac{1}{2}$ in). The hydraulic tubing cutter is driven by the piston of the hydraulic cylinder. When the piston pushes outward, the hydraulic cutting knives stretch out and are touched with the tubing. Conversely, the hydraulic cutting knives return back. The double acting circulating valve controls the position of the hydraulic centralizer by changing the flow of the circulating fluid.

In addition to these purposes, coiled tubing technology can also be used for logging operations in the completion process, perforation jobs, sand control for tubing, etc.; at the same time, in the process of drilling, it can also be used for the deepening of old wells, sidetracking, new well drilling and drilling open holes in marginal regions, etc. In addition, if the original production string cannot be used in ordinary jet pump wells, we can also adopt the coiled tubing jet pump technology. Thus, the popularization and application of coiled tubing technology should continue to be important in China.

PART VI THE FLOATING MULTIFUNCTIONAL PLATFORM FOR LNG PRODUCTION FOR DEVELOPMENT OF OFFSHORE GAS FIELDS

Traditional means of exploitation of offshore gas fields can be conducted, for example natural gas can be delivered through offshore pipelines to onshore natural gas liquefaction plant for liquefaction; or it can be delivered to pipe endpoints for storage, and then connected to onshore pipelines for

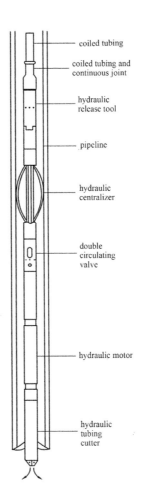

coiled tubing

coiled tubing and
continuous joint

hydraulic
release tool

pipeline

hydraulic
centralizer

double
circulating
valve

hydraulic motor

hydraulic
tubing
cutter

FIGURE 3-157

Combination of tools for coiled tubing cutting operation.

distribution. Conventional means not only increase the difficulty of construction of offshore pipelines, but also leads to higher costs and lower economic benefits. Currently, a new means has presented itself in many foreign countries, namely, conducting the process of liquefaction directly on the sea, while delivering liquefied natural gas (LNG) by ship. In doing so, burdens of pipeline construction are decreased and investment in a gas liquefaction plant is saved, which reduces costs and enhances economic benefits. However, such a working pattern requires a multifunctional platform for LNG production, which can float on the sea. Presently, there is only one platform of this kind being built on the globe, and it is expected to be completed in 2016. Therefore, this new kind of platform has been highlighted in discussion the exploitation of offshore gas fields. This section gives a brief introduction of the first multifunctional floating platform for LNG production, which is specifically detailed below.

3.23 **Composition of the Multifunctional Floating Platform for LNG Production**

The multifunctional floating platform for LNG production is called the FLNG (Floating Liquefied Natural Gas) system in foreign countries. With the help of single point mooring positioning system, the platform is fixed on operation waters, and is equipped with liquefaction devices that are equivalent to having a onshore liquefaction plant on board. However, the deck area is only 1/4 that of an onshore liquefaction plant, so the liquefaction process must be conducted in an orderly and efficient manner. On this floating platform, natural gas will be transformed into LNG, which can be stored and delivered by ships on a regular basis. The platform involves diversified functions, and is thus called the multifunctional floating platform for LNG production. Nevertheless, in comparison with onshore liquefaction plants of similar scale, the floating platform's investment could be reduced by 20%, and its construction period could be decreased by 25%. Additionally, during the process of liquefaction, the volume of gas shrinks by 600 times, which is beneficial to the storage and distribution of natural gas, yielding enormous economic returns. As a result, the multifunctional floating platform for LNG production should be highlighted in exploitation of offshore gas fields.

Figure 3-158 demonstrates the first and only multifunctional floating platform for LNG production built in the world. Sponsored by Royal Dutch Shell with 30 billion dollars, it will be the world's first floating platform, and is expected to be completed in 2016; will operate in the Prelude gas field, 200 miles away from West Australia.

The overall structure of the FLNG platform is presented in Figure 3-159, and it consists of a turret mooring system, a gas transmission system made up of risers, a condensate extraction device, a device

FIGURE 3-158

The first and only multifunctional floating platform for LNG production being built in the world.

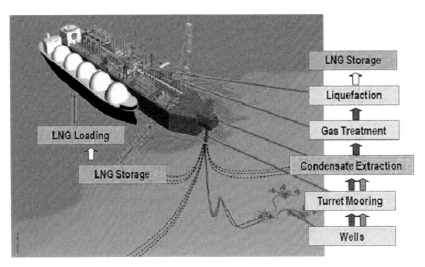

FIGURE 3-159

The structure and technological process of a marine floating LNG production multifunctional platform.

used for gas dehydration degasification, an LNG plant of liquid nitrogen, and an LNG tank, as well as a transportation and unloading system for LNG.

1. Turret mooring

   As is shown in Figure 3-159, through a certain connection method, turret mooring is used to fix the FLNG platform on the mooring points in the sea to make it rotate freely with the wind, waves, and current, thus reducing the environmental load and motion response of the FLNG system under external interference forces. It consists of turrets, a liquid transmission system, a rotating system, and an interface connection system. The turret is not only the mooring point of the FLNG system but also the channel for the stand pipe and umbilical cord system to reach the hull through the seafloor. Therefore, the turret fixed in the hull can ensure that the FLNG system can achieve continuous conveyance of gas under complex sea conditions.

   Turrets can be divided into internal turrets and external turrets based on their position as shown in Figure 3-160 and Figure 3-161. The turret device of the internal mooring system is located at the ship bow but external mooring overhangs on a cantilever.

2. Natural gas liquefaction production system

   There is a natural gas liquefaction system on the deck of the FLNG facility, which is equal to the arrangement of onshore natural gas liquefaction plants. But due to the limited deck area its technological process is very compact. The natural gas liquefaction production process is shown in Figure 3-159. It consists of condensate extraction, gas treatment, liquefaction, and so on. The liquefaction of natural gas usually uses nitrogen produced by a nitrogen expander cycle system as a freezing medium. The requirements of the whole system are compact, safe, and have a low sensitivity to ship motion.

FIGURE 3-160

The internal turret system of the FLNG platform.

3. LNG storage system

As is shown in Figure 3-159, an LNG storage system is used for storing qualified liquefaction gas. It consists of an LNG tank, LPG tank, and so on. Because the process of LNG storage is always under normal pressure and a low temperature of about -162°C, the materials and insulation of the tanks must meet these requirements. The LNG tanks can be MOSS type, SPB type, or GTT type. The structures of these three types are shown in Figures 3-162, 3-163, and 3-164, respectively.

FIGURE 3-161

External turret system of the FLNG facility.

FIGURE 3-162

MOSS type LNG tank.

4. LNG loading system

The LNG loading system is used for unloading reserved LNG to the transmission ship as shown in Figure 3-159. There are two unloading methods: tail unloading and side unloading.

The tail unloading method is shown in Figure 3-165. It transports the LNG to the ship bow to connect with the tail of the FLNG platform through a mooring line. Thus the LNG can be unloaded to the cargo ship through a flexible hose.

The side unloading method is shown in Figure 3-166. It is used to arrange the LNG cargo ship and FLNG platform side by side. Two ships are connected by a mooring line and fenders. The distance between the two ships is decided by the size of the fender.

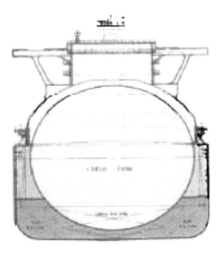

FIGURE 3-163

SPB type LNG tank.

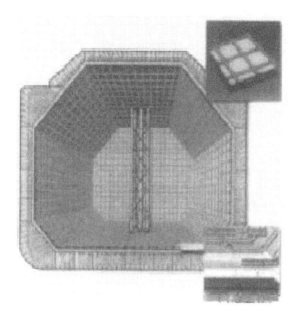

FIGURE 3-164

GTT type LNG tank.

FIGURE 3-165

LNG tail unloading mode.

3.24 **Key Technologies of the Offshore Floating Multifunctional Platform for Liquefied Natural Gas (LNG) Production**

Currently, the offshore floating multifunctional platform for LNG production is at the forefront of global offshore oil and gas projects, but it is only at a beginning throughout the world. The only floating platform, currently being built by the Royal Dutch Shell company, has not been put into

FIGURE 3-166

LNG side unloading mode.

production. Besides, more research is needed to solve some development problems. Some key technical problems are as follows:

1. Shake reduction technology for LNG container

   Because the liquidity of LNG in tanks is much higher than that of crude oil, the movement of the FLNG hull will trigger the LNG to slosh. LNG sloshing will cause great harm. It will in turn affect the overall movement of the FLNG hull. Therefore in the case of resonance, the hull will suffer fatigue damage. In addition, strenuous exercise of the liquefaction plant on the FLNG deck will cause LNG sloshing in the vessel, which will greatly reduce the production efficiency of liquidation. The following are some major available measures to reduce LNG sloshing in a container.

   a. Proper arrangement of liquefaction plant

      The position and direction of liquefaction plant on the FLNG is directly related with reduction of the movement response. The pitch is usually the most harmful. If liquefaction apparatus is arranged axially along the smallest pitch direction, movement can be reduced. Again, if the liquefied device is arranged as close as possible to the center of gravity position of the hull, this enable the response of heave motion caused by liquefaction plant moving along the vertical axis (Z-axis) to stay at a minimum.

   b. Proper design of liquefaction unit

      The overall size of the liquefier and geometry and size of its components should be properly designed. The diameter of the horizontal liquefier should be increased and the dimensions should be reduced as much as possible. As for the internal components, weir plates should be added but their number and geometry (such as meniscus, etc.) should be reasonably determined while optimizing the design. General liquefaction plant should be made to withstand 60° rolls and 30° pitching. The China University of Petroleum developed China patent No. 28626, "The resistance shaking device about reducing liquid sloshing in the internal container," which is based on optimal design theory, and presents an improvement of container productivity through repeated experiments.

2. Flexible technology for the tail input of LNG unloading

   FLNG platform tail unloading presents a process of connecting the shuttle tanker by one mooring cable and LNG hose. Usually, an offloading process lasts about 20 hours, and requires the delivery hose to float at sea level in the entire process. However, the temperature of LNG must be kept extremely low, $-162\,°C$. Therefore, not only must the material of the supplying hose withstand extremely low temperature, but also the hose itself should not be affected by water temperature for a long time, and a constant ultra-low temperature shall be maintained. In addition, the delivery hose also needs to overcome the impact of relative motion between the FLNG platform and shuttle tanker. At present, only foreign individual manufacturers are able to produce such high-end products.

3. Anti-collision technology in unloading

   When unloading leanly, the two close hulls will possess a strong nonlinear hydrodynamic influence, sometimes leading to a collision between two floating bodies. Therefore, we need to conduct research on the mutual hydrodynamic effect between two floating bodies and make an accurate prediction on the response of relative motion between the two bodies, especially the movement of the FLNG platform response. This involves not only nonlinear water dynamics research and prediction software, but also the development of anti-collision measures through experimental tank tests.

4. The improvement of liquefaction techniques

   Besides requiring the process to be very compact in order to suit the deck area, the liquefaction techniques also have requirements related to high adaptability to the different natural gas production areas and high thermal efficiency and also must be able to shut down immediately in extreme weather and start up rapidly when transferred to another producing location. There are requirements for the circulation mode, such as compact structure, good safety, high efficiency refrigerant with a small cooling chest, no need for fractionating tower, low sensitivity to ship motion, etc., to which we can apply optimization design theory to determine preferences.

5. Guaranteed safety of production

   The core part of the offshore floating LNG production multifunctional platform is similar to the onshore LNG plant. During the whole process of production, gas spilling or discharge will cause a fire. However, once there is a fire, the offshore platform will be more dangerous. Addressing and guaranteeing safety becomes the most important issue on an offshore floating LNG production multifunctional platform. So we should strictly follow the rules of the offshore oil operation safety management in all aspects including whole system design, building installation, usage, and maintenance.

3.25 The Main Applications of an Offshore Floating Multifunctional Platform for LNG Production

The application scope of an offshore floating LNG production multifunctional platform extensive; it can be used in both deep and shallow water, onshore and offshore. In detail, the main application are as follow:

1. The development of deep water gas fields

    Offshore floating LNG production multifunctional platforms are mainly used in the development of deep water gas fields; it can be combined with a subsea gas production system and LNG transport ship to form a complete system of gas production in deep water, treatment of oil and gas, natural gas liquefaction, and storage and loading of LNG, so that a perfect exploration of a deep water gas field can be achieved with high velocity, high quality, and high efficiency. This is because the offshore floating LNG production multifunctional platform has various capabilities, including adaptability of deep water gas recovery, strong wave resistibility in deep water areas, the processing capabilities for mass LNG liquefaction and oil, gas, and water production, and the ability to storage and load massive amounts of LNG.

2. The development of marginal gas fields

    The marginal gas field, measured from the area of economic benefits, refers to a field that can be developed with profit and also underdeveloped without profit. Compared with an onshore LNG factory in same scale, an offshore floating LNG production multifunctional platform helps lower investment 20% and shorten construction time 25%. With high flexibility, it can be moved to next gas field to reuse for high efficiency and it also can be combined with the wellhead platform as well as with jack-up or floating drilling and production platforms. It can be used in the exploration of deep and shallow water offshore fields and onshore marginal gas fields.

3. Early production of gas fields

    Early production refers that a portion of the gas field is firstly put into the production by making use of the advanced technique and facilities before a comprehensive exploration scheme is prepared well and the construction of natural gas production facilities are completed. The mode of early production can achieve economic profit as soon as possible. The offshore floating LNG production multifunctional platform not only combines with the jacket wellhead platform, but also with the jack up floating drilling platform or production platform, which can form an system integrated with gas production, oil-water processing, gas fluidification, LNG storage. So the multifunction platform is suitable for the early production, no matter the gas field is at the deep sea or at the shallow sea.

    To sum up the three aspects mentioned above, the floating offshore LNG production multifunctional platform is the latest highlight of offshore gas field development equipment; it has broad development prospects and will likely be used extensively in the future.

Special Problems of Deep-Sea Oil and Gas Engineering

In recent years, global offshore oil and gas development has been moving mainly toward the deep sea. So far, there are about 60 countries engaging in deep sea oil and gas exploration in the world. Seeking oil and gas from the deep sea has become a general trend. Although the development of China offshore oil fields has lasted for nearly 40 years, it has mostly occurred at the beach, in the shallow sea, and offshore. The area where the depth of the sea is more than 300 m in China is about $153.7 \times 10^4 \, km^2$, but it is basically unexplored. At present, China is the second largest oil consuming country all over the world. In 2014, China's foreign dependence has reached 58.8%, and it is expected to increase further in 2015. Facing the grim situation of Chinese oil and natural gas consumption growth, China Petroleum and Natural Gas Corporation, China Petrochemical Corp, and China National Offshore Oil Corporation realized the necessity and urgency of deep sea oil field exploration and development, thus putting a focus on seeking the oil and gas from the deep sea. So, in order to improve the degree of self-sufficiency of Chinese petroleum, apart from expanding the exploration and development of onshore oil and gas fields and fully tapping the potential of deep-sea oil and gas resources, going and seeking oil and gas from the deep sea is particularly important.

Compared with the offshore and shallow sea environments, the deep-water marine environment and marine engineering environment have their own unique aspects. Thus deep-sea oil and gas engineering technology and equipment have some different features compared with that used for offshore and shallow seas. So this chapter will discuss the special problems of deep-sea oil and gas engineering. This includes the combination of equipment for deep-water oil and gas collection and transportation, deep-sea mobile drilling and production platforms, deep-water oil and gas drilling technology and equipment, deep-water subsea production system problems, and so on.

4.1 Summary

4.1.1 The Definition of Deep Water

From the point of view of the oil industry, there is no exclusive definition for deep water in marine oil and gas engineering. The definition of deep water differs with time, region, and profession. As science and technology improves and the petroleum industry continues to develop, the definition of deep water also is changed constantly.

Offshore Operation Facilities. http://dx.doi.org/10.1016/B978-0-12-396977-4.00004-4

4.1.1.1 Different Definitions for Deep Water by Different Organizations

1. Definition by ABB (ABB Vector Gray)
   a. For drilling, water with a depth more than 3500 ft (1070 m) is categorized as deep water. In this area, semi-submerged platforms and floating drilling platforms are used.
   b. For oil extraction, water with a depth more than 1500ft (450m) is deep water. In this area, UWC (underwater completion), production and storage tankers, production, storage and offloading tankers, TLPs (tension leg platforms), and single leg platforms are used.
   Water with a depth more than 6000ft (1800 m) is ultra-deep water.
2. Definition by KMC (Kerr–McGee Corporation)
   a. For a production well, water with a depth more than 3000 ft (915 m) is deep water. In this area, UWC, TLPs, single leg production platforms, and production and storage tankers are used.
   b. For a drilling well, water with a depth more than 3500 ft (1070 m) is deep water. In this area, semi-submersible drilling platforms and floating drilling ships are used.

4.1.1.2 Different Definitions of Deep Water for Different Periods

Before 1990, water with a depth more than 1500 ft (450 m) was deep water.

In 1997, Serco defined that in general water with a depth more than 2000 ft (600 m) was deep water; but under some extreme environment conditions, water with a depth more than 1500 ft (450 m) was deep water.

In 1998, as published in *Offshore Magazine*, in order to distinguish deep-water floating drilling ships, water with a depth more than 3500 ft (1070 m) was defined as deep water.

In 1999, Hart (Hart's Deep-Water Technology) defined deep water as water with a depth more than 1300 ft (400 m).

In 2002, at the World Petroleum Congress held in Brazil, in order to divide sea oil and gas exploratory development by water depth of the water, the meeting reached a consensus that water with a depth less than 400 m was defined as normal water depth; water with a depth between 400 and 1500 m was defined as deep water; water with a depth greater than 1500 m was defined as ultra-deep water.

In general, at present, water with a depth more than 400 m is defined as deep water and water with a depth more than 1500 m is ultra-deep water.

4.1.2 Necessity of Deep-Water Oil and Gas Exploration and Development

4.1.2.1 Extent of Deep Water Area

The total area of seawater across the world occupies 78.9% of the earth's surface, the area of which is 5.11×108 km^2. The average depth of the global marine environment is 3730 m, and water with a depth of $3000 \sim 6000$ m is 73.83% of the total area of the ocean. However, the offshore continental shelf water region, with a depth less than 200 m, is only 7.49% of the total area of the ocean. Due to this existing disparity, we must vigorously search for resources from the deep sea.

4.1.2.2 Potential Oil and Gas Resources in the Deep Sea

The ocean is rich in oil and gas resources. At present, more than 1,600 offshore oil and gas fields are found all over the world, and only 200 of them have been put into production, yet the annual output has already reached $12 \times 10e^8$ t, accounting for about 1/3 of the world's total oil output. It should be pointed out that more than 56% of the ocean oil reserves that has been proven to exist are in sea areas with a water depth less than or equal to 500 m. However, 44% of global marine oil and gas resources are in water with a depth more than 500 m.

4.1.2.3 Development Trends for Global Offshore Oil Fields

In recent years, global offshore oil and gas development is mainly moving toward the deep sea. In 2001, more than 130 exploitation wells were built in sea with a depth more than 1000m. It is estimated that in 2001-2007, the number of offshore oil and gas field development projects worldwide was 43. Projects with a water depth of more than 500 m account for 48% of all such projects. In addition, 22% of projects involve a water depth deeper than 1200m. At present, more than 60 countries around the world have engaged in deep-sea oil and gas exploration, and more than 5×10^8 bbl from 33 giant oil and gas fields have been found. Therefore, seeking oil and gas from the deep sea is definitely the trend.

4.1.2.4 Advantageous Situations for Deep-Sea Oil and Gas in China

The South China Sea, with a water area up to 200×10^4 km^2, is one of the four major offshore oil and gas gathering centers in the world. The amount of oil reserves in the South China Sea are estimated to about $230 \times 10^8 \sim 300 \times 10^8$ t. And the amount of natural gas is about 338×10^{12} m^3, 70% of which is in deep water. In China, sea area with a water depth of more than 300 m is about 153.7×10^4 km^2, but only 16×10^4 km^2 of that area has been explored. So, China National Offshore Oil Corporation has realized the necessity and urgency of deep sea oil field exploration and development, and now is undertaking oil and natural gas exploration in 1.9×10^4 km^2 of deep water in the South China Sea. This situation shows that deep-sea oil and gas resource exploration and development must be the main focus for China offshore oil and gas in the future.

4.1.3 Characteristics of Deep-Water Oil and Gas Field Project Environment

At present, the depth of global deep-water drilling has reached 3095 m, and the maximum depth of oil and gas fields using underwater oil production technology has reached 2192 m. For oil and gas field development projects with such depth, the marine environment has presented a series of problems. The main differences from sea with a conventional depth are as follows.

4.1.3.1 Deep Working Depth

Due to the deep water working environment, the following issues are present in oil and gas field development engineering:

1. Difference equipment is demanded
   Compared with the conventional depth seas, the differences in equipment required are as follows:
   * Longer, larger drilling platform or ship
   * Higher, stronger derrick
   * Longer drilling riser
   * Greater tension and lifting capacity for the tensioned system
   * Greater driving force from the power equipment
   * Larger rig area
   * More and bigger drilling pump and bigger drilling fluid container
   * Winch with larger power and rotary table with larger opening diameter and power

- Bigger BOP stack and kill and choke line
- Larger storage capacity and more stable deck
- Stronger 10 or 12 point mooring system;
- Greater Dynamic Positioning (DP) redundancy and automatic exhaust pipe system
- Stronger Remotely Operated Vehicle (ROV) and operation system for it
- Wider facilities, including a wider moon pool for underwater tools and equipment
- Dynamic positioning systems

2. Different equipment capacities
Because of the deep water depth, equipment capacities are significantly different compared with conventional water depths. Taking the semi-submersible drilling platform or drilling floating ship as an example, the capacity of the main equipment is shown in Table 4-1.

3. Drilling cost difference
In deep water environments, drilling costs will increase. Taking the semi-submersible platform as an example, the comparison of cost between a water depth of 330 m (1000 ft) and 3300 m (10000 ft) is shown as Table 4-2.

4. Different well-construct cycle
Because of the increased water depth, more casing will be run, it will take long to trip the drilling string and the riser string, and time will increase for other nonproduction tasks; thus the whole well construction cycle will become longer.

4.1.3.2 Seabed Instability

The instability of the seabed in the deep sea, compared with conventional depth areas is another difference in the offshore engineering environment. It is caused by the rapid deposition of the landslide, as shown in Figure 4-1. With a lower depth than shallow water, soft and uncemented soils on the slope will quickly deposit in deep water, and form a thick, soft, high-water content, uncemented formation; thus an unstable seabed will form in the deep sea.

The influence of the instability of the seabed on deep-sea oil and gas engineering is as follows:

1. Instability of the borehole wall
Taking shale as an example, because of its instability and the inappropriateness of drilling fluid, which makes the pressure on the borehole wall unbalanced, the shale layer will break. The incrementing of formation pore pressure caused by the swabbing or fluctuating pressure can also make the shale rupture, as well as the invasion of drilling fluid, water stress, and drill string vibration. In short, the rupture of the unstable rock layer will result in borehole shrinkage or borehole collapse. And the drilling fluid pressure will fracture the formation and natural crushing zone, causing well leakage, which causes a lot of trouble for drilling engineering.

2. The bearing capacity of the foundation
Owing to the soft and uncemented soil of the unstable seabed, the static bearing capacity is low. But in deep-water oil and gas engineering, submarine drilling templates of multiple underwater wellheads and submarine foundations using tension leg platforms or spar platforms all need submarine soil with higher static bearing capacity, causing difficulty. When using a cylindrical base, not only is a higher static bearing capacity of the seabed soil required, but also friction between the cylinder wall and the soil. But for an unstable seabed, such requirements cannot be met.

Table 4-1 Deep-Sea Drilling Equipment Capacities

Type	Riser System		Hoisting System		Drilling Fluid Circulation System		
Semi-Submersible Platform or Drill Ship	Maximum Working Depth, ft	Riser Total Tension, klbf/in²	Maximum Drilling Depth, ft	Derrick Rated Weight, klbf/in²	Drilling Fluid Container, bbl	Pump	Pressure, kgf/in²
Second generation/A	600~1800	960	20000	1300	2500	2~3	5000
Third generation	1800~3000	1200	25000	1500	3~5000	2~3	5000
Fourth generation/B	3000~5000	1600	25000	1500~2000	10~15000	3	5000
Fifth generation/C	5000~10000	2500~4000	30000	2000	15000	4	75000

Note: The A, B, and C respectively represent the type of the drilling ship; the second generation, third generation, fourth generation, and fifth generation respectively represent the type of semi-submersible platform.

Table 4-2 Comparison of Semi-Submersible Drilling Cost at Different Water Depths

Project Cost	Working Depth, ft	
	1000	10000
Positioning system: anchor/winch/chain dynamic positioning/propeller fuel	$40 × 10^6 10~20 t/d	$70 × 10^6 10~60 t/d
Riser system	$2 × 10^6	$16 × 10^6
Tensioner system	$2 × 10^6	$8 × 10^6
BOP and Control system	$5 × 10^6	$10 × 10^6
Housing costs (size and variable load)	$90~110 × 10^6	$140~170 × 10^6
Monthly fee	$150~350 × 10^6	$350~650 × 10^6

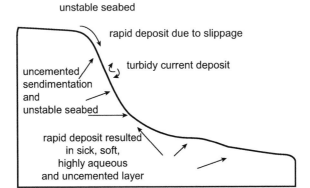

FIGURE 4-1

The formation of deep seabed instablity.

4.1.3.3 Low-Fracture Gradient

Fracture gradient here refers to the fracture pressure gradient, namely the minimum pressure gradient of the formation fracture. For formations with similar thick sedimentary layers, with increasing water depth, the fracture pressure gradient of the formation reduces, as shown in Figure 4-2.

The pore pressure gradient is the maximum pressure gradient of oil and gas fluid in the formation pore. In order to control the well and prevent blowout, we need to adjust the density of drilling fluid to make sure that the pressure of the bottom is higher than the maximum pore pressure gradient, and also control the density of drilling fluid to make the mud column pressure less than the minimum formation fracture pressure gradient. So, for a certain depth of well, the adjustable density range of drilling fluid can be formed between the formation fracture pressure gradient and the pore pressure gradient. It is usually called a window, namely the window between the formation fracture pressure gradient and the pore pressure gradient. Due to the increasing depth of the deep sea, and the decreasing fracture

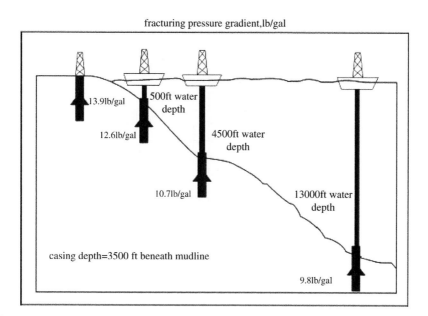

fracturing pressure gradient,lb/gal

13.9lb/gal

500ft water depth

12.6lb/gal

4500ft water depth

10.7lb/gal

13000ft water depth

casing depth=3500 ft beneath mudline

9.8lb/gal

FIGURE 4-2

The relationship of formation fracture pressure gradient and depth.

pressure gradient, the window between the formation fracture pressure gradient and the pore pressure gradient is very narrow. The relationship between the formation fracture pressure gradient and the pore pressure gradient changes with the variation of the depth of water and affects the window (Figure 4-3). The influence of the low fracture pressure gradient of deep water formations on deep-water oil and gas engineering is discussed in the following:

1. Easy circulation loss

   When using drilling fluid, after overcoming the wall mud cake, if the fluid pressure is transmitted to rock and it is larger than the minimum principal stress of the rock structure as well as the minimum fracture pressure, the formation will fracture and cracks will extend generally along a direction perpendicular to the minimum principal stress of the rock structure, resulting in circulation loss. In deep water, due to the narrow window between the formation fracture gradient and the pore pressure gradient, when adjusting the density of drilling fluid, the drilling fluid column pressure gradient can reach the cracking stress of rock structure easily, and give rise to circulation loss.

2. Increase of casing pipe layers

   In deep-water drilling, the depth of the casing shoe will be as deep as possible to reduce the number of layers of casing string. Due to the narrow window between pore pressure gradient and fracture pressure gradient in deep water, such a goal cannot usually be reached. In this way, compared with shallow waters or onshore oil wells with the same drilling depth, more casing layers are needed for the deep-water wells. Under some serious conditions, there is no casing string available for some oil and gas wells; therefore, the ultimate goal of drilling may not be achieved.

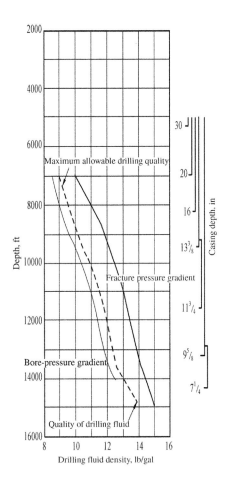

FIGURE 4-3

The window between pore and fracture pressure gradient.

3. The pressure management difficulties

When drilling in a marine environment, forecasting, monitoring, and controlling the pore and fracture pressure gradient is called pressure management. Pressure management is an important task for drilling and safe operation in a marine environment, especially in deep waters. But no matter if we predict conditions before drilling with geological interpretation, or monitor the drilling process through measurement while drilling, or interpret the obtained real-time data, or repeat the formation test after drilling, or even extend the test, it is more difficult to manage pressure in deep water than in shallow water. Especially, it is more difficult to adjust the density of drilling fluid to control the pore and fracture pressure gradient in deep-water drilling.

4.1.3.4 Seabed at Low Temperature

Under the sea surface, the water temperature decreases with increasing water depth, so in water the seabed temperature is low. The relationship between water depth and temperature below the sea surface is shown in Figure 4-4. The relationship between depth and temperature in the Gulf of Mexico is shown in Figure 4-4(b). Seen from this graph, near the surface, the temperature is 25~32°C, but with increasing water depth, the temperature decreases. When the water depth is 150 m, the temperature is reduced to 18~19°C. And when the water depth is 1000m, the temperature has dropped to 5°C. However, the change of temperature has a significant feature, that is, on the seabed, temperature decreases with the increase of depth, but below the seabed, the temperature rises with the increase of depth, as shown in Figure 4-5. Effects of the low temperature of the seabed are as follows:

1. Underwater appliances blocked
   Low temperature is one of the necessary conditions of natural gas hydrates generated in the deep sea. Under the conditions of low temperature and high pressure, natural gas hydrate will be generated easily, which will block the blowout preventer stack (BOP) pipeline, riser, and underwater casing head. These blocked appliances will bring risks to well control during the drilling process.
2. Difficult to assure flow
   Flow assurance ensures smooth flow of oil and gas in the pipeline. With the development of deep-water oil and gas fields, since the 1990s we have seen the growth of flow assurance technology. The low temperature of the seabed has become a major obstacle of flow assurance in deep-water oil and gas fields. For deep-water oil and gas transportation pipelines, the flow assurance difficulties caused by the low temperature are mainly:
   a. Deposited solid matter plugs the pipeline: Hydrate and wax deposition in the pipeline caused by low temperature is the main obstacle. In addition, there is asphalting, scaling, etc.
   b. After shutting down the transport of crude oil with a high condensation point and high viscosity, it's difficult to restart: When shutting down the transport of crude oil with high condensation point and high viscosity, because of the blockage of the pipeline caused by cooling crude oil, there will be big risk when restarting.

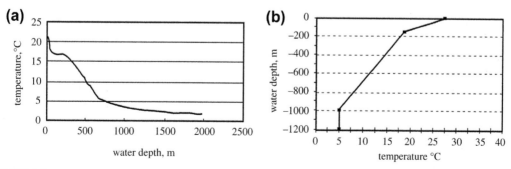

FIGURE 4-4

Relationship between water depth and temperature below the sea surface.

FIGURE 4-5

The sea water depth and temperature of up and down.

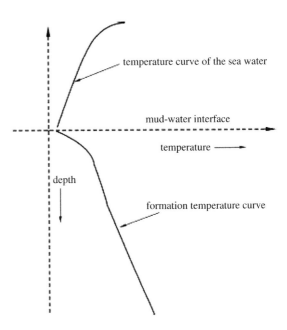

c. Slug flow in the riser pipe for gas-liquid mix conveyance: Due to the long length of the riser which transmits deep-water gas and liquid from the undersea to the surface, slug flow will be formed easily, resulting in foaming and emulsification of crude oil, which makes flow assurance difficult.

4.1.3.5 Existence of Gas Hydrate

Natural gas hydrates are also called clathrates. They are a cage-shape crystalline compound that are similar to ice. They are nonstoichiometric and composed of water and natural gas under certain conditions. As they can burn in a fire, they are also called "combustible ice." They can be expressed by $M \cdot nH_2O$, where M represents a the compound gas molecules and n represents the hydration index, which is the number of water molecules. The composition of natural gas, such as homologues CH_4, C_2H_6, C_3H_8, and C_4H_{10}, as well as CO_2, N_2, and H_2S, can form a single species or a variety of natural gas hydrates. The main gas resulting from the formation of natural gas hydrates is methane. And the natural gas hydrate where methane molecules account for more than 99% of the compositoin is usually called methane hydrate. As shown in Figure 4-6, natural gas hydrate is an ice-shaped gas and water combination. When it is on the ground, the volume can be expanded to $70 \sim 100$ times.

Formation conditions of natural gas hydrate mainly include the appropriate temperature, pressure, gas saturation, water salinity and pH. Due to the simultaneous presence of these necessary conditions, which includes a certain temperature, high pressure, water, and natural gas in the deep sea, the natural gas hydrates are easy togenerated. In the process of drilling in the deep sea, under the conditions of a certain temperature and pressure, gas impregnated drilling fluid may generate natural gas hydrate. Chinese and German scientists, in an expedition, found that there were natural gas hydrates existing in the South China Sea in June 2004.

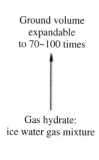

Ground volume
expandable
to 70~100 times

Gas hydrate:
ice water gas mixture

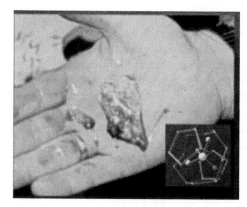

FIGURE 4-6

Shape and structure of gas hydrate.

For deep-water oil and gas engineering, potential serious damage caused by natural gas hydrates mainly include:

1. Ice jam
   a. Block, choke and kill pipelines, effecting well control
   b. Block BOP, resulting in disabling of ability to monitor subsurface pressure
   c. Block around the drill string, restricting operation of drilling appliances and easily causing sticking
   d. Block the production riser, making it difficult to convey gas-liquid from the bottom to the water surface
   e. Block the submarine pipeline and make flow assurance difficult.
2. Well control accidents
   a. Unable to close the BOP when you need to control the well
   b. After the BOP is closed, it cannot be opened
3. Production of gas
   The hydrate will generate a large amount of gas when it gasifies. These gases will affect the quality of drilling fluid seriously during the drilling process. At the same time, the thermal effect in the gasification process is also a big problem.
4. Environmental impact
   Natural gas hydrates affect the environment mainly though the following ways:
   a. The greenhouse effect: Methane is the main gas in a natural gas hydrate, but the greenhouse effect of methane is 21 times stronger than carbon dioxide. Methane's contribution to the greenhouse effect is about 15%.
   b. The ecological destruction: When a natural gas hydrate is formed, the production of methane (CH_4) will lead to the reduction of oxygen gas (O_2), thus the balance of the ecological environment is destroyed. This can be seen from the following chemical reaction:

$$CH_4 + 2O_2 = CO_2 + 2H_2O$$

$$CO_2 + CaCO_3 + H_2O = Ca(HCO_3)_2$$

4.1.3.6 Shallow Flow with High Pressure

So-called shallow flow includes shallow gas flow and shallow water flow. Shallow gas flow does not only existing in the deep sea, but the pressure of deep-sea shallow gas is usually higher, so it brings more serious harm to the deep-water oil and gas engineering. Shallow water flow refers to the shallow high-pressure water layer and it exists not only in the deep sea. Generally, the depth of the shallow high-pressure water layer is about 800 m below the mud line. In the drilling process, when caught in the shallow high-pressure water layer, the column pressure of drilling fluid cannot balance with the water pressure, which will produce the shallow water flow. For shallow water, deep sea pressure will be much higher. Therefore, if we can't handle it properly, it will bring adverse effects on deep-sea oil and gas engineering.

1. Effects of shallow gas layer

   When blowout occurs, in the shallow gas layer, the gas expands and diffuses rapidly in a funnel shape, and the scope of its impact is inclusive, as shown in Figure 4-7.

   If not handled properly, the consequences will be very serious. Once blowout occurs, fire disaster will follow, making the drilling ship or semi-submersible platform sink. These serious accidents have served as profound lessons in the history of offshore oil and gas drilling.

2. Effects of shallow water

   For offshore oil and gas engineering, the effects of shallow water flow are as follows:

   - Borehole collapse and well control problems caused by high-pressure shallow water flow in the process of drilling.
   - Failure of cementing operation caused by high-pressure shallow water flow in the formation in the process of well cementation.
   - After well cementation, if high-pressure shallow water flow happens, high-pressure water will flow from the catheter outside to the seabed. This may cause the catheter to sinks due to losing

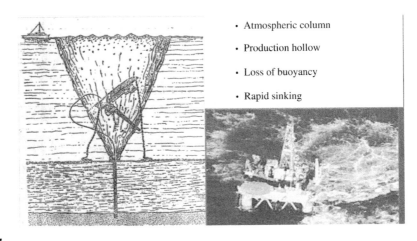

- Atmospheric column

- Production hollow

- Loss of buoyancy

- Rapid sinking

FIGURE 4-7

High-pressure shallow gas expansion and diffusion.

support surface, resulting in wellhead instability; we may even need for shifting and re-open wellbore.

In short, due to the series of problems above, the working hours an offshore drilling platform (ship) will be delayed for several months, and cost hundreds of millions dollars, creating a great impact on the economic benefits.

Generally, the reasons for shallow water accidents include the following aspects:

- Formation of fracture artificially. For example, the drilling fluid density is not appropriate and exceeds the corresponding formation fracture pressure gradient.
- Abnormal formation pressure caused by channeling cement during well cementation
- Fracturing shallow water layer due to abnormal pressure.

4.1.3.7 Internal Wave Activity

There is internal wave activity in the South China Sea. Internal waves act on offshore structures, resulting in large amplitude motion which brings serious harm to the deep-sea oil and gas engineering.

Due to some factors such as the sun's heat radiation, the density of seawater is affected by stratification effect. In the varying ocean layers, as long as there is a disturbance source, it will generate internal waves in the ocean. Therefore, an internal wave is a kind of dynamic process generally existing in the ocean.

1. Origin of internal wave activities in the South China Sea

   There are internal wave activities frequently in the South China Sea. In offshore oil and gas engineering, internal waves have become a critical disaster factor second only to the typhoon, It is an absolutely key environmental factor that should be considered in the design of offshore oil and gas engineering. In the northern South China Sea, the origin of internal waves are mainly:

   a. Vertical stratification effect of seawater

   In the northern part of the South China Sea, the water mass properties of the continental slope and shelf waters, affected by the Kuroshio and monsoon, are mainly governed by water of basin in the South China Sea and the Kuroshio water. Their flow pattern is changeable. With changing seasons, the complex stratification effect is also produced in the vertical direction for seawater.

   b. Effects of variation in bottom topography

   In the northern South China Sea, the water depth of the continental shelf changes sharply from 500 m to 1000 m and forms a steep slope. The terrain variation provides a natural marine environment for the production of internal waves.

2. The characteristics of internal waves in the South China Sea

   Active internal waves of the northern part of the South China Sea are formed by tides. They are formed by tides inspired by the sea-back, the edge of the continental shelf, and underwater terrain mutation. After the internal wave is formed stimulated by tide, it can simultaneously propagate a considerable distance towards the shallow water and deep water. Under certain conditions, it produces large amplitude internal solitary waves by fission in the dissemination process. The in-situ observation at sea and the marine remote sensing data show that in the northern South China Sea, internal solitary wave activities with large amplitude frequently occur.

   In the northern South China Sea, nonlinear internal waves with large amplitude originate in the Bash Channel. In the Lufeng oil field in the South China Sea, the maximum velocity of internal waves recorded is 149 cm/s, while in the Liuhua oil field in the South China Sea, the speed of the largest isolated wave observed at the depth of $20 \sim 100$ m is $50 \sim 150$ cm/s. In addition, the maximum velocity of solitary waves has exceeded 300 cm/s in some sea areas. Moreover, at the southern coast of the Dongsha Islands in China, solitary waves with amplitude near the 40 m have also been observed. In the northwest South China Sea, the maximum amplitude of semidiurnal tidal waves is 6 m, and the maximum amplitude of diurnal tide waves is 4 m.

3. Effects of internal waves on offshore oil and gas engineering

Internal waves are a disaster second only to typhoons, and its impact on offshore oil engineering mainly include following several aspects:

a. Large impulsive loads on offshore structures

Due to internal solitary waves with large amplitude, water which is close to a pycnocline in a shear state can induce a sudden strong current and produce a great impulsive load on offshore structures. Offshore platforms, ships, and offshore pipelines will be seriously harmed. In the Liuhua 11-1 oil field in the South China Sea, there has been an accident where unexpected strong flow of internal waves cut a ship cable out.

b. Production of large scale movement for floating structure

When nonlinear internal waves with large amplitude flow, they will encourage the offshore floating structure to move greatly, which poses a serious threat to the safety of the structure. For example, a floating drilling platform in the Andaman Sea was driven by internal waves to rotate $90°$ and move 30.48 m. This offset of well location causes serious damage in drilling engineering.

c. Large amplitude vibration for floating structure

Internal waves can bring great harm to floating offshore structure such as the floating drilling platform near the China Dongsha Islands; it's unable to operate normally when nonlinear internal waves with large amplitude pass by; the anchored tank in the sea swings $110°$ in less than 5 min. In addition, the offshore anchored observation research ship, when encountering a nonlinear internal wave with large amplitude, has also been forced to stop observation because of persistent sloshing.

4.2 Equipment Combination Programs for Acquisition, Processing, Storage, and Transportation of Deep-Sea Oil and Gas

Development projects for deep-sea oil fields involve collection of oil and gas in each wellhead, oil and gas separation, dehydration, and water purification and production processing; then the oil and gas is stored, and transported to shore in a timely manner. This involves an equipment combination program for acquisition, processing, storage, and transportation of deep-sea oil and gas, including what equipment is needed and how they should be combined in this process.

Abroad in the late 1980s, a deep-sea oil field development boom began. Over the past 20 years, the American Gulf of the Gulf of Mexico, Brazil, South America, and the West African waters, which is called "Golden Triangle," have obtained worldwide attention in the deep-sea oil field development and been leading the world in deep-water oil and gas engineering technology. They have also accumulated advanced experience that is suited to the characteristics of their equipment combination program for oil and gas acquisition, processing, storage, and transportation.

The South China Sea is abundant in deep-sea oil and gas resources, including the North Bay, YingGe Sea, Joan southeast is the South China Sea islands. These four basins have broad prospects for deep-sea oil and gas development. At present, China CNOOC, Petro China, and Sinopec are the three oil companies embarking on the study of the South China Sea and the potential deep-water oil and gas exploration and development. However, the South China Sea oil and gas engineering environment has different characteristics compared with that in foreign regions. In this case, which type of equipment combination program for oil and gas acquisition, processing, storage, and transportation China deep-sea development should choose is one of the focuses of China National Offshore Oil. This section is intended to introduce topics around this theme.

4.2.1 Several Foreign Existing Combination Programs

In 1996, Kerr-McGee installed the first single oil recovery column (SPAR) health production platform in the deep-water portion of the Gulf of Mexico. Since then, the United States has been committed to developing new deep-sea platform technology, mainly by studying an ultra-deep-water field that is 1666 m (5000 ft) to 3333 m(10000 ft) deep. Since 1986, Brazil, in order to establish independent deep-water engineering technology, has made a 15-year development plan, the third phase of which is ongoing, where the goal is to form deep-water field development capabilities of 2000 m deep. Countries, mainly including the United States and Brazil, have accumulated rich deep-sea oil and gas development experience and formed several distinctive options in equipment combinations for deep-sea oil and gas acquisition, processing, storage, and transportation. This is briefly introduced in the following:

4.2.1.1 The First Combination Program

This equipment combination program can be simplified as follows:

SWH+TLP/SPAR+FPSO/FSO+ST
(subsea wellhead) + (tension leg platform or single-pole platform) +(floating production, storage tankers or floating storage and offloading) + (shuttle tankers)

The equipment of this program adopts the underwater subsea wellhead (SWH); generally wet Christmas trees are used. Then, it collects each well's oil and gas, and transports it through the riser to the tension leg platform (TLP) or single (independent) column platform (SPAR) (drilling also uses the platform). Then it transmits the oil and gas to the floating production, storage, and offloading platform (FPSO) for oil and gas water treatment [or when the platform has its own production facilities for oil and gas water treatment, oil and gas is directly transmitted to the floating, storage, and offloading (FSO) vessel].When crude oil production reaches the quality standard ,it can be carried away by the shuttle tanker(ST) in a timely manner.

In this equipment combination, since the oil and gas water treatment and other production processes are carried out at sea, it is known as the all-sea type. It is characterized by drilling oil, the oil and water separation process, as well as storage and a variety of functions. We can use a well to drill a multi-well, offshore drilling chassis and satellite wells, suitable for transforming the original drilling platform into a drilling and production platform. However, TLP and SPAR types are new deep-sea platform types, and are difficult to design and construct.

Figure 4-8 shows a schematic view of the combinations of such equipment.

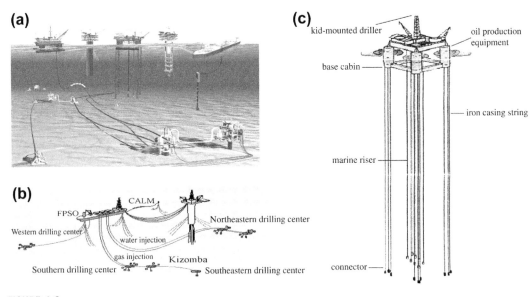

(a)

(b)

FPSO
Western drilling center
CALM
water injection
gas injection   Kizomba
Southern drilling center   Southeastern drilling center
Northeastern drilling center

(c) kid-mounted driller   oil production equipment

base cabin

iron casing string

marine riser

connector

FIGURE 4-8

Schematic diagrams of the equipment combinations.

4.2.1.2 The Second Combination Program

Combinations of such equipment can be simplified as follows:

SWH/SPS+SP+FPSO/FSO+ST
(subsea wellhead or subsea production systems) + (semi-submersible platform) + (production and storage tanker or floating storage and offloading platform) + (shuttle tankers)

This program is different from the first program, as here one can use an underwater subsea wellhead (usually a wet Christmas tree), or a subsea production system (SPS). The so-called subsea production system puts a Christmas tree, underwater pipe exchange center, oil storage, and oil and gas water separation processing equipment on the seabed. So when the subsea production system is adopted, we can no longer arrange production and processing facilities on the surface. The second difference is that the tension leg platform or single column platform is not applied on the surface of the water, but a semi-submersible platform is for drilling and production. When we use subsea production systems, the drilling platform can be enough, while if a subsea production system is not applied, the semi-submersible platform can be the drilling and production platform (oil production). Production and processing can be done in the floating production, storage, and offloading (FPSO) vessel. Among the equipment combinations, we can select FPSO, or storage and offloading (FSO), which depends on whether the subsea production systems or semi-submersible platform has the function of production and processing of oil and gas. Finally, products that are qualified after processing will be carried away by shuttle tankers. Thus, the second combination scheme belongs to the all-sea type, the same as the first scheme.

The features of the second program take full advantage of the underwater equipment, semi-submersible rigs, production storage and transportation round. There is flexibility and adaptability in moving from place to place whenever necessary; also, in terms of design and construction, the semi-submersible platform is more mature compared with the tension leg platform and single column (independent) platform. Of course, it has a variety of functions including drilling, production, storage, and offloading, so it can offer a complete all-sea development offshore oil field package and meet all the requirements.

From a foreign point of view, the second combination program has been an effective program for many years in Brazil. Brazil's deep-sea oil field development has basically used these equipment combinations; moreover, it has become the standard program for Brazil deep-sea oil field development and a typical equipment combination program globally for developing deep-water fields. By 2003, average daily production of crude oil in Brazil reached 153.5×10^4 bbl, natural gas per day was up to 4400×104 m3, and the oil self-sufficiency rate reached 80%, mainly through deep-sea oil field development and the second gear combination.

Figure 4-9 shows a schematic diagram of the second scheme, which is used in Brazil deep-sea oil field development. The underwater Christmas tree has three forms: wet, dry, and a combination of wet and dry.

4.2.1.3 The Third Combination Program
Combinations of such equipment can be simplified as follows:

SWH+FDPP+SPL
(subsea wellhead) + (floating drilling platform) + (subsea pipeline)

This program uses the following equipment: the subsea wellhead (SWH), floating, drilling, and production platform (FDPP) and submarine pipeline.

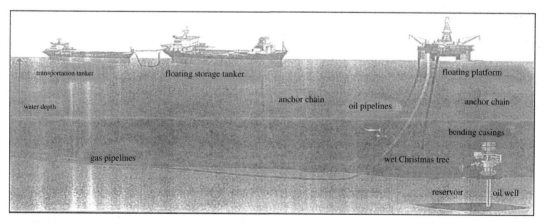

FIGURE 4-9

Brazil's second equipment combination.

The equipment combination of such programs is characterized by multiphase subsea wellhead oil and gas flowing to the floating platform. As the floating platform is a production platform, besides drilling operations, the floating platform can separate and process oil-gas-water. The processed qualified crude oil is sent away through a submarine pipeline to the shore for storage. Because there are no storage facilities at sea, So the Storage are carried out after the crude oil is transported to the shore. Therefore, we call it half-sea half-land, which is the main difference compared with the above two programs.

These equipment combinations are mainly used in United States, because the Ministry of Geology and Mineral Resources of the United States once had regulations in 2001 to restrict floating production, storage, and offloading (FPSO). Offshore oil and gas produced in the sea can be transmitted through a submarine pipeline to shore, which forms half-land half-sea type of offshore oil field development methods, such as the subsea pipeline in the U.S.Gulf of Mexico that has been extensively developed over the years, including a criss-cross large scale oil and gas pipeline network. That creates very convenient conditions for qualified oil and gas processing and external output on floating platform. It is easy to see from the above that this third combination is the product of specific historical conditions. In fact, the United States lifted the restrictions of the Ministry of Geology and Mineral Resources in 2001, which now allows the use of floating production, storage, and offloading (FPSO). In addition, there is another feature of this program, that is where the floating drilling platform adopts the single (independent) pillar (SPAR) platform. In the Gulf of Mexico, there are 15 deep-sea drilling single (single) columns. Figure 4-10 shows the deep-water Gulf of Mexico in the United States with a single-column (SPAR) platform used a floating drilling platform, as part of a schematic diagram of the third combination program. This third equipment combination used by the United States applies not only to single (independent) pillar (SPAR) platform, but also to the tension leg platform (TLP) and semi-submersible platform, as well as tower platforms as drilling level units; but the tower platform is used less frequently, as there are only four in the Gulf of Mexico, and the semi-submersible platform is used more (a total of more than 40 seats).

The problem of the third equipment combination is having to transport oil to shore via undersea pipeline for production and processing. Undersea pipeline transport is not as good as shuttle tanker transport; the latter is the superior means of water transport. Again, subsea pipelines used for mixed transportation of oil and gas that is not separated and processed will bring a series of complex equipment technology problems.

FIGURE 4-10

Third equipment combination in the deep-water Gulf of Mexico.

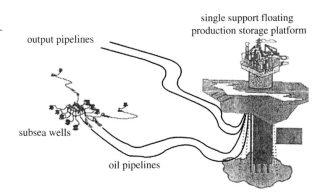

single support floating production storage platform

output pipelines

subsea wells

oil pipelines

4.2.1.4 The Fourth Combination Program

The item combination of such programs can be simplified as follows:

SDB+SOGSU+FPL+FSO/FPP+ST
(subsea drilling base) + (subsea oil and gas separating unit) + (flexible pipeline) + (storage and offloading or fixed oil platforms) + (shuttle tanker)

This combination scheme is equipped with the following equipment:

1. Subsea drilling base (SDB)
   Shown in Figure 4-11, this program installs a subsea drilling base in the bottom of the sea,which can drill a multi-port well. When a well has been completed, a submarine wet Christmas tree can be installed for early production of oil. At the same time, drilling can proceed. Of course, the drilling operations also need to have a drilling platform on the surface of the water.

2. Subsea oil and gas separating unit (SOGSU)
   A subsea oil and gas separating unit consists of an underwater oil and gas separation device and pump, as shown in Figure 4-12(a), in which the separate oil will deliver to the surface of the water by oil pump. Figure 4-12(b) shows the layout of the underwater oil and gas separation device at the bottom of the sea drilling chassis case.

3. Flexible pipeline (FPL)
   The fourth equipment combination program uses a flexible pipeline oil and gas separation device to separate oil and pump it to the surface. Thus, the flexible pipeline is an important piece of equipment for such programs.

4. FSO or fixed production platform (FPP)
   When the water depth is shallow, the fourth program can adopt a fixed oil platform for oil and gas processing and storage or just store at the origin. But when the water depth is large, the program uses the floating storage and offloading (FPO).

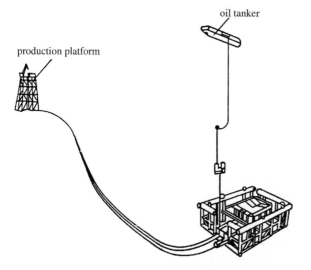

oil tanker

production platform

FIGURE 4-11

Schematic diagram of the fourth equipment combination.

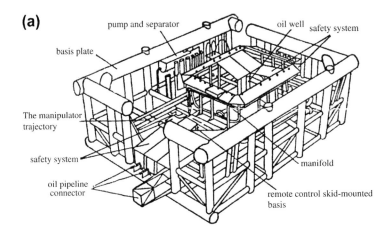

(a)

pump and separator

oil well safety system

basis plate

The manipulator trajectory

safety system

manifold

oil pipeline connector

remote control skid-mounted basis

FIGURE 4-12

Fourth equipment combination underwater oil and gas separation device.

5. Shuttle tanker (ST)

The fourth program uses shuttle tankers for crude oil transportation whether it adopts a floating storage and offloading (FSO) vessel or fixed oil platforms.

The fourth equipment combination program has the following features as it adopts an all-sea development program. Crude oil is produced in the sea and transmitted by shuttle tankers. Furthermore, it uses the submarine SDB to drill the chassis, while producing oil in a well. Multi-well drilling and oil production will be adopted in order to meet the requirements of early production. In addition, this program adopts a flexible pipeline to transport crude oil from the seabed to the floatign storage and offloading (FSO) vessel or fixed platform. The flexible pipeline presents complex technical issues under deep-water marine environment loads, so application of this program has been limited in deep water.

4.2.1.5 The Fifth Combination Program

Combinations of such equipment can be simplified a and expressed as follows:

WH + FPDSO (FPSO + TLD) + ST
(wellhead) + (floating production, drilling, storage, and offloading combination unit) + (shuttle tanker)

The core equipment of this program is a joint floating production, drilling, storage, and offloading device (FPDSO). This combination device is shown in Figure 4-13, which is developed on the basis of the floating production, storage, and offloading (FPSO) vessel apparatus. That is the FPSO is arranged at the center of a tension leg drilling deck (TLD). The tension leg platform deck adopts a similar technology of the tension leg platform and is tied with the drilling deck by a cable in the sea; the deck load is balanced through the outboard weight-block system. The weight block underwater is 100 m, in order to avoid wave action and reduce wobble, as shown in Figure4-13(b) below.

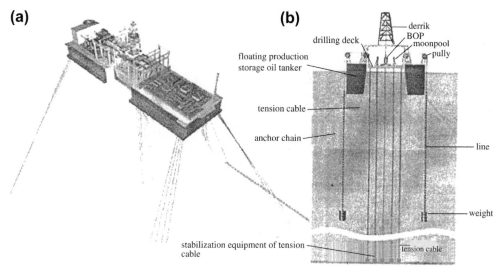

(a)

(b)

derrik
BOP
moonpool
pully

drilling deck

floating production
storage oil tanker

tension cable

anchor chain

line

weight

stabilization equipment of tension
cable

tension cable

FIGURE 4-13

Equipment combination with floating production, drilling, storage, and offloading system.

The main advantages of the fifth equipment combination program include:

- Adoption of the tension leg drilling deck. Thus the deck has almost no heave movement and heave compensation devices can be removed.
- Because it features an underwater weight balance system, the FPDSO has almost no effects on the drilling deck when it happens to heave, pitch, and roll. The anti-rolling device can be removed.
- It has no draught change limit.
- The wellhead BOP stack and Christmas tree can be easily installed on the drilling deck. This is very convenient in the FPDSO when in oil and gas processing and production. Due to the opening of the FPDSO at the center, when the FPDSO needs to move, the wellhead on the water will not be affected.

Since 2005, the fifth equipment combination plan, which provides the FPDSO as mentioned above, has also evolved a bit into new process. From 2009 to 2011, there were altogether three platforms with this new process around the globe. This kind of platform can be used for drilling, separation, and treatment of oil, gas, and water, as well as loading and unloading, due to the fact that it has adopted a semi-submersible or floating vessel with a dynamic positioning or anchoring system to locate it in the sea. Therefore, this platform may be called the floating drilling, production, storage, and offloading system, abbreviated as FDPSO. In the following, a brief introduction will be given to the three unique FDPSO platforms around the world.

1. In June 2009, the world's first cylindrical ultra-deep-water FDPSO was completed and put into operation.

    It is developed by the Norway Sevan Marine company [Figure 4-14(a)]. Some technical processes such as well drilling; separation; well maintenance; oil, gas and water treatment; and

FIGURE 4-14 (a)

Floating drilling, production, storage, offloading platform with deep-water cylindrical structure made in Norway.

storage offloading, can be performed on this floating platform. Also, the qualified crude oil can be shipped off by a shuttle tanker. Therefore, as noted above, it can be called an FDPSO platform. This platform operates at a depth of 3000 m, and is equipped with a DP3 dynamic positioning system. Adopting a cylindrical structure, its resistance against environmental loads is stronger than the ship type, and it is not sensitive to the direction of wind, waves, and flow. So the platform does not need to adjust its heading according to the flow direction. This quality is extremely important for a multifunctional platform which owns two moon pools and a drilling riser joint as well as a production riser, and can do well drilling and oil production at the same time. In addition, because the water plane area is comparatively large, oil storage and unloading has less impact on the platform.

2. In August 2009, the world's first ship-shaped deep-water FDPSO platform was completed and put into operation.

This platform is a floating vessel refitted by VLCC. After winning the contract, in November 2007, the Prosafe Company gave the project to Singapore's Keppel shipyard. Then in 2008, the Dalian COSCO shipping company was put in charge of the hull modification. After the modification, it became a FDPSO platform due to its capacity both in separation and treatment of oil, gas, and water as well as storage and unloading.

The capacity of oil storage is 1.3 MMb/d, and its processing capacity reaches 40,000 b/d. In August 2009, the platform entered the West African Republic of the Congo Oil Field—Azurite, and was put into use. The depth of the oil field is 1400 m. The offshore distance is 129 km and the working party is the Murphy Oil Company. Due to the small waves and light wind, a multi-spot mooring system localization is used instead of a DPS. Figure 4-14(b) presents us the working operation of the multifunctional platform.

3. In 2010, the world's first deep cylindrical semi-submersible ultra-deep-water multifunctional platform was completed by China Nantong COSCO Shipping Engineering Co. Ltd and Suzhou University.

It was finished at the COSCO Shipping Engineering Co. Ltd base. At present, the platform has been named the "SEVAN DRILLER" and will be consigned to the Brazil Oil Company and put into use there.

From the perspective of structure, the platform is made by using a specific semi-submersible cylindrical structure; see Figure 4-14(c). The stability of the platform is improved by utilizing a

FIGURE 4-14 (b)

Singapore Keppel shipyard converted a ship to a floating multifunctional platform.

FIGURE 4-14 (c)

China Nantong COSCO Shipping Engineering Co. Ltd made the "SEVAN DRILLER" floating multifunctional platform.

garden disc structure and automatic adjustment system; as a result, there is safety from abnormal waves. The platform utilizes a DP3 dynamic positioning system and utilizes a nonanchoring automatic positioning and resetting system.

As to the technological performance, the height of the platform is 135 m and the diameter is 84 m. The drillable well depth of the platform is 12000 m (world's deepest) and the loading capacity of the platform deck is 15000 tons. The platform can not only operate under extreme wind and wave conditions, but also can work with a temperature of 20 degrees below zero and ice loads.

It is easy to tell that the three global FDPSO platforms have the following advantages.

1. They are suitable for early production of oil fields.
Early production is a process to get access to oil earlier, recover funds, and put local small oil fields into production during the process of oil field exploration after finding the oil flow

without a comprehensive development plan and facility equipment. A FPDSO vessel can be equipped with a submarine Christmas tree which used for drilling and oil extraction. Meanwhile, due to the platform's combination device which has functions of separating oil/gas/water as well as storage and unloading, the qualified crude oil in early production can be exported by shuttle tankers. Thus, the purposes of early access to oil and gas and recovery of funds is achieved.

2. Suitable for the rolling development of adjacent oil fields

   Rolling development means that closely adjacent fields or block mobile platforms purely with drilling functions and floating multifunctional platforms are used interchangeably, so that a relay can be realized between the exploration and drilling of test wells from drilling platforms and the drilling of production wells and water injection wells and oil handling, storage, and export of multifunctional platforms. Thus, early production methods can constantly be rolled into neighboring fields through this method of expanding field development. Obviously, this method of field development can greatly reduce development costs and improve economic benefit. However, it requires a floating multifunctional platform, because only the floating multifunctional platform can achieve the functions of production and water injection well drilling and separation of oil/gas/water, as well as storage and exportation of qualified crude oil, after the exploration and appraisal well are drilled on the drilling platform. Moreover, after the floating multifunctional platform is transferred from the first oil field (block) to the second oil field (block), the oil and gas produced from the first oil field (block) can be delivered to the adjacent floating multifunctional platform through submarine pipelines for processing, storage, and export, in order to achieve the rolling development of oil fields.

3. Suitable to achieve the development of marginal fields

   Marginal fields refer to oil fields that on the subtle line of being profitable to develop or unprofitable in terms of economic benefits. Such oil fields often need to adopt advanced technologies and equipment, innovations, or other measures to cross over the "margin" and reach profit expectations for development. However, it can be seen from above that floating multifunctional platforms can achieve early oil production and also can collaborate with mobile drilling platforms for rolling development. Moreover, these two development methods can greatly reduce the costs and improve the economic benefits of oil fields drilling. Thus, using floating multifunctional platforms can turn the tide; in other words, marginal fields can be transformed into fields that can meet the profit expectations so that it is possible to develop marginal fields and utilize the idle oil and gas resources.

   To sum up, the emerging floating multifunctional platforms make it convenient to implement the fifth equipment combination program, which is conducive to the collection, transportation, and storage of deep-water oil and gas.

   The fifth equipment combination has some advantages, but it has not yet formally been put into production for applications. There was an industrial test in August 2001 in Brazilian waters in 1200 m deep water, but there are many issues that need further study and solution, such as FPDSO ports and multi-point mooring system mooring force, tension leg drilling, the gap between the deck and hull, risers and tension cables to withstand the dynamic load, and the increased restrictions on the wave pool and hull. In short, the fifth equipment combination is a new proposed program in recent years for the deep sea, which has a fresh concept and unique innovations. But as has not yet been put into operation, it still needs to be tested in practice.

4.2.2 The Choice of Deep-Sea Oil Field Development Equipment Combinations

4.2.2.1 Choice of Unit Equipment

1. Wellhead

   As can be seen from the analysis of foreign combination programs, all except the fifth kind of program, which adopts a wellhead on the surface of the water, adopt a subsea wellhead device for the wellhead unit. Thus, for deep-sea oil field development, the adoption of the subsea wellhead device is a near universal law, along with a wet underwater Christmas tree. Accordingly, the wellhead unit and equipment in the of the deep-sea oil field development equipment combination should use a wet Christmas tree underwater wellhead. Of course, the reason why the fifth combination adopts a surface wellhead is its drilling, oil production and oil storage special device. Therefore, if the fifth combination program is adopted, it should use the water wellhead devices.

2. Offshore platform

   It is easy to see from the introduction of the foreign portfolio of five equipment programs whether they use drilling platforms, or drilling and production platforms; though the forms of the semi-submersible platform, tension leg platform, and single (independent) column platforms are different, a type of floating platform is adopted. From this universality, China's deep-sea oil field can also select a floating platform for its combination of development and sea level equipment. And in view of that as the semi-submersible platform is more mature than several other floating platforms, China should give high priority to the use of the semi-submersible platform.

3. Production facilities

   Outside of the third combination which has choices due to special historical conditions, the programs are used in offshore oil and gas offshore production of crude oil processed are all-sea programs. Thus, although some production facilities are settled under the seabed, some will be installed on the platform, among which is the floating production, storage and offloading (FPSO) unit. It is commonly installed at sea instead on land. In accordance with this, China should adopt the all-sea type, while in view of that China subsea production systems have still lacked experience. Vessel construction technology in China is up to the level of the world's advanced technology, so it should choose the FPSO and install production and processing facilities at FPSO unit.

4. Transport and storage equipment

   Among these five equipment combinations, besides the third option which uses a subsea pipeline for transport due to special reasons, other programs use shuttle tankers to transport qualified crude oil and store oil in the floating production, storage, and offloading (FPSO) vessel or floating production and storage (FSO) vessel. According to this requirement, China has the ability in shipbuilding for the construction of a floating production, storage, and offloading (FPSO). According to this universal law, it is better to choose production, storage and offloading (FPSO) as crude oil equipment.

4.2.2.2 Choice of Equipment Combinations

1. China has the ability to construct major equipment for offshore oil field development

   a. China has the initial capacity to construct the semi-submersible drilling platform

   In the 1980s, China created its own design and constructed a semi-submersible drilling platform "Exploration III," which was drag-propelled, not self-propelled, not equipped with a dynamic positioning system, and put into use in 1984. Exploration III was manufactured by Shanghai Shipyard, with a working draft of 20 m, displacement of 21180 t, drilling depth of 6000 m, and working water depth of 200m. Since 1984, work on the development of the semi-submersible drilling platform was interrupted, and the current foreign semi-submersible drilling platform has been developed to the fifth generation, while the China Construction Exploration III was "made" just between the second and third generation, and the Exploration III working depth of only 200 m is still a conventional water depth. Therefore, for the deep sea, where there are new modern design and construction methods for semi-submersible drilling platforms, there is still a wide gap to cover that will require effort. But, in any case, China once constructed a semi-submersible drilling platform and has some experience with its design. Therefore, on this basis, there are more advantages in building this than tension leg platforms and single column platforms.

2. China has a strong capacity for the construction of floating production, storage and offloading (FPSO) units

There are 13 floating production, storage, and offloading (FPSO) units, of which there are eight belongs that are of China's own design and construction. The construction of a 30-ton FPSO is now underway by the Shanghai Takahashi Shipbuilding Co., LTD in cooperation with foreign countries, with a working depth up to 2000 m, daily processing of crude oil of 19×10^4 bbl and an oil storage capacity of 200×10^4 bbl. Thus, Chinese has strong capacity in designing and manufacturing vessels, and have been in the world's advanced ranks for the construction of deep-sea (water depth 2000m) FPSOs. In addition, since the 1980s China has adopted 13 Nanhai xiwang FPSO units in offshore oil fields, so China already has a wealth of experience. Accordingly, the use of FPSOs should be the first choice in China's deep-sea oil field development equipment portfolio.

3. China has the ability to manufacture offshore oil field wellheads

China has long been able to manufacture and supply offshore oil wells with a Christmas tree and drilling BOP group, for example a former Shanghai Second Petroleum Machinery Factory plant had supplied more than 170 sets of 21 MPa, 35 MPa, and 70 MPa Christmas trees. The Shanghai Petroleum Machinery Factory produced the Christmas trees for Penghu oil and gas field wells. As another example, as early as in the 1970s, full underwater equipment, including a seabed BOP stack, was designed by the former Ministry of Geology and Mineral Resources for the Shanghai Ocean Geological Survey and manufactured by the Shanghai prospecting machinery plant, and was used in the "Exploration One" catamaran floating drilling ship and the completed 9 wells petroleum geological survey and drilling in China Southern Yellow Sea. Thus it can be seen that China already has a manufacturing capacity for offshore oil drilling and oil production wellheads. However, for deep sea purposes there exist many gaps, such as the issue that technical performance is slow. The maximum diameter of the domestic BOP is 349.3 mm, with a maximum pressure of 70 MPa; foreign maximums are 476.3 mm and138 MPa. In addition, China can only manufacture wet Christmas trees, have yet to design and manufacture a casing head tieback system, and the entire subsea production system is empty. In summary, it is preferable to use underwater wellheads in China's deep-sea oil field development equipment combinations. China should make further efforts to improve the technical performance of the product using existing manufacturing experience, to make it fully capable.

4. Recent equipment combinations

   In the near term, in regards to oil in the deep sea, the main equipment combination can be considered as follows:

   SWH + SDP + FPSO + ST

   (subsea wellhead) + (semi-submersible drilling platform) + (floating production, storage, and offloading vessel) + (shuttle tankers)

   This equipment combination is based on the reality of existing major equipment design and manufacture capabilities. Design and construction for deep sea floating production, storage, and offloading (FPSO) vessels and shuttle tankers is not a problem. Just the new development of the fourth or fifth generation of semi-submersible drilling rigs for the deep sea and the fourth or fifth generation need to be manufactured. Wellhead devices also need to improve in technical performance, and some new products new to be developed. In short, there should be possible detours and a faster pace with smaller risk in this equipment combination.

5. Equipment combinations going forward

   From a long-term point of view, China should catch up with modern methods and select newer equipment combinations used in recent years abroad, that is, the earlier discussed fifth equipment combination.

   Obviously, after this program is selected, the focus will be on the development of a floating production, drilling, storage, and offloading combination (FPDSO) unit. But because China has strong design and construction capabilities for FPSOs, the time to use this program combination can be short as the real focus need only be on the tension leg drilling deck (TLD). As for the other components such as shuttle tankers (ST) and wellheads (WH), China does not have to spend a lot of time and energy. As a result, after a period of effort, China's deep-sea oil field development engineering technology can be expected to enter into the arena of the world's most advanced.

4.3 Deep-Water Offshore Floating Structures

From a global perspective, the Shell Oil Company took the lead in building the Cognac offshore platforms in the Gulf of Mexico at a water depth of 317 m (1040 ft) at the end of 1970. Since then, the United States has worked in deep-sea oil and gas exploration and development. Since then nearly 30 years, the enterprise of sea oil and gas exploration and development has increased rapidly around the world, and three "hot spot" areas have formed, commonly known as the "golden triangle," including the Gulf of Mexico, west Africa, and Brazil Campos basin. In recent years, China has also started deep-water oil and gas exploration and development. Demand for deep-sea oil and gas development will encourage the development of science and technology innovations, as progress of the deep-water offshore structures fully illustrates the issues that must be resolved.

Equipment used in deep-sea oil and gas field development has transformed from fixed platforms to the floating offshore structures gradually. As we know, fixed platforms recognized as the primary equipment that is widely used in sea oil and gas field development, but due to its weight, project cost dramatically increases with depth (as shown in Figure 4-15. This limits deep-water development. Take the Bullwinkle plinth-type jacket fixed production platform built by Shell Oil Company in 1988 in the Gulf of Mexico for example; the work depth is 412 m, there are 60 well slots in all, and the total weight of the structure is more than 70000 t, with a project investement of $500 million, which apparently was

FIGURE 4-15

The relationship between the depth
of the water and weight of fixed
platform's jacket.

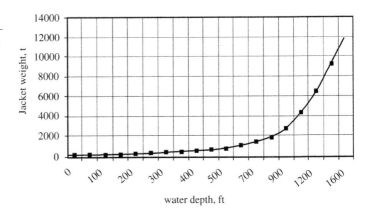

FIGURE 4-15

The relationship between the depth
of the water and weight of fixed
platform's jacket.

too high. So, the fixed platform is applied at 457 m (1500 ft) water depth. At the same time, another suit tower unit was developed. The first tower platform was built in a Gulf of Mexico lean oil field at a water depth of 305 m in 1983, then another four tower platforms were built around the world, as shown in Figure 4-16. The coordinate value on the right side of the figure where the unit is M represents the working depth. Let's use the Petronius platform in the Gulf of Mexico at 535 m water depth as an example; there are 10 production wells on the platform, and it has a weight of 43000 t and deck module weight of 3850 t, with a $500 million project investment. Similarly, the water depth for a tower machine is limited, and its practical application depth range is 305~610 m (1000 ft ~ 2000 ft). In short, although the pile foundation fixed jacket platform and the tower platform have advantages such as mature technology, convenient well repair, low operation and maintenance costs, and no mooring system, the main disadvantage is that they can't meet the needs of deep sea oil and gas exploration and development. So, the concept of development of the offshore floating structures came about for deeper waters. Since then, the tension leg platform (TLP), single (alone) pillar platform (SPAR), semi-submersible platforms and FPSO units that are used in deep-water offshore floating structures have appeared. This section will introduce some issues surrounding the use of different deep water offshore structure at different platforms.

FIGURE 4-16

Four compliant tower platforms around
the world.

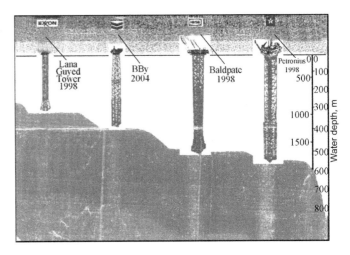

4.3.1 Types of Deep Water Offshore Floating Structures

In recent years, the different types of deep water offshore floating structures are as follows:

4.3.1.1 Tension Leg Platform (TLP)

The tension leg platform is used as a deep water mooring floating suit platform. Through nearly 20 years' development, a typical structure has formed, composed of five parts, including the upper part of the platform, column (including horizontal brace), lower part of the platform (the platform caisson) , anchoring foundation, and tension leg mooring system. The upper part of the platform, columns, and the lower part of the platform are in total called the platform ontology.

1. Tension leg platform overall structure types

There are 21 tension leg platforms under construction or in service in the world currently. The basic working principles of tension leg platforms are the same, but their structures and applications are different. According to the general structure of the tension leg platform at different evolutionary stages, we can divide structures into the first generation of tension leg platforms and the second generation.

a. The first generation tension leg platform

The first generation of tension leg platform appeared represents the largest number of tension leg platforms in the world today, as there are now 12 either in existence or under construction, which accounts for more than half of total (21). Its overall structure and composition is as shown in Figure 4-17. In order to distinguish it from other forms of tension leg platform structure developed later, it also is referred to a traditional type of tension leg platform.

When the first traditional type of tension leg platform was installed in the North Sea at a water depth of 147 m in Hutton, it has come into being production practice and development since 1984, its theoretical research and engineering application have been mature. For example, Unocal Company in build the West Seno tension leg platform in Indonesian waters east of Kalimantan; its weight reached 290,310 t in 2003, and it outputs 190,000 bbl of crude oil and 3.2×10^6 m^3 of natural gas per day, which is the world's largest tonnage and highest production capacity for a tension leg platform. The working depth breaks the 1000 m mark, holding the tension leg platform world record for depth of water in service.

Now, we look at the Auger platform tension leg platform built to a depth of 872 m in the Gulf of Mexico by Shell Oil Company in 1994 as an example, as shown in Figure 4-18; a brief introduction to its overall structure follows:

i. Main parts of the platform

As shown in Figure 4-18, the whole body consists of four cylindrical columns and four rectangular caissons, whose weight is 20,000 t. The diameter of each column is 22.6 m, and they are 48.5 m high. The caisson of a rectangular section is 10.7 m long and 8.5 m high, and it has a connection with the bottom of the bumper.

ii. The upper part of the platform

The upper part of the platform is an open truss structure: 100.6 m long, 88.4 m wide, and 21.3m high, with 10500 t steel load, and a full weight exceeding 23000 t. The life module on the upper deck is five layers of deck structure, including a control center and emergency processing center and places for 132 people to live. In addition, there is a drilling platform derrick installed on the upper deck which can complete well drilling,

FIGURE 4-17

The overall structure of the first generation of tension leg platform.

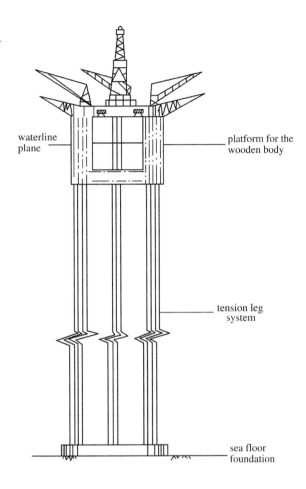

well completion and well workover. The drilling capacity of the drill is 7620 m deep. The wellhead area is rectangular, 61 m long, 23 m wide, and there are two slots on the well. Oil wells are arranged in one oval of 30 m x 63 m under the sea; the wells are about 5 m apart. The deck on the upper part of the platform is equipped with a full set of oil and gas separation, dehydration, and other processing equipment, processing 25,000 bbl of water, 72,000 bbl of oil, and 467×10^4 m^3 of gas per day.

iii. Tension leg mooring system

This is made up of four tension legs and each leg has three root tension units; each of the tension tendons is 841.2 m long and 66 cm in diameter, with a 33 mm wall for the hollow cylinder. The total weight of the 12 root tension tendons is 5800 t.

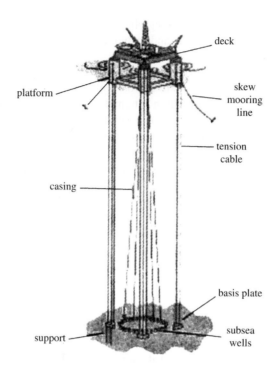

FIGURE 4-18

The architecture of the Auger traditional tension leg platform.

iv. Lateral mooring line system

This consists of eight slash mooring lines. The horizontal distribution radius is 2743 m, and each of the lines consist of cables of 2637 m long and chains 549 m long, which are supported by two buoys, and fixed by the holding power of an anchor; winch and tensioning devices are installed in the column on the platform. The total weight of the whole lateral oblique line mooring system is 6000 t, and the lateral stiffness of the system on the whole platform can be increased, reducing the deviation of the platform ontology caused by wind, waves, and flow, making the platform move within the permitted scope of the drilling and other work.

v. Anchor foundation

The Auger platform adopted a fixed pile foundation, and the pile foundation is composed of four independent pile foundation bases. The base is a steel frame structure, with dimension of 18 m×9 m×18 m where the total weight of the four bases is about 2440 t. Each base is fixed by four root piles into the bottom of pile, with a pile cap through the guide sleeve connected to the base; each root pile diameter is 1.83 m, and they are 131.1 m long. The 16 root piles of 4 bases have a gross weight is 3250 t.

For the first generation tension leg platform, the long service time, wide range of distribution, numerous platforms, and mature design theory all play an important role in the development process and usage of the tension leg platform. The production practice has further proved its performance and economy in the deep waters with semi-rigid and flexible movement. However, it still has the following disadvantages:

- Platform positioning is difficult in extremely dark water: when the water depth is over 1200 m, the length of the tendon increases due to tension. The weight of the tension leg upon itself is too much, and with the deterioration of tensile force of the tendon in deep waters, it's difficult to locate the platform.
- The gap between movement performance and economy: There is a deficiency in the design of the body structure of a traditional tension leg platform. This affects the platform cost, stress distribution and movement performance.
- The difference frequency load causes slow drifting movement: There is a slow change in the force of the difference frequency load, and this will also slowly change the leg platform plane motion as resonance occurs. In addition, wind vibration force also is in this difference frequency range, which will inevitably exacerbate this slow floating movement.
- Resonance is caused by waves of high frequency: High-frequency power and hydrodynamic components will cause tension leg platform resonance; this resonance is commonly called "springing" and "ringing." With the increasing water depth, these two problems will be more serious, and the impact on the safety of the platform structure will increase.
- Complex construction of deep-water anchor foundation: With increasing water depth, designation and construction in the undersea areas will be difficult.

b. The second generation tension leg platform

The second generation tension leg platform appeared in the early 1990s, and was developed on the basis of the first generation tension leg platform, inheriting the good movement performance and good economic efficiency that traditional tension leg platforms have, optimizing the structure to make the tension leg platform more suitable for deep water environments, and further reducing the construction cost. The second generation tension leg platform will be divided into three series, introduced in the following:

i. SeaStar series

This is designed by Atlantia Company; its outstanding characteristic is breaking the traditional tension leg platform column of three or four pillar-type units, adopting a unique single-column design scheme. Figure 4-19 is the SeaStar series tension leg platform structure.

It can be seen from Figure 4-19 that the main body of this kind of tension leg platform is made up of a vertically floating cylinder structure and a three rectangle cross section at the level of the buoys. The cylinder is located in the center of the main body of platform, called the central column. Near the bottom part of the central column, three rectangular cross sections of the buoy are integrated, and each of the buoys extends outward into a cantilever beam structure. The end of the section shrinks gradually; each of the horizontal plane angle is 120°. For a wellhead and Christmas tree on the water, it common to use the top tension riser through the central column center of sea wells and the central column. For the subsea wellhead and Christmas tree, a flexible production riser is used, and it is not run through the central column. Rather, a J-shaped channel is built on the outside of the body at the bottom,

(a)

(b)

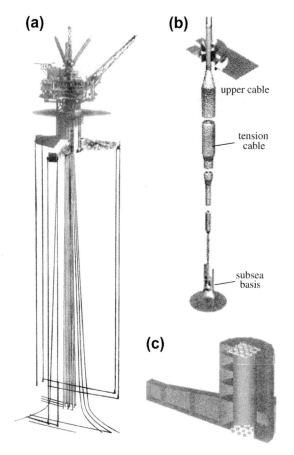

upper cable

tension cable

subsea basis

(c)

FIGURE 4-19

SeaStar series second generation tension leg platform.

with a lateral column into the central part of the hollow shaft, which is then connected to the upper production processing facilities platform. The buoyancy of the platform ontology is provided by three common buoys and the central column, the annular space in the central column and space in the cantilever is divided into separate compartments by using a buoy thick plate, so as to improve the platform's buoyancy; the internal structure is as shown in Figure 4-19(c). The platform's shell is welded by standard low carbon sheet steel.

It is furnished with an independent upper base near the top surface of the water in the central column. It is separated from the main body of platform by building an exoteric pipe. It can be used to support the platform deck, and make the surface wave go through the jacket directly, which reduces the wave load of the platform.

The upper part of the platform and traditional tension leg platform share a same basic structure, a skeleton module structure.

The SeaStar TLP tension tendon is assembled with high strength steel pipe, with mechanical connections on both ends of the steel pipe. Adopting new material and optimizing the structure that the tension leg uses makes the buoyancy equal or greater than the weight of the tendon. This method can solve the problem of the tension tendon's large weight. In addition, the tension tendon uses a hierarchical variable diameter design, as shown in Figure 4-19(b), which can give the tension leg system greater buoyancy, and reduce the load of the platform itself. A fairing is installed outside of the tension tendon, to reduce the vortex-induced vibration. The main body of the cantilever buoy tension tendon is connected with top through a link that makes examination and repair possible whether external or internal to the body.

The foundation of the SeaStar tension leg platform underwater is of three types, a suction type foundation, pile foundation, and foundation top to tendon interface, that forms an organic whole.

From 1998 to 2001, four SeaStar tension leg platforms were in service or under construction in the world, and their technology parameters are shown in Table 4-3.

ii. MOSES series tension leg platform

MOSES is short for Minimum Offshore Surface Equipment and Structure. It is the first tension leg of the Prince Project manufactured by MODEC Company and Japan's MITSUI Shipbuilding Company over five years, and was put into production in the Gulf of Mexico. The MOSES series of tension leg platform is designed with a displacement ranging from 3000 t to 50000 t, and the working depth ranges from 300 m to 1800 m. With its excellent workability and good economic benefit, it is seen as one of the new directions of tension leg platforms in the future. The major design changes are as follows:

- Platform main body

  The MOSES series platform main body is different from the SeaStar series. It still adopts the four pillar type main body structure of the traditional tension leg platform, as shown in Figure 4-20. The main body is made up of four long and thin prisms and a floating base structure underwater. The main body is focused in the center, and subject to the platform. Buoyancy is mainly provided by a underwater floating dock platform with a special base shape. The central part of it is a cube, and at four corners it has a diagonal outward extension, which forms four split hanging buoy structures, as shown in Figure 4-1 (b). A cantilever of longitudinal cross section is a right triangle, and sharp corners go outwards with tension tendons at the top of the interface. The floating dock is located under the surface of the water, and is affected less by surface waves, to improve the platform's water power performance. Four corner posts on the central base at the four corners of the cube cross the water to support the upper platform. The corner posts are thinner than the SeaStar tension leg platform center column, which can make the high-order wave force on the platform main body small, as wave interaction is small. The dead load of the platform above the water is small, which can reduce platform fatigue and improve the motion performance of the platform.

- Tension leg mooring system

  Tendons are composed of the standard parts of a rolling steel pipe assembly, and adopt low cost tendon tension end connection devices, which greatly reduces tension leg construction and installation cost. Due to the fact that the tension and the response in

Table 4-3 Parameters of SeaStar Series Tension Leg Platform

Name	Corporation	Water Depth, m	Year Finished	Location	Effective Load, MN	Central Rod		Quantity of Tension Legs
						Diameter m	Height m	
Morphett	British Borneo	518	1998	Ewing Bank Block 921	40	17.7	34.1	6
Allegheny	British Borneo	1006	1999	Green Canyon Block 254	40	17.7	34.1	6
Typhoon	Chevron	640	2001	Green Canyon Block 237	50	17.7	34.1	6
Matterhorn	Total Fina Elf	860	2004	Mississippi Canyon Block 243	110	25.6	38.1	6

(a) **(b)**

FIGURE 4-20

MOSES series Prince and Marco Polo tension leg platforms.

"fatigue area" has been minimized, the tension leg manufacturing technology requirements can be reduced greatly.

- Wellhead and riser system
The outstanding improvement in this aspect is the design of an eccentric wellhead where the wellhead device is located in the edge of the upper platform. Thus the production riser does no need to pass through the central platform, instead connecting with the lateral platform body at the end of the deck wellhead device directly. This design can place the wellhead area away from the work module, which improves safety to a degree and avoid repairing the facilitates outside the platform. Moreover we can avoid maintenance work within the central sea wells. The MOSES series tension leg platform riser system uses a system which is different from the traditional hydraulic or pneumatic tension system of the passive spring riser tensioned system. The riser trip is only 0.6 m less than the traditional $1.22 \sim 1.83$ m and because the spring group is always in a state of tension, it has automatic fault protection function. In addition, the riser pilot is on the platform keel and can transfer the horizontal motion of the stand pipe to vertical movement.

- Upper platform for MOSES series tension leg platform
This uses a light platform design similar to a jacket platform to optimize structure, reducing the dead weight of the upper platform significantly. And it reduces the displacement of the platform main body, the tension leg mooring loads of the system, and the construction cost of the upper platform, and shortens the construction period of the upper platform. It makes the effective load distribution of the platform more flexible; It simplifies the installation procedures of the upper platform, utilizing

one-time hanging with complete docking with the main body; this improves the efficiency of the whole platform construction project, and shortens the overall cycle of platform construction.

- Drilling machine

  The platform deck installs no permanent headframe. Drilling and workover operations can use a temporary light platform installation derrick, thus reducing the load on the platform.

- Platform installation

  The entire method of platform installation is flexible and convenient. Versatile ships and other kinds of ocean engineering ships can be qualified. The platform main body can be directly towed to the installation site, which reduces the platform installation cost greatly.

  Taking the Marco Polo tension leg platform as an example, it is the second series of MOSES force platform legs after the Prince tension leg platform. This platform was completed in July 2004, and put into operation in the Gulf of Mexico with 1310 m water depth. The Marco Polo tension leg platform's total displacement is 24879 t, and its payload weight is 129.84 MN (including the deck steel structure). The platform deck is a single mode block structure, with a total weight of 6300 t; the floating crane can be used once. The system adopts a surface wellhead and Christmas tree in the water. It can produce 12,000 bbl of crude oil, and 1132.68×10^4 m^3 of natural gas per day. The platform use six rigid riser tensioner to produce 6 wells locate on the subsea. The riser is a double shell structure; the outer diameter is 340 mm, and the inner diameter is 240 mm. The platform produces oil and gas, which can be transported through the steel catenaries riser. The oil pipeline diameter is 320 mm, while the gas pipeline diameter 460 mm.

iii. ETLP (Extended Tension Leg Platform)

This platform was first put forward by the ABB Company as the third series of the second generation of tension leg platform. The extended tension leg platform entered the building stage in 2003, and a total of three are under construction around the world, among which one sits in the Gulf of Mexico and will have a working after being put into production of 1433 m, which will create a world record for tension leg platform working depth.

Figure 4-21 shows an ETLP completed in June 2004 and put into production in the Gulf of Mexico, namely the Magnolia platform. We now take it as an example in this section to briefly introduce the structure of the ETLP.

The column cross section of the Magnolia platform is circular rather than rectangular, and has a 51 m length for the main body. The platform is positioned through eight tension tendon moorings, and two root tension units are distributed in the platform at the four corners of main body. The tension bar button at the top is attached to the cantilever on the tendon at the top of the buoy socket, about 18.3 m away from the water surface. The platform is built by two 7200 t, 67 m long modules. The platform has installed eight wellhead and Christmas tree sets, and oil production from the bottom of the sea goes through a rigid riser to the surface of the water. The oil is transported by pipeline of diameter 356 mm and the gas by pipeline of diameter 406 mm to land; it can produce 5×104 bbl oil, and 424.755×10^4 m^3 gas per day.

FIGURE 4-21

ETLP tension leg platform emanon bowel mimic diagram.

The ETLP is improving by correcting its design and its characteristics as follows:

- Bearing is more efficient

  Normally the carrying efficiency expression is as follows:

  The bearing efficiency of the platform = Platform total payload on the freeboard/total weight of main deck structure

  After improved design, the ETLP carrying efficiency is about 1.4, as seen from Table 4-4. Thus, the ETLP platform has obvious advantages in loading rate.

- Construction assembly is simple

  ETLP adopts modular construction, so construction is convenient. The modules of the main body of platforms are built first, and the main body of the platform consists of 24 super modules. Because its construction requirements are not high, it can be built in dry dock or on land. Modules are small, so they can be manufactured in a closed workshop directly, and then assembled at the construction zone on land or on a dock. The main body of the platform is first tested, and then transferred to the assembly area of the platform's upper part to be docked with the upper deck. After docking, with the aid of a floating crane, the assembly is loaded on the tugboat, which tugs it to the oil well location. Because the main body of the platform and the upper deck are built by dividing them into several modules respectively, then assembling the modules together, with docking at the end, it is convenient no matter how or where it is constructed, built, lifted, or transported.

Table 4-4 ETLP and TLP Carrying Efficiency in Different Waters

Platform/Location	Current Status	Loading Efficiency Range
ETLP/Southeast Asia	Conceptual design	1.4~1.5
ETLP/West Africa	Being built	1.3~1.4
ETLP/the Gulf of Mexico	Being built	1.1~1.2
TLP/the Gulf of Mexico	Finished	0.6~0.8

- Able to be installed as a whole

  Being able to be install as a whole means that the platform with the upper deck docked is an entire piece that is docked with the tension leg system and production riser system after being towed to the installation point. Therefore, because the whole installation function eliminates the assembly and construction work at sea and doesn't use a large floating crane, it shortens the offshore operation time, and reduces the risk and costs of the installation.

2. The advantages of the tension leg platform

   From what has been discussed above, we sum up the advantages of the various types of tension leg platform in the following points:

   a. Excellent athletic performance

   The structure of the tension leg platform makes it effectively avoid the effect of environmental forces, like storms, etc. It has almost no heave, roll, and pitch, as it stays stable because the tension legs restrain the platform's heave, pitching, and rolling motions effectively.

   b. High load-carrying efficiency

   The unique design of the main body of the tension leg platform carries with high efficiency. It is not hard to see from Table 4-4 that the second-generation ETLP's has reached 1.4 or so, which contributes to the economic benefits of the tension leg platform.

   c. Deeper operating water depth.

   From the current view of tension leg platforms producing oil and gas, the Magnolia ETLP working depth is the deepest, reaching to 1430 m. And in the future, it is possible that the working depth will break through 2000 m.

   d. Wide application range.

   Although usually a tension leg platform is more suitable for large offshore oil-gas field development, in recent years, the presence of economical, light, and small tension leg platforms has broaden their application range. The small tension leg platform is not only more and more popular with large oil-gas fields as it combines with FPSO, as shown in Figure 4-22, but also it has been well used offshore small oil-gas fields or offshore marginal oil-gas fields because it can solve the problem of multiple drilling centers and frequent oil well maintenance.

   e. Adopting the overwater wellhead

   An important advantage of the tension leg platform is that it can use the overwater wellhead and surface Christmas tree. In this way, they are more reliable than underwater trees, oil pipelines and risers in terms of technology, more economical and cheaper, and more convenient for repairing the well. There are great economic benefits, especially for the Micro Star tension leg, which is an unmanned tension leg wellhead.

   f. Easy to build and install

   From the view of ETLP construction and installation that has been introduced before, because the tension leg platform use modular construction technology, its construction is simple and the requirements for construction are not high. Not only are its main body and upper part respectively assembled by modules, but also they are docked with each other. Importantly, it can be used with a top tension type production riser and rigid catenaries riser at the same time, which can be more convenient.

FIGURE 4-22

Small tension leg platform and FPSO combination.

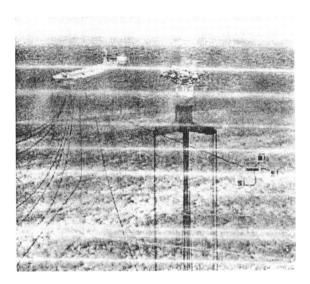

g. High economic efficiency.

From all of the above advantages we can see, due to the structure and optimization of the design of the tension leg platform, construction is simple, installation is convenient oil well repair is easy, and we can save steel, reduce cost, and save time, all of which can bring high economic benefits.

3. The shortcomings of the tension leg platform

The main disadvantages of the tension leg platform are:

a. Working depth restrictions

With increasing water depth the length of the tension leg itself will extend, and as the weight increases, tension increases, which affects the platform ontology. Therefore the working depth is restricted, and in current designs, the applicable depth is only 2000 m or less.

b. No storage capacity

The tension leg platform cannot store oil, therefore it can only combine with an FPSO unit or submarine oil pipeline, limiting applications.

c. Uncertainty of base

The anchor foundation of the tension leg platform depends on uncertain factors on the sea bottom, such as soil and slope.

d. Sensitive to the upper load

The change of tension leg platform variable load on the upper platform will directly affect the stability and raise the sensitivity of the tension leg platform mooring system.

e. Sensitive to the second order wave force

The natural frequency of the tension leg platform structure is low, and therefore outside the structure plane and plane within the frequency respectively and the wave sum frequency (the sum of the two frequencies is very large), difference frequency (two similar wave frequency, the difference between two frequency is very small) slow low frequency drift (missile drift) bounce (springing) response and high frequency response to motion platform is serious, it is worth attention to the two wave force sensitive theory response.

4. The development direction of the tension leg platform
   From the traditional tension leg platform (first generation) to the SeaStar, MOSES, and ETLP of the second generation of tension leg platform development, we can see the development direction of the tension leg platform:
   a. The platform main body is moving in the design direction of having a higher load-carrying efficiency and lower construction costs.
   b. The weight of the upper part is getting smaller, making construction easier.
   c. Further development of tension steel legs offer guaranteed strength, and more adaptability to the deep water environment.
   d. Smaller and lighter tension leg platforms are being developed.
      In recent years, a small tension leg platform (mini-TLP) able to adapt to extremely deep water environments has gradually become the norm of tension leg construction. Such as wellhead small tension leg platform is rising in recent years. A large tension leg flat machine costs $5 \sim 6 \times 10^8$, while the West Seno small tension leg wellhead platform and its combination with a floating production system (FPU) has a total investment of only 2.65×10^8. The wellhead small tension leg platform itself only accounts for the two-thirds the total investment cost, about $1.77 \times \$10^8$. Compared to the large Kizomba A tension leg wellhead platform that costs 6.5×10^8, the economic advantages of the small tension leg platform are very obvious.

4.3.1.2 Single Column Type (SPAR) Platform

The SPAR platform is a newer offshore floating structure that appeared in the 1980s that is used in deep sea oil and gas drilling, production, processing, storage and transportation, loading and unloading, and is seen as the development direction of the next generation deep-water platform. At present, there are 13 SPAR platforms all over the world and it has become the second main platform for deep-water oil field development just behind the TLP platform. The SPAR platform can be divided into three generations, as shown in Figure 4-23.

As shown in Figure 4-23, the three generations of SPAR platform are as follows: the first generation classic type single column (Classic SPAR) platform, the second generation (Truss SPAR) platform, and the third generation truss type single honeycomb (Cell SPAR) platform. Their primary difference is their different main structure which we introduce here:

1. The first generation SPAR platform: Classic SPAR platform
   The world's first Classic SPAR platform was the Neptune SPAR platform built in 1996, as shown in Figure 4-24. This platform in the Mexico Voice Knoll oil field has a water depth of 588 m. Its platform main body structure is characterized by a cylinder 215 m long, a 22 m diameter design draft of 198 m, a freeboard height of 17 m, a total weight of 129.06 MN, and a steel structure weight of 117.91 MN. The platform main body can bear the pressure load; the maximum dead weight is 95.14 MN. The middle of the platform well is in the main body epicenter, and its cross section is square, with a side length of 9.75 m; it can hold 16 production risers, and press 4×4 arrays. The wellhead is located in the central hole of the well mouth, between the workover decks and production deck. The upper body floating tank is 84 m long, and the bottom of the cabin is 30 m long, with several smaller watertight compartments in the cabin.

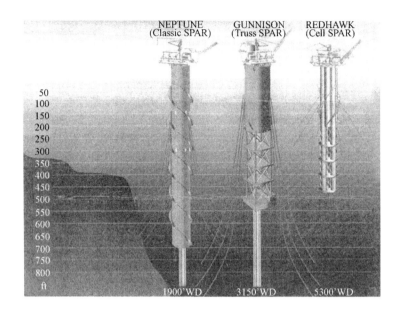

FIGURE 4-23

Comparison of three generations of SPAR platform.

FIGURE 4-24

The first generation SPAR platform: Classic Neptune SPAR platform.

The Neptune SPAR platform body is divided into three deck structures; from top to bottom there is the workover and well completion deck, middle deck, and production deck, respectively. Three deck and design on the surface of the water line distance is 35 m, 28 m, and 22 m respectively. The upper decks long 38 m and wide 25 m. The middle deck is 41 m long and 31 m wide; the scale of the production deck as same as the middle deck. The total weight is 59.93 MN. The steel deck structure weighs 12.43 MN, and the production facilities 33.55 MN. The deck equips with jacket light mast; so it cans completion wells and work over wells. The crude oil uses six units, with pump output of 6000 bbl, while gas uses two compressors with 85×104 m^3 output per day; the output pipe diameter 200 mm, and the submarine pipeline has a total length of 17.5 km. Three 900 kW turbine generators supply the power. The platform use four root columns with diameter 1.524 m that dock with the platform main body. The column base located on the vertical waterproof wall body and deep into a distance the subject.

The main body of the Neptune SPAR platform mooring system uses six 6root mooring lines, with a total length of 1125 m, each of them with a "chain-cable-chain" segmented catenaries structure. Its hypo mere is 73 m length chain into the underwater pile foundation; the middle is a 732 m spiral twisted wire, outer package fairing; the upper chain is 320 m long, and through the chock connect with a chain device. The cable diameter is 120 mm, with critical stress of 126.65 MPa; the wire diameter is 120 mm, with critical stress of 122.38 MPa. In the underwater pile foundation each root is 33 m long and 2 m in diameter, and goes through the steam hammer underwater.

The seabed wellhead of the Neptune SPAR platform is divided into two parallel line layouts, with spacing of 6 m; each governor is 32 m, and well spacing is 4.57 m. The platform doesn't have a drilling function, therefore before the platform is in place, the semi-submersible platform drills six wells first, Then the Neptune SPAR platform is installed after completion. The platform can use the chain machine's regulating system to keep the platform at the piles, and move within a certain range.

There are three Classic first-generation SPAR platforms around the world: the Kerr McGee Company's Neptune SPAR platform, Chevron's Genesis platform and Exxon's Hoover platform. Some parameters of Classic SPAR platforms are shown in Table 4-5.

From analysis of the Classic SPAR platform structure, it is not hard to see the overall structure of the SPAR platform component is as follows:

a. The upper platform

Figure 4-25 is a general structure diagram of the Classic SPAR platform. We can see from the figure that the SPAR platform body is generally composed of two or three deck modules, such as the drilling deck, production deck, and wellhead deck, etc., with each layer fixed by a column and brace structure. The upper main support columns dock directly to the platform main body at the top of the column base.

b. Platform

As shown in Figure 4-25, the main body of the SPAR platform is a vertical floating cylinder in water; the overall diameter is large, generally 20 ~ 40 m in length, the main draft is above 100 m, and the location of the center of gravity is below the waterline. The main body interior space divides into a multilayer, multiple tank structure, mainly composed of a hard capsule (hard tanks), middle area (mid-section), and soft capsule (soft tanks).

Table 4-5 First Generation SPAR Platform – Classic SPAR Platform Technology Parameters

Name	Main Body Scale, m		Main Body Weight, t	Design Load, kN	Mooring Line				Water Depth, m	Base	Finished
	Length	Diameter			Material	Quantity	Length, m	Diameter, m			
Neptune	215	22	12906	59300	Steel	6	1125	0.12	588	Pile foundation	1997
Genesis	215	37	26703	90000	Steel	14	1341	0.13	793	Pile foundation	1998
Hoover	215	37	35000	180000	Steel	12	2164	0.15	1463	Pile foundation	1999

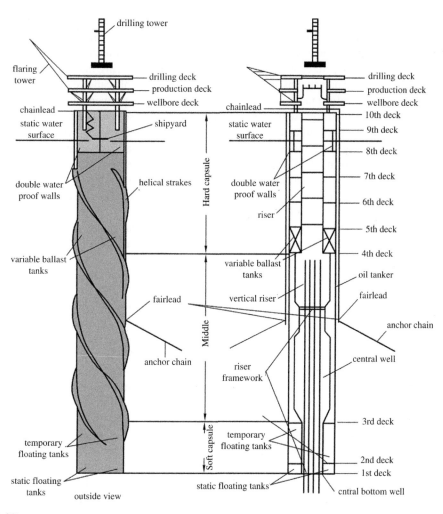

FIGURE 4-25

Classic SPAR general structure diagram.

The hard capsule located in the main body of the upper platform is the main source of buoyancy; the cabin is divided into a fixed floating tank and variable ballast tank. The fixed hard buoyancy tank located in the upper tank can support the subject itself and the upper deck and the upper load after filling. A variable pressure load tank is located at the tank bottom, with seawater ballast. We can adjust ballast and adjust the pressure loads of platform.

Midway through, as shown in Figure 4-25, from the variable pressure load tank at the bottom of the ballast tank to the temporary floating tank top deck, we have the middle part. The function of the middle part is to connect the hard capsule of the platform and soft capsule rigidly, and protect the central well production riser in the system from the current force at the same time. The middle part is composed of an outer shell and inner shell, the former protects the cabin, and the latter forms a central hole, where inside the shell ring there is a big area for an oil storage tank. The outer of the middle is assembled on the middle part outside the shell with a crown block structure, which is used to connect the mooring line.

Below the middle part, ballast is mainly provided by the soft capsule. The soft tank is divided into a fixed ballast tank and temporary ballast. For the permanent ballast space at the bottom of the main platform, seawater or specially designed solid ballast is available. In the upper part are a set of inflatable temporary floating tanks for water, which can make the platform change from a vertical suspended state to a horizontal floating state to facilitate transferring the platform when towing. When it arrives at a new location and is ready for placement in the sea again, and temporary ballast tank is filled with gas, which make the platform main body go into to vertical state of suspension; this process is called "erecting" or "upending" the platform main body.

A spiral side panel with two columns is installed on the platform body shell; this can has the effect of diverting water, thus reducing vortex-induced vibration of the platform.

c. The central shaft and riser system

The SPAR platform has a hole in the middle from bottom to top through the whole body, because of its opening bottom, so it filled by water. A production riser system is located in the central shaft, linking the upper platform and the upper of the production equipment together with the lower part in bottom of the sea, as shown in Figure 4-26. The draft for the platform is above 100 m, and is closed, so the production riser in the central well is almost completely not affected by waves and current, which is a great virtue of the SPAR platform design.

The production riser system is divided into a vertical tube and steel catenaries riser (SCR). In order to tension the riser, we use a riser buoy to fix the frame in the central borehole; it connects after filling the riser, and buoyancy will keep riser in tension. A tube buoy platform is set up separate from the platform main body, and it neither heaves nor collides with the platform main body.

The central shaft and central body at the bottom of the keel are equipped with guide framework with a function of transmitting horizontal load to the main body. The bending stress on the keel is high, and the riser through relative motion causes friction, so the strength of the keel joint design must meet the corner stress requirements, and should have enough resistance to wear.

d. Mooring system

SPAR platform is different from vertical tension mooring legs flat, it adopt the half tensioning of catenaries mooring system, which does not need to much tension, but at the state of tension and relaxation, rely on its own gravity, the natural formation of catenary. The mooring system of the SPAR platform is given in Figure 4-27.

It can be seen from Figure 4-27 that the SPAR platform mooring system has the following composition:

i. Mooring rope

The SPAR platform mooring line with subsection, its segment is composed of anchor chain, connect the chock and the upper machine of chain. The middle part is composed of spiral cables; in order to improve the stiffness some of them also add a ballast chain. The lower segment is under the bottom chain segment, the ends are linked with the sea bottom

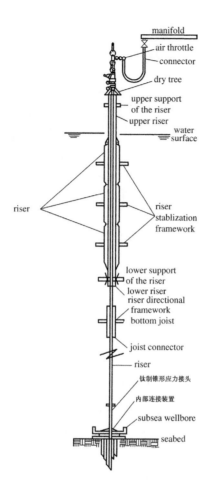

FIGURE 4-26

SPAR platform production riser system.

in the submarine base; some of them are set level, and some of them are vertical in the water in order to reduce the pulling force on the base.

ii. The chock

This is located near the center of gravity platform; in order to reduce dynamic load of the mooring cable, it is located in deep under the surface, reduce the influence of surface waves on the mooring cable. The chock is a slippery wheel mechanism, which is the join points of mooring cable and platform.

iii. The chain machine

This is located on the main roof deck platform to provide the mooring cable pretension; it is controlled by a computer system, which controls the length of the mooring line and pretension, in order to adjust the positioning of the platform.

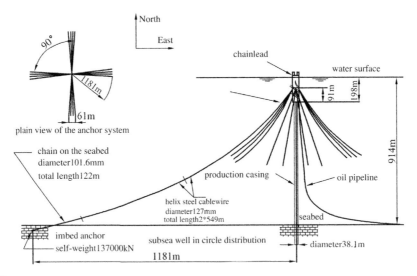

FIGURE 4-27

Mooring system of a SPAR platform.

    iv. The bottom of the sea

      The bottom of the SPAR platform mooring system adopts the traditional pile foundation and the suction type, etc.

2. The second generation SPAR platform: Truss SPAR platform

Compared with the first generation SPAR platform, the second SPAR platform adopts an open frame structure in the middle part of the main body of the platform, as shown in Figure 4-28. Although the improvements remove the enclosed cylinder structure formation of the oil storage tank, it has the following important advantages:

a. Improving the load and motion performance of the platform

Due to the platform main body in the middle being an open frame structure, so the horizontal bearing area is reduced, which reduce the water of the platform flat outside force and motion response of the horizontal platform.

b. Reduce vibration and improve the stability of the platform

Due to the middle platform using an open architecture, it can provide the ballast weight, and when there is heaving motion on the platform, it can produce larger viscous damping, increase the added mass, and has a lower natural frequency, thus reducing wave resonance. It also can reduce the vortex-induced vibration of platform, thus greatly improving the stability of the platform.

c. Save on steel quantity to reduce the cost of the platform

Due to the fact that the low-frequency wave response is small, which can reduce the main body's draft and reduce the length of the main body, and the fact that the open structure is lightweight, the platform can save a lot of steel, lowering the platform cost. For example, comparing the Truss SPAR platform at the Nansen platform with the first generation Enplane

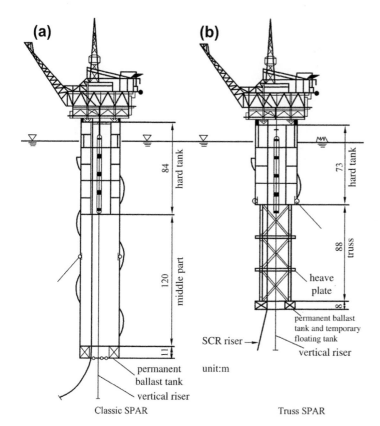

(a)

(b)

84 hard tank

120 middle part

11

Classic SPAR

73 hard tank

88 truss

heave plate

permanent ballast tank and temporary floating tank

vertical riser

SCR riser

8

unit:m

permanent ballast tank

vertical riser

Truss SPAR

FIGURE 4-28

Truss SPAR platform compared with Classic SPAR platform.

platform, the former main body length is only 166 m, while the latter is 215 m, and the payload of the Nansen platform was increased by 35% ~ 40%.

We take the Nansen Boomvang Gemini SPAR platform as an example to explain the technical parameters of the second generation SPAR platform. Figure 4-29(a) displays a general Truss SPAR platform, and Figure 4-29(b) is the Nansen Boomvang Gemini SPAR platform. The Nansen platform is the first Truss SPAR platform; it was put into production in January 2002 in the Gulf of Mexico oil fields. The working depth is 1120 m, and it is 165.5 m long, with a diameter of 27 m, and body weight of 12200 t, 87.50 MN. The upper sealed closed body has a length of 73 m, of which the freeboard height is 15 m; the bottom frame type main body has length 88 m. Heaving board structure, three layer to architecture so main body is divided into four layers, three layers above the height of 21 m, 24 m most lower level high, the bottom of the ballast tank is 5 m high. The Nansen SPAR has a body weight of 7800 t, the central shaft dimensions are 12 m × 12 m, and it can accommodate seven top tension type risers and eight steel catenaries. The platform adopts 9 mooring lines positioning, each divided into a set of three, and the total of three groups connect the platform to the underwater pile foundation.

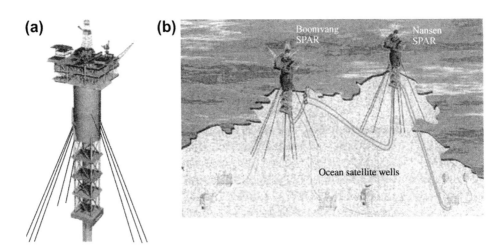

FIGURE 4-29

The second generation Trass SPAR platform.

The Boomvang SPAR platform is the sister of the Nansen SPAR platform, and as shown in Figure 4-29(b) it is located 14.5 km from the Nansen platform; it has a water depth of 1052 m, and its structure is almost exactly the same as Nansen. Two platform lightweight 170 mm, it the peak output capacity are 4×104 bbl of crude oil, natural gas, 566×104 m^3, each platform has 9 saliva on the wellhead and tree and three satellite wellhead and Christmas tree, the sea satellite well through the steel catenaries riser connects to SPAR production platform, and the surface wellhead tree connect with subsea wells by a vertical pipe.

The Boomvang SPAR platform was formally put into production in 2002; it is the second Truss SPAR platform in the world.

The Truss SPAR platform is one of the most active types of SPAR platform development; there are now a total of nine that have been built, and some of its technology parameters are in Table 4-6.

3. The third generation SPAR platform: Cell SPAR platform

The third generation SPAR platform is developed on basis of the first generation and second generation SPAR platforms. The first and second generation SPAR platforms had common disadvantages such as large volume and high cost. In order to overcome these shortcomings, the third generation SPAR platform looked to improve the platform main body structure; it adopts a new type of combined body structure. The use of "combined" refers to the fact that the main structure consists of a number a small lengths of hollow cylinder; each small cylinders can be built in a different location separately, and assembled together with a cylinder at the center. Other cylinders are bound to and surround the center column body to form the main structure in a honeycomb shape; therefore we call the third generation SPAR platform the Cell SPAR platform.

To this point, there is only a Cell SPAR platform in the world, called the Red Hawk Cell SPAR platform; it was completed and put into operation in 2004, and is located in the Gulf of Mexico at a water depth of 1500 m, as shown in Figure 4-30.

Table 4-6 Truss SPAR Platform Technical Parameters

| Name | Main Body Scale | | | Design Load, kN | Mooring Line | | Water Depth, m | Base | Finished |
	Length, m	Diameter, m	Main Body Weight, t		Material	Quantity			
Nansen	166	27	12000	87500	Steel	9	1120	Pile foundation	2001
Boomvang	166	27	12000	87500	Steel	9	1052	Pile foundation	2002
Horn Mountain	170	32	14600	86000	Steel	9	1646	Suction	2002
Medusa	168	29	11300	>48900	Steel	9	678	Suction	2003
Devil's Tower	179	29	<11300	>48900	Steel	9	1710	Suction	2004
Gunnison	167	30	13797	107700	Steel	9	960	Suction	2004
Holstein	227	45	37000	239900	Steel	16	1308	Suction	2004
Mad Dog	169	39	20000	163290	Nylon	11	1347	Suction	2004
Front Runner	168	29	>11300	>60000	Steel	9	1067	Suction	2004

The structure of the Cell SPAR platform is as follows:

a. Main body of the platform

The main body of Red Hawk Cell SPAR platform is made up of seven diameter are 6 m small cylinder division of four cylinders department of the length of 85 m, while the length of the three for 171 m, a 85 m short cylinder is located in central and six others are outlier, between each of cylinders is 0.6 m apart, they use fixed steel frame structure. Its effective diameter is 20 m, which is less than all current SPAR platform diameters. The weight of the platform main body is 7200 t. It consists of the following:

i. Hard capsule

Between the main roof deck to the short end of the cylinder in the division of amour plate is hard capsule. (Figure 4-30); a fixed floating platform tank and variable ballast tank both are located in this section.

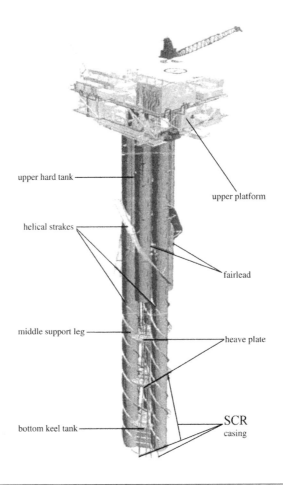

FIGURE 4-30

Red Hawk: the third generation Cell SPAR platform.

ii. Middle section

Since the hard tank below until midway through soft bottom deck is platform main body, the middle three is the principal part of the peripheral division of long cylinder. 3 root 171 m long cylinder division, stretching down to the bottom of the platform main body, and fixed the bottom and outside soft, serves as a support legs of rigid soft hard capsule.

iii. Heaving board

four floor heaving middle plate, the distance between each layer of 21 m, and it connect the three support legs under the hard tank. on the one hand, it can satisfy the main body's strength requirements, and on the other hand it also can increase the mass of platforms, and reduce the heaving motion of natural cycles.

iv. Spiral side panel

The platform main body hard capsule and three support legs are installed in the central part of the spiral side panel to reduce small eddy current effects on platform motion.

v. Soft capsule

A soft tank platform is located in the bottom of the main body and connected to the support legs; there are ballast tanks, including a 22 MN permanently fixed ballast to improve the stability of the platform, and another variable tank for seawater ballast.

b. The upper platform.

The Red Hawk SPAR platform has three layers of upper deck structure. The top is the main deck, located in the middle of production scale plate, they both are 34 m × 41 m, the lower level is wellhead deck, scale 23 m × 28 m, upper effectively load 47 MN.

c. The mooring system

The Red Hawk SPAR platform adopts six sectional fixed mooring ropes, with the mooring rope of each root system using nylon rope and two steel cables. The mooring rope has a 60° angle between departments. The underwater foundation uses six suction foundations, each base long 24 m, diameter 5.5 m.

d. Wellhead and Christmas tree

The Red Hawk SPAR platform adopts underwater wellheads a wet tree; there are two undersea wells, connected through a 100 mm diameter flexible flow line to the main body and with a hollow shaft connected to the upper production processing facilities. The platform outputs 850×104 m^3 of gas per day.

Throughout the above, we can see that as the Cell SPAR platform main body adopts a modular structure, it makes the body smaller and lighter. This reduces the installation cost. At the same time, due to the main body being divided into several parts, all of them can adjust to local conditions and be built independently; the site requirements are not high, which greatly reduces the overall cost.

4. SPAR platform construction and installation

The SPAR platform can be built after completion of the design and model testing in the pool. The main body of the SPAR platform is huge and there are high requirements for construction sites; construction should be carried out in a shipyard, such as the Finland Manty - Iouto shipyard that 13 SPAR platform 10 are built in it. During construction of the platform main body and the upper deck, the work of preinstalling the mooring system and riser system, such as installing steam hammer pile in the deep underwater foundation and laying submarine pipeline, can be done at the same time. After SPAR platform installation, four steps should be done as shown in Figure 4-31.

a. Towing platform: As shown in Figure 4-31(a), it can be towed [Figure 4-32(a)] or shipped [Figure 4-32(b)].

b. For the main body erection, as shown in Figure 4-31(b), we open a temporary floating tank in the water and the main body under the action of torque stands on its own [Figure 4-32(c)].

c. Offshore location: As shown in Figure 4-33(c), using winch the mooring cable is hoisted to the surface and connected to the plaform.

d. Lifting upper part of platform: As shown in Figure 4-31(d), the platform with a floating crane lifts the upper part [Figure 4-32(d)], then the upper parts docks with the subject [Figure 4-32(e)].

5. The advantages of the SPAR platform

Through the analysis of the third generation of SPAR platform, we can see its advantages:

a. Excellent stability

The SPAR platform has good motion performance; under the control of the mooring system and the buoyancy of its main body, the movement natural period of its six degrees of freedom is far away from the common marine energy band, and compared with several other types of deep-water offshore floating structures, its stability is excellent.

Figure 4-33 shows the amplitude of the heaving motion. It shows the difference between the SPAR platform and floating drilling ship and semi-submersible platforms. Table 4-7 shows a comparison of the maximum heaving motion and the maximum roll motion between the Classic type SPAR platform and semi-submersible platforms. Figure 4-33 and Table 4-7 shows the motion performance of the SPAR platform is superior to other deep-water offshore floating structures, as the heave and pitch and roll motions are small.

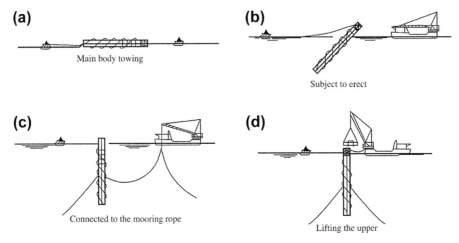

(a) Main body towing

(b) Subject to erect

(c) Connected to the mooring rope

(d) Lifting the upper

FIGURE 4-31

Installation steps and process of SPAR platform.

FIGURE 4-32

Towing and erection of the main body of the SPAR platform and lifting the upper part and docking it with the main body.

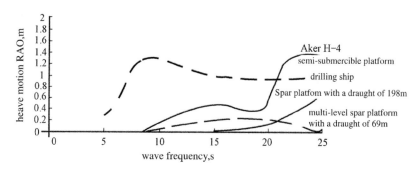

FIGURE 4-33

Heaving motion of SPAR platform and other deep-water offshore floating structures.

In ordinary usage, the draft of the main body of a SPAR platform is deeper, and the motion performance is better. Figure 4-34 shows the relationship curve between the SPAR platform heaving motion amplitude and its main draught. The figure shows two wave spectrum periods (14.9 s and 16.9 s). It shows that in different wave spectrum periods, the main draft of the platform has different effects on the first-order heaving motion. From

Table 4-7 Motion Performance of SPAR Platform and Semi-submersible Platform

	Semi-submersible Platform	Classic SPAR Platform
The maximum heaving motion/m	±10.668	±1.829
The maximum roll motion	20	4

FIGURE 4-34

The relationship of the SPAR platform main body draft and the amplitude of the heave motion.

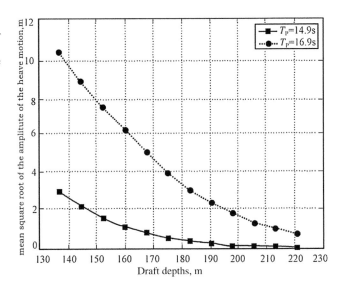

the SPAR platform in the Gulf of Mexico, the practice suggests that we choose a minimum value between 168 ~ 198 m water depth for the main draft of a platform to maximize the platform motion performance.

b. Excellent movement

During construction, the Cell SPAR platform consists of small cylinders, and it is easy to move. During manufacture, the platform uses a cable fixed mooring system, it is easy to tow and install, and we can dismantle the mooring system when necessary and directly transfer to the new location. So the SPAR platform is particularly applicable to well-dispersed, wide distribution offshore oil fields. In the state of the pile, we can adjust the length of mooring line, making platform on the surface of move a certain range, which is easy to dynamic position.

c. Economy

Compared with a fixed platform, as the SPAR platform is fixed by a mooring line, the cost will not increase sharply with an increase of the depth of the water. Compared with the tension leg platform (TLP), the cost is much lower. For example, the Harm Mountain SPAR platform's operating depth is 1646 m; the Mad Dog Truss SPAR platform's operating depth is 1646 m. And the cost of the former is 60 million dollars, while the latter only cost 35.5 million dollars.

The Auger tension leg platform cost 110 million dollars; its operating depth is just 872 m. The Brutus tension leg platform cost 75 million dollars; its operating depth is just 910 m. Clearly, the SPAR platform is more economical.

In a word, due to the three main advantages of SPAR platforms listed above, the SPAR platform is the type of platform most suitable for deep sea operation, and it has developed rapidly. It took four years to develop from the first generation to the second generation, but it only took only three years to go from the second generation to the third generation. It also continues to adopt breakthrough new technologies, moving toward greater water depths, better economic efficiency, and stronger adaptability. If we can grasp the opportunities offered by the platform, it will beneficial for deep-water oil exploration and development.

6. Problems of the SPAR platform

The SPAR platform has the following problems:

a. It is easy to induce vortex-induced vibration in the main body

The draft of the SPAR platform's main body is about 170-200 m. Underwater currents impacting the cylinder of the main body will create a vortex, thus causing vortex-induced vibration of the main body. Although after several generations of improvement of the SPAR platform designers have tried their best to reduce vortex-induced vibration, it is is still a problem.

b. Components are prone to fatigue failure

The vortex-induced vibration of the platform's main body makes the platform's main body and the supporting legs reach fatigue failure easily. At the same time the mooring system and riser connected with the main body are prone to fatigue failure. In addition, under the effect of the ocean current, the top tension riser (TTR) and its support riser under the main body are prone to fatigue failure. Therefore, we should do some further research to improve the fatigue strength of the SPAR platform.

4.3.1.3 Semi-submersible Platform

The architecture of the semi-submersible platform was introduced in the second chapter, so we won't repeat it here. But, the semi-submersible platform discussed here is not only a drilling semi-submersible platform; it refers to the multifunction deep-water semi-submersible platform. It means that it works in deep water with drilling, production, ,lifting, and pipeline installation functions.

The first semi-submersible platform came into service in 1963. According to the statistics at the end of 2002, 175 semi-submersible platforms are in service or under construction around the world. Forty-seven semi-submersible platforms are used in deep water, with 31 of them used below 1829 m (6000 ft) water depth, and 16 of them used below 2286 m (7500 ft) water depth; the deepest water depth has reached 3048 m (10000 ft).

Since the first semi-submersible platform launched in 1963, it has developed fast. There were 25 semi-submersible platforms in 1970 around the world, accounting for 13.15% of total mobile platforms around the world. In 1980, this number increased to 111, accounting for 24% of the total mobile platforms around the world. In 1985, it increased to 163 and it 1990, it increased to 171. By 2002 it had increased to 175. The semi-submersible platform has become the most promising floating platform for offshore oil and gas exploration and exploitation.

In the last 40 years, the semi-submersible platform has evolved from the first generation in 1963 to the sixth generation. We introduce some of their main features created during these development years.

1. The main characteristics of the semi-submersible platform
   The main characteristics of the semi-submersible platform from the first generation to the sixth generation are as follows:
   a. Improving the variable load capability
      The size of the variable load is an important index related to the performance of the semi-submersible platform. To improve the design of semi-submersible platforms, one of the development trends has been increasing the variable load at the same rate as the displacement and we can explain this by looking at the ratio of variable load and total displacement or the ratio of variable load and dead weight. From Table 4-8 we can see that during the development of the semi-submersible platform, the variable load has increased gradually. The ratio of variable load and total displacement has gradually increased from 0.1 to 0.175. This has been the trend of development.
   b. Increasing strength to ensure safety
      The strength of the semi-submersible platform's components has been developed to improve the platform's ability to resists storms. One purpose of the semi-submersible platform's development is to ensure the security of the platform over the years. For example, some platforms use high strength low carbon alloy steel or steel plate and sectional material with a Z shape; these platform have good welding performance and fatigue strength under normal temperatures. Double ship shells have been designed in the columns of semi-submersible platforms, and its stability requirements conform to internationally recognized standards of classification.
   c. Saving steel, reducing cost
      Trying to save steel, reduce cost, and improve the economic benefit of semi-submersible platforms are a major direction of the development of semi-submersible platforms. For example, the advanced semi-submersible platform uses a box body. Its shape is a flat plate, so the cost is effectively reduced and the strength of the structure is improved. And with efforts to reduce the number of solid components as far as possible, including reducing the column and brace, and minimizing the number of nodes (junction) and the number of poles (Table 4-8), we will save steel and reduce cost.
   d. Improving equipment performance
      The main equipment of the semi-submersible platform is developed for deep water. So we need to improve the technical performance as far as possible. For example, some semi-submersible platforms are equipped with a rig for which the drill and detection capability has reached 6000-9000 m. Some are equipped with three drilling pumps, each of them with power that can reach 2200 hp, and the blowout preventer's work pressure has reached 103.5 Mpa. Some are equipped with a top drive system, AC-VFD-DC drive, new type of drilling fluid, solid content control system, automatic discharge system for the drill pipe, etc. In addition, some equipment on semi-submerged platform develops to standardization, series and generation is also one of semi-submersible platform trend. For example, the windlass system, seawater pump, diesel power generation system and SCR systems, and fire rescue and desalination equipment are tending to adopt generalized equipment standards.
   e. Adding a dynamic positioning system
      This is a necessary technical measure for the semi-submersible platform to work in deep-water. A lot of semi-submersible platforms adopted the advanced dynamic positioning

Table 4-8 Comparison of Parameters between Several Developed Semi-submersibles

Platform Name	The Depth of Drilling Work, m	Total Displacement, t	Hull Weight, t	Variable Load, t	The Ratio of the Variable Load and the Weight	The Ratio of the Structure Contact Number and Extremely	The Ratio of Total Displacement and Weight	The Ratio of Variable Load and Total Displacement
BaiLong 2	210	16000	—	1600	—	—	—	0.1
BaiLong 3	300	21000	6500	2500	—	—	—	0.119
South China Sea 2	450	20630	—	2610	—	—	—	0.127
SEDCO 600	150	20000	—	3000	—	—	—	0.15
SEDCO 602	150	20000	—	3000	—	—	—	0.15
LIMCUN-NINGHAM	450	25000	9000	3000	0.33	—	2.77	0.12
ZAPATA ARCTIC	600	32000	9000	4500	0.5	—	3.55	0.14
Exploration 3	200	21180	—	1600	—	28/13	—	0.076
DSS 20	450	20000	7100	3500	0.493	4/2	2.82	0.175
PETROBRAS X V III	800	36000	10000	Deck load 9500	—	—	3.6	—

system in order to work in deep sea. For example, the DPS triple dynamic positioning system has four strong motors; each motor's power is 7MW.

f. Increasing operating depth

This is the general trend in recent years. From Table 4-8 we can see that the semi-submersible platform's developing trend is increasing operating depth. For example, eight semi-submersible platforms were built from 1988 to 1994, and six of the semi-submersible platforms' operating depths are below 914 m (3000 ft), with four of them working below 1500. In addition to this, 19 semi-submersible platforms were built from 1996 to 2001, and 18 of these semi-submersible platforms' operating depths are below 1500 m, with half of them work below a depth of 2000 m. It shows that a development trend of semi-submersible platforms is gradually increasing the working depth in the water.

2. Features of the fifth and sixth generation of semi-submersible platforms.

In recent years, "saike.quick" and "saike.energy" built by the French belong to the fifth generation of the semi-submersible platform. Figure 4-35 shows the sixth generation of the semi-submersible platform. These platforms were built up and put into production in 2001. These fifth and sixth generation semi-submersible platforms have some common characteristics.

a. Variable load

They all have large variable load decks; the variable load is at least 10×10^4 t, so the platforms have a large main scale and a big storage capacity for supplies used in drilling engineering. Furthermore, by improving the design, the ratio of variable load and the total displacement is more than 0.2, far surpassing the previous platform's 0.175. The ratio of total displacement and the weight of the platform is more than 4.0, far surpassing the previous platform's 3.6.

b. Increasing strength to ensure safety

The main structure of the platform uses high strength steel, good toughness, and good welding ability. The structure of the platform deck's life area is flat and short, and the hydraulic

FIGURE 4-35

Sixth generation of semi-submersible platform.

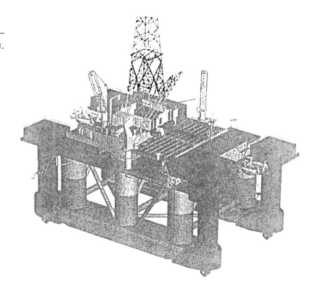

cylinder drilling rig has a cylindrical frame, thus reducing wind resistance. So the platform has an ability to adapt to big stormy waves. This improves safety and the ability to resist storms and with a longer term holding capacity the platform can adapt the open sea and deep water, in all weather conditions for an extended period.

c. Saving steel, reducing cost

They adopted a more concise structure. The sink pad and pillar sections are rectangular. The structure of the platform deck is a flat rectangular box. There are only two rectangular cross section columns connected with platform deck and a sink pad in each side. So except for four tie pieces on the left and right columns, it doesn't have any other nodes. So this reduces welding nodes and braces obviously, removing technical difficulties and saving construction costs. The platform is equipped with the newest cylinder type drilling rig (produced by Marine hydraulic company which is Library beauty companies in the United States joint venture with Norway). Its weight and power on per unit decreased obviously; this reduces drilling time by 30%, and construction costs are reduced by 30%. So, it saves steel and reduces cost.

d. Improving equipment performance

They are equipped with a new generation of advanced drilling equipment which has high power. They are equipped with a hydrostatic transmission rig controlled by a computer, the maximum power of the main winch is 2940 kw (4000 hp), the diameter of the rotary table is up to 1257 mm (49 ½ in), the drilling pump's delivery capacity is as high as 2500 L/min, and the working pressure is up to 34.5 Mpa (5000 psi). Some semi-submersible platforms are equipped with rigs where teh drilling depth can be 9000 m and drilling pump's power reaches up to 2200 hp. In addition, they are equipped with a drill pipe automatic discharge system and a new type of pumping system that combines the drilling fluid system and cementing system. All the above create favorable conditions for semi-submersible platforms working in deeper waters.

e. Dynamic positioning system

The fifth and sixth generation semi-submersible platforms are equipped with a new generation of dynamic positioning system.

f. Operation in deep sea

Taking the fifth generation semi-submersible platforms as an example, "saike.quick," "saike.energy," and "kajiang fast," all of them have an operating depth that has reached 2590 m. And four of the newly built "Bingo 9000" series have an operating depth of 3000 m (9843ft). It shows that the fifth and six generations semi-submersible platforms' operating depth is increasing. Thiss makes the development of deep sea oil and gas field easier.

3. The advantages of semi-submersible platform used in the deep sea

A semi-submersible platform is a kind of floating platform that is more suitable for the deep water. This is due to the following reasons:

a. Good stability

The semi-submersible platform uses extended anchoring; it doesn't need a special turret anchor system, but it is stable. Compared with other offshore floating structures such as an FPSO unit, movement is less. It can satisfy the platform motion performance requirements when drilling in deep water, and it is also easy to connect with steel catenaries risers.

b. Good movement

When semi-submersible platforms break away from the mooring system, we take the chain cable and mooring line back and pull the appliance under the water when it's time for the semi-submersible platform to move out of the well. The work is simple and the platform is easy to move.

c. Good depth adaptability

Semi-submersible platforms are adaptable to the water depth; they have a wide range of operating depths. They are strongly adaptable within the space

d. Function adaptability

Semi-submersible platforms can be used as drilling platforms and production platforms. They can complete deep-water drilling, oil production, production processing, lifting, laying and other functions.

e. Economy

Compared to other sea floating structures, a semi-submersible platform is easy to transform into a production platform, and it will not be influenced during the transformation. We can make full use of the offshore construction time and lower the development cost.

Of course, compared with the TLP and SPAR platform, the semi-submersible platform has its weaknesses, which have been introduced earlier.

4. A sixth-generation deep-water semi-submersible drilling platform has been put into operation by China.

On February 26, 2010, the world's most advanced sixth-generation deep-water semi-submersible drilling platform, towed by a total of eight tugboats, was successfully led out of dock from the China Shanghai Waigaoqiao Shipbuilding Co. Ltd, see Figure 4-36(a). The platform shown is named "Offshore Oil 981" the by China National Offshore Oil Corporation (CNOOC). The "Offshore Oil 981" semi-submersible drilling platform consists of two buoyancy tanks of 104 meters, four columns, and a box-shaped hull in the upper part. Its total weight is more than 30,000 tons, with the deck area equivalent to the area of a standard football field. The height from the bottom of the vessel to the top of the derrick is 136 meters, which amounts to a 45-story high rise,

FIGURE 4-36 (a)

(a)

The sixth generation deep-water semi-submersible drilling platform built by China in 2011.

and the length of the cable overall is over 800 kilometers. "Offshore Oil 981" has been used for deep-water drilling in the South China Sea since 2012, and as of 2013 there have been six deep water drillings with signs of oil and gas.

"Offshore Oil 981" can be called the world's most advanced sixth-generation deep-water semi-submersible drilling platform, not only because it is fully equipped with the six major characteristics of the fifth and sixth generation semi-submersible drilling platforms mentioned above, but also because there are some additional innovations and advances. The following will introduce its specific advances.

- As to improving the variable load, the "Offshore Oil 981" will break the world record for semi-submersible drilling platforms, reaching 9,000 tons. Thus it increases the reserve capacity of drilling equipment and daily necessities, and greatly improves the self-sustaining capabilities when operating in deep water.
- As to ensuring safety and security, "Offshore Oil 981" is the first in the world that is designed with wind and wave parameters for a 200-year storm, and also adds the unique internal waves in the South China Sea to its design. So the capacity of platform to resist disasters is largely improved, ensuring the safety and security of platform operations.
- As to saving the costs of steel, in order to build "Offshore Oil 981," China successfully developed the world's first ultra-high-intensity R5 anchor chain. This kind of chain not only leads to the addition of new content on R5 to the relevant international codes, but also saves lots of costs in building the platform because of the low cost of this steel, which lays a foundation for selling this domestic manufacturing abroad.
- As to improving the performance of equipment, a first-of-its-kind combination energy-efficient positioning system is applied to the design of "Offshore Oil 981." When the water is shallow, at a water depth within 1500 m, the conventional anchoring positioning system will be employed; while only in deep water areas, where the water depth is within 3000 m, will the DPS3 dynamic positioning system with large energy consumption be used. Obviously, this combination positioning system is an innovation creating an optimized energy-saving operation.
- From the perspective of expanding the water depth, the depth of the "Offshore Oil 981" reaches 3000 m, and it is capable of working in super depth water areas. It is deeper than fifth generation semi-submersible drilling platforms, such as "Sedco Fast" (operating water depth of 2590 m), and holds the line with sixth generation semi-submersible drilling vessels such as " the Bingo Liu 9000" series, which has a water depth of 3000 m.
- Taking improving security and environment protection into consideration, "Offshore Oil 981" has adopted the most advanced secure underwater blowout prevention system, and it can automatically close the wellhead through the system, so that it can prevent oil overflow, avoiding severe accidents caused by polluting the marine environment.
- As to scientific research, "Offshore Oil 981" has installed systematically an inspection system composed of sensors on key positions of the hull for the first time. It has established a systematic and comprehensive marine scientific platform for researching the distribution "stress" and the range of mooring strain of semi-submersible vessels on key structures during movement; such a submersible platform is unprecedented both at home and abroad. Evidently, based on proper arrangement, massive data gathered during working at the vessel can provide more key scientific design criteria for further developing the deep-water submersion vessels, which will have great importance going forward.

In addition to the above best in the world advancements, the "Offshore Oil 981" semi-submersion drilling vessel has 10 other units of advanced equipment. For example, it has installed a one-and-half derrick dual-activity rig. Figure 4-36(b) shows the dual-activity drilling derrick during the installation of the submersible drilling vessel: one is higher, and the other is shorter. The higher derrick is equipped with a standard drilling rig that accomplish the well drilling, well completion, and well repair work, while the shorter drilling rig tower, equipped with a lower power hoisting system, can do assisting work such connecting the drill column; therefore, both of the two work in parallel, saving marine operating time and improve drill performing efficiency. To be specific, for the exploration of a well we can save 15% of the total time, and for exploring a productive well we can save 30% of the total working time, reducing the cost of well drilling in general.

All in all, from the first application of those innovative projects, we can easily see the advanced level of the "Offshore Oil 981" semi-submersible drilling vessel.

4.3.1.4 Floating Production, Storage, and Offloading System

There is an important offshore facility named the floating production, storage, and offloading system, referred to as an FPSO.

1. The types of FPSO

   FPSOs can be divided into the following categories serving in the world currently.

   a. Floating production, storage, and offloading system

   This system is the most widely used. Nearly 100 FPSOs are in service around the world now; they are especially used in subsea production systems. It is equipped with a oil-gas-water disposal and production facility. This system floats on the sea through a single point moor. The system can do 360° full free rotation under the control of wind, waves, and current. Crude oil production can be stored in the cabin and then carried away by shuttle tanker at a fixed period.

FIGURE 4-36 (b)

The one-and-half derrick structure during construction of "Offshore Oil 981."

b. Floating storage and offloading system

This kind of offshore floating structure has the shape of a ship or barge. It has crude storage cabin and offloading equipment and doesn't have an oil-gas-water disposal and production facility. After the oil is stored, it will be carried away by a shuttle tanker.

c. Floating production system

This is an offshore floating structure equipped with an oil-gas-water disposal and production facility. It doesn't have a storage device. So the qualified oil needs to be transferred to the nearby shore or an FSO through a submarine pipeline in a timely manner.

d. Floating storage unit (FSU)

This is an offshore floating structure that can store oil. So the oil is fabricated after transporting it to shore through the offshore pipeline, where the oil is processed. If it is equipped with offloading equipment, then it is equal to an FSO.

2. Composition of an FPSO

The four kinds of FPSO above have the same structures:

a. Mooring system

An FPSO is an offshore floating structure. It is moored on the sea by the turret structure through the mooring system. The mooring system cable is connected to the rotary table of the turret structure. So the body can turn through external force. Therefore it is not only moored on the sea but also can do 360° full free rotation under the control of wind, waves, and current.

The mooring system is usually composed of chains, ropes, and anchor. In more than 300 m deep water, the suspended part of the chain is often made up of metal cord. In more than 2000 m deep water, a composite mooring system should be applied; the system should be a polyester tension leg system, as shown in Figure 4-37. As a result, we reduce the weight of the mooring system and enhance its stiffness, making it easy to apply in deep water.

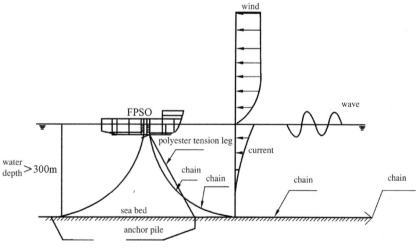

FIGURE 4-37

Synthesized mooring system in Brazil's deep waters.

   Figure 4-38 shows a mooring system with a submersible buoy that is a kind of mooring system used in deep water. It reduces the vertical load of the turret structure caused by the anchor chain through using an intermediate floating cylinder and upright buoy, thus increasing the stiffness of the mooring system.

b. Turret structure

   This is a kind of steel structure system. The upside is connected to the FPSO; the bottom is connected to underwater through the mooring system, and the top is equipped with rotary joints. The FPSO can move due to the wind, waves, and current. A flexible riser can be connected to rigid pipe in the bottom of the turntable through a flange.

   Turrets can be divided by their structure and location: overwater turret, aquatic turret, internal turret, and hinge/riser turret. The internal turret is a type of turret where the turret is located inside the hull and a mooring chain is fixed at the bottom of the turret, as shown in Figure 4-39. Figure 4-40 shows the structure of the hinge/riser turret, which is a cylindrical mooring riser; the riser is connected to the bottom of the turret through a universal hinged joint. The hull turns around the mooring riser, and the mooring system is fixed to the riser near the water surface.

c. Production and storage equipment on the water

   The production and storage equipment includes an oil tanker or oil storage inside a cabin and it has large storage capacity. The oil ship can provide a large deck area, so the production facility can be installed on the deck block to process the oil-gas-water; some other auxiliary help facilities are included. Some of them use a semi-submersible platform as production and oil storage equipment, but it should be equipped with a heave compensator to eliminate the effect of the semi-submersible platform's larger heave movement. Of course, whether using an oil tanker or semi-submersible platform,

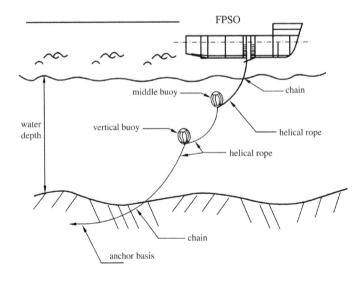

FIGURE 4-38

Submersible buoy mooring system used in deep water.

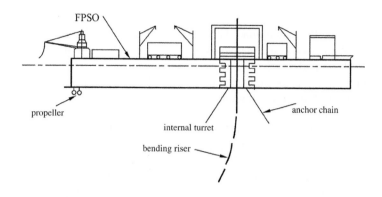

FIGURE 4-39

Structure diagram for the internal turret.

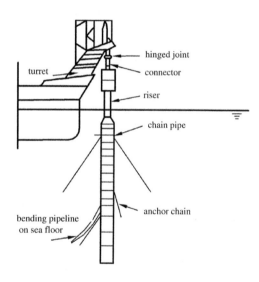

FIGURE 4-40

Structure diagram for hinge/riser turret.

in order to reduce the influence of the six degrees of freedom of movement on the production of oil and gas processing, the platform needs to install a dynamic positioning system.

d. Loading and unloading systems and shuttle tanker

The oil is processed by the offshore floating producing and storage system. If the oil is qualified, then it is carried away by shuttle tankers. Most shuttle tankers are equipped with a loading and unloading system in the cabin, and through hydraulic operation, oil is unloaded from the FPSO's tail to the shuttle tanker's stem through a flexible pipe connected to the shuttle tanker; unloading from the rear of the FPSO to a shuttle tanker bow, which is called "both loading and unloading."

3. FPSO use in foreign regions
   The FPSO has been used for more than 30 years around the world. It is mainly distributed in Australia, New Zealand, Pakistan, West Africa, and the North Sea of Norway. Because it can accommodate 20 to 2000 m water depth and a variety of marine environments, it is widely used all around the world. It was reported by a foreign source that there are 123 FPSOs in operation in 2005. Table 4-9 shows some FPSOs used in foreign regions.
4. Technical characteristics of FPSs in China
   After more than 20 years of effort, FPSOs have become an important facility for the development of offshore oil fields. Now there are 13 FPSOs used in 16 oil and gas fields in the Bohai Sea and the South China Sea with a total load capacity reaching 170×10^4 t. The characteristics are shown in Table 4-10. China is building the biggest and latest FPSO in China' its capacity has reached 30×10^4 t, and it will be used in the Penglai 19-3 oil field. In addition, there are two 10×10^4 t FPSO that will be built. Now, China has become one of the world's largest users of FPSOs and gradually is building its own technical ability and characteristics, as laid out in the following.
   a. Design capacity of and comprehensive evaluation of the overall plan
      When designing an overall exploitation program for an offshore oil field, oil and gas gathering, processing, production, transportation of equipment, reservoir type, marine environmental conditions, and crude oil transmission mode should be considered. Then we determine whether an FPSO is needed or not and determine the FPSO's technical performance and main technical parameters. At this time, we need to design and comprehensively evaluate the overall plan. The evaluation includes project investment, economic benefits, advancement, reliability, the safety of the operation, environmental protection, sustainable development, etc. China now has a complete comprehensive evaluation index system to evaluate the FPSO overall scheme and achieve rich experience from it. For example, in designing the overall development plan of the China Bohai sea oil field during the "the tenth five-year," when determining whether to use the pipe net or FPSO, the designer used a comprehensive evaluation system. Evaluation results showed that the network scheme would cost 25% more than the FPSO when comparing investments, with a total investment that was 5% ~ 8% more, and a yield rate 1% ~ 3% less. So the designer used an FPSO and determined the main technical parameters.
   b. Ability to optimize the constituent system
      China has the optimal design for a production process system, instrumentation and control system, power and electrical equipment system, public universal equipment system, and so on. And China has set up a full set of new methods to integrate design, and is accumulating experience. For example, a designer may undertake a consideration of the upper module production process system and hull of the ship, using integration design, isolating three crude oil processing tanks in the cabin class, and increasing the administrative levels of the oil/water separation. The oil in Bohai has high viscosity and low density which through a multistage process the oil. As another example, the upper and lower utility equipment such seawater cooling, diesel, fresh water, and fire protection systems, can be considered as an integrated design, allowing the sharing of resources, and efficient utilization, saving significant costs savings. As a result, on the No. 111 FPSO offshore oil platform seven less sea pumps were used. Similarly, for the electrical power equipment system we can take an integrated design approach, focused on power generation, district changing/distribution, the nearest power

Table 4-9 Overview of FPSOs Used in Foreign Regions

Owner	Project Name	Description of Project	Date Started	Date of Completion
Husky Energy	Sea Rose	14×10^4 t FPSO, east coast of Canada	–	2005
Trafigura	Jarnes Town	4×10^4 t FPSO, Nigeria	–	2004
Shell	Sea Eagle	20×10^4 t FPSO, Nigeria	–	2002
Husky Oil	Canada White Rose FPSO	FPSO of preliminary design and other shuttle set	1999	2000
Petro-Canada	Tema alova	20×10^4 t FPSO, east coast of Canada		1999
Ranger Oil	Kyle oil field development	BANFF oil field research on selection of floating structures	1997	1999
Hess	Triton FPSO	FPSO transformed from new oil tanker, oil production capacity of 105,000 bbl	1997	1999
Woodside Petroleum	Laminaria FPSO	Provided FPSO, underwater equipment for shuttle in north Australia, FPSO hull is the new shuttle	1997	1999
Petronas Carigali	Masa FPSO	Malaysian-built FPSO superstructure design, bidding and train build (EPC)	1997	1998
ESSO	Natuna barge development	Floating barge with concrete structure for development and production of oil and gas; every day oil and gas production are 2.72×10^8 m^3, it contains 71% CO_2	1997	1998
M.P. Zanrat	New Orleans	26×10^4 t FPSO, Tunisia	–	1998
Conaco	Independence	28×10^4 t FPSO, Nigeria	–	1997
Exxon	Ilyinskiy FSO development	Study FSO and other oil output optional equipment; oil capability is 200×10^4 bbl	1997	1997
AIOC	Caspian Sea PUQ Barge	Preliminary design of PUB barge	1997	1997
Apache Energy	Stag FSO	Research and preliminary design of FSO used in Australia	1996	1997
Texaco	Mariner FPSO	Design of FPSO submarine equipment, used in large oil field development; daily production capacity of 70000 bbl	1996	1998

Continued

Table 4-9 Overview of FPSOs Used in Foreign Regions—cont'd

Owner	Project Name	Description of Project	Date Started	Date of Completion
Kerr McGee	Gryphon operation and maintenance	FPSO operation and maintenance, including a renovation project	1994	1997
BP	Pierce FPSO	FPSO preliminary design, shuttle tanker, daily production capacity of 50,000 bbl	1996	1996
Shell Expro	Curlew FPSO	FPSO preliminary design, shuttle tanker, daily production capacity 60,000 bbl	1995	1996
Conoco	MacCulloch FPSO	FPSO preliminary design, transformed by oil tanker	1994	1995
Enterprise Oil	Sedgwick oil field development research	Sedgwick and Westbrae oil field offshore floating platform; system feasibility study	1994	1995
BP	Foinaven FPSO	FPSO; shuttle for Shetland oil fields in the west at the surface of the deep water area and semi-submersible structure research	1994	1994
Aran Energy	Gryphon FPSO examinination	Partner in oil field, the Gryphon FPSO technical inspection	1994	1994
Woodside Petroleum	Wanaea/Cossack FPSO	Preliminary design and construction, bidding and superstructure, construction management	1993	1994
Nova Scotia Resources	Penobscot Oil Field research	Oil field development research, review of alternative development equipment, including FPSO and semi-submersible structure	1993	1993
Shell Expro	Mallard/Teal upper structure research	Preliminary design of the upper structure, oil to oil wheel, oil and gas pipelines, output	1993	1993
Shell Expro	Shell Bonga	Oil field development research, preliminary design and construction	2003	2005
Chevron/Texaco	Sanha	Preliminary design and construction, bidding and superstructure, construction management	2004	2005

Table 4-9 Overview of FPSOs Used in Foreign Regions—cont'd

Owner	Project Name	Description of Project	Date Started	Date of Completion
Navis Explorer	FDPSO	Primary design of FPSO submarine equipment, used in large oil field development	1998	2000
Brazil Oil	N'Kossa	FPSO preliminary design, transformed by oil tanker	2008	2010

supply, and overall planning. For instrumentation and control system integrated design, we can look at centralized control, zoning implementation, and unified ordering, all of which can save significant costs.

c. Abilty to use joint design for on-site fuel and electric station

A large power station and heat station are generally set on an FPSO to supply electricity and heat for the whole oil field. We can save energy and improve the overall thermal efficiency by integrating the power station with the heat station, taking into consideration of the choice of fuel and waste heat utilization. The economic benefits will be more notable if we choose fuel reasonably, adopt the methods for optimizing design and use fuel on site. In recent years, China has gradually formed its own technical characteristics. It has established a complete method of power-heat station joint design and optimizing selection of fuel. For example, the Bohai heavy oil field took used heavy oil on site for the fuel of the power station. We can control the denseness, viscosity, and temperature of heavy crude oil in different heavy crude oil conversion modes. The dual fuel oil generator was adopted successfully in the Bohai Sea and Offshore Oil 111 FPSO, therefore greatly saving fuel for which the key is controlling oil temperature. The FPSO in Offshore Oil 113 using integrated design, equipped four joint supply devices, making full use of the waste heat produced by herculs 130 which had a combustion efficiency of only 25%–35%, and reduce the cost greatly.

d. Ability to design and build a new type of single point mooring positioning system.

At present, for a floating body such as an FPSO or FSO, the single point mooring system where mooring join point which is a single point has been widely used in offshore oil fields. China has used two forms of single point mooring system on eight FPSOs. They are the soft just arm single point mooring equipment and adduction tower single point mooring system. For example, the Bohai New Century FPSO on the Qinhuangdao 32-6 oil field uses a soft just arm single point mooring system. The Offshore Oil 111 FPSO on the Panyu oil field in the South China Sea uses the tower single point mooring system and has been tested of typhoon several times. All of these indicate that China has the ability to build these new types of single point mooring system and has accumulated certain experience in installation and connection construction technology.

e. Ability to design and build an FPSO which can work in ice areas on the sea

The China Bohai Sea is located in an ice area, and freezes in winter. In heavy ice years, more than 90% of the area is covered by ice, and the average thickness of ice is about 40 cm. China has researched the single point mooring system of the Suizhong 36-1 oil field on Bohai Sea since 1991 and adopted a series of measures for ice, so that the three FPSOs, "Bohaia

Table 4-10 An Overview of China's Floating Production and Storage Equipment

FPSO Name	Oil Field	Depth, m	Main Dimensions, m				Weight, 10⁴ t	Design Specification	Mooring Form	Date in Operation
			Length	Wide	Depth	Draft				
Bohai Friendship	Bozhong 28-1	23	210	31	17.6	10.5	5.2	ABS	Yoke SPM	1989
Bohai Changqin	Bozhong 34-2/4	21	210	31	17.6	10.5	5.2	ABS	Yoke SPM	1990
Bohai Jewel	Suizhong 36-1	32	210	32.8	18.2	11.7	5.8	DNV	Soft yoke SPM	1993
Bohai Century	Qinhuangdao 32-6	20	282	51	20.6	14.5	16	DNV	Soft yoke SPM	2001
Offshore Oil 112	Chaofeidian	24	262	51	23.6	15.6	15	DNV	Soft yoke SPM	2004
Offshore Oil 113	Bozhong 25-1	16.7	282	51	20.6	14.5	16	DNV	Soft yoke SPM	2004
NanhaiEndeavor	Wenchang 13-1/2	100	262	46	24.6	17.5	15	BV	Turret tower	2002
Offshore Oil 111	Fanyu 4-2/5-1	105	262	46	24.6	171	15	BV	Turret tower	2003
Nanhai Hope	Wei 10-3	40					9.3	–	Fixed tower	1986
Nanhai Discovery	Huizhou 21-1	116					25	–	BTM	1990
Nanhai development	Xijiang 24-3	100					15	–	BTM	1994
Nanhai Victory	Liuhua 11-1	304.8	285	58	26	90	15	–	BTM	1996
Nanhai Explorer	Lifeng 13-1	146	323	63	–	–	12.39	–	BTM	1993
Nanhai	LiuHua 11-1	–					23	–	BTM	2005
BoHai	PengLai 19-3						–	–	Soft yoke SPM	2009
Offshore Oil 116	NanhaiWen Chang 19-1	–	232.5	46	24.1	16	15	CNOOC	Soft yoke	2008

Friendship", "Bohai Changqing" and "Bohai Mingzhu" successfully work in ice areas. Practice shows that China has the ability to design and build FPSOs in the ice area on the sea.

f. Ability to design and build FPSOs on shallow waters

Normally, when designing a boat, to ensure safety, the depth of the water can't be less than 1.3 times of the draft of the ship in operation. The depth of water in China's Bohai Sea is shallow, but some of the oil fields in shallow waters require an FPSO. So FPSO application in shallow waters are required. China's CNOOC production and research center began to experiment in the late 1990s, putting forward the "shallow water effect," which is a new concept of the floating body. A set of optimum design and depth checking methods is established, which makes the floating body draft bigger than traditional requirements. It can increase the load weight by $6\% \sim 9\%$, which not only improves performance, but also reduces the volume of steel. This kind of design method was used for the FPSO of Offshore Oil 113 on the Bohai 25-1 oil field, making the draft increases 2 m beyond the traditional design. It was successfully to put into production in August 2005 in the sea area where the Bozhong 25-1 oil field was located, the depth of which is 16.7 m in extreme low water conditions. This shows that China has the ability to design and build FPSOs in shallow waters.

Despite the fifth and sixth abilities above having nothing to do with the deep water, in order to fully explain China's ability in design and construction of FPSOs they are listed.

4.3.2 Large-Scale Vertical Floating Columns and Wave Load of Column Group

The floating platform used in deep water, whether a TLP or SPAR platform, or semi-submersible platform, has a floating platform, and its main supporting wave load components are standing columns. However, the columns are large scale cylinders. Several columns constitute a column group. Here another special problem of deep water oversea floating structures is formed. It is the computation problem of wave load for the large-scale vertical floating columns and column group. The following proposes introducing the problem.

4.3.2.1 The Diffraction Theory of the Wave Action of Large-Scale Floating Upright Columns

1. Diffraction and radiation effect of wave on structure

For slight movement of waves in large-scale structures, the load effect of the wave on the structure can be decomposed into two parts, namely the diffraction and radiation of wave to the structure.

a. Diffraction effect

Diffraction refers to wave propagation on structures; moving forward over the surface of the structure will result in an outward scattering wave, and the superposition of the incident wave and scattering wave reaches the steady state and will form a new wave field. This wave field load structure is called diffraction.

b. Radiation effect

The radiation effect of a wave refers to a certain mode of small radial oscillations on the structure; the stability of the wave field generates an outward radiation wave field. Fluctuations in the field affecting the structure.

2. Theoretical foundation of diffraction waves on a structure
   a. Assume that the fluid is an incompressible, viscous, uniform ideal fluid, and argue that the wave field movement is irrational and influential.
   b. Set up a local coordinate system for a vertically floating cylinder group in deep water

      As shown in Figure 4-41(a), a tension leg platform is still in a depth d of a wave field, there is a column group composed of N vertical cylinders, each column with a draft of h and radius a. The coordinate system is established as shown in Figure 4-41, with the plane xoy located in the stillness of the free surface, oz the vertical axis, and the origin o located at the free surface of a stationary point. If the column group of a single cylinder in the center of the circle is o_j, with coordinates of (x_{oj}, y_{oj}, z) $(j = 1, 2.... N)$, then using this point as the origin, we can set up the j cylinder. In this way, for a local $o_j r_{j\theta}$ theta polar coordinate system, the k cylinder center o_j in j column local polar coordinates can be written as (R_{jk}, α_{jk}, z) $(j, k = 1, 2,..., N)$.
   c. Assume we are considering linear wave diffraction; the incident wave can be described by linear slight wave theory.
   d. Assume the hydrodynamic interactions between the cylinders in the column mainly come from the secondary diffraction caused by the secondary incident of every single cylinder to incident waves diffracted outside.
   e. For the secondary diffraction effect, introduce a maxim spacing hypothesis, namely that the column spacing R_{jk} is much bigger compared with radiation wave length L, namely $K_o R_{jk} >> 1$. So, by column in the group of cylinder under the action of incident wave diffracted wave produced by the other cylinder in water, can be approximately instead by the nonlinear equivalent plane wave correction.

3. Diffraction effect for an isolated vertically floating cylinder

   The single (independent) column type platform applied in deep water belongs to the category of isolated vertically floating cylinder, which has a certain draft depth. Considering the diffraction structure, the first problem we should solve is the velocity potential for solving problems under the diffraction. If we know the velocity potential, we can solve problems for the structure under the wave load computation.

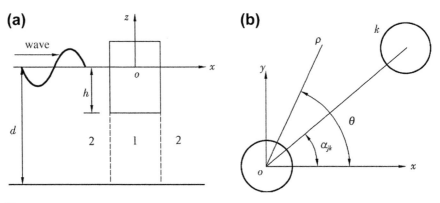

FIGURE 4-41

Coordinate system of floating upright columns.

a. Velocity potential of an isolated vertically floating cylinder

Consider a static isolated vertically floating cylinder j, and the division of its coordinate system and basin as shown in Figure 4-41(a), (b). From the figure we can divide the whole river basin into two parts: the domain (1) and outland (2). Domain 1 is located at the diameter of the cylindrical waters under the vertically floating cylinder; outland 2 is all the water area besides the interior domain and cylinder. In the two different water domains, the velocity potential is different. If we use the Φ_1 and Φ_2 expressions, they are:

i. The interior domain of the velocity potential Φ_1

There is no structure in this water area, that is, no diffraction. We only have incident wave velocity potential Φ_1 meaning

$$0 \leq r \leq a \quad -d \leq L \leq -h \tag{4-1}$$

ii. The external domain of the velocity potential Φ_2

There is structure in external domain 2, so diffraction exists. So the total velocity potential Φ_2 should be consist of the incident wave potential Φ_{21} and the diffracted wave potential Φ_{2D}. So the expression for Φ_2 should be the combination of these two parts as follows:

$$-h \leq r \leq \infty \quad -h \leq z \leq 0 \tag{4-2}$$

Set the cylinder j center coordinates to (x_j, y_j, o_j), define an angle with the x-axis β as the wave propagation direction, and set the outside of the incident wave velocity potential as Φ_{2I^j}. As covered by the slight wave theory in Chapter 1, cylindrical coordinates can be expressed as a series of Bessel functions and Φ_{2I}^j can be written as

$$\Phi_{2I}^j = \frac{gA \cdot chk_0(z+d)}{\omega \cdot chk_0 d} \cdot P_j \cdot \sum_{m=0}^{\infty} \varepsilon_m j^{m+1} j_m(k_0 r_j) \cos m(\theta_j - \theta) e^{i\omega t} \tag{4-3}$$

In the equation, A is the incident wave amplitude m, and with the x-axis is at angle β; Ω is the circular frequency of the incident wave, s^{-1}; P_j is the phase coherence; k_0 is the incident wave number; and m is the Bessel function order number.

In the same way, the incident wave potential of domain Φ_{1I}^j can be described by (4-3). Φ_{2D}^j is the diffracted wave potential in the outland. It can be calculated in accordance with the the first chapter, which discussed the solution for diffracted wave velocity potential for a large-scale structure. The expression is given as follows:

$$\Phi_{2D}^j = \frac{gA}{\omega} P_j \sum_{j,m=0}^{\infty} \Phi_{2D}^{(m)}(r_j, z) \cos m(\theta_j - \beta) e^{-i\omega t} \tag{4-4}$$

In equation (4-4) $\Phi_{2D}^{(m)}$ can be expressed in the first (Hankel) function and the second category of henkel function.

Thus we sum Φ_{2I}^j and Φ_{2D}^j together, and we get the total velocity potential Φ_2^j of the isolated vertically floating cylinder j:

$$\Phi_2^j = \Phi_{2I}^j + \Phi_{2D}^j \tag{4-5}$$

b. The wave load of an isolated vertically floating cylinder

    i. The pressure p_i of any point in the flow field

We get the velocity potential (i is the serial number of the water, $i = 1, 2$) through Equations (4-2) and (4-5). Then using the linear Bernoulli equation introduced in Chapter 1, we can get

$$P_i = -\rho \frac{\alpha \Phi_i^j}{\alpha t} \qquad (4\text{-}6)$$

    ii. The overall wave force F impacting the cylinder:

$$F = \iint_S p \cdot nds \qquad (4\text{-}7)$$

In Equation (4-7), n (n_x, n_y, n_z) is the unit external normal vector of the cylinder. Considering the geometric symmetry of the cylinder, the x (surge) and z (heaving) components of the linear wave forces impacting on the cylinder are not zero. So if $i = 1$, and order $z = -h$, then integrate along the surface S, and we get the z direction heaving wave force F_z of the floating vertical cylinder:

$$F_z = \iint_S p \cdot n_z ds = -\int_0^{2\pi} \int_{-h}^0 \rho \frac{\alpha \Phi_1^j}{\alpha t}\bigg|_{z=-h} \cdot rdrd\theta \qquad (4\text{-}8)$$

If we have order $r = a$, when $i = 2$, integrate along the surface S, and we get the x direction surge wave force F_x of the floating vertical cylinder:

$$F_x = \iint_S p \cdot n_z ds = -\int_0^{2\pi} \int_{-h}^0 \frac{\alpha \Phi_1^j}{\alpha t}\bigg|_{r=a} \cos\theta d\theta dz \qquad (4\text{-}9)$$

    iii. The overall wave moment of force M impacting the cylinder:

$$M = \iint_S p \cdot (r \times n)ds \qquad (4\text{-}10)$$

We suppose the cylinder water line's y-axis is the rocking-turn axis, and considering the geometric symmetry of the cylinder, the y direction of the linear wave forces impacting on the cylinder is not zero. So the pitch wave moment M_y is

$$M_y = \iint_S p(zn_x - xn_z)ds \qquad (4\text{-}11)$$

If $z = -h$, $r = a$, $i = 1, 2$, according to Equation (4-11) we can get

$$M_y = -\int_0^{2\pi}\int_{-h}^0 p \cdot z \frac{\alpha\Phi_2^j}{\alpha t}\Big|_{r=a} \cos\theta d\theta dz - \int_0^{2\pi}\int_0^a p\frac{\alpha\Phi_1^j}{\alpha t}\Big|_{z=-h} r^2 \cos\theta dr d\theta \qquad (4\text{-}12)$$

4. Diffraction of a vertically floating cylinder group

   For semi-submersible platforms and tension leg platforms used in deep water, their main bodies have several vertically floating cylinders, so we call them a cylinder group. The diffraction of the cylinder group is different than for a single cylinder. We should consider the effect of other vertically floating cylinder's diffraction on the cylinder. We introduce this problem as follows:

   a. The total incident wave velocity potential of the vertically floating cylinder group

      The total diffracted wave velocity potential of the cylinder j consists of three corresponding compositions:

      i. The impact of the external incident plane wave from the cylinder group on cylinder j

         We suppose the velocity potential can be expressed in local coordinates by the Equation (4-3).

      ii. The impact of the outside incident wave's diffraction of the other N-1 cylinders on cylinder j. Obviously, the impact is diffraction; it is a secondary diffracted wave caused by the diffraction of the other N-1 cylinders incident to cylinder j. The secondary diffracted wave can approximately be treated as an equivalent plane wave, so the velocity potential is

$$\Phi_{2I1}^j = \frac{gA}{\omega}\sum_{k=1,k\neq j}^N C_{jk}\exp\big[-ik_0\, r_j\cos m(\alpha_{jk}-\theta_j)\big]\cosh k_0(z+d)\cdot e^{-i\omega t} \qquad (4\text{-}13)$$

      iii. The plane correction term for the equivalent plane wave

         The diffracted wave of the cylinder is treated as an equivalent plane wave, so it can be revised into a correction term as; it includes 2 parts: one is the correction of N-1 cylinder to the diffraction plane wave of outside incident wave, the other is the correction of the N-1 cylinder to the secondary diffracted wave. Taking consideration of the two corrections overall, we can obtain the following equation:

$$\Phi_{2I2}^j = \frac{gA}{\omega}\sum_{k=1,k\neq j}^N \left\{P_k D\big(r_j,\theta_j,\beta\big) + \sum_{I=1,I\neq k} C_{kI}D\big(r_j,\theta_j,\alpha_{Ik}\big)\right\}\cosh k_0(z+d)\cdot e^{-i\omega t}$$

$$(4\text{-}14)$$

   So, the total velocity potential of the impact of the incident wave on cylinder j is the sum of those three together:

$$\Phi_{2I}^j = \Phi_{2I0}^j + \Phi_{2I1}^j + \Phi_{2I2}^j \qquad (4\text{-}15)$$

   b. Total diffracted wave velocity potential of an isolated vertically floating cylinder group

      In a similar way, the total diffracted wave velocity potential of cylinder j consists of three corresponding compositions. We suppose that the three compositions are: ϕ_{2D0}^j, ϕ_{2D1}^j, ϕ_{2D2}^j, So we can get

$$\Phi_{2D}^j = \Phi_{2D0}^j + \Phi_{2D1}^j + \Phi_{2D2}^j \qquad (4\text{-}16)$$

The three components in Equation (4-16) can be determined respectively from the next equations.

i. If we just consider the diffracted wave velocity potential caused by the external incident wave velocity potential to the cylinder j:

$$\Phi_{2D0}^j = \text{Re}\left[\frac{gA}{\omega}P_j\sum_{m=0}^{\infty}\Phi_{2D}^{(m)}(r_j, z)\cos m(\theta_j - \beta)e^{-i\omega t}\right] \qquad (4\text{-}17)$$

In Equation (4-17), the meaning of the symbols is as before. The function includes a Bessel function; it can be obtained by the next equation:

$$\Phi_{2D}^{(m)}(r_j, z) = -\varepsilon_m i^{m+1}\frac{\cosh k_0(z+j)J_m(k_0 a)}{\cosh k_0 d H_m(k_0 a)}H_m(k_0 r_j) + B_0^{(m)}\frac{H_m(k_0 r_j)}{H_m(k_0 a)}Z_0(z)$$

$$+ \sum_{q=1}^{\infty}B_q^{(m)}\frac{K_m(k_q r_j)}{K_m(k_q a)}Z_q(z) \qquad (4\text{-}18)$$

In Equation (4-18), is the first kind of m rank Bessel function. is the first kind of m rank Henkel function and is the second kind of m rank correct Bessel function. And and are functions derived from the independent variable of, while and and are plural coefficients; we can get them by taking the partial differential of r_j and then using the orthoginality of $Z_q(z)$. ε_m is the Neumann constant, $\varepsilon_0 = 1$, $\varepsilon_m = 2(m \geq 1)$.

ii. We consider the diffracted wave as a secondary incident wave; secondary diffracted wave velocity potential is caused under the action of the second incident wave. The velocity potential can be written as follows according to diffraction theory:

$$\Phi_{2D1}^j = \text{Re}\left\{\frac{gA}{\omega}P_j\left\{S_{kj}(\beta)\exp\left[-ik_0\cos(\alpha_{kj} - \theta_k)\right]\right\} + D_{kj}(\beta)\left[\cosh k_0(z+d)e^{-i\omega t}\right]\right\} \qquad (4\text{-}19)$$

iii. Correction for nonplanar wave

In Equation (4-19) is the correction for a nonplanar wave, its expression is

$$D_{kj}(\beta) = \frac{i}{k_0 R_{jk}}\left\{\begin{matrix}S_{kj}(\beta)\sum_{l=0}^{\infty}(-1)^l l^2\cdot J_1(k_0 r_k)\cos l(\alpha_{kj} - \theta_k) + T_{kj}(\beta)\sum_{l=0}^{\infty}\varepsilon_1(-i)^l l \\ J_1(k_0 r_k)\sin l(\alpha_{kj} - \theta_k)\cdot\end{matrix}\right\}$$

$$(4\text{-}20)$$

In Equation (4-19) and (4-20), is the amplitude equivalent plane wave. And the function of Hankel and Bessel function:

$$T_{kj}(\beta) = \sum_{m=0}^{\infty} \left\{ \frac{B_0^{(m)}\sqrt{2}}{H'_m(k_0 a)\left[1 + \frac{sinhzk_0 d}{zk_0 d}\right]^{1/2}} - \frac{\varepsilon_m i^{m+1}}{coshk_0 D} \cdot \frac{J'_m(k_0 a)}{H'_m(k_0 a)} \right\}$$
$$\cdot H_m(k_0 R_{jk})m\ sinm(\alpha_{jk} - \beta) \tag{4-21}$$

c. Wave load of vertically floating cylinder group

   i. The pressure p_i of any point in the flow field

   After we get the total incident wave velocity potential and total diffracted wave velocity potential from Equations (4-15) and (4-16), we can calculate the total velocity potential in the flow field of cylinder j in the vertically floating cylinder group from Equation (4-15):

$$P_j = -\rho \frac{\alpha \Phi_i^j}{\alpha t} \tag{4-22}$$

   ii. The total wave force of the cylinder j in the cylinder group:

$$F_j = \iint_S p_j \cdot nds \tag{4-23}$$

   iii. The total wave moment of the cylinder j in the cylinder group:

$$M_j = \iint_S p_j \cdot (r \times n)ds \tag{4-24}$$

   In Equation (4-24), r is the vector distance of any point in cylinder j to the torque point. S is the surface the cylinder. n (n_x, n_y, n_z) is the unit external normal vector of the cylinder j. We can refer to Figure 4-41.

   iv. The wave load impact on the cylinder group

   Suppose the cylinder group consist of N cylinders; sum the load impact on each cylinder and the moment point transform to (x_0, y_0, 0), where x_0, y_0 are the horizontal abscissa and longitudinal coordinate of the water plane center point of the column group under the global coordinate system, respectively. So we get the following:
   Total wave load:

$$F_j = \sum_{j=1}^{N} F_{ij} \tag{4-25}$$

In Equation (4-25), i (i = 1, 2, 3) means the different directions of the wave load; 1, 2, and 3 indicate the surging, swaying, and heaving directions respectively.

Total movement:

$$M_{x,y} = \sum_{j=1}^{N} M_{x,y}^{j} \qquad (4\text{-}26)$$

$$M_{z} = \sum_{j=1}^{N} \left(F_{x}^{j} y_{oj} + F_{y}^{j} x_{oj} \right) \qquad (4\text{-}27)$$

In Equations (4-26) and (4-27), x, y, and z are rolling, pitching, and yawing direction, respectively. and are the horizontal abscissa and longitudinal coordinate of local coordinate origin point for the overall coordinates, respectively.

It should be pointed out here that calculating Equations (4-26) and (4-27) above is very complicated and difficult. So we always use the simplified calculation method during a project. This method depends on the cylinder group and single cylinder, which are almost the same in the diffraction problem when looking at the external incident wave velocity potential. The other N-1 cylinders' plane wave potential component to cylinder j is similar to a plane wave. Outer plane correction of the diffraction plane wave can be expressed by the plane wave. So we don't need to calculate the total speed potential of the flow, we just calculate the expression of the isolated wave load on the vertically floating cylinder and then multiply by an appropriate constant. Thus the calculation of the cylinder group is greatly simplified. This simplified calculation method, due to the limited space, is not further introduced here.

4.3.2.2 Radiation Theory Wave Impact on a Large-Scale Floating Upright Cylindrical Group

Large-scale offshore floating structures are in a wave field. The interaction between waves and structures is decomposed into diffraction problems and radiation problems. The diffraction problems have been introduced earlier; we will introduce the radiation problems briefly now.

1. The definition of radiation problems

   When the structure has a slight oscillation motion in a certain mode, it produces an outward radiation wave field in a stable flow field, and the structure is impacted by the radiation wave; this is a radiation problem.

   From the definition of radiation problems, we know it is caused by slight oscillating motions, so the radiation wave load is connected to the structure's acceleration and velocity. We usually consider the acceleration and velocity separately. The radiation load coefficient related to the additional acceleration can be called a mass coefficient while the coefficient related to the additional velocity is called a damping coefficient. The two collectively are referred to as hydrodynamic coefficients. So the radiation problem becomes an analysis of the motion response of offshore floating structures and calculating the hydrodynamic coefficients.

2. The radiation wave velocity potential of an isolated vertically floating cylinder

To explore the floating cylinder radiation effect, we should know the exact solution of the flow field velocity potential. Usually we solve it by using the eigenfunction expansion and gradual matching method. Now, we assume a coordinate system and the division of the basin into a group of cylinders j (j= 1, 2......N). Thus, considering the symmetry of the geometrical axis of a cylinder, among the six modal movements, the rotation around the z-axis is zero, so the radiation movement potential of cylinder j can be written as (p = 1, 2,......5). Next we look at the methods and solving steps:

a. Write down an expression that satisfies the surface condition

For the j cylinder with wet surface conditions S_i, the different modal movements of velocity potential Φ_j^p have different forms, such as:

When $p = 1$

$$
\begin{cases}
\dfrac{\partial \Phi_j^1}{\partial r_j} = v_{1\,j} \cos \theta_j & \left(r_j = a, -h \le z \le 0\right) \\[3mm]
\dfrac{\partial \Phi_j^1}{\partial z_j} = 0 & \left(0 \le r_j \le a, z = -h\right)
\end{cases}
\tag{4-28}
$$

But when $p = 2$

$$
\begin{cases}
\dfrac{\partial \Phi_j^2}{\partial r_j} = v_{2j} \cos \theta_j & \left(r_j = a, -h \le z \le 0\right) \\[3mm]
\dfrac{\partial \Phi_j^2}{\partial z_j} = 0 & \left(0 \le r_j \le a, z = -h\right)
\end{cases}
\tag{4-29}
$$

b. Write down the expression that satisfies the matched condition

As shown in Figure 4-41, within the outland space radiation velocities are represented by Φ_{j1}^p and Φ_{j2}^p; within and outside the region at the interface, the outside the domain of the velocity potential Φ_{j1}^p and Φ_{j2}^p should meet the pressure, mass propagation conservation conditions, so we can write expressions as follows:

$$
\frac{\partial \Phi_{j1}^p}{\partial r} = \frac{\partial \Phi_{j2}^p}{\partial r} \qquad \left(r_j = a, \qquad -d \le z \le -h\right)
\tag{4-30}
$$

$$
\left(\Phi_{j1}^P = \Phi_{j2}^P\right)
\tag{4-31}
$$

c. Velocity potential for infinite series forms

In different basins l (l= 1, 2), corresponding to different control equations and boundary conditions, using the separation of variables method, the speed of the flow field potential $\Phi_{j1}^p(r_j, \theta_j, z)$ can become Φ_{j1}^p characteristics in a different function in the form of an infinite series, namely

$$
\Phi_{jl}^p(r_j, \theta_j, z) = f_j^p \cdot \psi_{jl}^p(r_j \cdot z) \cos r(\theta_j - \chi_p)
\tag{4-32}
$$

In the Equation, $l = 1, 2; p = 1, 2, ..., 5$; when $p = 3, r = 0$, for all others $r = 1$; and $f_j^1 = v_{1j} \cdot a, f_j^2 = v_{2j} \cdot a, f_j^3 = v_{3j} \cdot (d - h), f_j^4 = v_{4j} \cdot (d - h)^2, f_j^5 = v_{5j} \cdot (d - h)^2$, $X_1 = X_3 = X_5, X_2 = X_4 = \frac{\pi}{2}$.

d. Write down the characteristic function expression of the internal and external areas
As shown in Figure 4-41, the characteristic function ψ_{j1}^p of internal area 1 in the flow field should be written as

$$\psi_{j1}^p(r_j, z) = \frac{1}{2} A_{j0}^p \left(\frac{r_j}{a}\right)^r + \sum_{n=1}^{\infty} A_{jn}^p \frac{I_r\left(\frac{n\pi r}{d-h}\right)}{I_r\left(\frac{n\pi a}{d-h}\right)} \cos\left[\frac{n\pi(z+d)}{d-h}\right] + A_j^p \tag{4-33}$$

So the characteristic function ψ_{j2}^p of external area 2 in the flow field should be written as

$$\psi_{j2}^p(r_j, z) = B_{j0}^p \frac{H_r(k_0 r)}{H_r'(k_0 a)} Z_0(z) + B_{j1}^p \sum_{q=1}^{\infty} \frac{K_r(K_q r)}{K_r'(k_q a)} Z_q(z) \tag{4-34}$$

In the function, $H_r(x)$ is the first kind of r order Hankel function, $K_r(x)$ is the second kind of r order Bessel function. k_q is the positive root of the equation $\omega^2 + gk_q \tan k_q = 0$ $(q \neq 1)$, and $Z_q(z)$ is a function of orthogonalization in the interval $[-d, 0]$. It is defined as

$$\begin{cases} Z_0(z) = N_0^{-\frac{1}{2}} \cosh k_0(z + d) & (q = 0) \\ Z_q(z) = N_0^{-\frac{1}{2}} \cosh k_0(z + d) & (q \geq 1) \end{cases} \tag{4-35}$$

In the equation,

$$\begin{cases} N_0 = \left[1 + \frac{\sinh 2k_0 d}{2k_0 d}\right] / 2 \\ N_q = \left[1 + \frac{\sin 2k_q d}{2k_q d}\right] / 2 \end{cases} \tag{4-36}$$

e. Determinate the coefficient A_j^p and the expression of A_{jn}^p and B_{jl}^p
A_j^p is the particular solution of an inhomogeneous boundary condition of the outland radiation wave velocity potential for defined conditions, corresponding to different modes. It can be concluded that the corresponding solutions in order are

$$
\begin{cases}
A_j^1 = A_j^2 = 0 \\[2mm]
A_j^3 = \dfrac{1}{2(d-h)^2}\left[(z+d^2) - \dfrac{r_j^2}{2}\right] \\[4mm]
A_j^4 = A_j^5 = \dfrac{r_j}{2(d-h)^3}\left[(z+d^2) - \dfrac{r_j^2}{4}\right]
\end{cases}
\tag{4-37}
$$

If we set order $r = a$ in Equation (4-33), we can determinate the equation of the coefficient A_{jn}^p by using the orthogonality of the trigonometric function. It should be written as

$$
A_{jn}^p = \frac{2}{d-h}\int_{-d}^{-h}\left[\psi_{j1}^p(r_j,z) - A_j^p(r_j,z)\right]\Bigg|_{r_j=a} \cdot \cos\frac{\pi n(d+z)}{d-n}\,dz \qquad (n=0,1,2,\cdots)
\tag{4-38}
$$

In a similar way, we take the partial differential of r in Equation (4-34), $r = a$. We can determinate the equation of the coefficient B_{jl}^p by using the orthogonality of $Z_q(z)$. It should be written as

$$
B_{jq}^p = \frac{1}{k_q d}\int_{-d}^{0}\frac{\partial\psi_{j2}^p(r_j,z)}{\partial r_j}\Bigg|_{r_j=a} \cdot Z_q(z)\,dz \qquad (q=0,1,2,\cdots)
\tag{4-39}
$$

According to the matching conditions in (4-30) and (4-31), ψ_{j1}^p in Equation (4-38) and ψ_{j2}^p in Equation (4-39) shall meet the conditions as follows:

$$
\psi_{j1}^p = \psi_{j2}^p \qquad \left(r_j = a,\ -d \le z \le -h\right)
\tag{4-40}
$$

$$
\frac{\partial\psi_{j1}^p}{\partial r_j} = \frac{\partial\psi_{j2}^p}{\partial r_j} \qquad \left(r_j = a,\ -d \le z \le -h\right)
\tag{4-41}
$$

Putting Equation (4-40) into Equation (4-38), and integrating and consolidating, we get

$$
A_{jn}^p + \sum_{q=1}^{\infty} F_{nq}^p B_{jq}^p = R_{jn}^p \qquad (n=0,1,2,\cdots,\infty;\ p=1,\cdots,5)
\tag{4-42}
$$

In the same way, putting Equation (4-41) into Equation (4-39), we get

$$
B_{jq}^p + \sum_{n=1}^{\infty} G_{qn}^p A_{jn}^p = S_q^p \qquad (q=0,1,2,\cdots,\infty;\ p=1,\cdots,5)
\tag{4-43}
$$

In the equation above, R_{jn}^p is the scale function of a, d and h; F_{nq}^p is a function of $K_r(x)$, which is the second kind of r order transmutative Bessel function; G_{qn}^p is a function of $K_r(x)$, which is the first kind of r order correctional Bessel function; and S_q^p is a scale function of d, h, its sine trigonometric function, and hyperbolic sine and cosine function (their expression is not listed here).

f. The radiation wave potential of the motion modal

With simultaneous solution of (4-42) and (4-43), we can get the infinite order equations of the A_{jn}^p and $l = 1$, 2. If we consider the mathematical feature of the Bessel function in every equation, and guarantee the accuracy, we can solve the finite order equations after appropriately cutting off these infinite order equations, which is equivalent to the Bessel function oscillation failure reduction. Generally speaking, we choose $n, q \in [0, 20]$, which can satisfy the accuracy requirements. Through solving these finite order equations, we can calculate the coefficients A_{jn}^p and B_{jq}^p, and separately put them into Equations (4-33) and (4-34); we can get the characteristic functions ψ_{j1}^p and ψ_{j2}^p. Then we put ψ_{j1}^p and ψ_{j2}^p into Equation (4-32), and we can get radiation wave potential Φ_j^p of every motion modal $(p = 1, 2, \cdots, 5)$

3. Radiation wave load of the isolated vertically floating cylinder

When the column j makes slight periodic oscillatory motions on the p motion modal, it will cause the movement of the surrounding fluid, forming radiation waves that spread outside. This radiation wave acting on the columns will cause fluidic force and torque, which is the wave load of the radiation wave.

a. Any point pressure p_j^p in the flow field caused by the radiation wave

The radiation wave velocity potential $\Phi_{jl}^r(r_j, \theta_j, z)$ $(l = 1, 2)$ in the flow field space is calculated according to the above theory; we can get any point pressure p_j^p or $p_{jl}^p(r_j, \theta_j, z)$ $(l = 1, 2)$ in the flow field by using the Bernoulli equation as follows

$$p_{jl}^p = -\rho \frac{\partial \Phi_{jl}^p}{\partial t} = -i\rho\omega f_j^p \cdot \psi_{jl}^p(r_j, z) \cdot \cos r(\theta_j - \chi_p) e^{i\omega t} \tag{4-44}$$

b. The force of the radiation wave

Suppose that F_j^{pq} is the force of the radiation wave forming at the direction of the q modal when the column j has slight periodic oscillatory motion on the p motion modal. We have $p = 1, 2, 3, q = 1, 2, 3$, and when $l = 1, Z = -h$; when $l = 2, r = a$. We take the integral along the wet surface S on the cylinder, so we can get F_j^{pq} as

$$F_j^{pq} = \iint_{S_j} p_j^p \cdot n_j ds \qquad (p = 1, 2, 3, \quad q = 1, 2, 3) \tag{4-45}$$

c. The torque of the radiation wave.

Suppose M_j^{pq} is the torque of the radiation wave forming at the direction of the q modal when the column j makes slight periodic oscillatory motion on the p motion modal. According to the same method as above, we can get

$$M_j^{pq} = \iint_{S_j} p_j^p \left(n_i \times r_j\right) ds \quad (p = 1, 2, \cdots 5; \quad q = 4, 5 \ OR \ p = 4, 5; \quad q = 1, 2, 3) \tag{4-46}$$

In Equations (4-45) and (4-46), n_j is the unit inner normal vector of the column j; r_j is the vector distance between any point on the surface of the column j and action point of the torque; and S_j is the wet surface of column j.

4. Radiation wave velocity potential of the vertically floating cylinder group

The key to solving the problem of radiation wave velocity potential of the vertically floating cylinder group is how to determine the mutual hydrodynamism of the radiation wave between the various columns, so as to get the complete radiation wave velocity potential of the cylinder group. Moreover, this mutual hydrodynamism is mainly derived from the radiation wave caused by every modal motion of every column. To seize this key problem, first we must get a radiation wave caused by the single modal motion of a column; second, we introduce a large gap hypothesis that intercolumniation R_{jk} is larger enough than the wavelength of the radiation wave L. Then we can use the equivalent plane traveling wave to describe the interaction of radiation between each column. That is, the incidence function of the other round column basin caused by the radiation wave, which is caused by small amplitude periodic oscillatory motion of any cylinder with a frequency ω working the p motion modal in the cylindrical group, is replaced by the equivalent plane wave of a nonplanar correction term. So we can get the velocity potential of the incident wave. Last, considering the diffraction of the incident wave, we can get the velocity potential of the diffracted wave from the diffraction theory. Thus, the velocity potential of the incident wave and the velocity potential of the diffracted wave of the cylinder outside the column producing radiation wave are obtained in the cylindrical group. So we can obtain the total velocity potential. Based on the above ideas, the solution steps are as follows:

a. Radiation wave velocity potential produced by the column j

When the column j makes slight periodic oscillatory motion on the p motion modal, it will produce a corresponding radiation wave velocity potential acting on the outland 2. When it is expressed by Φ_{j2}^p, we can obtain

$$\Phi_{j2}^p = f_j^p \left[B_{j0}^p \frac{H_r(k_0 r_j)}{H'_r(k_0 a)} Z_0(z) + \sum_{q=1}^{\infty} B_{jq}^p \frac{K_r(k_q r_j)}{K'_r(k_q a)} Z_q(z) \right] \cdot \cos r\left(\theta_j - \chi_p e^{-i\omega t}\right) \tag{4-47}$$

Equation (4-47) is obtain through Equations (4-34) and (4-32).

b. The column j radiation wave as the velocity potential of the incident wave of the column k in the cylinder group.

We can transform the radiation wave velocity potential produced by column j into a local coordinate system $O_k r_k \theta_{kz}$ of the column k, which has the same radius a as the column j. The intercolumniation of the column k is R_{jk}. Using a large gap hypothesis, adopting the method of equivalent plane wave, and considering the nonplanar correction term, we can get the radiation wave produced by the p modal motion of the column j in the cylinder group. When the column k is the incident wave, its velocity potential $_j\Phi_{kl}^p$ is

$$
{}^{j}\Phi^{p}_{kl} = f_{i}\left[D^{p}_{kj} + \sum_{l=1,l\neq k}^{N} {}_{j}C^{p}_{kj}\cdot e^{-ik_{0}r_{k}\cos(\alpha_{kj}-\theta_{k})j}\right]\cdot \cosh k_{0}(z+d)\cdot e^{-i\omega t} \tag{4-48}
$$

In Equation (4-48), D^{p}_{kj} is the nonplanar wave amplitude; it consists of two parts. One part is the radiation wave amplitude caused by the p modal motion of column j in the basin of column l. The other is the equivalent plane wave amplitude of the incident wave caused by the diffracted wave of the column l produced by the radiation wave of the other column in the basin of column k. They can be obtained by using the theory that the radiation wave and the diffracted wave are the equivalent plane wave and the incident wave, respectively. f_{j} and f^{p}_{j} refer to the calculation method for the f^{p}_{j} in Equation (4-32).

c. ${}_{j}\Phi^{p}_{k1}$ and ${}_{j}\Phi^{p}_{k2}$, the diffracted wave velocity potential of the internal and external areas of column k in the cylinder group

In internal area 1:

$$
{}_{j}\Phi^{p}_{k1} = f^{p}_{j}\cdot e^{-i\omega t}\left\{A\delta_{jk}\psi^{p}_{k1}\cos r(\theta_{k}-\chi_{p}) + \sum_{l=1,l\neq k}\sum_{m=0}^{\infty}\left[\frac{1}{2}C^{k}_{0m}\left(\frac{r_{h}}{a}\right)^{m}\right.\right.
$$
$$
\left.\left. + \sum_{n=1}^{\infty} C^{k}_{nm}\frac{I_{m}[n\pi r_{k}/(d-h)]}{I'_{m}[n\pi r_{k}/(d-h)]}\cos\left(\frac{n\pi z}{d-h}\right)\right]{}_{j}V^{p}_{kl}(m)\right\}A \tag{4-49}
$$

In the external area 2:

$$
{}_{j}\Phi^{p}_{k2} = f^{p}_{j}\cdot e^{-i\omega t}\left\{A\delta_{jk}\psi^{p}_{k2}\cos r(\theta_{k}-\chi_{p}) + \sum_{l=1,l\neq k}\sum_{m=0}^{\infty}\left[\frac{1}{2}D^{k}_{0m}\cdot\frac{H_{m}(k_{0}r_{k})}{H'_{m}(k_{0}a)}\cdot Z_{0}(z)\right.\right.
$$
$$
\left.\left. + \sum_{n=1}^{\infty} D^{k}_{nm}\frac{K_{m}(k_{n}r_{k})}{K'_{m}(k_{n}a)}Z_{n}(z)\right]\cdot{}_{j}V^{p}_{kl}(m)\right\}A \tag{4-50}
$$

Equations (4-49) and (4-50) are obtained by using the methods of Equations (4-17) to (4-21), according to diffraction theory. The symbols in the equations have the same meaning. δ_{jk} is the Koneke constant in the equation. When $j\neq k$, $\delta_{jk}=0$; $j=k$, $\delta_{jk}=1$. ${}_{j}V^{p}_{kl}(m)$ can be obtained by

$$
{}_{j}V^{p}_{kl}(m) = {}_{j}C^{p}_{kl}\cos m(\theta_{k}-\alpha_{lk}) + \frac{i}{2k_{0}R_{kl}}\cdot[m^{2}\cos m(\theta_{k}-\alpha_{lk})]S^{p}_{kl}
$$
$$
+ 2m\sin m(\theta_{k}-\alpha_{lk})T^{p}_{kl} \tag{4-51}
$$

5. Radiation wave load of the vertically floating cylinder group

   Like the radiation effects of the single cylinder, when the column j makes slight periodic oscillatory motion on the p motion modal, it will lead to fluid motion in the flow field of the whole cylinder group. That is to say it will cause the outside spread function of the radiation wave. In this way, the radiation wave will produce force and torque acting on the moving column j and the other N-1 columns. That is the wave load of the radiation wave.

a. $_jp^p_{kl}$ or $_jp^p_{kl}(r_k, \theta_k, z, t)$, the dynamic pressure of any point in the flow field.
We can get the expression after ignoring the water pressure using the Bernoulli equation:

$$_jp^p_{kl} = -\rho \frac{\partial_j \Phi^p_{kl}}{\partial t} \tag{4-52}$$

In Equation (4-52), $l = 1, 2$; when $l = 1$, $z = -h$ and when $l = 2$, $r = a$.

b. The wave force $_jF^{pq}_k$ is the radiation wave of the column j making slight periodic oscillatory motion on the p motion modal acting on the q modal direction of the column k.
If we integrate the pressure $_jp^p_k$ integration along the wet surface S_k of the column k, we can get

$$_jF^{pq}_k = \iint_{S_k} {_jp^p_k} \cdot n_q ds \tag{4-53}$$

c. The wave torque $_jM^{pq}_k$ is the radiation wave of the column j making slight periodic oscillatory motion on the p motion modal acting on the q modal direction of the column k.
Make r_k the radius vector between any point on the surface of column k and action point of the torque. So we can write

$$_jM^{pq}_k = \iint_{S_k} {_jp^p_k} \cdot (n_k \times r_k) ds \tag{4-54}$$

In Equations (4-53) and (4-54), n_k is the unit inner normal vector of column k. S_k is the wet surface of column k.

4.3.2.3 Wave Load Calculation for the Four-Column Type Tension Leg Platform Body

The four-column type tension leg platform is shown in Figure 4-42, and the main geometrical and physical parameters are shown in Table 4-11. We take the wave incident angle β separately as

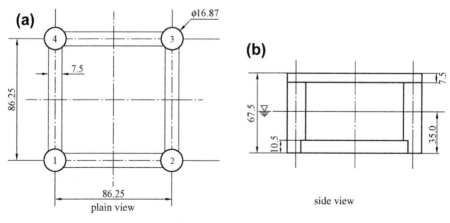

FIGURE 4-42

The mechanical structure of the four-column tension leg platform.

Table 4-11 The Main Geometrical and Physical Parameters of the Four-Column Tension Leg Platform Type

Parameter Name	Symbol	Unit	Value
The center distance between the columns	R	m	86.25
Column radius	α	m	8.435
Total height	H	m	67.50
Working depth of immersion	T_D	m	35.0
Displacement	Δ	t	54500
Platform weight	W	kN	405000
The total tension prestressing of the tension leg	T_0	kN	137340
Transverse metcentric height of the platform	GM_T	m	6.0
Longitudinal metcentric height of the platform	GM_L	m	6.0
The quality of the platform	M	t	4130
The transverse swing and longitudinal swing rotational inertia of the platform	I_x, I_y	$t \cdot m^2$	8.4×10^7
The vertical swing rotational inertia of the platform	I_z	$t \cdot m^2$	1.0×10^8
The height between the barycenter and the cellar deck	H_G	m	38.00
The initial length of the tension leg	l_0	m	415.0
The total vertical stiffness coefficient of the tension leg	K_{zz}	kN/m	8.044×10^5
The total transverse swing and longitudinal swing stiffness coefficient of the tension leg	$K_{\theta\theta}, K_{\Phi\Phi}$	kN·m/rad	1.5×10^9
The vertical swing stiffness coefficient of the tension leg	$K_{\psi\psi}$	kN·m/rad	2.12×10^9

$0°, 30°, 45°$. The frequency scope is $\omega = 0.2 \sim 1.7 rad/s$. We will calculate the maximum amplitude of the dimensionless wave load for different angles of incidence, and analyze the wave load at the angle of incidence $\beta = 0°$.

1. The maximum amplitude of the dimensionless wave load for different angles of incidence. According to diffraction theory, we calculated the maximum amplitude of the dimensionless wave load for different angles of incidence by using existing TWDF calculation software. For example, the computed result for column 1 is shown in Table 4-12.

Table 4-12 The Maximum Amplitude of the Dimensionless Wave Load for the Different Angles of Incidence of Column 1

	F_{xmax}	F_{ymax}	F_{zmax}	M_{xmax}	M_{ymax}
$\beta = 0°$	2.022	0.399	0.99	0.444	2.710
$\beta = 30°$	1.757	1.151	0.99	1.62	2.407
$\beta = 45°$	1.60	1.60	0.99	2.11	2.11

According to this calculated data, we can see the following:

a. For the double column against the waves in the anterior column (i.e., column 1), when the incident wave angle $\beta = 0°$, F_y and M_y are both maximum.

b. When the incident angle of the wave $\beta = 45°$, F_y and M_x are both maximum.

c. F_z is not affected by the incident angle of the wave.

2. The calculation of the dimensionless wave load for different angles of incidence on the four columns

Also using TWDF calculation software, we calculate the dimensionless wave load for different angles of incidence on the tension leg platform's four columns. Taking the incident angle $\beta = 0°$ as an example, its computation result is shown in Figure 4-43. From the Figure 4-43 amplitude-frequency curve, we can see:

a. When $\beta = 0°$, the four-column tension leg platform has a corresponding wave load at the modal direction of longitudinal oscillation, transverse oscillation, vertical oscillation, transverse swing, longitudinal swing, and vertical swing.

b. The change function of the wave load is different depending on the column because of the symmetry of structure. The change functions of the wave load on the first and fourth columns are the same. The change functions of the wave load for the second and third columns are the same.

c. The hydrodynamic interaction between each column can be divided into the interference effect and the shadowing effect obviously. The front column mainly has a shadowing effect on the back column, so as to make the wave load acting on the back a little less than the wave load of the single column. However, the back column mainly has an interference effect on the front column. The wave interference force and torque of the front column oscillate violently around the single column load curve.

d. When $\beta = 0°$, the wave load of the four column group combination has a significant interference effect and shadowing effect in the modal of longitudinal oscillation, transverse oscillation, transverse swing, and longitudinal swing. However, the vertical direction of the wave load is less affected by group of column.

4.3.3 Strength and Fatigue Life of Deep-Water Offshore Floating Structures

We mainly introduce the load under wave action for deep-water offshore floating structures above. The opposite of the load is the strength and fatigue life of the structure body. If the strength is greater than the load and the fatigue life is greater than accumulation of fatigue damage, the structures are safe and reliable. In order to ensure the safety and reliability of offshore floating structures during deep water

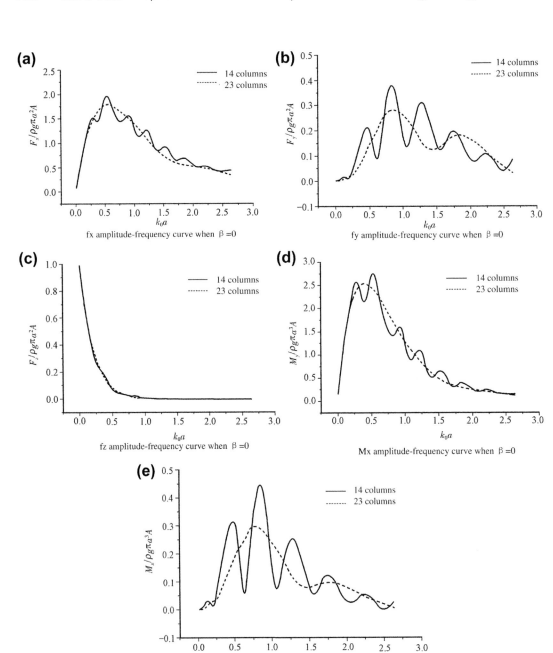

FIGURE 4-43

At $\beta = 0°$, the amplitude-frequency curve of the dimensionless wave load on the four columns.

work, we must make the structures' strength and fatigue life sufficient. However, in deep water, the traditional jacket platform and gravity platform are replaced by the tension leg platform, single column (SPAR) platform, and semi-submersible platform, which are more suitable for working in deep water. And the design of the strength and fatigue life has some different features from jacket platforms that use traditional methods. So we introduce below this special question regarding the strength and fatigue life.

4.3.3.1 The Strength and Fatigue Life of the Deep-Water TLP and SPAR Platforms

We consider some special problems as follows:

1. Dynamic response

   The dynamic response is the base of research of the fatigue life of the TLP and SPAR platforms. The dynamic response of the deep-water platform should consider three responses as follows:
   a. First-order wave frequency response

      This is the basic motion response, which relates to the sea stability of the deep-water floating structures. We can use radiation theory and an analytic method or boundary element method based on the Green function to get the first-order wave frequency response of the TLP and SPAR platform.
   b. The second-order difference frequency and sum frequency response

      The so-called difference frequency represents two kinds of waves with similar frequency whose frequency difference is very small, so as to produce low-frequency excitation, which leads to the slow drift motion of the platform. On the other hand, the sum frequency is a definition that represents two kinds of waves where the frequency difference can be quite large, so as to lead to vertical high-frequency jumping movement of the platform. A deep-water offshore floating structure produces obvious second-order nonlinear difference frequency and sum frequency movement, in addition to first-order wave frequency movement. These two kinds of movements both possibly take place for the TLP. But there is obvious difference frequency movement for the SPAR platform. They are both related to the fatigue life of the platform. Two methods are often used to solve this second-order nonlinear problem. We can use the direct solving method to get the second-order velocity potential. But the amount of calculation is large. Or we can use the indirect solving method where we get the second-order flow field velocity potential using a kind of adjuvant radiation wave potential. This method reduces the complexity of the problem, but brings the difficulties of a free surface integral.
   c. Higher frequency impulse response (ringing)

      The TLP will ring under a windstorm. It shows a sudden increase of tension leg tension, but a slow decay, like church bells, so it is called ringing. Because it belongs to the category of higher order nonlinear problems, it is more difficult to solve it.
2. The partial strength.

   TLP and SPAR platform in deep water is consisted by the stiffened column and board with larger diameter. This kind of structure is quite different from the jacket platform. Therefore, when analyzing partial strength, we should pay attention to the following problems:
   a. The shell unit instead of the space beam unit is used to analyze the partial strength
   b. The connection among the main modules of the platform often has a certain relative motion instead of a firm weld. Full attention should be paid to this feature of the compliant platform under the analysis of partial strength.

   c. The reinforcing rib of the noumenon structure makes the analysis of the TLP and SPAR platform become complicated. For example, the buckling problem of the stiffened cylindrical shell should be solved under the analysis of the partial strength.

   d. We should use an analytic method and finite element method to do fracture mechanics analysis for some key nodes.

3. The fatigue life.

The analysis of the fatigue life of the deep-water platform is different from the linear system analysis method of the jacket platform. Its special points are as follow:

   a. The fatigue problem caused by the second-order or higher order nonlinear motion can't be ignored.

   b. The fatigue reliability analysis method based on linear system transformation cannot be used for adeep-water platform.

   c. We must pay much attention to the fatigue problem caused by vortex-induced vibration. For example, the tension leg of the TLP platform can produce vortex-induced vibration. It could make both transverse vibration and line vibration. They should be used in the wake oscillator model or the corresponding model. They are based on the continuous function of the column amplitude to express the relevant lift amplitude acting on the columns, so as to make further fatigue analysis. The cylindrical floating body of the SPAR platform, which can produce vortex-induced vibration under certain conditions and its fatigue life should also be analyzed.

4. The coupling effect.

The wave load acting on the jacket platform can be determined by the wave field and has nothing to do with the platform response. However, the deep-water platform is different. The coupling effect between the load and the dynamic response of the TLP and SPAR platform must be considered. The existing analysis software now usually applies linear analysis and second-order frequency domain solutions to solve the coupling analysis problem. But this method is too conservative to design the platforms, as sometimes the safety coefficient is greater than 10. Therefore, fully considering the finite displacement, instantaneous wet surface, viscous force, etc., it is very necessary to develop a reasonable and economic kind of software which is based on an accurate nonlinear dynamic response and the analysis of the load for deep-water platform strength and fatigue life design.

4.3.3.2 The Strength and Fatigue Life of the Deep-Water Semi-submersible Platform

The deep-water semi-submersible platform updates faster to meet the needs of the deep-water development. At present, take the most advanced fifth generation semi-submersible platform in the world, the BINGO-9000, as an example; its working depth reaches 3000 m, and it can work under severe sea conditions with Low temperature of $-20°C$; wind speed 40 m/s; and wave height 32 m. Therefore, in order to guarantee safe and reliable operation of the platform, we need to correctly and reasonably design the strength and fatigue life of the platform. For the fifth generation semi-submersible platform, the design idea is that the platform structure strength design must be in line with the fatigue life analysis.

   1. Calculation of the platform load

The wave load is the main portion of the environmental load of the semi-submersible platform. It has the ability to test the total strength of the platform. It usually takes a once-in-a-century maximum wave to check the total strength of the platform, according to the regulations of the classification society in the relevant specification. So we no longer need consider the effect of the wind and the flow. There are two methods to deal with the wave load.

2. Wave design method

   We choose the maximum wave the platform might encounter as the design wave to calculate the platform movement, load, and stress components, according to the environment conditions and design requirements of the area where the platform is located. The specification usually states to use the once-in-a-century maximum wave.

3. The spectra analysis method.

   It regards the wave as a stationary Gaussian random process with zero mean. Its statistical properties can be determined by the power spectrum. According to the det norske veritas (DNV) rules, we can use the Pierson-Markowitz spectrum (P-M spectrum) as the wave spectrum, which applies to an infinite wind area and a fully developed wave. We make the significant wave height H_s and the average across zero cycle T_z parameters to the expression of the P-M spectrum $S(\omega)$:

$$S(\omega) = \frac{H_s^2 4\pi^3}{\omega^5 T_z^4} \exp\left(-\frac{16\pi^3}{\omega^4 T_z^4}\right) \tag{4-55}$$

If we take the platform as a linear system, its response spectrum can be equal to the wave spectrum multiplied by the system response amplitude operator (RAO). Therefore, the stress spectrum density $S_R(\omega)$ of structural elements can be written as

$$S_R(\omega) = S(\omega) \cdot |H(\omega)|^2 \tag{4-56}$$

In Equation (4-56), $H(\omega)$ is the stress response transfer function. $|H(\omega)|^2$ is the response amplitude operator. Suppose the n-order matrix of the stress spectrum is m_n. We can write

$$m_n = \int_0^\infty \omega^n S_R(\omega) d\omega \tag{4-57}$$

So we can get each order matrix of the stress spectrum from Equation (4-57). We can get the statistical characteristic value of the random stress as follows:

$$\left\{\begin{array}{ll} \text{the significant wave height} & H_s = 4\sqrt{m_0} \\[2em] \text{cross the zero rate} & f_0 = \dfrac{1}{2\pi} \\[2em] \text{bandwidth coefficient} & \varepsilon = \sqrt{1 - \dfrac{m_2^2}{m_0 m_4}} \end{array}\right. \tag{4-58}$$

It should also be noted that there is a big difference between the forces of the waves acting on the platform under different wave directions, different wave periods, and different wave phases. Therefore, when we calculate the wave load, we should select several different wave directions, cycles, and crest locations in different phase loads of the platform. And then choose one of the most unfavorable situations to analyze at the next step.

4. Analysis of the overall structure strength

The next step is to analyze the overall structure strength after the wave load calculation for the platform. We should consider the following problems at this time:

a. Working conditions

The working conditions of the semi-submersible platform can be divided into three kinds: drilling, self-storage, and towing. The drilling condition refers to the function of the hook load, i.e., the hook hanging over the entire drilling tool.

b. Bearing load status

This refers to the platform under load conditions, usually with the following:

i. Torsion status

This status occurs where the wave torque of the platform is the largest, when the oblique wave and wave enter the platform parallel to the diagonal direction of the platform. In this status, we should check the anti-shear strength of the platform floating body and columns and the total torsion deformation of the platform.

ii. Beam sea status

This is the status where the platform is bearing the maximum transverse force under the effect of a transverse wave. Then, when the peak is close to the center line of the platform, the floating bodies on both sides of the platform suffer the outward separation force. When the valley is close to the center line of the platform, the floating bodies on both sides of the platform suffer the inward extrusion force. Therefore, we should take the situation when the transverse separation force and the transverse extrusion force both reach the maximum to check the strength of the floating body and each column.

iii. Sag status

This is the status of maximum total longitudinal buckling. We should consider the status of the maximum hook load under the drilling operation, and the maximum total longitudinal buckling under the self-storage status and the towing status. When the platform bears the maximum total longitudinal buckling load, we mainly check the total longitudinal strength of the platform. It's based on the bending moment and shear force, to check the shear strength and bending ability of the platform floating bodies and columns. We should choose different wave periods and wave phases to calculate and select the working conditions of the maximum total bending moment for verification. According to the verification of the strength of the BINGO-9000 semi-submersible drilling platform, the partial stress produced by the outward plate of the floating body becomes larger under water stress and wave action, especially the floating body bone frame structure. In addition, the shear stress and bending stress is especially large through the entire deck and bulkhead structure where they intersect and connect. And the load on the diagonal strut resisting upper deck deformation is relatively large.

In a word, these parts are the important points of the strength verification for the sag status.

iv. Still water status

This is the status where the load of the platform is under the action of the hydrostatic pressure. Because this loading status is less serious than others, it is not commonly an important point of the strength verification.

v. Rising and sinking status

This is the status for reasons that make the platform rise and sink. Although the platform is in still water, when the emergency occurs like a surge or tsunami, it can lead the platform to rise and sink. At this time, the platform produces a certain acceleration in the upward direction, so as to suffer an inertia force action which is consistent with the direction of gravity; with the drilling operation and the hook load acting on the derrick, this inertia force is even more serious. Therefore, we must consider the inertia force when we check the strength under this bearing status. We can usually choose the acceleration value $1.5 \sim 2.5$ m/s^2 (vertical direction) for the calculation of the inertia force. Because the platform is in the status of speeding up and rising at this time, the pressure which the upper deck structure of the platform bears is very big, and the columns of the platform also bear greater pressure. So we should check the strength of these two parts emphatically.

Finally, it should be noted that when we analyze the overall structure strength as mentioned above we shall use practical engineering requirements to set up the drilling, self-storage and towing working conditions to calculate. However, each kind of condition also needs five kinds of bearing state to calculate, so we can make the total strength of the platform meet the requirements of all kinds of working and loading condition and loading conditions. We use the ANSYS finite element software to build a space plate girder composite structure mechanics model to analyze the platform main body. The model is shown in Figure 4-44.

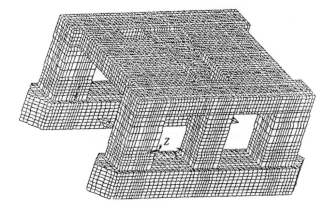

FIGURE 4-44

Finite element model of the BINGO-1000 semi-submersible drilling platform.

5. Analysis of the column buckling strength

The columns of the semi-submersible drilling platform mainly bear the gravity of all the structures and equipment above the deck, except the action of the static water pressure and wave force. Therefore, when we calculate the strength of the columns, we must consider the stability of the structure and analyze its buckling strength, and the important problems are as follows:

a. The working and load conditions and load condition.

Because the columns mainly bear the deck load and the deck has a heave motion, we usually select the load which the platform bears with the heave status of upward acceleration motion as the calculation basis for the drilling working condition when we analyze the strength of the columns. This is because the load the columns bear is the largest at this time.

b. The partial finite element model

We establish the finite element model for the main body of the middle column and back angle column on the right side and the partial part from the fourth deck to the seventh deck. The boundary conditions of the partial model are obtained from the result of the whole model calculation. That is, we make the displacements of all the boundary nodes of the partial model that are obtained from the result of the whole model calculation the boundary conditions of the partial model.

c. Buckling point analysis

According to the strength analysis of the columns of the BINGO-9000 semi-submersible drilling platform, the location of the large stress component amplitude focuses on the inboard angle columns close to the deck and platform. The vertical compressive stress in the columns, which have the largest influence on post buckling, is much larger than any other stress values. Therefore, the point of the buckling strength analysis is aiming at the maximum vertical pressure based on the relevant provisions for buckling strength analysis of the relevant standards group.

6. Analysis of the floating body ultimate strength

The so-called analysis of the floating body ultimate strength is a strength analysis when the structure reaches the ultimate state under extreme loading conditions. For a large semi-submersible platform working in the deep sea, sometimes the sea conditions are difficult, which makes the platform suffer extreme load caused by once-in-a-century sea conditions. At this time, the bending moment on the platform is particularly large. Therefore, the space steel frame structure composed by platform pole structure more easily leads to the destruction under the condition that some structures are instable or lose effectiveness. To this end, some classification society rules state that one must do a destructive strength verification and strength calculation under the damaged condition. Here, the strength calculation base on the damaged condition, which is the structure of the whole platform is damaged caused by ultimate strength. We mainly consider the following for the analysis of the damaged floating body ultimate strength:

a. Establishing the analysis model

According to the study of the floating body structure damage state, first we set up the requirements to meet the ultimate strength analysis, and then establish the mathematical mechanical model of the main structure of floating body and the main structure of floating body after being damaged.

b. Load calculation

When we analyze the ultimate strength, the load calculation of the main floating body should be focused on these two structure morphologies:

i. Normal working condition: The floating body is in its complete state.

ii. Unconventional working condition: The floating body is damaged, which leads to flooding and yielding water.

We should calculate the still water load and wave load for these two working conditions, calculate the most dangerous working condition, and get the corresponding maximum value of the still water load and wave load.

c. Bearing capacity calculation

If we calculate the ultimate bearing capacity of the structural beam, first we should use simplified analytic method to get the ultimate bending moment of the structural beam, and then use the ideal element method to analyze the nonlinear finite element.

d. The analysis of the ultimate strength.

When we analyze the ultimate strength of the floating body, we can use three-dimensional finite element analysis method to establish the 3D model of the floating body structure. We can also use the simplified progressive damage analysis method, which is often used in the ship domain; this method is especially suitable for analyzing the failure process of a ship body section step by step after the ship body beam suffers bending step by step.

7. Analysis of the tubular joint partial stress

The bracing tube structure is the weak link of the whole strength of the platform. Because we simplify the bracing tube to a beam element under the complete analysis, we cannot get a partial stress analysis of the joint part between the tubular joint and the bracing tube. So we need to analyze the partial stress of the tubular joint independently. The main methods are:

a. Building the partial model

When we analyze the partial stress of the tubular joint, we should build a quadratic partial model first. This model should contain the toggle plate and the axial stiffeners in the horizontal transverse bracing tube. It should use the four-point shell element for simulation. Near the intersecting line of the horizontal diagonal bracing tube and the horizontal transverse bracing tube, we should use unit partial encryption to get the accurate stress value of the part near the intersecting line. Figure 4-45 gives the finite element model of the horizontal bracing tube K-type tubular joint.

b. Determining boundary conditions

When we analyze the stress of the partial model, we should make the terminal displacement of the partial model obtained in the whole analysis the boundary conditions.

c. Calculating the load stress

Under the condition that the terminal displacement of the partial model obtained in the whole analysis is taken as the boundary condition, we should consider the self-weight of the tubular joint itself and the effect of the wave load to calculate the stress of the partial model.

8. Analysis of the floating body fatigue life

The fatigue analysis of the semi-submersible drilling platform mainly focuses on two aspects: on the one hand it is a floating body structure. It is similar to a main hull structure. On the other

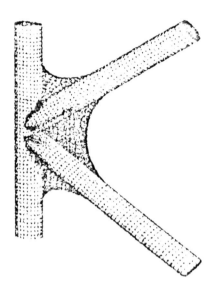

FIGURE 4-45

Finite element model of the horizontal bracing tube K-type tubular joint.

hand it is a cumulative fatigue damage calculation of the tubular joint in the special areas. Now, we research the first part as follows:

a. Selecting the fatigue "hotspot"

Verify the fatigue life of the whole floating body structure, and check the fatigue life of the under deck of large structure terminal. However, we must select the coupling end of the strength member in a large number of different parts to do the analysis of the fatigue life. Obviously, this requires a lot of work. Therefore, we must correctly choose the coupling end of the strength member of the most unfavorable position to check the fatigue life. This is the so-called the fatigue "hotspot." We usually selects three fatigue "hotspots" as the typical sample to analyze the fatigue life in the middle part of the floating body structure.

b. Structure calculation model

In order to calculate the response of the hull under the load, we must build the finite element calculation model of the whole structure. When building the model, ensure the boundary have little impact on the stress near the check point. And the structures near the check point should be a local network encryption. When we calculate the fatigue life of the structures, we select the cyclic principal stress range at the check point as the response stress. The cyclic principal stress range is usually determined by the rules and requirements of the classification society.

c. Calculation of the stress range

We often use the superposition calculation method to calculate of the floating body cabin dynamic load. Firstly, we make the stress and strain produced by the bending moment caused by the horizontal wave and the vertical wave the total stress and strain. Then, we make the stress and strain produced by the dynamic load caused by the wave acting on the floating body panels and the effect of the ballast water acting on the floating body compartment the partial stress and strain. When we calculate the stress range, first we superimpose all kinds of stresses; then we superimpose the stress of each class and between classes. So we can get the superposition range of the total stress and partial stress, and then get the stress range.

d. Fatigue analysis method

Analyzing the fatigue life is usually based on the S-N curve method and the Miner linear fatigue cumulative damage theory, and then we give the fatigue life of the "hotspot" selected in the floating body structure. The S-N curve is usually selected from the recommended curves in the rules of the classification society.

9. Analysis of the tubular joint fatigue life

The semi-submersible platform adopts lots of tubular joint structures, for example, the K-type tubular joint of the horizontal bracing tube and the tubular joint of the connection between the intermediate posts and the horizontal diagonal bracing. Because there is a high stress concentration at these tubular joints, their fatigue and fracture analysis is the important element in the analysis of the semi-submersible platform. We take the fatigue life analysis of the K-type tubular joint as an example to indicate the steps and methods as follows:

a. Building the meticulous finite element model

The K-type tubular joint usually has the minimum fatigue life where there is a stress concentration phenomenon where the chord member transversal surface changes. To take the South Sea No. 2 semi-submersible drilling platform as an example, its meticulous finite element model for the K-type tubular joint is shown in Figure 4-46. Figure 4-46(b) shows an elaborate picture of the K-type tubular joint chord member. Figure 4-46(a) shows the meticulous network finite element model of the whole K-type tubular joint.

b. Setting reasonable boundary conditions

When we calculate the stress concentration factor of the K-type tubular joint chord member, we make both ends of the pole rigid and both ends of the chord pole free in all degrees. When we establish the finite element model, in order to decrease the impact of the boundary condition acting on the stress concentration factor of the chord member "hotspot" areas, the pole and the chord member both extend a distance two times greater than the pipe diameter outward to the pole cross-section changing location. Thus we can get a real stress distribution.

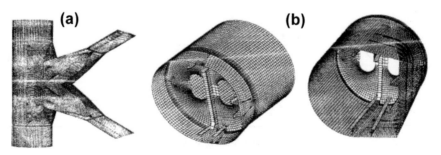

FIGURE 4-46

The meticulous finite element model of the K-type tubular joint.

c. Applying a unit load properly

We apply the unit axial load, the unit in-plane bending moment and the surface bending moment on the both ends of the chord member of the K-type tubular as shown in Figure 4-47. We should calculate the stress concentration factor separately under the three kinds of different external load effect conditions. The applied axial force and bending moment are a group, equal in size and opposite in direction.

d. Calculating the stress of the "hotspot" area

Through the finite element analysis, we can get the stress of the weld toe ends. We call this the stress geometric stress. The place that produces the maximum geometrical stress is called the "hotspot." The stress of the "hotspot" is called the "hotspot stress." The calculation of the "hotspot stress" usually reads the main stress value of each unit along the principal stress direction. Then, we get the stress of the "hotspot" by interpolation. Figure 4-48(a) gives the scope of the principal stress direction, and Figure 4-48(b) gives the stress interpolation curve.

e. Calculating the stress concentration factor

The positive stress σ_N is the stress in the section away from the tube nodes of the each tubular member, when the structure is under the ocean environment load. The stress concentration factor K_σ is the ratio of the "hotspot" stress σ_H and the nominal stress. So we can write this as

$$K_\sigma = \frac{\sigma_H}{\sigma_N} \tag{4-59}$$

There is usually a larger stress concentration phenomenon at the transversal surface at the welding line between the diagonal bracing connection side and the place close to the K-type tubular joint. Therefore, we take a number of points as the calculation points of the stress concentration factor at the weld line within an 180° arc of the chord member section mutational position. As shown in the Figure 4-48(c), we select five calculation points.

f. Calculating the stress transfer function

When we adopt the spectra analysis to calculate the fatigue life of the K-type tubular joint, we must get the stress transfer function of the K-type tubular joint under the action of the wave force first, so the calculation of the stress transfer function is an important step in the fatigue analysis of the K-type tubular joint. Usually under the calculation of a number of wave periods and the wave direction, in the finite element model of the whole structure, we get the nominal

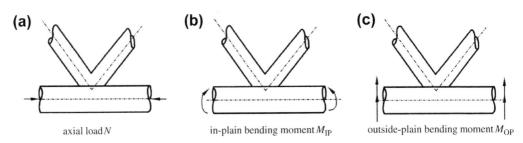

(a) axial load N

(b) in-plain bending moment M_{IP}

(c) outside-plain bending moment M_{OP}

FIGURE 4-47

Applied unit load of the K-type tubular.

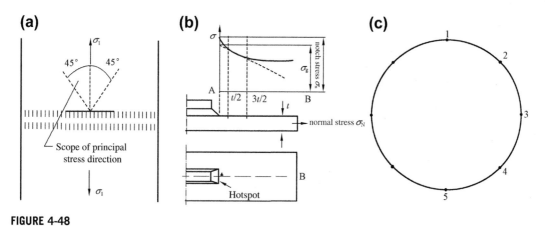

FIGURE 4-48

Calculation of the "hotspot stress."

stress of the boom head relevant parts of the K-type tubular joint under the three kinds of load (axial force, in-plane bending moment, external plane bending moment) action. Then we multiply the nominal stress by the relevant stress concentration factor obtained from the Equation (4-59), and then add the stress concentration factor together, so we can obtain the hotspot stress transfer function of the K type tubular joint at the abrupt change of cross section.

g. Giving the stress spectral density

The semi-submersible drilling platform can be regarded as a linear system. So its stress spectrum density function $S_R(\omega)$ is equal to the wave spectral density function $S(\omega)$ multiplied by the stress response transfer function $|H(\omega)|^2$, and we can get

$$S_R(\omega) = S(\omega) \cdot |H(\omega)|^2 \tag{4-60}$$

Therefore, if the stress transfer function $H(\omega)$ has been obtained, we can use Equation (4-55) to get the wave spectral density function $S(\omega)$ according to the South China Sea, so we can obtain the stress spectrum density function $S_R(\omega)$.

h. Giving the stress cycle times

Because of the different periods and different wave heights of the random wave, the magnitude of the alternating stress acting on the platform components is different. Therefore, when we calculate the degree of damage, we must know the stress cycle times of the alternating stress corresponding to different sizes. So we need the wave scatter diagram of the sea area where we are located. This wave scatter diagram can give the statistical data of the distribution of the given sea wave cycle times, such as shown in Table 4-13, which gives the statistical data for the Gulf of Mexico sea area. Its sea conditions are similar to the South China Sea, so we can use it in the South China Sea.

i. Selecting the S-N curve

The calculation of the fatigue life of the offshore oil platform components usually adopts the S-N curve recommended by the DNV RP 203 specification. The E curve is the sacrificial

Table 4-13 Wave Scatter Diagram of the Gulf of Mexico Sea Area

						T_z, s				
		5.6	6.3	7.7	9.1	10.5	11.9	13.3	14.7	16.1
H_s,ft	1.6	72	352	464	232	64	8	0	0	0
	4.9	24	368	1240	1336	704	224	48	8	0
	8.2	0	104	592	1040	824	384	120	32	8
	11.5	0	24	144	352	384	224	88	32	8
	14.8	0	0	32	88	120	80	40	8	0
	18.0	0	0	8	24	32	32	8	8	0
	21.3	0	0	0	0	8	8	0	0	0

Table 4-14 The E curve in the S-N Curve Recommended by the DNV

S-N curve	$N < 10^6$		$N > 10^6$		The modified
	lg a	m	lg a	m	thickness index k
E curve	11.610	3.0	15.350	5.0	0.20

anode protection in the seawater. It is consistent with the environmental conditions of the offshore platform underwater components. Table 4-14 gives the gradient m and the Y-axis intercept lg a of the E curve.

j. Calculating the fatigue cumulative damage

Usually when we calculate the fatigue life of the platform tube nodes, we use the Miner linear cumulative damage theory to calculate the fatigue cumulative damage of the structure under the action of alternating stress. That is, the total amount of damage D is the sum of the damage components D_i under the action of each different alternating stress. If the cycle times of the i class alternating stress S_i is n'_i, and the cycle times N_i are when the component reaches fatigue damage under the action of the S_i are obtained from the S-N curve, then from the $D_i = \frac{n_i}{N_i}$, we can get

$$D = \sum_{i=1}^{n} \frac{n_i}{N_i} \qquad (4\text{-}61)$$

k. Calculating the fatigue life of the tubular joint

According to the Miner rules, when we set $D = 1$, the tubular joint components can suffer fatigue failure. Therefore, the reciprocal of the total amount of damage D is the fatigue life of the tubular joint.

10. The strength design process of the semi-submersible platform.

To sum up, the process for the design calculation of the strength and fatigue life for the semi-submersible platform is shown in Figure 4-49.

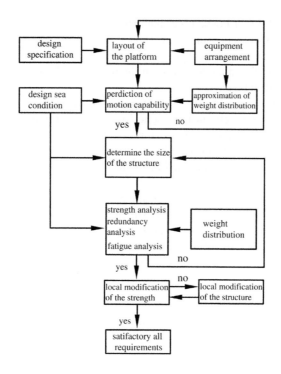

FIGURE 4-49

Strength design process of the semi-submersible platform.

4.3.3.3 Fatigue life analysis of an FPSO

1. Easy fatigue analysis areas of the FPSO

    For the FPSO, the easy fatigue failure areas are mainly:

    a. The hull center longitudinal framework structure

    When the hull of the FPSO is under the action of the wave load it can lead to the longitudinal framework producing alternating stress. Because the moment of the hull total longitudinal bending and the horizontal bending is at the maximum in the central hull areas, it can cause the longitudinal components to produce a higher total longitudinal bending force. So with the partial bending stress caused by the transverse load, a higher resulting stress will appear at the hull center longitudinal framework structure and join points at the ends of the transverse bulkhead as well as the ribbed plate and the connection points of the ribbed framework. That makes the hull center longitudinal framework nodes the important part of the fatigue check. So we must regard the transverse bulkhead at the ends of the middle part of the cabin and the adjacent strong ribbed framework as the fatigue check surface.

    b. Discontinuous place of the hull components

    The transverse components of the FPSO hull bear the lateral water pressure and the inertia force caused by the ballast water of the liquid cargo. The rib box girder bracket toe, transverse bulkhead horizontal girder bracket toe, inclined bottom tank roof, and inner bottom angle location are all high stress concentration areas under these load functions. These areas are usually selected as the fatigue check areas. The check points are usually selected at the middle of the cabin of the hull and a place close to the quarter of the stern.

c. Anchoring system structure

When the FPSO is working with a single point moored in a fixed sea area, the alternating anchor force acts on the anchoring system structure and the hull structure connected to the anchoring system structure, so the anchoring system structure must have enough fatigue life. The FPSO usually has two kinds of anchoring methods, that is, YOKE mode and center dropping anchor mode. They are shown as Figure 4-50(a) and (b) respectively.

For the YOKE anchoring mode, the alternate mooring force transmitted from the mooring system acts on the ship body of the forecastle deck, the main deck and the buttress bracket close to the longitudinal bulkhead at the bottom of the brace, and the strong transverse beams through the brace of the ship stem stents. These parts of the structure need to be checked for fatigue. Moreover, the stents and rocker arms through them transmit mooring force to the hull structure, which bears the alternating load uninterruptedly, so they need a fatigue check too.

For the center dropping anchor mode, the drum tower is usually installed in the hull bottom structure close to the stern. The mooring force transmitted from the mooring system acts on the neighboring inner bottom plating, external bottom plating, transverse bulkhead, ribbed plate, and the buttress bracket through the drum tower. The hull structures in these areas all need a fatigue check.

d. Torch tower structure

The torch tower is the jacket structure installed in the main deck of the FPSO and used to support the gas head. It is about $30 \sim 60$ m high. The structure produces alternating stress due to the inertia force under the motion of the FPSO hull. So we need to check the fatigue life of the tubular joint of its jacket.

2. Method of fatigue analysis

At present, for fatigue life analysis of the FPSO, we still use the S-N curve method and the Miner linear fatigue cumulative damage rule. The analysis steps and details are completely the same as

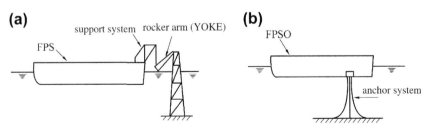

FIGURE 4-50

Two methods of anchoring the FPSO.

the fatigue analysis method of the tubular joint of semi-submersible platform described above, so we do not repeat them here.

3. Fatigue analysis process

Figure 4-51 gives the fatigue analysis flowchart of the FPSO.

As shown in the Figure, we can use the simplified analysis method to calculate the fatigue damage. Also, this kind of the simplified analysis method has been matured with software. We can use this kind of software directly and calculate the fatigue life. When we use this process to do the fatigue analysis, we must notice:

a. Fatigue check of the longitudinal framework nodes of the general ship

Because the stress concentration factor has been set to the default value of the software, we can use the A-B-C-D-E-F sequence in Figure 4-51 and rapid batch processing.

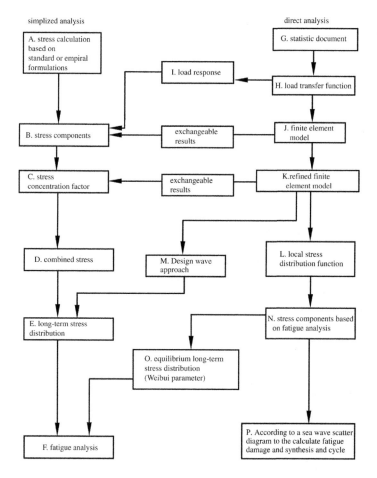

FIGURE 4-51

Fatigue analysis flow chart of the FPSO.

b. Fatigue check of the longitudinal framework nodes of the FPSO under different marine environments

Because the FPSO has a working condition and a moving condition, the marine environments are different, and we need the statistical information of the waves to do the simplified analysis. Therefore, its fatigue check should be based on the G-H-I-B-C-D-E-F sequence in Figure 4-51.

c. Other components beyond the longitudinal framework

This includes the rib box girder bracket toe, the transverse bulkhead stiffener bracket toe, the inclined bottom tank roof and inside bottom angle, the mooring system, the torch tower structure,etc.; they all need finite element analysis. We can do the calculation of the fatigue life after we get the stress of the check points. So we can carry it out according to the G-H-J-K-L-N-O-F sequence in Figure 4-51.

d. Fatigue cumulative damage calculation for the FPSO

Because the service period of the FPSO only includes two states, full load and pressure load, when we calculate the fatigue cumulative damage, we can separately calculate the fatigue cumulative damage under these two states according to the ratio 60%, 40%, or 50%.

4.4 Special Problems in the Deep-Water Drilling Process

From the perspective of deep-water drilling and oil production technology, besides the same problems we see for offshore drilling, there are several special problems that will be introduced below.

4.4.1 Deep-Water Drilling Design

There are some special problems that concern designing and planning deep-water drilling:

4.4.1.1 Choices of Drilling Platform

1. Deep-water drilling requirements for drilling platform

   There are some higher-level demands for the drilling platform for deep-water drilling:

   a. The platform design should meet the needs of the ultra-deep water.
   b. The platform design should meet the requirements of harsh conditions.
   c. The platform should have high mobility.
   d. The platform should have a high variable deck load and deck space.
   e. The platform should have the ability to use long-stroke motion.
   f. The platform should have natural oil storage capacity.
   g. The platform should have a dynamic positioning system.
   h. The platform should have both drilling and exploitation functions.
   i. The platform should efficiently handle operations and safety better.
   j. The platform design and construction shall conform to the specification requirements of the state and DNV, NPD, NMD, and IMO.

2. Drilling platform choices

   Deep-water drilling currently uses one the new type of drilling platforms, for instance:

   a. Semi-submersible drilling platforms: Generally, the fifth or sixth generation versions are used. The features of these platforms have been introduced in chapter III.

b. Floating drilling pontoon: The fifth generation of this drilling floating vessel is used for deep-water drilling currently, as shown in Figure 4-52. Its characteristics include a dynamic positioning system, a drilling rig with dual functions, a cuttings disposal system, a high variable deck load and deck space, sufficient storage for drilling fluid, and a polyester chain positioning system.

c. TLP and SPAR drilling platforms: These platforms were introduced in Section 4.3.

4.4.1.2 Drilling Rig Requirements

For deep-water drilling a dual operation drilling rig for drilling or completion or a hydraulic cylinder elevator drilling rig (ram type) is available.

1. Dual operation drilling rig

   A dual operation drilling rig is equipped with a double derrick system, as shown in Figure 4-53. It can reduce the production time with parallel operation for the purpose of saving drilling costs. Usually, drilling time for exploration wells can be reduced by 15%; for production wells we can reduce drilling time by 30%. Today, six dual operation drilling rigs exist around the world, assigned to the Enterprise, Deep seas, Deep Star, West Nation, West Future, and Navis Explorer. This kind of dual operation drilling rig is shown in the image. It shows the operation of the BOP and marine riser, and also, shows the order and path of the Christmas tree.

2. The hydraulic cylinder elevator drilling rig

   The main characteristics of this drilling rig is the use of the hydraulic cylinder elevator instead of the winch of the conventional drilling rig and hydraulic drive equipment instead of a wheel, which is used to make enough space for the drill floor and achieve the goal of a high platform deck space. Figure 4-54 shows the structure of this drilling rig, which can be dual operated, as shown in Figure 4-55. This kind of drilling rig is successfully applied to the semi-submersible platform and drilling floating ship, as shown in Figure 4-56(a) and Figure 4-56(b) respectively. Figure 4-56(c) shows the power source of the hydraulic pressure.

3. Technical performance of the drilling rig

   a. Winch unit: High power winch (5000 ~ 6000 hp) and AC motor applied.

   b. Rotary unit: Widening wheel with opening diameter of $60^{1}/_{2}$ in(1524 mm).

FIGURE 4-52

Fifth generation drilling floating vessel.

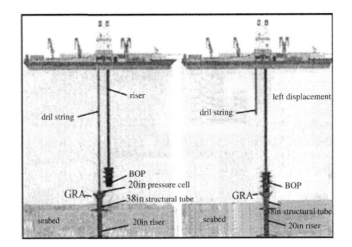

FIGURE 4-53

Dual operation drilling rig.

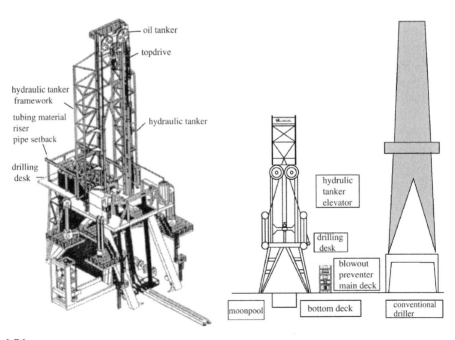

FIGURE 4-54

Hydraulic cylinder elevator drilling rig.

FIGURE 4-55

Hydraulic cylinder elevator type dual-function drill.

FIGURE 4-56

Hydraulic cylinder elevator drill equipped on a semi-submersible platform, and floating drilling ship.

c. Kill and choke lines: Diameter of $4 \sim 4^{1}/_{2}$ in ($101.6 \sim 114.3$ mm) instead of a tube diameter of $2^{3}/_{4} \sim 3$ in ($69.85 \sim 76.2$ mm), to further improve the choke and kill function.

d. Pipe discharge system: In order to improve the speed of the string, there should be an automatic pipe discharge system in deep-water drilling.

4.4.1.3 Improvement of the Riser System

In order to reduce production-time loss caused by bad weather, we should improve the drilling speed save drilling costs when drilling in the deep water. We need to improve the traditional riser to meet the requirements of deep-water drilling. Nowadays the independent water string (free-standing riser) is used broad. We will introduce some features below.

1. Structure of the independent marine riser

   Figure 4-57 presents the structure components of the marine riser. It is composed of two segments. The upper segment is a recyclable marine riser and the lower section is a freely independent marine riser. The two segments are connected by the upper marine riser package, shown in Figure 4-57(b). The marine riser package is located below $152.4 \sim 304.8$ m ($500 \sim 1000$ ft)

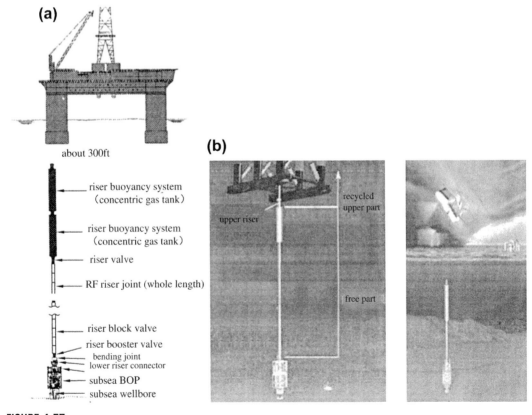

FIGURE 4-57

Structure of independent marine riser.

water depth, so the upper and lower section of the marine riser can be disassembled from this section underwater at about 91.44 m (300 ft). Below this point there is a freely independent marine riser section. The marine riser adopts the conventional riser style and the only improvement is that in order to provide buoyancy for the upper section, an upper string is made for an air tank with a 3.048 m (10 ft) concentric diameter, which is the buoyancy system of the marine riser column. It is located below the detachable connection and provides variable buoyancy for the lower marine riser column. The air buoyancy tank can hold the riser-containing mud column of 20.3704 kg/m^3 (17 1b/gal) and the drilling column, where we can adopt the conventional marine riser. There is a marine drilling riser fill-up valve between the buoyancy system and conventional marine riser, as shown in Figure 4-57(a). The bottom of the conventional riser consists of a dump valve for the underwater drilling riser, booster value, flexible spherical connector, hydraulic connectors, etc. The bottom parts connect with the subsea BOP and subsea wellhead. A detachable joint between the upper recycled pipe section and lower free independent pipe, which is made up of a ring sleeve connector and the unbind tube shear spindle. When a hurricane occurs, the connector can be dismantled with the help of the tubular spindle shear force. The lower portion can be raised to the upper section of the riser (about 150 ~ 300 m). The dismantled free independent marine riser at the bottom stays underwater through the buoyancy of the upper flotation system. So that it can eliminates the free and independent marine risers in this section, which reduce a large amount of nonproduction time and cost. It is confirmed that the free-standing drilling riser (FSDR) has been used successfully in deep-water drilling in the Gulf of Mexico.

2. Benefits of an independent marine riser in deep-water drilling
   a. The separate drilling string is in the buoyant marine riser and can be removed quickly.
   b. The connector between the upper part and lower part of the marine riser can be integrated to the air tank for the free and independent marine riser section.
   c. When it is needed for other operations for the well, the marine riser can be detached and discharged easily.
   d. When we carry on maintenance operations for the drill floor and tensioner, it is not necessary to take out the free and independent marine riser section.
   e. When the BOP stack is in use and the Christmas tree has been installed, the free and independent marine riser can be placed on the simulated wellhead.
   f. After the free and independent marine riser is installed, a smaller drilling rig is moved to the well location to continue work.
   g. For the second smaller drilling rig, just pull back this upper part of the riser, to restore the marine riser in place.
   h. The smaller drill is better for the load on the deck and is beneficial in reducing occupied deck space.
   i. When the marine riser is free and independent, it can withstand heavy hurricanes and sea currents according to the design.
   j. After the Christmas tree has been installed, the independent marine riser will be placed.

3. Economic of independent marine riser
   a. When drilling exploration well, it can greatly save on nonproduction time, reducing drilling cost. For example, for the conventional marine riser it will take 23 d to raise the underwater marine riser when a hurricane comes. But for the independent marine riser we just need 12h to raise the riser a length of 150 ~ 300 m above the opened place.

b. It can save additional operation time when drilling production wells. When the well is completed underwater, it is not necessary to raise the marine riser up before transferring Christmas tree down, as long as we place the free and independent marine riser on the simulating wellhead. This can save 6 days of raising the marine riser to reduce drilling cost.

4.4.1.4 Well Control System Design

1. Deep-water well control requirements
   a. Longer kill/block line
   b. Larger closing well pressure
   c. Longer marine riser
   d. Early prediction scheme for overflow
   e. Complete procedures for closing wells and the blowout preventer (BOP) group and its control system
   f. Enough circulating pressure and drilling fluid
2. BOP group choices
   There are two kinds of BOP group schemes for deep-water drilling well control to choose from.
   a. The BOP groups on water
   The BOP groups can be divided into those that are fixed and those that are unfettered in the horizontal direction, as shown in Figure 4-58. But because the unfettered BOP in the

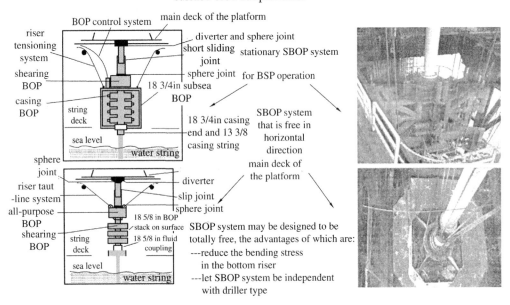

FIGURE 4-58

BOP groups on the surface of deep water. Note: The SBOP system will be designed to be unlimited in the future. The benefits are reducing the bending stress at the bottom of the BOP riser and making the BOP system have nothing to do with the rig type.

horizontal direction can reduce the bending stress of the lower marine riser, and make the BOP system have no connection with the type of drilling rig, we generally adopt unfettered BOP groups in the horizontal direction.

When we select a BOP group on the surface of water, it should be equipped with an underwater shutoff and departure device, as shown in Figure 4-59. This is the method of cutting and sealing to close the wellheads and separates the upper marine riser. The underwater shutoff and departure device is placed on the seabed with $13_{3/8}$ in marine riser, then place the 30 in conduit, drilling $17_{1/2}$ in well hole and place the $13_{3/8}$ surface casing and cementation. When the BOP group is installed on the water, pressure is tested by the marine riser. When the test pressure is qualified, the drilling rig is used for drilling.

b. Underwater BOP group

The underwater BOP group was introduced in the second chapter of this book, so it is unnecessary to repeat those discussions here. But we should note that for one well, if subsea wellhead equipment is adopted, its weight will reach up to 650,000 lb (244,125 kg); if BOP groups on the surface of the water are chosen, the weight, including the underwater shutoff and separation device, is only 200,000 lb (75,000 kg). Obviously, the latter has the advantages in weight. However, it has limitations, such as a maximum working depth only within 2133 m (7000 ft).

3. Preparation of the drilling fluid

We need to consider the following issues for drilling fluid for it to cooperate with well control for deep-water drilling:

a. How to inhibit hydrate formation

b. How to avoid and deal with the lost circulation

c. How to ensure the sand carrying capacity and good rheological properties

d. How to keep good rheological properties at low temperature

e. How to avoid or reduce the damage to the marine environment

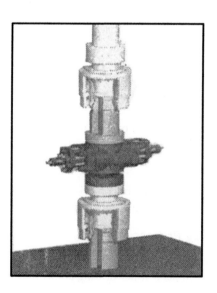

FIGURE 4-59

Deep-water drilling water shutoff and separation device.

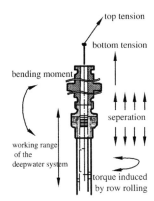

FIGURE 4-60

Deep-water drilling wellhead loads.

4.4.1.5 The Choice of Wellhead Device

1. Influence factors for deep water subsea wellhead
   For a subsea wellhead in deep water, the load bearing is very complex, as shown in Figure 4-60, where the loads mainly include:
   a. Tension from the top
   b. Tension from the bottom
   c. Bending moment
   d. Torque caused by drilling platform (ship) yawing
   e. The separation tendency
2. Deep-water drilling requirements for wellhead device selection
   a. Reliability
      In order to ensure safety and reliability for the wellhead device for deep-water drilling, the components of the seabed wellhead system, such as the flange of the wellhead BOP groups, wellhead connector, conductor pipe, and conductor pipe connector, should be able to withstand pressures of 103.42 ~ 137.103 MPa (15,000 ~ 20,000 psi) and a 9,490,600 N· m (7,000,000 ft · lb) bending moment.
   b. Stability
      The mud mat used on a 762 mm (30 in) or 914 mm (36 in) conductor line not only improves the stability of tubing string, but also prevents gas from directly impacting on the connector and the blowout preventer, etc.
   c. High efficiency
      If we use a guideline-less reentry assembly or guideline-less guide base to solve the problem of returning to the wellhead after completion, it will greatly improve work efficiency.

4.4.1.6 Design of Well Building Process

1. Existing problems in well building
   The problems of well building in deep-water drilling are as follows:
   a. Shallow water flow (SWF)
   b. Shallow gas
   c. The low stratum fracture pressure gradient

    d. Small annulus/casing clearance

    e. Lost circulation accident

2. Surface well building technology

  This mainly includes:

    a. Shallow geological engineering analysis

      This includes shallow seismic processing and analysis and drilling geological analysis.

    b. Evaluation of shallow geological disasters

      This requires the drilling of a pilot hole with MWD (measurement while drilling) technology.

    c. Down surface casing for operation

      This means taking the surface casing of 762 mm (30 in) or 914 mm (36 in) off for operation. Usually, in offshore shallow water areas first there is drilling, second the casing comes off and last there is well cementing to finish the job. But in deep water areas, because the seabed shallow strata is soft, the conventional borehole casing and cementing operation is difficult to achieve. Because of the long operation time, it is not available for the deep water drilling operation due to the expensive daily cost. A jetting-in mode is adopted at the surface layer of deep-water drilling abroad currently, namely taking the drill bit off in the conduit (762 mm or 762 mm). On the one hand, they use the weight of pipe column and the drill collar to press, and on the other hand open the pump for washing and take the conduit off simultaneously.

    d. Drilling operation for the second section

      In drilling operations for the second section, a reaming while drilling tool operated by a cam is used in deep water drilling, as shown in Figure 4-61.

    e. Evaluation for small pilot hole drilling

      For the evaluation of shallow geological disasters of deep water and determination of the depth of surface casing, a small pilot hole drilling well is used commonly, especially when drilling

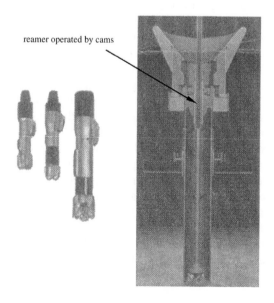

reamer operated by cams

FIGURE 4-61

Reaming while drilling (RWD) tools drill hole for the second well.

an exploratory well. We usually choose a 250.82 ~ 311.15 mm ($9^7/_8$ ~ $12^1/_4$ in) pilot, drilling down below the depth of the surface casing.

4.4.1.7 Well Cementing Technology Design

1. Difficult problems for deep-water well cementing
   a. Low temperature
      The temperature in 1000 m water depth is only 4°C and the temperature is lower in super deep water.
   b. Loose geological structure
      The shallow stratum in the deep seabed is relatively loose, which is adverse to well cementation.
   c. Pore pressure gradient under the seabed and the narrow window of stratum fracture pressure gradient.
      The adjustable drilling fluid density range is narrow to kill the well for blowout prevention, which make the fluid column pressure gradient exceed the stratum fracture pressure gradient and cause well collapse. This prejudices well cementation.
   d. Gas hydrate production in a high pressure and low temperature environment
      Gas hydrate are easily produced in deep water, which jams the cementing pipe and valve system causing cementation accidents.
   e. High pressure shallow water and gas flow in deep water
      Shallow water and gas flow in deep-water drilling will bring many difficulties to cementing technology.
   f. The strict standards of deep sea environmental protection
      These high standards and strict requirements, such as the cement slurry and the drilling fluid restrictions on marine pollution, etc., make it more difficult for the cementing operation.
2. Foam slurry cementing technology (foam cement system)
   For high temperature and pressure wells in complex strata abroad, foam slurry cementing technology, which is available for drilling deep-water cementing process design, has been adopted in recent years. Foam slurry is composed of basic cement dispersed by nitrogen. The cement slurry density depends on the content of nitrogen, and the amount of gas is decided by the basic slurry pumping rate. Therefore, the appropriate gas content can adjust the foam cement slurry pressure gradient between the fracture pressure gradient and the pore pressure gradient, so it can prevent leakage of liquid and reservoir fluid invasion effectively, which is good for well cementing.
   In the U.S. Gulf of Mexico, the problem of shallow water was solved by slurry successfully; in Mobile Bay, it was used to complete cementing of high temperature and pressure wells in complex strata. It showed that the cement ring reduces warp, rupture, and elongation of casing caused by stress. In short, practice proved the advantages of this kind of cement slurry:
   a. Helpful to control gas channeling
   b. Helpful for drilling fluid displacement
   c. Helpful for cementing annulus
      It is effective because the cement is sheathed by foam slurry that has resistance to the crushing stress on cement.

4.4.1.8 Structure Design of Slender Well

1. Slender well features
   The slender well structure has been considered abroard for deep water drilling, and it contains the follow characteristics:
   a. Small diameter riser
   The marine riser of 558 mm (22 in) outer diameter is replaced by a smaller one of 482 mm (16 in) for the general design, as shown in Figure 4-62. So we can adopt a drilling rig and drilling platform of the same capacity, but the work depth deepens greatly.
   b. Small wellhead
   A 346.075 mm ($13^5/_8$ in) wellhead are used in the common design, as shown in Figure 4-63, which can reduce the wellhead weight greatly.
   c. Small BOP stack
   As shown in Figure 4-62, the 346.075 mm ($13^5/_8$ in) BOP stack is used in the common design, which can reduce the cost of the BOP stack greatly.
   d. Reduce to 508 mm (20 in) casing
   d. Use small diameter chokes and kill lines
   Choke and kill lines of 114.3 mm ($4^1/_2$ in) inner diameter are used in common designs, which can reduce the weight.

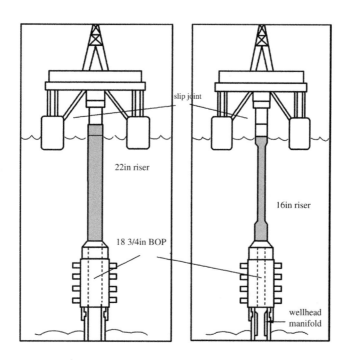

FIGURE 4-62

Comparison of two kinds of water string.

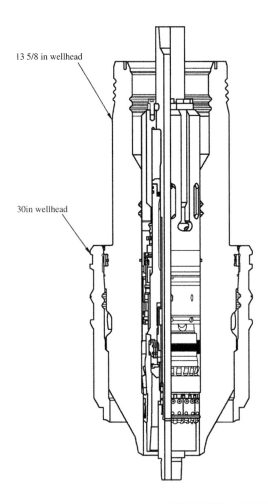

13 5/8 in wellhead

30in wellhead

FIGURE 4-63

Small-sized well 13 5/8 in well head; 30 in well head.

2. Slender well structure
   a. Casing sequence
   Surface casing: 762 mm (30 in)
   Intermediate casing: 346.075 mm ($13^5/_8$ in)
   Oil-string casing: 244.475 mm ($9^5/_8$ in)
   b. Tubing: 127 mm (5 in)
   c. Wellhead: 346.075 mm ($13^5/_8$ in)
   d. Christmas tree: 476.25 mm ($18^3/_4$ in). In this case, a 244.475 mm ($9^5/_8$ in) casing is allowed through the Christmas tree.

3. The advantages of the slender well structure
   a. Reduction of the total drilling cost
   i. Save steel to reduce cost: For example, the size and weight of the marine riser, casing pipe, and wellhead device are reduced, saving steel and reducing the cost.
   ii. Save materials to reduce cost: For example, the dosage of the drilling fluid and cement slurry is reduced.
   iii. Reduce the cost of the BOP stack.
   iv. Save drilling time to lower drilling cost.
   b. Reduce the cost of drilling rig and drilling platforms
   With a slender well structure, the second or third generation can reach the water depth of a fifth generation drilling rig. So, second or third generations Operating depth is broaden, which reduces the cost of the drilling platform and rig equipment. Figure 4-64 gives a comparison of conventional drilling with the fifth generation semi-submersible drilling platform and drilling well and using a third-generation semi-submersible drilling platform with a slender structure.
   c. Reduction of marine riser tension
   Due to the small diameter of the riser, its tension reduces, and thus the size of the marine riser tension rope is reduced.

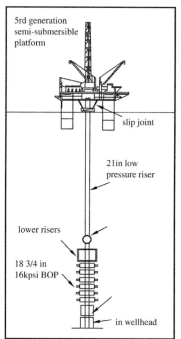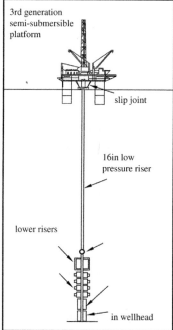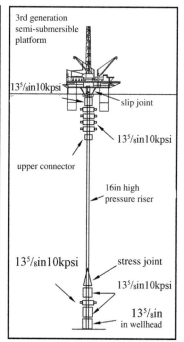

FIGURE 4-64

Comparison of fifth-generation semi-submersible drilling platform and third-generation platform with slender shaft structure.

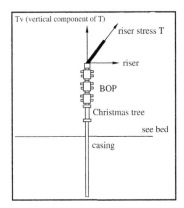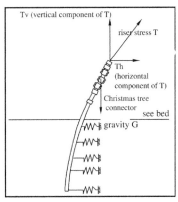

FIGURE 4-65

Slender well water string stress distribution for well bore structure.

4. Defects of slender well structure

As shown in Figure 4-65, when drilling a well with a slender structure, the whole BOP stack will bend, unlike what we see with a conventional well structure. So the vertical tension and the horizontal force of the marine riser will be different than the conventional well structure, which will worsen the stress state and bring a series of adverse effects.

4.4.2 Deep-Water Double-Gradient Drilling Technology

For deep-water drilling, the margin between a formation's pore pressure and fracture pressure gradient is small, but the mud column weight of the water riser column is large, which will cause problems for the deep-water drilling process. On the one hand, if the adjusted fluid density is small and makes the mud column pressure less than the formation pore pressure, it will lead to invasion of formation gas, which brings difficulties to well control. On the other hand, if the adjusted drilling fluid density is large and makes the mud column pressure greater than the formation pore pressure to prevent blowout, when the mud column pressure is greater than formation fracture pressure, it will lead to stratum fracture and brings serious problems, such as sticking, enlarge the well hole, drilling fluid leakage, washing well difficult. In the 1990s, investigation into deep-water double-gradient drilling technology started. Measures were taken to make the fluid density equal to the seawater. All pressure measures references the bottom of the sea, where the area between the fracture pressure and pore pressure gradient relatively widens, which solves the above contradiction. The following is an introduction to some of the double-gradient drilling technology.

4.4.2.1 The Basic Principles of Double-Gradient Drilling

Double gradient refers to two pressure gradients: one pressure gradient references the water surface, another references the bottom of the sea. The conventional drilling pressure gradient only has a reference point on the surface of the water, while double gradient drilling forms two gradients. The basic reason for that is that the drilling fluid density adjustable range expands when the double pressure gradient is introduced.

1. The adjustable range of conventional drilling technology drilling fluid density

   a. The minimum drilling fluid density ρ_{Nmin}.

   Mud column pressure usually can be expressed by the density of drilling fluid p_N. If the height of the mud column is h_N and the gravity acceleration is g, according to hydraulics, p_N for the mud column pressure can be expressed as

   $$p_N = \rho_N g h_N \tag{4-62}$$

   To prevent a blowout occurrence, mud column pressure should be greater than the pore pressure p_p. If the mud column pressure is equal to pore pressure p_p, the mud column pressure is at a minimum (no smaller or there is a blowout) p_{Nmin}. Set the density of mud column to ρ_{Nmin}, and then according to the Equation (4-62) it can be written

   $$p_{Nmin} = \rho_{Nmin} g h_N \tag{4-63}$$

   So we can conclude that the minimum density of drilling fluid in the conventional drilling process is

   $$\rho_{Nmin} = \frac{p_p}{g h_N} \tag{4-64}$$

   b. The maximum density of drilling fluid ρ_{Nmax}

   Usually the mud column maximum pressure is confined by the fracture pressure. Generally, it should not be more than the breakdown pressure of stratum p_f, otherwise it will cause hole wall collapse. Therefore, the maximum pressure of mud column p_{Nmax} is equal to p_f, and the drilling fluid density is the maximum density of drilling fluid ρ_{Nmax}; according to the Equation (4-62), we can write

   $$p_{Nmax} = \rho_{Nmax} g h_N = p_f \tag{4-65}$$

   that is,

   $$\rho_{Nmax} = \frac{p_f}{g h_N} \tag{4-66}$$

   c. The adjustable range of drilling fluid density R_N

   If we take Equation (4-66) minus (4-64), we can obtain

   $$R_N = \rho_{Nmax} - \rho_{Nmin} = \frac{p_f - p_p}{h_N g} \tag{4-67}$$

2. The drilling fluid density adjustable range for double-gradient drilling technology

   For double gradient drilling technology, the reference point for the pressure gradient should not be on the water surface but at the bottom of the sea, as shown in Figure 4-66. The height of the reference point of the mud column is different than for conventional drilling, as shown in Figure 4-66. There is a depth of 2736 m (9000 ft) from the bottom of the sea to the surface of the sea, but the actual total depth of the well is only 2438.4 m (8000 ft), and the height of the mud

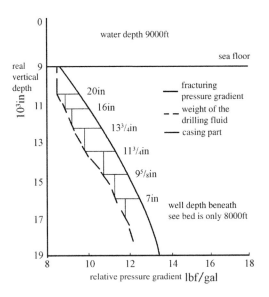

FIGURE 4-66

Pressure gradient for double-gradient drilling.

column h_a should be conformed to the actual total well depth. So, according to the same method as Equation (4-67), we can write the adjustable range of drilling fluid density for double-gradient drilling technology as follows:

$$R_d = \frac{p_f - p_p}{h_d g} \tag{4-68}$$

From Equation (4-68), we can see that the $h_d < h_n$, so $R_d > R_n$, which is the principle of the range of drilling fluid density for double-gradient drilling technology.

3. The influence factor of double gradient drilling technology η_d

If Equation (4-68) is divided by Equation (4-67) and the $\frac{R_d}{R_N}$ is defined as the η_d influence factors of the double-gradient drilling process, we can obtain

$$\eta_d = \frac{R_d}{R_N} = \frac{\frac{p_f - p_p}{h_d g}}{\frac{p_f - p_p}{h_N g}} = \frac{h_N}{h_d} = \frac{h_w + h_d}{h_d} = 1 + \frac{h_w}{h_d} \tag{4-69}$$

In Equation (4-69), h_w is the depth of the water. In the conventional drilling process, the surface of the water is a reference point for the pressure gradient, so the mud column height h_N should be the sum of water depth h_w and the mud column height h_d. Obviously, we can see the following from Equation (4-69):

a. With increasing water depth, the effect of the dual-gradient drilling technology is more obvious. As shown in Equation (4-69), when the well depth is constant ($h_d = \text{const}$), η becomes larger along with the increase of the value of h_w.

b. The effect of the double-gradient drilling technology performance better in shallow well. As shown in Equation (4-69), when the water depth is constant (h_w = const), η becomes larger along with the decrease of value h_d.

In super deep-water drilling, upper stratum and shallow water flow often occur. Usually, reservoir pore pressure is very high while the stratum fracture pressure is low. At this time, we are forced to use multilayer technology casing to pack the upper stratum part of the formation and the shallow water flow, but this will increase the drilling cost and the complexity of drilling operation. But double gradient drilling technology is better in shallow well sections with constant water depth, which solves the prominent contradiction of super deep-water drilling.

4.4.2.2 Implementation of the Double-Gradient Drilling Technology

According to the above basic principles, to become the ocean surface as the reference point of the pressure gradient, the marine riser above sea surface should be filled with water, instead of drilling fluid. The mud in drilling column of marine riser returns to the seabed wellhead from the bottom of well, then returns to the water surface platform by other pipelines instead of marine riser. This is how pressure gradient 1 and pressure gradient 2 are formed in Figure 4-67. Conventional drilling regards the water surface as reference point, and the mud column fills the marine riser from surface to the bottom of the sea and from the bottom of the sea to the bottom of well, so that there is only a single pressure gradient.

In order to achieve the double gradient, there are several plans, described in the following:

1. Drilling pump installed at the bottom of ocean
   The core of this scheme is that the drilling pump is installed in the bottom of the sea or near the bottom. In this way, we raise the annulus drilling fluid to the mud line through the underwater pump, and then further through the small diameter pipeline raise the drilling fluid to the surface of drilling platform (Figure 4-67). So the drilling fluid need does not need to go through the riser, and the riser annulus can be filled with water.
   In this underwater drilling fluid lift system, we can regard the underwater drilling pump as the main body; besides the drilling pump and powered equipment, there should be an underwater

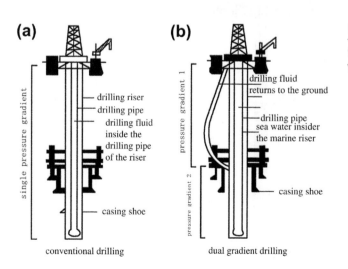

(a) conventional drilling

(b) dual gradient drilling

single pressure gradient

- drilling riser
- drilling pipe
- drilling fluid inside the drilling pipe of the riser
- casing shoe

pressure gradient 1

pressure gradient 2

- drilling fluid returns to the ground
- drilling pipe sea water insider the marine riser
- casing shoe

FIGURE 4-67

Comparison of double-gradient drilling with conventional drilling.

cuttings separation system and underwater well control system, etc.

In recent years, a drilling technique without a riser (riserless drilling, RD) was developed based on a submarine pump scheme. In the scheme, the riser column is canceled, and the drill column penetrates the water directly for drilling, which can achieve the double gradient.

The type of drilling pump installed on the seafloor can be divided into the following categories:

a. Subsea submersible pump scheme

This scheme was developed by Shell with the application stage launched at the end of 2002, and is called the Subsea Submersible Pump System (SSPS). This system is a subsea submersible pump installed underwater that is driven by cable. Drilling fluid is lifted to the drilling platform with a series of six underwater electric submersible pumps. The electric submersible pumps are similar to the electricity submersible pumps used in oil wells for lifting oil. This system includes underwater solid phase elimination equipment, for getting rid of cuttings in drilling fluid. At the bottom of the sea, it can cut pieces greater than 12.7 mm (0.25 in) before drilling fluid enters electric submersible pump. In this way, it makes the cuttings content in the drilling fluid lifted to the drilling platform less than 1%. For well control, separation equipment for drilling fluid and gas is installed. When kick occurs, the BOP groups will be closed first, then drilling fluid is transmitted with gas from the well bore to the sea bottom separation equipment. After gas is separated from the drilling fluid, drilling fluid is transferred to the tank, and lifted to the drilling platforms by underwater pump, while the isolated gas is discharged into the sea. Figure 4-68 shows this system developed by the Shell Company.

Because this system will discharge large pieces of debris into the bottom of the sea, which causes pollution, its application is limited.

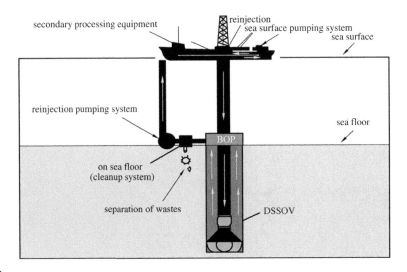

FIGURE 4-68

Submarine electric submersible pump system developed by the Shell Company.

b. Subsea diaphragm pump scheme

This solution was developed by 22 companies headed by Conoco Company, including the Hydril Company and BP, was put into application in 2002. The SML (subsea mud lift) underwater drilling fluid lift system with 3 ～ 6 underwater diaphragm pumps with 0.3028 m³ (80 US gal) of positive displacement is used to lift the drilling fluid. The pump is driven by high pressure water that reaches the bottom of the sea and goes through the lines within a 127 ～ 152.4 mm (5 ～ 6 in) in riser, and then is discharged into the bottom of the sea through the pump. In this way, it will save on the use of cables and winches on drilling platforms and underwater motors, reducing the cost. The application of the scheme shows that the diaphragm pump is suitable not only for circulating drilling fluid including the cuttings, but also for pumping gas mixture. A low-speed diaphragm pump (including large pump valve), allows for large suction and cutting off without stopping operation. The system has a rock crusher that can be used for crushing hard rock slices, clay, cement, etc., which become particles less than 38.1 mm (1.5 in) in size. So, they can go through the diaphragm pump, and the pump is not damaged. In the SML system, an underwater rotating separator transfers the drilling fluid to the subsea pump. A drill pipe and the revolving body are sealed by rotating rubber components, and the bearings are allowed to operate in the oil.

c. Undersea centrifugal pump scheme

This scheme was developed by Baker Hughes and Transocean Sedco Forex in 2001 and put into application in November 2001. This double-gradient drilling system system is called DV (Deep Vision), and it consists of 1 ～ 5 ocean centrifugal pumps developed by National Oil Well. One pump is an electric centrifugal pump, and a three-level pump is in the circulation system. Because the centrifugal pump is a chewing type, the blade can handle hard clay, sand, gravel, hard limestone, aluminum, cement, and rubber, preventing the pump from being damaged. According to the working depth and lift requirements of the pump, the DV system is equipped with between 1 and 5 submarine centrifugal pumps as necessary.

Table 4-15 shows the comparison of these three submarine pumps.

2. Injection of light hollow spheres

This scheme is characterized by injecting light hollow spheres to change the density of drilling fluid in the riser, achieving a double gradient. The material of the hollow spheres can be glass, plastic, composite material, and metal. They are pumped into the bottom of the sea from the water

Table 4-15 Comparison of Three Submarine Pumps for Double Gradient

Pump Type and Driven	The sea drilling fluid	Cuttings processing	BHA
diaphragm pump	separation		
Sea power-driven diaphragm pump	Rotary separatorRSU	Rock crusher Crushed particles less than 1.5	FSS
Electrically driven centrifugal pump	Sea water-drilling fluid separation system	Chomper The blade finely chopped particles are smaller than 0.5 in	FSS
Electric drive electric submersible pump	Separating means	Gumbo slide ilter out particles of 0.25 to the bottom of the sea and the exclusion of	DSSOV

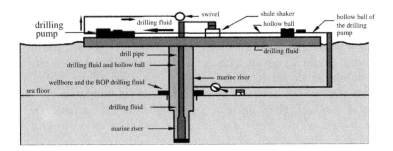

FIGURE 4-69

Hollow sphere double-gradient drilling system in Maurer.

surface and injected into the bottom of the riser through its control valve, which make the drilling fluid density of the riser annulus reduce. The hollow sphere double-gradient drilling system is shown in Figure 4-69.

There are two methods to transfer the hollow sphere from the surface to the bottom of the sea;

a. Transfer drilling fluid

This method mixes the drilling fluid and sphere on the platform first, transferring them into the bottom of the sea and injecting them into the bottom of the riser through the control valve after dilution. The advantage of this approach is that it doesn't need complex submarine equipment except for a control valve; while its drawback is that after the dilution, the hollow sphere concentration in the drilling fluid decreases, and it is difficult to form a high concentration of hollow sphere drilling fluid.

b. Transferring seawater method

We transfer the hollow sphere to the bottom of the sea, using seawater as the transmission fluid, separate the sphere from the seawater by separator, and then inject it into the riser with a pump. Obviously, the method has the following advantages:

i. The seawater transfer fluid can be discharged into the surrounding seawater without restriction.

ii. It's easy to achieve a high concentration of hollow spheres.

But the disadvantage is that it needs a subsea separator and bottom injection pump, and the device is relatively complicated. Therefore, the water transfer method is suitable for drilling fluid density more than 10lb/gal, and the drilling fluid transmission method suitable for drilling fluid density less than 10l/gal. From the above, we can see that the solution of installing a drilling pump on the bottom of the sea is not suitable for medium shallow water depth because of the expensive equipment and the high cost. But injection of lightweight hollow spheres is suitable for application at medium depth (600 ~ 1000 m). Especially when the well bore stability is poor, and there is shallow water flow and leakage of drilling fluid, this solution is better. The largest sphere content is between 35% ~ 50%. If the sphere content is 50%, we can replace drilling fluid of 1678 kg/m^3 with that of 1038 kg/m^3. Even with a hollow sphere volume content of 18%, we can reduce drilling fluid density from 1198 kg/m^3 down to the same as the seawater. Obviously, this method is suitable for all kinds of drilling fluid densities, and drilling can be carried out without a riser.

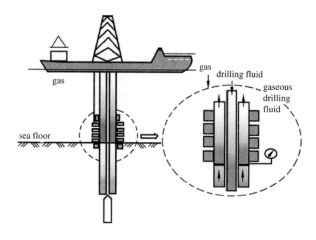

FIGURE 4-70

Gas lift dual-gradient drilling system schematic diagram.

3. Injecting gas

This scheme is characterized by gas injection to change the drilling fluid density in the annulus of risers; its principles are as shown in Figure 4-70. With compressor equipment, a certain amount of gas is injected in the bottom of the riser from the neighboring mud line; the commonly used gas is air and nitrogen, and seldom carbon dioxide or natural gas are used.

4. United program

This kind of union combines two of these three schemes. For example, injection of lightweight hollow spheres can be combined with installing an underwater pump, or the injection of gas can be combined with installing underwater pump schemes, as shown in Figure 4-71. In two of the schemes, gas

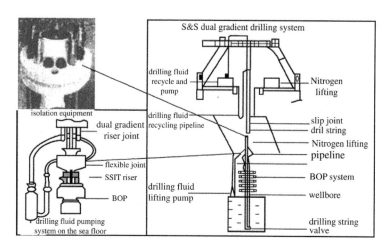

FIGURE 4-71

Injection of gas and submarine pump joint scheme.

injection and hollow sphere injection, the purpose is changing the density of drilling fluid in the annulus of the riser, so density from the sea surface to the seabed is different from the well bottom to the bottom of drilling fluid density. It is often regarded as double-density drilling technology.

4.4.2.3 Main Advantages of Double-Gradient Drilling Technology

As we can see from the above analysis, the main advantages of double-gradient drilling technology are as follows:

1. Due to the wider area between the fracture pressure and pore pressure, we can reduce kick, blowout, and well leakage accidents, saving time and processing costs.
2. By eliminating the need for multilevel intermediate casing, we save the time of casing and cementing, shorten the construction cycle, improve operational efficiency, reduce drilling costs, and improve economic efficiency.
3. By reducing the flow friction of the annulus of the riser column, we improve the speed of the drilling fluid returning to the water surface, which solves the problem of the drilling fluid carrying cuttings.
4. To reduce the requirements on drilling platforms and drilling rigs and other drilling equipment, smaller and cheaper drilling equipment can be realized by using a third-generation semi-submersible drilling platform to drill slender and deep wells.
5. The riser margin decreases, we have safer emergency evacuation, and the possibility of blowout and pollution of the marine environment are significantly reduced.
6. No depth restrictions; we can operate at any water depth.
7. The amount of drilling fluid is reduced greatly, saving costs.

Of course, the double-gradient drilling process can divide itself into different embodiments, which each have their own drawbacks. For example, the underwater pump installation program increases the need for submarine equipment and no matter what type of pump is used will inevitably increase cost and operational difficulties. The gas injection program will also bring well control and other security issues that need to be considered.

4.4.3 Deep-Water Jet Drilling Process

The hydraulic principles used in the jet drilling process create an effective technology that can increase mechanical drilling speed. However, this process presents some special problems in deep-water drilling.

4.4.3.1 Process Cycle Pressure Loss in Deep-Water Drilling

In accordance with the theory of high-pressure jet drilling, the pressure loss along the drill string Δp_P, in accordance with hydraulics principles, can be written as

$$\Delta p_p = \frac{B\rho^{0.8}\mu^{0.2}L_pQ^{1.8}}{d^{4.8}} \tag{4-70}$$

B - Constant, desirable domestic level for drill pipe is 0.51666
d - Drill pipe diameter, mm
ρ - Drilling fluid density, g/cm^3
L_p - Total length of the drill string, m
Q - Drilling fluid displacement, L/s

μ - Plastic viscosity of drilling fluid, Pa.s

ΔP_p - Process pressure loss along the drill string, MPa

In drilling deep-water wells, if hydraulic jet drilling technology is used, the following drilling process parameters for optimization design and drill pipe tool selection should be considered:

1. Displacement of the drilling pump should be as small as possible
   From Equation (4-70) we can see that the frictional pressure loss and displacement are proportional with a power ratio of 1.8. The more the displacement, the greater the frictional loss, and the lower the downhole hydraulic energy, so we should choose as small a displacement as possible.
2. Slim holes and small drill pipe diameters are not conducive to deep-sea wells
   Equation (4-70) shows that the pressure loss is inversely proportional to the drill pipe diameter with a 4.8 power ratio. Obviously, the smaller the drill pipe diameter, the higher the pressure loss; conversely, if the drill pipe inside diameter is slightly larger, the pressure loss will significantly decrease. For example, if a 127 mm (5 in) drill pipe is replaced by a 139.7 mm (5 in) pipe, increasing the drill pipe inside diameter 12.7 mm (in), the pressure loss will reduce 40%. Therefore, a slim drill pipe diameter is not conducive to deep-water drilling.
3. Deep-water drilling should minimize the use of drill pipe joints

With increasing water depth in deep-water drilling, the number of drill pipe joints will be greatly increased, and the cumulative result of local pressure loss of the joints is quite large, so deep-water drilling should minimize the use of drill pipe joints.

4.4.3.2 The Attenuation of Deep-Water Drilling Bit Nozzle Energy

1. The factors influencing the energy of the bit nozzle
   The use of jet drilling technology in practice shows that the major factors affecting the energy of the bit nozzle are the following:
   a. The distance from the bit nozzle exit to the drilling bottom
      Generally, the jet pressure loss is as high as 50% between the bit nozzle and the bottom section for a 215.9 mm (8 in) drill bit under normal pressure conditions.
   b. Confining pressure at the exit of the bit nozzle
      Practice shows that confining pressure is a powerful factor in jet pressure attenuation. When the confining pressure is small, the axis of jet dynamic pressure will fall with the increase of spraying distance; when the confining pressure is larger (such as 10 MPa), the axial dynamic pressure has a sharp drop at first and then slows to balance with the increase of spraying distance; if the confining pressure is increased from 0.2 MPa to 10 MPa for a spray distance of 45 mm, the axis of jet dynamic pressure will fall 75%, which is very serious.
2. The deep-water drilling hydraulic design parameters for the jet drilling process

The confining pressure is often as high as 50 MPa in deep-water drilling. Under such conditions, with the increase of spraying distance attenuation, we can foresee that the axis of jet dynamic pressure decreases sharply, causing great loss. Therefore, if we still adopt the nozzle exit hydraulic parameters to design the drilling hydraulic parameters when using the jet drilling technology, the error will be very large, so we must study the the flow pressure attenuation and design the hydraulic parameters based on the bottom of the borehole.

4.4.3.3 Application of Cavitation Jet Technology in Deep-Water Drilling

1. The cavitation phenomenon
   Cavitation is a phenomenon that generates cavities that grow and break up during the process of fluid flow when the local pressure is lower than the saturated vapor pressure.
2. Impact of damage from cavitation
   When cavitation occurs, cavitation collapse will generate significant damage. Calculations show that the instantaneous pressure of cavitation collapse is 8.6 ~ 124 times that of nozzle pressure drop, which is very powerful.
3. The problems in deep-water drilling that adopts cavitation technology
   To improve drilling speed in and bottom hole hydraulic energy in deep-water drilling, and further the role of the jet, cavitation technology is an important measure; cavitation technology is feasible, but the study of the dynamics characteristics of a cavitation jet is not enough, and if cavitation technology is used in deep-water drilling we need to answer the following questions:
   a. Cavitation jet dynamics problems under normal conditions
   b. Cavitation jet dynamics problems under high temperature and high pressure
   c. Cavitation jet mechanism of initiation, development, and collapse and problems with this mechanism under the conditions of high temperature and high pressure
   d. The rock fragmentation mechanism and application of the cavitation jet.

4.4.4 Deep-Water Subsea Production System

A subsea production system uses a subsea wellhead production system and underwater submarine tree as the core of an oil production system, The third chapter in this book introduced this system. Here, the application of a subsea production system in deep water will be discussed.

4.4.4.1 Subsea Production System in Deep Water

Here we look at how the subsea production system is applied in deep water.

1. It can adapt to different floating platforms
   A subsea production system for deep water can use a variety of different floating platforms, including semi-submersible platforms, or a new kind of tension leg platform, single pillar platform (independent), or floating production, storage, and offloading vessel, as shown in Figure 4-72.
2. Different water depths are available
   Deep-water drilling for oil and gas development has become an important area of the world's oil industry; the maximum depth of drilling has gone up to 3095 m. But, from the point of view of oil field development, subsea production systems can adapt to work with different depth from several meters to several km, as shown in Figure 4-73. In deep-water areas, we see subsea production systems working with four different floating platforms at four different water depths in Figure 4-74. For example, the one used in the Gulf of Mexico is a tension leg platform with a working depth of 1413 m; for Brazil's Seillean floating production, storage, and offloading (FPSO) vessel the operating depth is 1849 m. In addition, the Gulf of Mexico Na Kiki semi-submersible platform's operating depth is 1920 m; and the Mad Dog truss-type single (independent) column platform has an operating depth of 2073 m. This four different floating platforms are combined with a subsea production system, and present four kinds of different maximum working depths, shows the adaptability of a subsea production system.

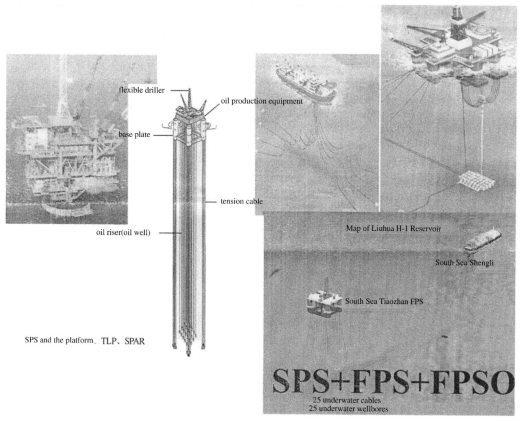

FIGURE 4-72

Subsea production systems are suitable for a variety of different floating platform.

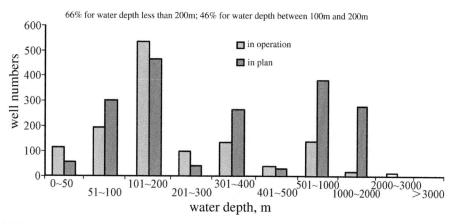

FIGURE 4-73

Global subsea production system applied at different water depths.

Magnolia tension leg platform in Gulf of Mexico
1434m

Mad Dog truss spar platform
2073m

Seillean floating production storage unit in Brazil
1849m

Na Kika semi-submersible oil platform in Gulf of Mexico
1920m

FIGURE 4-74

Subsea production system applied at different water depths in deep water.

3. Different types of oil field development are available

A subsea production system can be used for both large deep-water oil fields and marginal satellite oil fields; its adaptability for the development of deep-water oil fields is very strong. From the point of view of development, the number of subsea wellheads is increasing. Figure 4-75 clearly shows the trend of increasing subsea wellhead development in various parts of the world, including West Africa, Europe, the North Sea, the Mediterranean, the Gulf of Mexico, Brazil, and the Asian Pacific waters.

4. Adaptable to various sea conditions

Subsea production systems can be adapted to many sea areas, despite that large difference in this areas. As shown they can be applied in West Africa and the Asia-Pacific region, Brazil, Europe, the North Sea, the Mediterranean Sea, and the Gulf of Mexico. But, as is known to all, the conditions of the North

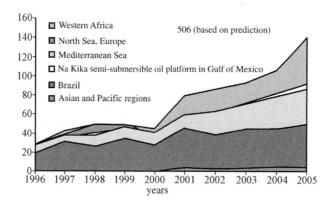

FIGURE 4-75

Increase in the global adoption of subsea wellheads.

Sea are relatively poor and those in the Gulf of Mexico are moderate, and random ocean wave spectrum can be quite different. Although the number of wellheads in Europe's North Sea that are currently in use is zero (0%), we know from Figure 4-75 that the North Sea subsea wellhead number will likely increase gradually, which indicates that subsea production systems are applicable for the North Sea.

4.4.4.2 Equipment Combinations for Subsea Production Systems

When we use subsea production system to explore deep-water oil fields, the exploration mode depends on whether there are surrounding facilities or we are in an independent development mode. No matter which solution is adopted, we can choose different solutions for combinations of equipment.

1. Equipment combination schemes
   Here several proposed equipment combinations refer to the subsea production system combination of SPS with surface production facilities. An existing subsea production system combined with equipment on the surface mainly has the following several of applications in deep water (Figure 4-72).
   a. SPS + SPAR
   subsea production system + single column (independent) production platform
   b. SPS + TLP
   subsea production system + tension leg platform
   c. SPS + FPSO
   subsea production system + floating production, storage, and offloading tanker
   d. SPS + FPS
   subsea production system + semi-submersible floating production platform
   e. SPS + (Fixed Platform)
   subsea production systems + fixed oil platforms
2. The distribution of several combinations in different years
   The proportions for the choice of the five equipment combinations above in various regions of the world for different years is shown in Table 4-16 below.

Table 4-16 The Proportion of Different Combinations and Equipment in the Global Subsea Production System (Unit %)

Equipment \ Years	1994	1995	1996	1997	1998	1999	2000	2001	2002	2003
FPSO	22	16	15	34	40	39	26	36	29	27
SPAR	0	0	0	0	0	0	2	5	6	5
TLP	0	5	1	8	2	3	4	9	7	8
FPS	25	27	34	27	23	29	31	16	18	32
Fixed	52	51	48	29	34	29	36	33	40	28

3. Subsea wellhead and equipment combinations trends

Figure 4-76 shows the development trends of the applications of deep-water subsea wellhead in the world. In Figures 4-75 and 4-76 and Table 4-16 we can see:

a. The number of applications of subsea wellheads varies with each year and continues to increase in every major region of the globe, and the number of subsea wellheads used in deep-water oil field is also increasing, from about 17% since the 1990s to 33% in the early part of this century.

b. Amongst subsea production system combinations, the FPSO vessel always occupies a large proportion through the years, as this advantageous scheme coincided with the growth in China's offshore oil field development.

c. The combination with subsea production systems with floating semi-submersible platforms also occupies a large proportion and it is related to the semisubmersible platform's comparatively mature design and construction. In turn, the new tension leg platform and single (independent) pillar platforms occupy a small proportion, because they are not mature.

FIGURE 4-76

Subsea wellhead quantity increasing over time.

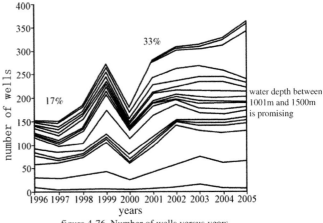

figure 4-76 Number of wells versus years

4. The combination with a fixed jacket platform also occupies a considerable proportion and had a larger percentage in the 1990s, which indicates that it is mainly applied to shallow water.

4.4.4.3 Major Components of a Conventional Subsea Production System

The major components of a conventional subsea production system are a subsea wellhead, Christmas tree, chassis manifold, and control system, as shown in Figure 4-77; Chapter 3 has introduced these components, so here we will give only a brief description of their use in deep water.

1. Christmas tree
A Christmas tree is a key piece of equipment for deep-water subsea production systems; Christmas tree for lift wells, flowing wells, or electric submersible pumps can be horizontal, as shown in Figure 4-78(a). They also can be vertical, as shown in Figure 4-78(b). They can be dry, semi-dry, or wet, with the pressure divided into 5,000 lbf/in^2, 10,000 lbf/in^2, and 15,000 lbf/in^2 that can be used when necessary.
2. Subsea wellhead manifold
Subsea wellhead manifolds used deep-water subsea production systems can be centralized, distributed, and folding, as shown in Figure 4-79.
3. Underwater control system

The underwater control system has two styles, hydraulic control and electro-hydraulic control, but the use of a composite electro-hydraulic control system is more common for deep-water depths; the control flowchart is shown in Figure 4-80 below. The equipment on the water is a wellhead platform top side device; underwater equipment consists of a downhole safety valve, production main valve, and

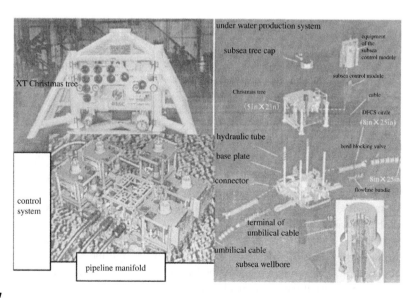

FIGURE 4-77

Components for deep-water subsea production system.

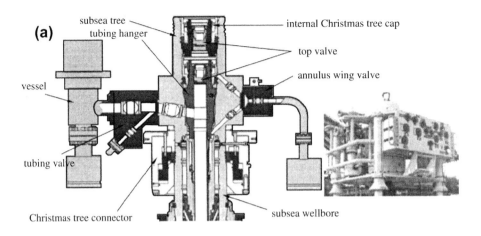

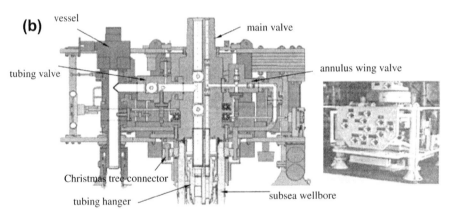

FIGURE 4-78

Deep-water subsea production system Christmas tree.

chemical injection control valve, as shown in Figure 4-81. The cross-sectional view of the underwater control tube bundle for the control system shown in Figure 4-82, consisting of the high hydraulic control tubes, low hydraulic control tubes and each spare pipeline, as well as a hydraulic return pipe, two chemical injection pipes and an alternate tube. Cable, compared with a 3×4-wire shielded cable, each Christmas tree should be a supply line, a signal lines, cables, Cable power line voltage DC $12 \sim 28V$, signal line was $4 \sim 20$ mA.

The work principles of the composite electro-hydraulic control system are as follows: A weak signal is sent to the subsea control module (SCM), and then through the MCU (Main Control Unit) control and electrically driven hydraulic system, the hydraulic high-pressure accumulator controls the valve block. Figure 4-83 present the complex electro-hydraulic control system working principles. In the figure, 1 is the subsea control module (SCM), 2 represents the subsea control module base, 3 is the power (signal) cross pipe, 4 is the five-line quick connector cross pipe, 5 is the pressure sensor cross

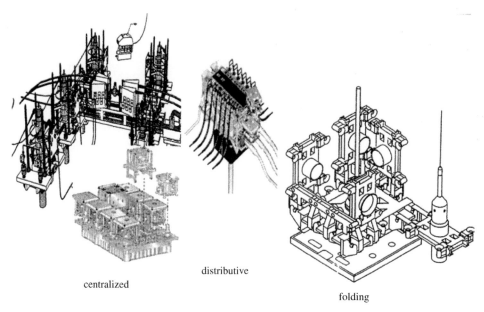

centralized

distributive

folding

FIGURE 4-79

Three kinds of subsea wellhead for deep-water subsea production system.

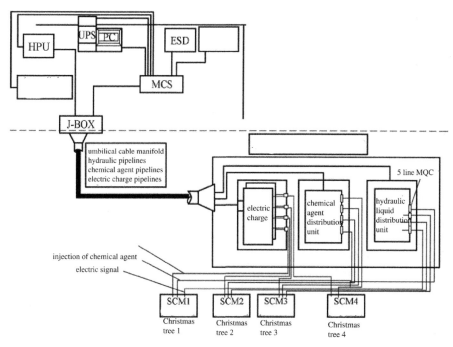

FIGURE 4-80

Underwater electro-hydraulic control flowchart.

FIGURE 4-81

Subsea control system surface and underwater equipment for deep water.

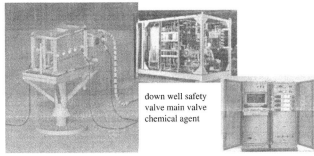

down well safety
valve main valve
chemical agent

UTH

FIGURE 4-82

Control tube bundle for deep-water subsea
control system.

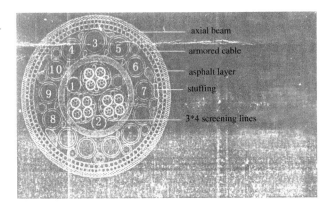

pipe, 6 is the temperature sensor cross pipe, 7 is the four-pin male buckle at the ROV wait, 8 is the 4-pin female buckle ROV tryout, 9 is the quick connector protection board, 10 is the quick connector plate hydraulic coupler, 11 is the quick connector waiting area, 12 is the single coupling circuit board part, 13 is a single-line coupler guard, 14 is the one-way coupler wait, and 15 is power (signal) EL short jumper tube.

As for the underwater equipment installation, maintenance equipment and tools, which is the basic composed of subsea production systems equipment. Such as deep water with the ROV, which has reached 3000M water depth, subsea production systems installation and maintenance and other operations, but chapter III of this book has introduced Related Content, so we will not repeat it here.

4.4.4.4 Deep-Water Long-Distance Underwater Multiphase Oil and Gas Flow Automatic Mining Technology

In deep-water oil and gas exploration, because of the long distance, it is appropriate to use remote automatic control technology. However, due to subsea wells' multiphase flow of oil, gas, and water, the implementation of remote automatic control mining for underwater multiphase flow of oil and gas has become a key technology for deep water oil and gas development. For example, as shown in Figure 4-84, in the Canyon deep-water fields, the tieback distance is up to 57 miles (91.713 km) with a

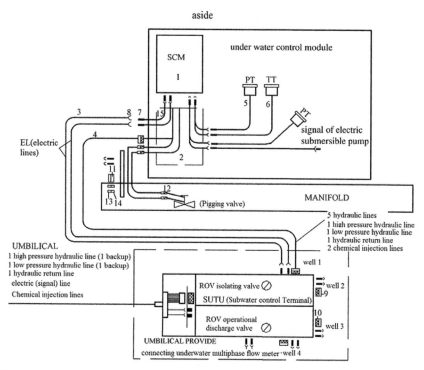

FIGURE 4-83

Composite electro-hydraulic control system working principles.

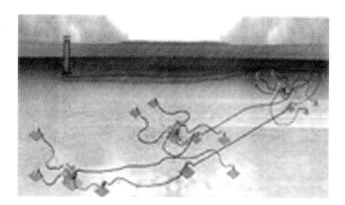

FIGURE 4-84

The mining of the Canyon oil field deep-water well over a long distance.

water depth of 7300 ft (2219.20 m). Obviously, automated mining techniques on the Sham distant wells must use pressurized lifting. Issues such as long-range submarine power supply technology, underwater modes of transmission and distribution systems, signal transmission system modes, and subsea pipeline flow security need to be addressed; when using pressurized lifting, the seabed booster pump installed outside the wellbore or wellhead electric submersible pump is one of the key pieces of equipment.

1. Subsea multiphase booster pump technology
   A multiphase booster pump under a rotating flow field makes vapor and liquid separate, for multiphase fluid pressurized conveying equipment. There are two types, water flooding and power driven. In 1994, foreign countries have developed 4 water flooding subsea multiphase booster pumps which tieback distance can reach 36.8 km; in 1996 an electrically driven multiphase booster pump came out; until 1999 two multiphase booster pumps made up of the electrically driven submarine multiphase booster pump station, for the 10km long distance subsea production systems oil and gas exploration technology. Figure 4-85 shows the station and multiphase booster pump and its cross-sectional view.
   Lucent field LF22-1 in the South China Sea is a typical marginal deep-water oil and gas development; its water depth is 333 m, and it uses the combination of the subsea production system and floating production, storage, and offloading (FPSO) vessel, with shuttle tankers taking the qualified crude oil, as shown in Figure 4-86. This oil field development adopts long-distance underwater automatic mining technology, and applies a multiphase booster pump that is installed in the seabed mud line. Figure 4-87 shows that the mud line booster pump is arranged at the mud line at the bottom of the sea. The subsea wellhead manifold is folding, the subsea production system uses the control system of the wet electrical connector to implement the electro-hydraulic control system, mud line booster pump and wellhead Christmas tree, all of them adopt the Complex electro-hydraulic control system. The subsea tree is a wet and horizontal tree.

2. Electrical submersible pump (ESP) technology
   Besides the seabed booster pump, the electrical submersible pump is another piece of equipment that is implemented in the deep-water and long-distance oil and gas subsea production systems

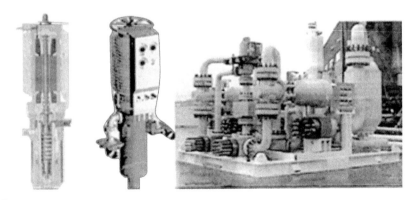

FIGURE 4-85

Electric subsea multiphase booster pump and pump station.

FIGURE 4-86

The South China Sea LF 22-1 oil field development equipment combination scheme.

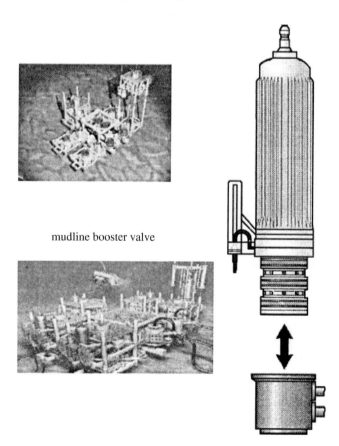

mudline booster valve

FIGURE 4-87

The South China Sea oil field development mud line booster pump.

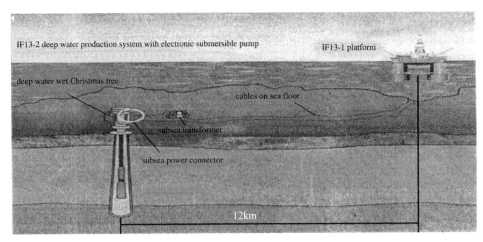

FIGURE 4-88

Subsea tieback electric submersible pump system.

that are automatically mining. The underwater electric submersible pump appears in 1996, and put into use in 1999. The electric submersible pump is installed at the seabed wellhead, as shown in Figure 4-88. It is installed below the deep-water horizontal Christmas tree, with supply of electricity for the electric submersible pump transmitted from the surface of the water on the semi-submersible production platform through undersea cables, and through a submarine power connector to the electric submersible pump. Figure 4-88 shows the 12 km long-distance oil and gas automation mining system at the Lucent oil field in China. Another is at the South China Sea Lihue oil field LH11-1; the 24-well subsea production system makes use of the underwater electric submersible pump technology. All 24 wells are horizontal wells, with a working depth of 310 m; it uses the following oil field development and equipment combination

SPS + FPS + FPSO

(subsea production system) + (semi-submersible production platform) + (floating production, storage, and offloading tanker)

As shown in Figure 4-89, in order to achieve the oil and gas mining automatic control technology for a long-distance subsea production system, we install the underwater electrical submersible pump shown in Figure 4-90. The underwater electric submersible pump uses a 1000V wet mutable electrical connector (WMEC) and is installed at the wellhead below the Christmas tree, with power transmitted long distance from the semi-submersible production platform. The Christmas tree and the electrical submersible pump (Figure 4-91) all use a hydraulic control system.

3. Long-distance submarine power supply technology

   Long-distance submarine power supply is also a key technology to achieve deep-water oil and gas long-distance subsea production systems for automatic mining. Because it is a long-distance transmission, we should solve the problem of low pressure to high-pressure transformation. Because it is a subsea production system, we also need to address the underwater wet connectivity issues. From 1996 until 1999, a 1000V wet mutable electrical connector (WMEC) and

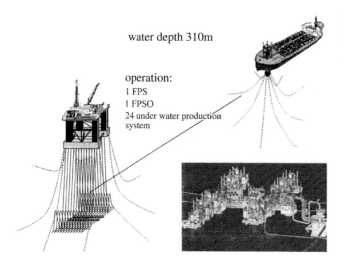

water depth 310m

operation:
1 FPS
1 FPSO
24 under water production system

FIGURE 4-89

South China Sea LH11-1 oil field development equipment combination plan.

FPS

Christmas tree cap

wet connector

WD-15 wellbore

tubing hanger

circle iv connector

FIGURE 4-90

Electric submersible pump adopted in China LH11-1 oil field.

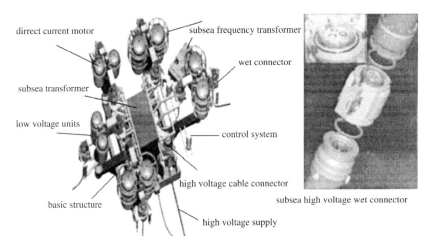

direct current motor

subsea frequency transformer

wet connector

subsea transformer

low voltage units

control system

high voltage cable connector

basic structure

subsea high voltage wet connector

high voltage supply

FIGURE 4-91

Long-distance submarine power supply device.

underwater transformer for subsea production systems was developed. Figure 4-91 shows the development of long-range submarine power supply unit and a new high-voltage submarine wet electrical connector (MECON) developed in recent years. Then the entire apparatus is installed on an infrastructure and the high-voltage power source is transmitted to the seabed. This means on the first connected to the high voltage cable connector (electrical connector), through the underwater transformer, into the low pressure and connected to the low pressure electrical connector transmit low pressure telex to the low-pressure unit, the power supply to the DC motor, the means to manipulate the control system, and underwater frequency converter is used as to change the motor frequency. Shows a flowchart of the subsea power distribution system. It supplies the power through the boost converter supercharging , underwater cable high-voltage power supplied to the underwater, and then through the buck converter to reduce voltage, thereby via the low-voltage electrical connector, frequency converter, motor, pumps and other implementing agencies in all processes.

4. Gas hydrate formation inhibition technology

For deep-water subsea production systems, subsea pipeline flow safety technology has developed in recent years in order to solve the current security issues of the subsea pipeline. Gas hydrate formation inhibition and prevention technology is one of the most important part. specially for the bottom of the sea Multiphase pipeline, this technology is more important to ensure the safety of the flow.

a. Method of inhibition.

Currently, in the production equipment and pipelines, thermodynamic inhibitors such as ethanol, ethylene glycol, sodium chloride, calcium chloride, and methanol are commonly used.

b. Mechanism of thermodynamic inhibition.

This method changes the thermodynamic hydrate formation conditions by injecting thermodynamic inhibitors, thus inhibiting or avoiding hydrate formation. Typically, the thermodynamic inhibitor concentration is generally from 15% to 50% to lower the hydrate formation temperature and make it below the operating temperature for transporting the multiphase flow

c. Dynamic suppression mechanism

Dynamic suppression refers to hydrate form as small crystals and present in the oil and gas pipelines, and without clogging the pipeline; or trying to slow down the formation speed of hydrate, so the generation time is shorter than the residence time of the water in the pipeline, thereby avoiding jamming the pipeline. Dynamic inhibitor concentration is generally less than 1%, making it more expensive than the thermodynamic inhibitors.

d. Dynamic inhibitor.

Currently, we have developed a third generation of dynamic inhibitor and the typical representatives are polyvinyl pyrrolidone (PVP), polyvinyl caprolactam (PVCAP), and a terpolymer mixture (VC-713). The third-generation products have been applied in the United States and the United Kingdom oil and gas fields within the range of 50 km pipeline transportation; practice in six oil and gas fields shows that obvious effect of dynamic inhibitors on the occurrence of gas hydrates.

4.4.5 Special Problems in Deep-Water Oil Production with Risers

Floating production systems and subsea production systems are used in the development of deep-water fields. An oil riser is a critical part in these systems. The deep-water riser fatigue loads caused by vortex-induced vibration is a prominent problem and the fatigue loads causing riser fatigue and fracture will bring serious harm to deep-water oil production, which is greatly significant to the deep-water oil production riser fatigue fracture failure reliability assessment. We will present a brief introduction here.

4.4.5.1 Two Types of Risers in Deep-Water Oil Production

There are two kinds of risers in deep-water field oil production development currently.

1. Steel catenaries deep-water riser

The riser is made of steel tubing. Because of the long distance from the seabed to the platform on the water, the riser maintains catenaries under its own weight; it is called the steel catenaries deep-water riser and it is used for water surface and deep-water semi-submersible platform floating production facilities connected to undersea pipelines to transporting oil and gas to the surface of the water.

2. Flexible deep-water riser

Such risers are made of titanium alloy or fiber-reinforced plastic composite material. Titanium alloy has a good strength-to-weight ratio, low elastic modulus, and good corrosion resistance and is suitable for deep-water riser material. Fiber-reinforced plastic composite is a new material for the manufacture of risers and its basic performance and fatigue properties meet the requirements of the deep-water riser and have been proved by the current practice to be suitable for use in 1000 m or deeper water fields.

4.4.5.2 Deep-Water Riser Fatigue and Fracture Analysis

1. Analysis of steel catenary deep-water riser fatigue

   C.A. Martins has analyzed the fatigue failure of steel catenary deep-water risers, outlined below:

   a. Location of motion

   Set S as the location of the starting point of the curve coordinate, and t as the time; the location of motion x(s, t) can be expressed as follows:

$$x(s,t) = \frac{1}{2}[1 + \sin \beta(s,t)]\{1 - \exp[e - \beta(s,t)]\} \, x_0 \left[1 + \frac{\tau(t)}{T_0}\right]^{-1} \tag{4-71}$$

   where

$$\beta(s,t) = \frac{S - x_0(t) + \lambda(t)}{\lambda(t)} \tag{4-72}$$

$$\lambda(t) = \lambda_0 \left[1 + \frac{\tau(t)}{T_0}\right]^{-1} \tag{4-73}$$

$$\lambda_0 = \sqrt{\frac{EJ}{T_0}} \tag{4-74}$$

   $X_0(t)$ - Location of movement;

   X_0 - Site of static catenaries curvature

   $\lambda(t)$ - Dynamic bending strength

   $\tau(t)$ - Point obtained by linear frequency range model, the dynamic tension

   T_0 - Location of static tension

   $\lambda(t)$- static bending strength

   E - Elastic modulus of steel

   J - Moment of inertia

   b. S-N curve expression

   The strain amplitude is set at a point $\Delta\varepsilon_i$; the S-N curve expression can be written as

$$N_i = a(\Delta\varepsilon_i)^{-b} \tag{4-75}$$

   In Equation (4-75), a and b are material constants and N_i is the number of cycles to failure.

   c. The accumulated fatigue damage

   According to the Miner rule, when the role of n_i times strain the fatigue accumulated losses D_i can be written as follows:

$$D_i = \frac{n_i}{N_i} \tag{4-76}$$

   Thus, the total cumulative damage D shall be as follows:

$$D = \sum_i D_i = \sum_i \frac{n_i}{N_i} \tag{4-77}$$

Usually when D = 1, representing the occurrence of fatigue failure, fatigue life can be obtained.

2. Fatigue analysis of flexible risers

Z. Chen and R.L. Reuben researched an analysis model to predict the fatigue life of flexible risers though theory and tests. They thought that the interlocking the (flexible risers connection) (Interlock) riser deformation, fatigue calculation, in consideration of the deformed shape of the interlock riser on the case of tension and bending, the critical strain can be expressed as follows:

a. Tensile case

The critical strain in a cross-sectional view of the flexible riser A at ε_A is

$$\varepsilon_A = 180tP\left[\frac{T}{G} - S_t(T)\right][\pi rA(90 - \beta)]^{-1} \tag{4-78}$$

In Equation (4-78), T is the tension of the riser; G is the axial stiffness of the riser; S_T (T) is slip function on the unit length; P, t, and r and beta is the dimensions after interlock, shown in Figure 4-92.

b. The bending conditions.

The flexible riser critical strain epsilon at cross section A is the following:

$$\varepsilon_A = 80tP\left[\frac{D_i + 2t - 2b}{2R + D_0 + 2b} - S_b(R)\right][\pi rA(90 - \beta)]^{-1} \tag{4-79}$$

In Equation (4-79), R is the radius of the bend of the elbow; D_i and D_0 are the riser inner and outer diameter, respectively; and $S_b(R)$, the slip coefficient, can be determined by experiment. From Equation (4-78) or (4-79) we obtain the strain ε_A in accordance with the S-N curve method and Miner fatigue cumulative damage rule, and Equations (4-75), (4-76), (4-77) above are similar to the calculations for the fatigue life of the flexible riser.

4.4.5.3 Deep-Water Riser Fatigue and Fracture Failure Reliability Analysis

1. Deep-water riser fracture failure assessment

Leiral gives the deep-water riser fracture failure assessment method and the assessment limit state function g(x):

$$g(x) = \min(\delta_{mat}f(L_r)\delta_1; L_{max} - L_r) \tag{4-80}$$

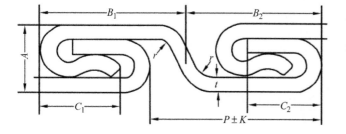

FIGURE 4-92

Flexible interlock shape dimensions of the stand pipe.

In Equation (4-80), the δ_{mat} and δ_1 are the critical and the actual load ratio, respectively; $f(L_T)$ is an influence function which reflect the stability of the fracture and plastic impact; and $f(L_r)$ is determined by the following equation:

$$f(L_r) = \frac{1}{I_\delta}\left(1 - 0.14L_r^2\right)\left[0.3 + 0.7\exp\left(-0.65L_r^6\right) - \rho\right]^{-2} \tag{4-81}$$

In Equation (4-81), ρ is the plastic correction factor and I_δ is the model correction factor; when calculating plane stress, take $I_\delta = 1$, and when calculating plane strain, take $I_\delta = 2$.

2. Reliability assessment of the fatigue failure of catenaries steel riser

Souza and Conclaves studied the deep-sea rigid riser fatigue life and fatigue reliability assessment with statistical methods.

a. Basic assumptions

i. S-N curve under constant amplitude loading fatigue:

$$S^b N = K_s \tag{4-82}$$

In Equation (4-82), b and K_s are obtained by testing coefficients.

ii. Applying Palmer-Miner's rule, ignore the influence of the load order on the fatigue process; the fatigue damage in each structure is constant. With δ as the injury, we can write

$$\delta = \frac{1}{K_s}S_{ai}^b \tag{4-83}$$

When the stress amplitude is S_{ai}, the total injure $D\delta$ in n time cycles is

$$D\delta = \sum_{i=0}^{n-1}\delta_i = \frac{1}{K_n}\sum_{i=0}^{n-1}S_{ai}^b \tag{4-84}$$

b. Regarding the stress amplitude as random fatigue damage $D\delta$

$E(n)$ is the mathematical expectation of riser on n cyclamen (S_{ai}^b) is the mathematical expectation of S_{ai}^b; so we can obtain the total injury $D\delta$ as follows:

$$D\delta = \frac{1}{K_s}E(n)E\left(S_{ai}^b\right) \tag{4-85}$$

c. The long-term stress amplitude on the riser fatigue caused by environmental loads can be expressed the total amount damage as D.

If the probability of occurrence in the environment during the life of the structural design load is Pj; [E (n)] j for j at ambient conditions in the riser on the mathematical expectation of the number of stress cycles; [E(S_{ai}^b)] environmental conditions stress amplitude S_{ai}^b mathematical expectation, we have

$$D = \frac{1}{K_s}\sum_{j=1}^{k}P_j[E(n)]_j\left[E\left(S_{ai}^b\right)\right]_j \tag{4-86}$$

Equation (4-86) can be written as

$$D = \frac{1}{K_s} T \sum_{j=1}^{k} P_j f_j \left[E(S_{ai}^b) \right]_j \tag{4-87}$$

In Equation (4-87), T is the fatigue life of the riser; and f_j is stress amplitude frequency in the j sea conditions. Therefore Tf_j replaces $[E(n)]$ j.

d. The prediction of rigid deep-water riser service life.

According to the Miner rule, if D > delta (delta is the critical cumulative damage, usually delta = 1), fatigue failure occurs. Therefore if we make D = delta, (4-87) can be written as

$$T = \Delta \cdot K_s \left\{ \sum_{j=1}^{k} P_j f_j \left[E(S_{ai}^b) \right]_j \right\}^{-1} \tag{4-88}$$

In Equation (4-88), T is the prediction of service life expressed by time.

e. The reliability analysis of fatigue failure in a deep-water rigid riser

For the limitation function for riser fatigue failure, we can obtain

$$g_s(X) = \Delta - \frac{1}{K_s} \left\{ \sum_{j=1}^{k} P_j f_j \left[E(S_{ai}^b) \right]_j \right\}^{-1} \tag{4-89}$$

In Equation (4-89), Δ is the cumulative damage that has occurred when the fatigue failure is critical; K_S is the constant factor for determining the fatigue strength of the S-N curve (S-N curve in the ordinate axis intercept); and $\left[\sum S_{ai}^b \right]_j$ is effect for each sea condition j on the riser and stress amplitude. In a given sea state condition, the number of stress cycles is huge, and the coefficient b is a fixed value of the normal distribution to approximately describe the probability density function $\left[\sum S_{ai}^b \right]_j$. Thus, for a normal probability density function of gamma stress amplitude $\left[\sum S_{ai}^b \right]_j$ the mathematical expectation $\left[\sum S_{ai}^b \right]_j$ can be written as like this:

$$E\left[\sum S_{ai}^b \right]_j = E\left[S_{ai}^b \right]_j T \frac{n_s}{t_s} \tag{4-90}$$

In Equation (4-90), T is the estimated life of the riser; and n_s is the stress amplitude observed over cumulative time:

$$E\left[S_{ai}^b \right]_j = \frac{1}{\lambda^b} \Gamma\left(m + \frac{b}{c} \right) \left[\Gamma(m) \right]^{-1} \tag{4-91}$$

In Equation (4-91), λ, m, and c are normality gamma parameters; Γ is a gamma function.

3. Titanium alloy flexible riser local buckling and reliability analysis

Bordet applied Von-Misses yield criteria and the nonlinear finite element method to calculate flexible deep-water riser local buckling, and based on numerical analysis, analyze the reliability.

g(x) represents the maximum bending moment limit state function in a given external pressure; x is the yield strength and Young's modulus, so the limit state function $\widehat{g}(x)$ that is associated with the response surface function can be written as the following:

$$\widehat{g}(x) = g(x) - c \tag{4-92}$$

In Equation (4-92), c is a specific lower limit of the response amount g(x); if it exceeds the limit, then $\widehat{g}(x)<0$, and fatigue failure will occur. Accordingly, we can obtain the deep-water flexible riser fatigue failure probability P_t:

$$P_f = P[\widehat{g}(x) \leq 0] = \int_\Omega f_x(x)dx \tag{4-93}$$

In Equation (4-93), the function $f_x(x)$ is the joint probability density function that determines the parameters.

4.4.5.4 Deep-Water Riser Vortex-Induced Vibration

In recent years, many overseas scholars have studied deep-water riser vortex-induced vibration problems, and given some software for analysis.

1. The influence of vortex-induced vibration on deep-water risers
   There are two aspects of influence that vortex-induced vibration has on deep-water risers:
   a. Increasing the deformation of the riser
      When the vortex is distributed, the drag force acting on the deep-water riser will increase causing increased deformation of the deep water riser.
   b. Riser fatigue damage
      Because the vortex distribution arouses the vibration of the riser, the effects of the dynamic stress amplitude cycle vibration will lead riser fatigue damage.
2. Fatigue analysis software for steel catenary deep-water riser vortex-induced vibration
   Maher and Finn forecasted steel catenary deep-water riser fatigue damage, giving the vortex-induced vibration fatigue time domain and frequency domain calculation procedures ABAVIV and CATVI. The two programs are the foundation calculations that can predict the deep-water riser's (rigid) fatigue life.
3. Flexible deep-water riser strength and fatigue life evaluation software
   R.Gant and R.Litton gave a vortex-induced vibration procedure in time domain for the flexible deep-water riser. It can be used to calculate the deep-water riser vortex-induced vibration, and we can compare the calculated results with large-size simulation test data. They used this software to adopt a time domain method, simulate any tube structure, and analyze the dynamic response of wave-induced motion. The results of the analysis can assess the stand pipe strength and fatigue cumulative damage, and predict the strength and fatigue life of the flexible riser in deep water.

Special Problems in Sea Petroleum Engineering for Beaches and Shallow Sea Areas

The shoreline of China's Bohai Bay is 1150 km long. Around the Bohai Bay there is the Liaohe oil field, Jidong oil field, Dagang oil field and Shengli oil field. According to the second national oil and gas resource evaluation, in the beach/shallow sea area between land and sea and where the water depth is $0 \sim 5m$, the underground oil and gas resources are so rich that the divinable prospective oil resources are $(30 \sim 40) \times 108$ t. So this area is one of the important oil and gas reserve replacement areas in eastern China, and also one of the major areas of recent oil and gas exploration and development in eastern China. Very recently, the Nanpu oil field was found in Caofeidian, Tangshan, China in the beach/shallow sea area. In order to rapidly improve oil and gas production in China, this beach oil fields are a key project. But this broad beach/shallow sea area not only provides a very promising prospect for oil and gas reserves, but also provides severe challenges for oil and gas field exploration and development because of the difficult environmental conditions in these areas.

The beach/shallow sea area includes beaches and extremely shallow sea of $0 \sim 5m$ water depth. In this area, oil and gas exploration and development is different from inland, general sea, and the desert. Because of the special working environment, engineering geology, and special technical equipment, the existing mature experience of offshore and onshore oil fields including desert oil field exploration are not appropriate for beach/shallow sea oil and gas fields. While there is some beach/shallow sea oil and gas field exploration and development experience overseas, the specific conditions are much different. So the experience used in the Beaufort Sea in Canada, the Amazon rain forest of South America, and offshore Indonesia is difficult to use for reference. In this case, China beach/shallow sea oil and gas field exploration and development has developed independently in recent years, and a new approach using the Chinese characteristics of offshore oil and gas field exploration and development has preliminarily grown over a long period. This chapter proposes to introduce the experience of China beach/shallow sea oil and gas field exploration and development according to the characteristics of China beach/shallow sea oil projects. Content involving onshore and offshore oil and gas fields will not be introduced in this chapter.

5.1 Overview
5.1.1 Division of Beach/Shallow Sea Area

Beach/shallow sea is a general term for the $0 \sim 5m$ depth water and beach area in China. The beach/shallow sea area can be divided into the following three areas based on different technical characteristics:

1. Intertidal zone (beach/shallow sea);
2. Amphibious zone ($0 \sim 2m$ depth water area)
3. Extreme beach/shallow sea ($2 \sim 5m$ depth water area)

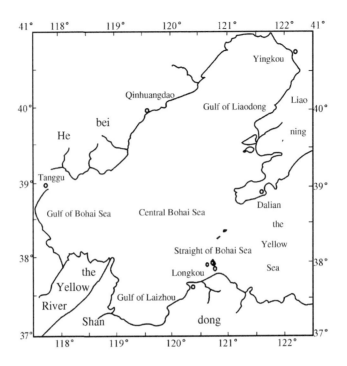

FIGURE 5-1

Bohai Sea area in China.

The Bohai Sea area in China includes Liaodong Bay, Bohai Bay, Laizhou Bay and the Bohai Sea central basin, as shown in Figure 5-1. Among them, the Liaodong Bay is the area from the mouth of the Daqinghe River in Hebei province to north of the Old Iron Mountain in the southern tip of the Liaodong Peninsula; Bohai bay is located on the west side of the Bohai Sea, in the area from the Daqinghe River mouth in Hebei province to the Yellow River Estuary in Shandong province; Laizhou Bay is located on the northern shore of Shandong Peninsula in the Bohai Sea, in the area from the Yellow River Estuary to the Jimudao Peninsula of Longkou.

The Bhallow sea of Bohai extends from Bayuquan in Liaoning province to Laizhou Bay in Shandong province. The coastline is 1150 km long. The total exploration area is $1 \sim 14 \times 10^4$ km^2. Among them, the Liaohe oil field beach/shallow sea area is 3412 km^2, the Jidong oil field beach/shallow sea area is 1500 km^2, the Dagang oil field beach/shallow sea area is 2257 km^2, and the Shengli oil field beach/shallow sea area is 4229 km^2.

The water depth near the coast of the Bohai Sea is shallow, generally within 10 m. The area near the river is much shallower, usually less than 5 m. Shallow places such as the river estuary are only 0.5m. There is a stretch of beach. The intertidal zone is relatively wide, and some zones are more than ten kilometers wide. The oil and gas exploration area of beach/shallow sea of the Bohai Sea is mainly distributed along the zone.

5.1.2 The Characteristics of the Marine Environment

In this chapter, the marine environment mainly refers to the environmental conditions of the offshore oil engineering. It mainly includes the climatic characteristics and marine hydrology environment characteristics. The hydrological and meteorological data for the Bohai Sea area is presented in Table 5-1; it describes the marine environment of the five components of the Bohai Sea area.

With regard to the marine environment, the Bohai Sea area mainly has the following features:

1. Windy area

   The Bohai Sea area is windy, and the seasonal variation is significant. Monsoons are strongest in winter and spring, and weakest in autumn. So autumn is the best season for an oil offshore engineering operation.

2. Storm surge

   Because the Bohai Sea area is surrounded by land on three sides and faces the Yellow Sea on the east, the easterly winds will cause the addition of water at the windward beach/shallow sea area and then affect the sea level. This combination between wind increasing water and the astronomical tide together will form a storm surge whose tide height is much higher than that of the normal tide. This storm surge will bring a serious threat to beach/shallow sea area oil field operations.

3. Stormy waves are dominant

   Because the Bohai area is surrounded by land on three sides and the strait is narrow, the Bohai Sea has the characteristics of a closed inland sea area and it is not easy for the surge to come in. So the stormy waves are the main force. So when the main wind is the offshore wind, the beach/shallow sea area does not raise huge waves. But when the main wind is the onshore wind, the beach/shallow sea area will raise huge waves that will reach a maximum in the winter, especially in the beach/shallow sea area of the southern Bohai Sea, due to the open sea, long wind zone, and bigger waves.

4. Primary tide

   Tide is given priority in the trend of the Bohai Sea current, but in strong winds, especially the onshore wind, the wind current combined with the tide will make the current velocity of the beach\shallow sea area be the vector sum of both their velocities, increasing it significantly. As shown in Table 5-1, the maximum flow rate can be up to 5 knots.

5. Ice heavier

   The Bohai Sea belongs to the ice zone. The ice conditions of the beach/shallow sea area are shown in Figure 5-2. The ice conditions of the Bohai Sea are heavier, and each beach/shallow sea area is different. The most serious area is Liaodong Bay. The coastal fixed ice age is about two and a half months, and the fixed width is 5 km, with a maximum up to 10 km. The average ice thickness is $25 \sim 30$ cm, and the maximum ice thickness is 100 cm. In estuaries and fords ice stack height is up to $2 \sim 3$ m on average. The ice conditions of the Bohai Bay and Laizhou Bay are weaker than Liaodong Bay, and the data related to ice conditions can be found in Table 5-1 and Figure 5-2.

5.1.3 Characteristics of Engineering Geology

The beach/shallow sea area of the Bohai Sea has many rivers from the surrounding land that go into the sea. The Liao River, Hai River, and Yellow River carry large amounts of sediment into the sea for a

Table 5.1 The Main Hydrometeorological Data for the Bohai Sea

Sea Area	Project	Area km2	Maximum wind speed m/s	Monthly average maximum wind speed m/s	Monthly average minimum wind speed m/s	Often wind direction	Extreme maximum temperature °C	Extreme minimum temperature °C	The main rivers into the sea	Maximum depth m	Tidal nature	Maximum average tidal range m	Maximum possible tidal range m	Average wave height m	Maximum wave height m	Maximum wave period s	Maximum velocity section	Maximum water temperature °C	Minimum water temperature °C	Maximum salinity %	Minimum salinity %	Beginning date of ice	Ice date	End date of ice
Bohai Strait	Dalian city, Dalu island, Ocean island	28000	40	8.8	4.4	N NE	33.5	-12.7	—	78	—											—	—	—
Liaodong Bay	Yingkou, Hulu Island	28000	40	5.1 - 6.8	3.5	SW ENE	34	-23.6	Liao River Luan River	32	Irregular semidiurnal											Last third of November	105 - 121	Last third of March
Bohai Bay	Tanggu	14000	28	6.7	3.7	SW SE	38.5	-18.3	The Yellow River Haihe River Ji River	28	Irregular semidiurnal											Middle third of December	90 - 110	Last third of March
Central Bohai	Qinhuangdao	29800	32			SW NE	33.5	-12.7	—	30	Regular diurnal											—	—	—
Laizhou Bai	Longkou	11000	40	7.1	3.1	SSE	39.8	-21.3	The Yellow River Wei River	18	Regular semidiurnal											Last third of November	100 - 121	Last third of March

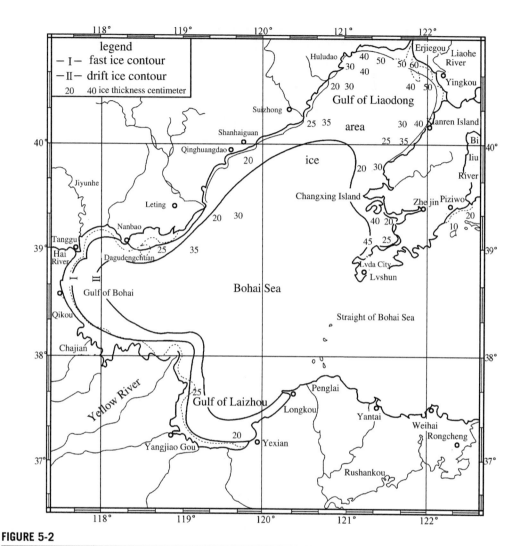

FIGURE 5-2

Overview of beach/shallow ice conditions of the Bohai Sea area.

long time, which constantly deposit, thus forming the geology characteristics of alluvial fan, flat seabed, and multilayer deposition in the beach/shallow sea area. The specifics are discussed in the following sections.

5.1.3.1 Engineering Geological Profile Varies with Area Location

Because of the difference of the river alluvial conditions and the influence of waves and currents, the engineering geological condition of the beach/shallow sea areas of the Bohai Bay, Laizhou Bay, and Liaodong Bay are different.

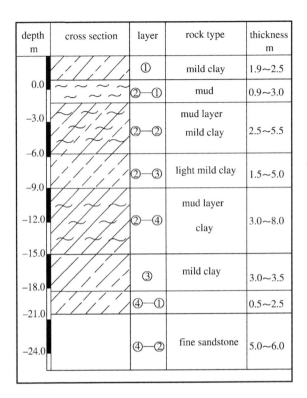

depth m	cross section	layer	rock type	thickness m
		①	mild clay	1.9~2.5
0.0		②—①	mud	0.9~3.0
−3.0		②—②	mud layer mild clay	2.5~5.5
−6.0		②—③	light mild clay	1.5~5.0
−9.0				
−12.0		②—④	mud layer clay	3.0~8.0
−15.0				
−18.0		③	mild clay	3.0~3.5
−21.0		④—①		0.5~2.5
−24.0		④—②	fine sandstone	5.0~6.0

FIGURE 5-3

Engineering geological section of the Bohai Bay beach area.

1. Beach/shallow sea area of the western Bohai Sea (Bohai Bay)

   This mainly refers to the Bohai Bay area, and the beach and neritic seabed of this area is very smooth. The gradient of the intertidal zone is 0.5% ~ 1%, and the width is 2.5 ~ 3.5 km. The gradient of the amphibious zone seabed is 0.77% ~ 1.5%, and the width is 4.5 ~ 7.0 km. The width of the 0 ~ 5m extreme beach/shallow sea is 20.4 ~ 115 km, and the maximum gradient is 2%. The typical engineering geological profile of this beach/shallow sea area is shown in Figure 5-3.

2. Beach/shallow sea area of Laizhou Bay

   This area is different from Bohai Bay. Taking the Chengdao oil field developed for many years as an example, the stratum of the Laizhou Bay area can be divided into modern Yellow River delta front deposit, the holocene deposits, and the pleistocene deposits from top to bottom. Table 5-2 shows the data of a north Chengdao 12-hole engineering geological section located in the Laizhou Bay waters. This table reflects the engineering geological characteristics of the beach/shallow sea area.

Table 5.2 Engineering Geological Profile Data for the Laizhou Bay Chenbei 12-Hole Section

Horizon	Soil Name	Soil Depth m	Thick Layer m	Moisture %	Density g/cm3	Relative Density	Void Ratio	Compression Factor cm2/kg	Quick Shear	Consolidation	Quick	Allow Carrying Capacity	Characteristic
1	Silty sand	0 - 2.8	2.80										Lithology is uneven, with clumps of clay
2	Clayey silt	2.8 - 4.9	2.10										Lithology is uneven, with clumps of clay, low plasticity soil
3	Mild clay	4.9 - 8.9	4.00										Med plastic soil, shows soft plastic
4	Clayey silt	8.9 - 10.5	1.60										—
5	Mucky clay	10.5 - 13.4	2.90										High plastic soil, shows soft plastic - flow plastic silty sand mass
6	Mucky mild clay	13.4 - 15.6	2.20										Med plastic soil, shows flow plastic
7	Mild clay	15.6 - 19.21	3.61										Med plastic soil, composition is uneven, more silt sandwich
8	Clayey silt	19.21 - 26.72	7.51	—	—	—	—	—	—	—	—	—	Low plasticity soil
9	Silty sand	26.72 - 28.2	1.48	—	—	—	—	—	—	—	—	12.3	
10	Clayey silt	28.2 - 35.67	7.47	—	—	—	—	—	—	—	—	14	—
11	Silty sand	35.67 - 40.32	4.65	—	—	—	—	—	—	—	—	14	—
12	Mild clay	40.32 - 43.64	3.31	—	—	—	—	—	—	—	—	12	—
13	Clayey silt	43.63 - 49.04	5.41	—	—	—	—	—	—	—	—	14	—
14	Silty sand	49.04 - 52.72	3.68	—	—	—	—	—	—	—	—	17	—

3. Beach/shallow sea area of Liaodong Bay

The coarsening impact of estuarine change and wind power has little effect on the engineering geological conditions of the Liaodong Bay beach/shallow sea area because its strata are formed by the alluvial fan. So its engineering geologic conditions are between the western Bohai Sea and north Chengdao beach/shallow sea areas in general. Figure 5-4 shows the engineering geological drilling profile of the LH-1 hole in the south sea area of the Shuangtaizi estuary of northern Liaodong Bay. This reflects the engineering geological characteristics of this beach/shallow sea area.

5.1.3.2 Deep Silt Layer with Poor Engineering Properties

Another feature of the Bohai Sea area engineering geology is that there are deep silt layers in the entire Bohai Sea area. Among them, those in the western Bohai Sea area are the most prominent. Generally there is a plastic mud flow of about 0.5m to just more than 1 m thickness on the seabed surface. But in some sea areas of the Dagang oil field, the mucky clay will stretch down more than 11 m. The engineering properties of silt layers of this depth are poor, and have a great influence on petroleum engineering.

1. Low carrying capacity and prone to slip

This silt layer is not only difficult to pass, but also causes gravity-type equipment or engineering structures to slip because of the low carrying capacity.

2. Unable to use a gravity-type pile foundation

With external disturbances this silt layer can easily lose strength and generate creep, and it cannot serve as a gravity-type pile foundation.

3. Not suitable for opening a channel for dredging method

The silt layer under the action of currents and tide will produce a strong back silting effect, and back silt fills the trench. So, a seabed where back silting is serious is not suitable for opening a channel for the dredging method.

5.1.3.3 Easily Liquefied and Scoured Silty Sand Layers

In the entire Bohai Sea area there are silty sand layers all around, especially in the Chengdao area of Laizhou Bay. In the Chengdao area there is a 2.8 ~ 5.2 m depth silty sand layer on the surface of the seabed. Clay grain is contained in the silty sand layer. It has a bearing capacity of about 9.2 ~ 12 t/m^2, and can be used as a supporting layer for the submersible drilling platform. But there are two major drawbacks:

1. The soil structure can be easily destroyed and then liquefied

When the silty sand layer is shocked, the structure is easily destroyed and then liquefied. This will make pile foundation or gravity structures unstable.

2. The soil structure is loose and vulnerable to scouring

This kind of silty sand layer structure is relatively loose, therefore it's vulnerable to scouring due to current effects, so the bottom of the submersible drilling platform can be hollowed out and then the platform slips.

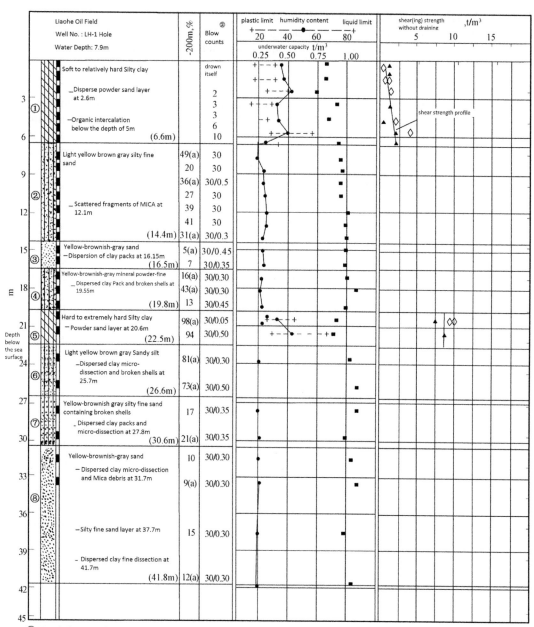

FIGURE 5-4

Engineering geological profile of the LH-1 hole in the Liaodong Bay sea area.

5.1.3.4 Complex Unstable Seabed Landform

Because of the dynamic effect of the estuary changes, stormy waves, seabed, etc., the seabed landform of the Bohai Sea beach/shallow sea area is complicated and unstable, which is most evident in Liaodong Bay and Laizhou Bay.

1. Liaodong Bay
   The Liaodong Bay beach/shallow sea area has formed a seabed where the landform moves up and down and the changes are complex due to the cutting function of the river and the scouring effect of waves and currents. For example, in the Shuangtaizi River and Liaohe River estuary, there are some independent sand and mud flats not connected to land and beach. They sink in the water at high tide and emerge from the water at low tide, and the water depth changes a lot. There are some underwater beaches, sand ridges, and underwater channels, which increases the complexity of the seabed.
2. Laizhou Bay
   In the Chengdao Sea area, for example, the coast is erosive muddy beach in the Yellow River delta. Because of the Yellow River diversion and its strong erosion, the coastline is falling back to land at a faster speed; water depth is on the increase. The original mouth spit has gradually transformed into a spit-type sandbank that extends along the coast. Many unstable landforms such as potholes, collapsed depressions, submarine landslides, and mud flow tongues have developed. This has formed complex terrain, with an unstable settled layer area on the submerged delta. In the area, smooth and stable sea only accounts for about 35%. The unstable seabed with a strong disturbance traverses across the whole area from west to east, about 10 ~ 12 km in width.

5.1.3.5 The Seismic Region Influenced by Fault Movement

The North China Plain is one of the seismic active regions. The main active fault zones are the three large fault zones as shown in Figure 5-5.

The Eastern Zhangjuhe River district of Bohai Bay is located on the edge of the three active fault intersections. Any of the three faults' tectonic movements will directly affect the surface activity of this region. The largest seismic belt, the Tanlu seismic belt, is in the eastern part of the western Bohai beach/shallow sea area in eastern China, and several earthquakes have occurred in the past. In the north of the beach/shallow sea area is the Yanshan Bohai fault belt, and the tectonic movement there is also very strong. In the past 1000 years, 22 earthquakes greater than 6 magnitude have occurred. In the west of the beach/shallow sea area is the Hebei Plain fault zone, and both the groove seismic frequency and intensity are stronger. In history, there have been 13 earthquakes of magnitude 6 ~ 6.9, 5 earthquakes of magnitude 7 ~ 7.9 and 1 earthquake of magnitude greater than 8. The effect of the strong earthquakes of this beach/shallow sea area are shown in Table 5-3.

1. Periodicity of seismic activity in the western Bohai beach/shallow sea area
   The seismic period is around 300 years in Northern China. The seismic activity of thewestern Bohai Sea area is almost synchronous with the North China Plain over time. Now the seismic activity in Northern China is in the tail of the summit. In the next 50 ~100 years a strong

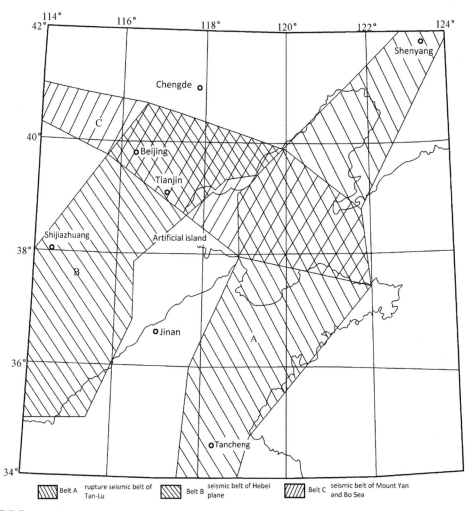

FIGURE 5-5

Distribution of the three major fault zones in the North China Plain.

earthquake may still occur, but the intensity and frequency will be lower than the level of 1966-1985 in Northern China.

2. Repeatability of seismic activity in the western Bohai beach/shallow sea area
   Seismic activity is repeatable. The weaker earthquakes are, the higher the repetition rate will be; on the other hand, the stronger earthquakes are, the lower the repetition rate will be.

Table 5.3 Effect of Strong Earthquakes in the Western Bohai Beach/Shallow Sea Region in History

Time	Location	Scale	Epicentral intensity	Regional influence degree of crack
1548.9.13	Bohai	7	> VIII	IV – VI
1568.4.25	Laizhouwan	6	> VII	IV – V
1597.10.26	Bohai	7	> VIII	VI – VII
1668.7.25	Tancheng	8.5	XII	VI – VIII
1679.9.2	Pinggu	8	XI	V
1829.11.19	Yidu	6	VIII	IV
1888.6.18	Heze	7.5	> VIII	VI – VIII
1937.8.1	Heze	7	IX	IV
1966.3.22	Heze	7.2	VIII	V
1969.7.18	Xingtai	7.4	VIII	VI – VIII
1976.7.28	Tangshan	7.8	XI	VI – VII

Darthquakes of a magnitude equal or greater than 7 haven't been repeated in the original area. In 1969 a 7.4 magnitude earthquake occurred in the western Bohai Sea, but magnitude 6 aftershocks has been rare in the aftershock series. So in the coming decades the possibility of late aftershocks in this area is very large.

In any event, the Crustal Stress Institute of the National Seismological Bureau has concluded that the largest earthquake intensity of the eastern part of the western Bohai Zhangjuhe area in history is VI degrees. So, in the next 100 years the biggest effect from the surrounding area is not greater than VI degree earthquake intensity. This region does not have the ability to see earthquakes equal to or greater than 6 magnitude in the region, but there may be magnitude 5 earthquakes, and the epicenter intensity is a maximum of VI degrees.

5.1.4 Characteristics of Technical Equipment

In beach/shallow sea area oil field development, both drilling and production operations, in terms of the technology, are not much different from onshore and offshore development. But the technology and equipment has its own characteristics.

5.1.4.1 Special Requirements for Operating Equipment

In drilling operations, for example, the mobile drilling platform is used, and the special requirements of the platform are "ease of getting in (transportability), and standing firm." In sea areas with depth greater than 2.5 m there are some mature technologies such as the bottom-supported platform and jack-up drilling platform, etc. Their movement and positioning technology is mature. So using them in the beach/shallow sea area of 2 ~ 5 m depth is not difficult. But in the $0 \sim 2$ m depth amphibious zones and beach areas, because the mud is deep and the water is shallow, the bottom-supported or jack-up platforms are very difficult to "get in" and "stand firm." Special requirements exist for drilling equipment transport and location that requires them to be able to "get in" and "stand firm" in amphibious zones and beach areas. It is one of characteristics that what kind of solution and method should be used to solve the special problems.

5.1.4.2 Special Requirements for Vehicles

In addition to the operating equipment, in order to carry out drilling, production and other operations, there must be unobstructed vehicles for the supply of equipment, transport, and subsistence. In an extreme beach/shallow sea area a ship can be used as transport. But in the amphibious area and the beach, the silt conditions are serious; it is difficult for ships. The method of opening up a canal is not only difficult to achieve, but also has high costs. Building a long canal to stretch into the beach area can only be used in in-shore areas and the cost is high. Foreign scholars have developed a variety of beach sea area vehicles, such as tire vehicles, tracked vehicles, metal roller wheeled vehicles and hovercraft and air boats, but they are required to adapt to different environmental conditions. Therefore, an effective vehicle should be developed for the characteristics of the beach of the Bohai Sea region and the conditions of different regions.

5.1.4.3 Special Requirements for Controlling Oil Spills

With the development of oil fields in beach/shallow sea areas, environmental pollution and the prevention and treatment of such are inevitable. There is a potential for accidents and causing massive oil spills in the process of drilling, production, and transportation. For general pollution from oil spills on the sea there have been some mature prevention and control measures. However, in particular areas such as extreme beach/shallow sea, amphibious zones, and beaches, large-scale oil spill prevention and treatment is inadequate, and must be developed further. Oil spill prevention and control for the beach/shallow sea area has special difficulties including the following:

1. Complexity of the mixture of sediment and impurities
   In the beaches and amphibious zones there are large amounts of sediments. If and oil spill accident, a large amount of sand will be mixed into the oil spill and will cause special difficulties for clearing the oil spill.
2. Complex impact of the terrain and the tidal current
   Compared with the inshore area, the terrain of the beach/shallow sea area is more complex. For example, the silt swamp of the beach area is not very conducive to clearing an oil spill; the rise and fall in a wavelike manner of the tidal current in the amphibious zone also increases the difficulty of the cleanup operation. In a word, these complex effects all make oil spill prevention and clearance more difficult than for coastal waters.
3. Difficulties of beach/shallow sea and beach cleanup
   The depth of the extreme beach/shallow sea is low. If an oil spill occurs, the cleanup is also more difficult than for the inshore area. Similarly, we have the difficulty with beach cleanup

described above. These are the characteristics that must be considered for oil spill prevention in beach/shallow sea areas.

5.2 Seawalls and Artificial Islands

The construction of seawalls in beach/shallow sea areas is commonly used in the exploration and development of oil and gas at home and abroad. Along with the deepening of exploration and development of oil and gas resources, the method of construction of an artificial island in beach/shallow sea areas has developed. With regard to the exploration and development of sea oil and gas in beach/shallow sea areas in China, the construction of seawalls began in 1965. In Dagang oil field, the first artificial island was built. Years of practice has proved that these two methods are effective in the exploration and development of sea oil and gas in beach/shallow sea areas under certain conditions. This section will present a brief introduction that focuses on the valuable experience of China in this regard.

5.2.1 Seawalls

China's 30 years of practice has shown that building seawalls, depending on the land, is not only successful but also economic for the beach/shallow sea oil and gas exploration. However, this method also has its limitations. The limited position of seawall construction is 1 to 1.5 km inside the 0 m depth line in the sea chart. Using a seawall to build a well site and a test well and then using directional drilling technology to drill large displacement directional wells can solve exploration problems in beach/shallow sea areas. For example, for the Dagang oil field in China, since 1965, 11 seawalls have been built, as briefly outlined in Table 5-4.

Table 5.4 Dagang Oil Field Profiles of Seawalls Built before 1993

Seawall name	Year	Length,km	Natural ground elevation,m
25 Wells	1965	1	1.8
69 Wells(3 Wells)	1968	1.8	0.62
21 –66 Wells	1968	1.8	1.343
77 Wells	1969	1.8	0.48
18 Wells	1969	1.8	0.5
80 Wells	1970	2.28	0.96
24 –66 Wells	1970	1.37	2.47
9 –69 Wells (539 Wells)	1972	0.65	0.88
25 –66 Wells	1985	9.01	2.08
Gangdong seawall	1988	4.85	−0.2
Zhang Don seawall	1993	2.2	−0.08
Total		28.56	11

5.2.1.1 Seawall Classification

1. Seawall categories

   They can be divided into two types: causeway and jetty.

   a. Causeway

   A seawall is used to block the water, and there is a surrounding land area with no invasion by seawater. The causeway trap area increases the area of oil and gas exploration, and can be the site of other comprehensive development. However, its disadvantage is the long duration and high cost, especially for a causeway parallel to the sea side, which is not easy to build and brings high risk.

   b. Jetty

   It extends from the shore into the sea, and sets a well field on its head. This jetty is just a connection of land and sea, so the advantage is short duration and low cost. But its drawback is less exploration area with a certain risk.

2. Seawall section categories

   It can be divided into four forms: vertical, sloping, hybrid, and composite slope types. At home and abroad, the sloping and vertical seawalls are commonly used. But taking the characteristics of the weak foundation of the China beach area into account, the composite slope type should be adopted due to its simple structure, convenient operation, suitability for soft foundations, higher overall stability, use of local materials, and ease of repair after destruction. The so-called composite slope type refers to a seawall which is sloping overall, but has composites of different section forms along the length of the different segments. For example, the Zhangdong embankment built in the Dagang oil field in 1992 has a 2.2 km long jetty. This jetty was made by different cross sections for the characteristics of different wave actions on all aspects of the jetty. In the 0 ~ 1800 m section it uses a single slope structure, and in the 1800 ~ 2200 m section and the well site at the end it uses the method of adding a "stage" (platform).

3. Seawall position

   There are two positions: the overwater seawall and underwater seawall. An underwater seawall refers to the usual sea wall. And an overwater seawall is to allow sea water swept over a certain height of the wall. A beach/shallow sea area seldom uses an overwater seawall.

5.2.1.2 Seawall Design

A seawall project is a large investment with high risk. Therefore, before building the seawall, a careful site selection should be done and it should be well designed.

1. Basic principles of design

   a. Try to consider the circumstances of oil and gas production

   For example, if the oil and gas production is unknown, the overwater seawall or a lower standard (such as the once every 20 years standard) ordinary seawall can be used. On the contrary, if the oil and gas production is expected to be relatively higher, a higher standard (such as the once every 50 years standard) seawall can be used.

   b. Try to use the jetty form to design the seawall

   Because of the good flexibility of the jetty, when the oil area discovered is large, the jetty can be connected to form a causeway at any time, and some offshore operating conditions can convert to land operating conditions to reduce the cost of exploration and development.

c. Try to shorten the distance to the well site

When using seawalls to solve drilling problems for a beach/shallow sea exploration well, it's necessary to try to shorten the length of the associated seawall from the shore to the well site. When building a seawall, the distance from the seawall to the shore is generally in the range of 1 ~ 6 km.

d. Try to use a nearby building materials

For things such as embankment material, try to obtain raw materials locally such as beach shoals soil, so that you can reduce the cost of the seawall.

e. Try to shorten the operating time in the tide period

For example, the appropriate use of new technologies and new materials can shorten the operating time in the tidal zone, thereby reducing the amount of engineering and cost. In short, more economical, better, and faster construction of the seawall are the basic principles in seawall design.

2. Determination of the tide elevation

When the seawall is designed, the tide elevation is generally determined by the following formula:

$$Z_p = h_p + R_{F\%} + \Delta h \tag{5-1}$$

In the formula, Z_p means the design tide elevation of the design annual frequency, in m. h_p means the design high tide value of the design annual frequency, in m. $R_{F\%}$ means higher climb value with an accumulation (F%) of the design wave, for example, the Dagang oil field beach/shallow sea region often has an $R_{F\%}$ equal to 0.8 ~ 2.9 m. Its units is m. Δh means ultra-high safety value and we usually take Δh equal to 0.5 ~ 1.0 *m*. Its unit is m

The design standard of the common seawall is based on conditions that happen once in 50 years, so the design annual frequency can be determined.

Usually on the top of a seawall there are anti-wave parapets, and the elevation of the top of the wall is generally 1.2 to 1.4 times the wave height safety high level, which is the height of the 1% highest wave at the safety high level.

3. Choice of the form of seawall

The choice of the form of seawall includes the use of a causeway or jetty and whether the form of seawall is sloping, vertical, composite sloping or mixed type. In addition, whether the seawall is over the water also needs to be considered. For the Dagang oil field beach/shallow sea region, the construction of seawalls mainly selects the overwater composite sloping jetty. If necessary, the jetty can be expanded to a causeway.

4. Ground treatment program

The surface silt in the beach of the Bohai Sea area is very soft. Therefore, a treatment program against the soft foundation must be given in the design. Usually treatment methods against the soft foundation are the desilting and changing the soil, the sand pile method, the gravel pile compaction method, bagged manhole and plastic (paper) plate vertical consolidation drainage method, the throwing rubble method. and the laying of geotextile. But years of practice shows that among these six methods, the laying of geotextile to deal with soft foundation, with easy construction, a short period, and less investment, is most suitable for building seawall in the Beach Sea region in China. Now some details of laying geotextile are described below:

a. The choice of geotextile

Geotextile is a nonwoven needle. Due to its ability to adapt to large ground deformation and good permeability with small holes in it, it can be used as a reinforcing and isolating seawall mud foundation. Currently, CT-300S acupuncture nonwoven fabric is used in the China beach/shallow sea area, and its technical performance parameters are shown in Table 5-5.

b. Geotextile mechanism

i. As the drainage blanket of a foundation surface layer, geotextile has strong permeability, forming a well-drained surface, which can accelerate the foundation soil pore water discharge to increase the strength of the foundation.

ii. Under body load, there will be compression for the foundation, and the geotextile will be under the action of tension. However, due to the tensile strength of geotextile and the shear resistance between it and the foundation, it can constrain the lateral displacement of the foundation. In addition, the joint action of the geotextile and Jing Ballet piece forms a cushion with different stiffness with the seawall soft ground (see Figure 5-6), which also makes it a flexible piece of the seawall foundation. Therefore, through this cushion, it can not only pass the levee load to the soft ground, but also isolate the tensile failure zone of the seawall from the shear plastic zone of the foundation of the seawall. As a result, it can maintain the overall stability of the seawall and improve the bearing capacity of the foundation.

c. Ultimate bearing capacity of the foundation

When geotextile is used, the ultimate bearing capacity of the foundation P_{stc} can be calculated as follows:

$$P_{stc} = \alpha C N_c + 2P \sin \theta + \beta N_q \cdot p/r \qquad (5-2)$$

In this formula, α, β are the shape parameters of the foundation soil. C means the cohesion of soil, kPa. N_p, N_c are the bearing capacity parameters related to the internal friction angle of the soil. θ means the tilt angle of the geotextile at the edge of the foundation (°). r means the radius of the imaginary circle, m. P means the tensile strength of the geotextile (checked from Table 3-3), kN/m.

α, β, C and N_p, N_c can be checked from soil mechanics reference books. θ and r are given in the design of seawalls.

The result calculated shows that after the geotextile method is used, the bearing capacity of the foundation is 12% higher than the natural foundation.

Table 5.5 Technical Performance of the CT-300S Geotextile

Mass per unit area, g/m^2	Zonal strength, N/5 cm	Tear strength, N/5 cm	Tear strength,N	Permeability coefficient,cm/s	Effective aperture, mm
300	514	578	553	1.8×10^{-2}	0.053

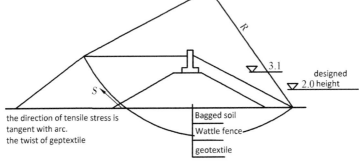

FIGURE 5-6

Analysis of seawall after foundation reinforcement with geotextile.

d. Overall stability

When using the geotextile method, the safety factor of the overall stability of the seawall should be raised. Usually the overall stability of the seawall is expressed by safety factor K, that is

$$K = \frac{M_R}{M_S} \tag{5-3}$$

In this formula, M_R means the torque against dike displacement, and M_S means the torque against levee sliding.

After using geotextile, in the calculation of the overall stability of the levee, there are two models as follows:

i. Sweden calculation model

Set the safety coefficient after geotextile use as K', then

$$K' = [M_R + S(a + b \cdot \tan \phi_1)]/M_S \tag{5-4}$$

In Equation (5-4), $S \cdot a$ is the sliding skid torque of the center of the sliding circle generated by the pulling force of the geotextile along the horizontal direction, as shown in Figure 5-6(a). And $S \cdot b \cdot \tan \phi_1$, when there is inner vertical cracking in the embankment, is the friction torque generated by the geotextile, as shown in Figure 5-6(a), where φ_1 is the internal friction angle of the embankment. As seen from Equation (5-4), the two skid torques are increased after the use of geotextile; therefore, the stability safety factor is raised.

ii. Netherlands calculation model

This calculation model assumes that at the cutting of the sliding arc, there is a distortion that adapts to the sliding arc in the geotextile, and that the tension of the geotextile S is directly tangent to the sliding arc, as shown in Figure 5-6(b). A skid torque $S \cdot R$ will be generated; thus after the use of geotextile, the safety factor K' of overall stability of the seawall increases, and

$$\Delta K' = \frac{S \cdot R}{M_S} \tag{5-5}$$

In the formula, R means the radius of the sliding arc as shown in Figure 5-6(b); M_S means the torque force against levee sliding.

Calculations show that $\Delta K'$ increased by about 4%.

5.2.1.3 Seawall Construction

A whole seawall construction project is divided into two stages in accordance with experience in the beach area.

1. First stage

   The low tide period runs from November to March for China's Bohai Sea, so the first stage of construction is in general scheduled in winter. Due to the reduction in influence of sea conditions on the construction and increased operating time, the seawall construction can be easily achieved. The work of the first phase is as follows:

   a. Basic construction

   Foundation construction is mainly a treatment for the soft foundation. When using geotextile, we should lay the geotextile first, and each geotextile should be vertical to the embankment and be laid in turn extending along the embankment. The width of two laps among them is 0.1 to 0.15 m. Then lay Jing Ballet piece on geotextile in order to increase flexibility and porosity. Soil bags or rubble are used to fix Jing Ballet piece to prevent floating. In this way, the geotextile and Jing Ballet piece together constitute the drainage layer, which can effectively control the uneven settlement of foundation. On the Jing Ballet piece, it should masonry with the bagged soil (see Figure 5-6). It is with the requirement of pocket binding tightly, occlusal masonry, cracks staggered to form one. The masonry height should be 0.5 m higher or lower than the tide level in each intertidal tide, in order to reduce wave erosion. Foundation construction should precede continuing the heightening of the seawall, preferably by 4 to 5 days, so that there will be a pre-press process for the foundation and the bearing capacity is improved.

   b. Raising the seawall

   The seawall height is about 1 m above the high tide. The filling process is that 'low elevation filling, focusing on extending, front and back the cross-operation, follow-up synchronization heightening'. That is, the key job is to extend the length and width of embankment to open it to traffic so bulldozers can roll during low tide to fill soil to raise the embankment above the construction water level before the high tide. With the extension of the seawall, the influence of the tide increases. In order to avoid building the embankment directly in the water, we can use bagged soil on the foundation to form a small dam so that there is no or less water inside, thereby effectively controlling the moisture content of the filling and reducing the loss of earthwork.

c. Temporary slope protection

Due to the fact that the direct propulsion method is used in the construction of the seawall and a dam is not built, during construction, the bagged soil and filling cannot withstand the long-period impacts of the wave, so we must build slope protection during the construction to ensure safety. This slope protection is temporary surface protection during construction. It is divided into inner, middle, and outer.

Middle surface protection is slope protection for the interim period in front of the rubble, covering the inner surface protection. Use the method of multilayer masonry of geotextile bags in order to resist the loss of filling from the tide. The outer surface protection is the rubble slope, and geotextile is an inverted filter layer which must be strictly constructed according to design requirements. For the geotextile layer along the embankment, the lower take the pressure of the upper, with laying from the bottom up. Meanwhile, we complete the rubble slope, and promote up layer by layer, but attention should be paid to prevent damage to the geotextile caused by the rubble slope. The inner layer is built by smaller weight rubble, using the method of inserting laying rubble and embedding and fixing it, in order to increase the overall stability of the levee.

2. Second stage

The first phase of construction is completed, which basically reaches the land construction stage. The second phase of construction is as on the land. The work in this stage includes:

a. Continue layer heightening

In order to control the rate of increasing load, we should heighten the thickness of each layer 30 cm and ram each layer down to meet the design requirements. We should reserve settlement height according to the design elevation, and the covering area of each heightening layer should be as large as possible, in order to avoid local higher segments and soil slip.

b. Improve the body quality

The temporary retaining surface must be checked after a few months of seawater erosion before construction, and comprehensive maintenance must be conducted. The emphasis is on checking the inverted filter layer of the geotextile, to eliminate hidden dangers. Following is the inspection of the rubble surface protection, including the quality of rubble, the mass of rubble and the quality of inserted layering; all of this should meet quality standards and make the outer surface protection both strong and neat.

c. Placement of prefabricated grille

The prefabricated grille should be settled from the bottom to the top. The placement of the seat slurry at the bottom should be accurate and the error should be less than 2 cm. The horizontal and vertical lines should be aligned and both it and the outer surface protection are the first stage of wave dissipation facilities covering the seawall.

5.2.2 **Artificial Island**

The construction of the artificial island has a rather long history. An artificial island was used for petroleum exploration and development in the beginning of the 20th century. However, the development is very quick, and until 1990, 35 artificial islands have been built only in a certain sea area. More artificial islands have been built overseas. Artificial island construction technology research has taken place mainly in western oceanic countries, such Holland, who has constructed a large number of artificial islands, America, Britain, and the Middle East; Japan has also built some artificial islands. In order to meet the need of industrial land, since the beginning of the 20th century, Japan has built about 50 artificial islands. In Alaska and the Canadian Beaufort Sea, more than 20 artificial island were built in the 70s. American and Canadian oil companies, for the development of Beaufort Sea Basin oil and gas resources, have since 1972, have developed practices for island structure according to the island site depth, offshore distance, sources of construction materials, local topography, geomorphology and other factors. China's construction technology for artificial islands gradually developed in the mid-20th century. Before 1990, there were no artificial islands used in domestic oil and gas exploration and development. In 1988-1992, Dagang oil field built the first artificial island in China. Afterwards, in 1994 Liaohe oil field began the construction of the second artificial island for the exploration and development of offshore oil and gas. And then Dagang oil field completed the Chenghai artificial island by the end of 2006, and it was put into use at the beginning of 2007. Through the construction of these three artificial islands, China has gradually accumulated experience and created a new way suitable for the features of China Bohai Sea to build artificial islands. The following is brief introduction of the experiences in the construction of artificial islands of China for oil and gas exploration and development.

5.2.2.1 *Artificial Island Categories*

At home and abroad, there are five methods to build an artificial island as follows:

1. Filling type

   This was the early main method for foreign islands, accounting for more than 40% of the total number of artificial islands. First, build the surrounding embankment, usually called a berm or revetment. Then mountain sand, dredging sediment, or industrial waste is used for internal filling, so it is called a filling type of island. This is the original method to build an artificial island. This construction method is simple and low cost, but generally it is only used for beach/shallow sea and beach area adjacent to the coast. If this method is used for an artificial island used in oil drilling, in addition to the above-mentioned construction procedure, some procedures should be added, such as drainage consolidation on the island, piling and concrete steel structure installation, and equipment installation on the island.

   The key process to building this kind of artificial island is the construction of the berm. Methods for the construction of the berm are many, but commonly used methods are the two following:

   a. Caisson type

      This method is to build a berm with caissons. The caisson material can be steel or reinforced concrete. Caisson is first built on the shore or in plant, and then it is tugged to the towing place to sink down. A certain number of individual caissons are made based on the ability of the tugboat and marine environmental conditions. The height of the caisson should generally meet the design height for one-time production, but can also be produced in segments, then dragged to the sea and connected. After all caissons are dragged to the sea and sank down, it can be connected with molding in the field.

b. Steel sheet pile type

In this method, the steel sheet pile will be inserted into the mud, or piled. Then we connect the steel sheet into a berm, and the seawater is isolated. Steel sheet pile can be constructed by dividing into inner and outer rings, using construction equipment.

2. The pile type

The typical case of the pile artificial island is the jacket artificial island. This jacket is the same as the jacket in a fixed platform for offshore oil and gas drilling and production. Thus, its construction and the procedures and methods of construction are the same as the jacket introduced in Chapters 2 and 3 in this book, and not restated here. Obviously, this kind of artificial island is not suitable for beach/shallow sea and beach sea areas, and it is generally used for sea area with a water depth of 5 to 20 m.

3. Bottom-supported type

This form of artificial islands is similar to the bottom drilling platform. The entire artificial island is built and shaped in the factory, and then hauled to position by the floating craft; after positioning, other operations can be carried out on the island. Thus, 60% of the workload of the artificial island construction can be done on land, which can not only reduce the risk of construction, but also reduce the project cost. It can be used in beach/shallow sea and beach sea areas.

Obviously, the key technology to the construction of this artificial island is floating the entire island into position. For a soft foundation, usually the method of gravity sink is used. And for sandy foundation, the combination of a gravity-type sinking method and flush can be used, and sinking depth can be determined based on the foundation stability. If necessary, we can use piling or another load-bearing method. During the sinking, the load can be added or decreased at different orientations to ensure the accuracy of the position.

4. Jack-up type

This artificial island is similar to a jack-up drilling platform that can move. But it can only be used in beach/shallow seas, not the extreme beach/shallow seas or the beach/shallow sea areas with amphibious beach.

5. Floating

This is similar to the semi-submersible platform. The entire island is a semi-submersible floating body which can move, and it is also suitable for shallow and deep water areas, not just the beach sea region.

The three artificial islands that were built in China for oil exploration and development since 1992 are similar to the first in the six types above. But there is another form of independent innovation in China, the relevant details of which will be described in detail below.

The types of foreign artificial island are natural revetment artificial island, artificial island in winter, sandbag or stone revetment artificial island, and slope protection plate embankment artificial island, the floating trap artificial island, the ship skin trap artificial island, cross linked gel material trap artificial island, steel plate (pipe) pile trap artificial island, and caisson type artificial island.

5.2.2.2 Design of Artificial Islands Built by China

China has built three artificial islands for oil exploration and development, that is, China's first artificial island in the Dagang oil field, "Liaohai Sunflower 1" artificial island in the Liaohe oil field, and the Chenghai artificial island in the Dagang oil field. The type of the former two artificial islands is double wall steel caisson pile single ring up a whole pad.

1. Basic principles of design

   Artificial island construction comes with high investment and risk. Therefore, the design should be very careful to follow a few basic design principles:
   * Comply with the requirements of the relevant state laws and regulations.
   * Comply with the inspection standards of the National Bureau of Shipping.
   * Take the requirements of the drilling and production process into account.
   * Meet the engineering and construction economic requirements.

   The economic requirements mentioned here are to minimize the difficulty and workload of offshore construction, to shorten the time of offshore operations, and to make full use of the adjacent construction conditions.

2. The choice of the structure

   In accordance with the principal materials used in the construction of artificial islands, the structure of artificial island can be divided into the following two kinds:

   a. Gravel artificial island

   This is an artificial island filled by throwing or blowing the gravel. In the 1980s in north Dagang reservoir an artificial island was filled with soil, and belongs to the gravel artificial island category. Its method of construction is the same as with the construction of seawalls, and not repeated here.

   b. Steel-concrete artificial island

   This is mainly built by sinking steel molds and pouring concrete, and this method was used in "the first man-made island" built by China in Dagang oil field beach sea area. The selection of the structure is shown in Figure 5-7.

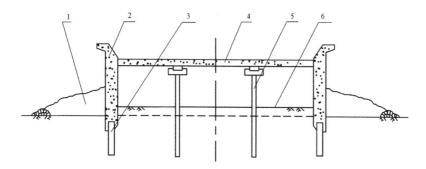

FIGURE 5-7

Ring wall and steel mold: structure of the concrete artificial island: 1 - river bottom protection; 2 - island wall; 3 - pile foundation; 4 - the island surface 5 - island core pile 6 – blanket.

As shown in Figure 5-7, the periphery of the artificial island is a ring wall. The outer diameter of the ring wall is approximately 63.6 m, and the inner diameter of the upper and lower portions is different, respectively 61.6 m and 60 m. The total height of the island walls is 12.5 m, buried about 2.8 m. There are 36 cast-in-situ piles in the island wall, which have with a diameter of 1.3 m and a length of 22.4 m. The entire island is built by pouring reinforced concrete, of which the weight is almost 13000 t. In order to prevent erosion from ocean currents on subsea island wall, the wall around the island has riprap bottom protection along the direction of NE, SE with a range of 120°, 20 m outward from the wall. The elevation of the wave wall is 7.4 m, which extends outwardly 50 cm in order to prevent splashing waves. The core structure of the steel-concrete artificial island, shown in Figure 5-7, is divided into upper and lower layers. The upper surface of the island is 9.6 m and the lower island surface is 5.5 m high. The clearance of the lower island body is 4.5 m, of which the base is backfilled by mountain surface soil. The upper island surface is a high pile cap structure, and the designed surface load is 20 kN/m^2. The upper island surface is supported by 207 island core piles, which are 400 mm × 400 mm × 25.2 m prefabricated reinforced precast concrete square piles. On the top of the pile, there are pile caps, of size 1000 mm × 1000 mm (shown in Figure 5-7). There are embedded parts located on the pile cap for connecting with the island surface beams. The island surface beam has a height of 900 mm and width of 500 mm, prefabricated by C25 concrete on the shore. Generally the length of each beam is 5.0 m and the width is 2.0 m. There are embedded parts at both ends of the beam which has to be welded onto the pile cap when lifting. There is a crane set up at the northeast corner and the southwest corner of the island. The suspended deck of the crane has the same elevation as the upper island surface. The thickness of it is 1300 mm; the length is 6 m and the width is 6 m. The suspended deck is supported by four reinforced concrete piles with a diameter of 1300 mm. There are two wellhead grooves on the island, located in the south and the west of the island; the two grooves' top surface is at the same level as the elevation of the upper island surface. The height of the beam is 3.8 m, the length is 33.2 m, and the width is 3.65 m. In order to move the derrick, there are four tracks on the wellhead beam, which are supported by 400 mm × 400 mm precast concrete square piles. Thirty wells can be drilled for two well slots.

The living area in the artificial island is 409 m^2, and can contain 68 people. It is a four-story steel frame structure. All the buildings are 16.1 m high. The layers below the upper island surface are the drinking water tank zone, and the upper part is the residential area. On the living roof there is a small stop platform. The basis of the entire life building is a pile cap structure, where the piles are cast-in-place piles with a diameter of 800 mm and the suspended deck is a reinforced concrete suspended deck with the thickness of 800 mm.

3. The design of the steel mold

The design of the steel mold is important in the design of artificial islands; the structure of a steel mold is shown in Figure 5-8. The design of the overall structure shall be calculated in accordance with "Design of Steel Structures" and "Grid Structure Design and Construction Procedures."

a. Determine the distance from the top of the steel mold to the blade foot surface

As shown in Figure 5-8, setting the distance H as the design height of the top of the steel mold to the foot blade, you can write:

$$H = h + D + \Delta h \tag{5-6}$$

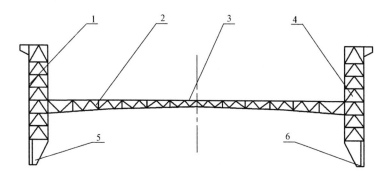

FIGURE 5-8

Ring wall steel mold for concrete-steel structure: 1 – rampart; 2 - large roof truss; 3 - large top panel; 4 - level of ring beam; 5 - pile positions; 6 - blade feet.

where h means design embedded depth, D means design high water level corresponding to the actual water depth, and Δh means surplus height (generally $\Delta h = 1.0$ m).

b. Calculate pad weight of steel model pad system

By using the cushion pads and towing, the island wall steel mold can be placed in the setting area of the beach. The pad system generally consists of power equipment, fans, island, and aprons, as shown in Figure 5-9. After the fan (2) presses air, under the action of the inner cavity pressure p_c, the apron (3) gets off the ground together with the steel mold (1), held by the air cushion. Due to the fact that the fan and power equipment depends on the weight of the entire pad, the design should calculate the pad weight W first as follows:

$$W = p_c A_c \qquad (5\text{-}7)$$

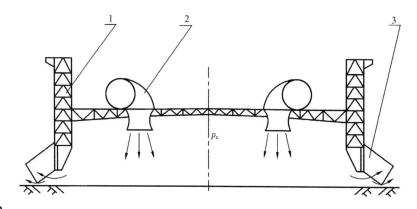

FIGURE 5-9

The composition of the artificial island steel die cushion pad lifting system: 1 - steel mold; 2 – fan; 3 – apron; p_c - inner chamber pressure.

In this formula, p_c means the cavity pad pressure, A_c means the area of the air cushion, and W means lifting weight.

c. Cushion pad flow calculation

Set the total flow rate of the cushion nozzle jet as $Q(m^3/s)$, and it can be written as

$$Q = \Phi L H_c \sqrt{\frac{2p_c}{\rho_a}} \tag{5-8}$$

where Φ means flow coefficient (generally take $\Phi = 0.6$), L means the length of the discharge, H_c means average flying height, p_c means lifting pressure, and ρ_a means the density of the air.

d. Stability calculation for cushion towing

During the cushion towing, the static pad stability and towing stability should be considered. The towage stability, influenced by wind and waves, can be determined by experiment and calculation, and specific requirements are as follows:

i. Lateral stability

Set the width of cushion to B_c and the height of initial metacenter to h_c:

$$h_c/B_c \geq 0.3 - 0.4 \tag{5-9}$$

ii. Longitudinal stability

Set the height of initial longitudinal metacenter to h_ϕ and the the taly tion of structure,tificial island length of cushion to L_c, then

$$\frac{h_\phi}{L_c} \geq 1.0 - 1.5 \tag{5-10}$$

4. The design of the safety system for the artificial island

For an artificial island, not only should the high investment and high requirements for the prevention of pollution of the marine environment be considered, but also the safety of maritime staff. Therefore, strong safety protection measures must be designed to ensure the safety and reliability of offshore oil and gas production and to prevent accidents. Usually we can be divide it into active and passive protection measures.

a. Active protection measures

Fire and combustible gas detection systems can alarm and automatically deal with dangerous situations in a timely manner; emergency shutdown systems may turn the subsurface safety valve off automatically to prevent accidents. Pressure relief and decompression systems should be provided, so the gas can be guided to a safe place in both normal and abnormal conditions, the gas can be guided to a safe place; also the commonly used fire protection systems and ventilation systems should be set up.

b. Passive protection measures

All kinds of facilities in the artificial island should be strictly arranged in accordance with the normal partition layout, and also should be divided into danger zones and nonhazardous areas. And fire-resisting walls and blast walls should be set at open flame devices. In addition, the escape channel and life-saving equipment as well as the broadcasting and communication facilities should be smooth and easy to use.

5.2.2.3 *Construction of the Artificial Islands Built by China*

Artificial island construction technology, which is based on civil engineering technology, is also closely related to marine engineering technology. The artificial islands built in China, of the "pile foundation, single-ring, double-walled, steel caisson, overall pad cushioning" style, are now briefly introduced in the following:

1. Initial preparation for construction

    Before the construction of the artificial island, the construction design should be well prepared. Using construction organization design books, materials, personnel, equipment, funding, and quality assurance, should be planned with as much detail as possible. Also, the precast plant and assembly plant for onshore construction should be ready to use. The construction site should be "leveled," with the island location and channel obstacles cleaned up to ensure smooth progress.

2. Onshore construction

    Artificial island shore construction is a combination of multiple high-tech processes, including the following:

    a. Construction site

    The artificial island is sunk down after being assembled onshore, so steel construction can only be along the beach for easy towage. In this way, the reclamation dam should be constructed around the assembly venue, and the road construction, steel mold assembly foundation, and basic cycle road construction should also be completed.

    b. Construction of base factory building

    This includes the construction of housing, warehouses, depots, and other premises.

    c. Performance test of pile foundation

    Artificial islands can be located in soft ground, using cast-in-place piles. Therefore, before construction on the sea, performance tests of piles should be carried out onshore to obtain data such as single pile load-bearing shear resistance.

    d. Steel field assembly

    The steel mold of the artificial island is divided into two parts, the wall of the island and the island core. The wall of the island is assembled by 36 blocks, and the core of the island can also be lifted and divided into a number of blocks. The assembly program is:

    Partly prefabricate in factory → assembly of the wall of island (leaving the last two sections for import and export) → assembly of the island core → close the wall of island → remove the supporting piles → quality inspection

    e. Air cushion system debugging

    For the artificial island air cushion system, the installation and commissioning procedures are: The installation and commissioning of wind turbine → the installation and debugging of aprons → the installation of the steel mold monitoring system → the installation of ballast leveling system

    f. The towage program

    The towage program of artificial island is the following:

    i. High strength power anchors + large deck barge + winch

    ii. Amphibious operation truck + shallow draft barge

    iii. (Hovercrafts + winch) + cable + artificial island steel mold + amphibious operation truck

Due to the specific circumstances of Dagang oil field beach/shallow sea region, the third option is appropriate.

g. Preparation of construction team

The team for the offshore construction of the artificial island should be well organized.

3. Offshore construction

Offshore construction jobs mainly include the following:

a. Towing in place

Towage should generally be carried out at high tide, and the placement and ballast and destruction of aprons should be carried out during the day.

First, the towage winch and initial tension should be lifted on a hovercraft after tensile testing. As shown in Figure 5-10, next to the refuge islands, rope should be laid between the half pile and coast, about 4125 m long. The end of the wire rope near the shore is fixed with the pile foundation onshore through the rope after crossing the winch on the hovercraft or tensing by the bulldozer. There are three other wire ropes on the bottom of the winch connecting to the steel mold of the artificial island, so that towage can be achieved.

After towing in place, the artificial island can be settled by a four-point positioning method.

b. Ballast sinking

After the artificial island steel mold is in place, it is necessary to do the first sinking using the ballast sinking method. It shall promptly release the cushion air to pour water into the wall of the island. Also, the handling equipment on the steel mold should be used to stack the construction materials (cement, sand, etc.) on the top of steel mold to increase the weight as soon as possible and sink down 1.2 m. Then, the first sinking procedure is realized.

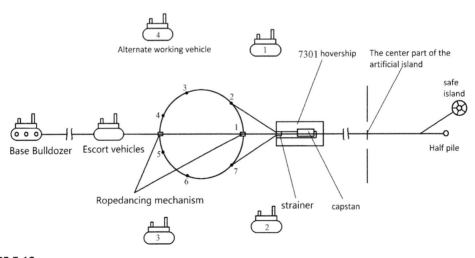

FIGURE 5-10

The artificial island towage program.

c. Island wall casting

After the first sinking is realized, concrete should be pouring into the 36 positions in the wall of the island immediately. The construction method of pouring concrete in a two-tier diagonal style should be used in order to uniformly sink the island body.

d. Pile foundation construction

The weight of the artificial island body must be carried by the pile. The pile foundation uses cast-in-place piles with sink pipe, which fuse with the body of the island. There are 36 cast-in-place piles, with a diameter of 1.3 m, length of 22.4 m and into-mud depth of 22 m.

e. Construction of the upper wall on the island

The upper wall on the island is constructed using steel mold material as the construction site. Jobs such as stirring together material, banding up the island wall steel, supporting, and concreting progress in turns until the casting of the upper wall on the island is complete.

f. Water backfill

After the construction of the upper wall on the island is complete, the steel mold, which is used for cushion and as a construction site, should be the first to be cut off. Then, drainage and consolidation of the core of the island should be done. After consolidation, a bearing ability of 40 to 60 km/m^2 could be reached by the method of backfilling soil (or sand). This is to create good conditions for further construction.

The first artificial island was designed and constructed by China after 16 months of construction, and completed in August 1995. It completed the drilling of three wells in 1996, and obtains oil and gas flow with high production and remains in operation. This fact indicates that China's independent innovation in artificial island engineering technology was successful.

5.3 Drilling Equipment and Its Movement in Beach/Shallow Seas

As for drilling technology, there are not many special points when drilling in beach/shallow sea region. It can basically take advantage of the existing onshore drilling and offshore drilling technology. In addition, in the beach/shallow sea areas where the water depth is deeper than 2.5m, mature drilling equipment, including drilling rig movement and location operations, with the bottom-sitting drilling rigs and jack-up drilling platforms. However, the beach area's special point is the beach and amphibious area. Because of the drilling equipment and its special movement difficulty, we must solve the "easily in and out" and "stand firm" problem. For example, a new directional drilling technology in the Shengli oil field using an adjustable AGS stabilizer and FEWD wireless MWD, with the new black positive gel silicon polymer drilling fluid, effectively solved technical problems such as friction and torque. Judging from exploration tasks in the beach area of the Bohai Sea, about 60% of of the exploration wells and 70% of appraisal wells are located in beach and amphibious areas. The recent discovered oil field in Tangshan Nanpu in China is mostly situated on the beach and amphibious areas. Problems with the drilling equipment and its movement have to be solved. Furthermore, taking into account the high risk of exploration drilling, oil fields have generally used mobile drilling equipment, and rarely involve the construction of fixed permanent buildings. Therefore, mobile drilling equipment in the beach area is generally appropriate. In short, the solution for movement of drilling equipment in the beach and amphibious areas for drilling work is the principal issue. This section is intended to discuss some aspects of this particular issue.

5.3.1 Drilling Equipment Overall Movement Program

5.1.3.1 Some Programs Existing at Home and Abroad

At present, for exploration drilling at beach or amphibious areas at home and abroad, there are some solutions for drilling equipment and movement programs. Some have been successfully used in a large number of drilling operations, and they are summed up in the following:

1. Construction of seawall stretching from the beach
    This program will build seawalls that extend into the beach to form areas of land, so we can take advantage of the seawall as a channel to apply mature transport techniques to move drilling equipment easily. Moreover by using the land we can make drilling equipment onshore drilling. Regarding the engineering problems of seawall construction, the last section has an introduction, and it is not repeated here. But what should be noted is that this solution to drilling equipment and movement is limited, because this can be only used for the drilling locations away from the shore where seawall construction conditions exist.

2. Dredge the waterway for drilling barge
    A dredger, which can be used to dredge the waterway, is applied in this program. Until the waterway reaches the requirements for shipping, sail the drilling barge to the well location, and then sit it in place, and finally drill. This program has been successfully carried out into the beach sea area of the Mississippi River in the United States for drilling operations for years, and has been successfully used in oil field development and production with the use of the seaplane as a means of delivery, which is not only required to transport the drilling equipment, but also for the establishment of the oil platforms, transport staff, and maintenance machines.
    A drilling barge is a mature piece of beach/shallow sea drilling equipment, whose main feature is solving beach/shallow sea access problems. Moreover, in recent years, it weight engineering has been used, the concept of rational arrangement of the maximum load which minimizes the weight of the barge and reduces barge draft greatly. The 0.9 m drilling barge is conducive to the beach/shallow sea area. However, with the use of barges, difficulties include wind resistance, poor stability, slipping on a layer of silt covering the seabed, and the limited ability to adapt to the water depth. Therefore, this program is mainly applied in the beach/shallow sea, inland river, and internal lakes.

3. The construction of the platform using module rigs
    This program is to build a small platform in the beach/shallow sea, and then use a rig that is modular and easy to move to move each module rig through small delivery means to the beach/shallow sea on a small platform. The last process is drilling. The MSH-2000 modular rig designed and manufactured in the USA can drill to a depth of 6000 m. The rig can be broken down into modules, which include a module for a winch drum and brake weighing 1.86 t and other modules that do not exceed 8 t. These modules are transported by helicopter. The derrick is divided into eight modules in transportation. The rig is 6.7 m and is easy to transport and install. This scheme has been successfully used in floating platforms on inland lakes in the United States as well as Africa and Pakistan in some difficult transit areas. The modular rig has also successfully been used in the swamps of South America's Amazon rainforest zone. However, this program requires the conditions allowing the building of a shallow small platform, or the modular rig is useless.

4. Module platform delivery system

   Such a program goes a step further than the third option. It not only makes the rig modular, but also makes the drilling platform modular. Then, the use of small vehicles solves the problem of drilling platform and drilling rig movement. For example, a fixed drilling platform is assembled by piles (used in the Peru Amazon rainforest swamp in South America). The platform is assembled with a light steel pile structure; the diameter of the pipe pile is only 12 3/4 in (323.85 mm) with a deck area of 79 m × 26 m. The rig is installed in the center of the deck floor. The residence building is at the end portion and a helicopter pad extends from a surface. All the components and modules of the platform and rig are transported by helicopter. When the platform is being assembled, we must first build a small temporary structure to install piling on it. This temporary structure can be used for piling for the assembled drilling fixed platform and lifting module. Obviously, the requirements of this program must include the necessary conditions to set up a small temporary structure.

5. Use and amphibious walk drilling barge

   This solution was invented by the Chinese, independently innovated and designed by the heart of the Chinese Academy of Engineering, Gu Yi, chief engineer of the China Shengli oil field. Amphibious walk means that when the drilling barge has the ability to be carried by sea, it can be carried from the sea to the well location. But when there are not conditions to carry by sea, it can also use the crawling step-by-step approach to shift from land to the well located at the beach. The ship's scale is 75.24 m × 43.14 m × 12. 6 m, which can be at a depth of 0 to 6.8 m during drilling operations, with a 4500 m drilling depth of the electric drive rig. The basic principle of the walk is that it adopts the mats of the body boats and the composition of the outer body and uses a vertical and horizontal hydraulic cylinder for the alternating movement of the mat of the inner body and outer body. The inner body moves one step with the body and then the outer moves one step. Each step is a stroke of the hydraulic cylinder, which is 10 m. In this way, you can crawl step by step, and the mat moves forward until you reach the beach well located in a difficult position. If the well is located in a navigable position, the barge can float by a tugboat from the water and then be dragged to the well locations. Obviously, this program is a solution for the beach sea region, including beach/shallow sea and the amphibious belt. The adaptability of the beach drilling equipment and its movement makes this a good program, but for areas such as the China Dagang oil field with a thick silt layer surface environment, there are still difficulties.

6. Using a variety of air-cushioned drilling equipment

   Such a program applies the mature air-cushion technology to beach sea drilling equipment, and takes drilling equipment (though hovercraft) to difficult to access amphibious areas. Then drilling equipment can be moved to regions with depth of less than 2 m with intertidal mud.

   Abroad, when using this program, the large cushion means of delivery is used for the entire migration of drilling equipment, and generally an independent air cushion barge is used. It doesn't install a propeller, but utilizes tractor towing with a full-cushion hovercraft. Such a non-self-propelled air cushion barge has high carrying capacity and low speed, suitable for the overall migration of the rig. For example the Russian 7500 t cushion drilling platform (total weight of 7500 t, cushion pressure 8000 Pa) can be loaded with a full set of drilling equipment. It is the bottom platform ride in the tide room with 0 to 2m of the beach-shallow seas of the drilling oil and gas wells. Dagang oil field of China and Institute 708 jointly developed a self-propulsion air-cushion barge whose load is 35 t, and has been put into use as a delivery vehicle for many years.

In addition, the Dagang oil field cooperated with some research institutes, and utilized air-cushion equipment in the beach sea region suitable for a jack-up drilling platform.

It can be seen from the above six programs that based on the drilling equipment movement, beach sea drilling equipment can be divided into two categories, namely the movement of the completed beach sea drilling equipment together and the modular transport of beach sea drilling equipment, which will be described later.

5.3.2 Movement of Complete Beach Sea Drilling Equipment

For the complete movement of drilling equipment (except for extreme beach/shallow sea) in beach/shallow sea areas we have mature experience to review in the harbor area drilling platforms, and there are beach/shallow sea drilling barges, air-cushion drilling equipment, and walking-type drilling equipment.

5.3.2.1 Beach/Shallow Sea Drilling Barge

The beach/shallow sea drilling barge is a bottom-drilling platform, and it involves the overall movement of drilling equipment which can adapt to a working depth of 2 m in very shallow beach sea areas. Because its structure and composition have been introduced in Chapter 2, it is not repeated here.

1. Basic form

   A drilling barge's (in beach/shallow sea) basic form is divided into the following two types:

   a. Well trough type

   The well slot of the beach/shallow sea barge is at one end, and it can be divided into a single-deck and deck post support structure. The former one, which only has simple mat-drilling equipment, is arranged in the mat deck surface, the so-called single deck; the latter one is above the mat, but also uses a number of pillars to support the high-level deck set drilling equipment, and is called post support deck type. Shengli oil field's Victory No. 1, No. 2, No. 3 and No. 4 belong to the latter. For example with "Victory 4," the total length and width is 64 m and 24 m. The mat depth is 4.3 m and the pillar height is 4.3 m. The deck's height is 9.1 m, the well groove is 3 m $\times$ 12 m (width and length), the maximum water depth is 7.6 m, the light-load draft is 1.8m, the maximum drilling depth is 9144 m, the lightship weight is 1680 t; when working each side has 8 to 10 root piles.

   b. Cantilever type

   This shallow-water drilling barge is different from the above, and uses a cantilever structure. The structure makes the rig out overboard, in order to facilitate a small wellhead platform with the default drill well, as shown in Figure 5-11. A beach/shallow sea drilling barge in this form is especially well suited for small wellhead platforms, to establish an early production system or marginal oil field production system.

2. Nonslip caisson

   A beach/shallow sea drilling barge sits on the bottom of the sea. Therefore, it is very important to prevent slippage and silt seabed deposits in the area. In recent years, there has been use of a movable caisson base station, which is very suitable for solving the problem of preventing slipping undersea. The caisson is removable. When drilling is completed and after the removal of the drilling barge, a dredger would move gravel so the caisson can be used again. Rollers are on both sides of the caisson. After dragging a drum water ballast to a predetermined position, you

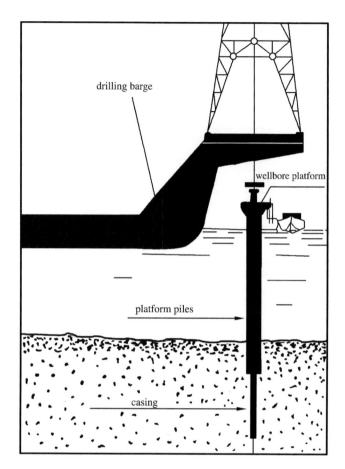

FIGURE 5-11

Cantilever drilling barge with wellhead platform.

can make the caisson sink underwater, and caissons will be filled with gravel and built into a sitting drilling barge underwater base station. This process can greatly reduce the necessary filling material and time. Operating procedures for a caisson base station operating procedures are shown in Figure 5-12.

3. Channel dredging

Limited by the minimum draft of the beach/shallow sea drilling barge, it is generally difficult to enter water with a water depth less than 2m, even with weight engineering concepts to minimize the weight of the hull structure. The applicable water depth is 0.9 to 3.7 m, which cannot meet amphibious areas and on the beach movement needs. Therefore, dredging waterways is often used to expand the operating range of the beach/shallow sea drilling barge abroad. The following problems should be solved before dredging:

a. The choice of dredger

The dredger has a wide range of types: suction Yang style, chain bucket, grab and raise style and shovel field type. As for the Bohai beach/shallow sea region, the beach is sand based, so the selection of suction style is appropriate. A cutter suction and two trailing suction hopper

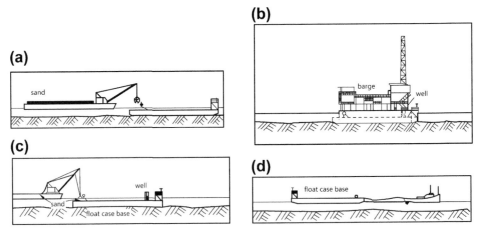

FIGURE 5-12

Movable caisson base station operating procedures. (a) Pull the caisson base stations to the location; sink ballast and fill in the sand; (b) Drilling barge sitting on the caisson foundation; (c) After drilling; remove all caisson base stations in the sand and base station wellhead; (d) Pull the caisson base station to the new base station location.

suction styles, due to the cutter suction mud pipe mud can have continuous operation, high production efficiency to open the dredging, sand and sandy loam-based, it is also suitable for small waves, low flow rate region. So applying the cutter suction dredger in the Bohai sea amphibious belt and beach is appropriate.

b. Estimate the amount of earthwork

The amount of dredging earthwork is a key factor for the assessment of dredging feasibility. To estimate earthwork, we must first consider the ultra-deep, to ensure the design water depth and check the unevenness of the bottom of the tank, with the ultra-deep reaching 30 to 40 cm. Second, we must consider the width, which should sufficiently allow the dredger to turn around in case of emergency evacuation. Again, it is crucial to consider the slope to prevent the end groove side from collapsing to ensure the stability of the slope. And generally for dense sand stone, it is desirable to have a slope gradient of 1:2 to 1:3; soft clay and silt prefers 1:3 to 1:5; and for soft mud 1:10 or more desirable. Siltation is a restricting factor which determines whether dredging is feasible or not. The siltation rate tends to be siltation thickness (mm) created per unit of time (days), and is used for the determination of the actual bottom of the tank.

The siltation for earthwork V can be calculated as follows:

$$V = \frac{1}{2} \times \delta_b \times B \times L/1000 \tag{5-11}$$

where δ_b is bottom of the tank siltation rate, mm/d; L is trench length; B is trench average width; and T is construction period.

Take $\frac{1}{2} \times \delta_b$ as the average siltation rate, because the siltation rate increases with groove depth. For example, in the beach/shallow sea in Dagang oil field, a serious siltation rate can reach 114 mm/d. Therefore, the dredging program must be paid full attention, generally only in regions where siltation is serious. When the well location is closed to the main channel or deep water area it is fitted to use dredged drilling barge in dredged shorter.

5.3.2.2 Cushion Drilling Equipment

Air cushion technology at home and abroad has been applied for many years; air cushion technology can be used for the overall movement of drilling equipment. It can be adopted using a non-self-propelled loaded large air cushion platform, and then matched with the appropriate running gear and a large air cushion platform supporting smooth movement for all drilling equipment in beach and amphibious areas. In addition, in recent years there has been a jack-up cushion drilling platform. It can be non-self-propelled, but it can also use a walking mechanism for movement. The following is the introduction to some of this equipment.

1. Large truck air cushion platform
    The load capacity of air cushion platform is very large, and it can be used to move the entire drilling equipment. For example, Russia has developed an air cushion platform with a set weight of 2000 t to completely move large equipment. This air cushion platform is scaled 20 m × 60 m and fans are 36 m diameter stainless steel blades, with a wind flow of $40 m^3/s$ and cushion pressure 800 kgf/m^2. Out skirt with capsule refers to the type which has two-point touchdown with our fingertips and can reduce the splash with the bag. The apron is mounted on a removable frame whose size is 3 m × 6 m, a total of 25 pieces. It can be used on land and in water, dragged by cushion traction sheaves with a force of 150 t. The Dagang oil field in China with domestic scientific research units was developing a dead-weight 100 t passive air cushion transport platform in 1997, which can load 35 t air cushion transport platforms and the hovercraft with 33 seats. The transport platform has been successfully used in the Dagang oil field beach-shallow sea region for many years.
2. Running gear of air cushion platform
    The air-cushion platform with load capacity is divided into self-propelled and non-self-propelled types. Ancillary non-self-propelled types must be matched with running gear. For movement, there are crawler running gears, tires, a tire/paddle wheel combo and other forms existing in domestic use, but wheeled running gear has been fitted for the China Bohai silt beach area. The 708 Research Institute of China has developed a propulsion wheel walking advancing mechanism similar to an agricultural tractor. It is applied in the China Dagang oil field beach area. The mechanism can be used with 35 t cushion support platform. This running gear has a primary width of 400 mm, stab wheels with 1800 mm diameter which is driven by a hydraulic motor, and a buried depth in the 0 to 500 mm range. Adjusted by hydraulic arms, steering can also be achieved through the manipulation of rigid wheel traction. The walking mechanism can be used in depth ranging from 0 to 500 mm of the beach and intertidal zone, and the rigid wheel thrust is 7 kg/hp. It is applied not only to the marshy surface, but can also paddle forward in plastic flow like mud, to provide thrust for a general advancing mechanism.
3. A variety of air-cushion drilling platforms
    There are several different air-cushion styles used for movement of whole drilling platforms.

a. Bottom-sitting air-cushion drilling platforms

Bottom-drilling platform have been made with air cushion, for example the UK Mackace cushion drilling equipment system is made with a 6000 t total weight non-self-propelled bottom-sitting cushion drilling barge and a 160 t load self-propelled air cushion loading platform. The bottom-sitting cushion drilling platform is 91.4 m long, 87 m wide; the cushion pressure is 10560 Pa and the cushion height 2.44 m. Another cushion load platform is self-propelled by propeller, with a load of 160 t, and speed of up to 15 km/s; it is mainly used as a transport for emergency supply and services for the sitting cushion platform.

b. Jack-up air-cushion drilling platform

This form makes jack-up drilling platform using an air cushion. For example, Japan has constructed a non-self-propelled jack-up cushion platform, which can be used for the amphibious belt and beach areas for geological drilling. The leg of the jack-up platform adopts liquid pressure as a lifting mechanism. The pile speed is 15 m/h, and the speed to raise the platform is 8 m/h. Each leg has a 100 t lifting capacity. Jack-up platform's scale is 256 m × 19 m, the platform weight is 310 t, with pile leg diameter of 0.5 m, pile leg support leg spacing of 224 m × 144 m, and pile leg length of 20 m. The platform cushion has a pressure of 602 kgf/m^2 and can load 80 t. The platform has five hidden water compartments. There are four balanced water tanks in the four corners set to adjust the load on the platform.

Dagang oil field in 1996 studied a new cushion jack-up drilling platform suited for beach/shallow sea drilling, and the features are the following:

- Jack-up platform can be lifted off the surface 11.3 m and line to deeper bear stratum with four piles leg-line, which can effectively solve the problems of silt layer slip and underrun, thus achieve "stability".
- The platform adopts a 6032 Pa pressure cushion to pad surface of the ground (water), and uses tractor towing for amphibious performance and traction to effectively solve the 0 to 2 m depth of the beach and the problems of "getting in" and "out" in very beach/shallow sea areas.
- The platform adopts plant solutions that use a machine with two diesel engine fan units for each series. A generation machine can supply the power to improve the efficiency of equipment when in drilling operations.
- Platform deck facilities use a skid-mounted modular structure which is easily disassembled and relocated to balance load.
- Hull structure and rig are high strength alloy steel construction, and the weight engineering method is used to optimize the design, so as to make the platform sufficiently strong and stiff, and keep the weight at a minimum.
- Platform adopts air-cushion technology for delivery and platform support which ensures the supply of the platform's necessarily equipment and personnel transport, including the use of 100 t load passive air cushion transport platform and amphibious traction tractor and air-cushion boat.

5.3.2.3 Walking Drilling Equipment

Walking drilling equipment is located at the end of the high deck of a pillar supporting a drilling barge, and can move by land if necessary to the beach/shallow sea region to drill into the sea. When the wells are located in the tidal flat areas, not fit for sailing, the equipment goes by land. When the wells are located in beach-shallow sea area that can be sailed, you can go from the sea to the well location. The "Victory II" Chinese design is an example of this walking drilling equipment, is put into use in 1988. That walking drilling platform was first proposed by with a conceptual design by Yi, chief engineer of the Chinese Academy of Engineering, and patented by the Shengli oil field Drilling Research Institute and Shanghai Jiao Tong University Design and Development. This walk-up drilling platform works for water depth less than 8 m, has a drilling depth of 7620 m, scale of 75.24 m × 43.14 m × 12.6 m, and a 4500 m electric drive moves the rig. Walk type platform composes of the inner body and the outer body, and the outer body is mat and can sit on the bottom. There are vertical and horizontal directions for the cylinders on the platform. When the inner body and the outer body is fixed to the inner body or outside the body in turn in the fluid, cylinder is driven to the vertical or horizontal direction step line, with each walking stroke of 10 m. In this way, you can crawl in the beach area.

5.3.3 Modular Beach/Shallow Sea Transport Drilling Equipment

This drilling equipment is based on drilling rigs and drilling platforms adopting a modular-based style to transport drilling rigs and the platform to well sites. First the drilling platform is assembled; the modular rig will be broken down again and assembled on the platform, and then drilling operations can occur. Integrally movable platform is similar to other alternative means which may also be used in carrying rig light load Conditions, the shift chasing place will then be decomposed rig movement to be assembled on the platform and drilling operations.

5.3.3.1 The Advantages of the Program

The advantages of the program that modularizes beach/shallow sea drilling equipment and movement are:

1. Reducing the difficulty of transporting equipment. Drilling rigs and platforms are modular and lighter, so it can greatly reduce the difficulty of movement.
2. Saving on the initial investment in the movement. Due to the decreasing of movement expenses, the initial investment of the project is lowered.
3. Flexibility to adapt to a variety of environmental conditions flexibility. This style can be adapted to the shallow sea, the amphibious zone, and the beach and the beach-shallow, and therefore it has a greater flexibility and adaptability.
4. Platform and rig can be utilized separately. After completion, the rig moves to the new well location for reuse. The drilling platform can be either removed or continue to be used in the original well location.

5.3.3.2 Modular Rigs

The modularity of the rig allows achievement of rapid decomposition and assembling, which is the basis of the transportation of beach/shallow sea drilling equipment. In order to adapt to the marsh, desert, hills, and other areas which are difficult to, easy transport modular rigs have been developed to meet the needs for drilling operations over the years.

1. The TBA-2000 modular rig
   The TBA (Transportable by Anything)-2000 rig was developed by the United States for tool transport and to allow modular rigs to access areas of the South American swamp.

a. Power systems

It is either diesel engine driven or electrically driven; the former one is made by four diesel engines that have a power of 375 hp, with a drive dial, and the other three driven with winches; the latter is made using a 2550 hp motor-driven winch, a 350 hp motor drive turntable, and a GPO-500 triplex drilling pump with a winch common drive.

b. Technical performance.

The drill pipe has drilling depth of 6000 m, the derrick height is 42 m, the derrick carrying capacity is 465 t, the large hook holds 304 t weight, the winch drum is 66 cm × 122 cm, and the brake is 117 cm × 30 cm.

c. Transport

The TBA-2000 rig has total weight of 227 t, which can be decomposed into 286 modules, among which the largest module weighs no more than 1.8 t. The winch can be broken down into seven modules, and the maximum does not exceed 1.8 t; the drilling pump can be decomposed into three modules. The derrick adopts a special structure and lifting mast. As the derrick goes up, the rig is mounted vertically to reduce the space required. The entire process besides assembly of the rig derrick is in a narrow (30 ft × 30 ft) platform area. When transported by helicopter or small boats, the four 30.5 m long, 12.2 m wide barges with rig and drill pipe and drilling materials require timely completion.

2. Modular rig made by Simmons in Canada

This rig can drill to a depth of 1372m (4500ft), and is a light rig. In the rig modular structure, the general module has a weight of 1.7 t, and the maximum module weight is 1.9 t, which is suitable for transport by helicopter. The underlying structure of such a rig is divided into five basic parts: racking, winches, winch engine fixed on two ski racks, water tank, and turntable. The derrick adopts the TBA-2000 portable derrick by the United States introduced above. The rig machinery and equipment transportation module is as follows:

a. Winch is divided into three modules, namely the roller, roller shaft and clutch input shaft and gearbox, and the engine is mounted on a skid frame.

b. The triplex drilling pump engine is divided into five lifting segments.

c. The mud tanks are divided into four modules; each of them is mounted on the skid frame. When adopting this drilling rig, you can also add a spare base, which can start at the location of drilling wells, and be assembled by helicopter transport. After the well drilling, the rig can be directly transported to assembly using the alternate base, which can greatly reduce machine downtime for the rig.

This modular rig is fit for drilling in everglades. According to Canadian experience, when the relocation distance ranges from 3.5 km to 8 km, the relocation is completed approximately within 30 h.

3. U.S. MSH-2000 modular rig

Such a modular rig has a drilling depth of 6000 m, and has been used in Indonesia. It also uses helicopter transport. When transporting, the weight of the module besides the weight of the winch drum and brake module is 18 t; other modules do not exceed 1.8 t. The entire consignment is 151 segments. When transported, the derrick is divided into eight modules. The height of the rig is 6.7 m, and all control systems are housed in a ski rack.

5.3.3.3 Assembled Platforms

To solve drilling equipment's transportation problems in area which is difficult to be accessed including amphibious, beach/shallow sea and compete with the modular rig, assembled drilling platforms have been developed by foreign companies. It can be classified similarly with the common offshore drilling platforms, and there are mainly four types.

1. Fixed drilling platform assembly

   The assembled fixed drilling platform has been used in Peru's Amazon rainforest swamp area. At the beginning, it often adopts a stake structure, later switching to steel piles and concrete pile structures to protect corrosion.

   This lightweight steel piles structure is assembled with a platform with steel piles with a diameter of 323.1 mm ($12\frac{3}{4}$ in), additional steel beams, and wooden deck. The deck area is 79 m × 26 m.The rig is installed in the central location of the deck and living space is stretched out at the end, with the helicopter pad on the side. All of the components of the platform are transported by helicopter.

   When being installed, first a small temporary structure has to be built for the installation of a modular piling machine to hit a few steel piles. After having installed the few steel piles on the steel beams and deck, which results in the basic platform, the piling machine would be moved to a basic level stage. Then piling is extended and the platform moves outward to the desired scale. When the platform assembly is complete, drilling holes can be assembled.

   Water transportation can be used where heavier concrete piles are used instead of steel piles. Concrete pile diameter is 1371.6 mm (54 in), and they can be mounted on a barge piling machine. The assembled fixed platform can be demolished, reused, or left in place for the production platform.

   For the fixed platform components, in addition to the helicopter and barge, one also can use a hovercraft, amphibious vehicles, and other transport tools.

2. Bottom-sitting drilling platform assembly

   Bottom-sitting drilling platforms have the following two assembly forms abroad:

   a. The use of mud sled on the drilling platform.

   This does not take advantage of the waterway in the beach/shallow sea region, as it adopts the mud sled to transport and drill.

   The size of the mud sled made in the United States and used in the swamp area of Indonesia is 8.1 m × 1.8 m × 0.9 m, and each mud sled is equipped with a diesel engine-driven winch, which can walk by winch traction on mud skid with the use of pitching of pile. Mud sleds can be prefabricated and sized based on the type of platform, and mud sleds moved to the well location may be different according to the size of the carriage and equipment, and by the number of modules assembled, i.e., they can be assembled into rigs and the rigs mounted on the spot. After completion, it can be also disassembled and transported to a new well location to be reused. Completed drilling equipment in general has nine mud skid modules including one mounted winch, one mounted drilling pump and diesel engine, and one mounted fitted frame along with other ancillary equipment from the remaining six mud skid modules. Other suitable traction tools can also be used with such mud sled towing.

b. Use of barges by bottom-sitting drilling rigs

This assembled drilling platform can only be applied to water with certain tidal ranges and a seabed that is suitable for bottom-sitting platform operations. It's just decomposition barge on drilling equipment drilling platform open. In certain tidal waters, the barge rig is not installed and not carrying supplies and equipment, so the draft is very shallow, and the platform can take advantage of the high tide into the well. Then, high tide can be used again when the drilling rig and other equipment are to be transported to the well location, mounted on a barge. Before installing the rig, the ballast water needs to be injected to let the barge sit on the bottom, and smooth the seabed before sitting.

Such an assembled sitting platform has been applied in the mouth of the Elbe in Germany, by two 91 m × 27 m × 6. 5 m flat hopper barges built into the sitting drilling platform.

3. Assembled drilling jack-up platform

There are two main forms assembled jack-up drilling platform:

a. Installation of moving rig lightweight jack-up platform

This is actually a beach/shallow sea drilling barge with a solid leg formation, which will hold a lightweight drilling platform. It is a beach/shallow sea barge with its own lifting pile leg, called a jack-up assembled platform. The legs are a solid formation with carrying capacity and stability advantages. It also uses the beach/shallow sea barge flexibility, that is, sitting on the bottom and rising, so it is very suitable for serious silt beach-shallow sea regions aboard. Figure 5-13 shows a schematic view of such a platform.

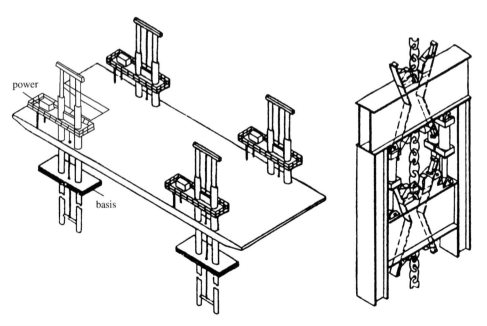

FIGURE 5-13

A schematic diagram of the jack-up lightweight assembly platform.

This platform water depth is 3.5 m to 5.5 m, and if the beach/shallow sea barge is raised, the maximum working depth is up to 10.7 m. While working, first, the mat is dragged from the platform to well, then staked as shown in Figure 5-13. Note the water ballast falls to sufficient depth underwater, and then after the beach-shallow sea barge sit the end of the mat when needed. Water would be discharged in the mat to reach the desired height, thus ensuring the drilling barge carried out drilling operations. This platform load capacity is 2000 t and 3000 t, on which you can put a 64 m × 18 m drilling barge for the beach/shallow sea area.

The movement of this platform is generally takes advantage of the high tide when dragged to the well bit, because of the fact that there is no installation of drilling equipment, so the drafts of the general jack-up platform are shallow. After being staked and raising the barge, modular drilling rigs and cushion equipment will be transported to installation by amphibious means of delivery. After completion of the drilling operations, such a platform can be moved to new wells.

This platform has no load from drilling rigs and other equipment, so the requirements for the pile leg lifting mechanism is lower than the average jack-up platform.

Obviously, if such a platform is equipped with an air-cushion pad lifting system, it can be changed to air-cushion lightweight jack-up platform; its effective range will be further expanded. The Dagang oil field described earlier researched a jack-up cushion combination drilling platform, which is based on this concept.

b. Modular rig and platform jack-up platform assembly

This platform is designed specifically for conditions which are very difficult in swampy areas, and this style has been successfully used in swampy areas in the Amazon rainforest of South America.

The platform drilling rigs and platforms are assembled in a modular fashion, and the module weight does not exceed 1.8 t; they are transported by helicopter to the site, and have doubled buoyancy in weight, so they can float on water for storage. In the case of fairway, bundles of rafts group are compiled to haul to the present field. These components can be assembled in a floating state, with the longitudinal pin connector connecting each component with vertical and horizontal cable tensioning. This can form a temporary water raft. When in the installation preparation stage, the first stage is composed of six types of rafts: the helicopter down raft, assembly working raft, equipment raft, crane assembly rafts, leg assembly rafts, and auxiliary work raft. The first modular piling machine crane is assembled on the tops of the rafts, which serves as the basic platform in the water to maintain position and anchoring. Then with the leg assembly rafts and the use of a crane four good legs are assembled good legs, using a leg guide sleeve along the base platform into the seabed. Level to be the basis of positioning instrumentation, each pile legs are fitted with hydraulic with lifting the locking mechanism on the leg guide tube, with the oil pump on the hydraulic power basic platform is raised, and then lock it generally to rise the basic platforms from the water surface 3.65 m.

After the raising of the basic platform, the crane can be used to expand the platform along the horizontal and vertical in the same way until the work area is as shown in Figure 5-14.

After the completion of the platform assembly, the modular rig is transported for installation and drilling. After completion, the rig station and platform are broken down into modules, for movement to new wells. Abroad, it usually takes about 20 working days to assemble a 99 m × 21.6 m platform. The platform is capped with steel deck and can be equipped a drilling rig with a drilling depth of 5400 m.

(a)   **(b)**

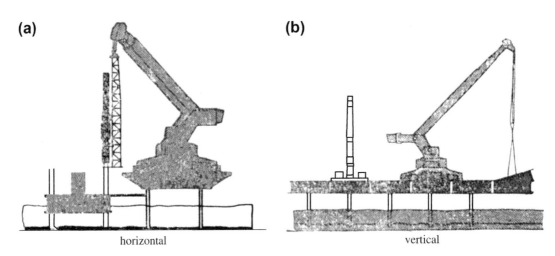

horizontal   vertical

FIGURE 5-14

Platform assembly with expansion of the horizontal and vertical direction.

This platform can also be equipped with a multifunctional auxiliary floating platform, outsourcing steel shell sealed polystyrene foam feed members held together. The scale of the individual component is 1.2 m × 10.8 m × 0.75 m. When 65% of the volume of member is submerged to carry about 5 t, this platform uses an auxiliary wooden deck for storage of supplies and equipment, or for other purposes.

In short, both the modular rig and the assembly jack-up platform are not only easy to transport and install, but also easy to prefabricate, and can be put into use in a very short time to reduce the preparation time for exploration. In addition, because of demolition and reuse, we can also avoid the use of fixed engineering facilities, and the risk of significant investment in the exploration stage, so it is suitable for an economic beach/shallow sea platform.

C. Floating drilling platform assembly

This platform is for beach-shallow seas with less severe storms or inland lakes using a floating platform, known as the FFS (Flexible Floating System), which can be used for certain China beach-shallow sea regions, in particular, extreme beach/shallow seas and lakes.

The platform structure is a modular rig and platform, but the platform module is a standardized self-floating module, and for easy transportation, the module size is limited; the module is divided into three sizes, 3.0 m × 3.0 m × 2.1 m; 6.1 m × 3.0 m × 2.1 m; and 12.2 m × 3.0 m × 2.1 m. The module is a watertight steel plate welded structure, with internal ribs to ensure proper bearing and connection. The maximum weight of a single module is 16.5 t and other modules can be transported by helicopter. The module can carry 16.36 t (36,000 lb). The modules are assembled into a platform; there are two scales, namely, 54.9 m × 27.4 m × 2.1 m and 48.8 m × 21.3 m × 2.1 m. The platform installation uses a modular production TBA rig, with a drilling depth of up to 6000 m.

The positioning of this platform uses an eight-point mooring system and you can also add a number of stakes in order to prevent loss of position. In order to prevent equipment which is located on the platform from slipping due to wind and waves, spot welding can be used on the deck.

Connection between the modules of the platform assembled in the water is very convenient; in order to ensure the interchangeability of the modules, strict manufacturing is used to control the quality. In addition, in order to prevent the corrosion of the modules in the water, the platform is recoated with epoxy resins yearly on average.

In short, the floating assembly drilling platforms and rigs are modular in design, and assembled according to the needs of platform into different scales and for different purposes; they are highly flexible and adaptable. In addition, the platform is easily transported either by land, air, and water, with hauling using a series of rafts or mud sled movement. Therefore, it has been successfully applied to the Great Salt Lake in the United States and Green Lake inland lakes, and can also be adapted to certain local conditions in the China beach/shallow sea region.

5.4 **Special Vehicles in Beach/Shallow Seas**

Besides serious silt in beach areas, there are changes in land and water with tidal transition zones, and very shallow areas in beach/shallow seas, so generally used vehicles are difficult to adapt to the characteristics of these environments. In order to solve the problem of vehicles for these scenarios, many kinds of vehicles have been developed both domestically and overseas, which mainly consist of wheeled and tracked amphibious vehicles, hovercraft and air-cushioned transportation platforms, and mud pry and ships for extremely shallow sea areas. In the following sections let's introduce some of these vehicles.

5.4.1 **Wheeled and Tracked Vehicles**

Both wheeled and tracked vehicles are used in beach/shallow sea areas.

5.4.1.1 *Wheeled Vehicles*

The characteristics of such vehicles mainly are large low-pressure tires, which not only guarantees very low ground pressure while running on the ground, but also can float in the water, relying on the buoyancy of the tire and amphibious functions. Examples include the K14 and K1 × 6 transport vehicle with wide rubber tires, the wide tire four-wheel drive of the U.S. Rolligon 4460 and six-wheel drive of the Rolligon 6660 and Rolligon 6 × 6 Challenger amphibious vehicles, etc. Now we take the U.S. Rolligon amphibious vehicle as an example and a detailed introduction follows.

1. Rolligon amphibious vehicle
   The technical performance of several U.S. Rolligon amphibious vehicles are shown in Table 5-6. From the Rolligon's technical performance, it is easy to see:
   a. The high ground pressure is suitable for sandy beaches
   You can see from Table 5-6 that the pressure ratio is $0.14 \sim 0.31 \ kgf/cm^2$ when it's at full load, so it is applicable for sandy beaches, such as Shengli oil field. However, it is not appropriate for muddy beaches, such as Dagang oil field, where the ground specific pressure is too high, coupled with easy slipping when the tire grooves are on the sludge.

Table 5-6 Technical Performance of Several Rolligon Amphibious Vehicles

Vehicle Model Item	Rolligon 4460	Rolligon 6660	Rolligon 6 × 6 Challenger
Number of wheels for drive	4	6	6
Engine power, hp	115	145	210
Tire size (D × B × d), cm × cm × cm	137 × 172 × 46	137 × 172 × 46	183 × 172 × 71
Maximum load on ground, kgf	4536	9072	13608
Maximum load for floating, kg	2268	4536	4536
Maximum speed on land, km/h	38.6	38.6	–
Maximum load floating in static water, km/h	4.8	4.8	–
Ground pressure with no load, kgf/cm^2	0.093	0.117	0.21
Ground pressure with full load, kgf/cm^2	0.14	0.273	0.31
Overall dimensions (L × W × H), m × m × m	6.1 × 4.0 × 2.4	7.3 × 4.0 × 2.4	10.36 × 4.57 × 3.84

b. Poor airworthiness is suitable mostly for ground driving

The Rolligon amphibious vehicle can float and move on the water with the buoyancy of the tire. However, by relying on tire paddling, thrust is very small. So the speed of floating is very low, only 4.8 km/h. In addition, because its overall profile is still car-shaped, as shown in Figure 5-15, it will come across too much resistance when floating. Even if we use the additional propeller propulsion, the airworthiness is also very poor. Therefore, it is mainly applied on the ground rather than floating in the beach area, which is very restrictive.

c. Limited wading depth exerts an effect on operating range

When the actual depth reaches 0.8 ~ 1. 0 m, the Rolligon amphibious vehicle enters a floating state, thus limiting the effective operation to the scope of the amphibious zone and intertidal zone. For the 6×6 Challenger, the chassis has been changed from the original

FIGURE 5-15

U.S. Rolligon amphibious vehicle appearance.

Table 5-7 Delta and MOL Wading Vehicle Technological Parameters

Vehicle Model Item	Delta 2B	Delta 2C	Delta 100	Delta K1244
Engine power, hp	197	175	136	127
Load, kgf	9030	6810	5400	6000
Ground pressure, kgf/cm^2	1.27	1.2	0.41	
Tire size (D × B × d), cm × cm × cm	168 × 109 × 64	168 × 109 × 64	168 × 109 × 64	168 × 109 × 64
Fording depth, m	1.37	1.52	1.27	

63.5 cm to 76 cm, the diameter of the tire has increased, and the wading depth has been move up to 1.22 m. Still, its wading depth is limited, which impacts the effective operating range of such vehicles.

2. Delta and MOL wading vehicle

This vehicle is equipped with wide tires with lower specific pressure. It does not have the ability to float in amphibious zones, but with a lower ground pressure and larger diameter of the tire it has a certain driving capability. The technical performance of different variations is given in Table 5-7. Such a vehicle is suitable for beach/shallow seas and better-quality beach areas.

In addition to the above two types of vehicle, there are ordinary truck tires, which are mounted with rubber crawlers in order to increase the ground contact area and decrease the ground pressure. For example, TRUCK TRACK is installed with a 0.61 m long rubber crawler, thus ground pressure of ordinary trucks goes from 4.9 kgf/cm^2 to 0.42 kgf/cm^2, which significantly improves the capacity of the truck in the tidal flats. Its speed can reach 40 km/h under trackless conditions, and disassembly is very convenient.

5.4.1.2 Tracked Vehicles

The basic characteristic of a tracked vehicle is relying on a wide track to increase the ground area, decrease ground pressure (about $0.15\ kgf/cm^2$), and obtain the necessary thrust so as to run on soft surfaces like silt beach. A crawler vehicle usually has pontoon bodywork, thus once a certain depth is exceeded it can rely on the buoyancy of the floating box to float the whole tracked vehicle and travel on the water, which can ensure amphibious capability.

In China, there are some tracked amphibious operation vehicles in oil fields in the beach/shallow sea regions, such as the foreign Quality, Kori, and Hagglunds. Besides, there are GJH 12 beach vehicles and a 10 t traction amphibious tractor self-developed by Dagang oil field. Now we present the main details briefly as follows.

1. Several foreign tracked amphibious vehicles

The technical performance of several foreign crawler amphibious vehicles is shown in Table 5-8. The Quality 104 T-82 and Kori 4526 vehicle from United States are what our country uses more. Using a metal caterpillar or crawler, these vehicles are often badly worn with a low life of only 200 ~ 300h and high operating maintenance cost. The Hagglunds BV-206 D from Sweden replaces the metal caterpillar with rubber crawler. Thus the landing speed is improved, while it maintains the same floating speed in water as the metal caterpillar (Table 5-8). In addition, it

Table 5-8 Technical Performance of Several Foreign Crawler Amphibious Vehicles

Vehicle Model Item	Quality 104T-82	Kori 4526	Hagglunds BV-206D
Engine power, hp	210	210	136
Dead weight, t	8.17	18.2	Limber 2.67 Rear 1.67
Load, tf	4.54	2.5	Limber 0.5 Rear 1.6
Land speed, km/h	12.9	4.02	50
Floating travel speed, km/h	4.8	3.2	5
Ground pressure, kgf/cm^2	0.10	0.12	0.13

increases the hydraulic position pile, and adds a pusher in the tail, so as to improve the performance of floating operations in water, as shown in Figure 5-16.

Several of the above amphibious tracked vehicles can apply to the general silt beach. But in areas with serious mud, trapping is also prone to occur. To solve this problem, the German company GST produces an amphibious vehicle with variable crawler chassis height, as shown in Figure 5-17, for which the height of the chassis and track contact length can be changed by hydraulics; thereby the ground pressure ratio is changed. On good ground, raising the chassis can reduce the track's ground contact length and improve the speed, as shown in Figure 5-17(a).

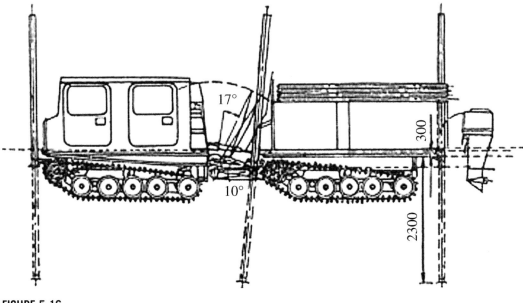

FIGURE 5-16

Hagglunds BV-206 D tracked amphibious vehicle.

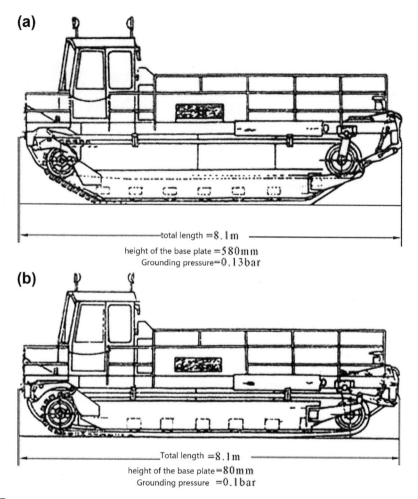

(a)

total length = 8.1 m
height of the base plate = 580mm
Grounding pressure = 0.13bar

(b)

Total length = 8.1 m
height of the base plate = 80mm
Grounding pressure = 0.1bar

FIGURE 5-17

German GST tracked amphibious vehicle with variable ride height.

Conversely, when traveling on weak ground, put down the chassis to increase the ground contact area with a minimum ground pressure, and improve traffic capacity, as shown in Figure 5-17(b). This GST NB/S-1 crawler amphibious vehicle, when traveling and floating in the water, relies on jet propulsion, and the maximum speed can reach 10km / h with a rubber track 9 m wide. In addition, the Canada Transport Development Centre (TDC) developed an air-cushion tracked vehicle, as shown in Figure 5-18. By means of the air cushion with 1 m wide rubber, it makes a track pad 0.6m long and decreased ground pressure approximately 4/5, thereby increasing the capacity on harsh surfaces. The vehicle's engine power is 415 hp and air-cushion pressure is 0.0063 kgf/cm^2. In addition, its track pressure on the ground reaches 0.356 kgf/cm^2 and its travel speed is 30 km/h.

FIGURE 5-18

Air-cushioned tracked vehicle developed by Germany.

Also, the French AMG Company developed metal floating wheel amphibious vehicles, as shown in Figure 5-19. Instead of the low-pressure rubber tires and tracks, it uses large diameter round metal floating wheels, which float through buoyancy. Meanwhile, the metal outer edge of the roller wheel is equipped with a water paddle advance program or similar structure, thus the roller wheel goes forward like paddle. Generally such a vehicle is driven by four hydraulic wheels alone and it has better capacity on a weak ground beach. Meanwhile, due to the large diameter wheel, it can also wade in deeper water. Its speed reaches 8 km/h on land, but when it is floating and driving, its speed is low, only 5 km/h.

FIGURE 5-19

Amphibious wheeled vehicles with metal buoy car by AMG in France.

2. GJH12 beach operation truck developed by China Dagang oil field

The GJH12 beach operation truck is developed by Dagang oil field, and is mainly used for transport and applied to very shallow sea areas with a depth of less than 3 m, intertidal zones, and marsh areas. It also improves the local structure design, and introduces a new type of hydraulics available for drilling. A brief introduction will be given in the following section.

a. The main technical parameters

   i. Dimensions are as follows: length is 83 m; width is 72 m; height is 84 m; dead weight is 9500 kg; load is 2500 kg; and ground pressure is 10 KPa.

   ii. Walking parameters (full load) are as follows: maximum speed is 10 km/h; hydrostatic maximum speed is 4km/ h; climbing angle is 30°; the buoyancy reserve is 30%; and turning radius is 4 m.

   iii. Engine parameters are as follows: air-cooled diesel engine of is model F6L913; the maximum transfer is 2800 r/min; the maximum output torque is 384N m (1600 r / min); and the maximum output power is 95.6 kW.

b. Structure composition

As shown in Figure 5-20, it is composed of the engine (1), the engine cover (2), the cab (3), the active chain wheel chassis (5), the pontoon connection box (6), the floating box (7), the drive chain (8), the driven chain wheel shaft (9), and other components.

c. Transmission system.

Its drive system is shown in Figure 5-21.The power output from the engine crankshaft is delivered by clutch, gearbox, drive shaft, intermediate gearbox, drive shaft, differential gearbox, the saline shaft, and the left and right chain box. Then the power is spread around the pontoon on the active chain wheel. Finally, a caterpillar chain is driven by the sprocket to drive the overall unit.

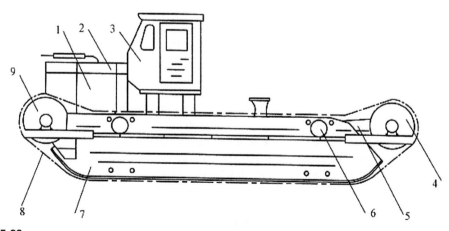

FIGURE 5-20

Structure composition of GJH12 beach working vehicle: 1 - engine; 2 - hood; 3 - cab; 4 - driving sprocket shaft; 5 - chassis; 6 - pontoon connecting shaft; 7 - pontoon; 8 - drive chain; 9 - driven sprocket.

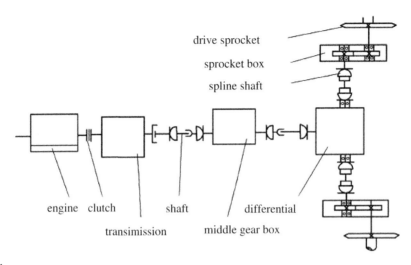

drive sprocket

sprocket box

spline shaft

engine   clutch   shaft   differential

transimission   middle gear box

FIGURE 5-21

Drive system of CJH12 beach working vehicle.

There are five gears available in the gearbox, including four forward gears and one reverse gear. In addition there are two blocks ready for use in the intermediate gearbox, in which one is used for the direct drive and another is the deceleration afterburner block. Steering action is accomplished by two pneumatic valve operating handles to control the differential left and right steering brake band. When the side of the brake makes the output shaft decelerate, the running speed of the left and right pontoon will be different, so that the job of vehicle steering is accomplished.

3. 10 t traction amphibious tractor developed by Dagang oil field

This is an amphibious tractor directly from the land into the sea which can go through land, beach, and sea areas and can produce large traction in mud areas. Besides it is self-designed and developed to solve the problem of Dagang Oil field beach, and air cushion drilling platform is required by very shallow sea area of 0 ~ 2.5 m depth and traction of passive air cushion transport platform.

a. Main technical parameters
   i. Major scale and performance features: length is 10.78 m, width is 6.21 m, height is 4.66m; ground pressure is 13 kPa, and weight is 32500 kg.
   ii. Traction parameters for the turtle board beach are: max traction is 174 kN; max speed is 6.0 km/h. Beach mud flats: max traction is 101kN; max speed is 6.0 km/h. Transition zone: max traction is not less than 50 KN. The travel speed on the road and in the still water is not less than 10 km/ h; the land turning radius is less than 30 m.
   iii. Wind and wave adaptation parameters (in floating traction conditions) are that wind speed is less than 8.0 m/s, and wave height is less than 1.2 m. Buoyancy reserve is 125% and, continuous working time is 8 h.iv. Engine parameters are that two TBD234V6-1261A turbocharged diesel engine are used; the rated speed is 2100 r/min; rated power is 250 kW, and max torque is 1492 N • m (1600 r/min).

b. Structure composition.

As shown in Figure 5-22, this is mainly composed of the vehicle body (1), the engine (2), the transmission device (3), the mobile device (4), appliance apparatus (5), water propulsion mean (6), communication device (7), marine habitat attachments, and other components. Among them, the transmission device is mechanical, and is composed of a gimbal gearbox, main clutch and its operating mechanism, gearbox and its operating mechanism, steering clutch, brake and its control mechanism, as well as couplings and an end reducer. Its function is to pass and cut off the driving force of engine to the capstan, change the running speed of the vehicle, which makes the vehicle steer and brake, and provide power output for the vehicle's auxiliary parts. The mobile device is composed of a track roller frame, inducer wheel, driving wheel, supporting wheel and final reducer as well as track, with a set on the left and right, which is capable of carrying the weight of the whole vehicle

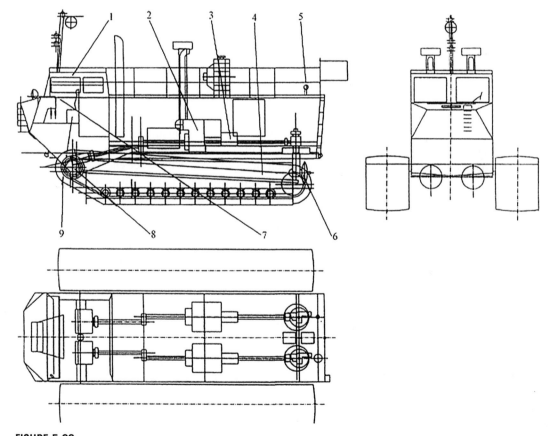

FIGURE 5-22

Structure of 10 t traction amphibious tractor: 1 - vehicle body; 2 - engine; 3 - gear; 4 - action device; 5 - electrical equipment; 6 - water propulsion unit; 7 - communications equipment; 8 - drainage; 9 - marine habitat attachment.

and accomplishing walking and towing operations in the turtle board beach, silt, and transition zone. A water propulsion unit in the rear of the tractor is the general term from the front of the engine to the connection among the parts and the steering oar, which consists of toothed rubber coupling, steering oar clutch, propeller shaft, and a rudder propeller and its operating mechanism. Its function is to transfer at sea or cut off the power output of the engine, and change the tractor's sailing speed on the sea and direction of travel.

c. Action principles
Engine will output power with the crankshaft, deliver power to the left and right end reducer to drive the capstan by the gearbox and main clutch, and gearbox and then drive the track by the capstan, so that the machine runs. On the other hand, its steering is controlled by the operating handle of the engine throttle. While sailing on water, it is controlled by the left and right rudder propeller's dynamic change around the backend of the vehicle, which can change the traveling speed and direction of the vehicle.

5.4.2 Air-Cushion Vehicles

5.4.2.1 Hovercraft Working Principles

A hovercraft is generally divided into the sidewall and cushion. Cushion hovercraft has been applied for development in Dagang oil field in China; therefore, the principles of work presented here is mainly focused on cushion hovercraft.

The biggest difference between all hovercrafts and ships is that there is a pad lifting system in this hovercraft which enables the boat to go out of the water (or onto ground) and is composed of pad blower, aprons, and airway. Cushion hovercraft are equipped with one or more fans, around which there is a circle of flexible apron made of rubber nylon and below which the space enclosed by apron is called the air cushion chamber. Pressure air discharged by the pad fan goes through the airway, pouring into the inside of the cushion. Due to the seal action around the apron, only a small amount of air escapes from the gap in the low portion of the apron, forming a certain pressure cushion below the hull. Although the cushion pressure is not large (generally about 1000 to 6000 Pa), it is enough to support the weight of the ship, so that the hovercraft is off the ground or water surface (bearing surface) in a floating state, as shown in Figure 5-23. In this case, with the help of thrust of an air propeller mounted

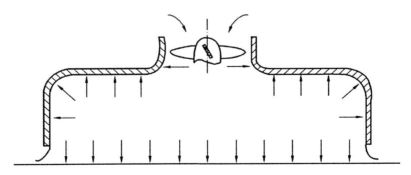

FIGURE 5-23

Operating principle of hovercraft.

on a hovercraft, a hovercraft can sail on the surface of the water or on the ground with relatively high speed. Being able to be completely out of the water or on the ground, the cushion hovercraft has good amphibious performance. Thus in the area where ordinary vessels and land transport can't move, it can sail with a higher speed.

5.4.2.2 The Structural Composition of Hovercrafts

A hovercraft is mainly composed of the pad system, hull, power units, the manipulation device, and outfitting and electrical equipment. Now the different main parts of the general hovercraft will be introduced as follows.

1. Pad lifting system

   The pad lifting system is the most critical device, and is unique to the hovercraft. It is comprised of:

   a. Lift fan

   This is a centrifugal blower, and is the gas source of pad lifting system. It should have a large flow and a certain pressure so as to provide the desired pressure for the air cushion.

   b. Apron

   This has a special significance on hovercraft, directly influencing the performance of the hovercraft. To make hovercraft not destroy the gas seal of the entire cushion apron when itShow jumpingShow jumping surmounts an obstacle, the general solution should be to make the hovercraft barrier height not be greater than 70% ~ 75% of the height of the apron. However, in order to ensure the stability of the ship, the apron's height is not greater than one-sixth of the breadth of the ship. In operation, if a part of the apron is damaged, the system can automatically compensate. Even when the damage extent of the apron reaches 10%, the system can still maintain the air cushion effect. The aprons are generally made of high-strength nylon rubber cloth, among which sac apron is most widely used.

2. Propulsion system

   Because a sidewall hovercraft is not out of the water, the majority use a water propeller or water jet propulsion unit. But a self-propelled jack-up hovercraft propeller is used by air cushion types, and most are ducted propellers. Thrust generated by the propeller must overcome various obstacles of the hovercraft's voyage. The rudder force generated by the air rudder installed at the rear of the catheter can change the running direction of the hovercraft.

   Non-self-propelled hovercraft do not have a propulsion system, and may be dragged to sail by a self-propelled hovercraft, tugboat, tractor, or winch external traction system.

3. The hull

   Because a hovercraft needs to support the full weight of the ship under the fairly low pressure of the air cushion and can fully float in the water on cushion pressure, a lightweight hull, relatively large bottom area, and sufficient buoyancy are required. Usually small hovercraft hulls are welded or riveted aluminum alloy plate. The lower load-bearing part of a large hovercraft generally uses welded steel plate, with the upper part made of multipurpose aluminum alloy plate. Alternately, the hull is a box-shaped structure which is composed of a number of water-sealed pontoons. Compared to other ships, a hovercraft has a large hull width, small aspect ratio, and small thickness.

4. Power plant

The power plant of a hovercraft is mainly for propulsion and pad lifting system power. Thus large power, light weight, small size, and corrosion resistance are required. Generally it uses a high-power high-speed diesel engine, especially an air-cooled diesel engine.

5.4.2.3 35 t Self-Propelled Air Cushion Transport Platform Developed by China Dagang oil field

This air cushion transport platform is developed jointly by Dagang oil field and the relevant mainland authorities. Its load is 35 t, it can accomplish self-propelled action, wand is named 7301(code) by Dagang oil field. The transport platform has a steel main hull, aluminum superstructure, and using light 12V180 diesel engines for power can accomplish self-propelled action with a catheter air propeller; it is mainly for the beach, intertidal areas of 0 ~ 2 m deep, and the oil fields of muddy swamp whose surface can carry 10000 Pa. Usually it is used to supply necessary equipment and tools to artificial islands or platforms.

1. Main technical performance parameters
    a. The total length is 19.2 m; total width is 12.2 m; and total height is 6.9 m when in the air cushion.
    b. Air cushion pressure is 4900 Pa.
    c. Navigation capability is that when in full load, hydrostatic speed is 10 km/h, speed at the beach mud flat is 10 km/h, and endurance is 6 h.
    d. Load capacity is that maximum load is 35 t, and full load displacement is 80 t.
2. Structural composition
    The structure of the 35 t self-propelled air cushion transport platform is shown in Figure 5-24. It mainly consists of the propulsion system (1), pad lifting system (2), power plant (3), operating device (4), hull (5), and outfitting and electrical (6), etc.
    a. Pad lifting system
        The pad lifting fans are G4-73-11 single-inlet centrifugal fans whose diameter is 4 m. It uses a capsule type apron, and the apron materials are the English "4275" fabric, double-sided coated rubber.
    b. Propulsion system
        This consists of shafting, thrust bearings, and two air ducted propellers. The diameter of an air ducted propeller is 3 m.
    c. Power plant
        The main engine is a light 12V180 diesel engine, the maximum power is 661.5 kw, and the speed is 1600 r/min; auxiliary is a 24 kw generator of T2HS double sub driven by a F4L912 air-cooled engine, of which the power is 31.6 kW, and the speed is 1500 r/min.
    d. Control device
        This mainly consists of host manipulation and a propeller pitch steering system. Host manipulation can drive a screw to control the host speed using the reversible ZD2530 motor. The propeller pitch steering system, with two CB-40 pumps as a power source, uses a hydraulic steering gear and the solenoid valve in two ways to steer and allows the two rudder blade groups not mechanically connected to achieve synchronization by reversible shunt.

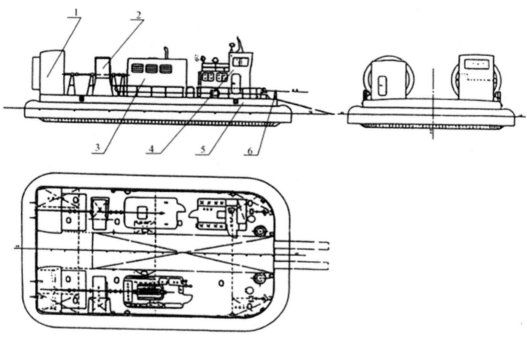

FIGURE 5-24

35 t self-propelled air cushion transport platform: 1 - propulsion system; 2 - pad lifting system; 3 - power device; 4 - manipulation device; 5 - hull; 6 - outfitting and electrical equipment.

e. Main hull

The main hull of the platform is a box-shaped water-sealed welded-steel structure, which is composed of a roof rack, floor rack, side plating, and vertical and horizontal bulkhead. In order to ensure that a damaged compartment doesn't sink the hull, the main hull is divided into several watertight areas. The superstructure is an aluminum alloy riveted structure.

f. Outfitting equipment

The platform is equipped with two 35 kg high power anchors; and four streamlined rudders that are placed in the rear of the two propellers. The left and right of the platform's head respectively has an electric harmonic winch of 25 KN maximum pulling force.

g. Powered device

The F4L912 diesel engine drives a T2HS three-phase AC generator that provides electricity for the platform. On the other hand, two F-1500J host engines provide electricity for the total distribution pane by driving 24 V DC generators.

5.4.2.4 100 t Passive Air Cushion Transport Platform Developed by China Dagang oil field

The platform, which is named "7303" (code) by Dagang oil field, is mainly used in the beach, very shallow sea (0 ~ 5 m) and amphibious oil fields as transportation for equipment, apparatus, and other components in oil exploration and development. The air cushion transport platform is designed as a passive type without a traction device. In order to solve the problem of traction for short-distance transport in the fixed waterway between the artificial island and base, people establish an electric, endless rope winch traction device on the platform. Thus, the platform can also be self-propelled on a fixed channel.

1. Main technical performance parameters
   a. The main dimensions are a total length of 28.3 m; total width of 18.8 m; and overall height of 8.8 m when using the air cushion.
   b. Load capacity is a rated load of 100 t, and maximum load of 130t.
   c. Design pressure of air cushion is 6027 Pa.
   d. Navigation capability is that towage speed in hydrostatic conditions is 10 km/h; towage speed in the beach mud area is 10 ~ 20 km/h, and maximum dragging is 3 km/ h.
   e. When in extremely shallow beach area, if the wind is below 8 magnitudes and wave height is not less than 1.2m, the platform can realize the pad liter state-hauling air. When in very shallow sea area and in drainage air state, if the wind is below 8 magnitudes and wave height is not greater than 1.2m, the safety of the platform can be ensured.

2. Structural composition of platform
   The structure of the passive air cushion transport platform of 100 t load is shown in Figure 5-25. It predominantly consists of the pad lifting system (1), traction device (2), power unit (3), manipulation device (4), hull (5), and outfitting and electrical equipment (6).
   a. Pad lifting system
      This is composed of the air inlet, three lift fans, airway, and aprons. The lift fan is aluminum centrifugal fan single inlet, DS-2, with an impeller diameter is 1.12 m; the apron is capsule-type apron with low sac pressure. The bow and stern are two low resistance cystic aprons, whose material is composed of double-sided coated cotton fabric rubber.
   b. New traction device of the transport platform
      This includes a 60 kN traction electric endless rope winch and two electric initial tension devices, as well as four guide ropes etc. When the traction cable ends are fixed on the artificial island and the shore side of the base of the pier, the platform can be moved round trip with a 50m/min maximum speed with the traction device, and we can also regulate the speed of movement through the inverter device.
   c. Power plant
      The power plant of the platform uses 3 BF12L513C air-cooled diesel engines made in Germany as pad lifting engines; also the platform is equipped with an F8L413F air-cooled diesel engine, which comprises an 100 kW auxiliary generating set.
   d. Manipulation device
      The main and auxiliary engines are manipulated through a centralized flexible shaft by a cab; the ballast balance system is manipulated through a solenoid valve; and the traction winch of the platform is controlled and manipulated jointly by cab.
   e. Hull
      The platform body is equipped with longitudinal framing, whose material is marine Class B steel welding, and the deckhouse and cabin shed uses riveted LY12CE aluminum.

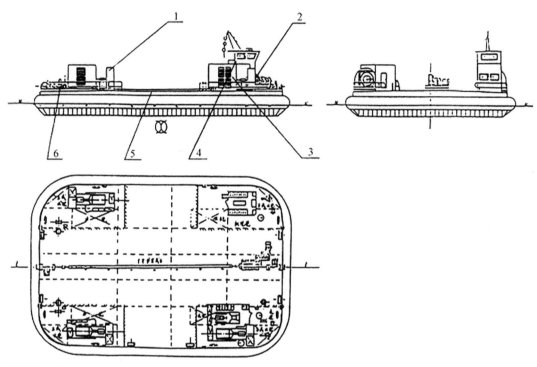

FIGURE 5-25

Passive air cushion transport platform of 100 t load: 1 - pad lifting system; 2 - traction device; 3 - power plant; 4 - manipulation device; 5 - hull; 6 - outfitting and electrical equipment.

f. Outfitting

The platform is equipped with two 50 kg high power anchors on the bow deck and a steel cable guide clamp and bollard. The dragging device is equipped with 3 sets of 50 kN pull winches for electric harmonic.

g. Electrical equipment

This mainly includes a 55 kw traction motor (Y280M-6-H), two sets of XWJD4-4 initial tension motors of 4 kw, and three sets of winch motors of 50 kw.

The Navigation of air cushion transport platforms of 35 t and 100 t load is shown in Figure 5-26(a) and (b).

5.4.3 Mud Sled Means of Delivery

Mud sleds have already been mentioned earlier in this chapter; they are a boat-like structure used as a means of delivery of the ship. Mud sleds can not only float in the water at sea, and taxi on mud shoals, but can also skid on the ice. All in all, it is a simple, economic, and practical amphibious carrier.

(a)

(b)

FIGURE 5-26

Navigation of air-cushion transport platforms of 35 t and 100 t load.

The TH4001 mud sled structure with 100 t load capacity, constructed by Dagang oil field in China in the 1990s, was successfully used in the oil field transport operations of a muddy beach area. A brief introduction of the TH4001 mud sledding system follows.

5.4.3.1 Main Features
There are mainly three characteristics of the TH4001 mud skid.

1. Barge has a water floating air system as a base
   The TH400 mud sled system is a large-scale amphibious vehicle, based on the conversion of a flat deck barge. So it has a water floating air system that can support certain weight loads and drainage.
2. Using the sludge characteristics for delivery on the beach
   Because of 90% of the relative content of clay mineral in the mud on the beach of Dagang oil field and about 70% of dispersed clay minerals, sludge can make hydrated clay mineral dispersed, forming suspension of colloidal liquid or vermicular colloidal under the condition of the water infiltration, coupled with the disturbance of sludge. And it can form a thin film between the moving objects and the soil surface to reduce the friction coefficient (generally not greater than 0.10), thus contributing to the effectiveness of mud skidding on the muddy beach.

3. Being able to transport in a fixed waterway with a traction cable

   When a traction cable has two ends fixed at both ends of the shipping route, the mud sled can accomplish transport in the fixed waterway if diesel generator sets, an electric endless rope winch, and an initial tension device are installed. In this case, the rope enters one end of the mud sled and twines 3 ~ 4 circles on the friction wheel of the winch, then through initial tension, leaves the other end of the mud sled. So when the motor drives winch friction wheel rotation and the initial tension at the end of rope is working at the same time, a winch rope end generates the initial tension. When the tension reaches a certain predetermined value, the frictional force generated by the friction between the wheels of the rope winch enables the friction wheel to scroll forward along the rope, through the winch base, driving the mud sled forward along the cable. Conversely, when the motor rotates in reverse, the winch friction wheel rotates in reverse, driving the mud sled towards the opposite direction. Thus, by adjusting the positive and negative going motion of motor, the mud sled can run along the cable round trip, and carry out transport tasks in a fixed channel.

5.4.3.2 Technical Performance

1. Size parameters

   The sled body's length is 27.52 m; width is 5.53 m; and molded depth is 63 m.
2. Load capacity

   Maximum load capacity of the mud sled is 80 t.
3. Weight of the mud sled

   Including the weight of the equipment, the mud sled weight is 72 t.
4. Movement speed

   Speed of mud sled movement is 20 m/min.
5. Dynamic power

   Power of the mud sled generator is 200 kW.
6. Performance of winch

   Motor power of the winch is 90 kW and Max traction is 250 kN.
7. Initial tension power

   Initial tension motor power is 1.5 kW.
8. Structural compositionThe TH4001 mud sled mainly consists of sled body, power systems, traction systems and living facilities, as shown in Figure 5-27.

   a. Mud sled body

   In Figure 5-27, the sled body (2) is the hull of flat-bottomed barge, which is divided into five watertight bulkheads. Among them the bow is the water tank, which can accommodate up to $10m^3$ water for delivery of domestic water. The front is the machine cabin, in which the traction winch (6) is equipped. The middle and rear of the tank is empty, being used for storage room. The fore part is the fuel tank, with a 2.5 m^3 fuel tank. In the middle of the deck above, there is a 15 m cargo loading area, which is approximately 80 m^3.

   b. Power system

   The power system consists of a 200 Cummins diesel generator set and a 2.5 m^3 fuel tank, as shown in Figure 5-27. It mainly provides electricity for the traction system, auxiliary systems, and living systems.

   c. Traction system

   As is shown in Figure 5-27, the traction system consists of a 2JDM-25B Rally for the 250 kN electric endless rope winch, winch control cabinet (7), two initial tension pieces (8), traction

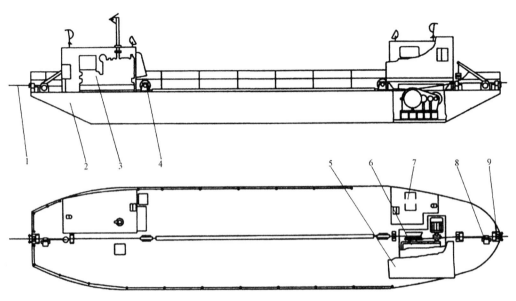

FIGURE 5-27

Structural composition of the TH4001 mud sled: I - traction cable; 2 - sled body; 3 - diesel generator sets; 4 - pulley; 5 - lounge; 6 - winch; 7 - cabinet; 8 - initial tension; 9 – fairlead.

cable, (1) and lead rope and pulley (4), etc. The winch consists of a 90 kW squirrel cage motor, an arc gear III gearbox, and two double-cone friction wheels. When the motor drives the rotation of the friction wheel through a gearbox, and the friction wheel is wound around the rope end with a certain initial tension, the generated friction force between the friction wheel and rope is called winch traction.

The initial tension device consists of a gear reducer motor, active sheaves, roller, lever, and pressure iron. An active sheave is mounted on the motor shaft, and a cable is embedded in the active sheave groove, above which is placed the pressure roller. Through the pressure iron and lever, the pressure roller has some positive pressure on the cable. Thus, when the motor rotates, rope is dragged by the generated frictional force between the capstan and pressure roller, which makes the rope end of the winch friction wheel produce a certain initial tension.

d. Living facilities

The living facilities are mainly composed of the lounge (5) shown in Figure 5-27.

5.4.4 Ship-Type Delivery Vehicles

Generally speaking, delivery using a ship in beach area is not significantly different from the ordinary ship delivery. But the technical performance must meet the needs of exploration and development of oil field in the beach sea region. Therefore, I do not intend to occupy a lot of space, but only to introduce the main aspects of some ships for beach/shallow sea region oil fields self-designed and built by China.

5.4.4.1 Shallow Draft Dragging Supply Vessels

Take the shallow draft 2400 kW dragging supply vessels as an example here for description; they are designed and built by Dagang oil field.

1. The main purposes are to:
   a. Drag the offshore oil drilling platform in place;
   b. Deliver fuel, fresh water, cement, barite equipment and other equipment and materials for the platform
   c. Undertake the mission of guarding and rescuing the shallow drilling platform
   d. Weigh anchor and drop anchor for the drilling platform of beach/shallow sea areas
   e. Supply an external fire capacity of 1200 m^3/h of water, foam
   f. Tow barges
2. Size parameters
   The total length of the ship is 59.4 m; the length of the design waterline is 57.0 m; the length between perpendiculars is 53.7 m; the width of the beam is 12.0 m; and the depth is 4.0 m.
3. Technical performance
   a. Draft and displacement
   Draft of a light load is 9 m; displacement is 940 t; heavy draft is 2.6 m; and displacement is 1388 t.
   b. Navigation
   In deepwater navigation area the draught is 1.9 m, and the maximum speed is 13.5 kn; in the light-load condition in a deep-water navigation area, wind is 13.8 m/s, and the speed of dragging the platform is 5.0 kn. In the condition of overloaded navigation service speed, endurance is 1000 n mile; the self-sustaining force is not less than 20 d.
   c. Bollard pulls
   The bollard pull of the sham air district is 400 kN; and the bollard pull of beach/shallow sea trade area is 250 kN.
   d. Cargo capacity
   Cargo capacity of overloaded deck is 250 t, and amount of bulk cement (canned) is 100 t.
   e. Dynamic power
   3 MWMTBD620V8 diesel engines made in Germany are the host engines, with a rated speed of 1500 r/min. Rated power is 829 kW. Generator sets are 275M × DFCC made in United States, with NTA855G3 and HCM534C engines, and a total of three units. Their speed is 1500 r/min and capacity is 275 kW. The emergent generator is CCFJ50Y-1500, whose engine is the 4135AD; there is only one unit generator, the TFHX with a speed of 1500 r/min and capacity of 50 kW.
4. Weight
   The total weight of the entire ship is 902.1 t; and the net weight is 270.6 t.
5. Structural composition
   The structure of the shallow draft 2400 kW dragging supply vessel is shown in Figure 5-28. The ship is a beach/shallow sea supply vessel which has a long side, single continuous deck, single bottom, closed cabin; forward side column, Transom stern, and three paddle machines and twin rudders.
   a. The hull
   The hull has a welded steel structure, single bottom, single deck, and transverse framing. The carrying capacity of the cargo deck is 50 kN/m^2.

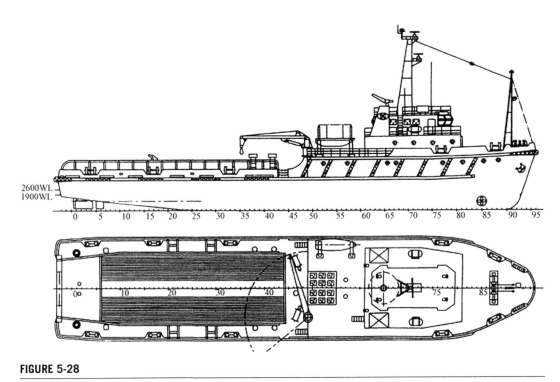

FIGURE 5-28

2400 kW shallow draft ship structure.

b. The turbine system

The turbine system includes three main propulsion units, three propulsion shafts, and a set of bow thrusters. The main propulsion unit consists of three diesel engines. Propulsion shafts are arranged in parallel with the ship's three machines propeller, ducted propeller, propeller rotation direction of internal rotation and external rotation. Thus, the middle propulsion shaft is arranged in the midline plane. But the port and starboard propulsion shafting is arranged 3.5 m away from the midline surface, three promote axis are parallel, 1.2 m away from the baseline. The bow thruster is in its own compartment, and is furnished with a lateral conduit propeller by catheter, gearbox, motor, control cabinet, and so on. Other equipment, such as generator sets, deck machinery, life-saving equipment, cabin oily wastewater treatment systems, sewage treatment equipment, ballast water, fresh water and oil fuel transfer systems, powder pneumatic conveying systems, fire systems, and electrical and communications equipment, are the same as for the ordinary ship, so it will not be repeated here.

5.4.4.2 Dayun 202 Multipurpose Shallow Tugboat

The Dayun 202 multipurpose shallow tugboat was self-designed by the Dagang oil field, and it has a long forecastle, single deck, single bottom, double machine, sculls, double tunnel, and shallow draft; it is shallow multipurpose tug. The function of the ship is the same as the above-described shallow draft 2400 kW dragging supply vessel with low displacement and cargo capacity compared with that vessel,

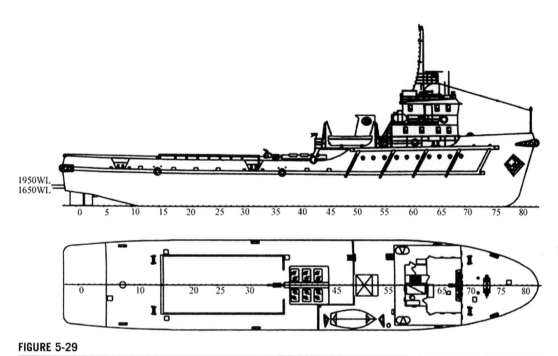

FIGURE 5-29

Dayun 202 multipurpose shallow sea tug contour map.

with a max 80 t cargo capacity. The host power is only 2×260 kw, but max speed is 9.5 kn. A drawing of the ship is shown in Figure 5-29.

5.5 **Problems in Beach Oil Field Development and Production**

For many years China's onshore oil field development and production, and offshore oil field development and production have formed a set of mature models with accumulated experience. But for beach/shallow sea areas, whether in scale; or in technical, environmental, and development issues; or in economic evaluation standards and methods; or in some of the production technology and process flow, there are differences with onshore and offshore techniques. Especially in the China Bohai area, there are vast areas with dissimilar environmental conditions. Thus, the mode of development and production system must take into account the specific circumstances and analysis. This section is about special problems of oil field development and production for beach/shallow sea areas.

5.5.1 **System Choice for Oil Field Production Facilities**

The oil field development system is composed of production facilities and equipment, which are called oil field production facility systems.

5.5.1.1 Categories of Oil Field Production Facilities

From the domestic and foreign point of view, there are two types of oil production facility systems: early production systems and conventional production systems.

1. Early production systems
   In the early period of production in oil field development, some shifting production equipment is used to support oil production systems before fixed production facilities have been built. Therefore, the core of the early production system is use of removable and repeatedly reusable oil field production equipment.
2. Conventional production systems
   Conventional production system are fixed production systems, with the fixed production platforms and oil storage platform at the center; this is the mining oil production system commonly used for offshore large oil field development.

5.5.1.2 Choice of Beach/Shallow Sea Production Facilities

1. Production oil fields with certain reserves
   Similar to the development of offshore fields, for beach/shallow sea oil field development with certain reserves, a production oil field can choose a fixed conventional production system. The large-scale stationary production facilities allowing for driller maintenance and operation and convergence as well as the production process of oil extraction work well, and also facilitate the establishment of a compact and complete production flow process, centralized management, and high efficiency production. In addition, the concentration of the production process is conducive to address issues such as environmental protection. Although the use of large-scale stationary production facilities requires significant investment and a long construction period, the mass production and high efficiency can achieve good economic returns due to the advantages mentioned above.
2. Harsh environmental conditions and low recoverable reserves of small- and medium-sized oil fields
   Once a fixed production system is used in the beach sea area of these types of fields, capital costs are expensive, even with full oil field production during the development period. It is difficult to recover the investment or recycle the investment, and little profit is produced. In this case, such an oil field from the economic perspective will not be worth developing, or there must be cost-effective production and technical measures for what are commonly referred to as marginal fields. Obviously, for marginal fields such as beach/shallow sea areas, the choice of an early production system is an important technical measure to improve the economic benefits. Because the core of the early production system is removable and can be used repeatedly as production facilities and equipment, this greatly reduces the burden of a field, which requires only the burden of production systems and equipment rental, but not all of the cost of investment, which often makes the development of marginal fields become economically feasible.
   However, we should add that, early production system does not apply to marginal oil fields, some ready oil field development in early or even the entire development period. Early production system can make these large oil fields develop early with recovery of funds and benefits before uncompleted fixed production systems.

5.5.2 The Problems of Early Production Systems

If an early production system has been chosen for a beach/shallow sea are, the following issues need to be considered.

5.5.2.1 Wellhead Platforms and Completion Methods

Wellhead platforms and completion of system facilities is the main part of the early production system which need to pay a fixed facility investment points, and it's rationality is directly related to the economic viability of marginal field development. Completion system facilities and completion style are inseparable. The beach/shallow sea region has favorable conditions for simple completion methods to reduce the cost of the completion system facilities.

Simple wellhead platforms should be used for beach/shallow sea region, here are several beach/shallow sea wellhead platforms.

1. Single pile or piles of concrete wellhead platforms

    A concrete platform was developed by Lagoven in Venezuela and used for the depth of the water of Lake Maracaibo. Now we look at the single pile concrete wellhead platform as an example to illustrate this. The concrete pile wellhead platform consists of four prefabricated groups as shown in Figure 5-30.

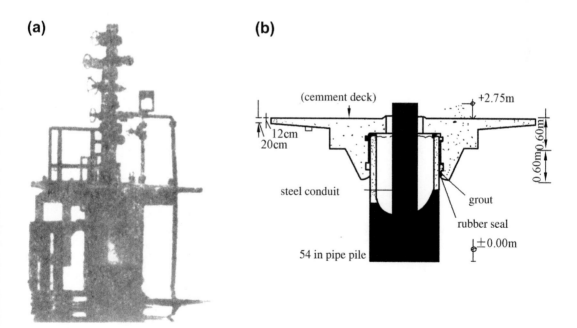

FIGURE 5-30

Single pile concrete wellhead platform.

a. Concrete piles

As shown in Figure 5-30(b), a reinforced concrete caisson centrifugal pouring is made, whose length and diameter depend on water depth and engineering geological conditions. For example, in Lake Maracaibo, when the water depth is 6 m, this pile outer diameter is 914.4 mm (36 in), and in water depth of 18 m, this pile diameter is 1625.6mm (64 in). Pile piling machine play into the sea bed. As for double-pile wellhead platform, it is strengthened by a small pile on the diagonal. Figure 5-30(a) shows a single pile wellhead platform.

b. Steel pipe

Steel pipe is used, as shown in Figure 5-30(b), and first goes into the pile, and then into the catheter.

c. Concrete deck

As Figure 5-30(b) shows, it is first prefabricated and after the pipe pile and catheter are laid, this deck is loaded.

d. By berth

It is installed finally to be used for stopping after the deck is inaugurated.

The scale of the pile and catheter diameter, length, and thickness, should be based on wind, waves, and current and depend on the type of the vessel as well as the process geological conditions. Lagoven has built more than 350 single pile wellhead platform for beach/shallow sea; the pipe pile diameter is 1676.4 mm (66 in), wall thickness is 152 mm (6 in) and the length is 56 m.

The economy of such small wellhead platforms makes them suitable for simple oil production single well platform-based production systems; in Figure 5-31, we see a bird's-eye view of the Venezuelan single-well platform-based production system used in Lake Maracaibo. The use of

FIGURE 5-31

Production system based on a single wellhead platform.

single pile, double pile, or three concrete wellhead platforms cantilevered drilling barges overall migration associated with drilling and completion operations. In this way, we can not only rig the overall relocation, but also accelerate the speed of the drill production wells, and the platform is no longer withstand the weight and drilling rig hook load, so the size and weight of the wellhead platform can greatly reduce then in order to achieve a simple structure. Using this single well platform for oil production, generally through the underwater pipeline, the oil and gas can be gathered to a number of production platforms, to processing and output. This single well platform oil field development and production plan is used in China's Daqing oil field reservoirs and the Huang Jue, Jiangsu oil field. But it is also used in China in the very shallow seas of the Bohai area. The Shengli oil field in Chengbei oil field, which adopts the left on the water with a diameter of of 508 mm (20 in) casing pipes, around additional the casing welded into the well head protection platform to anti-ice into the catheter and sleeve; In Liaohe Oil field, the method that through injecting into the catheter and cannula the annular space is affused by cement is used to against ice-resistant. In short, the above shows taht the use of a simple single wellhead platform and water completion methods in the Bohai Sea, China beach/shallow sea area is a viable oil production program.

2. The oil production workover integration platform
   An oil production workover integration platform for the Shengli oil field in China was independently designed and built for very shallow sea. Such a platform solves the single wellhead platform workover problem. It mainly consists of a fixed oil platform portion (the structural base part) and repair mining module part (moving parts) as shown in Figure 5-32.

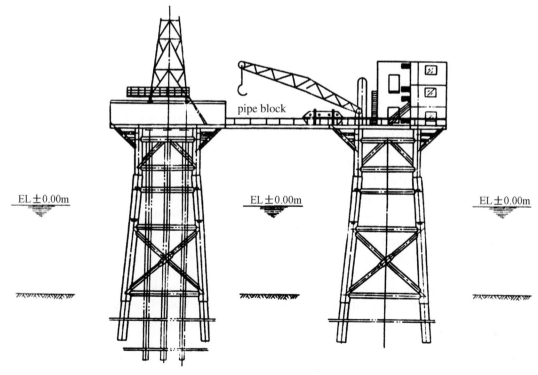

FIGURE 5-32

Integration of drilling and workover platform.

a. Structure

    i. The base structural part

        The fixed oil production platform section is shown in Figure 5-32. Other fixed oil platforms include the measurement platform workover platform and trestle. The measurement platform (shown in Figure 5-32 left) is a measured jacket, the upper measurement platforms, and equipment room; the workover platform is a workover jacket (shown in Figure 5-32 right), upper workover platform; trestle part different pipes, profiles welded space truss structure, is mainly used for process piping, cable across, and doubles as a disposal site and test site work over operations tubing.

    ii. Repair mining part of the module

        This is an integrated platform of moving parts, which can be flexible for lifting and handling, mainly for resettlement workover with a variety of special equipment. This module can connect the wellhead platform and mobile operations, which can solve workover issues with a wellhead platform production of single well.

b. Main features

    i. Relatively fixed position of well

        The oil well workover jacket (shown in Figure 5-32 right) within the framework of the two fixed relative position, size, angle deviation requirements are small, which will not affect the structure of the upper workover platform wellhead oil trees, electricity pump and sucker rods.

    ii. The fixed length of trestle

        Trestle used for stacking tubing is required to have a fixed length, and the length of error should be from 0 m to 1 m. Otherwise, additional length requires the strengthening of the trestle structure, which when too short will fail to fulfill the main role of the trestle.

    iii. Groove mosaic structure lap

        For the overlapping end of the trestle and workover platform, a groove mosaic structure is used. This is because if the orientation of the workover platform jacket (shown in Figure 5-32 right) has been determined, the orientation of the trestle will also be able to be determined. Then the relative orientation and required deviation between the measurement platform jackets and workover platform jackets is very small, so a groove mosaic structure lap is used. Otherwise, if the measurement platform is at the end of the trestle, the lateral offset is too large, and it will affect the structural strength of the platform and the measurement of the oil production platform.

        In the above if the reference workover well integrated platform concept is introduced to a simple wellhead platform, with the water well completion method it can satisfactorily solved an issue of single well oil production wellhead platforms.

5.5.2.2 Mobile Production Storage Equipment

The beach/shallow sea region in addition having a well close to the shore can take oil and gas output to onshore processing. Most of the wellhead platforms are required to first gather oil and gas and transport them to production at the oil storage facilities, where they are handled by centralized processing. An important feature of the early production system is lower investment and space than occupied by large production processing facilities and mobile storage and offloading facilities that can be reused. Therefore, the following will make a brief introduction on the mobile production storage equipment for beach/shallow sea regions.

1. Floating production storage barge

   This is similar to a barge for offshore floating production, storage, and offloading (FPSO), and applies to very shallow beach sea areas. The installation of a set of oil and gas processing facilities and storage and offloading facilities in smaller barges, and then an appropriate mooring in water, allows them to be used as a floating production storage and offloading platform. For example, Arco (United States) uses a 3000 t load production, storage and offloading barge for Indonesia oil production, storage, and offloading, with a working depth of 3 m to 90 m, anti-storm wave height 2 m, and scale of 64 m × 18 m × 3.7 m. The front deck on barge is equipped with oil and gas processing equipment, including two horizontal three-phase with a daily processing capacity of 2000 t. Part of the gas is separated for power generation fuel, and the excess is sent through the underwater pipeline to flare. Separation of oil and water is metered, and pumped to treatment not far from the fixed production platform. Accordingly, the barges on the water do not have treatment devices; the end of the barge to the production area can accommodate 32 people. Sometimes when the barge need not to be used in conjunction this barge will be used with the separate installation of storage and offloading barge.

2. Bottom-sitting production storage platform

   For very shallow sea and beach areas, if a floating production storage barge is not suitable, a removable bottom-sitting platform can be used.

   This platform is a sink pad barge, on which, according to the size of the water depth and water level fluctuation, mounting pillars are installed to support the upper deck, and oil and gas processing, storage, and offloading facilities. This bottom-sitting production storage platform can take the tide into the berth, and then sit at the end with a fixed pile as the production, storage, and offloading platform. When moving, the ballast water in the mat is discharged, it can float and be dragged to a new berth, and can be reused. Here are its several forms:

   a. Simple form of oil and gas processing

      Shown in Figure 5-33, this is only a cement barge converted into oil and gas processing movable platform, and is equipped with an oil well on the high deck and well processing device.

   b. Both production and offloading and storage facilities

      As shown in Figure 5-34, there is both the reservoir removable bottom-sitting production storage platform and offloading and storage. It is also a cement barge, but in addition to the high deck equipped with production and processing equipment, there are offloading facilities, and oil storage with 1400 t oil storage tanks in the hull of the mat.

      The sitting cement barge-based mobile production storage platform, saves steel, has a low cost, and is easily constructed; there are many years of construction experience with this platform.

   c. Simple form oil reservoir

      Without a storage facility a bottom-supported production processing platform can group and add cement barge tanks to a mobile bottom type oil storage platform for use. This is a simple oil storage with removable bottom-supported storage platform, with a cement barge

FIGURE 5-33

Removable bottom-type oil and gas processing platform.

FIGURE 5-34

Removable bottom production platform with storage function.

that has two 140 t oil tanks, the other in the hull and the teams from four storage 140t oil storage tanks. For safety, the pile of set bits should be added around.

d. Anti-ice features form

With the China Bohai area freezing in winter, we need to address the issue of anti-ice beach/ shallow sea removable bottom production storage platforms. Figure 5-35 shows a 45° sloping segment side-heated barge, which has been successfully used at water depths from 2 to 9 m, with ice thickness of 2.1 m and an accumulation of ice as thick as 3.9 m sitting on the barge. This barge has heat pipes in the side of the ship, using the waste heat of the engine to cool water, and heat the ice in the ship's side section. In this way, the ice can produce a layer of melted water film in contact with the side of the ship, which is conducive to bending and

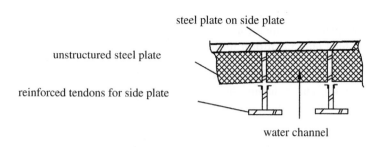

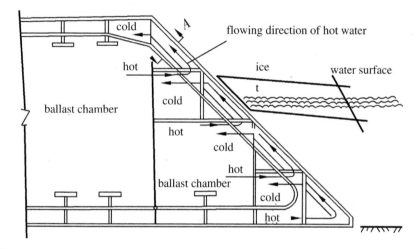

FIGURE 5-35

Anti-ice heated side for barge. Side of the side plate, No structural steel plate, Lateral side stiffening rib, water passes, ballast tank, cold, warm, Hot water flow direction, ice cake, water surface.

breaking the ice along the 45° sloping side. Because the flexural strength of the ice is much lower than the compressive strength, the ship's side is made with a 45° slope type to easily damage the ice.

3. Movable pile production storage platform

The removable bottom-sitting production storage platform, which is suitable for load carrying capacities for sandy beach areas, and for some deep mud beach sea area with available foundation of reinforced silt seabed, may not be economically viable due to the high cost; we also need to consider the mobile pile production storage platform.

The pile foundation of this platform should be modular and able to be transported and assembled in extremely shallow sea water, and the production processing platforms and oil storage facilities should also be modular in design to facilitate delivery vehicle movement as well as platform installation. Modular jack-up drilling platforms and pile modular assembly platforms have been introduced in the third main section of this chapter, and will not be repeated here.

It should be pointed out that this assembled movable platform can be used not only as a production storage platform, which can also be assembled into a small wellhead platform, but also can be assembled into a platform for a number of cluster wells.

5.5.2.3 Simplify the Processes of Production and Processing

Simplifying the processes, and reducing the platform area and the cost for the beach sea region development is of great significance. The beach/shallow sea oil field first takes advantage of the favorable conditions of being close to shore, with the output of oil and gas to the onshore production facilities, thus eliminating the need for production and oil storage platforms. Or there may only be a simple device for preliminary separation on the platform, with output and main processing of the oil and gas and water purification placed onshore that can greatly simplify and reduce the required production platform. Simplifying and reducing the production platform is another important way to simplify and shorten the process of production and processing, and the following will put forward some measures in this regard.

1. Short flow injection
   The traditional oil field water injection process is like this:
   water → centrifugal pump for water → water storage tank → centrifugal pump upgrade → sand off → deoxy → storage tank → the pump → high-pressure injection pump → injection wells
   Obviously, this injection process has a lot of processes and equipment ranges, accounting for a very large area, and the existence of both deoxidation and repeated aeration raises complex problems, increasing the complexity of the process. However, the following simplified injection process can be considered according to the experience of some of the oil fields in China:
   water wells → centrifugal pump water → closed pressure desanding → high-pressure injection pump → modular water distribution stations → injection wells
   In this process, a modular water distribution station is used, and there is a switch to small-size valves and throttling device, which reduces the water distribution station space about 50%.

2. Short process of oil and gas gathering and transportation
   The traditional oil gathering and transportation process is a large-scale process that has various equipment, a large occupied area, and is not suitable for dispersing and low yield oil beach areas. Therefore, some of China's oil field reference experience can be considered to take measures to simplify oil and gas gathering processes; these features are:
   a. With a full sealed jet mixing device, oil and gas will be delivered directly to the production processing station on the land.
      This process can not only simplify and reduce the complexity of the gas and oil transmission process, but also can reduce the wellhead back pressure, increasing the oil yield and associated gas recovery. This scheme can be used in the common Y type oil pump at output liquid as the power fluid, a jet pump feeding device, the cost compared with ordinary transfer station, can be decreased by about 90%.
   b. Multifunctional storage tank to complete storage and loading operation of crude oil
      This program uses fire tube heating, and handles heating, dewatering, loading, and shipping of oil and other functions in one multifunctional energy storage tank. In this way, not only can we greatly reduce the flow, but also because of the modular unit and flexible assembly,

we can use different scale crude oil loading platforms and adjust the tank unit number, thus the occupied space and cost have been greatly reduced.

c. Direct implementation of power fluid concentration separation, dehydration, pressure, and other full circulation system technology in the production platform

A hydraulic piston pump is usually used in the traditional power fluid system process: crude oil $\rightarrow$ metering station $\rightarrow$ union station $\rightarrow$ oil production separator $\rightarrow$ a settling tank $\rightarrow$ pump $\rightarrow$ hydraulic power station $\rightarrow$ pump and high pressure pipeline $\rightarrow$ power fluid pressure distribution valve group $\rightarrow$ dynamic fluid pipeline $\rightarrow$ wellhead metering station input production system

However, a production platform realizing a dynamic fluid treatment scheme has been put forward according to some of China's oil field experience. It will produce liquid on the platform directly into a multifunctional device; in this device, oil and gas separation, dehydration, and oil heating technique will be done in conjunction. Then, through the dynamic liquid booster pump, power is distributed to the liquid column of each well, so not only can we reduce the flow and occupied space, but also the cost is low, and it is convenient to move. Obviously, it has vital significance to the production platform for early production systems. Of course, this process is also applicable to conventional production systems.

5.5.2.4 Suitable Programs for Oil and Gas Transmission

Crude oil and gas transportation solutions in offshore oil fields generally use shuttle tankers of crude oil and associated gas in addition to the flare method to make fuel for power generation. But for beach/shallow sea regions, you need to distinguish between different situations and evaluate the options.

1. Close to shore to beach/shallow sea oil field

In this case, we should take full advantage of the near shore favorable conditions and use oil and gas pipelines directly sent to the shore production facilities for processing. And in order to simplify the construction and reduce corrosion on the beach or very shallow sea areas, we can also consider a simple trestle to set up the pipeline, but if in ice, such as in the Bohai beach/shallow sea in winter which often has ice floes and pack ice, the trestle structure should be resistant to ice; otherwise, it should be a buried pipelines.

2. Far away from shore beach/shallow sea oil field

In this case, the following programs are available:

a. Low-yield dispersed well

By a production storage barge turns separate wells, periodic intermittent mine oil, direct oil mined and transported, and the so-called "Bee oil production".

b. Remote and isolated well

If the isolated well has plenty of storage, a suitable form of movable production storage facility can be installed near the wellhead platform, such as the earlier-described modular assembly pile-type mobile production storage platform for the preliminary processing of oil and gas; the shuttle tankers carry it away on a regular basis. Well that are low yield are treated like dispersed wells.

c. Intensive low-yield wells

At this time, you can moderate position, production, storage, and installation of movable facilities, and then through the trestle or pipeline, well output of oil and gas gathering and transportation to the facility, forming qualified products which can be transported by shuttle tankers or pipeline.

The barge should be suitable for beach area application; a barge or air cushion transport platform can be used.

d. Small marginal fields

This also comprises a considerable amount of oil field associated gas that can be utilized, but with the laying of pipelines it is not feasible economically. This case can be considered a shipping method using compressed natural gas (CNG), chasing after further processing and utilization of the method to the shore. Using this scheme, the shuttle transport ship moors at the wellhead. Processing associated gas and natural gas will not be handled when loading at the wellhead, and will be compressed to about 2400psi to fill the gas tank on board. Then when transported to the purpose, after unloading the gas will be handled with measurement, and dehydration. A single boat can be used for intermittent transportation, or a few boats can be used to ensure continuous production.

5.5.3 **Problems in the Conventional Production System**

It has been mentioned that so-called conventional production system is a production system which regards fixed production storage facilities as the center when developing a large oil field.

For large oil offshore oil field development, such a production system is often used. Conventional production systems, due to the large fixed production storage facilities, facilitate convergence of the production process of drilling and well maintenance operations; the establishment of a compact and complete production process, centralized management, and efficient production has a lot of advantages. However, large fixed production storage facilities require a large investment and long construction period.

It has been pointed out that, beach sea area oil fields of considerable size using conventional production systems need a viable field development pathway. The beach/shallow sea oil field conventional production system uses large-scale stationary production storage facilities with artificial islands or pile fixed platforms. These fixed production facilities can be a cluster of oil wells, and the completion of satellite wells from single well wellhead platform water with the composition of the centralized production system, the output of crude oil by pipeline or tanker ship out.

For oil fields near the beach/shallow sea, shore production oil storage facilities can also be used instead of the beach offshore fixed production storage facilities; oil and gas well output is input to the onshore processing and transport system.

In the beach/shallow sea region, the basic production system and transport method used for conventional crude oil gathering are underwater pipelines, bridges, or shipping similar to that used with an early production system.

There are two types of conventional production systems in beach/shallow sea oil field completions: one is cluster oil wells on an artificial island or fixed production platform, similar to land-based method of completion; another is when satellite wells are included in the production system, they can use the same single wellhead platform as early production systems with water completion.

From the above, we that for a beach/shallow sea region using conventional production systems, an artificial island or fixed production platform and other facilities make up the core of the production system; we intend to introduce them below.

5.5.3.1 Choice of the Type of Artificial Island for Production

At home and abroad there are dozens of artificial islands with diverse forms used in petroleum mining, and among those forms the following forms are more suitable for the environmental conditions of beach sea.

1. Filled artificial island

    This type is based on the artificial island with gravel or sand filling, which applies to shallows as it is easy to obtain filling materials such as sand and stone in the beach sea area. It is easy to make an excellent filling.

    Filled artificial island construction methods are different depending on the source of the filling material. For example, we can use a dredging boat to reclaim material directly from the seabed or shipped to the island and dump filling. However, if sand and gravel must be transported from land, depending on the distance and transport conditions, the use of different methods of transport such as cars, boats, or conveyors can be used for filling. Such as, ice can take advantage of glacial transport on ice or ice-breaking filling underwater slope protection and other methods, to be thawed before. It should be noted that sand bags plus geotextile to prevent island slope erosion in the sea is a commonly used method.

    Jiangsu, China Huangjue field is used with this type of production system; there is a storage facility and artificial island with a single wellhead platform that operates with good results. We need the installation of pumping units, a single wellhead platform, and an artificial island equipped with production processing equipment and oil storage facilities and an external input device. Each single wellhead platform produces oil and gas output to the underwater pipeline to the artificial island for centralized processing, and transported by tanker. Figure 5-36(a) shows the single-well oil field wellhead platform, and Figure 5-36(b) shows the oil fields of artificial islands and foreign transportation facilities. Compared with the filling-type artificial island, the steel platform is about six times higher in cost, so in suitable conditions, the choice should be the filled artificial island instead of the steel platform. For a filled artificial island, the wellhead is generally set in a concrete basement to prevent the risk of fire, explosions, and other accidents caused by crude oil pollution in the beach/shallow sea.

(a) **(b)**

FIGURE 5-36

Chinese Huangjue oil field filling-type artificial island and wellhead platform.

2. Caisson trap artificial island

This type of artificial island has a deeper applicable depth than the filled type; it is generally used for an unstable sea state or the situation that gravel filling material source is insufficient, and the filling earthwork need to be reduced. It involves prefabricated concrete caissons or steel caissons hauled to the island site, then the caisson sits, composed of traps, with closed circle of space, and is filled with sand and gravel to build an artificial island.

Caisson combination traps have a prefabricated whole steel shell diameter, which is hauled to the island and seated at the end, to be built into a concrete wall, and then filled with sand and gravel to build mature artificial islands. The Dagang oil field discussed earlier in this chapter describes the construction of an artificial island that is of this type.

The caisson trap artificial island is filled under the conditions of the traps, and without making the slope the Diaspora loss, reducing more filling than filling artificial island. In addition, since we can use prefabricated caisson applications, we can also reduce the amount of construction in the sea and beach area, which are advantages.

There are two caisson trap artificial island used in foreign areas, namely:

a. The concrete caissons trap artificial island

This artificial island caisson is made of prefabricated and prestressed concrete, transported by barge to the island site, seated at the end of the installation, and then filled. For example, Canada constructed the Tarsuit artificial island; the plane and the cross-section are in Figure 5-37. The Tarsuit artificial island is four caissons of 69 m × 15 m × 11.5 m, and each caisson weighting 5000t flexible connection between the caissons, so that each caisson may pass the standing wind, wave and ice load directly to the heart part of island. In order to reduce the ice load and effectively use the carrying capacity of the island embankment, the peripheral is designed with four octagonal caissons, which can reduce the cost 20% from a rectangular caisson. The required sand filling for the island can be reduced to about 3/5. The concrete caisson when necessary can be removed from floating and then dragged to other places for repeated use.

For the Tarsuit artificial island, before the installation of caissons, a dredger should first be used to dredge the island site, and the sand should be replaced by fine sand. Then build a circular mound at the top, and fill the sand into underwater base station (shown in Figure 5-37). The caisson is 25000t barge to island site, the anchor winch location, after water caisson bottom, and then the intermediate filling.

b. The steel caisson trap artificial island

This caisson used for an artificial island is made of steel, so it can be designed into a detachable type, and can be transferred for utilization elsewhere; this is the main advantage of it.

Figure 5-38 shows the plan of an artificial island with steel caisson trap design in the United States of America. It is composed of eight trapezoidal caisson composition traps; each caisson is 43 m long, 13.1 m wide, and weighs 1500 t. Caissons are held together with two wire hoops, each of which has a bundle of cables composed of eight steel wire ropes of 76 mm diameter, with a bearing capacity of 400 t. After connecting the caisson, composed of an octagonal trap, its external diameter is 117 m (shown in Figure 5-38). The caisson can float with no load, and shallow draft. So, it can be transported directly in the water (also it can be transported by barge to the island site), and located with an anchor winch. When assembling and being injected with ballast water, caisson sits on the bottom, then the heart part of the island in a trap can be filled with earth. The quantity of earthwork saves 80% than filling type.

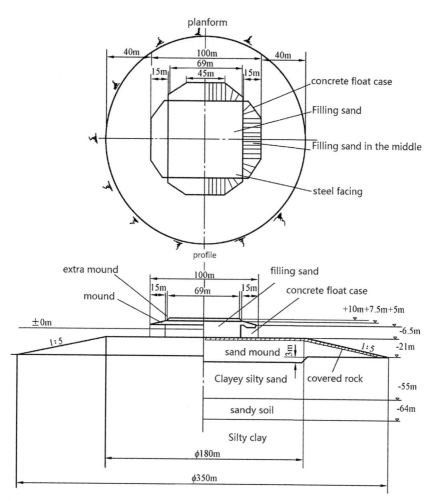

FIGURE 5-37

Canadian Tarsuit artificial island.

"China's first artificial island" in the Dagang oil field also belongs to the steel caisson trap artificial island type, the difference is that there is an overall steel wall concrete caisson instead of the prefabricated steel caisson structure. In addition, it is also a soft foundation with pile foundation reinforcement; this structure can also reduce the filling quantity of sand, but needs concrete pouring on the seabed. The details in the second section of this chapter present a detailed introduction, and it is not repeated here.

3. Concrete gravity artificial island

This kind of concrete gravity artificial island is applicable to beach/shallow seas, and it is different than the concrete gravity platform in deep water. It uses a caisson or stacked structure simple

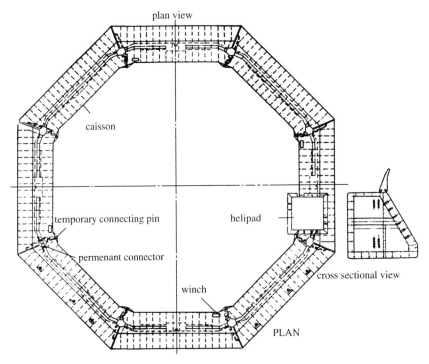

FIGURE 5-38

Plan for a steel caisson trap artificial island in the United States.

bottom to form a production platform, so it can be called a bottom-sitting concrete platform. It has the two classes:

a. Concrete caisson type artificial island

Brazil had built three such concrete caisson type artificial island; the caisson is a rectangular structure, length 52.3 m, width 45.3 m, height 25.7 m, with 42 separate compartments, and 20 compartments in the middle. Peripheral compartments can be installed with ballast sand and ballast water, as shown in Figure 5-39(a). The caisson is a prestressed concrete structure, plate is 0.8 m thick, with a peripheral side wall thickness of 0.4 m. There is construction of a prebuilt sand bed 1.5 m high, after the caisson is transported to the cabin, water is put into the caisson and it sinks, and the bottom is located in the sand bed. After that, loading sand can be throw into the cabin and riprap used for protection is throw around the caisson.

During drilling, the artificial island is equipped with a full set of drilling equipment, as shown in Figure 5-39 (b). When in production, it will drill out, installed production and

(a)

main direction of wave propagation

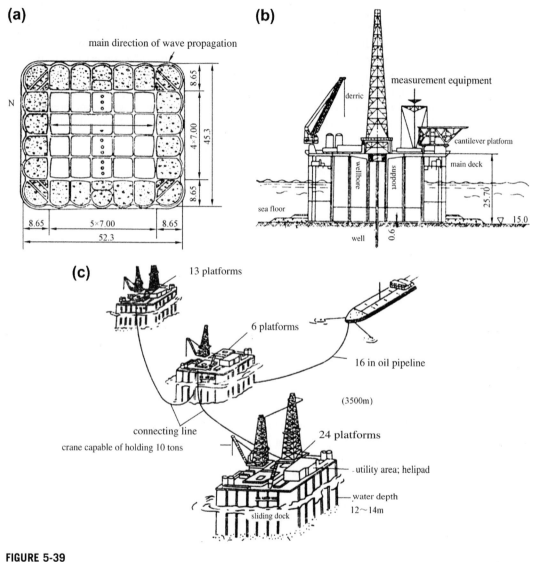

N

8.65
4×7.00 45.3
8.65

8.65
5×7.00
8.65
52.3

(b)

measurement equipment

derric

cantilever platform

main deck

wellbore
support

sea floor

25.70

15.0

well
0.6

(c)

13 platforms

6 platforms

16 in oil pipeline

(3500m)

connecting line

crane capable of holding 10 tons

24 platforms

utility area; helipad

water depth
12~14m

sliding dock

FIGURE 5-39

Concrete caisson type artificial island built in Brazil.

processing equipment, and storage is taking oil-water replacement method, the direct use of caisson in oil storage tank, the replacement of the sewage water separator, the sewage discharged into the sea. The crude oil hose to the mooring buoy in the tankers Sinotrans, as shown in Figure 5-39(c).

b. Stacked concrete artificial island.

The stacked artificial island is combined with three layers, namely, the metal base, a concrete island located in the middle of the body, and the surface island drilling barge. When working, these three parts are stacked together, sitting at the end of the cornerstone of the seabed; when work is completed, it can be recycled.

The role of the metal base lies in the expansion bearing area to reduce pressure and prevent slip, and it can be separated into a number of holds for the purpose of setting the pressure. The middle of the concrete island honeycomb body is made of one or two layers of lightweight concrete caisson, which has good anti-ice performance.

When working, the base and the island body are stacked in the water together first, sinking into deep water. Then, its support of the drilling barge onto the top of it, to be positioned, and the sink on the base of the water and the island of body float to engagement with the bottom of the drilling barge, the appropriate method to be fixed, together onto the island site at the sinking sit at the end of drilling. The whole process of this assembly is shown in Figure 5-40. Japan NKK Shipyard, which is built with this artificial island body, is octagonal, 29 m high, 95 m long, 89 m wide, with atotal weight of 56000 t; the stacked concrete artificial island has the following features:

   i. Stratified island body structure, with layers that can be changed with depth and adapted to a larger depth range

   ii. The weight and height of each layer are greatly reduced, reducing the difficulty of construction and workplace depth.

   iii. Because the weight of each layer is light and there is a shallow draft, it's easy to haul from the beach/shallow sea-to-water assembly place.

   iv. The concrete island body is central, and thus has good anti-ice, especially for polar ice.

   v. It can be layered to floating, so it is easy to extricate adsorption.

   vi. Because the top adopts the overall barge form, it is not only conducive to the construction and assembly, but it can also utilize the drilling barge to transform it to a production barge, which has both drilling and production functions.

4. Steel caisson gravity artificial island

Such an artificial island is characterized by the use of the steel caissons themselves as an artificial island. Therefore, it is not like the closed circle artificial island, with filling used for the heart of the island and the formation of the operating area. Different caissons circumstances have different forms, such as the use of old vessels as restructuring caissons for an artificial island. For example, Japan has old tankers with a length of 202 m, width of 53 m, and deck area of 10700 m^2; the inner wall adopts a thick 1 m concrete reinforcement. It is hauled to the site of the island and put on the underwater abutment built earlier to make an artificial island. Later, the old

FIGURE 5-40

Assembly process of stacked type concrete artificial island.

supertanker is transformed, splitting the hull, and assembled into a large-scale oil extraction artificial island, as shown in Figure 5-41.

Obviously, it is feasible for shallow draft or small vessels be converted into artificial islands in extremely beach/shallow sea areas.

5. Steel caisson artificial island.

This artificial island is made of steel caissons. Molikpaq is an artificial island in Canada, which consists of eight assembled right angle trapezoidal steel caissons, with the caissons installed on a fully enclosed deck, as shown in Figure 5-42. In order to maintain the stability of the island, from the steel caisson bottom to the caisson center chamber part is about 11×10^4 m^3 of dense sand. When migration after unloading sand. Its greatest feature is the overall ease of migration, especially due to the fact that a steel caisson is lighter than the weight of the concrete platform and has a shallow draft, so it is appropriate to haul in beach/shallow sea areas. In addition, due to the overall structure, installation is greatly simplified, which allows reducing the workload of offshore construction. For example, in the harbor area, after reducing the weight of caisson, it can be towed with a no load condition without installed equipment, so the draft is greatly reduced.

5.5.3.2 Fixed Production Platform Innovation

The concrete gravity fixed production platform in deep water is not suitable for using in beach/shallow seas, and the caissons concrete gravity structure which was introduced above, so as for fixed production platform, Jacket platform will be mainly introduced here. However, pile foundation jacket steel fixed platform design and construction work for offshore oil engineering has accumulated a lot of experience, but this experience cannot be used for the specific conditions of the beach/shallow sea. For example, environmental conditions are not often as harsh as offshore, and the load of the platform and the strength requirements are not as rigorous as offshore. Moreover, a steel fixed offshore platform jacket is hauled in its entirety, but this is difficult to implement in the beach/shallow sea area. The pile jacket steel must be designed pursuant to the specific circumstances of beach area oil field conventional production, and the following example will illustrate it.

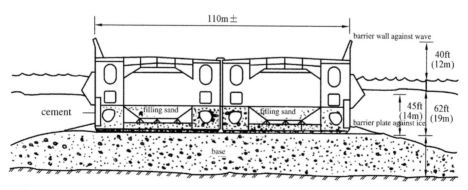

FIGURE 5-41

Production of artificial island from a hull assembly.

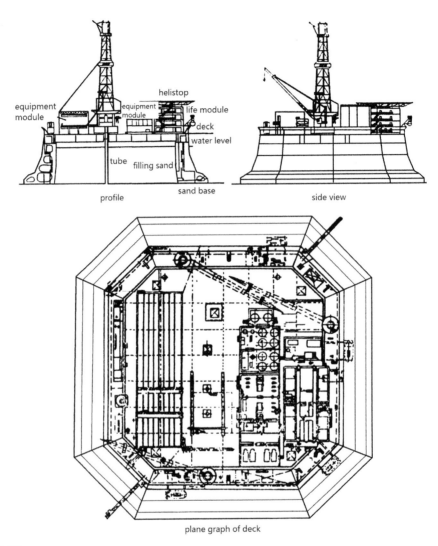

profile

side view

plane graph of deck

FIGURE 5-42

Molikapaq steel caisson artificial island.

1. Segmented pile jacket steel fixed platform assembly

   In order to solve the problem of haulage of the entire jacket not applying in the beach/shallow sea, we cite an innovative subassembly of the jacket platform structure, as shown in Figure 5-43. The Blue Water Maintenance Company in the United Sates designed it and built it for Gulf of Mexico beach/shallow sea marginal field development. The innovation of this platform is the use of a

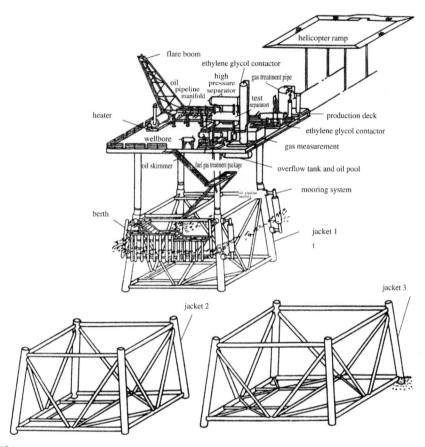

FIGURE 5-43

Segmented jacket steel fixed platform.

subassembly jacket, the jacket consists of section I, section II, and section III. In this way, the jacket sections are either prefabricated at the factory, or the volume decreases because of the weight, and depending on the water depth, we can adopt suitable beach sea haulage; there is agreat deal of flexibility and adaptability.

The platform has a 18.3 m × 21.3 m production deck for nine oil wells, treatment facilities, and processing capacity of 142×104 m^3 for natural gas, and 398 m^3 for condensate water; all equipment is skid mounted, which is convenient for transportation, installation and replacement. The platform can be unmanned.

2. Bucket foundation jacket steel fixed platform

This platform is located in the China Shengli oil field area, in the Chengdao development; it is an innovative fixed steel platform jacket for very shallow water/beach areas. It is mainly used to solve the hard construction problem of piling pile foundation of jacket fixed platforms at sea and mobility problem of jacket fixed platform, and it is an innovation based on the jacket steel fixed platforms. The following is an introduction to installation of the technology built around the Chengdao oil field for well group CB20B.

a. Structural features

i. Platform deck and jacket

The CB20B platform is a four-barrel base jacket platform; the platform deck is a square structure, deck water surface elevation for 9m jacket total height of 14m, from the surface to the top 5.5m, and set up a total of three horizon support structure heights of 4.00m, -1.6 m and -6.9m, between -1.6m and -6.9m, without oblique support. In order to reduce the force of the sea ice platform in the jacket surface of the water over the ice area without ramp support.

ii. Bucket foundation

Each of the four platform jacket legs has a bucket foundation. So it is called a bucket foundation, because it is shaped like inverted steel vats. Barrel roof top plate with a certain inclination of the disk contour has a certain depth of the barrel skirt and the tip of the barrel skirt open. A valve is connected between the lower ends of each leg of the platform jacket with the bucket foundation. When the barrel skirt is pressed deeper and mud poured, you can close this exhaust gas valve, thus the suction with a pump, in order to enable that your sealed 's outside the barrel phase respect to the top for rent, the formation of the negative pressure.

iii. Ballast tank

A ballast water tank is installed in the upper part of the platform deck; the main dimensions are 14.5m × 6.5 m × 5.0 m, which are divided into four tank areas without cover, and a box board, separated the walls of the box and the bottom box board, the side wall of the tank and the bottom box panels are slotted plate girder structure tank to separate the walls of the box at a level support for supporting the box girder, from the bottom 2.5m.

b. Bucket foundation in place work

When installed in place, a bucket foundation through its own quality (including attached superstructure quality), makes the tip of the barrel skirt pressure poured into the mud a certain depth, then coupled with a certain quality ballast pressure is poured into the barrel skirt mud deeper, to ensure that the tip of the skirt barrel can form a reliable seal. Meanwhile, it adjusts the jacket so that the tilt is minimized. Then, the exhaust valve should be closed by pumping; the sealed barrel, with respect to the outside of the top plate, forms negative pressure. With negative pressure and self-respect, the barrel skirt penetrates into the mud to a predetermined depth. After installation, suction barrel negative pressure will gradually disappear due to soil seepage water flows.

In the penetration process, we need to have sufficient negative pressure to overcome the barrel skirt tip resistance and bucket friction inside and outside the skirt wall to prevent the pressure difference inside and outside from being too large, leading to barrel skirt buckling. Therefore, in addition to the appropriate thickness and strength, we need to control the maximum negative value.

The bucket foundation must withstand the marine environment load (including wind, waves, currents, and ice) the uplift force, in addition to the barrel skirt wall friction, the main uplift force generated in the barrel of huge short-term negative pressure on the pull force balance, and the negative pressure on and pulling force increases this basis compared to pile rely on the friction-supported, have a distinct advantage.

c. Bucket foundation platform offshore installation

i. Sea preparation

The whole platform is hauled by pontoon hauling to the platform workplace, before setting sail first the catheter rack, pump module and its cable, ballast tank, control box, measurement and control sensor and its signal line, water lines, etc. are assembled on the fixed pontoon.

ii. In place in the sea

When the platform is in place, a lifting floating crane is used to lift the platform with four equal-length wire ropes lifting; the shipment platform pontoon is removed and it is ready to sink. Sinking before commissioning, the platform is into the water with speed. In this process you should debug suction pump and off the working condition of the non-return valve and ultrasonic depth sounder, pressure transmitters and other instruments.

iii. Platform sinking

This is the most critical step in the installation process. First the weight is sunk at a speed of 0.3 m/min until touchdown, making sure the weight goes into the mud 0.5 m; if it not reached, the pump discharge should be lessened and the amount of ballast increased. Then there is negative pressure sink, with manual control and negative pressure suction. The suction pump speed is generally controlled at 15Hz. If it goes into the mud no more than 1 m, negative values should be controlled at 0.02MPa; after sinking to 2 m, it will change automatically to control the sink, which should be controlled throughout the sinking process with an inclination within 0.5 degree of the platform. The final tilt error should not exceed 1 degree.

iv. Completion

After the end of sinking, the diver goes into the water to close the valve, and open the quick connector of suction piping and ballast piping. At the same time, the water ballast bottom valve will open, draining away the water. And then, the ballast tank, the pump group module, as well as some of the hanging control instruments will go back into the floating box, and at last we can return.

d. Bucket foundation jacket steel platform advantages

i. Eliminating the need for piling and pile construction

It eliminates the need for the piling of a traditional pile jacket fixed platform at sea, which removes the troublesome construction work for a pile, thus saving construction time, reducing construction costs, and reducing the risk of offshore construction.

ii. Can be migrated and reused

By compressing air or using water, the bucket foundation platform can be pulled out in its entirety and removed for repeated use, which not only improves the utilization but also protection of the marine environment.

iii. Improve the quality of offshore construction

Because we can use measurement and control technology to control the posture of the platform in the penetration process, we can avoid the various environmental effects cause by pile platform construction and avoid platform tilt, which greatly improves the quality of offshore construction.

iv. Significant savings in the amount of steel

Mainly through negative pressure from the pulling force caused by the load of the marine environment, there is a large diameter barrel to withstand lateral loads, and therefore the system does not require elongated structures of steel piles, thus saving on the amount of steel, which decreases approximately 20%–40% than pile foundation jacket platform.

5.5.4 Beach Sea Region Oil Spill Cleanup

Oil density is smaller than the sea water, so, once spilled into the sea, it floats on the water surface and extends into a smooth film, drifting along with the wind and the direction of the surface current. With film expansion and drifting process, the light components of the oil evaporate quickly, usually within 2 or 3 days about 25% to 30% evaporates. Oil-soluble component is soluble in water, while the part of the recombinant component will deposit or adhesion on a suspension of solid particles in seawater, and then sink to the seabed. When heavy crude oil is spilled into the sea, within a few hours emulsification occurs. Then oil-in-water is formed, and its degradation speed is much faster than oil degradation. The density of oil-in-water, sedimentation on the seabed, microbial degradation phenomenon can't be observed after 1 to 2 years; oil-in-water is easily on the adhesion of the solid, and thus adhesion of oil on the coast of the package more and more water. Consume a lot of oil spilled into the ocean, oxygen, general oil complete oxidation of 40×10^4 L oxygen consumption which makes marine animals to death. Oil was rushed to the coast, then the serious pollution of the beach makes wasteland of beach, destruction of Mari culture and Salt pan production, and destruction of coastal tourist areas. In short, the harm caused by an oil spill is serious. Therefore, the international community is becoming increasingly sensitive and placing constraints and strict requirements on the offshore oil industry.

After years of continuous efforts at home and abroad for oil pollution marine oil spill prevention there is more understanding of the appropriate system and control measures. However, for the special circumstances of beach/shallow sea areas there is less knowledge on prevention measures. Therefore, the following is intended to introduce some of the oil spill cleanup technology and equipment for beach/shallow sea areas.

5.5.4.1 Mechanical Methods for Beach/Shallow Sea Area Oil Spill Cleanup

Here is some technology and equipment for Chinese beach/shallow sea oil field mechanical methods of cleaning up.

1. Oleophilic disc skimmer system

This system uses solid floating booms to limit the drift and diffusion of the oil spill, and and oleophilic disc skimmer to recover the oil spill; oil-water separation is achieved through, the separated oil is recovered, and the separated water discharged directly, as shown in Figure 5-44. It can be seen from Figure 5-44 that the system consists of booms (1), the oleophilic disc skimmer (2), hose (3), self-priming pump (4), gravity oil-water separator (5), and storage tank (6). The oleophilic disc skimmer, as shown in Figure 5-45, is composed of body, power source, oil cars, and accessories. Electric hydraulic pump for liquid motor power, the liquid motor drive pro

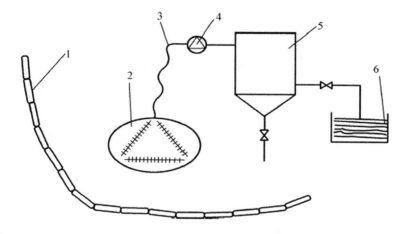

FIGURE 5-44

Disc collection system for oil: 1 - oil containment boom; 2 - Oil wet disc type receiving machine; 3 - flexible hose; 4 – self-priming pump; 5 - gravity type oil water separator; 6 - oil storage barrels.

oil pan (4) to (3) on the drive in the spindle. The disc has a lipophilic, floating on the surface of the oil spill to the disc, and then scratch component (5) blew into the storage tank (6). Storage tanks contains free water oil spill, which is pumped to gravity oil-water separator (see Figure 5-44) by screw pump. The separated oil is collected in drums, and water is emitted, then the oil spill recovery will be completed. The support assembly (7) in Figure 5-45 includes a support spindle, and the floating body (1) supports the entire skimmer.

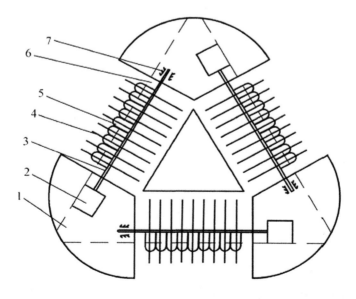

FIGURE 5-45

Oleophilic disc structure for oil recovery: 1 - floating body; 2 - hydraulic motor components; 3 - principal axis; 4 - oil wet disc component; 5 - scraping components; 6 - oil collector; 7 - supporting components.

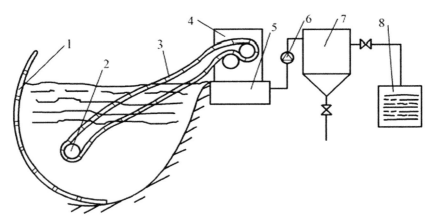

FIGURE 5-46

Crawler suction pillow—roller extruder overflow oil collection system: 1 - floating body; 2 - guide pulley; 3 - crawler type oil absorption pillow; 4 – roller-type extrusion machine; 5 - oil sump tank; 6 - self-priming pump; 7 - gravity type oil-water separator; 8 - oil tanker.

The oleophilic disc skimmer in the beach/shallow sea are in the Dagang oil field has effectively recovered a liquid high- and medium-viscosity oil spill.

2. Crawler oil pillow: roller-style extruder oil spill recovery system

The composition of such a system is shown in Figure 5-46, which includes booms (1), steering wheel (2), crawler oil pillow (3), roller-type extrusion machine (4), oil sump tank (5), pump (6), gravity oil and water separator (7), and storage tanks (8). The crawler oil pillow face uses knotless polyethylene mesh cloth, oleophilic hydrophobic treatment of polyethylene foam, and polypropylene oil-absorbing material. Due to the oil pillow floating in the water, it can absorb a variety of different viscosities of crude oil or refined oil. Roll extruder selection of chain drive mechanism, sprocket driven by two parallel gum elastic roll rotation, roll rotating extrusion friction, affecting crawler oil pillow campaigns. When the oil absorbency pillow past the roll, the oil is extrusion, then Oil absorbency pillow after the oil spilling was brought to the surface oil to absorb oil again.

In this way, through the "range - extrusion - range" process continuously repeating many times, almost all of the surface oil spill can be sucked up. The oil spill is extruded into the oil sump tank, and then pumped to the gravity water separator for oil-water separation.

This system is convenient to use, and easy to operate and recycle, etc. Various viscosity liquid crude oil can be efficiently recovered at the beach/shallow sea region.

3. Pump suction recovery system

The structure of the system shown in Figure 5-47 consists of booms (1), oil head (2), inlet pipe (3), pump suction machine (4), the main pump (5), gravity oil-water separator (6), and storage tanks (7). The suction head is supported by a buoy floating in the water, through the pump suction and self-priming main pump installed in a small car, and not need water and direct suction. The suctioned oil spill is sent to the oil-water separator for oil-water separation. This system has been in used in China for various viscosities of crude oil or refined oil floating on the water and has been effective for recovery.

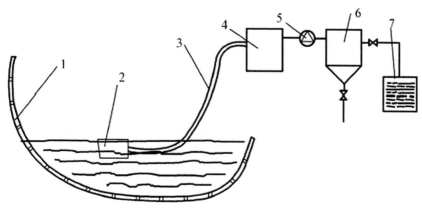

FIGURE 5-47

Pump suction machine oil spill recovery system: 1 - floating body; 2 - suction oil tanker; 3 - oil inlet; 4 – suction pump of oil suction machine; 5 - main pump; 6 - gravity type oil-water separator; 7- oil collector.

4. Oil trawl oil spill recovery system

The system shown in Figure 5-48 consists of hauling (1), wing nets (2), floating body components (3), the main framework (4), the main network (5), a connection box (6), bag nets (7), and a floater (8).

Figure 5-48 shows the BTC1 solid oil spill recovery system for the China Dagang oil field, whose trawl can be used with single vessel towing. A two-boat V-shaped towing solid form oil spill

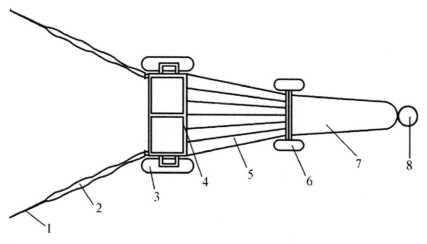

FIGURE 5-48

BTC1 solid oil spill recovery system: 1 - guide rope; 2 - wing net; 3 - floating body components; 4 - main body frame; 5 - main network; 6 - connection box; 7 - landing net; 8 – floater.

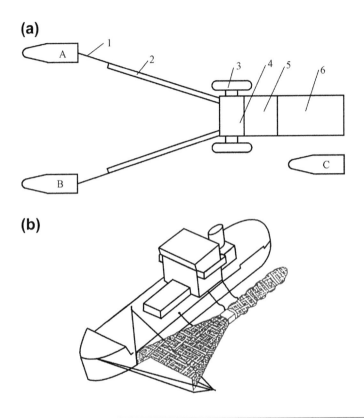

FIGURE 5-49

Trawl oil spill recovery system: 1 - guide rope; 2 - oil containment boom; 3 - floating body; 4 - main body frame; 5 - main network; 6 - landing net.

recovery system is shown in Figure 5-49(a); a single-boat arm hanging net solid form oil spill recovery system is shown in Figure 5-49(b), which can also be a single ship arms hanging on the oil spill recovery.

The trawl oil spill recovery system is suitable for recycling solid form oil spill in wide beach sea waters, such as the Dagang oil field in China, Most of the waxy oil has a higher freezing point, and it will become a solid block in normal temperatures. So BTC1-a solid-like oil spill recovery system has a multiplier effect ever.

5. Crawler absorption felt: roller extruder oil spill recovery system

This system consists of a crawler absorption felt, roller extruder, steering wheel, fuel tank, and gravity oil-water separator, with a system structure and method of operation that is basically as same as the crawler oil pillow system with roller extruder suction. It has wear-resistant polypropylene and polyethylene materials through the oleophilic hydrophobic treatment made by linoleum mat core material, thereby enhancing the strength of the crawler suction linoleum. The absorption felt through the extruder roller still works in the oil spill even if sand, wood chips, broken plastic, and other debris exist.

Tests in Dagang oil field in China show that this oil spill system is suitable to use in the quagmire of the beach/shallow sea areas for recycling various viscosity liquid oil spills.

5.5.4.2 Chemical Methods for Beach/Shallow Sea Area Oil Spill Cleanup

Chemical methods are important for cleaning up oil spills. Dispersion and condensation are often used. Measures to combat oil pollution are recovery, sedimentation, diffusion, and condensation. Sedimentation and diffusion are achieved by spraying oil spill dispersants. In recent years, the method of using condensate oil spill cohesion piles, then recycling with an oil trawl had drawn more and more attention. From the point of view of the beach/shallow sea area, chemical methods to clean up the oil spill by using oil-absorbing material, spraying dispersants, and using condensate oil are now briefly described as follows:

1. Oil absorbent material

   The oil spill is adsorbed into the oil-absorbing material, which is a physical process. The suction is affected by the oil-absorbing material itself, the physical and chemical properties, the characteristics of the oil spill, as well the adsorption process. After oil spills into the water, due to the different densities and the surface tension between them, oil and water are immiscible, and when the lipophilic oil is in contact with the oil-water, it i mainly absorbs the oil, not water. For beach/shallow sea oil spill cleanup, the oil-absorbing materials can be divided into the following two categories:

   a. Natural materials

      The advantage of natural materials is cost, and that they can be adapted to local conditions. However, their oil absorption capacity is limited and deployment and recovery is not easy. If any oil absorption naturally settles to the bottom of the sea, it is easy to cause secondary pollution, so generally it should not be used.

   b. Synthetic oil absorbent material

      Foam and some artificial materials such as synthetic fibers, due to a chemical synthesis process, can change its physical properties and chemical properties. Therefore, it is possible to obtain absorbent material with excellent absorption properties. Moreover, the shape of the foam can be artificially controlled, to facilitate mechanized deployment and recovery, which is especially suitable for recycling and reuse.

      Tests showed that polyurethane and polypropylene synthetic materials, after a lipophilic hydrophobic and oil absorption, can absorb the liquid viscosity of crude oil and refined oil, and be used by extrusion, re-use, as a crawler oil absorption pillow or the suction felt core material.

2. Oil spill dispersants

   The role of il spill dispersants, commonly known as degreasing agents, is to assist the cleanup oil spill recovery, and do final cleaning of the sea. The oil spill dispersants can be used to deal with a film thickness of 1 mm, and it's difficult to play a role on the recovery unit. Therefore, in certain circumstances, such as adverse sea conditions, fire happens and causes serious harm to the marine environment, so the area is sprayed, sometimes with an aircraft, to eliminate the oil slick.

   Oil spill dispersants have been developed to a third generation of new products, agricultural by three pears, sugar alcohols, fatty acid as the raw material, using the synthesis of ethylene glycol as the solvent concentrated oil spill dispersants, whose toxic 1000 times less than the first generation. Chinese production of the "double elephant" brand oil spill dispersant has reached the level of foreign products, and needs the approval of the China Bureau of Shipping for use in an offshore oil spill event.

The main problem of oil spill dispersants in beach/shallow sea areas is that it's toxic. Surfactant-based oil spill dispersant toxicity has decreased, and is basically harmless to fish and shellfish, but it may be harmful to juvenile fish, algae, and fish eggs. Therefore, some countries prohibit or ban the use of it in offshore marine areas. In the Bohai beach/shallow sea region the survival of fish, shrimp, crabs, and other marine life is critical, and it a good place for migration and breeding and an important part of the Bohai Sea fishing grounds; thus outside of the event of a fire emergency with an oil spill, oil spill dispersants must be approved by the state. Otherwise generally they cannot be used.

3. Oil coagulating agent

The oil gelling agent controls the diffusion effect in 1 cm to 1.5 cm thick oil film. On 0.3 cm to 0.5 cm thick oil film after being sprayed with condensate oil, there is a cross-linking reaction, which can cause rapid solidification of the oil spill so as to facilitate recovery.

Foreign countries have been developing condensate oil to deal with an oil spill. The main raw material of such oil is sorbitol. Chinese raw materials are ample, and it's a promising oil spill chemical treating agent. It is particularly suitable for the beach/shallow sea region, where there can be very effective use of condensate oil, and then trawling for recycling.

5.5.4.3 Biodegradable Oil Technique

This technique uses marine microorganisms to degrade the oil and clean up oil spills; it is a specific application of a bioengineering technique for cleaning up marine oil spills, and a combination of traditional microbiological methods with modern biological engineering technology to provide new and effective means for environmental protection. Abroad, the United States and Japan are conducting experimental research, and Russia has developed a bio-oil scavenger, which has opened up new avenues for use of biodegradable oil.

1. The mechanism of the biodegradable oil

The biodegradable oil mechanism is oil oxidation and decomposition by the microorganisms naturally present in the ocean or the soil to remove the oil spilled into the ocean.

Microbial degradation of petroleum hydrocarbon components is a biological oxidation process. Hydrocarbons are being degraded under the action of enzymes secreted by the microorganisms. In this process, part of the hydrocarbons is decomposed into carbon dioxide and water, and another portion is decomposed by microorganism use, constituting the biological extracellular matrix components.

The enzymes secreted by microorganisms are a complex protein, which can make the organic molecule or substrate fixed on it, and for the decomposition of organic matter, they play the role of catalytic substance, greatly improving the speed of the decomposition reaction.

Different species of bacteria generate the optimum temperature of the enzyme type, and thus the microorganisms involved in the biodegradation processes can be used under various temperature conditions.

2. Microorganisms to clean up oil spill

Microbial treatment of oil spill: First is the isolation of the contaminated area and the blooming or enhancement of microorganisms in the vicinity; second is the introduction of a new microbial

Table 5-9 China Bohai Oil Microbial Test Results

Sample Category	Surface Water Samples			Surface Sediment Samples		
	Number of samples	Number of detections	%	Number of samples	Number of detections	%
Oil bacteria	124	111	89.5	97	62	63.9
Oil yeast	93	59	63.4			
Oil mold	86	80	93.0	65	46	70.8

Table 5-10 Bohai Beach/Shallow Sea Area Oil Spill Cleanup Technology and Equipment Recommendations

Oil Spill Scenario	Recommend Oil Spill Recovery Equipment
Shore mud flats with a small amount of liquid oil spill ($<$ 1 ton each time)	Simple suction synthetic oil absorbent material or natural oil absorbent material
Shore mud flats with a medium amount of liquid oil spill (1 ~ 10 t/time)	Oil-absorbing oil pillow, roller extrusion type, or XB-2, XB-3 type self-priming skimmer
Shore mud flats with large liquid oil spills ($>$ 10 t/time)	Booms to block the spread of the oil spill into the sea, and then use a lot of oil pillows with absorption felt and roller extrusion technique; or more than one XB-2, XB-3 self-priming oil machines
Shore mud flats, a small amount of solid-state oil spill ($<$10 t/time)	Artificial small push net or kegs, teaspoon of oil revenues
Shore mud flats with a large solid-state oil spill ($>$ 10 t/time)	Block the spread of the oil spill into the sea, and then use a large push net collection of solid-state oil spill booms
Beach/shallow sea, a small amount of liquid oil spill ($<$1 t /time)	Natural oil-absorbing materials or synthetic oil-absorbing material suction and collection of oil-absorbing material with a small network
Beach/shallow sea, a medium amount of liquid oil spill (1~10 t /time)	Small oleophilic disc skimmer with oil-absorbing pillow using roller extrusion
Beach/shallow sea, a large amount of liquid oil spill	Surround the oil spill with booms simultaneously with use of oil pillow using roller extrusion; small oleophilic disc skimmer; and a natural oil-absorbing material to absorb the small amount of remaining oil spill
Beach/shallow sea ($<$ 0.5 m) with a small and medium amount of solid-state oil spill	Closed fishing with small mesh
Beach/shallow sea ($<$ 0.5 m) with a large amount of solid-state oil spill	Use two U-shaped towing recycling solid oil spill system
Beach/shallow sea (0.5-5 m) with a large amount of liquid-state oil spill	Pump-priming oil spill recovery; oleophilic disc skimmer; recovery of oil spill with absorption felt using roller extrusion
Beach/shallow sea (0.5-5 m) with a large amount of solid-state oil spill	Trawling with solid oil spill recovery to recycle oil spill

genetic improvement to the contaminated areas, used to clean up pollutants. Currently, 70 genera and 200 species of microorganisms that are capable of oxidizing one or more of petroleum hydrocarbons are known, including 28 genera of bacteria, 30 genera of filamentous fungi, and 12 genera yeast.

The United States during the marine oil pollution research projects with yeast found that clearing marine oil pollution with yeast compared to other microorganisms such as bacteria has the following advantages:

a. A strong bactericidal effect of sun and seawater osmotic resistance.

The sun can kill a virus, and the osmotic pressure of seawater can damage the cell walls of bacteria, which interfere with the function of bacterial decomposition of oil. Yeast has strong resistance, which can quickly break down the oil, or drill into the oil droplets to reproduce.

b. Factors that affect the microbial degradation of oil are relatively insiginifcant

The following affect the microbial degradation of the oil: temperature, dissolved oxygen, pH value, the physical energy, the diffusion speed of the oil, the concentration of nutrients, and animals feeding on microorganisms. Bacteria is impacted greatly by these factors, while the yeast is relatively small.

There are 700 kinds of yeast strains that can be used in oil-contaminated beaches, estuaries, and rivers, which creates favorable conditions for the use of biodegradable techniques for cleaning up oil spill pollution.

3. The China Bohai petroleum microorganism situation

China's State Oceanic Administration tested Bohai Sea surface water samples and surface sediment samples, and the test results are shown in Table 5-9. It indicates that oil microorganisms exist in the Bohai Sea water samples and mud samples, which opens the way for the governance of biodegradable oil spill technology in the Bohai beach/shallow sea region. Therefore, it is urgent to vigorously carry out research in this regard.

5.5.4.4 The Choice of Beach/Shallow Sea Region Oil Spill Technology and Equipment

The China beach/shallow sea area has a complex and changing environment, diverse varieties of crude oil, and there are liquid and solid states under normal temperature and sea temperature. Therefore, according to the depths of water in different geographic locations at seashores and beaches as well as different oil spills, we need different oil spill cleanup technology and equipment. Table 5-10 has recommended technology and equipment that can be used for the Bohai beach/shallow sea area.

Submarine Pipelines and Pipeline Cable Engineering

CHAPTER

Note: This chapter is available in its entirety online at http://store.elsevier.com/product.jsp?isbn=9780123969774

Offshore Operation Facilities: Equipment and Procedures. http://dx.doi.org/10.1016/B978-0-12-396977-4.00006-8

Safety System Engineering for Offshore Oil

Note: This chapter is available in its entirety online at http://store.elsevier.com/product.jsp? isbn=9780123969774

Offshore Operation Facilities: Equipment and Procedures. http://dx.doi.org/10.1016/B978-0-12-396977-4.00007-X

Appendix

APPENDIX 1: NUMERICAL TABLE OF NORMAL DISTRIBUTION

$$\Phi(Z) = P(Z \leqslant x)$$

Z	0.00	0.01	0.02	0.03	0.04	0.05	0.06	0.07	0.08	0.09	Z
−0.0	0.5000	0.4960	0.4920	0.4880	0.4840	0.4801	0.4761	0.4721	0.4681	0.4641	−0.0
−0.1	0.4602	0.4562	0.4522	0.4483	0.4443	0.4404	0.4364	0.4325	0.4286	0.4247	−0.1
−0.2	0.4207	0.4168	0.4129	0.4090	0.4052	0.4013	0.3974	0.3936	0.3897	0.3859	−0.2
−0.3	0.3821	0.3783	0.3745	0.3707	0.3669	0.3632	0.3594	0.3557	0.3520	0.3483	−0.3
−0.4	0.3446	0.3409	0.3372	0.3336	0.3300	0.3264	0.3228	0.3192	0.3156	0.3121	−0.4
−0.5	0.3085	0.3050	0.3015	0.2981	0.2946	0.2912	0.2877	0.2843	0.2810	0.2776	−0.5
−0.6	0.2743	0.2709	0.2676	0.1543	0.2611	0.2578	0.2546	0.2514	0.2486	0.2451	−0.6
−0.7	0.2420	0.2389	0.2358	0.2327	0.2297	0.2266	0.2236	0.2206	0.2177	0.2148	−0.7
−0.8	0.2119	0.2090	0.2061	0.2033	0.2005	0.1977	0.1949	0.1922	0.1894	0.1867	−0.8
−0.9	0.1841	0.1814	0.1788	0.1762	0.1736	0.1711	0.1685	0.1660	0.1635	0.1611	−0.9
−1.0	0.1587	0.1562	0.1539	0.1515	0.1492	0.1469	0.1446	0.1423	0.1401	0.1379	−1.0
−1.1	0.1357	0.1335	0.1314	0.1292	0.1271	0.1251	0.1230	0.1210	0.1190	0.1170	−1.1
−1.2	0.1151	0.1131	0.1112	0.1093	0.1075	0.1056	0.1038	0.1020	0.1003	0.09853	−1.2
−1.3	0.09680	0.09510	0.09342	0.09176	0.09012	0.03851	0.08391	0.08534	0.08379	0.08226	−1.3
−1.4	0.08076	0.07927	0.07780	0.07636	0.07493	0.07353	0.07215	0.07078	0.06944	0.06811	−1.4
−1.5	0.06681	0.06552	0.06426	0.06301	0.06178	0.06057	0.05938	0.05821	0.05705	0.05592	−1.5
−1.6	0.05480	0.05370	0.05262	0.05155	0.05050	0.04947	0.04846	0.04746	0.04648	0.04551	−1.6
−1.7	0.04457	0.04363	0.04272	0.04182	0.04093	0.04006	0.03920	0.03836	0.03754	0.03676	−1.7
−1.8	0.03593	0.03515	0.03438	0.03362	0.03288	0.03216	0.03144	0.03074	0.03005	0.02938	−1.8
−1.9	0.02872	0.02807	0.02743	0.02680	0.02619	0.02559	0.02500	0.02442	0.02385	0.02330	−1.9
−2.0	0.02275	0.02222	0.02169	0.02118	0.02068	0.02018	0.01970	0.01923	0.01876	0.01831	−2.0
−2.1	0.01786	0.01743	0.01700	0.01659	0.01618	0.01578	0.01539	0.01500	0.01463	0.01426	−2.1
−2.2	0.01390	0.01355	0.01321	0.01287	0.01255	0.01222	0.01191	0.01160	0.01130	0.01101	−2.2
−2.3	0.01072	0.01044	0.01017	$0.0^{2}9903$	$0.0^{2}9642$	$0.0^{2}9387$	$0.0^{2}9137$	$0.0^{2}8894$	$0.0^{2}8656$	$0.0^{2}8424$	−2.3
−2.4	$0.0^{2}8198$	$0.0^{2}7976$	$0.0^{2}7760$	$0.0^{2}7549$	$0.0^{2}7344$	$0.0^{2}7143$	$0.0^{2}6947$	$0.0^{2}6755$	$0.0^{2}6569$	$0.0^{2}6387$	−2.4
−2.5	$0.0^{2}6210$	$0.0^{2}6037$	$0.0^{2}5868$	$0.0^{2}5703$	$0.0^{2}5543$	$0.0^{2}5386$	$0.0^{2}5234$	$0.0^{2}5085$	$0.0^{2}4940$	$0.0^{2}7499$	−2.5
−2.6	$0.0^{2}1661$	$0.0^{2}4527$	$0.0^{2}4396$	$0.0^{2}4269$	$0.0^{2}4145$	$0.0^{2}4025$	$0.0^{2}3907$	$0.0^{2}3793$	$0.0^{2}3681$	$0.0^{2}3573$	−2.6
−2.7	$0.0^{2}3467$	$0.0^{2}3364$	$0.0^{2}3264$	$0.0^{2}3167$	$0.0^{2}3072$	$0.0^{2}2930$	$0.0^{2}2890$	$0.0^{2}2803$	$0.0^{2}2718$	$0.0^{2}2635$	−2.7
−2.8	$0.0^{2}2555$	$0.0^{2}2477$	$0.0^{2}2401$	$0.0^{2}2327$	$0.0^{2}2256$	$0.0^{2}2186$	$0.0^{2}2118$	$0.0^{2}2052$	$0.0^{2}1938$	$0.0^{2}1926$	−2.8
−2.9	$0.0^{2}1866$	$0.0^{2}1807$	$0.0^{2}1750$	$0.0^{2}1695$	$0.0^{2}1641$	$0.0^{2}1589$	$0.0^{2}1538$	$0.0^{2}1489$	$0.0^{2}1441$	$0.0^{2}1395$	−2.9

Z	0.00	0.01	0.02	0.03	0.04	0.05	0.06	0.07	0.08	0.09	Z
-3.0	$0.0^2 1350$	$0.0^2 1306$	$0.0^2 1264$	$0.0^2 1223$	$0.0^2 1183$	$0.0^2 1144$	$0.0^2 1107$	$0.0^2 1070$	$0.0^2 1035$	$0.0^2 1001$	-3.0
-3.1	$0.0^3 9676$	$0.0^3 9354$	$0.0^3 9043$	$0.0^3 8740$	$0.0^3 8447$	$0.0^3 8164$	$0.0^3 7888$	$0.0^3 7622$	$0.0^3 7364$	$0.0^3 7114$	-3.1
-3.2	$0.0^3 6871$	$0.0^3 6637$	$0.0^3 6410$	$0.0^3 6190$	$0.0^3 5976$	$0.0^3 5770$	$0.0^3 5571$	$0.0^2 5377$	$0.0^3 5190$	$0.0^3 5009$	-3.2
-3.3	$0.0^3 4834$	$0.0^3 4665$	$0.0^3 4501$	$0.0^3 4342$	$0.0^3 4189$	$0.0^3 4041$	$0.0^3 3897$	$0.0^3 3758$	$0.0^3 3624$	$0.0^3 3495$	-3.3
-3.4	$0.0^3 3369$	$0.0^3 3248$	$0.0^3 3131$	$0.0^3 3018$	$0.0^3 2909$	$0.0^3 2803$	$0.0^3 2701$	$0.0^3 2602$	$0.0^3 2507$	$0.0^3 2415$	-3.4
-3.5	$0.0^3 2326$	$0.0^3 2241$	$0.0^3 2158$	$0.0^3 2078$	$0.0^3 2001$	$0.0^3 1926$	$0.0^3 1854$	$0.0^3 1735$	$0.0^3 1718$	$0.0^3 1653$	-3.5
-3.6	$0.0^3 1591$	$0.0^3 1531$	$0.0^3 1473$	$0.0^3 1417$	$0.0^3 1363$	$0.0^3 1311$	$0.0^3 1261$	$0.0^3 1213$	$0.0^3 1166$	$0.0^3 1121$	-3.6
-3.7	$0.0^3 1078$	$0.0^3 1036$	$0.0^4 9961$	$0.0^4 9574$	$0.0^4 9201$	$0.0^4 3842$	$0.0^4 8496$	$0.0^4 8162$	$0.0^4 7841$	$0.0^4 7532$	-3.7
-3.8	$0.0^4 7235$	$0.0^4 6948$	$0.0^4 6673$	$0.0^4 6407$	$0.0^4 6152$	$0.0^4 5906$	$0.0^4 5669$	$0.0^4 5442$	$0.0^4 5223$	$0.0^4 5012$	-3.8
-3.9	$0.0^4 4810$	$0.0^4 4615$	$0.0^4 4247$	$0.0^4 4427$	$0.0^4 4074$	$0.0^4 3908$	$0.0^4 3747$	$0.0^4 3594$	$0.0^4 3446$	$0.0^4 3304$	-3.9
-4.0	$0.0^4 3167$	$0.0^4 3036$	$0.0^4 2910$	$0.0^4 2789$	$0.0^4 2673$	$0.0^4 2561$	$0.0^4 2454$	$0.0^4 2351$	$0.0^4 2252$	$0.0^4 2157$	-4.0
-4.1	$0.0^4 2066$	$0.0^4 1978$	$0.0^4 1894$	$0.0^4 1814$	$0.0^4 1737$	$0.0^4 1662$	$0.0^4 1591$	$0.0^4 1523$	$0.0^4 1458$	$0.0^4 1395$	-4.1
-4.2	$0.0^4 1385$	$0.0^4 1277$	$0.0^4 1222$	$0.0^4 1168$	$0.0^4 1118$	$0.0^4 1069$	$0.0^4 1022$	$0.0^5 9774$	$0.0^5 9345$	$0.0^5 8934$	-4.2
-4.3	$0.0^5 8540$	$0.0^5 8163$	$0.0^5 7801$	$0.0^5 7455$	$0.0^5 7124$	$0.0^5 6807$	$0.0^5 6503$	$0.0^5 6212$	$0.0^5 5934$	$0.0^5 5668$	-4.3
-4.4	$0.0^5 5418$	$0.0^5 5169$	$0.0^5 4985$	$0.0^5 4712$	$0.0^5 4498$	$0.0^5 4294$	$0.0^5 4093$	$0.0^5 3911$	$0.0^5 3732$	$0.0^5 3561$	-4.4
-4.5	$0.0^5 3398$	$0.0^5 3241$	$0.0^5 3092$	$0.0^5 2949$	$0.0^5 2813$	$0.0^5 2682$	$0.0^5 2558$	$0.0^5 2439$	$0.0^5 2325$	$0.0^5 2216$	-4.5
-4.6	$0.0^5 2112$	$0.0^5 2013$	$0.0^5 1919$	$0.0^5 1828$	$0.0^5 1742$	$0.0^5 1660$	$0.0^5 1581$	$0.0^5 1506$	$0.0^5 1434$	$0.0^5 1366$	-4.6
-4.7	$0.0^5 1301$	$0.0^5 1239$	$0.0^5 1179$	$0.0^5 1123$	$0.0^5 1069$	$0.0^5 1077$	$0.0^6 9680$	$0.0^6 9211$	$0.0^6 8765$	$0.0^6 8339$	-4.7
-4.8	$0.0^6 7933$	$0.0^6 7547$	$0.0^6 7178$	$0.0^6 6827$	$0.0^6 6492$	$0.0^6 6173$	$0.0^6 5869$	$0.0^6 5580$	$0.0^6 5304$	$0.0^6 5042$	-4.8
-4.9	$0.0^6 4792$	$0.0^6 4554$	$0.0^6 4327$	$0.0^6 4111$	$0.0^6 3906$	$0.0^6 3771$	$0.0^3 3525$	$0.0^6 3348$	$0.0^6 3179$	$0.0^6 3019$	-4.9

APPENDIX 2: UNIT CONVERSION TABLE

$1\,\text{ft} = 0.3048\,\text{m}$

$1\,\text{in} = 25.4\,\text{mm}$

$1\,\text{ha} = 10^4\,\text{m}^2$

$1\,\text{acre} = 4.046873 \times 10^3\,\text{m}^2$

$1\,\text{mD} = 10^{-3}\,\mu\text{m}^2$

$(^\circ\text{F} - 32)/1.8 = 1\,^\circ\text{C}$

$1\,\text{bbl} = 0.1589873\,\text{m}^3$

$1\,\text{psi} = 6.894757 \times 10^{-3}\,\text{MPa}$

$1\,\text{mile} = 1.609344\,\text{km}$

$1\,\text{cP} = 1\,\text{mPa} \cdot \text{s}$

$1\,\text{gal} = 3.785412 \times 10^{-3}\,\text{m}^3$

$1\,\text{St} = 10^4\,\text{m}^2/\text{s}$

Bibliography

[1] Guoting A, Peiqiong L. Offshore oil development technology and equipment. Tianjin: Tianjin University Press; 2001.

[2] Dagang Oilfield Technology Books editorial board. Beijing harbor engineering. Beijing: Petroleum Industry Press; 1999.

[3] Yanqiu D. Deep-sea oil platforms wave loads and response. Tianjin: Tianjin University Press; 2005.

[4] Menglan D, Yongning W, Zhaojie G. High fixed oil platforms of ice-induced vibration in the time domain and frequency domain response. Mechanical Strength 1999;21(89):14–7.

[5] Huacan Fang, Guoming C. Fuzzy probabilistic fracture mechanics. Dongying: China University of Petroleum Press; 1999.

[6] Huacan Fang, Guoming C. Ice offshore structures for reliability analysis. Beijing: Petroleum Industry Press; 2000.

[7] Huacan Fang, Wei Z. Reliability analysis of offshore structures in ice. Beijing: Petroleum Industry Press; 1995.

[8] Huacan Fang. On problems concerning China's deep-sea oilfield development project. China Offshore Platform 2006;21(4):1–8.

[9] Huacan Fang. Theoretical basis for offshore oil drilling equipment. Beijing: Petroleum Industry Press; 1984.

[10] Huacan Fang. Offshore oil drilling and production equipment and structure. Beijing: Petroleum Industry Press; 1990.

[11] Huacan Fang. Ice ocean petroleum steel structure engineering mechanics. Dongying: China University of Petroleum Press; 1996.

[12] Huacan Fang. The fatigue life of offshore petroleum steel structures. Dongying: China University of Petroleum Press; 1990.

[13] Shuchu F, Xuemin W, Kuichang G. Oil and gas gathering and transferring. Dongying: China University of Petroleum Press; 1988.

[14] Pengcheng S, Baozhi C, Xu S. Principium of safety. Beijing: Chemical Industry Press; 2005.

[15] Haibo X, Jinzhou Z, Yongqing W. Technology and management for offshore oil and gas field development engineering. Beijing: Petroleum Industry Press; 2005.

[16] Qianjin Y. Ice load—interaction between sea ice and offshore structures. International Academic Dynamic 1999;5(104).

[17] Jianbo Z. Overview of technical inspection and maintenance of underwater structures. Beijing: Petroleum Industry Press; 2005.

[18] China Classification Society. Submarine pipeline system specification. Beijing: China Communications Press; 1992.

[19] Marine Steel Structure Branch Secretariat of China Steel Structure Association. Steel structure, 21. Academic conference proceedings for Marine Steel Structure Branch meeting of China Steel Structure Association in 2006; 2006 (90).

[20] Shouwei Z. High-tech and practice for China national offshore oil. Beijing: Geological Publishing House; 2005.

[21] Donghong Z, Kaiqing G. Safety system engineering. Beijing: Chemical Industry Press; 2004.

[22] Huacan Fang. New progress of oil equipment used for deepwater platform. China Offshore Platform 2010;25(2):1–7.

[23] Huacan Fang. Reflections on the China's construction and application of floating rig. Petroleum and Petrochemical Equipment Information 2011;504/505(5/6):10–4.

[24] Huacan Fang. On the platform and engineering ship used for deep water in the South China Sea. China Offshore Platform 2012;27(3):4–8.

[25] Huacan Fang. Floating LNG production and multi-function platform for offshore field development. China Offshore Platform 2013;28(2):1–5.

[26] Huacan Fang. Foreground of marine floating drilling rig. Oil and Equipment 2013;50(3):52–4.

Index

X

Printed and bound by CPI Group (UK) Ltd, Croydon, CR0 4YY

08/05/2025

01864894-0001